MIDDLE AMERICAN TERRANES, POTENTIAL CORRELATIVES, and OROGENIC PROCESSES

MIDDLE AMERICAN TERRANES, POTENTIAL CORRELATIVES, and OROGENIC PROCESSES

Edited by

J. Duncan Keppie • J. Brendan Murphy

F. Ortega-Gutiérrez • W. G. Ernst

CRC Press is an imprint of the
Taylor & Francis Group, an **informa** business

CRC Press
Taylor & Francis Group
6000 Broken Sound Parkway NW, Suite 300
Boca Raton, FL 33487-2742

First issued in paperback 2019

ISBN 13: 978-0-367-45266-7 (pbk)
ISBN 13: 978-1-4200-7370-6 (hbk)

Library of Congress Cataloging-in-Publication Data

Middle American terranes, potential correlatives, and orogenic processes / editors, J. Duncan Keppie ... [et al.].
p. cm.
Includes bibliographical references and index.
ISBN-13: 978-1-4200-7370-6 (alk. paper)
ISBN-10: 1-4200-7370-2 (alk. paper)
1. Orogeny--Mexico. 2. Orogeny--Appalachian Mountains. 3. Subduction zones--Mexico. 4. Subduction zones--Appalachian Mountains. 5. Geology, Structural--Mexico. 6. Geology, Structural--Appalachian Mountains. 7. Geology, Stratigraphic. I. Keppie, J. Duncan.

QE621.5.M6.M53 2008
551.8'20972--dc22 2007046783

Visit the Taylor & Francis Web site at
http://www.taylorandfrancis.com

and the CRC Press Web site at
http://www.crcpress.com

MIDDLE AMERICAN TERRANES, POTENTIAL CORRELATIVES, AND OROGENIC PROCESSES

International Geological Correlation Program Project No. 453: Comparison of Modern and Ancient Orogens—Uniformitarianism Revisited

J. Duncan Keppie, J. Brendan Murphy, F. Ortega-Gutiérrez, and W. G. Ernst, eds.

Contents

Potential Correlatives

Orogenic Processes

EDITORS

Duncan Keppie has been an investigator in the Instituto de Geología at the Universidad Nacional Autónoma de México since 1995. Prior to then, he worked for 20 years in the Department of Natural Resources in Nova Scotia (Canada), one year as a sabbatical replacement in the Geology Department at Acadia University (Canada), 3 years (1970–1973) with the Zambian Geological Survey under a British Aid scheme, and 3 years (1967–1970) at Bryn Mawr College (USA) as a visiting professor. He earned his B.Sc. and Ph.D. degrees from Glasgow University, Scotland. He and his coauthors have published over 200 scientific papers, geological maps and field guides, and co-edited five books. He has also been a co-leader of five projects under the auspices of the International Geological Correlation Program (UNESCO).

Brendan Murphy is a professor in the Department of Earth Sciences at St. Francis Xavier University, Canada, where he has been since 1982. He earned his B.Sc. degree from the University College Dublin, Ireland, before emigrating to Canada in 1975. He acquired an M.Sc. from Acadia University in 1977 and a Ph.D. from McGill University in 1982. He teaches courses in structural geology, tectonics and the evolution of the Earth. He has published over 160 scientific articles in academic journals, book chapters, monographs, or geological field guidebooks, and has authored or coauthored more than 200 conference presentations. He is currently editor of the *Geological Society of America Bulletin.*

Fernando Ortega Gutiérrez is currently full professor at the Instituto de Geología, Universidad Nacional Autónoma de México and was formerly director of the Institute from 1986 to 1994. He received his Ph.D. from the University of Leeds, U.K., in 1975 and has worked since then in the areas of geology and petrology of basement terranes and tectonics of Mexico. He served as councilor of the Geological Society of America from 1992 to 1994, has been an adjunct professor at the University of Arizona, Tucson, and has been an invited lecturer in several universities of the United States, Canada, and Guatemala. His long career in scientific research is documented in nearly 100 papers published in fully refereed national and international journals and books. He was the main author for the 1992 Geological Map of Mexico at the scale of 1:2,000,000.

W. Gary Ernst joined the University of California at Los Angeles faculty in 1960. He was chair of the Geology Department (1970–1974), chair of the Earth and Space Sciences Department (1978–1982), and director of the UCLA Institute of Geophysics & Planetary Physics (1987–1989). He moved to Stanford University for a 5-year term as dean of the School of Earth Sciences (1989–1994). Named a professor in the Department of Geological and Environmental Sciences (1999–2004), Ernst became the Benjamin M. Page professor emeritus in 2004. A member of the National Academy of Sciences, American Academy of Arts & Sciences, and American Philosophical Society, Ernst served as president of the Mineralogical Society of America (1980–1981) and the Geological Society of America (1985–1986). He received the MSA Award (1969), the Geological Society of Japan Medal (1998), the Penrose Medal of the GSA (2004), and the Roebling Medal of the MSA (2006). Author of 6 books and research memoirs, editor of 14 others, Ernst has also authored more than 200 papers dealing with: physical chemistry of rocks and minerals; Phanerozoic interactions of lithospheric plates and orogenic belts, especially in central Asia, the Circumpacific and the western Alps; early Precambrian petrotectonic evolution; ultrahigh-pressure subduction-zone metamorphism and tectonics; geobotanical studies/remote sensing; and geology and human health.

CONTRIBUTORS

E. Belousova
GEMOC National Key Centre
Department of Earth and Planetary Sciences
Macquarie University
North Ryde, Australia

Jamie F. Braid
Department of Earth Sciences
St. Francis Xavier University
Antigonish, Nova Scotia, Canada

Kenneth L. Cameron
Earth Sciences Division
University of California at Santa Cruz
Santa Cruz, California

E. Centeno-García
Instituto de Geología
Universidad Nacional Autónoma de México
México D.F., México

B. N. Church
Geological Services
Victoria, British Columbia, Canada

J. Dostal
Department of Geology
St. Mary's University
Halifax, Nova Scotia, Canada

David S. Dowe
Department of Geological Sciences
Clippinger Laboratories
Ohio University
Athens, Ohio

Kerstin Drost
Staatliche Naturhistorische Sammlungen Dresden
Forschungsmuseum für GeoBio Wissenschaften
Dresden, Germany

Joseph M. English
School of Earth and Ocean Sciences
University of Victoria
Victoria, British Columbia, Canada

Javier Fernández-Suárez
Departamento de Petrología y Geoquímica
Universidad Complutense
Madrid, Spain

Michael Gehmlich
Staatliche Naturhistorische Sammlungen Dresden
Forschungsmuseum für GeoBio Wissenschaften
Dresden, Germany

Arturo Gomez-Tuena
Centro de Geociencias
Universidad Nacional Autónoma de México
Querétaro, México

W. E. Hames
Department of Geology
Auburn University
Auburn, Alabama

T.S. Hamilton
Camosun College
Victoria, British Columbia, Canada

Andrew Hynes
Department of Earth and Planetary Science
McGill University
Montreal, Quebec, Canada

Teresa E. Jeffries
Natural History Museum
London, United Kingdom

G. A. Jenner
Department of Earth Sciences
Memorial University
St. John's, Newfoundland, Canada

Stephen T. Johnston
School of Earth and Ocean Sciences
University of Victoria
Victoria, British Columbia, Canada

J. Duncan Keppie
Instituto de Geología
Universidad Nacional Autónoma de México
México D.F., México

J. W. K. Lee
Department of Geology
Queens University
Kingston, Ontario, Canada

R. Lopez
Geology Department
West Valley College
Saratoga, California

H. Macdonald
Department of Geology
St. Mary's University
Halifax, Nova Scotia, Canada

C. Maciás-Romo
Instituto de Geología
Universidad Nacional Autónoma de México
México D.F., México

Uwe Martens
Universidad de San Carlos de Guatemala–
Centro Universitario del Norte
Cobán, Guatemala

B. V. Miller
Department of Geological Sciences
University of North Carolina at Chapel Hill
Chapel Hill, North Carolina

Sergio Morán-Ical
Universidad de San Carlos de Guatemala–
Centro Universitario del Norte
Cobán, Guatemala

Dante J. Morán-Zenteno
Instituto de Geología
Universidad Nacional Autónoma de México
México D.F., México

J. Brendan Murphy
Department of Earth Sciences
St. Francis Xavier University
Antigonish, Nova Scotia, Canada

R. Damian Nance
Department of Geological Sciences
Clippinger Laboratories
Ohio University
Athens, Ohio

H. Ochoa-Camarillo
Procesos Analíticos Informáticos
México D.F., México

Fernando Ortega-Gutiérrez
Instituto de Geología
Universidad Nacional Autónoma de México
México D.F., México

Carlos Ortega-Obregon
Instituto de Geología
Universidad Nacional Autónoma de México
México D.F., México

A. Ortega-Rivera
Centro de Geociencias
Universidad Nacional Autónoma de México
Centro Querétaro, Qro., México

Myriam Osorio
Instituto de Geofísica
Universidad Nacional Autónoma de México
Ciudad Universitaria
Coyoacan D.F., México

J. V. Owen
Department of Geology
St. Mary's University
Halifax, Nova Scotia, Canada

Forrest G. Poole
U.S. Geological Survey
Federal Center
Denver, Colorado

S. Quiroz-Barroso
Museo de Paleontología
Facultad de Ciencias
Universidad Nacional Autónoma de México
México D.F., México

Margarita Reyes-Salas
Instituto de Geología
Universidad Nacional Autónoma de México
México D.F., México

Jaime Roldán-Quintana
Instituto de Geología
Universidad Nacional Autónoma de México
Hermosillo, Sonora, México

L. Rosales-Lagarde
Instituto de Geología
Universidad Nacional Autónoma de México
México D.F., México

J. L. Sánchez-Zavala
Instituto de Geología
Universidad Nacional Autónoma de México
México D.F., México

Charles A. Sandberg
U.S. Geological Survey
Federal Center
Denver, Colorado

Peter Schaaf
Instituto de Geofísica
Universidad Nacional Autónoma de México
Ciudad Universitaria
Coyoacan D.F., México

Luigi A. Solari
Instituto de Geología
Universidad Nacional Autónoma de México
México D.F., México

Jesús Solé
Instituto de Geología
Universidad Nacional Autónoma de México
México D.F., México

F. Sour-Tovar
Museo de Paleontología
Instituto de Geología
Universidad Nacional Autónoma de México
México D.F., México

Craig D. Storey
Department of Earth Sciences
The Open University
Milton Keynes, United Kingdom

Bodo Weber
División Ciencias de la Tierra
Centro de Investigación Cientifica y de Educación Superior
Ensenada B.C., México

Actualistic models for Paleozoic convergence and orogenesis require consideration of the changing profiles of Andean-type subduction zones with time. Modern subduction zone profiles change with variations in convergence rate, slab age, and motion of the upper plate, and range from flat-slab to angles of up to 45°. The Late Mesozoic–Early Cenozoic Laramide orogeny in the southwestern United States is often held to be an example of orogenesis driven by flat-slab subduction, and provides a model for similar-aged orogenesis in Mexico and the Acadian orogeny in Maritime Canada. English and Johnson point out that none of the proposed mechanisms for driving Laramide orogenesis satisfactorily explain the geometry, timing, or extent of the orogeny. Inasmuch as Laramide-age deformation occurred along the entire length of North America regardless of which oceanic plate was subducting beneath it, it seems unlikely that slab age or convergence rate were the principal driving force of orogenesis. Although eastward migration of arc magmatic belts in western Canada and Mexico during Laramide time may reflect a shallowing of the subducting slab, the forces that drive foreland deformation are not apparent. English and Johnson consider the absolute motion of the overriding plate (driven by ridge-push from the Atlantic spreading center) and basal drag to be unlikely mechanisms.

Paleozoic orogenesis in the Middle America terranes was related to convergence between Laurentia and Gondwana, culminating in the amalgamation of the supercontinent Pangea. Although the existence of Pangea is a cornerstone in the understanding of Phanerozoic tectonics, mechanisms responsible for its amalgamation are controversial. Moreover, it is clear from the geological record that repeated supercontinent amalgamation and dispersal has occurred since the Archean. Murphy and Nance point out that following supercontinent breakup, there are two geodynamically distinct oceans: an *interior ocean* formed between the dispersing continents whose lithosphere is younger than the time of supercontinent breakup, and an *exterior ocean* that surrounded the supercontinent prior to breakup and consequently is dominated by lithosphere that is older than the time of breakup. Sm-Nd isotopic analyses of mafic complexes accreted to continental margins prior to terminal collision can reveal whether the oceanic lithosphere consumed by subduction was part of an interior or exterior ocean. They find that Pangea was formed by closure of interior oceans, whereas the Late Neoproterozoic supercontinent Pannotia was formed by closure of an exterior ocean, implying that supercontinents can be formed by fundamentally distinct geodynamic processes.

—J. Duncan Keppie, J. Brendan Murphy,
F. Ortega-Gutiérrez, and W. G. Ernst

MIDDLE AMERICAN TERRANES

Terranes of Mexico Revisited: A 1.3 Billion Year Odyssey

J. DUNCAN KEPPIE[1]

Instituto de Geología, Universidad Nacional Autónoma de México, 04510 México D.F., México

Abstract

During the Precambrian and Paleozoic, Mexican terranes were either part of or proximal to Laurentia and Middle America (basements of Mesozoic Maya, Oaxaquia, and Chortis terranes that bordered Amazonia). Obduction of the Sierra Madre proximal terrane in the Late Ordovician was followed by Permo-Carboniferous amalgamation of all proximal terranes into Pangea. Middle Jurassic breakup of Pangea resulted in two continental terranes, Maya and Chortis, which were surrounded by small ocean-basin/arc terranes: Gulf of México, Caribbean Sea, Juarez, Motagua terranes, and the Guerrero composite terrane. All of these terranes were obducted onto North America during the Late Cretaceous–Early Cenozoic, Laramide orogeny. Neogene propagation of the East Pacific Rise into the North American margin has led to separation and northwest translation of the Baja California terrane.

Introduction

TERRANE MAPPING was first applied to Mexico by Campa and Coney (1983) and Coney and Campa (1984) as part of projects in the North American Cordillera (Silberling et al., 1992) and around the Pacific Ocean (Howell, 1985), with its primary purpose being an understanding of the plate tectonic evolution (Coney, 1983) and metallic mineral and energy resource distributions. Applying the principles of terrane analysis (Jones and Silberling, 1979; Howell et al., 1985), Campa and Coney (1983) recognized 12 terranes (Fig. 1A), of which 7 were considered to be composite (C), because existing geological maps did not allow subdivision. They grouped the terranes into three categories depending on their provenance: (1) North American provenance—two Precambrian–Mesozoic terranes (Chihuahua and Caborca [C]); (2) Gondwanan provenance—three Paleozoic terranes accreted to North America during the latest Paleozoic Ouachita-Marathon orogeny (Coahuila, Maya, and Sierra Madre, all composite); and (3) Pacific provenance—seven Mesozoic (–Precambrian) terranes accreted to western Mexico in the Late Cretaceous (Alisitos, Vizcaino [C], Guerrero [C], Juarez, Mixteca [C], Oaxaca, and Xolapa).

The next comprehensive terrane analysis of the whole of México appeared 10 years later by Sedlock et al. (1993) who outlined 16 terranes (Fig. 1B), 2 of North American provenance, 7 of Gondwanan provenance, and 7 of Pacific provenance, named after the various indigenous cultures of Mexico. Although the basic outlines of the terranes are similar to those of Campa and Coney (1983), the following changes occurred: (a) in the Gondwanan terranes, the Coahuila terrane and Sierra Madre terranes were each split into two, and the Mixteca and Oaxaca terrane were recognized as being of Gondwanan (rather than Pacific) provenance; (b) the Alisitos and Guerrero terranes were each split into two; and (c) the Vizcaino terrane was subdivided into four subterranes. Sedlock et al. (1993) went on to propose a tectonic evolution for Mexico.

Since 1993, further studies have allowed many of the composite terranes to be subdivided. Thus the Guerrero composite terrane on mainland Mexico has been subdivided into five terranes (Centeno-García et al., 1993, 2000, 2003; Talavera-Mendoza and Suastegui, 2000; Freydier et al., 2000). The ~1 Ga basement of the Oaxaca terrane has been traced into the Sierra Madre terrane, thereby giving rise to the Oaxaquia microcontinent (Ortega-Gutiérrez et al., 1995). Subsequently, Ortega-Gutiérrez et al. (1999) subdivided the Mixteca terrane into ophiolitic and sedimentary units that were inferred to have been juxtaposed in the Late Ordovician–Early Silurian. In 2001, Dickinson and Lawton classified the Mexican terranes in terms of eight, internally coherent Permian–Cretaceous crustal blocks with a detailed explanation of how they relate to previously defined terranes (Table 1 and Fig. 1 of Dickinson and Lawton, 2001), which they used to develop a Carboniferous–Cretaceous plate tectonic model. Again,

[1]Email: duncan@servidor.unam.mx

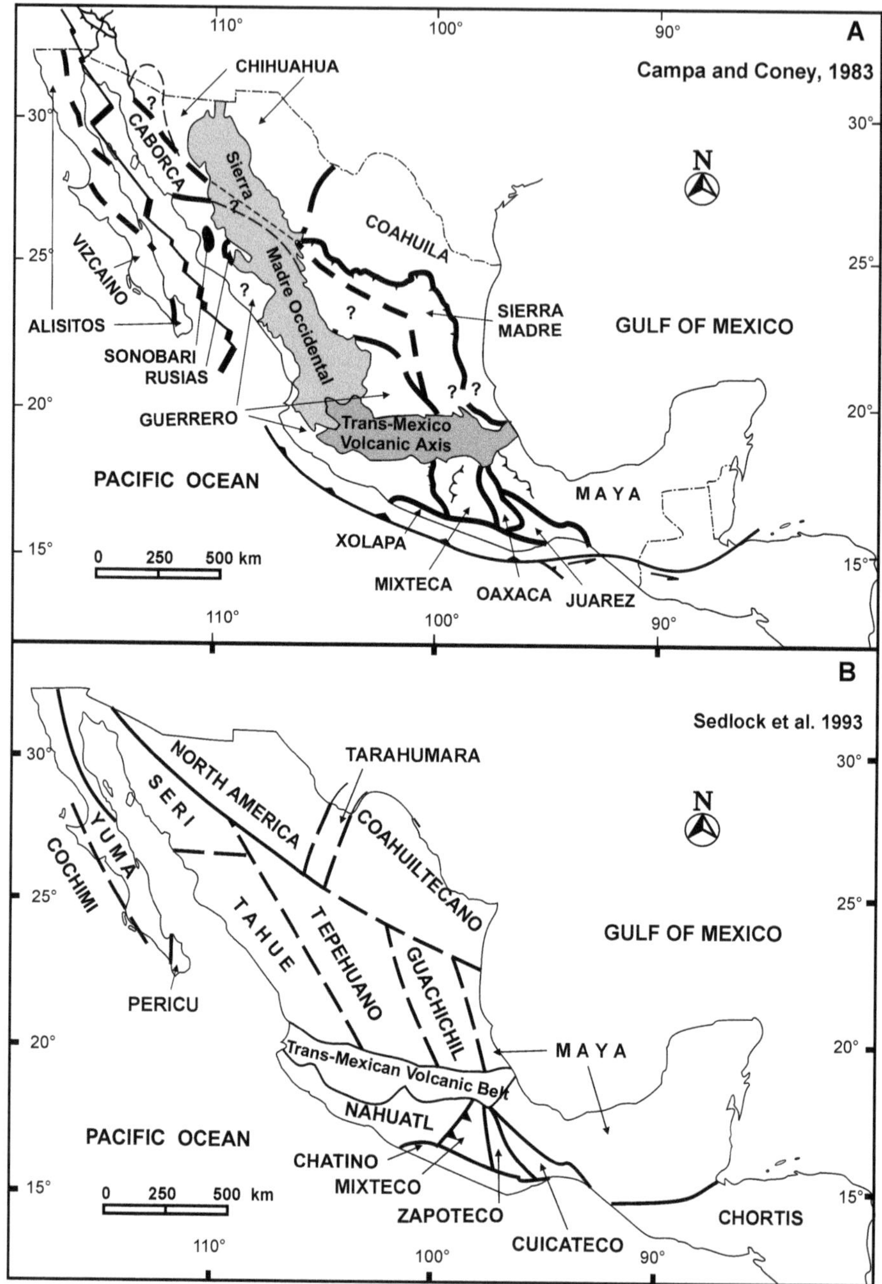

FIG. 1. Terrane maps of (A) Campa and Coney (1983), (B) Sedlock et al. (1993), and (C) this paper (facing page).

these crustal blocks generally build on earlier terrane analyses after the amalgamation of Pangea. The main changes are: (a) the amalgamation of Gondwanan terranes in the Coahuila, Tampico, and Del Sur blocks; (b) the elevation of the Guerrero terrane to an undivided superterrane; and (c) the recognition of two collisional zones—a Permian–Jurassic subduction complex on the western margin of the Gondwanan terranes, and a middle Cretaceous suture zone (closed ocean basin) between the

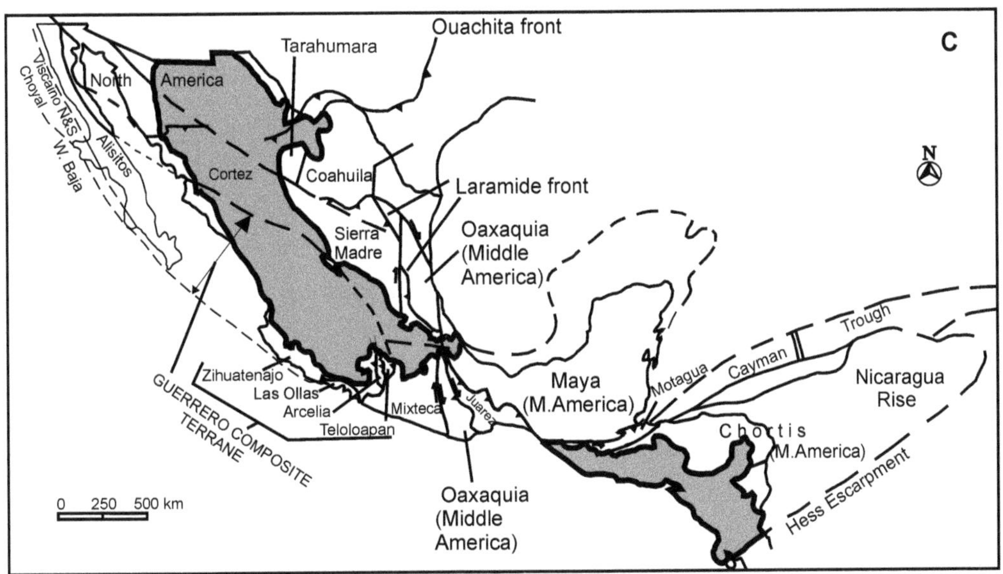

Guerrero superterrane and North America/Gondwana terranes. Several new terranes have been introduced in abstracts; however, the lack of published maps and tectonostratigraphy makes evaluation difficult.

The present analysis has allowed the following advances: (1) subdivision of the composite terranes; (2) delimitation of the terranes in time; (3) tectonic interpretation of the terrane geological records; (4) provision of better constraints on the provenance; and (5) reconstruction of actualistic palinspastic maps from the Mesoproterozoic to the present. The present analysis (Fig. 1C) also builds on the earlier terrane analyses in: (a) accepting the subdivision of the Guerrero composite terrane (Centeno-García et al., 2000); (b) introducing the Cretaceous, Motagua oceanic terrane between the Maya and Chortis terranes; (c) acquitting four terranes: (i) the Caborca terrane because it is merely an offset part of North America (Dickinson and Lawton, 2001); (ii) the Las Delicias terrane (McKee et al., 1999) as it appears to have been an arc developed on Pangea; (iii) the Xolapa terrane because it has been shown to be a Mesozoic overstep sequence passing from continental on the Acatlán Complex to shallow marine towards the coast (Ortega-Gutiérrez and Elías-Herrera, 2003); and (iv) the Juchatengo terrane, as new data shows the lavas, rather than of oceanic affinity, are continental tholeiites probably similar to those in the Acatlán Complex (Grahales-Nishimura et al., 1999); (d) accepting terrane subdivisions of the Coahuila composite terrane (Sedlock et al., 1993); (e) recognizing peri-North American and peri-Gondwanan elements as terranes in their own right that originated as continental rise, trench complexes, or suture zones, e.g. eugeoclinal rocks of the Cortez terrane bordering the North American craton, and Sierra Madre and Mixteca terranes bordering Middle America (Dickinson and Lawton, 2001); and (f) more clearly defining the birth, life, and death of individual terranes. Given the extensive descriptions of the geological records of the various terranes in earlier works, their geological records are only briefly summarized in the Appendix. In order to provide a stepping stone toward tectonic models, these geological records are interpreted in terms of tectonic settings in three time-and-space diagrams: Figure 2 summarizes terrane evolution, whereas Figures 3 and 4 show transects across northern and southern Mexico.

Terminology

Terrane terminology requires their definition in space and time. Although the geographical extent of the Mexican terranes has been relatively clearly defined, time constraints on their existence were generally loosely defined. For example, terranes in an oceanic realm may amalgamate to form a composite terrane before being accreted to a craton—e.g., individual Mesozoic terranes originating in the paleo-Pacific Ocean appear to have

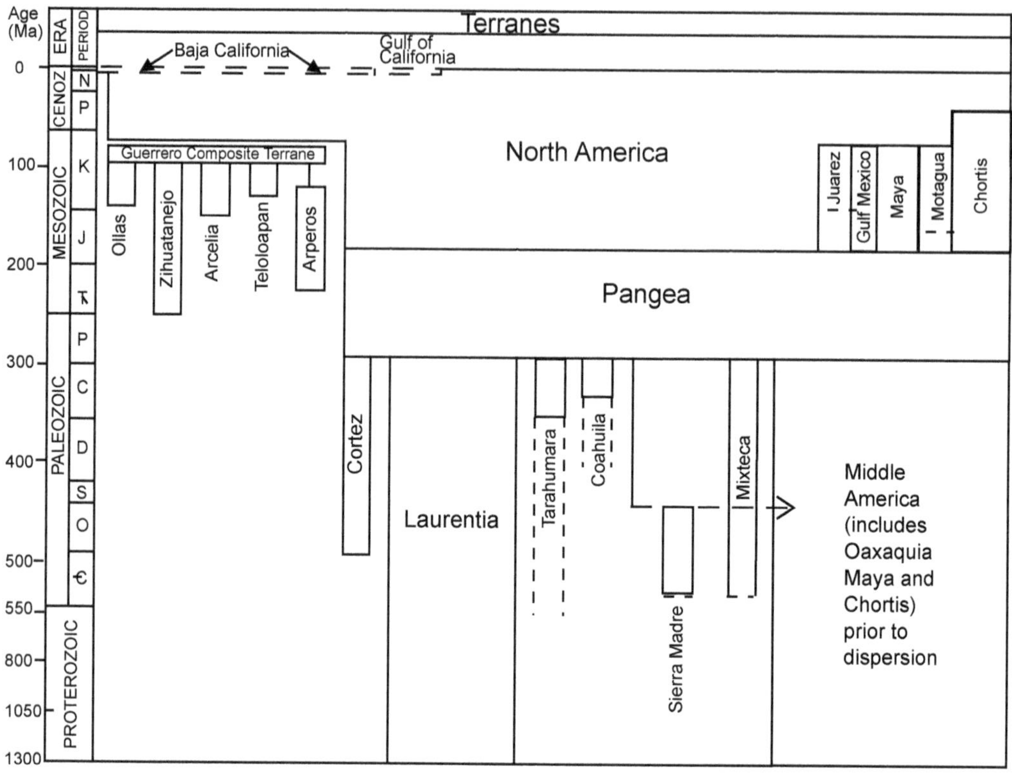

FIG. 2. Time-space summary diagram of terranes identified in this paper. Abbreviations: Є = Cambrian; O = Ordovician; S = Silurian; D = Devonian, C = Carboniferous; P = Permian; T = Traissic; J = Jurassic; K = Cretaceous; P = Paleocene; N = Neogene.

been amalgamated into the Guerrero composite terrane before being accreted to the North American craton in the latest Mesozoic–Early Tertiary (e.g., Centeno-García et al., 1993), and these distinctions need to be reflected in the terminology. A more complex example is the Maya terrane, which has a complex history: (a) during the Precambrian and Paleozoic, it appear to have formed part of a single Middle America terrane on the margin of Amazonia (e.g., Keppie and Ramos, 1999); (b) in the Permo-Triassic, it formed part of Pangea; (c) in the Mesozoic, opening of the Gulf of Mexico separated the Yucatan block from Permo-Triassic Pangea, forming the Maya terrane (e.g., Marton and Buffler, 1994); and (d) during the latest Cretaceous–Early Cenozoic, it was reamalgamated with mainland Mexico by the Laramide orogeny (e.g., Sedlock et al., 1993). However, only the term Maya terrane has been applied to this block and applies only to Mesozoic time. A partial solution to this problem was the application of the term Oaxaquia to the ~1 Ga Precambrian rocks of nuclear Mexico (Ortega-Gutiérrez et al., 1995), however, this included the inferred subsurface extent, and the term has gradually been expanded to include all or parts of the Mixteca, Maya, or Chortis terranes (e.g., Cameron et al., 2004). In as much as a terrane map shows their surface distribution, the term Oaxaquia as originally used must be redefined for use on a terrane map. In this synthesis the term is restricted to the Oaxaca, Huiznopala, and Novillo complexes, the latter extracted from the original Sierra Madre terrane because it is in tectonic contact with ophiolitic mélange of the Granjeno Schist (Carrillo-Bravo, 1991). The ~1 Ga basement of Oaxaquia is generally correlated with that in the Maya and Chortis, for which the term Middle America terrane is introduced in this synthesis. Should this correlation prove robust, the term Oaxaquia would be acquitted, as it would be part of Middle America in the

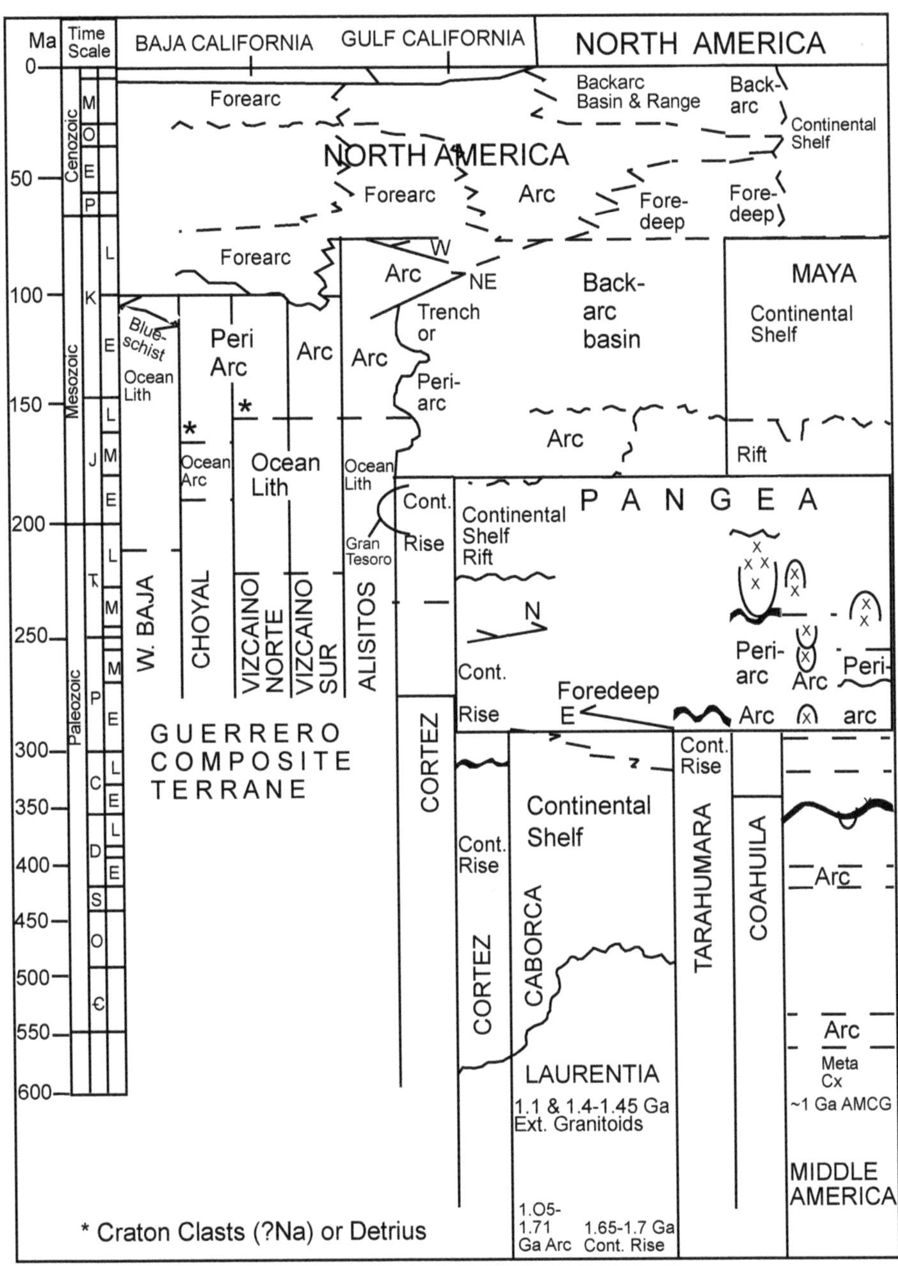

FIG. 3. Time-space diagram for northern Mexico. Abbreviations are the same as in Figure 2, plus X = plutons; E = Early; M = Middle; L = Late, N = North; NE = northeast; W = West; E = east; AMCG = anorthosite-mangerite-charnockite-granite; blues = blueschist; Cont. = continental; Cx = complex; Ext = extensional; Lith = lithosphere; Meta = Metamorphic; O.L. = oceanic lithosphere.

Precambrian and Paleozoic, and part of North America in the Mesozoic and Cenozoic. Individual Mexican terranes have received different names (Campa and Coney, 1983; Sedlock et al., 1993; Dickinson and Lawton, 2001), leading to confusion and/or double-barreled names. In order to simplify

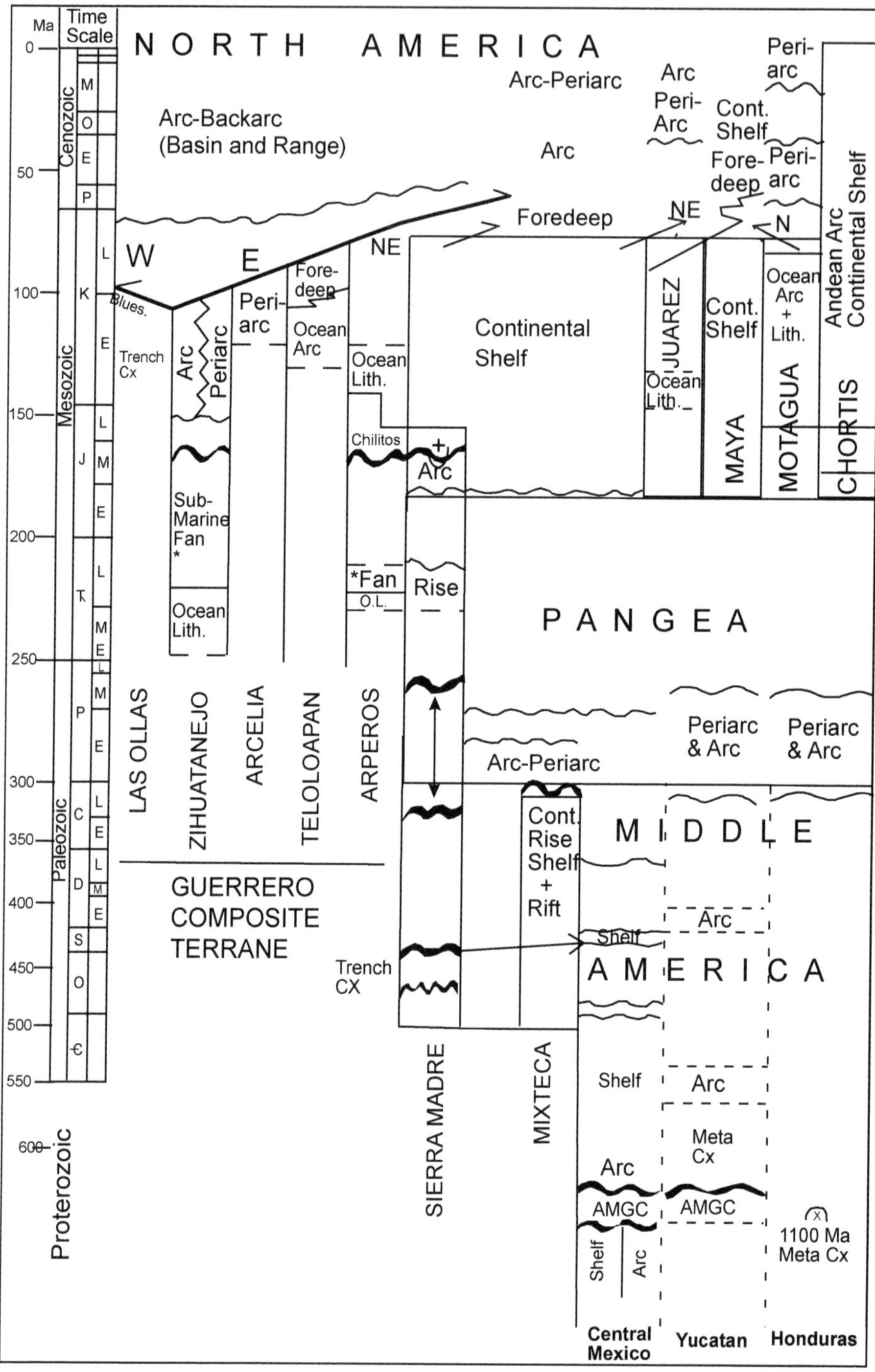

FIG. 4. Time-space diagram for southern Mexico. For abbreviations, see Figures 2 and 3.

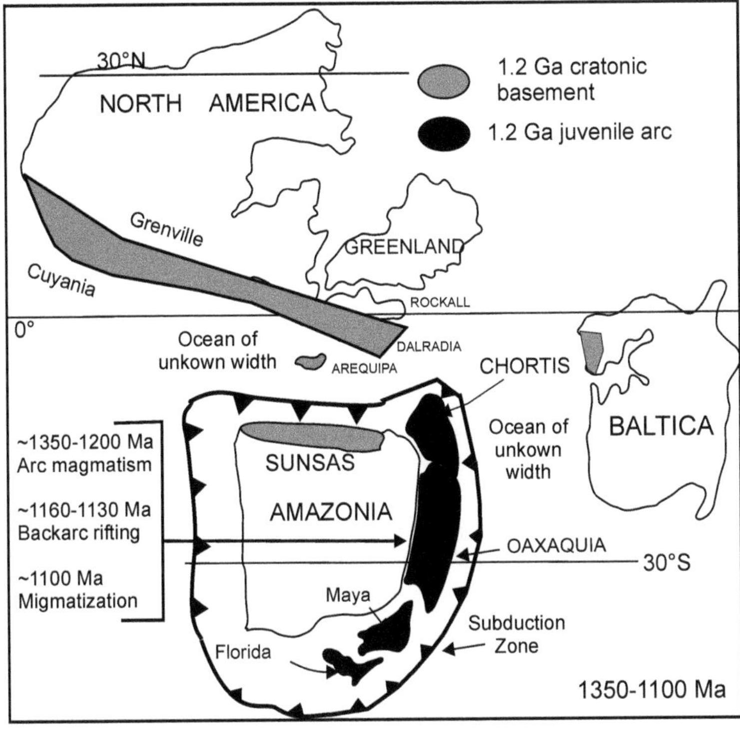

FIG. 5. 1350–1100 Ma reconstruction showing the distribution of juvenile and cratonic 1.2 Ga basement (modified after Dostal et al., 2004). Barbed line is a subduction zone with teeth oriented toward the overriding slab.

the nomenclature, previous names are retained, giving priority to the earliest terms, and only introducing new names where absolutely required (Figs. 2–4).

Cratons, Cratonic Blocks, and Their Edges

A first-order analysis requires recognition of the edges of the North and Middle America–Amazon cratons beyond which lie the accreted terranes. The Oaxaquia, Maya, and Chortis terranes form the basement of Middle America, and are inferred to have formed a belt along the northwestern margin of the Amazon craton of South America (Keppie and Ramos, 1999). In practice, the continent-ocean boundary is covered by a passive margin sequence that may be subdivided into continental shelf and continental rise prism, and the boundary between them is generally close to the oceanic-continental lithospheric transition. This buried boundary may also be identified using Sr, Nd, and Pb isotopes in cross-cutting igneous rocks (Keppie and Ortega-Gutiérrez, 1995). However, this boundary may be significantly offset at the surface by subsequent thrusting.

North American craton

Precambrian terranes in the Laurentian carton are beyond the scope of this paper. The shelf-rise transition in latest Proterozoic–Paleozoic rocks around the *southwestern margin* of North America and the Caborca terrane was documented by Stewart (1988); however, the contact is either a Late Permian–Middle Triassic, north-vergent thrust or a high-angle, Cenozoic, normal fault (Stewart et al., 1990). The 700–950 km, intracratonic, sinistral displacement of the Caborca block relative to North America has been interpreted in terms of transcurrent and transform movements (Anderson et al., 1991; Dickinson and Lawton, 2001); thus the

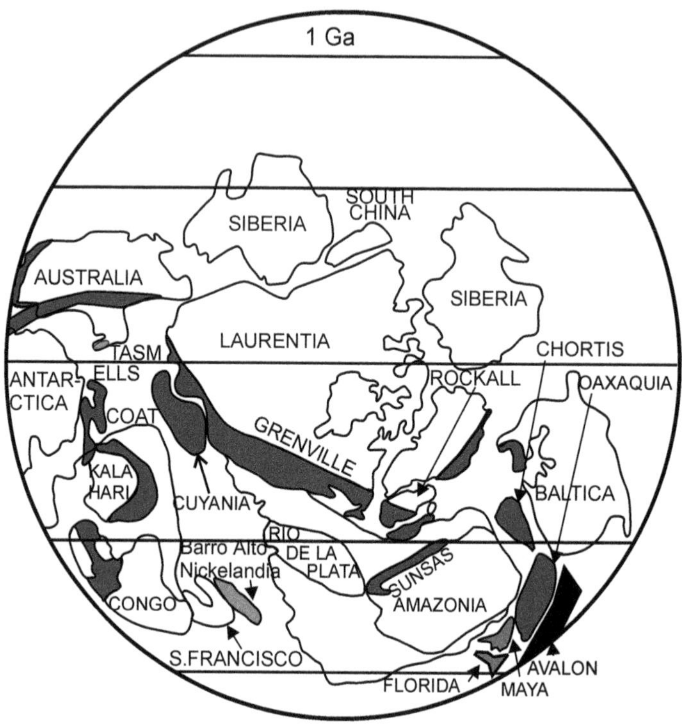

FIG. 6. ~1 Ga reconstruction of Rodinia (modified after Keppie et al., 2003a).

Caborca terrane is acquitted. The ocean-craton boundary off southwestern North America in mainland Mexico is covered in general by Mesozoic and Tertiary rocks. In Baja California and southern California, although the boundary is obscured by intrusion of the Peninsular Ranges batholith and associated high-grade metamorphism, it may be traced in the following changes in its geochemistry (west to east): (i) increasing initial $^{87}Sr/^{86}Sr$ ratios, (ii) increasing $\delta^{18}O$ values, (iii) increasing fractionation of rare-earth elements (REEs), (iv) increasing radiogenic Pb isotopes, and (iv) decreasing ε_{Nd} values of +7.9 to –2.5 (Silver et al., 1979; Hill et al., 1986; Silver and Chappell, 1988).

The *southeastern cratonic* boundary is inferred beneath allochthonous rocks of the Ouachitan orogen (Thomas, 1989). These rocks have been traced into Mexico in the Tarahumara terrane (Sedlock et al., 1993); however, it is not known whether these rocks represent the continental rise prism bordering either North or Middle America, or an intervening oceanic assemblage.

Middle America terrane (i.e., Oaxaquia-Maya-Chortis prior to Jurassic dispersion)

In contrast, the edge of the Middle America terrane has not been so clearly established, partly due to extensive Mesozoic–Cenozoic cover, the similarity of craton and craton-derived sediment, isotopic signatures, metamorphic overprint, and a lack of geochronology. Lower Paleozoic rocks resting unconformably upon ~1 Ga Oaxaquia are interpreted as shelf sequences with Gondwanan faunal affinities (Robison and Pantoja-Alor, 1968; Shergold, 1975; Boucot et al., 1997). The ubiquitous presence in all the ~1 Ga basement inliers in Mexico of ~1008 Ma anorthosite-mangerite-charnockite-granite association (Keppie et al., 2003a), and 1000–980 Ma granulite-facies polyphase deformation (Solari et al., 2003) implies that they are part of one terrane. However, pre-1100 Ma rocks represent either a rift-shelf sequence (northern Oaxacan Complex: Ortega-Gutiérrez, 1984) or an arc sequence (Lawlor et al., 1999; Keppie et al., 2001; Cameron et

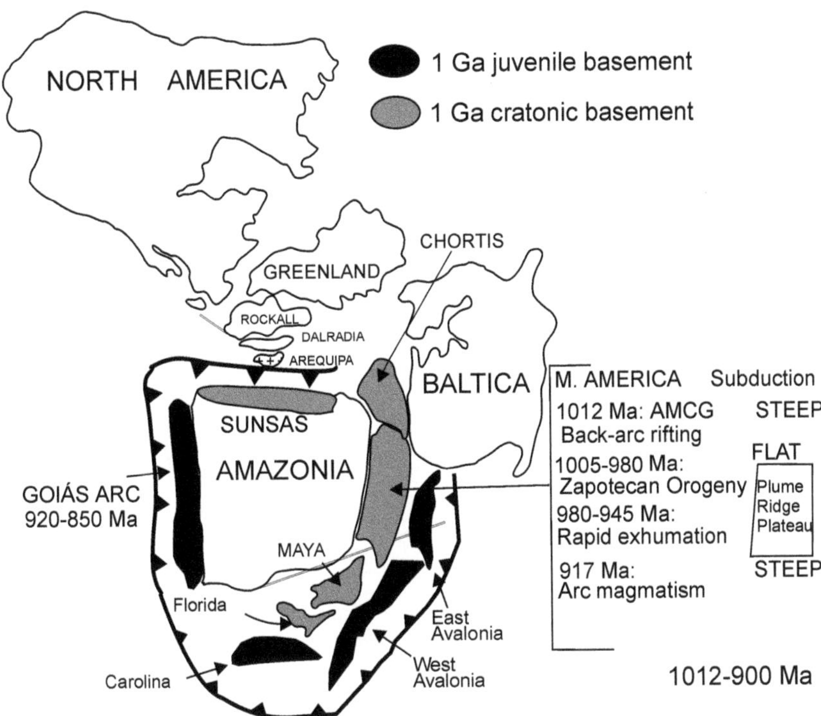

FIG. 7. 1012–900 Ma reconstruction showing ~1 Ga juvenile and cratonic basement surrounding Amazonia (modified after Ortega-Obregón et al., 2003; Solari et al., 2003; and Keppie et al., 2004d).

al., 2004; Dostal et al., 2004) that may, or may not, indicate more than one terrane. On the other hand, the presence of similar metasedimentary rocks in all Mexican, ~1 Ga inliers suggests that the arc may have been developed on one terrane.

During the Paleozoic, the western border of Middle America is best exposed in southern Mexico where Elías-Herrera and Ortega-Gutiérrez (2002) have documented that it is a Permian dextral flower structure. To the west of this boundary, psammitic and pelitic rocks of the Paleozoic Acatlán Complex are inferred to represent either trench-forearc deposits (Ortega-Gutiérrez et al., 1999) or a continental rise prism bordering Middle America (Ramírez-Espinoza et al., 2002); the latter is a conclusion consistent with its comparable Nd signature (Yañez et al., 1991).

The western boundary of Middle America is also exposed near Ciudad Victoria, where Paleozoic ophiolitic mélange (Granjeno Formation interpreted as trench complex) is juxtaposed against ~1 Ga rocks (Novillo Complex) along a dextral fault: a minimum age for the Granjeno is given by 320–260 Ma K-Ar ages on metamorphic mica (references in Sedlock et al., 1993). However, Ordovician obduction is indicated by the presence of Granjeno pebbles in the Wenlockian sediments unconformably overlying the ~1 Ga Novillo Complex (Fries et al., 1962; De Cserna et al., 1977; De Cserna and Ortega-Gutiérrez, 1978). In this region, the Granjeno and Novillo belts crop out in NNW-trending horsts and grabens. Granulite-facies xenoliths of presumed Middle American provenance have been recorded as far west as the edge of the Guerrero composite terrane (references in Sedlock et al., 1993; Ortega-Gutiérrez et al., 1995). This implies a west-dipping thrust contact between oceanic and cratonic rocks that was subsequently displaced by NNW-trending vertical faults.

The *southern margin of Pangea* is separated from the amphibolite-greenschist–facies rocks of the Xolapa terrane by a sinistral-normal fault with relatively little lateral displacement (Ratchbacher et al., 1991; Meschede et al., 1997; Tolson-Jones, 1998) (Fig. 1). The Xolapa Complex consists of quartzo-feldspathic schists and gneisses migma-

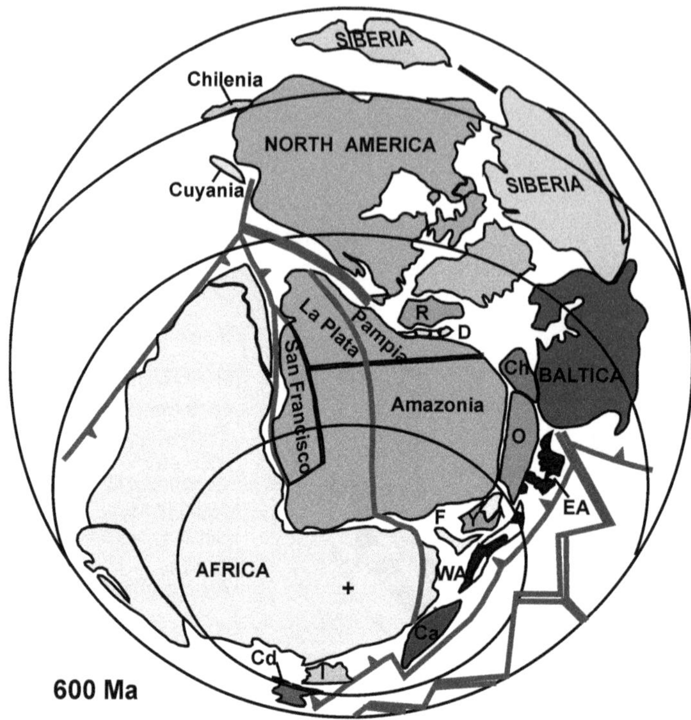

FIG. 8. 600 Ma reconstruction modified after Keppie and Ramos (1999) and Keppie et al. (2003b). Abbreviations: Ca = Carolina; Cd = Cadomia; Ch = Chortis; D = Dalradia; EA = East Avalonia; F = Florida; I = Iberia; O = Oaxaquia; R = Rockall; WA = West Avalonia; Y = Yucatan. Barbed line is a subduction zone with teeth oriented toward the overriding slab; double line is a mid-ocean ridge, and single line is a transform fault.

tized at ~132 ± 2 Ma (nearly concordant, U-Pb, zircon, lower intercept; Herrmann et al., 1994) intruded by 35 to 27 Ma, calcalkaline plutons (Herrmann et al., 1994). Ortega-Gutiérrez and Elías-Herrera (2003) have traced the Xolapa Complex into the Jurassic rocks overstepping the Acatlán Complex, a discovery that acquits the Xolapa terrane.

The Terranes and Constraints on Amalgamation

The geological record and the tectonic interpretation of each terrane are briefly summarized in the Appendix. Constraints on the time of amalgamation include overstep (or overlap) sequences, stitching plutons, exotic pebbles, deformation, and metamorphism (Howell, 1985; Keppie, 1989). They provide a younger limit on the life of a terrane and will now be outlined in historical order.

Paleozoic terranes: Most of Gondwana's provenance

Paleozoic Mixteca terrane. This terrane was defined by Campa and Coney (1983), and subsequently renamed the Mixteco terrane by Sedlock et al. (1993). It consists of two sequences that are tectonically juxtaposed against one another: the low-grade Petlalcingo Group and the eclogitic Piaxtla Group. The Petlalcingo Group consists of a polydeformed sequence of graywackes (Chazumba Formation), variably metamorphosed to psammitic and pelitic schists and migmatized in the Jurassic (Keppie et al., 2004b), overlain by slates and phyllites (Cosoltepec Formation) that are interpreted to represent either trench and forearc deposits (Ortega-Gutiérrez et al., 1999), or a continental rise prism adjacent to Middle America (Ramírez-Espinoza, 2001). The Cosoltepec Formation is tectonically interleaved with the eclogitic Piaxtla Group (Meza-Figueroa et al., 2003) containing mafic rocks of

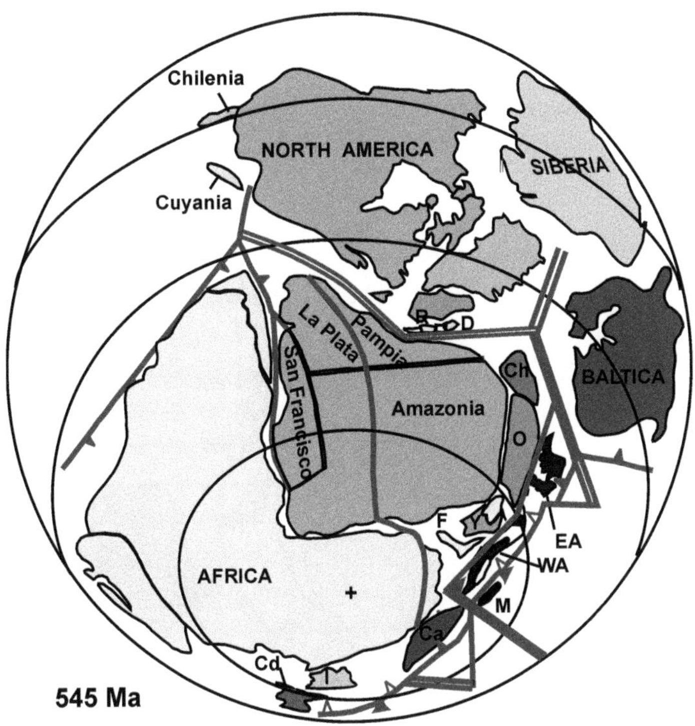

FIG. 9. 545 Ma reconstruction modified after Keppie and Ramos (1999) and Keppie et al. (2003b). Abbreviations and symbols are the same as those in Figure 8, plus M = Meguma.

inferred oceanic affinity that have yielded Sm-Nd garnet-whole rock ages of 388 ± 44 Ma (Yañez et al., 1991). These units are unconformably overlain by the Tecomate Formation.

The general lack of well-constrained fossils in these units has led to inferences based on geochronology: namely that syntectonic granitoid intrusion occurred during the Late Ordovician Acatecan orogeny and the Devonian Mixtecan orogeny (Ortega-Gutiérrez et al., 1999). By extension, this provided a younger age limit of Late Ordovician on the Petlalcingo and Piaxtla groups and Late Devonian on the Tecomate Formation. However, new data is pointing to significant revisions. The Esperanza granitoid, part of the Piaxtla Group, has yielded Ordovician–Silurian protolith ages (concordant U-Pb age of 478 ± 5 Ma: Campa-Uranga et al., 2002; and concordant U-Pb ages of 480–460 Ma: Keppie et al., 2004a). ~1 Ga upper intercept, U-Pb zircon ages in the Esperanza granitoid indicate the presence of such a basement beneath the Acatlán Complex during granite genesis: this is consistent with the similar Nd model ages in these granitoids (Yañez et al., 1991). Granulite-facies xenoliths probably derived from buried Oaxacan Complex have been recorded as far west as the eastern Guerrero composite terrane (Elías-Herrera and Ortega-Gutiérrez, 1997), indicating that the Lower Paleozoic continent-ocean boundary is a west-dipping thrust that was subsequently displaced by a Permian dextral shear zone. The eclogite-facies tectonothermal event has been dated at 346 ± 3 Ma, followed by migmatization at ~350–330 Ma (Keppie et al., 2004a). Emerging geochemical data indicate that parts of the Piaxtla Group consist of craton-derived metasediments and continental rift tholeiites rather than oceanic lithosphere (Keppie et al., 2003c). Furthermore, new U-Pb geochronology and well-dated fossils in the Tecomate Formation indicate a Pennsylvanian–Middle Permian age, which therefore is a facies equivalent of both the Matzitzi and Patlanoaya formations (Malone et al., 2002; Keppie et al., 2004c). Deposition of these units appears to have been synchronous with both an Early–Middle

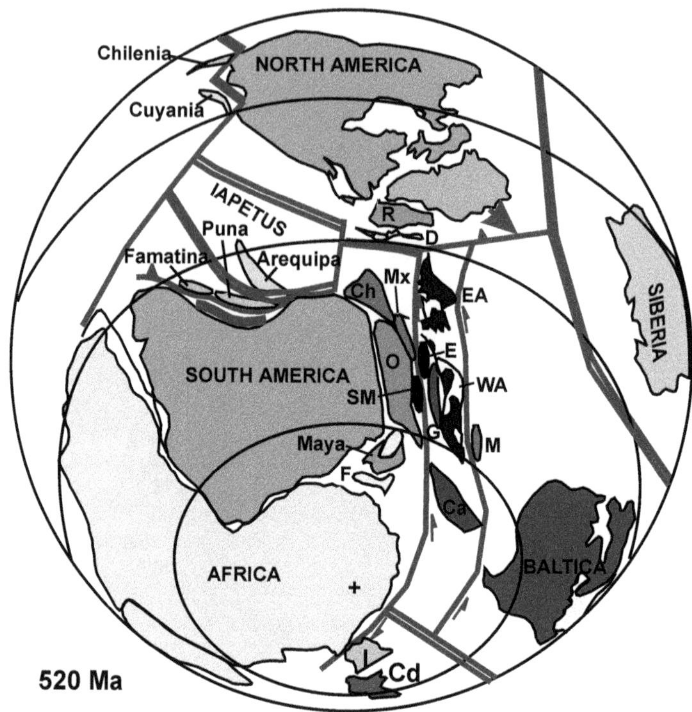

FIG. 10. 520 Ma reconstruction modified after Keppie and Ramos (1999) and Keppie et al. (2003b). Abbreviations and symbols are the same as in Figures 8–9, plus E = Exploits; G = Gander; Mx = Mixteca; SM = Sierra Madre.

Permian tectonothermal event and the arc magmatism along the margin of Pangea (Torres et al., 1999).

The presence of similar, shallow marine–continental Permo-Carboniferous rocks on top of both the Oaxacan and Acatlán complexes (Keppie et al., 2003c) implies that the Mixteca and Oaxaquia terranes were close, if not adjacent, at this time. This is consistent with the observation that the dextral transcurrent boundary between the Acatlán and Oaxacan complexes is overstepped by the Leonardian (Early Permian) Matzitzi Formation (Elías-Herrera and Ortega-Gutiérrez, 2002; Malone et al., 2002).

During Mesozoic time, the continent-ocean boundary on the western side of the Mixteca terrane may be placed at the eastern margin of the oceanic, Early Cretaceous Arperos terrane (Freydier et al., 2000), and on the eastern side of the Triassic Zacatecas Formation (interpreted as continental rise on oceanic crust by Centeno-García and Silva-Romo, 1997).

Paleozoic Sierra Madre terrane. This terrane formerly included both the ~1 Ga basement (Novillo Complex) and the Paleozoic Granjeno Formation, with an unconformity separating the units (Campa and Coney, 1983). However, all such contacts are tectonic and so the ~1 Ga basement is included in the Oaxaquia terrane (Ortega-Gutiérrez et al., 1995), and the Sierra Madre terrane is restricted to the Granjeno Formation and equivalents. Thrusting of the Sierra Madre terrane, which consists of an ophiolitic mélange (Granjeno Formation), over the ~1 Ga Novillo Complex in the Ordovician–Early Silurian is documented by the presence of Granjeno pebbles in Wenlockian sediments unconformably overlying the Novillo Complex (Fries et al., 1962). In the Permian, this thrust boundary was cut by a steeply dipping dextral shear zone (Garrison et al., 1980). The Granjeno Formation is inferred to represent a Paleozoic trench complex. Triassic rocks overstep the boundary between these two terranes.

(?)Paleozoic Tarahumara and Coahuila terranes. The Tarahumara terrane consists of low-grade metasedimentary rocks metamorphosed in the late Paleozoic (Sedlock et al., 1993, and references therein). The adjacent Coahuila terrane is similar except that it also contains Permo-Carboniferous flysch with synchronous calc-alkaline volcanic

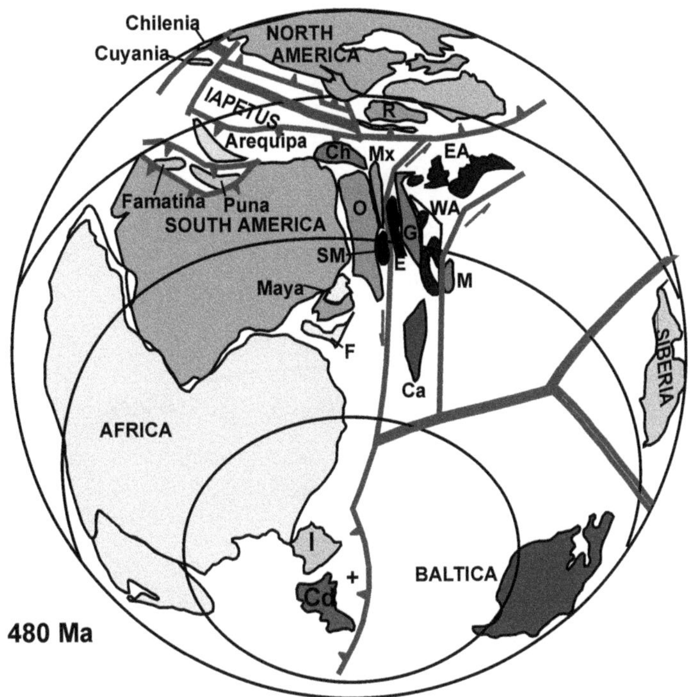

FIG. 11. 480 Ma reconstruction modified after Keppie and Ramos (1999) and Keppie et al. (2003b). Abbreviations and symbols are the same as in Figures 8–10.

detritus and Neoproterozoic, Mesoproterozoic, and Paleoproterozoic boulders deposited in a periarc basin (López et al., 2001). The paleogeographic location of these two terranes is uncertain—adjacent to either North or Middle America or in the intervening ocean basin.

Permo-Triassic Pangea

All of the above terranes were involved in the collision between North and Middle/South America during the assembly of Pangea. The collisional zone is represented by the Ouachita orogen, which terminates in northeastern Mexico against the paleo-Pacific Ocean (Fig. 1C). Proximity of Middle and North America by Missisippian times is indicated by the presence of Midcontinent bachiopod fauna in both regions (Navarro-Santillán et al., 2002).

The Pangea-Pacific boundary appears to have been an active margin at this time, as documented by the presence of a Permo-Triassic arc that may be traced from California through Mexico (McKee et al., 1999; Torres et al., 1999), and Permo-Triassic folding and N-vergent thrusting in the Caborca block (Stewart et al., 1990). A Late Triassic–Jurassic overstep sequence is present in many parts of Mexico and the adjacent United States, which is generally synchronous with the breakup of Pangea.

Mesozoic terranes of Pangean provenance

The fragmentation of Pangea spalled off two large cratonic terranes, Maya and Chortis, which appear to have separated sequentially, the Chortis terrane separating from southern Mexico before mid-late Jurassic times, followed by separation of the Maya terrane from the southern margin of North America in the Callovian–Late Jurassic (see below). Nuclear Mexico (Oaxaquia) remained attached to the North American craton. The Maya terrane is surrounded by rhombochasmic-shaped oceanic lithosphere—the Gulf of Mexico, the proto-Caribbean, and Juarez and Motagua terranes. Paleomagnetic data indicate that the Chortis terrane rotated ≥100° clockwise during the Mesozoic (Gose, 1985), and was also surrounded by oceanic lithosphere—Motagua, proto-Caribbean, and Pacific.

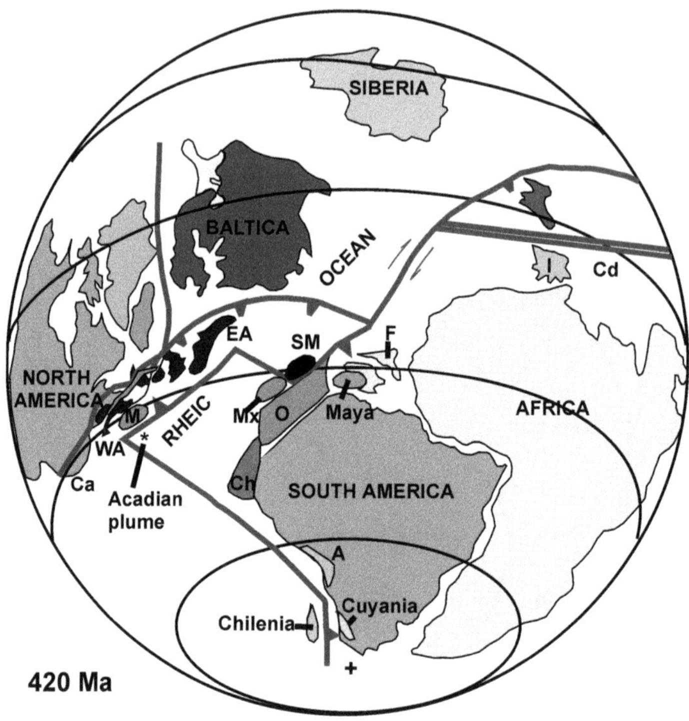

FIG. 12. 420 Ma reconstruction modified after Keppie and Ramos (1999) and Keppie et al. (2003b). Abbreviations and symbols are the same as in Figures 8–10.

Chortis terrane. Many authors have placed the Chortis terrane adjacent to southern Mexico and northwestern South America in pre-Tertiary reconstructions (e.g., Anderson and Schmidt, 1983; Pindell et al., 1988; Ross and Scotese, 1988; Schaaf et al., 1995; Keppie and Ramos, 1999). However, an alternative reconstruction using the Cayman transform results in back-rotation of the Chortis terrane into the Pacific Ocean, so that in the Eocene the southern Mexican coast would have faced an open ocean (Keppie and Moran-Zenteno, 2004). The latter is consistent with the proposals that part of the southern margin of Mexico has been removed by subduction erosion (Moran-Zenteno et al., 1996). On the other hand, there appear to be pre-Cenozoic connections between southern Mexico and the Chortis terrane. Thus, a Precambrian connection between the Chortis and Oaxaquia terranes is suggested by the presence of a ~1 Ga granitoid in northeastern Honduras (Manton, 1996) that is synchronous with similar plutons in Oaxaquia (Keppie et al., 2003a). However, in Honduras, the country rocks are amphibolite facies, whereas they are granulite facies in Mexico, possibly reflecting a metamorphic facies change. Such a change in P-T conditions is also suggested by the presence of lower-greenschist facies, Mesoproterozoic boulders in a Carboniferous conglomerate in the Coahuila terrane (López et al., 2001). A Permo-Carboniferous, tectonomagmatic event has been recorded in the basement rocks of northern Honduras (305 ± 12 Ma Rb-Sr isochron on orthogneisses: Horne et al., 1990), and may be correlated with a similar event in the Acatlán Complex of southern Mexico. Subsequent separation of the Chortis terrane is possibly recorded in southern Mexico if the Xolapa Complex represents a continental rise prism of mid-Jurassic age as suggested by Ortega-Gutiérrez and Elías-Herrera (2003). Such a separation is consistent with the apparent absence in the Chortis terrane of the high-grade, Early Cretaceous tectonothermal event in the Xolapa Complex of southern Mexico. The northern edge of the Chortis terrane is the steeply dipping Jocotan-Chamelecon fault zone, which forms the southern border of the Motagua ophiolitic terrane (Guinta et al., 2001). These ophiolitic

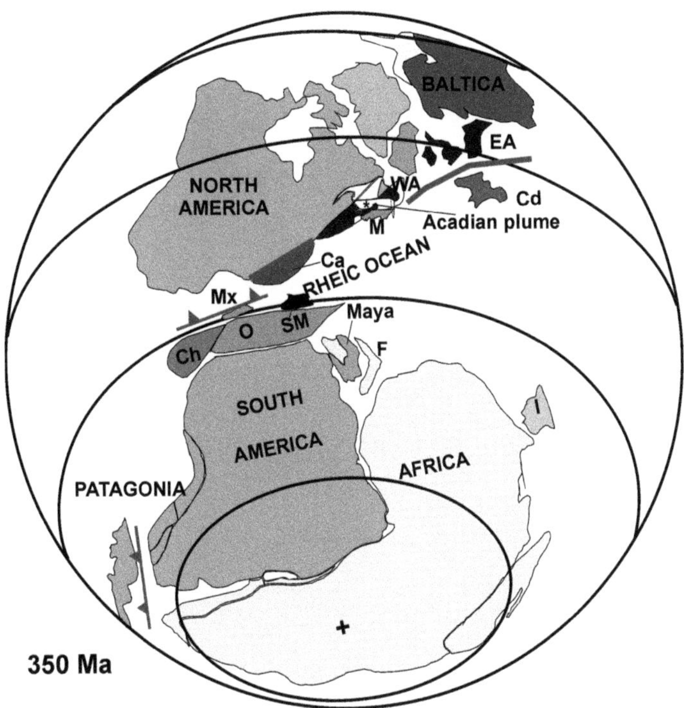

FIG. 13. 350 Ma reconstruction modified after Keppie and Ramos (1999) and Keppie et al. (2003b). Abbreviations and symbols are the same as in Figures 8–10. Asterisk indicates location of the Acadian plume.

massifs are not thrust over the Chortis terrane, an observation consistent with the absence of foredeep deposits.

Given the existence of the oceanic, Middle Jurassic–Late Cretaceous, Motagua terrane between the Chortis and Maya terranes, direct continuity of Mesozoic strata between the latter is improbable, a conclusion similar to that reached by Horne et al. (1990). This ocean was probably relatively narrow because paleomagnetic data indicate that the Chortis terrane lay at a similar paleolatitude relative to southern Mexico during the Mesozoic (Gose, 1985). Final closure of the Motagua ocean during the Late Cretaceous–Paleogene Laramide orogeny is recorded by the latest Cretaceous–Early Eocene flysch on the Maya terrane. The western edge of the Chortis terrane is covered by the Cenozoic arc, the southern margin is inferred to be a continuation of the Hess Escarpment, and the eastern part includes the Nicaragua Rise that is submerged beneath the Caribbean Sea.

Maya terrane. Paleomagnetic data indicate ~42° anticlockwise rotation of the Maya terrane about an Euler pole near the present northeast corner of the Yucatan Peninsula in the Callovian–Late Jurassic (Molina-Garza et al., 1992) to which Dickinson and Lawton (2001) add a further 18° anticlockwise rotation in Late Triassic–Middle Jurassic times.[2] This places the southern side of the Maya terrane (Guatemala-Chiapas massif) on the eastern side of the Coahuila terrane along the northern continuation of the Granjeno-Novillo fault boundary. Recent work in the Chiapas massif indicates a significant event at 260–250 Ma involving intrusion of the Chiapas batholith and high-grade deformation (Weber and Cameron, 2003) indicating that it was part of the magmatic event along the border of Middle-South America. This appears to be approximately synchronous with Permo-Carboniferous migmatization in the Chaucus Formation in Guatemala (Ortega-Gutiérrez et al., 2004).

[2]The Maya terrane is here regarded as one block rather than splitting it into two pieces as done by Dickinson and Lawton (2001).

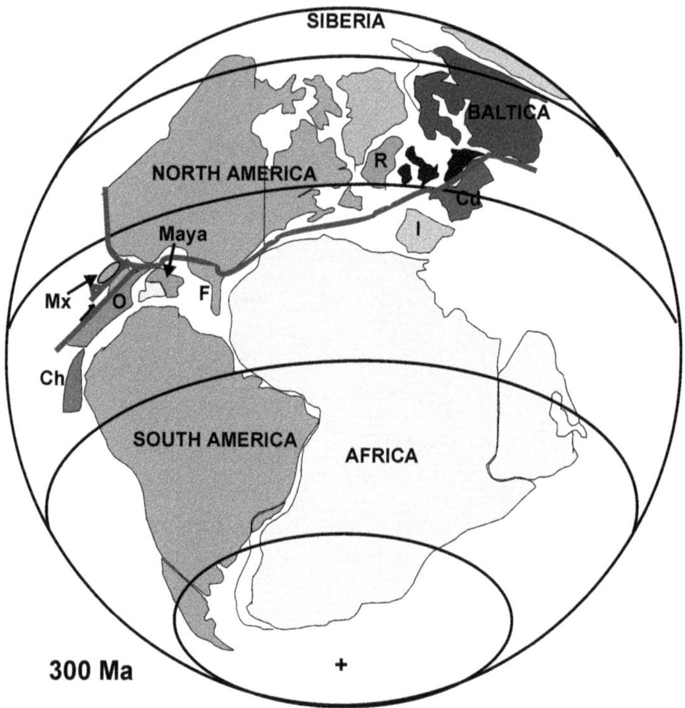

FIG. 14. 300 Ma reconstruction modified after Keppie and Ramos (1999) and Keppie et al. (2003b).

Schouten and Klitgord (1994) suggested that the Maya terrane acted as an independently rotating microplate during the separation of North and South America. Given its rectangular shape, rhombochasms opened along its sides, whereas the corners became compressive zones. This is consistent with the wedge shape of the Gulf of Mexico (Marton and Buffler, 1994), and with the triangular shape of both the eastern Maya margin (Dickinson and Lawton, 2001), and the Juarez terrane (Fig. 1). Such a reconstruction provides a proximal source in the southern Maya terrane for the Mesoproterozoic and Neoproterozoic boulders found in a Carboniferous conglomerate in the Coahuila terrane (López et al., 2001). This reconstruction also suggests that prior to the Mesozoic, the Maya terrane was continuous with Oaxaquia.

Juarez terrane. The Juarez terrane is a southward-widening terrane consisting of poorly dated, (Jurassic-) Cretaceous ophiolitic rocks (gabbro, serpentinite, mafic lava, tuff, greywacke, quartzite, slate, limestone, conglomerate containing pebbles of granulite and phyllite) metamorphosed under greenschist-facies metamorphic conditions at or before ~82 Ma (K-Ar on phyllite: Carfantan, 1983). Its western boundary is a major mylonite zone that has a complex history involving: (1) E-vergent thrusting; (2) Jurassic dextral shear; and (3) Cenozoic listric normal faulting (Alaniz-Alvarez et al., 1996). The eastern boundary is an E-vergent thrust of Late Cretaceous–Early Paleogene age, a younger limit being provided by the presence of undeformed Oligocene–Miocene rocks that form an overstep sequence (references in Sedlock et al., 1993). The triangular shape of the terrane suggests that it opened southwards, whereas it pinches out northwards into a transcurrent shear zone.

Motagua terrane. The Motagua terrane consists of Jurassic–Cretaceous ophiolites of arc and MORB affinities (Guinta et al., 2001), and both high- and low-grade, metasedimentary and meta-igneous rocks that include eclogites. It is inferred that this assemblage represents oceanic lithosphere, oceanic arc, associated sediments, and subducted oceanic lithosphere. The southern edge of the Motagua terrane is the steeply dipping Jocotan-Chamelecon fault zone. Eclogite-facies metamorphism in ophiolitic bodies along this southern margin of the

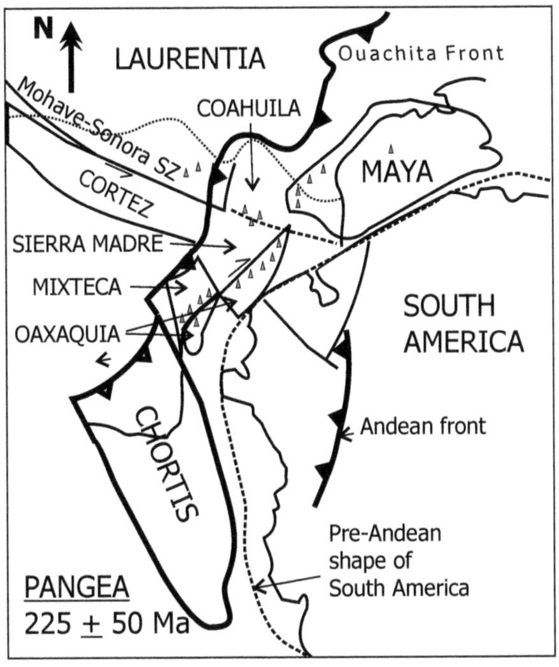

FIG. 15. 225 ± 50 Ma reconstruction of Pangea modified after Dickinson and Lawton (2001) showing magmatic arc (triangles) and Ouachita and Andean thrust fronts.

Motagua terrane formed at 161 ± 20 Ma (Nd model age: Sisson et al., 2003) cooling through ~350°C at 125–113 Ma ($^{40}Ar/^{39}Ar$ age on phengite: Harlow et al., 2004). On the other hand, north of the Motagua fault, phengites yielded $^{40}Ar/^{39}Ar$ ages of 77–65 Ma (Harlow et al., 2004), which are similar to the ~72–48 Ma, K-Ar cooling ages on hornblende and mica recorded by Ortega-Gutiérrez et al. (2004) along the northern margin of the Motagua fault zone. These data imply: (1) separation of the Maya and Chortis terranes with the formation of oceanic lithosphere in the Motagua terrane prior to the mid-Late Jurassic; (2) subduction of the southern margin of the Motagua terrane in the mid-Late Jurassic followed by obduction and exhumation during the Early Cretaceous; and (3) obduction of the Sierra Santa Marta ophiolitic massif onto the southern margin of the Maya terrane during the latest Cretaceous–Paleocene as recorded by the ophiolitic detritus in the Sepur Formation (interpreted as a foredeep deposit by Guinta et al., 2001). Note that the Jurassic subduction is synchronous with the opening of the Gulf of Mexico, suggesting that the two events may be connected. Cenozoic dextral movements on the Motagua fault zone totals ~170 km (Donnelly et al., 1990 and references therein). Tertiary arc volcanic rocks represent an overstep sequence.

Mesozoic terranes of Pacific provenance

Guerrero composite terrane. Much has been written about the Guerrero composite terrane (Sedlock et al., 1993; Centeno-García et al., 2000, 2003, and references therein), and will not be repeated here. Suffice it to say that south of the Trans-Mexican Volcanic Belt it has been subdivided into five subterranes (elevated to terranes herein). From west to east, these are the: (1) Las Ollas (?) Lower Cretaceous blueschist mélange; (2) Zhihuatanejo mid-Triassic to mid-Jurassic oceanic lithosphere (Arteaga Complex) overlain unconformably by a mid-Jurassic to mid-Cretaceous arc-periarc sequence; (3) Arcelia late Lower Cretaceous oceanic periarc assemblage; (4) Teloloapan Lower Cretaceous oceanic arc overlain by Albian-Cenomanian flysch (Talavera-Mendoza and Suastegui, 2000); and (5) Arperos Lower Cretaceous oceanic lithosphere (Figs. 1C and 3). Mid-Cretaceous thrusting of the Zihuatenajo and Arcelia terranes over the Teloloapan terrane is recorded by the Albian–Cenomanian flysch, suggesting intra-oceanic amalgamation that predates by

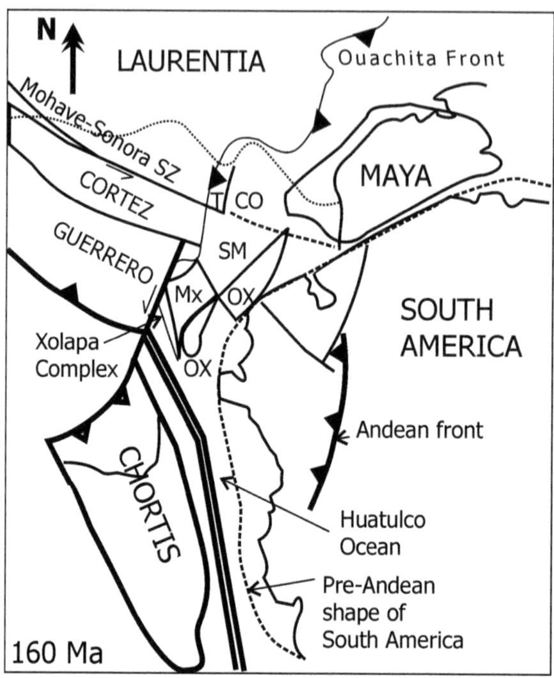

FIG. 16. 160 Ma reconstruction showing the separation of the Chortis terrane. Abbreviations: CO = Coahuila; Mx = Mixteca; OX = Oaxaquia; SM = Sierra Madre; T = Tarahumara.

≥20 m.y. accretion of the Guerrero composite terrane to North America recorded by the late Upper Cretaceous Mexcala flysch.

These terranes have been traced through isolated inliers north of the Trans-Mexican Volcanic Belt; however, their connection with those recorded in Baja California is not clear. Prior to opening of the Gulf of California (Dickinson and Butler, 1998; Keppie and Dostal, 2001), it appears that the mid-Cretaceous blueschist mélanges of Las Ollas and the Western Baja terrane line up. Similarly, the eastern boundary of the Guerrero composite terrane may be traced directly across the future Gulf of California; however, several differences are apparent in passing from Baja California to the mainland (compare Figs. 3 and 4): (1) the age of the ocean floor is ~25 m.y. older in the north; (2) the eastern Guerrero arc began ~20 m.y. earlier in the north; (3) the arc straddles the ocean-continent boundary in the north, passing southward into a predominantly oceanic arc; and (5) beneath the western Guerrero arc, the oceanic lithosphere is ~30 m.y. younger in the north. The first three observations may be explained in terms of either a southward-migrating triple point (Dickinson and Lawton, 2001), or the existence of two subparallel arcs (Moores, 1998). It is significant that the western arc terranes (Zihuatenajo and Choyal) contain continent-derived detritus (Boles and Landis, 1984; Centeno-García et al., 1993). In the case of the Choyal terrane, Boles and Landis (1984) suggested that the continental detritus was derived from North America, a conclusion consistent with deposition of mid-Jurassic arc volcanic rocks upon, and intrusion of mid-Jurassic parts of the Peninsular Ranges batholith into, eugeoclinal rocks of the Cortez terrane bordering the Caborca block (Sedlock et al., 1993). Correlation of the continentally derived, Upper Triassic–Lower Jurassic sediments of the Arteaga Complex at the base of the Zihuatenajo terrane with the Zacatecas sandstone along the periphery of North-Middle America supports the idea that the Arteaga Complex was rifted off western North America (Centeno-García and Silva-Romo, 1997).

Accretion of the Guerrero composite terrane onto western North America is recorded by the uppermost Cretaceous–Lower Cenozoic Mexcala flysch in the foreland basin in front of the advancing Laramide nappes, which eventually advanced within 100 km of the present coast of the Gulf of Mexico. The

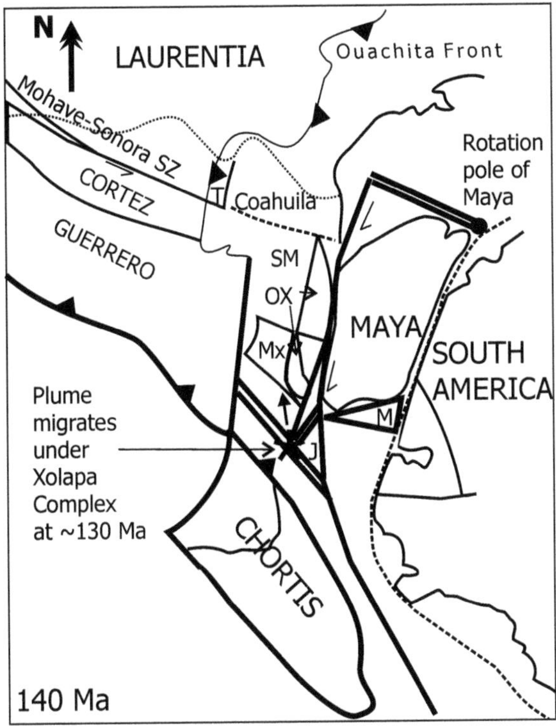

FIG. 17. 140 Ma reconstruction showing rotation of the Maya terrane. Abbreviations are the same as in Figure 16, plus J = Juarez; Mo = Motagua.

eastward migration of the Laramide deformation front is synchronous with eastward migration of the magmatic arc, and both have been related to flattening of the subducting slab in response to increased convergence rates and/or subduction of progressively younger oceanic lithosphere (Clark et al., 1982; Coney, 1983). Subsequently, these Laramide structures were overstepped by Cenozoic rocks (Figs. 3 and 4) that define the present extent of North America.

Baja California terrane. Impingement of the East Pacific Rise with the trench off western North America produced a T-R-F triple point that migrated southwards (Lonsdale, 1989; Atwater and Stock, 1998). At ~13 Ma the triple point reached the mouth of the Gulf of California, following which the East Pacific Rise extended inland leading to the separation of Baja California at ~6 Ma, which now rides northwestward on the Pacific plate. Thus Baja California has become a terrane.

Plate Tectonic Reconstructions

Advances in terrane mapping in Mexico have generally gone hand-in-hand with development of plate tectonic models. Thus, Coney (1983) modeled the Mesozoic and Cenozoic using the terrane subdivision of Campa and Coney (1983). Sedlock et al. (1993) combined their terrane map of Mexico with a series of paleogeographic maps from 600 Ma to the present. Dickinson and Lawton (2001) subdivided Mexico into a series of Permian-Cretaceous blocks (terranes) as the basis for palinspastic, plate tectonic maps. Based on the terrane map of Mexico presented in this paper, a series of paleogeographic maps for the Mesoproterozoic–Present tectonic evolution of Mexico is constructed (Figs. 5–19). These draw heavily upon previous maps modified, where needed, by data presented in the present terrane analysis. This is a two-stage process involving determining the provenance of a terrane followed by

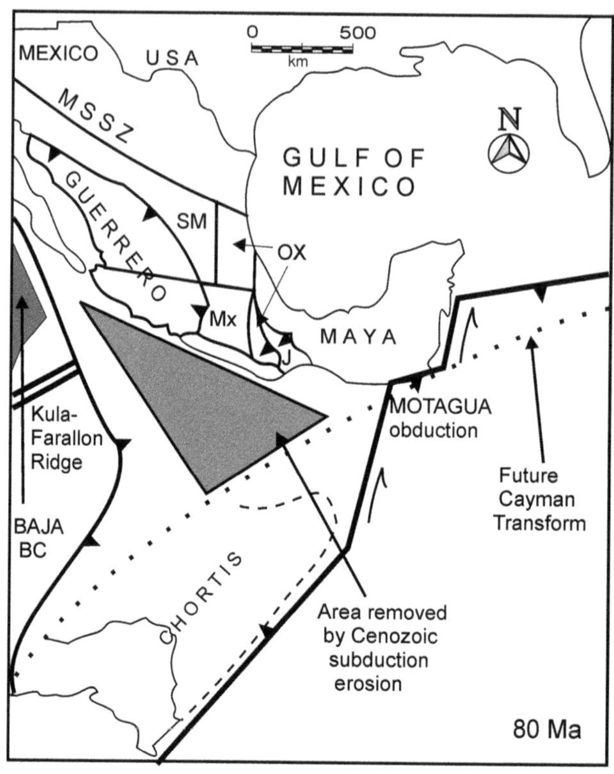

FIG. 18. 80 Ma reconstruction showing obduction of the Guerrero composite terrane during the Laramide Orogeny. Abbreviations are the same as in Figure 16.

devising a plate tectonic model for its history. The reader is also referred to more detailed palinspastic models for: (1) the Mesozoic evolution of the Guerrero composite terrane presented by Dickinson and Lawton (2001); and (2) the Cenozoic locations of the Chortis terrane (Keppie and Moran-Zenteno, 2004).

Precambrian–Paleozoic

The Gondwanan provenance of the Middle America terrane was initially based upon its Ordovician and Silurian fauna (Robison and Pantoja-Alor, 1968; Boucot et al., 1997); however, its location on this margin has varied from Venezuela to Bolivia. Cocks and Fortey (1988) identified two Ordovician facies zones surrounding Gondwana. However, a gap opposite Venezuela and Colombia is neatly filled by the Middle America terrane (combined Maya-Oaxaquia-Chortis) represented by the shelf facies of the Tremadocian Tiñu Formation (lying unconformably upon the Oaxacan Complex; c.f. Cocks and Torsvik, 2002, Fortey and Cocks, 2003). This location is consistent with the correlation between the Silurian rocks at Ciudad Victoria and Venezuela (Boucot et al., 1997).

The presence of ~1 Ga basement throughout Middle America (Maya-Oaxaquia-Chortis) is also consistent with an origin off Venezuela-Colombia as part of the circum-Amazonian ~1 Ga orogens, which includes the Sunsas orogen, the Andean massifs, and the Tocantins Province (Figs. 5–7; Restrepo-Pace et al., 1997; Ramos and Aleman, 2000; Pimentel et al., 2000; Keppie et al., 2001, 2003a). In this location, Middle America experienced several stages.

1. At ~1.3–1.2 Ga, development arc magmatism in a primitive island-arc system (Fig. 5; Lawlor et al., 1999; Keppie et al., 2001; Dostal et al., 2004).

2. At ~1.16–1.13 Ga, intrusion of rift-related plutons, followed at ~1.1 Ga by migmatization, and at ~1012 Ma by renewed AMCG, rift-related plutons, all possibly occurring in a backarc setting as the trench migrated seaward leading to the birth of the Avalonian primitive island arc outboard of Middle America (Figs. 5–7; Keppie et al., 2003a).

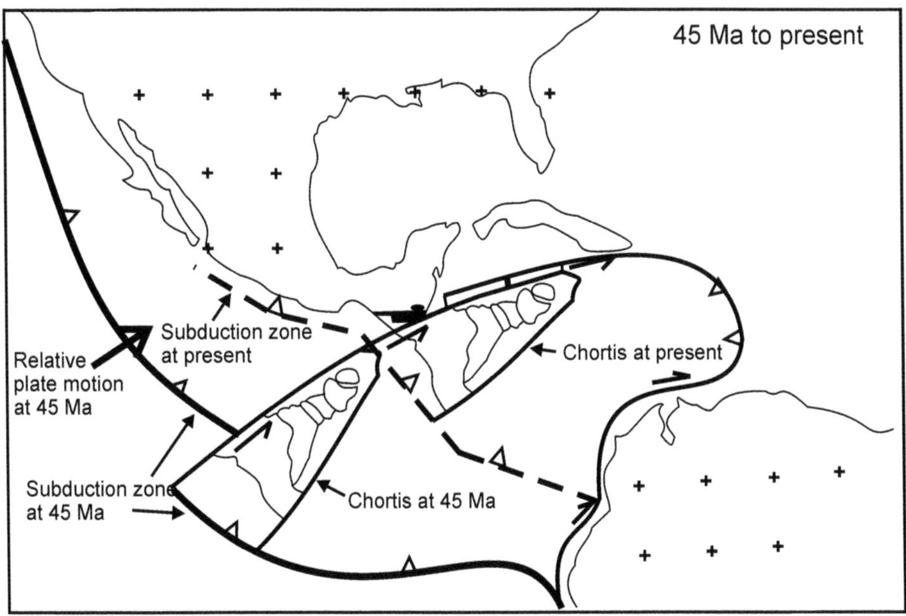

FIG. 19. 45 Ma to Present reconstruction showing rotation of the Chortis terrane to its present position contemporaneous with NE migration of the subduction zone, and tectonic erosion of the southern coast of Mexico (modified after Keppie and Moran-Zenteno, 2004).

3. At ~1,005–980 Ma, polyphase deformation at granulite facies, followed by rapid exhumation between 980 and 945 Ma, possibly related to flat-slab subduction as a result of collision of a ridge, a plume, or an oceanic plateau with the trench (Fig. 7; Ortega-Obregón et al., 2003; Solari et al., 2003; Keppie et al., 2004c).

4. At ~917 Ma, intrusion of an arc-related pluton accompanied by extensive hydration, possibly due to steepening of the subduction zone (Fig. 7; Keppie et al., 2001; Ortega-Obregón et al., 2003). The main ~1 Ga juvenile arc may be represented by the Avalonian basement, which is inferred to have lain outboard of Middle America, and appears to pass laterally into similar juvenile basement in the 900–850 Ma, Goiás magmatic arc (Arenópolis and Mara Rosa arcs: Pimentel et al., 2000) on the eastern side of the Amazon craton.

5. Between 700 and 600–550 Ma, subduction beneath Avalonia and the Yucatan produced voluminous arc magmatism (Keppie et al., 2003b) synchronous with a tectonothermal event that has also been recorded in the southern Oaxacan Complex (Krogh et al., 1993a, 1993b; Schulze et al., 2004): diachronous switching of arc to rift magmatism in Avalonia has been related to ridge-trench collision (Figs. 8–9; Nance et al., 2002; Keppie et al., 2003b).

6. Separation of Avalonia from Middle America is inferred to have been a two-stage process: in the latest Precambrian–earliest Cambrian rifting is inferred to have taken place by a mechanism analogous with separation of Baja California or Baja British Columbia (i.e., extension of the ridge into the continental margin), and this was followed in latest Cambrian–earliest Ordovician by a rift-drift transition (Figs. 10–11; Keppie et al., 2003b). As a consequence, continental rise prisms developed on the margins of Avalonia and Middle America—the Gander and Mixteca terranes, respectively. The intervening Rheic Ocean is represented by the Exploits oceanic terrane. The Tarahumara and Coahuila terranes also represent parts of the intervening area between North and Middle America.

7. During the Late Ordovician, Avalonia and the Gander terrane were accreted to North America (Fig. 12; Keppie et al., 2003b). At the same time, the Sierra Madre terrane was obducted onto Oaxaquia.

8. During the Mississippian, the Mixteca terrane was subducted, followed by Early–Middle Permian,

tectonic imbrication (Figs. 12–14). This appears to have been synchronous with amalgamation of Pangea and subduction along the western margin of Pangea (Fig. 15).

Throughout the Paleozoic, the Middle America terrane is inferred to have traveled passively with South America until it collided with the southern margin of Laurentia during the Carboniferous and Permian to form Pangea (Keppie and Ramos, 1999).

Mesozoic–Cenozoic

The breakup of Pangea led to the birth of several terranes that were subsequently accreted to North America:

1. Prior to the mid-Jurassic, the Chortis terrane separated from southern Mexico as recorded by the southward-thickening passive margin sequence (Tecocoyunca Group–Xolapa Complex; Ortega-Gutiérrez and Elías-Herrera, 2003) passing into the Huatulco Ocean (Figs. 15–16).

2. Callovian–Late Jurassic separation of the Maya terrane led to opening of the Gulf of Mexico, and the Juarez and Motagua terranes, all floored by oceanic lithosphere: it is inferred that a plume may have formed at the R-R-R triple point at the junction of the Juarez and Huatulco oceans (Fig. 17). This was synchronous with backarc rifting that separated the Zihuatanejo terrane (Guerrero composite terrane) from Pangea (Centeno-García and Silva-Romo, 1997)

3. Late Jurassic–Early Cretaceous development of the Guerrero composite arc terrane (Centeno-García et al., 1993, 2000, 2003) and subduction along the margin of the Chortis block. At ~130 Ma, the southern margin of Mexico overrode a plume, producing high-temperature/low-pressure metamorphism in the Xolapa Complex (Fig. 17).

4. In the mid-Cretaceous (~90 Ma), Baja British Columbia collided with the Guerrero composite terrane (Keppie and Dostal., 2001), which initiated amalgamation of the Guerrero composite terrane (Fig. 18). This was followed by birth of the Kula plate, which led to northward transport of Baja British Columbia.

5. Late Cretaceous–Paleocene flattening of the subducting slab, possibly the result of Kula ridge/trench collision, and/or increased convergence rates, led to accretion of the Guerrero composite terrane to North America during the Laramide orogeny and obduction of the Juarez and Motagua terranes onto the Maya terrane (Fig. 18).

6. Sinistral Cenozoic rotation of the Chortis terrane ~1100 km along the Cayman tranform fault occurred concurrently with subduction erosion of a triangular area south of Mexico and flattening of the subduction zone, possibly due to collision of the Tehuantepec Ridge with the subduction zone, anticlockwise rotation of the volcanic arc to its present position along the Trans-Mexican Volcanic Belt, and development of the Chiapas foldbelt (Fig. 19; Keppie and Moran-Zenteno, 2004).

7. In the Late Neogene, propagation of the East Pacific Rise into the continental margin led to the transfer of Baja California from the North American plate to the Pacific plate.

Acknowledgments

I thank Dr. Zoltan de Cserna for reviewing the manuscript, and Dr. Fernando Ortega-Gutiérrez for extensive discussion of terranes in Mexico. Funding for this project was provided by PAPIIT project # IN103003. This paper represents a contribution to IGCP projects 453 and 497.

REFERENCES

Alaniz-Alvarez, S. A., van der Heyden, P., Nieto-Samaniego, A. F., and Ortega-Gutiérrez, F., 1996, Radiometric and kinematic evidence for Middle Jurassic strike-slip faulting in southern Mexico related to the opening of the Gulf of Mexico: Geology, v. 24, p. 443–446.

Anderson, T. H., McKee, J. W., and Jones, N. W., 1991, A northwest trending, Jurassic fold nappe, northernmost Zacatecas, Mexico: Tectonics, v. 10, p. 383–401.

Anderson, T. H., and Schmidt, V. A., 1983, The evolution of Middle America and the Gulf of Mexico-Caribbean Sea region during Mesozoic time: Geological Society of America Bulletin, v. 94, p. 941–966.

Atwater, T., and Stock, J., 1998, Pacific-North America plate tectonics of the Neogene southwestern United States—an update: International Geology Review, v. 40, p. 375–402.

Blum, J. D., Chamberlain, C. P., Hingston, M. P., Koeberl, C., Marin, L. E., Schuraytz, B. C., and Sharpton, V. L., 1993, Isotopic comparison of K/T boundary impact glass with melt rocks from the Chicxulub and Manson impact structures: Nature, v. 364, p. 325–327.

Boles, J. R., and Landis, C. A., 1984, Jurassic sedimentary mélange and associated facies, Baja California, Mexico: Geological Society of America Bulletin, v. 95, p. 513–521.

Boucot, A. J., Blodgett, R. B., and Stewart, J. H., 1997, European Province Late Silurian bachiopods from the

Ciudad Victoria area, Tamaulipas, northeastern Mexico, *in* Klapper, G., Murphy, M. A., and Talent, J. A. eds., Paloeozoic sequence stratigraphy, biostratigraphy, and biogeography: Studies in honour of J. Granville ("Jess") Johnson: Geological Society of America Special Paper 321, p. 273–293.

Cameron, K. L., López, R., Ortega-Gutiérrez, F., Solari, L., Keppie, J. D., and Schulze, C., 2004, U-Pb geochronology and common Pb isotope compositions of the Grenvillian rocks of eastern and southern Mexico, *in* Tollo, R. P., Corriveau, L., McLelland, J. B., and Bartholemew, G., eds., Proterozoic tectonic evolution of the Grenville Orogen in North America: Geological Society of America Memoir, in press.

Campa, M. F., and Coney, P. J., 1983, Tectono-stratigraphic terranes and mineral resource distributions in Mexico: Canadian Journal of Earth Science, v. 20, p. 1040–1051.

Campa-Uranga, M. F., Gehrels, G., and Torres de Leon, R., 2002, Nuevas edades de granitoides metamorfizados del Complejo Acatlán en el Estado de Guerrero: Actas Instituto Nacional de Geoquímica, v. 8, No. 1, p. 248.

Carfantan, J.-C., 1983, Les ensembles géologiques du Mexique meridional: Evolution géodynamique durante le Mésozoïque et le Cénozoique: Geofisica Internacional, v. 22, p. 9–37.

Carrillo-Bravo, J., 1991, Geología del anticlinorio de Huizachal-Peregrina al N-W de Ciudad Victoria, Tamaulipas: Asociación Mexicana de Geólogos Petroleros Boletín, v. 13, p. 1–98.

Castillo-Rodríguez, H., 1988, Zur des kristallinen Grundgebirges der Sierra Madre Oriental-insbesondere des Granjero-Schiefer-Komplexes-im Sudteil des Huizchal-Peregrina-Antiklinoriums (Raum Ciudad Victoria, Bundestaat Tamaulipas, Mexiko): Unpubl. Ph.D. thesis, Unversität Münster, Germany, 138 p.

Centeno-García, E., Corona-Chávez, P., Talavera-Mendoza, O., and Iriondo, A., 2003, Geology and evolution of the western Guerrero terrane—a transect from Puerto Vallarta to Zihuatanejo, Mexico: Guidebook for field trips of the 99th Annual Meeting of the Cordilleran Section of the Geological Society of America, Publ. Esp. 1, Instituto de Geología, Universidad Nacional Autónoma de México, p. 201–228.

Centeno-García, E., Lomnitz, C., and Ramírez-Espinosa, J., eds., 2000, Geologic evolution of the Guerrero Terrane, western Mexico: Journal of South American Earth Science (special issue), v. 13, p. 293–476.

Centeno-García, E., Ruiz, K., Coney, P. J., Patchett, P. J., and Ortega-Gutiérrez, F., 1993, Guerrero terrane of Mexico: Its role in the Southern Cordillera from new geochemical data: Geology, v. 21, p. 419–422.

Centeno-García, E., and Silva-Romo, G., 1997, Petrogénesis and tectonic evolution of central Mexico during Triassic-Jurassic time: Revista Mexicana de Ciencias Geologicos, v. 14, p. 244–260.

Clark, K. F., Foster, C. T., and Damon, P. E., 1982, Cenozoic mineral deposits and subducted-related magmatic arcs in Mexico: Geological Society of America Bulletin, v. 93, p. 533–544.

Cocks, L. R. M., and Fortey, R. A., 1988, Lower Paleozoic facies and faunas around Gondwana, *in* Audley-Charles, M. G., and Hallam, A., eds., Gondwana and Tethys: Geological Society of London Special Publication no. 37, p. 183–200.

Cocks, L. R. M., and Torsvik, T. H., 2002, Earth geography from 500 to 400 million years ago: a faunal and palaeomagnetic review: Journal of the Geological Society of London, v. 159, p. 631–644.

Coney, P. J., 1983, Un modelo tectónico de Mexico y sus relaciones con América del Norte, América del Sur y el Caribe: Revista del Instituto Mexicano del Petroleo, v. 15, p. 6–15.

Coney, P. J., and Campa, M. F., 1984, Lithotectonic terrane map of Mexico (west of the 91st meridian): U.S. Geological Survey Miscellaneous Field Studies Map MF-1874-D, scale 1:2,500,000.

de Cserna, Z., 1989, An outline of the geology of Mexico, *in* Bally, A. W., and Palmer, A. R., eds., The geology of North America—an overview: The Geology of North America, v. A: Boulder, CO, Geological Society of America, p. 233–264.

de Cserna, Z., Graf, J. L., Jr., and Ortega-Gutiérrez, F., 1977, Alóctono paleozoico inferior en al región de Ciudad Victoira, Estado de Tamaulipas: Revista Instituto de Geología, Universidad Nacional Autónoma de México, v. 1, p. 33–43.

de Cserna, Z., and Ortega-Gutiérrez, F., 1978, Reinterpretation of isotopic age data from the Granjeno Schist, Ciudad Victoria, Tamaulipas; y reinterpretación tectónica del Esquisto Granjeno, Tamaulipas; Contestación: Revista Instituto de Geología, Universidad Nacional Autónoma de México, v. 2, p. 212–215.

Dickinson, W. R., and Butler, R. F., 1998, Coastal and Baja California paleomagnetism reconsidered: Geological Society of America Bulletin, v. 110, p. 1268–1280.

Dickinson, W. R., and Lawton, T. F., 2001, Carboniferous to Cretaceous assembly and fragmentation of Mexico: Geological Society of America Bulletin, v. 113, p. 1142–1160.

Donnelly, T. W., Horne, G. S., Finch, R. C., and López-Ramos, E., 1990, Northern Central America: The Maya and Chortis blocks, *in* Dengo, G., and Case, J. E., eds., Decade of North American geology, v. H, the Caribbean region: Boulder, CO, Geological Society of America, p. 37–76.

Dostal, J., Keppie, J. D., Macdonald, H., and Ortega-Gutiérrez, F., 2004, Sedimentary origin of calcareous intrusions in the ~ 1 Ga Oaxacan Complex, Southern Mexico: Tectonic implications: International Geology Review, v. 46, p. 528–541.

Elías-Herrera, M., and Ortega-Gutiérrez, F., 1997, Petrology of high-grade xenoliths in an Oligocene rhyodacite plug—Precambrian crust beneath the southern Guerrero Terrane, Mexico?: Revista Mexicana de Ciencias Geologicas, v. 14, p. 101–109.

______, 2002, Caltepec fault zone: An Early Permian dextral transpressional boundary between the Proterozoic Oaxacan and Paleozoic Acatlán complexes, southern Mexico, and regional implications: Tectonics, v. 21, No. 3 [10.1029/200TC001278].

Elías-Herrera, M., Sánchez-Zavala, J. L., and Macías-Romo, C., 2000, Geologic and geochronologic data from the Guerrero terrane in the Tejupilco area, southern Mexico: New constraints on its tectonic interpretation: Journal of South American Earth Science, v. 13, p. 355–376.

Fortey, R. A., and Cocks, L. R. M., 2003, Palaeontological evidence bearing on global Ordovician–Silurian continental reconstructions: Earth Science Review, v. 61, p. 245–307.

Freydier, C., LaPierre, H., Ruiz, J., Tardy, M., Martínez-R. J., and Coulon, C., 2000, The Early Cretaceous Arperos basin: An oceanic domain dividing the Guerrero arc from nuclear Mexico evidenced by the geochemistry of the lavas and sediments: Journal of South American Earth Science, v. 13, p. 325–336.

Fries, C., Jr., Schmitter, E., Damon, P. E., Livingston, D. E., and Erickson, R., 1962, Edad de las rocas metamórficas en los Cañones de La Peregrina y de Caballeros, parte centro-occidental de Tamaulipas: Universidad Nacional Autónoma de México, Instituto de Geología, Bolletin, v, 64, p. 55–69.

Garrison, J. R., Jr., Ramírez-Ramírez, C., and Long, L. E., 1980, Rb-Sr isotopic study of the ages and provenance of Precambrian granulite and Paleozoic greenschist near Ciudad Victoira, Mexico, *in* Pilger, R. H., Jr., ed., The origin of the Gulf of Mexico and the early opening of the central North Atlantic Ocean: Baton Rouge, LA, Louisiana State University, p. 37–49.

Gose, W. A., 1985, Paleomagnetic results from Honduras and their bearing on Caribbean tectonics: Tectonics, v. 4, p. 565–585.

Grahales-Nishimura, J. M., Centeno-García, E., Keppie, J. D., and Dostal, J., 1999, Geochemistry of Lower Permian or older basalts from the Juchatengo complex of southern Mexico: Tectonic implications: Journal of South American Earth Science, v. 12, p. 537–544.

Guinta, G., Beccaluva, L., Coltorti, M., Sienna, F., Mortellaro, D., and Cutrupia, D., 2001, The peri-Caribbean ophiolites: Structure, tectono-magmatic significance and geodynamic implications: Caribbean Journal of Earth Science, v. 34, p. 12–27.

Harlow, G. E., Hemming, S. R., Avé Lallemant, H. G., Sisson, V. B., and Sorensen, S. A., 2004, Two high-pressure/low-temperature serpentine-matrix mélange belts, Motagua fault zone, Guatemala: A record of Aptian and Maastrichtian collisions: Geology, v. 32, p. 17–20.

Herrmann, U. R., Nelson, B. K., and Ratschbacher, L., 1994, The origin of a terrane: U/Pb geochronology and tectonic evolution of the Xolapa complex (southern Mexico): Tectonics, v. 13, p. 455–474.

Hill, R. I., Silver, L. T., and Taylor, H. P., Jr., 1986, Coupled Sr-O isotope variations as an indicator of source heterogeneity for the northern Peninsular Ranges batholith: Contributions to Mineralogy and Petrology, v. 92, p. 351–361.

Horne, G. S., Finch, R. C., and Donnelly, T. W., 1990, The Chortis block, *in* Dengo, G., and Case, J. E., eds., The Caribbean region: Boulder, CO, Geological Society of America, Geology of North America, v. H, p. 55–76.

Howell, D. G., ed., 1985, Tectonostratigraphic terranes of the Circum-Pacific region: Circum-Pacific Council for Energy and Mineral Resources, Earth Science Series, no. 1.

Howell, D. G., Jones, D. L., and Schermer, E. R., 1985, Tectonostratigraphic terranes of the Circum-Pacific region, *in* Howell, D.G., ed., Tectonostratigraphic terranes of the Circum-Pacific region: Circum-Pacific Council for Energy and Mineral Resources, Earth Science Series, no. 1, p. 3–30.

Jones, D. L., and Silberling, N. J., 1979, Mesozoic stratigraphy—the key to tectonic analysis of southern and central Alaska: U.S. Geological Survey Open File Report 81-792, 20 p.

Kamo, S. L., and Krogh, T. E., 1995, Chicxulub crater source for shocked zircon crystals from the Cretaceous–Tertiary boundary layer, Saskatchewan: Evidence from new U-Pb data: Geology, v. 23, p. 281–284.

Keppie, J. D., 1989, Northern Appalachian terranes and their accretionary history: Geological Society of America, Special Paper 230, p. 159–192.

Keppie, J. D., and Dostal, J., 2001, Evaluation of the Baja controversy using paleomagnetic and faunal data, plume magmatism and piercing points: Tectonophysics, v. 339, nos. 3-4, p. 427–442.

Keppie, J. D., Dostal, J., Cameron, K. L., Solari, L. A., Ortega-Gutiérrez, F., and López, R., 2003a, Geochronology and geochemistry of Grenvillian igneous suites in the northern Oaxacan Complex, southern Mexico: Tectonic implications: Precambrian Research, v. 120, p. 365–389.

Keppie, J. D., Dostal, J., Ortega-Gutiérrez, F., and López, R., 2001, A Grenvillian arc on the margin of Amazonia: evidence from the southern Oaxacan Complex, southern Mexico: Precambrian Research, v. 112, nos. 3–4, p. 165–181.

Keppie, J. D., Miller, B. V., Nance, R. D., Murphy, J. B., and Dostal, J., 2004a, New U-Pb zircon dates from the Acatlán Complex, Mexico: Implications for the ages of tectonostratigraphic units and orogenic events [abs.]: Geological Society of America Abstracts with Program.

Keppie, J. D., and Moran-Zenteno, D., 2004, Tectonic implications of alternative Cenozoic reconstructions for southern Mexico and the Chortis block: International Geology Review, v. 46, in press.

Keppie, J. D., Nance, R. D., Murphy, J. B., and Dostal, J., 2003b, Tethyan, Mediterranean, and Pacific analogues for the Neoproterozoic–Paleozoic birth and development of peri-Gondwanan terranes and their transfer to Laurentia and Laurussia: Tectonophysics, v. 365, nos. 1–4, p. 195–219.

Keppie, J. D., Nance, R. D., Powell, J. T., Mumma, S. A., Dostal, J., Fox, D., Muise, J., Ortega-Rivera, A., Miller, B. V., and Lee, J. W. K., 2004b, Mid-Jurassic tectonothermal event superposed on a Paleozoic geological record in the Acatlán Complex of southern Mexico: Hotspot activity during the breakup of Pangea: Gondwana Research, v. 7, no.1, p. 239–260.

Keppie, J. D., and Ortega-Gutiérrez, F., 1995, Provenance of Mexican terranes: Isotopic constraints: International Geology Review, v. 37, p. 813–824.

Keppie, J. D., and Ramos, V. S., 1999, Odyssey of terranes in the Iapetus and Rheic Oceans during the Paleozoic, *in* Ramos, V. S., and Keppie, J. D., eds., Laurentia-Gondwana connections before Pangea: Boulder, CO, Geological Society of America Special Paper 336, p. 267–276.

Keppie, J. D., Sandberg, C. A., Miller, B. V., Sánchez-Zavala, J. L., Nance, R. D., and Poole, F. G., 2004c, Implications of latest Pennsylvanian to Middle Permian paleontological and U-Pb SHRIMP data from the Tecomate Formation to re-dating tectonothermal events in the Acatlán Complex, southern Mexico: International Geology Review, v. 46, in press

Keppie, J. D., Solari, L. A., Ortega-Gutiérrez, F., Elías-Herrera, M., and Nance, R. D., 2003c, Paleozoic and Precambrian rocks of southern Mexico—Acatlán and Oaxacan complexes, *in* Geologic transects across Cordilleran Mexico, Guidebook for the field trip of the 99th Geological Society of America Cordilleran Section Annual Meeting, Puerto Vallarta, Jalisco, Mexico, April 4–10, 2003: Mexico, DF, Instituto de Geología, Universidad Nacional Autónoma de México, Publicacion Especial 1, Field trip 12, p. 281–314.

Keppie, J. D., Solari, L. A., Ortega-Gutiérrez, F., Ortega-Rivera, A., Lee, J. W. K., and Hames, W. E., 2004d, U-Pb and $^{40}Ar/^{39}Ar$ constraints on the cooling history of the northern Oaxacan Complex, southern Mexico: Tectonic implications, *in* Tollo, R. P., Corriveau, L., McLelland, J. B., and Bartholemew, G., eds., Proterozoic tectonic evolution of the Grenville Orogen in North America: Geological Society of America Memoir, in press.

Kettrup, B., Deutsch, A., Ostermann, M., and Agrinier, P., 2000, Chicxulub impactites: Geochemical clues to the precursor rocks: Meteorites and Planetary Science, v. 35, p. 1229–1238.

Krogh, T. E., Kamo, S. L., and Bohor, B. F., 1993a, Fingerprinting the K/T impact site and determining the time of impact by U/Pb dating of single shocked zircons from distal ejecta: Earth and Planetary Science Letters, v. 119, p. 425–429.

Krogh, T. E., Kamo, S. L., Sharpton, B., Marin, L., and Hildebrand, A. R., 1993b, U-Pb ages of single shocked zircons linking distal K/T ejecta to the Chicxulub crater: Nature, v. 366, p. 232–236.

Lawlor, P. J., Ortega-Gutiérrez, F., Cameron, K. L., Ochoa-Camarillo, H., López, R., and Sampson, D. E., 1999, U-Pb geochronology, geochemistry, and provenance of the Grenvillian Huiznopala gneiss of eastern Mexico: Precambrian Research, v. 94, p. 73–99.

Lonsdale, P., 1989, Geology and tectonic history of the Gulf of California, *in* Winterer, E. L., Hussong, D. M., and Decker, R. W., eds., The eastern Pacific Ocean and Hawaii: Boulder, CO, Geological Society of America, The geology of North America, v. N, p. 499–521.

López, R. L., Cameron, K. L., and Jones, N. W., 2001, Evidence for Paleoporterozoic, Grenvillian, and Pan-African age crust beneath northeastern Mexico: Precambrian Research, v. 107, p. 195–214.

Malone, J. W., Nance, R. D., Keppie, J. D., and Dostal, J., 2002, Deformational history of part of the Acatlán Complex: Late Ordovician–Early Silurian and Early Permian orogenesis in southern Mexico: Journal of South American Earth Science, v. 15, p. 511–524.

Manton, W. L., 1996, The Grenville in Honduras [abs.]: Geological Society of America Abstracts with Program, v. 28, no. 7, p. A493.

Marton, G., and Buffler, R. T., 1994, Jurassic rconstruction of the Gulf of Mexico basin: International Geology Review, v. 36, p. 545–586.

McKee, J. W., Jones, N. W., and Anderson, T. H., 1999, Late Paleozoic and early Mesozoic history of the Las Delicias terrane, Coahuila, Mexico, *in* Bartolini, C., Wilson, J. L., and Lawton, T. F., eds., Mesozoic sedimentary and tectonic history of north-central Mexico: Geological Society of America Special Paper 340, p. 161–189.

Meschede, M., Frisch, W., Herrmann, U. R., and Ratschbacher, L., 1997, Stress transmission across an active plate boundary: An example from southern Mexico: Tectonophysics, v. 266, p. 81–100.

Meza-Figueroa, D., Ruiz, J., Talavera-Mendoza, O., and Ortega-Gutiérrez, F., 2003, Tectonometamorphic evolution of the Acatlán Complex eclogites (southern Mexico): Canadian Journal of Earth Science, v. 40, p. 27–44.

Molina-Garza, R. S., van der Voo, R., and Urrutia-Fucagauchi, J., 1992, Paleomagnetism of the Chiapas massif, southern Mexico: Evidence for rotation of the Maya block and implications for the opening of the Gulf of Mexico: Geological Society of America Bulletin, v. 104, p. 1156–1168.

Moores, E. M., 1998, Ophiolites, the Sierra Nevada, "Cordilleria," and orogeny along the Pacific and Caribbean of North and South America: International Geology Review, v. 40, p. 40–54.

Morán-Zenteno, D. J., Corona-Chávez, P., and Tolson, G., 1996, Uplift and subduction erosion in southwestern Mexico since the Oligocene: Pluton geobarometry constraints: Earth and Planetary Science Letters, v. 141, p. 51–65.

Nance, R. D., Murphy, J. B., and Keppie, J. D., 2002, A Cordilleran model for the evolution of Avalonia: Tectonophysics, v. 352, p. 11–31.

Navarro-Santillán, D., Sour-Tovar, F., and Centeno-García, E., 2002, Lower Mississippian (Osagean) brachiopods from the Santiago Formation, Oaxaca, Mexico: Stratigraphic and tectonic implications: Journal of South American Earth Science, v. 15, p. 327–336.

Ortega-Gutiérrez, F., 1984, Evidence of Precambrian evaporites in the Oaxacan granulite complex of southern Mexico: Precambrian Research, v. 23, p. 377–393.

Ortega-Gutiérrez, F., and Elías-Herrera, M., 2003, Wholesale melting of the southern Mixteco terrane and origin of the Xolapa Complex [abs.]: Geological Society of America Abstracts with Program, v. 35, No. 4, p. 66.

Ortega-Gutiérrez, F., Elías-Herrera, M., Reyes-Salas, M., Macías-Romo, C., and López, R., 1999, Late Ordovician–Early Silurian continental collision orogeny in southern Mexico and its bearing on Gondwana-Laurentia connections: Geology, v. 27, p. 719–722.

Ortega-Gutiérrez, F., Ruiz, J., and Centeno-García, E., 1995, Oaxaquia, a Proterozoic microcontinent accreted to North America during the late Paleozoic: Geology, v. 23, p. 1127–1130.

Ortega-Gutiérrez, F., Solari, L. A., Solé, J., Martens, U., Gómez-Tuena, A., Morán-Ical, S., Reyes-Salas, M., and Ortega-Obregón, C., 2004, Polyphase, high temperature eclogite-facies metamorphism in the Chuacús Complex, central Guatemala: Petrology, geochronology, and tectonic implications: International Geology Review, v. 46, p. 445–470.

Ortega-Obregón, C., Keppie, J. D., Solari, L. A., Ortega-Gutiérrez, F., Dostal, J., López, R., Ortega-Rivera, A., and Lee, J. W. K., 2003, Geochronology and geochemistry of the ~917 Ma, calc-alkaline Etla granitoid pluton (Oaxaca, southern Mexico): Evidence of post-Grenvillian subduction along the northern margin of Amazonia: International Geology Review, v. 45, p. 596–610.

Pimentel, M. M., Fuck, R. A., Jost, H., Ferreira Filho, C. F., and de Araújo, S. M., 2000, The basement of the Brasilia fold belt and the Goiás magmatic arc, *in* Cordani, U. G., Thomaz Filho, A., and Campos, D. A., eds., Tectonic evolution of South America: 31st International Geological Congress, Rio de Janeiro, Brasil, p. 195–230.

Pindell, J. L., Cande, S. C., Pitman, W. C., III, Rowley, D. B., Dewey, J. F., Labreque, J., and Haxby, W., 1988, A plate-kinematic framework for models of Caribbean evolution: Tectonophysics, v. 155, p. 121–138.

Ramírez-Espinosa, J., 2001, Tectono-magmatic evolution of the Paleozoic Acatlán Complex in southern Mexico, and its correlation with the Appalachian system: Unpubl. Ph.D. thesis, University of Arizona, 170 p.

Ramírez-Espinoza, J., Ruiz, J., and Gehrels, G., 2002, Procedencia Pan-Africana en la sedimentación de la Formación Cosoltepec del Complejo Acatlán: Evidencia del margen pasivo oriental del Iapetus en el sur de Mexico: Actas Instituto Nacional de Geoquímica, v. 8, no. 1, p. 181–182.

Ramos, V. A., and Aleman, A., 2000, Tectonic evolution of the Andes, *in* Cordani, U. G., Milani, E. J., Thomaz Filo, A., and Campos, D. A., eds., Tectonic evolution of South America: 31st International Geological Congress, Rio de Janeiro, Brasil, p. 635–685.

Ratschbacher, L., Riller, U., Meschede, M., Herrmann, U., and Frisch, W., 1991, Second look at suspect terranes in southern Mexico: Geology, v. 19, p. 1233–1236.

Restrepo-Pace, P. A., Ruiz, J., Gehrels, G. E., and Cosca, M., 1997, Geochronology and Nd isotopic data of Grenville-age rocks in the Colombian Andes: New constraints for Late Proterozoic–Early Paleozoic paleocontinental reconstructions of the Americas: Earth and Planetary Science Letters, v. 150, p. 437–441.

Robison, R., and Pantoja-Alor, J., 1968, Tremadocian trilobites from Nochixtlan region, Oaxaca, Mexico: Journal of Paleontology, v. 42, p. 767–800.

Ross, M. I., and Scotese, C. R., 1988, A hierarchical tectonic model of the Gulf of Mexico and Caribbean region: Tectonophysics, v. 155, p. 139–168.

Ruiz, J., Tosdal, R. M., Restrepo, P. A., and Marillo-Muñeton, G., 1999, Pb isotope evidence for Colombia-southern Mexico connections in the Proterozoic, *in* Ramos, V. A., and Keppie, J. D., eds., Laurentia-Gondwana connections before Pangea: Geological Society of America Special Paper 336, p. 183–198.

Sánchez-Zavala, J. L., Centeno-García, E., and Ortega-Gutiérrez, F., 1999, Review of Paleozoic stratigraphy of Mexico and its role in the Gondwana-Laurentia connections, *in* Ramos, V. S., and Keppie, J. D., eds., Laurentia-Gondwana connections before Pangea: Geological Society of America Special Paper 336, p. 211–226.

Schaaf, P., Morán-Zenteno, D., del Sol Hernández-Bernal, M., Solís-Pichardo, G., Tolson, G., and Köhler, H., 1995, Paleogene continental margin truncation in southwestern Mexico: Geochemical evidence: Tectonics, v. 14, p. 1339–1350.

Schouten, H., and Klitgord, K. D., 1994, Mechanistic solution to the opening of the Gulf of Mexico: Geology, v. 22, p. 507–510.

Schulze, C. H., Keppie, J. D., Ortega-Rivera, A., Ortega-Gutiérrez, F., and Lee, J. K. W., 2004, Mid-Tertiary cooling ages in the Precambrian Oaxacan Complex of southern Mexico: Indication of exhumation and inland arc migration: Revista Mexicana de Ciencias Geologicos, in press.

Sedlock, R. L., Ortega-Gutiérrez, F., and Speed, R. C., 1993, Tectonostratigraphic terranes and tectonic evolution of Mexico: Geological Society of America Special Paper 278, 153 p.

Shergold, J. H., 1975, Late Cambrian and Early Ordovician trilobites from the Burke River structural belt, western Queensland, Australia: Department of Minerals and Energy, Bureau of Mineral Resources, Geology, and Geophysics, Bulletin 153, 221 p.

Silberling, N. J., Jones, D. L., Monger, J. M., and Coney, P. J., 1992, Lithotectonic terrane map of the North American Cordillera: U.S. Geological Survey Miscellaneous Investigation Series Map I-2176, scale 1:5,000,000.

Silver, L. T., and Chappell, B. W., 1988, The Peninsular Ranges Batholith: An insight into the evolution of the Cordilleran batholiths of southwestern North America: Transactions of the Royal Society of Edinburgh, Earth Science, v. 79, p. 105–121.

Silver, L. T., Taylor, H. P., Jr., and Chappell, B. W., 1979, Peninsular Ranges Batholith, San Diego and Imperial counties, *in* Mesozoic crystalline rocks: Guidebook for Geological Society of America Meeting: San Diego, CA, Department of Geological Science, San Diego State University, p. 83–110.

Sisson, V. B., Harlow, G. E., Sorensen, S. S., Brueckner, H. K., Sahn, E., Hemming, S. R., and Ave Lallemant, H. G., 2003, Lawsonite eclogite and other high-pressure assemblages in the southern Motagua fault zone, Guatemala: Implications for Chortis collision and subduction zones [abs.]: Geological Society of America Abstracts with Program.

Solari, L. A., Keppie, J. D., Ortega-Gutiérrez, F., Cameron, K. L., López, R., and Hames, W. E., 2003, 990 Ma and 1,100 Ma Grenvillian tectonothermal events in the northern Oaxacan Complex, southern Mexico: Roots of an orogen: Tectonophysics, v. 365, p. 257–282.

Steiner, J. B., and Walker, J. D., 1996, Late Silurian plutons in Yucatan: Journal of Geophysical Research, v. 101, no. B8, p. 17,727–17,735.

Stewart, J. H., 1988, Latest Proterozoic and Paleozoic continental margin of North America and the accretion of Mexico: Geology, v. 16, p. 186–189.

Stewart, J. H., Poole, F. G., Ketner, K. B., Madrid, R. J., Roldán-Quintana, J., and Amaya-Martínez, R., 1990, Tectonics and stratigraphy of the Paleozoic and Triassic southern margin of North America, Sonora, Mexico, *in* Gehrels, G. E., and Spencer, J. E. eds., Geologic excursions through the Sonoran Desert region, Arizona and Sonora: Arizona Geological Survey Special Paper 7, p. 183–202.

Talavera-Mendoza, O., and Suastegui, M. G., 2000, Geochemistry and isotopic composition of the Guerrero terrane (western Mexico): Implications for the tectono-magmatic evolution of southwestern North America during the late Mesozoic: Journal of South American Earth Science, v. 13, p. 297–324.

Thomas, W. A., 1989, The Appalachian-Ouachitan belt beneath the Gulf Coastal Plain between the outcrops in the Appalachian and Ouachita Mountains, *in* Hatcher, R. D., Jr., Thomas, W. A., and Viele, G. W., eds., Decade of North American geology, v. F-2, The Appalachian and Ouachitan orogen in the United States: Boulder, CO, Geological Society of America, p. 537–553.

Tolson-Jones, G., 1998, Deformación, exhumación y neotectónica de la margen continental de Oaxaca: Datos estructurales, petrológicos y geotermobarométricas: Unpubl. Ph.D. thesis, Universidad Nacional Autónoma de México, 98 p.

Torres, R., Ruiz, J., Patchett, P. J., and Grajales, J. M., 1999, A Permo-Triassic continental arc in eastern Mexico: Tectonic implications for reconstructions of southern North America, *in* Bartolini, C., et al., eds., Mesozoic sedimentary and tectonic history of north-central Mexico: Geological Society of America Special Paper 340, p. 191–196.

Vachard, D., and Flores de Dios, A., 2002, Discovery of latest Devonian/earliest Mississippian microfossils in San Salvador Patlanoaya (Puebla, Mexico); biogeographic and geodynamic consequences: Compte Rendu Geoscience, v. 334, (2002), p. 1095–1101.

Vachard, D., Flores de Dios, A., Buitron, B. E., and Grahales, M., 2000, Biostratigraphie par fusulines des calcaires Carboniferes et Permienes de San Salvador Patlanoaya (Puebla, Mexique): Geobios, v. 33, p. 5–33.

Vera-Sánchez, P., 2000, Caracterización geoquímica de las unidades basales del bloque de Yucatan y su afinidad con unidades similares en el Golfo de Mexico: Unpubl. M.Sc. thesis, Instituto Geofisica, Universidad Nacional Autónoma de México, 105 p.

Weber, B., and Cameron, K. L., 2003, U-Pb dating of metamorphic rocks from the Chiapas massif: Evidence for Grenville crust and a Late Permian orogeny in the southern Maya terrane [abs.]: Geological Society of America Abstracts with Program, v. 35, no. 4, p. 65.

Weber, B., and Hecht, L., 2003, Petrology and geochemistry of metaigneous rocks from a Grenvillian basement fragment in the Maya block: the Guichicovi complex, Oaxaca, southern Mexico: Precambrian Research, v. 124, p. 41–67.

Weber, B., and Köhler, H., 1999, Sm/Nd, Rb/Sr, and U-Pb geochronology of a Grenville terrane in southern Mexico: Origin and geologic history of the Guichicovi complex: Precambrian Research, v. 96, p. 245–262.

Yañez, P., Ruiz, J., Patchett, P. J., Ortega-Gutiérrez, F., and Gehrels, G., 1991, Isotopic studies of the Acatlán Complex, southern Mexico: Implications for Paleozoic North American tectonics: Geological Society of America Bulletin, v. 103, p. 817–828.

Appendix. Terrane Descriptions

Chortis terrane: (A) Mesoproterozoic amphibolite facies para- and ortho-gneisses yielded concordant U-Pb SHRIMP zircon ages of 1074 ± 10 Ma (magmatic age) and 1017 ± 20 Ma (metamorphic age; Manton, pers. comm.); (B) greenschist–lower amphibolite facies metasediments and metavolcanics of unknown age; intruded by (C) deformed granitoid plutons with early Mesozoic Rb-Sr ages; (D) nonconformably overlain by Mesozoic sedimentary and volcanic rocks (Middle Jurassic–Lower Cretaceous siliciclastic rocks and volcanic rocks, Lower–Upper Cretaceous carbonates and andesitic volcanic rocks, and Upper Cretaceous–Paleogene redbeds) that were folded and eroded before deposition of (E) overstepping mid-Tertiary to Recent volcanic arc rocks (Donnelly et al., 1990).

Motagua terrane: (A) Mantle peridotites (metamorphosed to eclogites, jadeitites, amphibolites, and serpentinized peridotites); (B) Upper Jurassic–Lower Cretaceous gabbro, amphibolite, pillow basalt (MORB, OIB, and island-arc tholeiitic [IAT] affinities), and radiolarian chert; (C) Mid-Cretaceous eclogite-facies metamorphism and obduction in the southern Motagua terrane (Harlow et al., 2004); (D) Upper Cretaceous, calc-alkaline, island-arc basalt-andesite, radiolarian chert, limestone, and phyllite synchronous with eclogite-facies metamorphism and exhumation in the northern Motgua terrane (Harlow et al., 2004); unconformably overlain by (D) overstepping Eocene molasses and volcanicalstics; unconformably overlain by (E) Miocene-Quaternary sediments (Guinta et al., 2001).

Maya terrane: (A) metasedimentary rocks and ~1238 Ma AMCG suite metamorphosed to granulite facies at ~990-975 Ma (Weber and Köhler, 1997; Ruiz et al., 1999; Weber and Hecht, 2003); (B) Basement clasts in the Chicxulub crater include quartzite, quartz-mica schist, felsic-intermediate granitic gneiss, volcanic arc granitoids (quartz diorite, granodiorite, and tonalite with a SiO_2 range of 48–68%), and mafic, arc volcanic rocks (basaltic andesite and olivine tholeiite; Vera-Sánchez, 2000): granitic gneiss clasts yielded a depleted mantle model Nd age of 1.2–1.4 Ga (Kettrup et al., 2000); melt rocks have yielded depleted mantle Nd model ages of 1,060 ± 20 Ma (Blum et al., 1993) to 1.1–1.2 Ga (Kettrup et al., 2000); U-Pb analyses from Chicxulub breccia have yielded ages of 2,725 ± 57 Ma (one zircon), 550 ± 15 Ma (predominant: 6 zircons), 286 ± 14 Ma (1 titanite), and from distal ejecta of 544 ± 5 Ma and 559 ± 5 Ma (Colorado), 418 ± 6 Ma (Haiti and Chicxulub—3 zircons), 320 ± 31 Ma (Colorado; Krogh et al., 1993a, 1993b; Kamo and Krogh, 1995). Magmatic arc diorite-granodiorite-granite in the Maya Mountains of Belize yielded an intrusive age of 418 ± 4 Ma (upper intercept, U-Pb zircon data) and an inheritance age of 1210 ± 136 Ma (upper intercept, U-Pb zircon data: Steiner and Walker, 1996); (C) Upper Pennsylvanian–mid Permian (late Leonardian) shelf clastic, carbonate, and volcanic rocks displaying folds accompanied by lower greenschist-facies metamorphism (Steiner and Walker, 1996), and migmatization of the Chaucus Group (Ortega-Gutiérrez et al., 2004); unconformably overlain by (D) overstepping Upper Jurassic–lowermost Cretaceous continental sediments overlain by Cretaceous and Lower Tertiary marine carbonates: upper Campanian–lower Eocene turbiditic flysch records thrusting of Motagua ophiolites onto the southern margin of the Maya terrane (Guinta et al., 2001), followed by Miocene folds and SW-vergent thrusts in the Chiapas foldbelt (de Cserna, 1989) that are unconformably overlain by (E) Plio-Pleistocene clastic and volcanic rocks.

Juárez terrane: (A) metamorphosed serpentinite, gabbro, mafic volcanics, felsic tuff, greywacke; (B) Lower Cretaceous andesite, volcaniclastic rocks, tuff, flysch, and schist, and Berriasian–Valanginian (Lower Cretaceous) flysch, slate, and limestone deformed and metamorphosed at ~131–137 Ma ($^{40}Ar/^{39}Ar$ hornblende plateau ages; Sedlock et al., 1993); unconformably overlain by (C) overstepping Campanian–Maastrichtian flysch deformed in latest Cretaceous–Paleogene Laramide Orogeny; unconformably overlain by Cenozoic clastic and volcanic rocks (Sedlock et al., 1993 and references therein).

Oaxaquia terrane: (A) paragneisses, arc volcanic rocks, intruded by within-plate, rift-related, ≥1140 Ma orthogneisses; (B) deformed during the ~1100 Ma Olmecan migmatitic, tectonothermal event; (C) intruded by 1012 ± 12 Ma, anorthosite-charnockite- granite suite; (D) deformed by 1004 ± 3 to 979 ± 3 Ma Zapotecan orogeny under granulite-facies metamorphic conditions (Keppie et al., 2003a; Solari et al., 2003); (E) intrusion of 917 ± 6 Ma, arc-related, granitoid pluton (Ortega-Obregón et al., 2003); (F) unconformably overlain by Tremadocian clastic and carbonate rocks containing a

Gondwanan faunal assemblage in the south and Silurian, shallow-marine clastic rocks with Gondwanan fauna in the north (Robison and Pantoja-Alor, 1968; Boucot et al., 1997); (G) unconformably overlain by Carboniferous–Permian clastic and carbonate rocks; (H) overstepped by Upper Jurassic and Cretaceous continental-shallow marine clastic and carbonate rocks; (I) Cenozoic red beds and volcanic arc rocks.

Mixteca terrane: (A) metamorphosed psammites and pelites of uncertain age containing detrital zircons as young as Ordovician (Ramírez-Espinoza et al. 2002); (B) thrust slices of psammitic and pelitic metasediments intruded by mafic-felsic igneous rocks (~478–440 Ma U-Pb zircon ages; Ortega-Gutiérrez et al., 1999; Campa et al., 2002; Keppie et al., 2004a); (C) Mississippian eclogite-facies metamorphism and exhumation (346 ± 3 Ma U-Pb zircon age; Keppie et al., 2004a); (D) deposition of Upper Devonian–Middle Permian, shallow-marine clastic and carbonate rocks and arc-backarc volcanic rocks (Vachard et al., 2000; Vachard and Flores de Dios, 2002; Keppie et al., 2004c); (F) deformed and metamorphosed at greenschist (-amphibolite) facies during the Early–Middle Permian with synchronous intrusion of arc plutons; overstepped by (G) unconformably overlain Lower Jurassic–Cretaceous, continental-marine clastic and carbonate rocks; (H) deformed by Late Cretaceous–Eocene Laramide Orogeny; (I) unconformably overlain by Cenozoic continental clastic and volcanic-arc rocks.

Sierra Madre terrane (redefined to exclude the ~1 Ga Novillo gneiss and overlying Siluro-Devonian rocks): (A) pelitic and psammitic schist with lenses of serpentinite, metagabbro, metabasalt, and metachert of the Granjeno Formation (Carrillo-Bravo, 1991; Castillo-Rodríguez, 1988); (B) polyphase deformation accompanied by greenschist-facies metamorphism prior to deposition of pebbles in Wenlockian sediments unconformably overlying the Novillo Complex (Fries et al., 1962; de Cserna et al., 1977; de Cserna and Ortega-Gutiérrez, 1978); unconformably overlain by (C) overstepping Triassic–Lower Jurassic redbeds, and shallow-marine clastic rocks, with minor volcanic rocks, that were locally folded before deposition of (D) Middle Jurassic–Cretaceous redbeds, evaporites, shallow-marine clastic and carbonate rocks, felsic volcanic rocks, that were deformed by the Laramide orogeny before deposition of (E) the unconformably overlying Cenozoic, continental rocks.

Coahuila terrane: (A) Upper Pennsylvanian–Permian, low-grade, volcaniclastic flysch, calc-alkaline volcanic rocks, and clastic and carbonate rocks, intruded by Triassic granitoids that were deformed prior to deposition of (B) overstepping and unconformably overlying Upper Jurassic–Cretaceous, shallow-marine limestone, shale, evaporite, siltstone, sandstone, and local coal, overlain by (C) Paleocene–Miocene continental-shallow marine clastic rocks and Oligocene-Quaternary felsic and alkaline volcanic rocks (Sedlock et al., 1993, and references therein).

Tarahumara terrane: (A) basinal sedimentary rocks similar to those in the Ouachita orogenic belt that were deformed and metamorphosed at greenschist facies in the Permian before being unconformably overlain by (B) overstepping Upper Jurassic–Cretaceous clastic, carbonate, and evaporitic rocks, which were deformed during the Laramide orogeny before being unconformably overlain by (C) Cenozoic, calcalkaline volcanic rocks (Sedlock et al., 1993, and references therein).

Cortez terrane: (A) Upper Ordovician, quartzite, carbonates and chert; (B) Devonian, Carboniferous and Permian, psammitic and pelitic rocks, rare chert and limestone; overstepped by (C) Upper Jurassic–Lower Cretaceous, magmatic arc rocks and associated sedimentary rocks that were affected by the Laramide orogeny before being unconformably overlain by (D) Upper Cretaceous–Quaternary, andesitic-rhyolitic volcanic rocks (associated with plutons) and continental rocks (Sedlock et al., 1993; Sánchez-Zavala et al., 1999).

The Guerrero Composite Terrane is characterized by Upper Jurassic–early Upper Cretaceous, submarine (-subaerial) volcanic and sedimentary rocks that were accreted to cratonic Mexico in the Late Cretaceous, producing the Turonian-Maastrichtian foreland basin deposits (Centeno-García et al., 2000, 2003). The terranes are best defined south of the Trans-Mexican Volcanic Belt (TMVB)—potential equivalents north of the TMVB are shown in brackets. Terranes 6–10 occur in Baja California and have been defined by Sedlock et al. (1993).

1. Arperos terrane (shown as a suture on Fig. 1C): (A) Lower Cretaceous, basalts produced by mixing of OIB and N-MORB siliceous sediments, pelagic carbonates, and turbidites (Freydier et al., 2000).

2. Teloloapan terrane: (A) Lower Jurassic, andesitic-dacitic, volcanic rocks, phyllite, and sericitic tuff contemporaneous with granite that yielded U-Pb zircon data with intercepts at 186 ± 7 Ma and 1242 ± 126 Ma (Elías-Herrera et al., 2000); (B) Neocomian–Albian, volcanic-arc rocks, limesones, shale, and sandstone that are thrust eastward over Aptian–Turonian rocks of the Mixteca terrane during the Laramide orogeny.

3. Arcelia(-Guanajuato) terrane: (A) Albian–Cenomanian, primitive island arc, back-arc basin, OIB, and MORB basalts, ultramafic rocks, pelagic limestone, radiolarian chert, and black shales that are thrust eastward over the Teloloapan terrane during the Laramide orogeny.

4. Zihuatenejo(-San José de Gracia) terrane: (A) Triassic, siliceous continent–derived sediments with ε_{Ndi} = –6 to –7 and T_{DM} ages = 1.3–1.4 Ga, and basalts, that were deformed and metamorphosed in the Early–Middle Jurassic and intruded by Middle Jurassic granitoids before being unconformably overlain by (B) Lower Cretaceous (Neocomian–Albian) arc-volcanic rocks (andesitic-dacitic flows) and associated sedimentary rocks that are thrust westward over the Las Ollas terrane during the Laramide orogeny, and overstepped on the east by (C) mid-Tertiary ignimbrites. Paleozoic rocks appear to underlie the Cretaceous arc rocks in the San Jose de Gracia terrane.

5. Las Ollas terrane: (A) probable Lower Cretaceous, ophiolitic mélange consisting of blocks of ultramafic rocks, immature island-arc tholeiitic gabbro, basalt, amphibolite, dolerite, limestone, quartzite, and chert in a matrix of flysch and serpentine with blueschist metamorphic minerals.

6. Alisitos terrane: (A) Upper Jurassic–late Lower Cretaceous, calc-alkaline, arc-volcanic and volcanogenic rocks, and limestone that are coeval with older parts of the Peninsular Ranges batholith that apparently straddled the continent-ocean boundary: the arc passes east and west into periarc sandstones; unconformably overlain by (B) Upper Cretaceous–Eocene marine clastic rocks and minor tuff overlain by mid-Miocene marine clastic rocks and calc-alkaline volcanic rocks passing upwards into upper Miocene–Cenozoic alkalic-tholeiitic volcanic and sedimentary rocks.

7. Vizcaino Sur terrane: (A) Upper Triassic ophiolite, chert, limestone, breccia, and sandstone; (B) Lower Jurassic volcanic and volcaniclastic rocks.

8. Vizcaino Norte terrane: (A) Upper Triassic ophiolite and tuffaceous sediments; (B) Upper Jurassic–Upper Cretaceous volcanogenic rocks containing granitoid clasts with discordia intercepts of 1,340 ± 3 Ma and 150 ± 3 Ma.

9. Choyal terrane: (A) Middle Jurassic, mafic-felsic volcanic, volcaniclastic clastic, and ophiolitic rocks intruded by granitoids, (B) Middle and Upper Jurassic clastic rocks of continental derivation (including Pennsylvanian limestone and quartzite clasts). Terranes 7, 8, and 9 are overstepped by Albian–Campanian siliclastic turbidites unconformably overlain by Miocene–Pliocene shallow-marine strata.

10. Western Baja terrane: (A) Upper Triassic to mid-Cretaceous, ocean-floor basalt, siliciclastic metasedimentary rocks, chert, and rare limestone affected by deformation and blueschist-facies metamorphism. The boundary between this terrane and nos. 7–9 is a serpentinite-matrix mélange containing blocks of orthogneiss, eclogite, ultramafic rocks, blueschist, amphibolite, and greenschist that have yielded ages ranging from Middle Jurassic to mid-Cretaceous.

Baja California terrane: (A) Mesozoic rocks of the Guerrero composite terrane (nos. 6–10); (B) Middle Miocene–Holocene, rift–passive margin, volcanic, and associated sedimentary rocks.

Oaxaquia Terrane

Sedimentary Origin of Calcareous Intrusions in the ~1 Ga Oaxacan Complex, Southern Mexico: Tectonic Implications

J. DOSTAL,[1]

Department of Geology, Saint Mary's University, Halifax, Nova Scotia, B3H 3C3, Canada

J. D. KEPPIE,

Instituto de Geologia, Universidad Nacional Autonoma de Mexico, 04510 Mexico D.F., Mexico

H. MACDONALD,

Department of Geology, Saint Mary's University, Halifax, Nova Scotia, B3H 3C3, Canada

AND F. ORTEGA-GUTIÉRREZ

Instituto de Geologia, Universidad Nacional Autonoma de Mexico, 04510 Mexico D.F., Mexico

Abstract

Intrusive calcareous bodies, marbles and calc-silicate rocks, are a distinctive feature of the high-grade metamorphic suites of the ~1 Ga northern Oaxacan Complex. They typically form dike-like intrusions up to 4 m thick which cut across the surrounding high-grade granulite- and upper-amphibolite facies metamorphic rocks. Various protoliths are possible for these carbonate bodies: (1) sediments including evaporites; (2) metasomatic skarns; and (3) carbonatites. An evaporitic protolith is supported by the predominance of scapolite, low abundances of incompatible trace elements (including Nb and rare-earth elements) relative to carbonatites, and the presence of a sharp contact with host rocks without a significant contact metamorphic aureole or fenitization. It is inferred that limestones and related rocks were remobilized under granulite-facies conditions and intruded into the host rocks. The widespread distribution of such evaporites in the Oaxacan Complex is consistent with deposition after the worldwide ~1.3 Ga oxygenation event that increased the marine sulfate reservoir. Intrusion of rift-related plutons into the sediments at ~1157–1130 Ma provides a younger limit on the age of protholiths of the metamorphic suites. Modern analogues for such evaporites are rifts associated with passive margins (e.g., Red Sea) and active margins (e.g., Gulf of California). The presence of evaporites implies a paleolatitude of 10–35°, a conclusion consistent with a paleogeographic provenence for the Oaxacan Complex adjacent to either Amazonia or eastern Laurentia in Rodinia reconstructions.

Introduction

THE COMPOSITION OF metasedimentary rocks can provide important clues to the provenance and tectonic settings of their deposition. It is particularly important in deciphering the origin of high-grade metamorphic rocks in Precambrian areas where deformation and metamorphism have destroyed most of the features used to constrain their genesis. Such is the case for the calcareous rocks in the ~1 Ga Oaxacan Complex of southern Mexico. Two origins have been proposed for these rocks: (1) a sedimentary origin as part of an evaporite sequence deposited in a passive-margin tectonic setting (Ortega-Gutiérrez, 1984); and (2) an igneous origin as carbonatites (Melgarejo and Prol-Ledesma, 1999). The rocks could also have originated through metasomatic processes. The geochemical and mineralogical signatures and field relations of such processes are distinctive. To determine these relations and constrain the origin of the Oaxacan calcareous rocks, they were studied along the two highways connecting Nochixtlan and Oaxaca in southern Mexico (Figs. 1–3).

[1]Corresponding author; email: jarda.dostal@stmarys.ca

Geological Setting

The Oaxacan calcareous rocks (containing 10 to >90 vol% carbonates) have been subdivided according their modal composition by Ortega-Gutiérrez (1984) into marbles (> 50 vol% carbon-

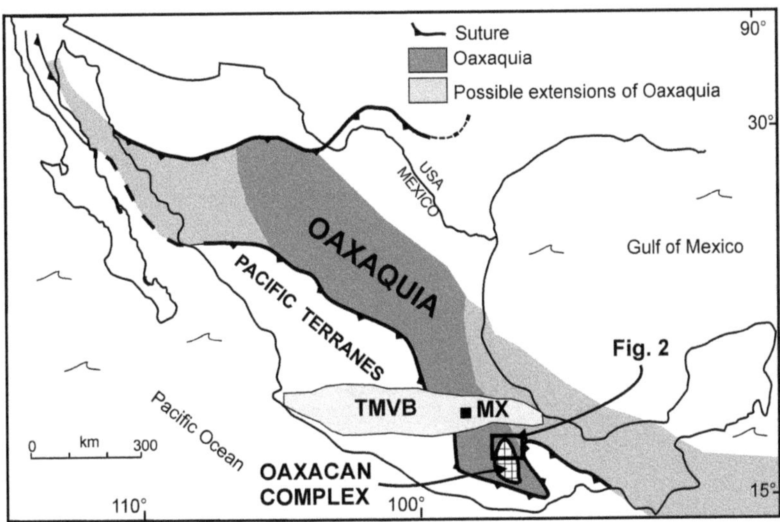

FIG. 1. Simplified geological map of Mexico with inset showing the location of Figure 2. Oaxaquia and its possible extension is after Ortega-Gutiérrez et al (1995). TMVB–Trans-Mexican volcanic belt; MX–Mexico City.

ates) and calc-silicates (10–50 vol% carbonates). They form concordant layers (<200 m thick), and discordant bodies up to 4 m thick, that are an integral part of the high-grade metamorphic sequence (Ortega-Gutiérrez, 1984). The discordant calcareous rocks are the dike-like intrusions, which cut across the surrounding metamorphic rocks. These discordant bodies are mostly marbles with xenoliths of calc-silicate and/or host rocks, whereas the concordant bodies are both calc-silicates and marbles.

The calcareous bodies occur in the upper "paragneiss" thrust slice of the northern Oaxacan Complex (Figs. 2–3). This slice was intruded by within-plate, ≥1140 Ma charnockite and meta-syenite (Keppie et al., 2003) before being involved in two tectonothermal events, the ~1100 Ma Olmecan event, and the ~1005–980 Ma Zapotecan event (Solari et al., 2003). The latter involved polyphase deformation under granulite-facies metamorphic conditions (Solari et al., 2003). Peak granulite-facies metamorphism reached 700°–750°C and 7.2–8.2 kbar (Mora et al., 1986). The concordant calcareous lenses contain Zapotecan foliations and folds, and although the discordant bodies cut across these ductile structures, they record the granulite-facies metamorphism, indicating that the high-grade metamorphism outlasted the deformation. Titanite and phlogopite from these discordant calcareous bodies have yielded ages of 968 ± 9 Ma (concordant U-Pb age) and 945 ± 10 Ma ($^{40}Ar/^{39}Ar$ laser fusion analyses), respectively, which have been interpreted as dating cooling through 660–700°C and ~450°C (Keppie et al., in press).

The host rocks range in compositions from mafic to felsic. The main mineral assemblages of most of the thrust slice are of granulite facies. The assemblages of the rocks are mostly anhydrous and include quartz, plagioclase (commonly antiperthitic), hypersthene, clinopyroxene, garnet, alkali feldspar (perthite and mesoperthite), opaque mineral (ilmenite), and a variable amount of titaniferous hornblende and biotite. Hydrous phases, including hornblende and biotite, are in textural equilibrium with the anhydrous minerals (Mora et al., 1986). In places, the rocks are composed of mineral assemblages containing biotite, amphibole, plagioclase and opaques, indicative of only upper amphibolite-facies–grade metamorphism. Phase equilibria were reported by Prakash et al. (1991) for some of these calc-silicates with emphasis on the rare assemblage of wollastonite-quartz-graphite.

Petrography and Mineral Chemistry

The calcareous rocks, both marbles and calc-silicate rocks, typically contain calcite, diopside, scapolite, and phlogopite as essential minerals with minor to accessory amounts of dolomite, forsterite, wollastonite, pargasitic amphibole, spinel, garnet, titanite, sulfides, Fe-Ti oxides, clinozoisite, and

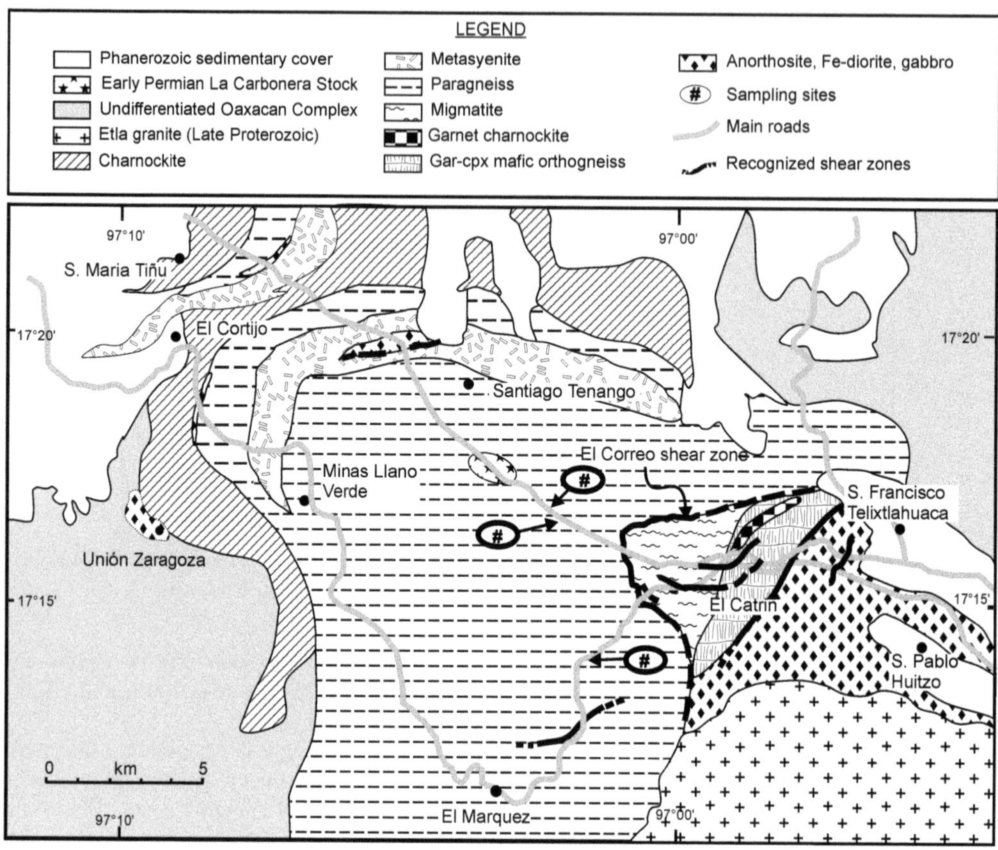

FIG. 2. Geological map of the northern Oaxacan Complex (modified after Solari et al., 2003) showing the locations of sampled calcareous intrusions.

humite minerals, such as chondrodite. Quartz, plagioclase and alkali feldspar occur in variable amounts. In addition, accessory apatite, vivianite, iron phosphate, cancrinite, and anhydrite (mostly pseudomorphed to gypsum) occur in some of these rocks. Retrograde minerals include numerous hydrated phases such as chondrodite, serpentine, talc, brucite, zoisite/clinozoisite, epidote, tremolite, gypsum, and Mg-Fe chlorite, which indicate a high flux of water-rich fluids during retrogression.

Marble

The marble is white to light grey, massive, and composed principally of equant calcite. In addition to carbonates (>50 vol%), the marbles typically contain forsterite, diopside, phlogopite, scapolite, rare amphibole, chondrodite, spinel, apatite, sulfides, titanite, quartz, and feldspars (Tables 1 and 2). The marble without dolomite also contains wollastonite. Texturally, all of these minerals seem to represent an equilibrium assemblage.

Carbonate minerals have a grain size typically ranging between 0.2 and 1 mm, and exceptionally reaching up to 1.5 cm in diameter. Dolomite typically accounts for <10 % of the modal composition. Clinopyroxene is slightly pleochroic, light green diopside (Table 1), which forms prismatic grains ranging in size from ~1 to 10 mm. Olivine (~Fo_{92}), partially serpentinized, was found only in marbles. Brownish, pleochroic amphibole (Table 2) occurs in marbles in association with clinopyroxene. Phlogopite (with low ~1 wt% TiO_2) is present in some marbles as a major constituent (5–10 vol%). Minerals present in minor to accessory amounts include: (1) spinel, which occurs only in marbles with dolomite; (2) chondrodite, which forms along the grain boundaries of the dolomite as an

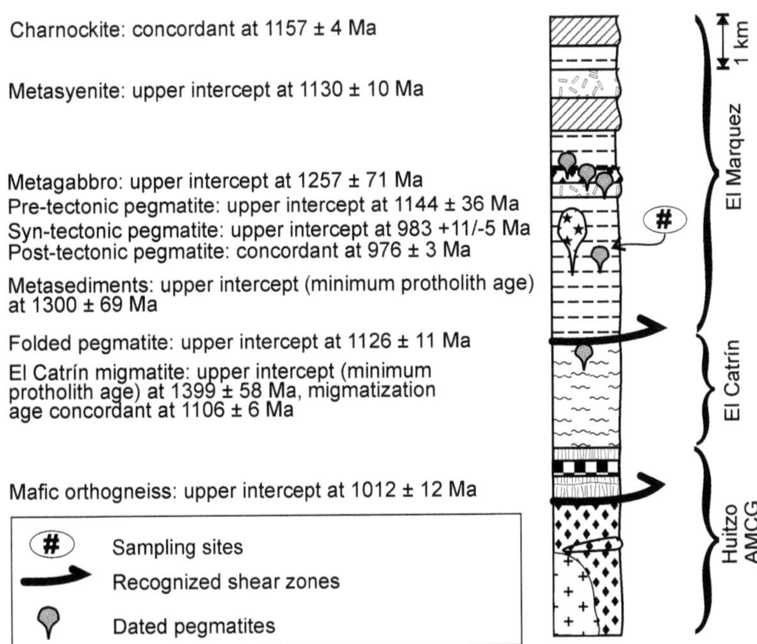

FIG. 3. Structural section of the northern Oaxacan Complex (modified after Solari et al., 2003) showing the location of the sampled calcareous intrusions. Symbols are the same as in Figure 2. AMCG = anorthosite-mangerite-charnockite-granite suite; Huitzo, El Catrin and El Marquez refer to three thrust sheets of the Oaxacan Complex (Solari et al., 2003).

alteration feature; (3) feldspars, including calcic plagioclase, nearly pure albite An_2 and K-feldspar (Table 2); (4) scapolite; and (5) titanite. Rare anhydrite, mostly pseudomorphed to gypsum, was likely in textural equilibrium with the other mineral phases. Grains of sulfides, apatite, and other phosphate minerals (probably vivianite and heterosite) occur in trace proportions.

Calc-silicate rocks

The calc-silicate rocks are medium to coarse grained (mostly < ~2 mm in size) and consist predominantly of a granoblastic-polygonal mineral assemblage composed of significant amounts of diopsidic clinopyroxene (~15–90 vol%) and scapolite (up to 45 vol%). Scapolite is Ca-rich with the meionite component ranging from 0.53 to 0.97. The contents of carbonates, principally calcite, are highly variable, averaging around 15 vol%. These rocks may also contain graphite, quartz, plagioclase, alkali feldspars (K-feldspar and albite), wollastonite, hedenbergite, andradite and phlogopite. In calc-silicate samples, amphibole (magnesiohastingsite; Table 2) typically occurs in association with scapolite. Titanite is evenly distributed as an accessory mineral, but in some cases forms prismatic crystals > 2.5 cm in size. Clinozoisite might reach up to 5 vol%. Apatite, iron phosphate, zircon, Fe-Ti oxides, and sulfides are other accessory phases.

Contact aureole

The amphibole-biotite mafic gneisses that generally host the carbonates show a narrow, typically <5 cm wide contact aureole. The host gneisses are medium grained (0.5–1 mm in size) composed predominantly of brown amphibole, biotite, and plagioclase ($An_{\sim 30}$). In the 2–3 cm wide contact zone surrounding the cross-cutting carbonate intrusions, the host gneisses were converted into fine-grained rocks (typically 0.3–0.5 mm in grain size) containing mainly plagioclase, K-feldspar, quartz, strongly altered mafic minerals (chiefly biotite and clinopyroxene), opaques, and sulfides. Mafic minerals, which are less abundant than in the mafic gneisses, were replaced by a mixture of fibrous minerals and Ti-magnetite.

Xenoliths

Xenoliths in the marble intrusions are typically cms to dms in size and are composed of amphibole-

TABLE 1. Representative Analyses of Minerals from the Calcareous Intrusions[1]

	Clinopyroxene					Olivine		
	Marble			Calc-silicate		Marble		
Sample	OX-5	OX-5	OX-50	OX-3	OX-3	OX-45	OX-45	OX-47
SiO_2 (wt%)	53.19	52.84	51.57	52.91	52.88	41.83	41.2	40.75
TiO_2	0.32	0.39	0.66	0.37	0.31			
Al_2O_3	2.52	2.3	5.22	1.94	2.31			
FeO	1.93	2.8	3.18	7.39	7.18	6.9	7.06	8.67
MnO			0.27		0.37			0.42
MgO	16.45	15.78	14.4	12.7	13.35	51.68	51.8	49.86
CaO	25.02	24.81	24.64	24.14	23.05			
Na_2O	0.38	0.46	0.74	0.61	0.73			
Σ	99.81	99.38	100.68	100.06	100.18	100.41	100.06	99.7
X_{Mg}	0.94	0.91	0.89	0.75	0.77	0.93	0.93	0.91
Wo	0.51	0.51	0.52	0.51	0.49			
En	0.46	0.45	0.43	0.37	0.39			
Fs	0.03	0.04	0.05	0.12	0.12			

[1]End-member components: Wo = wollastonite; En = enstatite; Fs = ferrosilite; mineral compositions were determined using a JEOL Superprobe 733 at Dalhousie University (Halifax, Nova Scotia). Data were reduced using ZAF corrections.

biotite gneisses, amphibole-clinopyroxene granulites, and calc-silicate rocks. Amphibole-biotite gneiss xenoliths are rimmed by a thin contact zone composed of plagioclase-clinopyroxene-biotite rock, whereas the granulite xenoliths are surrounded by clinopyroxene-plagioclase-titanite rims.

Analytical Methods

Samples were analyzed for major and some trace (Rb, Sr, Ba, Zr, Nb, Y, Cr, Ni, Sc, V, Ga, and Zn) elements by X-ray fluorescence spectrometry at the Regional Geochemical Center at Saint Mary's University, Halifax, Nova Scotia. The precision and accuracy of the data have been reported by Dostal et al. (1986). Representative samples were then chosen for analyses of rare-earth elements (REE), Th, Ta, Nb, and Hf (Table 3) by inductively coupled plasma-mass spectrometry (ICP-MS) at the Geoscience Laboratories of the Ontario Geological Survey. Precision and accuracy are given by Ayer and Davis (1997) and are generally within 5%. Mineral compositions were determined at the Department of Earth Sciences of Dalhousie University (Halifax, Nova Scotia) using a JEOL Superprobe 733 with a wavelength-dispersive spectrometer and operated with a beam current of 15 kV at 5nA.

Geochemistry

The calcareous rocks show a spectrum of mineral and element concentrations ranging from relatively pure marbles to calc-silicate rocks. The two groups of the Oaxacan calcareous rocks (marbles and calc-silicates) defined by Ortega-Gutiérrez (1984) are distinct not only in modal proportions of forsterite, diopside, scapolite, feldspars, and carbonates, but also on the basis of their chemical compositions (Table 3).

Marbles

The marbles have high contents of CaO (>30 wt%) and loss on ignition (LOI) but low SiO_2 (<30 wt%), reflecting the dominance of calcite in these

TABLE 2. Representative Analyses of Minerals from the Calcareous Intrusions[1]

	Spinel		Phlogopite		Amphibole		Feldspar			Scapolite	
	Marble		Marble		Marble	CS	Marble			Marble	CS
Sample	OX-45	OX-45	OX-5	OX-5	OX-5	LS-53	OX-7	OX-7	OX-49	OX-1	LS-62
SiO_2 (wt%)			41.00	40.89	44.16	43.76	62.78	63.82	68.89	50.10	50.21
TiO_2			1.25	1.10	1.05	1.38					
Al_2O_3	69.24	68.60	13.60	13.65	13.67	15.29	19.05	18.84	19.85	25.00	25.34
FeO	6.65	6.52	2.52	2.38	2.58	1.53					
MnO											
MgO	23.23	22.80	24.51	24.64	17.94	19.73					
CaO					12.87	13.13			0.35	13.25	13.37
Na_2O			0.41	0.38	2.07	2.23	0.75	0.24	10.34	5.77	5.81
K_2O			10.41	10.33	1.96	2.01	15.65	16.30		0.93	0.89
BaO							1.06				
ZnO	1.17	1.31									
Cl										1.32	1.23
Σ	100.29	99.23	93.70	93.37	96.30	99.06	99.29	99.20	99.43	96.37	96.85
X_{Mg}	0.86	0.86	0.95	0.95	0.93	0.96					
Me										0.53	0.54
Or							0.93	0.98			
Ab							0.07	0.02	0.98		
An									0.02		

[1]Abbreviations: CS = calc-silicate rock. End-member component: Or = orthoclase; Ab = albite; An = anorthite; Me = meionite.

TABLE 3. Representative Analyses of Calcareous Rocks from the Oaxaca Complex[1]

	Marble					Calc-silicate				Gneiss	
	OX-45	OX-47	OX-49	OX-50	98-2-2	98-3-1	S-2-2	98-2-1	S-2-9	98-1-4	98-1-1
SiO_2(%)	11.98	7.84	7.53	14.66	26.20	49.07	44.56	51.79	50.14	50.46	68.76
TiO_2	0.01	0.12	0.12	0.17	0.20	0.63	1.01	0.45	0.40	1.43	0.74
Al_20_3	1.75	1.17	1.28	1.76	4.86	12.48	13.89	6.63	7.70	17.12	13.93
Fe_20_3	3.38	1.11	1.05	1.49	2.72	7.41	4.40	4.34	7.98	9.85	4.71
MnO	0.19	0.16	0.14	0.10	0.13	0.17	0.12	0.34	0.20	0.14	0.06
MgO	15.20	7.13	2.12	5.05	3.08	5.88	8.98	11.75	9.16	6.49	0.73
CaO	33.54	44.24	47.37	44.17	35.96	14.13	22.09	21.92	21.84	6.10	1.95
Na_2O	0.38	0.25	0.01	0.33	0.99	2.24	0.77	1.59	1.41	3.90	2.51
K_2O	1.01	0.42	0.33	0.98	1.64	1.93	1.05	0.29	0.33	2.58	5.57
P_2O_5	0.02	0.02	0.05	0.02	0.06	0.20	0.01	0.03	0.02	0.40	0.18
LOI	33.70	37.20	39.20	31.10	23.58	6.12	3.46	1.52	1.26	1.70	1.28
Σ	101.16	99.66	99.20	99.83	99.42	100.26	100.34	100.65	100.44	100.17	100.42
Cr (ppm)	19	9	13	25	4	117	32	14	27	115	10
Ni	16	3	4	6	4	21	19	3	6	97	7
V	10	17	24	28	42	107	141	80	80	183	80
Zn	246	23	41	34	57	127	56	165	217	139	44
Rb	4	12	8	5	50	59	52	7	9	90	80
Ba	29	50	89	50	653	742	84	28	25	506	1464
Sr	220	116	702	404	762	497	126	138	124	583	206
Ga	2	3	0	6	5	22	13	12	15	18	22
Ta	0.02	0.05	0.41	0.04	0.58	0.65	0.79	0.73	0.55	0.99	1.56
Nb	0.4	1.1	1.2	0.3	4.4	7.0	1.2	7.5	5.9	12.0	30.4
Hf	0.03	0.50	0.51	1.67	1.15	4.49	2.74	7.06	5.57	5.82	9.68
Zr	1	25	17	66	39	170	128	223	186	228	453
Y	40	14	22	56	27	34	7	12	15	31	32
Th	0.22	0.36	1.39	0.15	1.80	3.91	0.32	4.33	1.86	1.30	0.17
La	75.64	8.08	24.52	102.2	35.26	24.34	6.23	5.57	9.03	26.22	17.64
Ce	132.9	14.73	38.91	189.5	59.01	53.42	13.40	13.78	22.15	57.05	39.03
Pr	14.70	1.97	5.03	22.24	7.03	7.36	1.70	2.09	3.29	7.64	5.79
Nd	49.40	8.01	18.54	77.47	25.59	31.21	6.65	8.57	13.60	31.71	26.88
Sm	7.83	1.74	3.22	12.88	4.44	6.58	1.31	1.99	3.10	6.65	6.33
Eu	1.20	0.41	0.80	1.74	1.17	2.28	0.46	0.48	0.55	2.17	4.70
Gd	6.67	2.02	3.06	10.83	3.84	6.16	1.12	1.88	2.66	6.35	6.46
Tb	0.98	0.30	0.47	1.53	0.58	0.93	0.17	0.32	0.42	0.95	0.96
Dy	5.85	1.85	2.75	9.21	3.55	5.48	1.07	1.97	2.49	5.66	5.78
Ho	1.25	0.39	0.57	1.86	0.74	1.11	0.22	0.44	0.50	1.11	1.17
Er	3.59	1.22	1.77	5.31	2.08	3.27	0.60	1.27	1.47	3.21	3.29
Tm	0.52	0.16	0.24	0.77	0.31	0.48	0.08	0.21	0.23	0.48	0.48
Yb	3.37	1.04	1.45	4.76	1.94	3.09	0.50	1.54	1.96	2.90	3.24
Lu	0.56	0.16	0.22	0.72	0.32	0.55	0.06	0.30	0.36	0.42	0.50

[1]Sample 98-1-1 = host rock at the contact with discordant carbonate intrusion; 98-1-4 = host rock ~5 cm away from the contact and from sample 98-1-1.

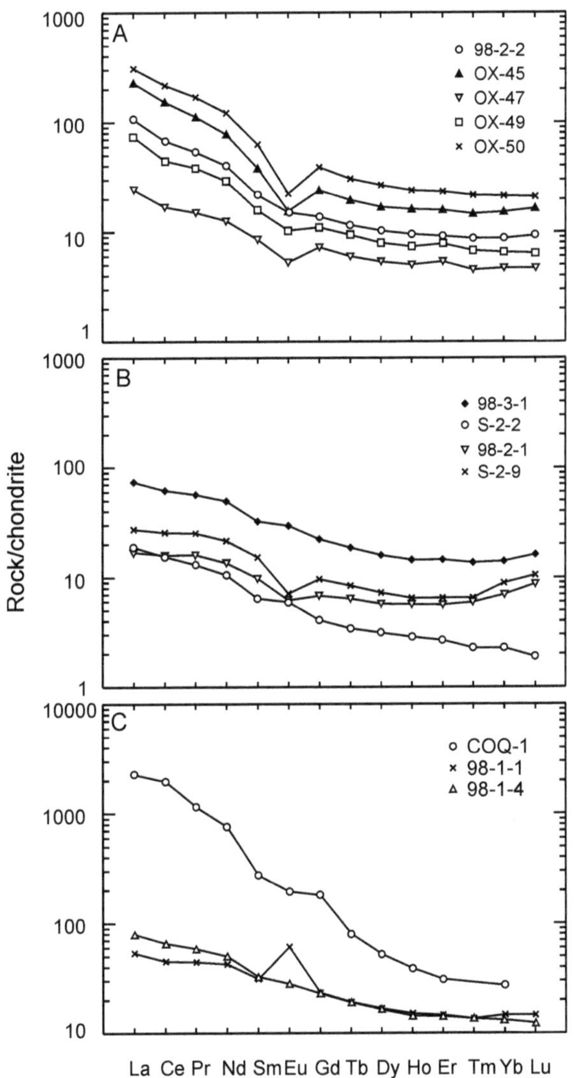

FIG. 4. Chondrite-normalized rare-earth element abundances in calcareous intrusions and related rocks. Normalizing values after Sun and McDonough (1989). A. Marbles. B. Calc-silicate rocks. C. Host rocks 98-1-4 (~5 cm away from the contact) and 98-1-1 (at the contact). Carbonatite from Oka Complex (USGS geochemical reference standard rock COQ-1) is shown for comparison.

rocks. As with most carbonate rocks, CaO correlates negatively with SiO_2 and Al_2O_3, suggesting that the rocks are admixtures of carbonate and silicate components. Correlations between modal abundances of minerals and chemical compositions are in agreement with observations of other carbonate rocks (e.g., Condie et al., 1991; Rock et al., 1987), which show that most major and trace elements are contained in non-carbonate, chiefly silicate phases. The marbles are poor in iron with respect to magnesium, and MgO/FeO ratios display a negative correlation with SiO_2. The rocks also have a wide range of the Mg/Ca ratio from 0.01 to 0.9, reflecting changes in the proportions of dolomite and Mg-rich silicates relative to calcite. Titanium, Zr, and REE are variable, and Ti shows a positive correlation with Zr. The varying concentrations of these elements may indicate an irregular distribution of heavy minerals

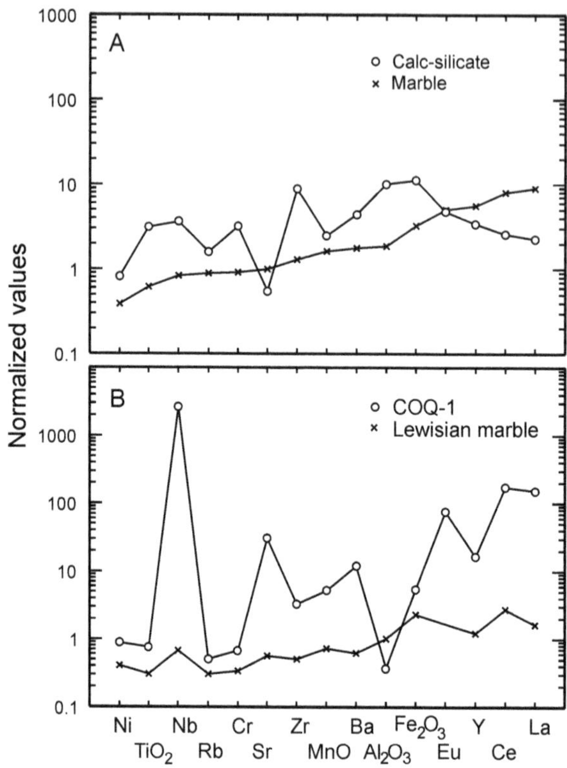

FIG. 5. Normalized element distributions in (A) average marble ("marble"; average of six samples) and average calc-silicate rock ("calc-silicate"; average of four samples) from the Oaxacan Complex. B. Carbonatite from the Oka Complex (USGS geochemical reference standard rock COQ-1) and average of Lewisian marble (Rock et al., 1987) are shown for comprison. Normalizing values for average Phanerozoic marine limestone are from Condie et al. (1991) and are as follows (in wt%): MnO = 0.084, Fe_2O_3 (total) = 0.54, Al_2O_3 = 1.0, TiO_2 = 0.2; (in ppm): Rb = 20, Cr = 15, Ni = 15, Y = 5, Zr = 20, Eu = 0.2, La = 5, Ba = 85, Sr = 400, Nb = 1.5.

in the original protolith. A wide range of Sr/Y ratios (4–35) in the marbles supports the important role of accessory phases in the distribution of trace elements in these rocks. The chondrite-normalized REE patterns of the marbles (Fig. 4) are similar; they are light REE-enriched, have a relatively flat heavy REE segment with $(Gd/Yb)_n$ between 1.6 and 1.9. The absolute concentrations are variable with La_n ranging between 24 and 310 × chondrites (Fig. 4). Some patterns exhibit Eu anomalies.

The abundances of some critical elements in the Oaxacan marbles and calc-silicate rocks normalized to an average Phanerozoic marine limestone of Condie et al. (1991) are shown in Figure 5. A sequence of elements in the graph is arranged according to increasing concentrations in the average Oaxacan marble. In general, the Oaxacan marbles resemble Phanerozoic limestones. Relative to average Phanerozoic marine limestone, the elements Ni through Rb are depleted while Sr through La are progressively enriched in the marble (Fig. 5). The pattern is also similar to those of other marbles from high-grade terranes, including those of the Precambrian Lewisian marbles (Fig. 5) from Scotland (Rock et al., 1987), which are of sedimentary origin. The patterns differ, however, from those of carbonatites, which are distinctly enriched in Nb, Sr and REE (Fig. 5).

Calc-silicate rocks

The chemical composition of the calc-silicate rocks reflects an admixture of carbonate and silicate components. Compared with the marbles, they contain a significantly higher proportion of silicate components and higher concentrations of oxides of most major elements including SiO_2, Al_2O_3, FeO_{tot},

TiO_2, Na_2O, but lower CaO (Table 3). Higher concentrations of some trace elements in the average of the Oaxacan calc-silicate rocks compared to the marbles and Phanerozoic limestones (Fig. 5) confirm the presence of these elements mainly in silicates. Negative anomalies of Mn and Sr in the patterns of the calc-silicates indicate that these elements occur chiefly in carbonates. The chondrite-normalized REE patterns of the calc-silicate rocks show an enrichment of light REE but are highly variable in their shapes (Fig. 4). They also have variable REE concentrations with La_n ~5–90 but, in general, lower than those of marbles. Inasmuch as metamorphic minerals are similar in both calc-silicates and marbles, it appears that the differences in REE patterns were inherited from the protoliths, which compositionally are similar to those of sedimentary rocks (Taylor and McLennan, 1985). The REE patterns resemble patterns of various clastic sedimentary rocks (Lentz, 2003).

Contact aureole

The host rocks at the immediate contact with the carbonate intrusions have significantly higher contents of SiO_2, K_2O, Ba, Zr, Hf, Nb, and Ta, but are lower in Al_2O_3, FeO_{tot}, MgO, MnO, CaO, Na_2O, TiO_2, Sr, Th, Cr, Ni, Zn, and light REE than the host rocks farther away from the contact (Table 3). The REE pattern of the contact rock (98-1-1) is similar to those of the gneiss but also shows a positive Eu anomaly (Fig. 4), consistent with the enrichment of secondary feldspars observed in the contact rocks. However, the rocks at a distance of about 5 cm from the contact (98-1-4) do not show an Eu anomaly.

Discussion

Various hypotheses have been invoked to explain the origin of calcareous rocks in high-grade metamorphic terranes. They include derivation from: (1) metasomatic calcareous skarns; (2) carbonatites; and (3) metamorphosed sediments including evaporites. The available data for the Oaxacan carbonates are now evaluated in the context of these different hypotheses.

Skarn origin

This hypothesis essentially assumes that calc-silicate rocks and impure marbles were produced by metasomatic introduction of silica and other elements into pure carbonate rocks. Metasomatic calcareous skarns are granoblastic rocks that generally occur at or near an intrusive igneous body and are typically characterized by the presence of calc-silicate minerals such as wollastonite, garnet (andradite or grossularite), epidote, and vesuviantite, with, or without calcite (Blatt and Tracy, 1996; Easton, 1995; Lentz, 1998). In the Oaxacan Complex, the absence of associated intrusive bodies and the absence of zonations in the calcareous bodies are inconsistent with a skarn genesis. The intrusive and crosscutting nature of some of the calcareous bodies, and the relatively sharp contacts between the calcareous intrusions and host rocks, are also incompatible with metasomatism.

Carbonatite origin

Melgarejo and Prol-Ledesma (1999) suggested that the calcareous bodies may be carbonatite intrusions. Several features are consistent with this model: (1) their intrusive nature; and (2) their post-peak of metamorphism emplacement. However, carbonatite intrusions are typically characterized by: (1) extensive fenitization of the surrounding rocks marked by K-metasomatism; (2) an unusual mineralogy including monazite, perovskite, scapolite, apatite, and zircon; and (3) very high contents of REE, as well as highly fractionated patterns (Fig. 4) and high-field-strength elements (HFSE), particularly Nb (Wall and Mariano, 1996; Wyllie et al., 1996; Hornig-Kjarsgaard, 1998). The Oaxacan calcareous bodies differ from carbonatites in having low contents of HFSE and REE (Fig. 5). Fenitization is absent in the host rocks, and several key minerals of carbonatites such as perovskite, zircon, barite, and magnetite are absent or present only in trace amounts.

Sedimentary origin

This model postulates that the intrusive calcareous rocks are isochemically metamorphosed limestones and related rocks remobilized under granulite facies and intruded into the host rocks (Ortega-Gutiérrez, 1984). However, a question arises about the nature of the protolith. The carbonates could be either marine limestones or a part of an evaporite sequence.

The Oaxacan calcareous bodies differ from Phanerozoic marine limestones (Taylor and McLennan, 1985), due to the presence of significant amounts of scapolite (Cl and Na rich), alkali-rich minerals including alkali feldspar and phlogopite, and Mg-rich minerals such as clinopyroxene, forsterite, dolomite, and phlogopite. These features,

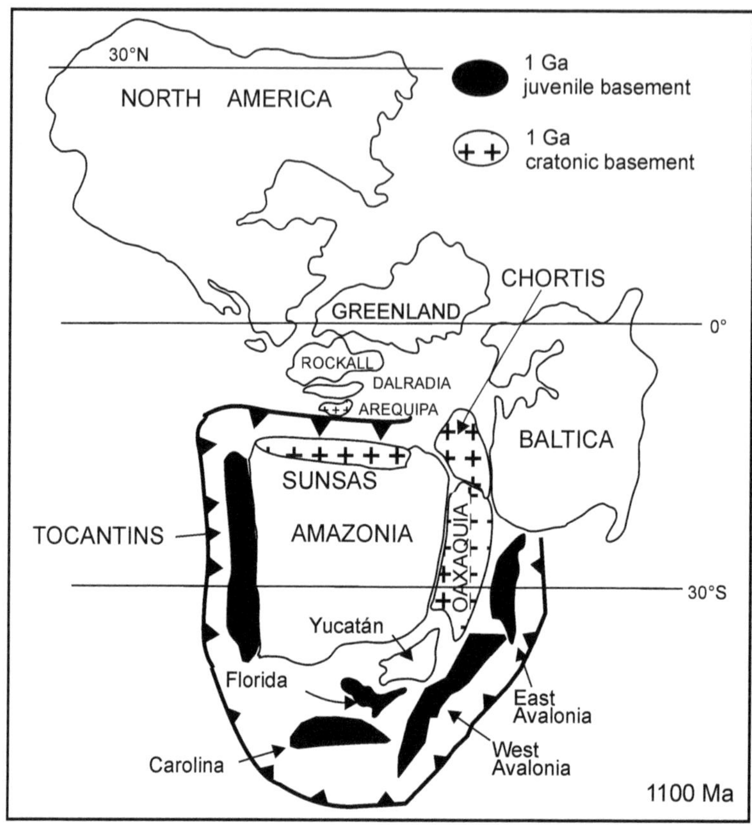

FIG. 6. 1100 Ma reconstruction of Rodinia showing the location of Oaxaquia, Mexico as part of a 1 Ga belt on the margin of Amazonia (modified after Keppie et al., 2003). Note that carbonates in Oaxaquia lie between 10° and 35°S, which are the limits for the formation of Phanerozoic evaporites. The barbed line is a subduction zone with triangles on the upper slab.

particularly the presence of scapolite (Moine et al., 1981; Tysdal and Desborough, 1997), suggest an evaporite precursor. Elevated concentrations of chlorine, alkalis, calcium, magnesium, and sulfur are principal characteristics of sedimentary rocks associated with evaporites (Moine et al., 1981). In fact, Ortega-Gutiérrez (1984) suggested that the rocks are metamorphosed carbonates from an evaporite sequence that originally formed in a passive-margin environment.

Evaporites are produced by the extensive if not complete evaporation of water from saline solutions. Precipitated minerals include calcite, dolomite, halite, sulfates (gypsum and anhydrite), and borax. Inasmuch as high-grade metamorphism destroys most of the evaporite minerals, scapolite and other minerals containing Na, Cl, S, F, and B (Hietanen, 1967; Serdyuchenko, 1975) such as tourmaline (Abraham et al., 1972), alkali feldspars, and Mg-rich carbonates have been used as indicators inherited from ancient evaporite piles. High abundance of meionitic scapolite, dolomite, phlogopite, albite, and K-feldspar in the Oaxacan carbonate bodies supports this genesis. The marialitic nature of the scapolite minerals implies the presence of Cl and Na in the system. Albite, K-feldspar, phlogopite, and cancrinite contain elevated concentrations of alkalis, whereas, dolomite, forsterite, clinopyroxene, and amphibole are rich in Mg, and sulfides are rich in S. On the other hand, tourmaline, which is also considered to be an important indicator of the evaporite precursor, is absent in Oaxacan rocks. As borates are some of the last minerals to precipitate from sea water, it is possible that the evaporites associated with these rocks never reached the stage of precipitating borates. An alternative explanation for the absence of tourmaline in the Oaxacan Complex may be thermal instability of the tourmaline under the low water pressures characteristic of granulite facies. Boron, if originally present, probably migrated with the fluid and melt phases to lower-temperature and wetter zones of the orogen. Low contents of boron may also indicate that the source of the Oaxacan carbonate rocks was non-granitic.

The marbles were probably derived from impure limestones and evaporites whereas the protholith of calc-silicate rocks were calcareous sediments with higher contents of silicates. The common presence of spinel may indicate some silica-poor, bauxitic component in the clay fraction, and thus probably highly weathered source rocks.

Conclusions

Based on the evidence presented here, it is concluded that the carbonate rocks of the Oaxacan Complex best fit the model of an evaporite-limestone precursor (Ortega-Gutiérrez, 1984). Precambrian evaporites became widespread after the ~1.3 Ga limited oxygenation event, which increased the marine sulfate reservoir (Kah et al., 2001). This together with the intrusion of rift-related plutons into the sediments at ~1157–1130 Ma (Keppie et al., 2003) brackets deposition of the Oaxacan sediments between ~1.3 and 1.16 Ga.

In the Phanerozoic, evaporites generally occur between 10° and 35° (McKerrow et al., 1992). Similar paleolatitudinal constraints appear to apply to ~1.3–1.0 Ga evaporites, which have been recorded: (1) in the 1.3–1.15 Ga evaporites at Balmat in the Adirondack Lowlands (Whelan et al., 1990); (2) in the 1.2 Ga carbonate-evaporite of Baffin and Bylot Islands (Kah et al., 2001), (3) at McArthur River in Australia (Williams and Ray, 1974); (4) at Outokumpo in Finland (Makela, 1974); and (5) in the ~1.1 Ga Upper Roan Group in Zambia (Strauss, 1993). The Oaxacan evaporites would also lie within these paleolatitudes if Oaxaquia is placed near either Amazonia or eastern Laurentia in ~1 Ga reconstructions of Rodinia (Fig. 6; e.g., Keppie et al., 2003). The apparent absence of similar-aged evaporites in Siberia (Bartley et al., 2001) may support the hypothesis of Sears and Price (2000) that Rodinia lay off western Laurentia at 1 Ga.

Evaporites are typically found in rifts associated with passive margins, such as the Red Sea, and rifts associated with active margins, such as the Gulf of California, and intra-arc rifts and backarc basins. Potentially correlative, ~1.3–1.15 Ga metasediments in the southern Oaxacan Complex are associated with arc volcanic rocks (Keppie et al., 2001), and the ~1157–1130 Ma rift-related plutons that may have been intruded during associated rifting (Keppie et al., 2003).

Acknowledgments

Funding for various aspects of this project was provided by CONACyT grants (0255P-T9506 and 25705-T), PAPIIT grants (IN116999 and IN10799) to JDK and FOG, and a NSERC Discovery grant to JD. We are grateful to A.K. Chatterjee for enlightening discussions and Drs. Brian Fryer and John Greenough for their constructive reviews.

REFERENCES

Abraham, K., Mielke, H., and Povondra, P., 1972, On the enrichment of tourmaline in metamorphic sediments of the Arzberg Series, N.E. Bavaria: Neues Jahrbuch fur Mineralogie, Monatshefte, v. 5, p. 14.

Ayer, J. A., and Davis, D. W., 1997, Neoarchean evolution of differing convergent margin assemblages in the Wabigoon Subprovince: Geochemical and geochronological evidence from the Lake of the Woods greenstone belt, Superior Province, northwestern Ontario: Precambrian Research. v. 8, p. 155–178.

Bartley, J. K., Kaufman, A. J., Semikhatov, M. A., Knoll, A. H., Pope, M. C., and Jacobsen, S. B., 2001, Global events across the Mesoproterozoic–Neoproterozoic boundary: C and Sr isotopic evidence from Siberia: Precambrian Research, v. 111, p. 165–202.

Blatt, H., and Tracy, R. J., 1996, Petrology: Igneous, sedimentary, and metamorphic: San Francisco, CA, W. H. Freeman and Company, 529 p.

Condie, K. C., Wilks, M., Rosen, D.M., and Zlobin, V. L., 1991, Geochemistry of metasediments from the Precambrian Hapschan Series, eastern Anabar Shield, Siberia: Precambrian Research, v. 50, p. 37–47.

Dostal, J., Baragar, W. R. A., and Dupuy, C., 1986, Petrogenesis of the Natkusiak continental basalts, Victoria Island, N.W.T.: Canadian Journal of Earth Sciences, v. 23, p. 622–632.

Easton, R. M., 1995, Regional geochemical variation in Grenvillian carbonate rocks: Implications for mineral exploration, *in* Summary of field work and other activities 1995: Ontario Geological Survey, Miscellaneous Paper 164, p. 6–18.

Hietanen, A., 1967, Scapolite in the Belt Series in St. Joe–Clearwater region, Idaho: Geological Society of America, Special Paper 86, 1–56.

Hornig-Kjarsgaard, I., 1998, Rare earth elements in sovitic carbonatites and their mineral phases: Journal of Petrology, v. 39, p. 2105–2120.

Kah, L. C., Lyons, T. W., and Chelsey, J. T., 2001, Geochemistry of a 1.2 Ga carbonate-evaporite succession, northern Baffin and Bylot Islands: Implications for Mesoproterozoic marine evolution: Precambrian Research, v. 111, p. 203–234.

Keppie, J. D., Dostal, J., Cameron, K. L., Solari, L. A., Ortega-Gutiérrez, F., and Lopez, R., 2003, Geochronology and geochemistry of Grenvillian igneous suites in the northern Oaxacan Complex, southern México: Tectonic implications: Precambrian Research, v. 120, p. 365–389.

Keppie, J. D., Dostal, J., Ortega-Gutierrez, F., and Lopez, R., 2001, A Grenvillian arc on the margin of Amazonia: Evidence from the southern Oaxacan Complex, southern Mexico: Precambrian Research, v. 112, p. 165–181.

Keppie, J. D., Solari, L. A., Ortega-Gutiérrez, F., Ortega-Rivera, A., Lee, J. W. K., and Hames, W. E., in press, U-Pb and $^{40}Ar/^{39}Ar$ constraints on the cooling history of the northern Oaxacan Complex, southern Mexico: Tectonic implications, *in* Tollo, R. P., Corriveau, L., McLelland, J. B., and Bartholemew, G., eds., Proterozoic tectonic evolution of the Grenville Orogen in North America: Geological Society of America Memoir.

Lentz, D. R., 1998, Mineralized intrusion-related skarn systems: Mineralogical Association of Canada, Short Course 26, 664 p.

______, 2003, Geochemistry of sediments and sedimentary rocks: evolutionary considerations to mineral deposit-forming environments: Geological Association of Canada, Geotext 4, 184 p.

Makela, M., 1974, A study of sulfur isotopes in the Outokumpo ore deposit, Finland: Geological Survey of Finland Bulletin 267, 45 p.

McKerrow, W. S., Scotese, C. R., and Brasier, M. D., 1992, Early Cambrian continental reconstructions: Journal Geological Society of London, v. 149, p. 599–606.

Melgarejo, J. C., and Prol-Ledesma, R. M., 1999, Th and REE deposits in the Oaxaca Complex in southern Mexico, *in* Stanley, C. J., Mineral deposits: Processes to precessing: Rotterdam, Netherlands, Balkema, p. 389–392.

Moine, B., Sauvan, P., and Jarousse, J., 1981, Geochemistry of evaporite-bearing series: A tentative guide for the identification of metaevaporites: Contributions to Mineralogy and Petrology, v. 76, p. 401–412.

Mora, C. I., Valley, J. W., and Ortega-Gutiérrez, F., 1986, The temperature and pressure conditions of Grenville-age granulite-facies metamorphism of the Oaxacan Complex, southern México: Revista del Instituto Mexicano del Petroleo, v. 5, p. 222–242.

Ortega-Gutiérrez, F., 1984, Evidence of Precambrian evaporites in the Oaxacan granulite complex of southern México: Precambrian Research, v. 23, p. 377–393.

Ortega-Gutiérrez, F., Ruiz, J., and Centeno-Garcia, E., 1995, Oaxaquia, a Proterozoic microcontinent accreted to North America during the late Paleozoic: Geology, v. 23, p. 1127–1130.

Prakash, G. O., Murillo, M. G., Grajales, N. J. M., Torres, V. R., and Bosh, G. P., 1991, A rare wollastonite-quartz-graphite assemblage from a high-grade regional metamorphic terrain of late Precambrian age in Oaxaca, Mexico: Revista del Instituto Mexicano del Petroleo, v. 13, p. 5–13.

Rock, N. M. S., Davis, A. E., Hutchison, D., Joseph, M., and Smith, T. K., 1987, The geochemistry of Lewisian marbles, *in* Park, R. G., and Tarney, J., eds., Evolution of the Lewisian and comparable Precambrian high grade terrains: Geological Society (London) Special Publication 27, p. 109–126.

Sears, J. W., and Price, R. A., 2000, New look at the Siberian connection: No SWEAT: Geology, v. 28, p. 423-426.

Serdyuchenko, D. P., 1975, Some scapolite-bearing rocks evolved from evaporites: Lithos, v. 8, p. 1–7.

Solari, L. A., Keppie, J. D., Ortega-Gutiérrez, F., Cameron, K. L., Lopez, R., and Hames, W. E., 2003, 990 Ma and 1,100 Ma Grenvillian tectonothermal events in the northern Oaxacan Complex, southern Mexico: Roots of an orogen: Tectonophysics, v. 365, p. 257–282.

Strauss, H., 1993, The sulfur isotopic record of Precambrian sulfates: New data and a critical evaluation of the existing record: Precambrian Research, v. 63, p. 225–246.

Sun, S. S., and McDonough, W. F., 1989, Chemical and isotopic systematics of oceanic basalts: Implications for mantle composition and processes, *in* Saunders, A. D., and Norry, M. J., Magmatism in the ocean basins: Geological Society (London) Special Publication 42, p. 313–345.

Taylor, S. R., and McLennan, S. M., 1985, The continental crust: Its composition and evolution: Oxford, UK, Blackwell Scientific, 328 p.

Tysdal, R. G., and Desborough, G. A., 1997, Scapolitic metaevaporite and carbonate rocks of Proterozoic Yellowjacket Formation, Moyer Creek, Salmon River Mountains, central Idaho: U.S. Department of the Interior, U. S. Geological Survey, Open File Report 97-268.

Wall, F., and Mariano, A. N., 1996, Rare earth minerals in carbonatites: A discussion centred on Kangankunde carbonatites, Malawi., *in* Jones, A. P., Wall, F., and Williams, C. T., eds., Rare earth minerals: Chemistry, origin, and ore deposits: London, UK, Chapman and Hall, p. 193–225.

Whelan, J. F., Rye, R. O., deLorraine, W., and Ohmotot, H., 1990, Isotopic geochemistry of a mid-Proterozoic evaporite basin: American Journal of Science, v. 290, p. 396–424.

Williams, N., and Ray, D. M., 1974, Alternative interpretation of sulphur isotope ratios in the McAuthur lead-zinc-silver deposit: Nature, v. 247, p. 535–537.

Wyllie, P. J., Jones, A. P., and Deng, J., 1996, Rare earth elements in carbonate-rich melts from mantle to crust, *in* Jones, A. P., Wall, F., and Williams, C. T., eds., Rare earth minerals: Chemistry, origin, and ore deposits: London, UK, Chapman and Hall, p. 77–103.

Geochronology and Geochemistry of the ~917 Ma, Calc-alkaline Etla Granitoid Pluton (Oaxaca, Southern Mexico): Evidence of Post-Grenvillian Subduction along the Northern Margin of Amazonia

C. ORTEGA-OBREGON, J. D. KEPPIE,[1] L. A. SOLARI, F. ORTEGA-GUTIÉRREZ,

Instituto de Geología, Universidad Nacional Autónoma de México (UNAM), 04510 México D.F., México

J. DOSTAL,

Department of Geology, St. Mary's University, Halifax, Nova Scotia B3H 3C3, Canada

R. LOPEZ,

Geology Department, West Valley College, Saratoga, California 95070

A. ORTEGA-RIVERA,

Centro de Geociencias, Campus Juriquilla, Universidad Nacional Autónoma de México (UNAM), Apdo. Postal 1-742, Centro Querétaro, Qro. 76001, Mexico

AND J. W. K. LEE

Department of Geology, Queens University, Kingston, Ontario, Canada, K7L 3NG

Abstract

The post-tectonic Etla pluton intrudes the ~1 Ga granulitic Oaxacan Complex that cooled through 450°C by ~945 Ma. The Etla pluton consists of massive, coarse, porphyritic granodiorite-monzogranite (plagioclase, K-feldspar, quartz, biotite ± hornblende) with fine-grained felsic rocks along the margin. Geochemistry indicates that it is a peraluminous, I-type, medium-K, calc-alkaline, volcanic-arc granite-trondjemite with relatively low contents of high-field-strength elements and flat REE patterns. U-Pb zircon isotopic analyses fall on a chord with intercepts at 180 ± 50 Ma and 920 ± 25 Ma: the latter is similar to the $^{207}Pb/^{206}Pb$ age of 917 ± 6 Ma of the least discordant (1%) analysis and is inferred to date the time of intrusion. This pluton is synchronous with similar igneous activity in Avalonia (eastern Appalachians) and in Tocantins Province of central Brazil, which may form parts of a peri-Amazonian magmatic arc. $^{40}Ar/^{39}Ar$ laser step-heating analyses of biotite and K-feldspar yielded plateau ages of 207 ±5 Ma and 221 ± 3 Ma, respectively, that may be related to Phanerozoic reheating.

Introduction

THE ETLA PLUTON intrudes the ~1 Ga Oaxacan Complex about 20 km NNW of Oaxaca in southern Mexico (Fig. 1A). Previous geochronology produced a whole-rock Rb-Sr isochron of 272 ± 8 Ma, with an initial $^{87}Sr/^{86}Sr$ ratio of 0.7047 ± 0.0005 (Ruiz-Castellenos, 1979). This led Torres et al. (1999) to include it in a Permo-Triassic magmatic arc that extended along the length of Mexico. However, in the course of mapping the northern Oaxacan Complex, the La Carbonera stock located about 10 km northwest of the Etla pluton yielded an intrusive age of 275 ± 4 Ma (lower intercept zircon age: Solari et al., 2001). Contrasts in rock type and structure between the Etla pluton and La Carbonera stock led us to further examine the Etla pluton.

[1]Corresponding author; email: duncan@servidor.unam.mx

Geological Setting

The Oaxacan Complex that hosts the Etla pluton has been interpreted as a >1160 Ma arc intruded by two rift-related plutonic suites at ~1160–1130 Ma and ~1012 Ma, which were deformed during two tectonothermal events (Olmecan and Zapotecan) dated at ~1100 Ma and ~1004–980 Ma, respectively, the latter involving granulite-facies metamorphism (Keppie et al., 2001, 2003a; Solari et al.,

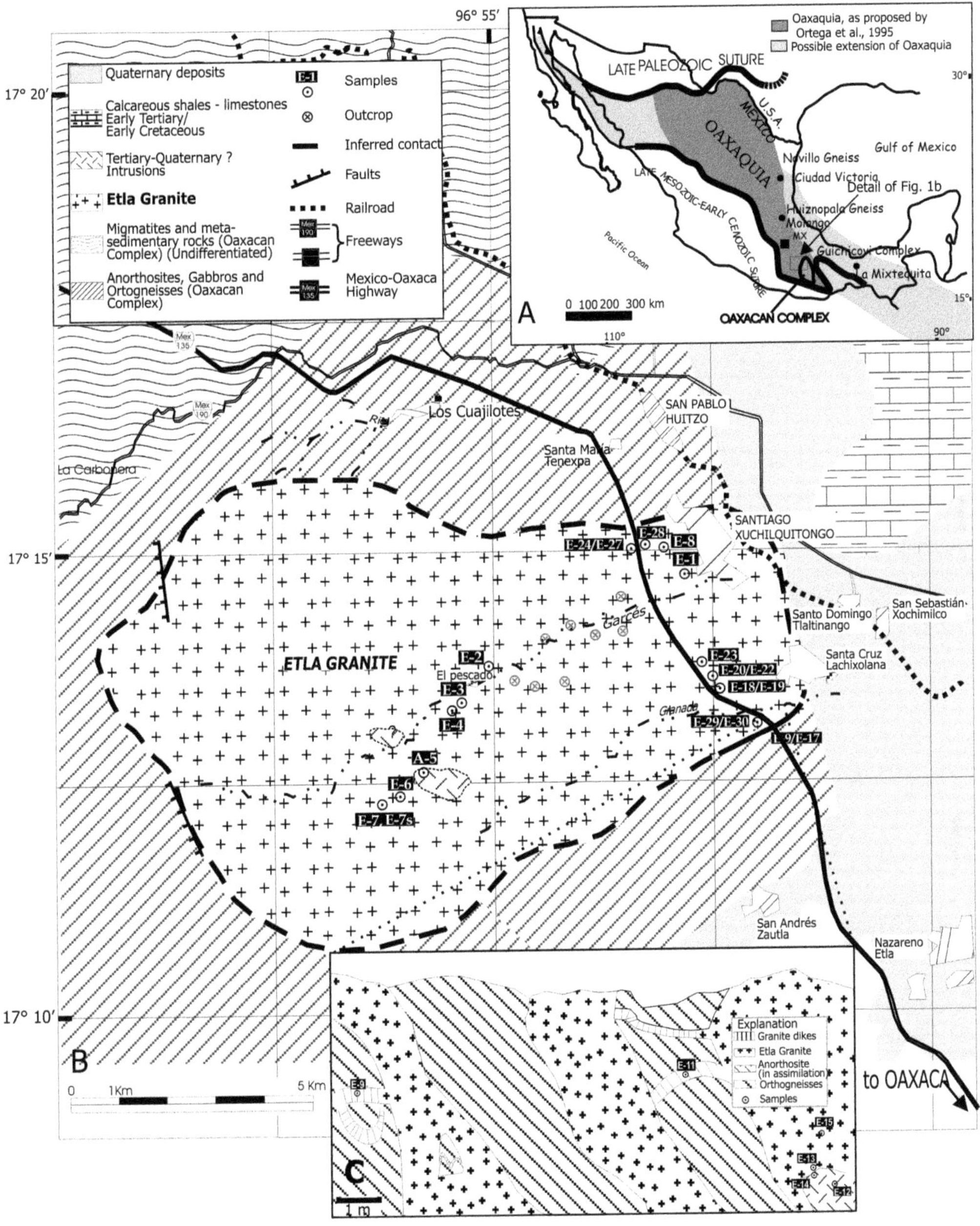

FIG. 1. A. Location of Oaxacan Complex within Oaxaquia, modified after Keppie et al., 2001. B. Geological map of the Etla pluton and surrounding area. C. Sketch maps of southern contact of the pluton showing deformed felsic sheets cut by undeformed sheets.

2003). Cooling through ~450°C and ~300°C is recorded by $^{40}Ar/^{39}Ar$ ages on phlogopite and biotite at 945 ± 10 Ma and 856 ± 10 Ma, respectively (Keppie et al., in press). These and similar rocks exposed in inliers underlie the backbone of Mexico and constitute the Oaxaquia microcontinental terrane (Fig. 1B) (Ortega-Gutierrez et al., 1995). Although traditionally regarded as a southern continuation of the

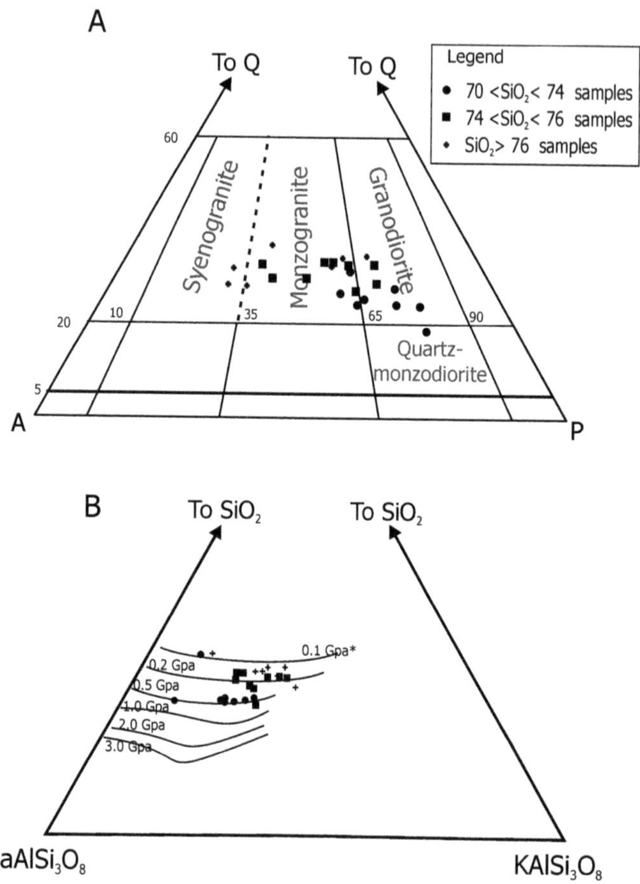

FIG. 2. Modal analyses of the Etla samples plotted on (A) quartz–alkali feldspar–plagioclase diagram (after Streckeisen, 1976), and (B) quartz–albite–orthoclase diagram, showing minimum melt pressures.

Grenville orogen of eastern Laurentia (de Cserna, 1971; Shurbert and Cebull, 1987; Dalziel, 1992; Karlstrom et al., 1999; Burrett and Berry, 2000), paleomagnetic data indicate that it is an allochthonous terrane (Ballard et al., 1989) probably derived from Amazonia based on comparable geological records (Keppie and Ortega-Gutiérrez, 1995, 1999; Keppie et al., 2001, 2003b; Solari et al., 2003) and Lower Paleozoic faunal provinciality (Robison and Pantoja-Alor, 1968: faunal nomenclature revised by Shergold, 1975; Boucot et al., 1997). Most current reconstructions of the 1 Ga Rodinia juxtapose eastern Laurentia and western South America (Dalziel et al., 2000; Cawood et al., 2001), and the 1190–980 Ma Grenvillian orogeny has generally been attributed to arc-continent followed by continent-continent collision (Starmer, 1996; Sadowski and Bettencourt, 1996; Rivers, 1997). In such a scenario, the nature of the Etla pluton and its bearing on the location of Oaxaquia are critical.

The Etla pluton is an oval shaped pluton, ~80 km^2 in area, that intrudes the Oaxacan Complex (Fig. 1A). It varies from coarsely crystalline, porphyritic granodiorite and granite to fine-grained aplites on the margin. Pegmatites also occur sporadically. The intrusive contact exposed on Toll Highway 135 shows that it cuts across all the structural fabrics in the anorthosites and orthogneisses. Some intrusive sheets emanating from the pluton are deformed by E-trending, subhorizontal, upright folds with an axial planar fabric developed under lower greenschist-facies metamorphism. On the other hand, some of the granite sheets cut across these folds, suggesting that deformation was synchronous with

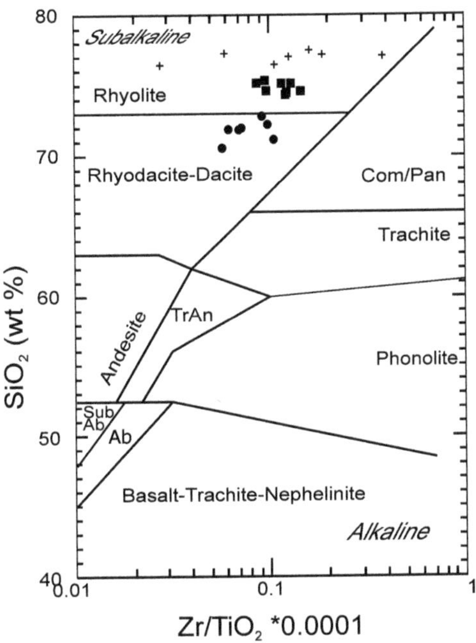

FIG. 3. Etla granitoid analyses plotted on SiO_2 versus Zr/TiO_2 diagram, after Winchester and Floyd (1977).

intrusion (Fig. 1C). The orientation of these structures is nearly perpendicular to the structures in the Oaxacan Complex, which are generally NNW-trending and gently dipping (Solari et al., 2003). The pluton is unconformably overlain by Tertiary and Quaternary deposits along its eastern margin (Fig. 1A).

Petrography

The Etla granitoid varies from megacrystic to aplitic, and from quartz-porphyritic to K-feldspar megacrystic (up to 12 cm in size). The rock is essentially unfoliated except along certain parts of its margins, where it appears strongly cleaved and locally mylonitic. The primary, magmatic mineralogy consists essentially of quartz + K-feldspar + oligoclase + magnetite ± biotite ± prismatic, green-brown hornblende, with euhedral titanite, euhedral zircon, and apatite as common accessories, and rare allanite and xenocrystic garnet. Modal analyses of the coarse Etla rocks indicate that they are mainly granodiorite and monzogranite with a few rocks falling into the syenogranite field (Fig. 2). The presence of hornblende-titanite-magnetite suggests an I-type granitoid. The fine-grained marginal facies has a similar mineralogy, but generally lacks ferromagnesian minerals and contains secondary muscovite.

Metamorphism accompanied by distinct deformation affected most rocks, but some remained undeformed and preserve fresh igneous textures. Reddish-pale brown biotite is pervasively altered to green biotite and in a few cases to chlorite. Epidote/clinozoisite and white mica are common hydrothermal minerals, and most alkaline feldspar was converted to microcline and microcline perthite. The effects of deformation include undulose extinction of plagioclase, lobate to serrated margins of all quartz grains and some plagioclase, subgrain recrystallization, checker patterns in quartz, mortar structure in feldspar, and ubiquitous myrmeckite. Plagioclase shows mild to moderate oscillatory zoning and is partly sericitized, whereas white mica, typically in combination with magnetite, replaced oligoclase, suggesting the reaction magnetite + oligoclase = paragonitic phengite + epidote.

Evidence of rapid magmatic crystallization and high vapor pressure is the common presence of alkaline feldspar megacrysts, acicular apatite, inequigranular textures, feldspar zoning, and complex replacement textures, including abundant myrmeckite (Pitcher, 1993). This is consistent with the 6–15 km depth deduced from the minimum melt graph (Fig. 2B). On the other hand, development of granoblastic textures (blastesis) and metamorphic minerals such as white mica, biotite, and microcline clearly indicate subsequent burial metamorphism of the granite.

Analytical Methods

Twenty-six samples were selected for chemical analyses from ~100 samples collected during the mapping of the pluton. The major and trace elements Rb, Ba, Sr, Zr, Nb, Y, Ga, Cr, Ni, Cu, and Zn were analyzed by X-ray fluorescence methods at the Geochemical Centre at Saint Mary's University, Halifax, Nova Scotia. Precision is generally better than 5% for the major oxides and between 5 and 10 % for trace elements (Dostal et al., 1986). Subsequently, three samples were selected for the determination of rare-earth elements (REE), Th, U, Ta, Zr, Nb, and Y by an inductively coupled plasma–mass spectrometry at the Geoscience Laboratories of the Ontario Geological Survey at Sudbury, Ontario. This method was by described by Ayer and Davis (1997). The

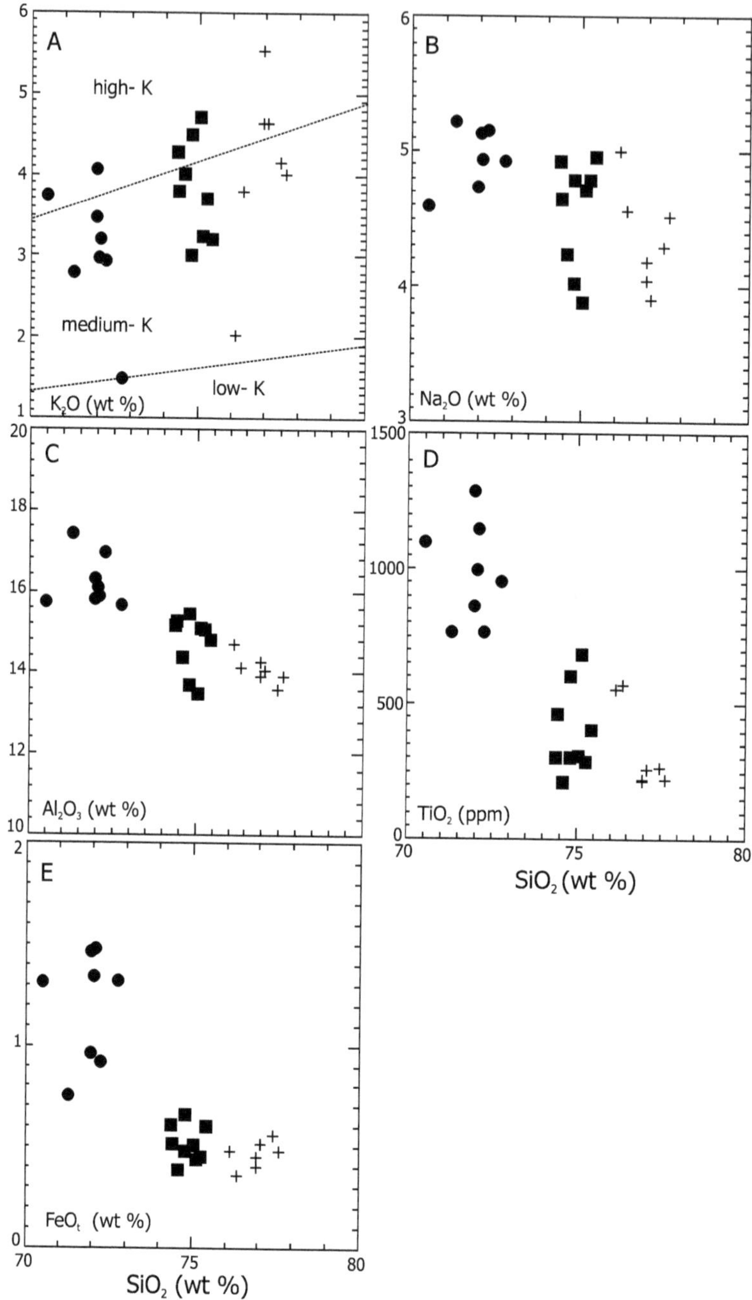

FIG. 4. Etla granitoid analyses plotted on various diagrams. A. K_2O versus SiO_2. B. Na_2O versus SiO_2. C. Al_2O_3 versus SiO_2. D. TiO_2 versus SiO_2. E. FeOt versus SiO_2.

TABLE 1. Chemical Analysis of Rocks from the Etla Pluton

Sample:	E-1	E-2	E-3	E-4	E-4P	A-5	E-6	E-7	E-8	E-9	E-10	E-11
SiO_2, wt%	75.04	72.03	74.61	71.95	76.94	68.06	75.28	71.94	77.45	75.42	76.34	70.49
TiO_2	0.051	0.166	0.035	0.143	0.037	0.647	0.048	0.214	0.044	0.067	0.095	0.183
Al_2O_3	13.49	16.09	14.37	16.3	14.26	15.83	15.05	15.81	13.59	14.81	14.13	15.75
Fe_2O_3	0.46	1.31	0.32	0.95	0.33	2.62	0.43	1.42	0.54	0.59	0.35	1.28
FeO	0.1	0.16	0.1	0.11	0.1	0.32	0.06	0.18	0.07	0.07	0.05	0.16
FeOT	0.51	1.34	0.39	0.96	0.4	2.68	0.45	1.46	0.56	0.6	0.36	1.31
Fe_2O_3T	0.57	1.49	0.43	1.07	0.44	2.98	0.5	1.62	0.62	0.67	0.41	1.46
MnO	0.001	0.034	0.001	0.019	0.002	0.048	0.001	0.035	0.004	0.003	0.002	0.042
MgO	0.01	0.37	0.01	0.31	0.01	2.13	0.02	0.52	0.02	0.02	0.03	0.21
CaO	0.82	2.39	1.16	1.52	0.85	0.83	1.23	2.34	0.86	1.36	0.68	1.55
Na_2O	3.88	5.13	4.24	4.73	4.18	5.85	4.79	4.73	4.29	4.96	4.56	4.59
K_2O	4.71	2.97	4.02	4.06	4.63	1.45	3.71	3.47	4.15	3.21	3.8	3.74
P_2O_5	0.015	0.066	0.015	0.024	0.013	0.24	0.013	0.077	0.013	0.012	0.016	0.067
LOI[1]	0.09	0.17	0.19	0.93	0.09	2.1	0.3	0.28	0.1	0.4	0.3	0.56
Total	98.66	100.91	99.09	101.06	101.46	100.17	100.94	101.04	101.14	100.93	100.36	98.64
ACNK[2]	1.040	1.006	1.070	1.091	1.062	1.246	1.065	1.001	1.036	1.050	1.100	1.092
Cr, ppm	<4	<4	<4	<4	<4	27	<4	4	<4	5	<4	<4
Ni	<3	<3	<3	<3	<3	19	3	<3	<3	<3	<3	<3
Co	<5	<5	<5	<5	<5	10	<5	<5	<5	<5	<5	<5
Cu	<5	<5	<5	<5	<5	28	<5	<5	<5	<5	<5	<5
Pb	12	10	16	13	14	13	11	8	13	13	16	18
V	11	29	11	25	13	78	15	36	12	17	<4	23
Zn	20	60	16	45	16	74	19	58	22	23	18	98
Rb	116	81	144	94	128	63	106	84	121	98	113	167
Ba	305	759	449	1220	354	454	525	1244	<5	333	479	602
Sr	176	708	410	674	355	439	421	774	72	328	267	414
Ga	28	24	20	21	24	23	22	24	26	24	19	28
Nb	8	3	3	3	4	5	3	5	6	3	6	13
Zr	60	119	51	123	48	183	63	134	26	65	27	107
Y	<3	<3	<3	<3	<3	<3	<3	<3	<3	<3	<3	3
Th	8	6	6	7	7	9	6	6	6	7	8	8
Nd	6	<5	6	<5	<5	31	<5	<5	6	<5	<5	5
U	2	<1	<1	<1	1	<1	<1	<1	3	1	1	1
La	10	8.36	<5	<5	9	22	<5	11	<5	<5	<5	2.22
Ce	–	16.03	–	–	–	–	–	–	–	–	–	5.42
Pr	–	1.95	–	–	–	–	–	–	–	–	–	0.64
Nd	–	7.49	–	–	–	–	–	–	–	–	–	2.91
Sm	–	1.37	–	–	–	–	–	–	–	–	–	0.80
Eu	–	0.45	–	–	–	–	–	–	–	–	–	0.23
Gd	–	1.00	–	–	–	–	–	–	–	–	–	0.95
Tb	–	0.13	–	–	–	–	–	–	–	–	–	0.15
Dy	–	0.69	–	–	–	–	–	–	–	–	–	0.88
Ho	–	0.15	–	–	–	–	–	–	–	–	–	0.20
Er	–	0.41	–	–	–	–	–	–	–	–	–	0.59
Tm	–	0.06	–	–	–	–	–	–	–	–	–	0.10
Yb	–	0.40	–	–	–	–	–	–	–	–	–	0.73
Lu	–	0.08	–	–	–	–	–	–	–	–	–	0.13

[1]Loss on ignition.

[2]Modal $Al_2O_3/(CaO + Na_2O + K_2O)$.

E-12A	E-12B	E-15	E-16	E-17	E-18	E-19	E-20	E-21	E-22	E-23	E-24	E-28
72.77	76.95	72.24	71.3	77.08	74.8	75.15	76.13	74.42	77.61	74.81	74.39	72.06
0.159	0.036	0.127	0.127	0.043	0.05	0.114	0.092	0.077	0.037	0.1	0.05	0.191
15.67	13.92	16.94	17.42	14.05	13.68	15.1	14.71	15.28	13.9	15.44	15.16	15.87
1.29	0.44	0.9	0.73	0.51	0.47	0.43	0.47	0.51	0.47	0.65	0.6	1.44
0.16	0.05	0.11	0.09	0.06	0.059	0.05	0.058	0.06	0.058	0.08	0.075	0.18
1.32	0.45	0.92	0.75	0.52	0.48	0.44	0.48	0.52	0.48	0.66	0.61	1.48
1.47	0.5	1.02	0.83	0.58	0.54	0.49	0.53	0.58	0.53	0.74	0.68	1.64
0.035	0.004	0.028	0.013	0.004	0.007	0.005	0.008	0.011	0.011	0.009	0.004	0.037
0.28	0.01	0.16	0.15	0.01	0.01	0.09	0.09	0.06	0.01	0.14	0.09	0.5
2.51	0.36	2.31	2.23	1.05	0.79	1.34	1.68	1.37	0.92	1.57	1.03	2.38
4.93	4.04	5.15	5.21	3.9	4.02	4.71	5	4.65	4.52	4.79	4.92	4.94
1.49	5.54	2.93	2.79	4.64	4.49	3.24	2.02	3.8	4.02	3	4.28	3.21
0.068	0.013	0.048	0.033	0.015	0.014	0.014	0.016	0.013	0.013	0.033	0.021	0.068
1.1	0.19	0.59	0.89	0.1	0.2	0.58	0.57	0.49	0	0.8	0.39	0.36
100.48	101.55	101.54	100.99	101.46	98.60	100.83	100.85	100.75	101.56	101.43	101.02	101.26
1.097	1.047	1.069	1.113	1.053	1.060	1.103	1.092	1.072	1.033	1.104	1.038	0.996
14	<4	<4	<4	<4	<4	<4	<4	<4	<4	<4	<4	<4
52	<3	<3	<3	<3	<3	<3	<3	<3	<3	<3	<3	<3
<5	<5	<5	<5	<5	<5	<5	<5	<5	<5	<5	<5	<5
<5	<5	<5	<5	<5	<5	<5	<5	<5	<5	<5	<5	<5
14	20	14	12	11	17	11	13	15	13	14	11	11
25	11	24	22	13	12	19	16	16	12	18	13	32
79	18	70	51	26	30	31	39	39	26	38	19	65
90	251	115	105	145	149	101	80	105	111	92	100	80
684	120	1497	1353	561	623	738	226	647	100	395	300	830
587	32	729	813	284	277	564	463	446	99	498	388	710
23	28	27	24	18	25	22	25	24	29	23	28	24
9	5	10	7	3	5	4	6	3	4	7	5	4
147	140	127	136	81	59	101	101	90	59	107	65	139
<3	15	<3	<3	<3	<3	<3	<3	<3	<3	<3	<3	<3
8	9	8	8	8	6	6	6	5	8	7	7	6
14	11	<5	<5	<5	8	9	7	9	6	6	<5	<5
<1	8	<1	<1	1	2	<1	<1	<1	4	<1	1	<1
9	13	1.99	12	<5	5	<5	12	<5	<5	5	6	10
–	–	3.50	–	–	–	–	–	–	–	–	–	–
–	–	0.46	–	–	–	–	–	–	–	–	–	–
–	–	1.88	–	–	–	–	–	–	–	–	–	–
–	–	0.39	–	–	–	–	–	–	–	–	–	–
–	–	0.26	–	–	–	–	–	–	–	–	–	–
–	–	0.40	–	–	–	–	–	–	–	–	–	–
–	–	0.06	–	–	–	–	–	–	–	–	–	–
–	–	0.31	–	–	–	–	–	–	–	–	–	–
–	–	0.06	–	–	–	–	–	–	–	–	–	–
–	–	0.20	–	–	–	–	–	–	–	–	–	–
–	–	0.03	–	–	–	–	–	–	–	–	–	–
–	–	0.23	–	–	–	–	–	–	–	–	–	–
–	–	0.04	–	–	–	–	–	–	–	–	–	–

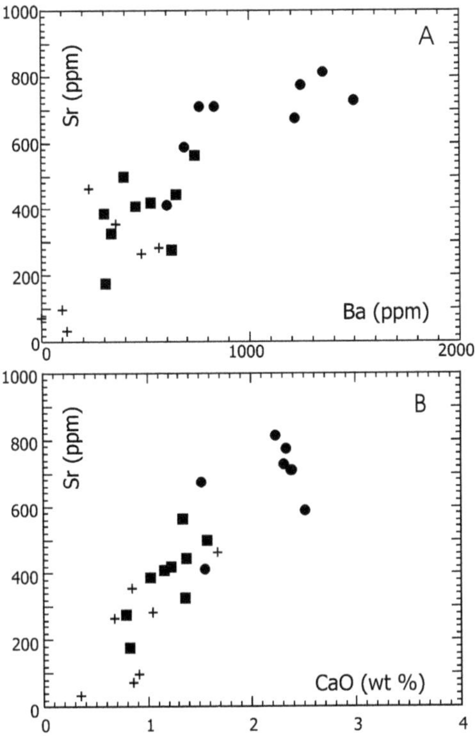

FIG. 5. Etla granitoid trace element analyses plotted on (A) Sr versus Ba diagrams; and (B) Sr versus CaO diagram.

analytical error of the trace element determinations is 2–10%. Analyses are compiled in Table 1.

All the rocks have been affected by secondary processes and low-grade metamorphism. However, as will be shown below, the chemical composition of these rocks does not appear to have been significantly modified by these processes.

Geochemistry

The rocks of the Etla pluton have SiO_2 contents ranging from 70 to 77 wt% (Figs. 3A and 3B); they are slightly peraluminous (Shand, 1951) with Al_2O_3/($CaO + Na_2O + K_2O$) > 1. Most samples have A/CNK between 1.05 and 1.1. The value of 1.1 was used as the dividing line between I-type and S-type granites (Chappel and White, 1974) and thus by this definition, most of these rocks can be classified as I-type granites. This is consistent with the low initial $^{87}Sr/^{86}Sr$ ratio of 0.7047 ± 0.0005 recorded by Ruiz-Castellanos (1979). Major-element oxides show correlations with SiO_2, which are typical of calc-alkaline suites (Rollinson, 1993; Fig. 4). The plot of K_2O versus SiO_2 shows a shallow sloping trend typical of a medium-K calc-alkaline suite. In terms of normative An-Ab-Or components, the rocks plot into the granite and trondhjemite fields. The rocks have relatively low contents of high-field-strength elements such as Nb and Zr. In particular, Nb contents of 3–13 ppm are low compared to typical granite values of ~20 ppm (Table 1). Zr generally decreases with increasing SiO_2, presumably reflecting zircon fractionation. Plots of Sr versus Ba and Sr versus CaO (Fig. 5) show trends indicative of feldspar fractionation.

The REE patterns (Fig. 6) are variable but generally flat, with (La/Yb)n ranging from 2 to 13. The analyzed samples cluster in the volcanic-arc granite or

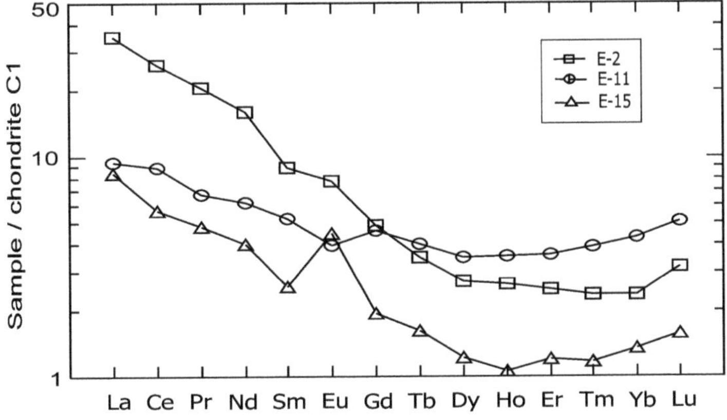

FIG. 6. Etla granitoid rare-earth analyses compared with chondrites, after Sun and McDonough (1989).

Table 2. U-Pb Isotopic Analyses of Zircons, Etla Granite, Mexico[1]

Fraction[2]	Weight mg	U ppm	Total Pb ppm	Com. Pb pg	$^{206}Pb/^{204}Pb$	$^{207}Pb/^{206}Pb$	$^{208}Pb/^{206}Pb$	$^{206}Pb^*/^{238}U$	$^{207}Pb^*/^{235}U$	$^{207}Pb^*/^{206}Pb^*$	$^{206}Pb^*/^{238}U$	$^{207}Pb^*/^{235}U$	$^{207}Pb^*/^{206}Pb^*$	% Disc.
					Observed ratios[3]			Atomic ratios			Age (Ma)[4]			
Etla 1, abr, stby, multif., 5 grn	0.11	101	10	110	355	0.00986	0.00281	0.08699	0.79102	0.06595	538	592	805 ± 6.5	33
Etla 2, stby, multif, transp, no abr, 14 grns	0.1	91	11	200	185	0.01096	0.00540	0.09141	0.84604	0.06713	564	622	842 ± 12	33
Etla 3, prism, abr, 12 grns	0.15	162	14	8	1159	0.00616	0.00086	0.08597	0.79427	0.06701	532	594	838 ± 19	37
Etla 4, bip, elg, no abr	0.26	67	6	25	1071	0.01482	0.00093	0.08439	0.76695	0.06591	522	578	804 ± 12	35
Etla 5, 1 grn, abr	0.06	38	6	16	1338	0.08022	0.17904	0.15098	1.44971	0.06964	906	910	917 ± 6	1

[1]Asterisk symbol (*) denotes radiogenic Pb. Zircon sample dissolution and ion-exchange chemistry modified after Krogh (1973) and Mattinson (1987) in Parrish (1987) -type microcapsules.
[2]Abbreviations: abr = abraded; grn = grains; prism =prismatic; elg = elongate; multif= multifaceted; stby= stubby; transp= transparent; bip= bypiramidal terminations.
[3]Observed isotopic ratios are corrected for mass fractionation of 1‰ for both ^{208}Pb and ^{205}Pb spiked fractions. Fractions spiked with the mixed $^{235}U/^{205}Pb$ tracer are also corrected for spike and blank relative contributions. Uncertainties in the $^{206}Pb/^{204}Pb$ ratio vary from 0.1% to 2.4%.
[4]Decay constants used: $^{238}U = 1.55125 \times 10^{-10}$; $^{235}U = 9.48485 \times 10^{-10}$; $^{238}U/^{235}U = 137.88$. Estimated uncertainties of the U/Pb ratio are ± 0.4 based on replicate analyses of a single zircon standard fraction (see Lopez et al., 2001). $^{207}Pb^*/^{206}Pb^*$ age uncertainties are 2 sigma and from the data reduction program PBDAT of K. Ludwig (1991). Total processing Pb blank amount varied between 2 pg and 30 pg, generally averaging <10 pg at UCSC. At LUGIS, total analytical blanks ranged between 40 and 200 pg. Initial Pb composition are from isotopic analysis of feldspar separates. Isotopic data were measured on a VG 54-30 sector multicollector mass spectrometer with a pulse counting Daly detector at UC Santa Cruz, whereas at LUGIS was used a Finnigan MAT 262 multicollector mass spectrometer, with a SEM multiplier.

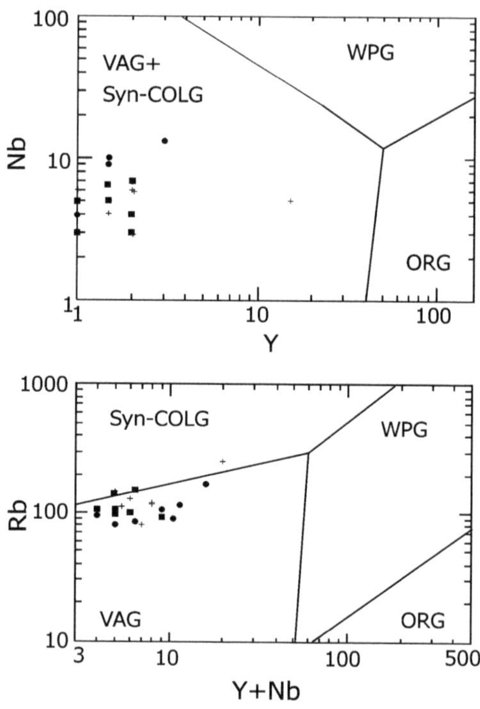

FIG. 7. Etla granitoid analyses plotted on: (A) Nb versus Y diagram, and (B) Rb versus Y + Nb diagram, after Pearce et al. (1984).

syncollisional granite fields on the Nb-Y diagram, and mainly in the volcanic-arc field on the Rb-(Y+Nb) tectonic setting discrimination diagram (Fig. 7).

Geochronology

U-Pb data

Two samples were collected for U-Pb zircon analyses. One sample (OX-Gr) was analyzed at the University of California at Santa Cruz using methodology described by Lopez et al. (2001). The other sample (E-2) was analyzed at the Laboratorio Universitario de Geoquimica Isotópica (LUGIS) at the Institutos de Geologia and Geofisica of the Universidad Nacional Autónoma de México using the procedures outlined by Solari et al. (2001). The data are given in Table 2 (see also Fig. 10). All stated errors are 2σ.

Most of the data are highly discordant. However, one abraded, single-grain analysis yielded nearly concordant data (1% discordant) with a $^{207}Pb/^{206}Pb$ age of 917 ± 6 Ma (Table 2). A chord drawn through all the data intersects concordia at 920 ± 25 Ma and

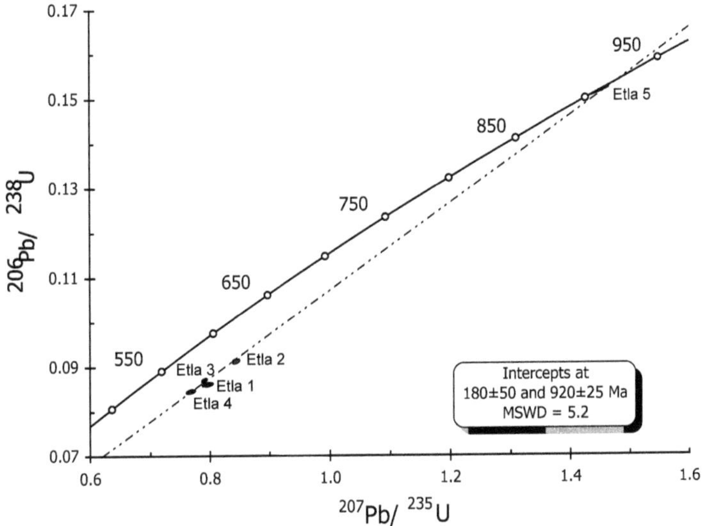

FIG. 8. Analytical data from the Etla pluton plotted on the U-Pb concordia diagram.

180 ± 50 Ma (Fig. 8). These ages are inferred to date the time of intrusion and a later Pb-loss event, respectively. The possibility that the ~917 Ma age represents inheritance is considered unlikely because: (1) ages in the host rocks are >978 Ma (Keppie et al., 2003a; Solari et al., 2003) and so cannot be a source of the ~917 Ma zircon; (2) inclusion of a zircon in the magma would probably lead to significant overprinting, such that the data would no longer be almost concordant near the upper intercept; and (3) greenschist-facies thermal overprinting during the Phanerozoic is least likely to be recorded by low-U zircon (~917 Ma zircon), whereas higher-U zircons are more affected by the thermal overprint (Table 2).

$^{40}Ar/^{39}Ar$ data

$^{40}Ar/^{39}Ar$ laser step-heating analyses were performed at the Queen's University $^{40}Ar/^{39}Ar$ Geochronology Laboratory on biotite and K-feldspar using the procedure described by Clark et al. (1998). The analyses are given in Table 3 (see also Fig. 10). All stated errors are 2σ.

Biotite from sample E-15 yielded discordant data with a plateau age of 207 ± 5 Ma (Fig. 9A). Biotite from sample E-2 shows a saddle-shaped spectrum, typically indicating the presence of excess argon (Fig. 9B). On the other hand, K-feldspar from the same sample, except for the first two steps, generally shows progressively older ages with increasing temperature, with dates from 169 ± 3 Ma to a plateau at 221 ± 3 Ma for 48.2% of the gas released at the high-temperature steps (Fig. 9C, Table 3). This spectrum is typical of <150°C reheating, which may be extrapolated to ~169 Ma.

Interpretation

Estimated blocking temperatures for $^{40}Ar/^{39}Ar$ in biotite and K-feldspar are ~280°C and ~150°C, respectively (Harrison and McDougall, 1982; Harrison et al., 1985). Given the fact that the surrounding Oaxacan Complex appears to have cooled through ~300 °C by ~856 Ma (Keppie et al., in press), and that the Oaxacan Complex is unconformably overlain by lowermost Ordovician rocks (Centeno and Keppie, 1999), it is suggested that the 220–200 Ma ages in biotite and K-feldspar record cooling during the Triassic following a reheating event, possibly associated with intrusion of the nearby La Carbonera pluton dated at 275 ± 4 Ma (Solari et al., 2001). The K-feldspar spectrum (Fig. 9C) appears to record partial resetting at ~169 Ma in the Late Jurassic. This latter reheating event also appears to be recorded in the Pb-loss event at 180 ± 50 Ma.

Tectonic Implications

The 917 ± 6 Ma Etla pluton presently appears to be unique in Mexico. However, this event may be more widespread, because the lower amphibolite-

TABLE 3. Ar-Ar Analyses, Etla Granite, Southern Mexico

Sample E-15 biotite[1]

J value:	Mass:	Volume 39K:	Integrated age:
–0.007425 ±–0.000056	16.5 mg	5.48 × 1E-10 cm3-NTP	195.86 ± 4.70 Ma

Power	$^{36}Ar/^{40}Ar$		$^{39}Ar/^{40}Ar$		r	Ca/K	%40Atm	%^{39}Ar	$^{40}Ar^{*}/^{39}K$		Age
1.00	0.002525	±0.000100	0.020596	±0.000239	0.51	0.481	74.57	23.26	12.323	±1.055	157.96 ± 12.95
<2.00>	0.001569	0.000085	0.03237	0.000265	0.359	0.246	46.31	29.55	16.571	0.595	209.35 ± 7.1
<2.00>	0.001834	0.000086	0.027418	0.000232	0.318	0.07	54.16	12.71	16.704	0.773	210.93 ± 9.21
<4.00>	0.001973	0.000089	0.025118	0.000231	0.423	0.149	58.26	11.32	16.599	0.810	209.68 ± 9.66
<5.00>	0.001615	0.000153	0.033895	0.000374	0.334	0.201	47.68	7.75	15.423	1.116	195.6 ± 13.41
<6.00>	0.001555	0.000273	0.035655	0.000623	0.427	0.542	45.89	4.22	15.162	1.784	192.46 ± 21.48
<7.00>	0.00163	0.000421	0.031536	0.000982	0.514	1.431	48.11	2.42	16.44	2.883	207.79 ± 34.42
<8.00>	0.001557	0.000129	0.032729	0.000334	0.399	2.152	45.98	8.76	16.492	0.909	208.41 ± 10.85

Sample E-2 K-Feldspar[2]

J value:	Mass:	Volume 39K:	Integrated age:
0.007395 ±–0.000056	17.0 mg	23.78 × 1E-10 cm3 NTP	212.64 ± 2.07 Ma

Power	36Ar/40Ar		39Ar/40Ar		r	Ca/K	%40Atm	%^{39}Ar	$^{40}Ar^{*}/^{39}K$		Age
1.00>	0.00147	±0.000514	0.003278	±0.001294	0.988	6.275	43.41	0.4	172.637	±8.568	1484.23 ± 50.21
2.00>	0.000688	0.000081	0.039351	0.000282	0.256	0.953	20.29	4.03	20.251	0.51	251.75 ± 5.91
[3.00>	0.000156	0.000055	0.072127	0.000369	0.068	0.19	4.61	8.95	13.224	0.216	168.32 ± 2.63
[4.00	0.000091	0.000092	0.072784	0.00128	0.032	0.107	2.69	9.08	13.369	0.414	170.09 ± 5.03
5.00	0.0001	0.00005	0.067742	0.00062	0.036	0.392	2.93	9.28	14.328	0.238	181.69 ± 2.87
5.50	0.000076	0.00008	0.06455	0.000384	0.084	0.212	2.2	5.74	15.151	0.346	191.59 ± 4.15
6.00	0.000084	0.000119	0.062825	0.000495	0.119	0.335	2.44	3.69	15.528	0.516	196.11 ± 6.17
7.00	0.000116	0.000104	0.060324	0.000397	0.17	0.296	3.42	3.99	16.009	0.449	201.85 ± 5.36
7.50	0.000156	0.000124	0.058321	0.000422	0.123	0.114	4.57	3.38	16.361	0.585	206.04 ± 6.96
8.00	0.000052	0.000136	0.057452	0.000393	0.176	0.108	1.49	3.25	17.146	0.634	215.37 ± 7.5
<8.50	0.0001	0.000102	0.05532	0.000403	0.002	0.001	2.93	6.26	17.546	0.56	220.1 ± 6.61
<9.00	0.000158	0.000038	0.054212	0.000304	0.056	0.186	4.67	35.54	17.583	0.213	220.54 ± 2.51
<10.00	0.000182	0.000098	0.052822	0.0004	0.004	0.001	5.38	6.41	17.912	0.565	224.41 ± 6.65

Sample E-2 Biotite[3]

J value:	Mass:	Volume 39K:	Integrated age:
–0.007405 ±–0.000056	22.0 mg	43.57 × 1E-10 cm3 NTP	240.37 ± 2.12 Ma

Power	$^{36}Ar/^{40}Ar$		$^{39}Ar/^{40}Ar$		r	Ca/K	%40Atm	%39Ar	$^{40}Ar^{*}/^{39}K$		Age
1.00	0.002901	±0.000072	0.018222	±0.000149	0.275	0.283	85.68	5.15	7.831	±1.040	101.7 ± 13.13
2.00	0.000578	0.000034	0.046366	0.000232	0.096	0.094	17.05	19.92	17.884	0.211	224.37 ± 2.49
3.50	0.000236	0.000034	0.04511	0.000245	0.077	0.153	6.97	22.04	20.621	0.229	256.37 ± 2.66
4.50	0.000257	0.000036	0.045973	0.000237	0.112	0.092	7.58	12.36	20.1	0.224	250.32 ± 2.6
6.00	0.000273	0.000031	0.046557	0.000246	0.094	0.078	8.06	16	19.746	0.195	246.2 ± 2.28
7.00	0.000217	0.000043	0.046307	0.000254	0.109	0.094	6.41	7.76	20.208	0.261	251.58 ± 3.03
<8.00	0.000146	0.000036	0.045544	0.00024	0.093	0.509	4.29	16.78	21.013	0.23	260.91 ± 2.66

[1]Initial 40/36: 322.46 ± 149.80 (MSWD = 0.64, isochron between 0.37 and 2.26). Correlation age: 187.15 ± 41.40 Ma (76.7% of ^{39}Ar, steps marked by >). Plateau age: 207.19 ± 4.53 Ma (76.7% of ^{39}Ar, steps marked by <).

[2]Initial 40/36: 740.23 ± 508.57 (MSWD = 0.50, isochron between -0.41 and 3.83). Correlation Age: 156.57 ± 15.99 Ma (13.4% of ^{39}Ar, steps marked by >). Plateau age: 221.00 ± 2.73 Ma (48.2% of ^{39}Ar, steps marked by <); 169.21 ± 3.10 Ma (18.0% of ^{39}Ar, steps marked by [).

[3]Plateau age: 260.91 ± 3.23 Ma (16.8% of ^{39}Ar, steps marked by <).

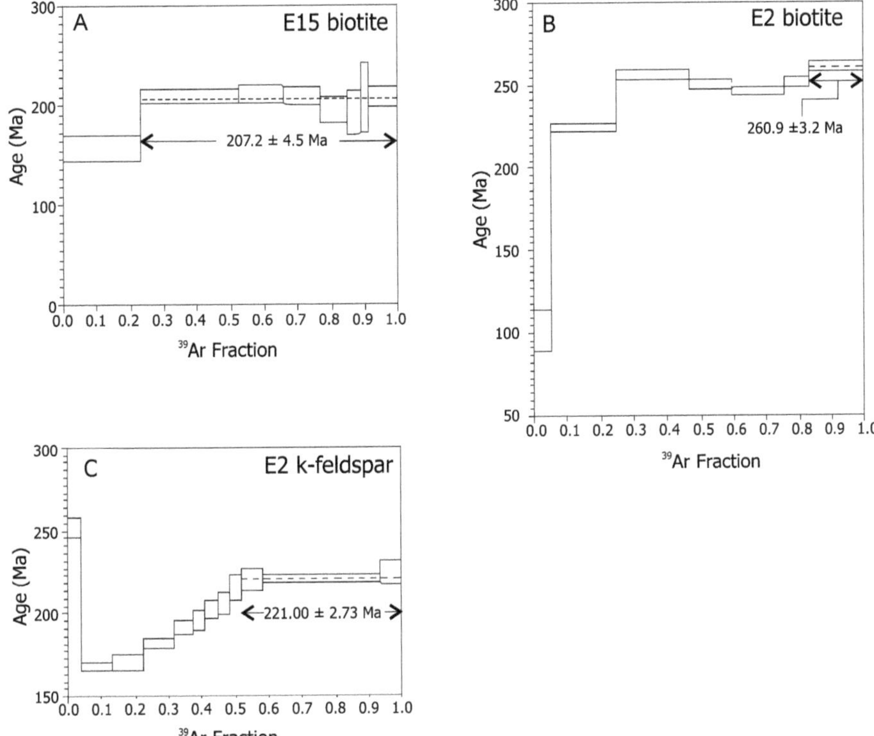

FIG. 9. $^{40}Ar/^{39}Ar$ laser step-heating mineral analyses from the Etla pluton plotted on graphs of calculated age versus ^{39}Ar released. A. Biotite from sample E-15. B. Biotite from sample E-2. C. K-feldspar from sample E-2.

facies hydration of the Zapotecan granulites is widespread within Oaxaquia (Keppie et al., 2001, 2003a). Both the Etla calc-alkaline magmatism and this hydration can be related to subduction beneath Oaxaquia. This implies that oceanic lithosphere lay on one side of Oaxaquia at ~917 Ma, and precludes a palinspastic location for Oaxaquia in the ≥980 Ma continent-continent collision zone between eastern Laurentia and Amazonia (Sadowski and Bettencourt, 1996; Starmer, 1996; Rivers, 1997). However, this collision is presently in doubt because Ramos and Aleman (2000) showed a Late Neoproterozoic Brasiliano orogen between the Peruvian Arequipa massif and the Amazon craton, and Loewy et al. (2000) suggested that the Arequipa masif is allochthonous. This would allow one to place Oaxaquia anywhere around the northern, western, or southern periphery of Amazonia. However, a northern location is favored by the occurrence of contemporaneous, 950–900 Ma arc magmatism in Tocantins Province, central Brasil and in proto-Avalonia that is inferred to have lain along the northern margin of Amazonia (Fig. 10) (Pimental et al., 2000; Keppie et al., 2003b). Such a location was also favored by Keppie and Ortega-Gutiérrez (1995, 1999) and Keppie et al. (2001) based upon comparable P-T-t data and Ordovician faunal affinities.

The 220–200 Ma age recorded in the $^{40}Ar/^{39}Ar$ biotite and K-feldspar data in the Etla pluton are younger than those recorded in the ~275 Ma La Carbonera pluton (Solari et al., 2001) 10 km north of the Etla pluton, and are similar to those recorded in the Cosoltepec Formation in the neighboring Acatlán Complex (Keppie et al., in press). Thus these ages may be interpreted as cooling following a Permian tectonothermal event (Keppie et al., in press). Such a reheating event may also explain the 272 ± 8 Ma, Rb-Sr whole-rock isochron produced by Ruiz-Castellenos (1979) in the Etla pluton. This Permo-Triassic event appears to be part of an extensive magmatic arc that extended along the length of Mexico (Torres et al., 1999).

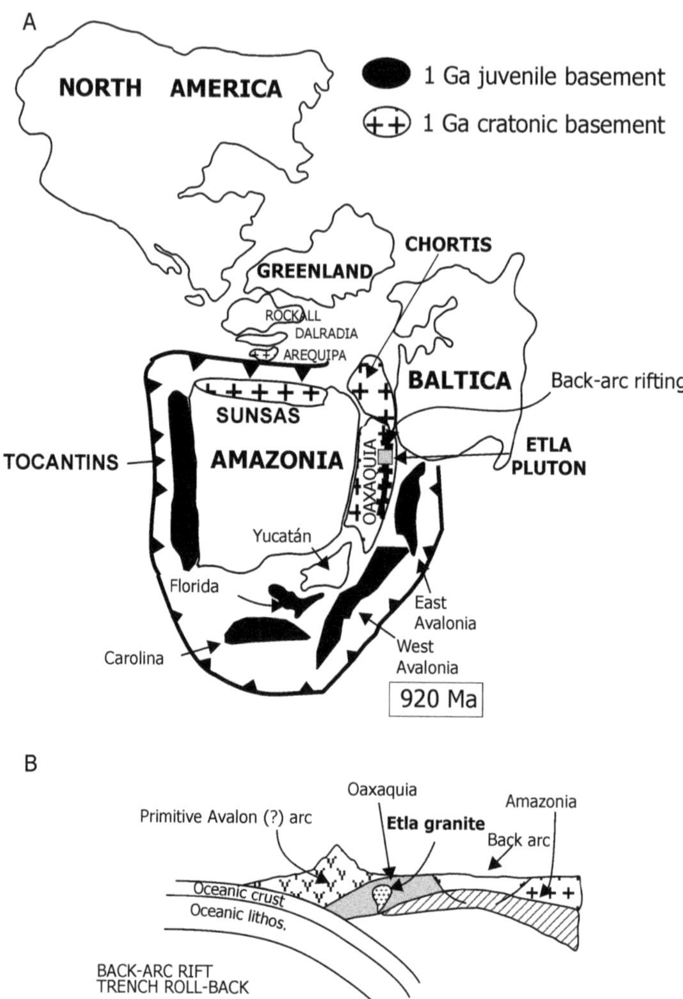

FIG. 10. Palinspastic reconstruction and cross-section at ~917 Ma, showing the arc-backarc location of the Etla pluton in Oaxaquia and its correlatives in Tocantins Province, central Brasil and Avalonia, modified after Keppie et al. (in press).

The ~170 Ma age recorded in the $^{40}Ar/^{39}Ar$ and U-Pb data from the Etla pluton is also similar to those recorded in the eastern Acatlán Complex, which have been related to the overriding of a mid-Jurassic plume (Keppie et al., in press). Such reheating appears to have also affected the Etla pluton, and may also be related to gravitational unroofing of the Paleozoic rocks that unconformably overlie the Oaxacan Complex (Centeno-Garcia and Keppie, 1999; Keppie et al., in press).

Acknowledgments

JDK would like to thank PAPIIT and CONACyT for providing funds for the project. JD acknowledges the Natural Sciences and Engineering Research Council (NSERC) for a grant that supported the geochemistry. The $^{40}Ar/^{39}Ar$ analyses were supported by a NSERC grants to JKWL. AOR received support from grants from the Instituto de Geologia (UNAM) and CONACyT grant 33100T. D. M.

Martinez and J. T. Vasquez-R provided valuable help in sample preparation at UNAM. Gabriela Solis-Pichardo, Juan Julio Morales, and Peter Schaaf are also thanked for analytical help and TIMS supervision at LUGIS, UNAM.

REFERENCES

Ayer, J. A., and Davis, D. W., 1997, Neoarchean evolution of differing convergent margin assemblages in the Wabigoon Subprovince: Geochemical and geochronological evidence from the Lake of the Woods greenstone belt, Superior Province, Northwestern Ontario: Precambrian Research, v. 81, p. 155–178.

Ballard, M. M., Van der Voo, R., and Urrutia-Fucugauchi, J., 1989, Paleomagnetic results from Grenville-aged rocks from Oaxaca, Mexico: Evidence for a displaced terrane: Precambrian Research, v. 42, p. 343–352.

Boucot, A. J., Blodgett, R. B., and Stewart, J. H., 1997, European Province Late Silurian brachiopods from the Ciudad Victoria area, Tamaulipas, northeastern Mexico, *in* Klapper, G., Murphy, M. A., and Talent, J. A., eds., Paleozoic sequence stratigraphy, biostratigraphy, and biogeography: Studies in honor of J. Granville ("Jess") Johnson: Geological Society of America Special Paper 321, p. 273–293.

Burrett, C., and Berry, R., 2000, Proterozoic Australia–Western United States (AUSWUS) fit between Laurentia and Australia: Geology, v. 28, p. 103–106.

Cawood, P. A., McCausland, P. J. A., and Dunning, G. R., 2001, Opening Iapetus: Constraints from the Laurentian margin in Newfoundland: Geological Society of America Bulletin, v. 113, p. 443–453.

Centeno-Garcia, E., and Keppie, J. D., 1999, Latest Paleozoic–early Mesozoic structures in the central Oaxaca terrane of southern Mexico: Deformation near a triple junction: Tectonophysics, v. 301, p. 231–242.

Chappel, B. W., and White, A. J. R., 1974, Two contrasting granite types: Pacific Geology, v. 8, p. 173–174.

Clark, A. H., Archibald, D. A., Lee, A. W., Farrar, E., and Hodgson, C. J., 1998, Laser probe $^{40}Ar/^{39}Ar$ ages of early- and Late-stage alteration assemblages, Rosario porphyry copper-molybdenum deposit, Collahuasi District, I Region, Chile: Economic Geology, v. 93, pp. 326–337.

Dalziel, I. W. D., 1992, On the organization of American plates in the Neoproterozoic and the breakout of Laurentia: GSA Today, v. 2, no. 11, p. 1–2.

Dalziel, I. W. D., Mosher, S., and Gahagan, L. M., 2000, Laurentia-Kalahari collision and the assembly of Rodinia: Journal of Geology, v. 108, p. 499–513.

De Cserna, Z., 1971, Precambrian sedimentation, tectonics, and magmatism in México: Geologische Rundschau, v. 60, p. 1488–1513.

Dostal, J., Baragar, W. R. A., and Dupuy, C., 1986, Petrogenesis of the Natkusiak continental basalts, Victoria Island, N.W.T.: Canadian Journal of Earth Sciences, v. 23, p. 622–632.

Harrison, T. M., Duncan, I., and McDougall, I., 1985, Diffusion of ^{40}Ar in biotite: Temperature, pressure, and compositional effects: Geochimica et Cosmochimica Acta, v. 49, p. 2461–2468.

Harrison, T. M., and McDougall, I., 1982, The thermal significance of potassium feldspar K-Ar ages inferred from $^{40}Ar/^{39}Ar$ age spectrum results: Geochimica et Cosmochimica Acta, v. 45, p. 2513–2517.

Karlstrom, K. E., Williams, M. L., McLelland, J., Geissman, J. W., and Ahäll, K-I., 1999, Refining Rodinia: Geologic evidence for the Australia–western U.S. connection in the Proterozoic: GSA Today, v. 9, No. 10, p. 1–7.

Keppie, J. D., Dostal, J., Ortega-Gutierrez, F., and Lopez, R., 2001, A Grenvillian arc on the margin of Amazonia: Evidence from the southern Oaxacan Complex, southern Mexico: Precambrian Research, v. 112, Nos. 3-4, p. 165–181.

Keppie, J. D., Dostal, J., Cameron, K. L., Solari, L. A., Ortega-Gutiérrez, F., and Lopez, R., 2003a, Geochronology and geochemistry of Grenvillian igneous suites in the northern Oaxacan Complex, southern México: Tectonic implications: Precambrian Research, v. 112, No.3-4, p. 165–181.

Keppie, J. D., Nance, R. D., Murphy, J. B., and Dostal, J., 2003b, Tethyan, Mediterranean, and Pacific analogues for the Neoproterozoic–Paleozoic birth and development of peri-Gondwanan terranes and their transfer to Laurentia and Laurussia: Tectonophysics, v. 365, p. 195–219.

Keppie, J. D., Nance, R. D., Powell, J. T., Mumma, S. A., Dostal, J., Fox, D., Muise, J., Ortega-Rivera, A., Miller, B. V., and Lee, J. W. K., in press, Mid-Jurassic tectonothermal event superposed on a Paleozoic geological record in the Acatlán Complex of southern Mexico: Hotspot activity during the breakup of Pangea: Gondwana Research.

Keppie, J. D., and Ortega-Gutiérrez, F., 1995, Provenance of Mexican terranes: Isotopic constraints: International Geology Review, v. 37, 1995, p. 813–824.

______, 1999, Middle American Precambrian basement: A missing piece of the reconstructed 1 Ga orogen, *in* Ramos, V.A., and Keppie, J. D., eds., Laurentia-Gondwana connections before Pangea: Boulder, CO, Geological Society of America Special Paper 336, p. 199–210.

Keppie, J. D., Solari, L. A., Ortega-Gutiérrez, F., Ortega-Rivera, A., Lee, J. W. K., and Hames, W. E., in press, U-Pb and $^{40}Ar/^{39}Ar$ constraints on the cooling history of the northern Oaxacan Complex, southern Mexico: Tectonic implications, *in* Tollo, R. P., Corriveau, L., McLelland, J. B., and Bartholomew, M. J., eds., Proterozoic tectonic evolution of the Grenville Orogen in North America: Boulder, CO, Geological Society of America Memoir.

Krogh, T. E., 1973, A low-contamination method for hydrothermal decomposition of zircon and extraction of U and Pb for isotopic age determinations: Geochimica et Cosmochimica Acta, v. 27, p. 485–494.

Lopez, R. L., Cameron, K. L., and Jones, N. W., 2001, Evidence for Paleoporterozoic, Grenvillian, and Pan-African age crust beneath northeastern Mexico: Precambrian Research, v. 107, p. 195–214.

Loewy, S., Connelly, J. N., Dalziel, I. W. D., Gower, C. F., and Cawood, P. A., 2000, Testing a propsed Rodinia reconstruction using Pb isotopes and U-Pb geochronology [abs.]: Geological Society of America Abstacts with Program v. 32, No. 7, p. A455.

Ludwig, K. R., 1991, PdCat: A computer program for processing Pb-U-Th isotope data, version 1.24: U.S. Geological Survey Open-File Report 88–542.

Mattinson, J. M., 1987, U-Pb ages of zircons: A basic examination of error propagation: Chemical Geology, v. 66, p. 151–162.

Ortega-Gutiérrez, F., Ruiz, J., and Centeno-Garcia, E., 1995, Oaxaquia, a Proterozoic microcontinent accreted to North America during the late Paleozoic: Geology, v. 23, p. 1127–1130.

Parrish, R. R., 1987, An improved micro-capsule for zircon dissolution in U-Pb geochronology: Chemical Geology, v. 66, p. 99–102.

Pearce, J. A., Harris, N. B., and Tindle, A. G., 1984, Trace element discrimination diagrams for the tectonic interpretation of granitic rocks: Journal of Petrology, v. 25, p. 956–983.

Pimentel, M. M., Fuck, R. A., Jost, H., Ferreira Filho, C. F., and de Araújo, S. M., 2000, The basement of the Brasilia fold belt and the Goiás magmatic arc, *in* Cordani, U.G., Thomaz Filho, A., and Campos, D.A., eds., Tectonic evolution of South America: 31st International Geological Congress, Rio de Janeiro, Brasil, p. 195–230.

Pitcher, W. S., 1993, The nature and origin of granite: London, Blackie Academic & Professional, UK.

Ramos, V. A., and Aleman, A., 2000, Tectonic evolution of the Andes, *in* Cordani, U. G., Milani, E. J., Thomaz Fihlo, A., and Campos, D. A., eds., Tectonic evolution of South America: 31st Int. Geol. Cong., Rio de Janeiro, Brasil, p. 635–685.

Rivers, T., 1997, Lithotectonic elements of the Grenville Province: Review and tectonic implications: Precambrian Research, v. 86, p. 117–154.

Robison, R., and Pantoja-Alor, J., 1968, Tremadocian trilobites from Nochixtlan region, Mexico: Journal of Paleontology, v. 42, p. 767–800.

Rollinson, H., 1993, Using geochemical data: Evaluation, presentation, interpretation. Essex, UK: Addison Wesley Longman Ltd.

Ruiz-Castellanos, M., 1979, Rubidium-strontium geochronology of the Oaxaca and Acatlán metamorphic areas of Southern Mexico. Unpubl. Ph.D. thesis, University of Texas, Dallas, 188 p.

Sadowski, G. R., and Bettencourt, J. S., 1996, Mesoproterozoic tectonic correlations between eastern Laurentia and the western border of the Amazon craton: Precambrian Research, v. 76, p. 213–227.

Shand, S. J., 1951, The study of rocks: London, UK: Thomas Murby and Co., 236 p.

Shergold, J. H., 1975, Late Cambrian and Early Ordovician trilobites from the Burke River structural belt, western Queensland, Australia: Department of Minerals and Energy, Bureau of Mineral Resources, Geology and Geophysics, Bulletin 153, 221 p.

Shurbert, D. H., and Cebull, S. E., 1987, Tectonic interpretation of the westernmost part of the Ouachita-Marathon (Hercynian) orogenic belt, west Texas–México: Geology, v. 15, p. 458–461.

Solari, L., Dostal, J., Ortega-Gutiérrez, F., and Keppie, J. D., 2001, The 275 Ma Carbonera stock in the northern Oaxacan Complex of southern Mexico: U-Pb geochronology and geochemistry: Revista Mexicana Ciencias de Geologia, v. 18, p. 149–161.

Solari, L. A., Keppie, J. D., Ortega-Gutiérrez, F., Cameron, K. L., Lopez, R., and Hames, W. E., 2003, 990 Ma and 1,100 Ma Grenvillian tectonothermal events in the northern Oaxacan Complex, southern Mexico: Roots of an orogen: Tectonophysics, v. 365, p. 257–282.

Starmer, I. C., 1996, Accretion, rifting, rotation and collision in the North Atlantic supercontinent, 1700–950 Ma, *in* Brewer, T. S., ed., Precambrian crustal evolution in the North Atlantic region: Geological Society of London Special Publication 112, p. 219–248.

Steckeisen, A. L., 1976, To each plutonic rock its proper name. Earth Sciences Review, v. 12, p. 1–33.

Sun, S. S., and McDonough, W. F., 1989, Chemical and isotopic systematics of oceanic basalts: Implications for mantle composition and processes, *in* Saunders, A. D. and Norry, M. J., eds., Magmatism in the ocean basins: Geological Society of London Special Publication 42, p. 313–345.|

Torres, R., Ruíz, J., Patchett, P. J. and Grajales-Nishimura, J. M., 1999, Permo-Triassic continental arc in eastern Mexico; tectonic implications for reconstructions of southern North America, *in* Bartolini, C., Wilson, J. L., and Lawton, T. F. eds., Mesozoic sedimentary and tectonic history of north-central Mexico: Geological Society of America Special Paper 340, p. 191–196.

Winchester, J. A., and Floyd, P. A., 1977, Geochemical discrimination of different magma series and their differentiation products using immobile elements: Chemical Geology, v. 20, p. 325–343.

Geochemistry of the Tremadocian Tiñu Formation (Southern Mexico): Provenance in the Underlying ~1 Ga Oaxacan Complex on the Southern Margin of the Rheic Ocean

J. BRENDAN MURPHY,[1]

Department of Earth Sciences, St. Francis Xavier University, Antigonish, Nova Scotia, B2G 2W5, Canada

J. DUNCAN KEPPIE,

Instituto de Geología, Universidad Nacional Autónoma de México, México D.F. 04510, México

JAMIE F. BRAID,

Department of Earth Sciences, St. Francis Xavier University, Antigonish, Nova Scotia, B2G 2W5, Canada

AND R. DAMIAN NANCE

Department of Geological Sciences, Clippinger Laboratories, Ohio University, Athens, Ohio 45701

Abstract

In southern Mexico, the Tremadocian Tiñu Formation with a Gondwanan fauna rests unconformably upon the ~1 Ga Oaxacan Complex, and is unconformably overlain by Carboniferous sedimentary rocks containing a Laurentian fauna. The Tiñu Formation is inferred to represent deposition along the southern margin of the Rheic Ocean in its early stages of development. The geochemistry of sandstones and shales of the formation show: (1) high iron, magnesium, and titanium, indicating chemical immaturity in some samples; (2) a positive correlation between Zr and SiO_2; (3) derivation from crust with mafic and felsic components; (4) LREE enrichment with a moderate Eu anomaly; (5) low $\varepsilon_{Nd(t)}$ values (–7.0 to 7.8, calculated for t = 485 Ma); and (vi) Mesoproterozoic T_{DM} ages (1.5–1.83 Ga). Taken together, these data indicate a proximal source with two main components, an ancient, felsic-intermediate, moderately differentiated continental component, and a minor mafic component. These data are consistent with derivation from the underlying Oaxacan Complex, with little or no input from distal or juvenile sources. This is in accord with: (1) published Nd data from the Oaxacan Complex that yielded T_{DM} ages of 1.5–1.8 Ga; (2) detrital zircon data from the Tiñu Formation, which gave 990–1200 Ma ages; and (3) the Gondwanan affinity of Silurian fauna from northeastern Mexico. The lack of a juvenile, Cambrian signature in these geochemical data is inconsistent with an origin in the Iapetus Ocean, and is compatible with a rift-to-drift model involving opening of the Rheic Ocean.

Introduction

THE PALEOZOIC HISTORY of southern Mexico is central to an understanding of geodynamic linkages between the closure of Paleozoic oceans (such as the Rheic and Iapetus oceans) and the amalgamation of Pangea. Northeast of the city of Oaxaca in southern Mexico (Fig. 1), an unmetamorphosed Lower Ordovician (Tremadocian) cover sequence containing a Gondwanan fauna (Robison and Pantoja-Alor, 1968; Shergold, 1975) unconformably overlies granulite-facies gneisses consisting of calc-alkaline and rift-related igneous rocks and clastic-carbonate-evaporitic sedimentary rocks dated between ~990 and ~1300 Ma, known as the Oaxacan Complex (Keppie et al., 2001, 2003, Dostal et al., 2004). The Tiñu Formation is tectonically overlain by Mississippian rocks containing a Laurentian fauna (Navarro-Santillan et al., 2002): the tectonic contact is inferred to be a sheared unconformity (Centeno-García and Keppie, 1999). These faunal data indicate that Oaxaquia, the ~1 Ga crustal block of which the Oaxacan Complex is part, was transferred from Gondwana to Laurentia between the Tremadocian and the Mississippian. According to Ortega-Gutiérrez et al. (1999), Oaxaquia formed the

[1]Corresponding author; email: bmurphy@stfx.ca

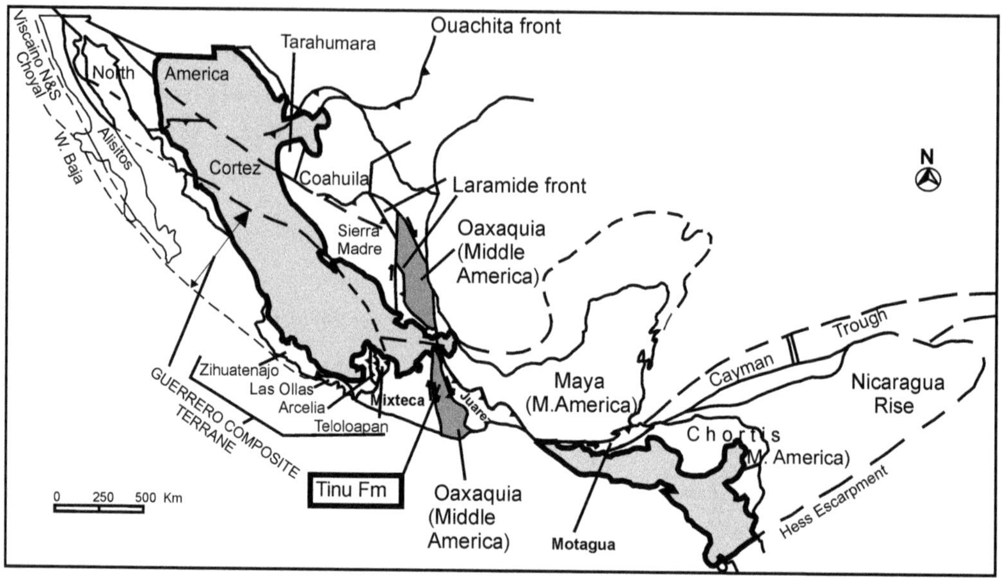

FIG. 1. Summary terrane map of Mexico, showing the location of Oaxaquia in Mexico (modified after Keppie, 2004).

southern margin of the Iapetus Ocean and preserves a Neoproterozoic–Cambrian, rift-drift record, followed by subduction and closure by the end of the Ordovician. However, Keppie and Ramos (1999) proposed that Oaxaquia formed the southern flank of the Rheic Ocean, which opened during the Ordovician and closed during the Devonian and Carboniferous. An origin along the southern flank of the Rheic Ocean is consistent with the observation that the Silurian fauna with Gondwanan affinity occur in rocks unconformably overlying ~1 Ga rocks in the northern Oaxaquia terrane (Boucot et al., 1997), and with recent data from the Acatlán Complex (which lies to the west of the Oaxacan Complex across a Permian flower structure; Elías-Herrera and Ortega-Gutiérrez, 2002) that indicate orogenic events occurred in the Carboniferous and Permian (Malone et al., 2002; Keppie et al., 2004) rather than Ordovician and Devonian (Ortega-Gutiérrez et al., 1999).

These recent interpretations suggest that the Lower Ordovician Tiñu Formation may provide constraints on the tectonic evolution of the southern (South American) flank of the Rheic Ocean during the early stages of ocean development. Many studies have shown that geochemical analyses of siliciclastic rock units can constrain the provenance and tectonic environment at the time of deposition (e.g., McLennan et al., 1990). This paper presents lithogeochemical and Sm-Nd data from 16 representative samples of the Tiñu Formation. These data provide some broad constraints on tectonic and sedimentary processes along a portion of the southern flank of the Rheic Ocean at the time of deposition of the formation. The abundance of labile oxides indicates a chemical immaturity that suggests a provenance proximal to the depositional basin, and the geochemical signature of the sediments suggests a mixed felsic-intermediate-mafic source for the detritus. The presence of 1.0 Ga detrital zircons (Gillis et al., 2001) and a T_{DM} age of 1.4–1.6 Ga (Patchett and Ruiz, 1987) indicate a Mesoproterozoic cratonic source, probably in the underlying Oaxacan Complex.

General Geology

The Tiñu Formation occurs in two outliers in southern Mexico (Fig. 1), where it rests unconformably upon granulitic gneisses of the ~1.0 Ga Oaxacan Complex. The northern Oaxacan Complex essentially consists of two thrust slices: (1) a lower, ~1012 Ma anorthosite-mangerite-charnockite-granite-gabbro suite; and (2) an upper metasedimentary suite with clastic-carbonate and evaporitic protoliths (<1300 Ma) intruded by ~1140 Ma gabbroic, syenitic, and granitoid plutons (Keppie et al., 2003; Dostal et al., 2004). These rocks were deformed during the ~1100 Ma Olmecan tectonothermal event and the ~1104–980 Ma Zapotecan orogeny (Solari et al., 2003) before being intruded by the calc-alkaline Etla pluton (Ortega-Obregon et al., 2003). The southern Oaxacan Complex also contains pre-1100 Ma calc-alkaline rocks of inferred volcanic origin (Keppie et al., 2001). The northern Oaxacan Complex has yielded ε_{Nd} values of 1.4–1.6 Ga (Patchett and Ruiz, 1987).

The time of deposition of the Tiñu Formation is tightly constrained in the Tremadocian by fossils, such as trilobites, brachiopods, and conodonts, which have close affinities with the South American margin of Gondwana (Robison and Pantoja-Alor, 1968; Shergold, 1975). The total thickness of the Tiñu Formation ranges from 23–60 m in the Santiago Ixaltepec outlier to 100 m in the type section at Rio Salinas (Robison and Pantoja-Alor, 1968): the latter, more complete section was chosen for sampling in this study. At Santiago Ixtaltepec, the Tiñu Formation is separated from overlying Mississippian rocks by a listric normal fault that is inferred to be a fault-modified unconformity (Centeno-García and Keppie, 1999). At Rio Salinas, the Tiñu Formation is unconformably overlain by Tertiary rocks. The formation at the type section consists of interbedded marine limestone and shale that grade upward into predominantly shale and siltstone (Robison and Pantoja-Alor, 1968; Fig. 2). Based on this change in lithology, the formation has been subdivided into two members: a lower member consisting of conglomerate, sandstone, limestone, and shale, and an upper member dominated by shale with a few thin limestone and sandstone lenses. The lower unit is dominated by imbricated tabular clasts, which are composed predominantly of fossil fragments (to 90%). These fragments are composed of Tremadocian trilobites, brachiopods, gastropods, conodonts, nautiloids, and minor fragments of echinoderms and algae (Robison and Pantoja-Alor, 1968; Fig. 2). The depositional environment of this limestone has been interpreted as a shelf environment (Centeno-Garcia and Keppie, 1999). The other detrital, fine-grained limestone units, which grade upward, have been interpreted as channel deposits. The percentage of limestone decreases upward in the sequence as the lithology changes to predominantly black shale and a few lenticular layers of fine-grained sandstone that are interbedded in the upper parts of the section. These sandstone layers are rich in quartz and mica, and contain abundant trace fossils. A thick calcareous red siltstone layer containing an abundance of well-preserved brachiopod shells dominates the top of the sequence. Mafic sills and dikes of unknown age intrude the unit. The Tiñu Formation has yielded detrital zircons with common ages of ~992, 1079, and 1155 Ma that are inferred to have been derived from the underlying Oaxacan Complex (Gillis et al., 2001).

The Tiñu Formation at Rio Salinas is deformed into a regional syncline associated with an upright slaty cleavage (Fig. 2B). Bedding/cleavage intersection lineations are parallel to minor fold axes and plunge gently to the northeast, implying a genetic relationship between regional and outcrop-scale structures. Centeno-García and Keppie (1999) suggested that these structures are related to Permo-Triassic E-vergent thrusting along the eastern boundary of the Oaxacan Complex.

Geochemistry

Methods

In order to characterize the clastic rocks according to their chemical composition and as a means of identifying the possible source regions from which the sediments were derived, 13 representative samples were taken from shale and fine-grained sandstone (TU-A to P) and were analyzed for major and selected trace elements by X-Ray Fluorescence at St. Mary's University, Halifax (Table 1). Precision and accuracy are generally within 5% for most major elements, and within 5–10% for minor and trace elements. Analytical methods are detailed in Dostal et al. (1986, 1994).

Three representative samples (TU-D, E, I) were analyzed for rare-earth elements (REE) and Sm-Nd isotopic compositions. REE were analyzed by inductively coupled plasma mass spectrometry (ICP-MS) at Memorial University, Newfoundland (Table 2). These analyses have an accuracy and precision of better than 10%; all procedures are detailed in Longerich et al (1990). Sm-Nd isotopes were analyzed at the Atlantic Universities Regional Isotopic Facility (AURIF) in Memorial University, Newfoundland (Table 3). Nd and Sm were measured by isotope dilution. $^{143}Nd/^{144}Nd$ ratios were then calculated by thermal ionization mass spectrometry, following the separation of Nd from Sm and other REEs by ion exchange chemistry (Kerr et al., 1995). $^{147}Sm/^{144}Nd$ ratios were calculated with data reduction software developed at Memorial University. Depleted mantle model ages (T_{DM}) are calculated assuming a modern depleted mantle of $^{143}Nd/^{144}Nd$ =

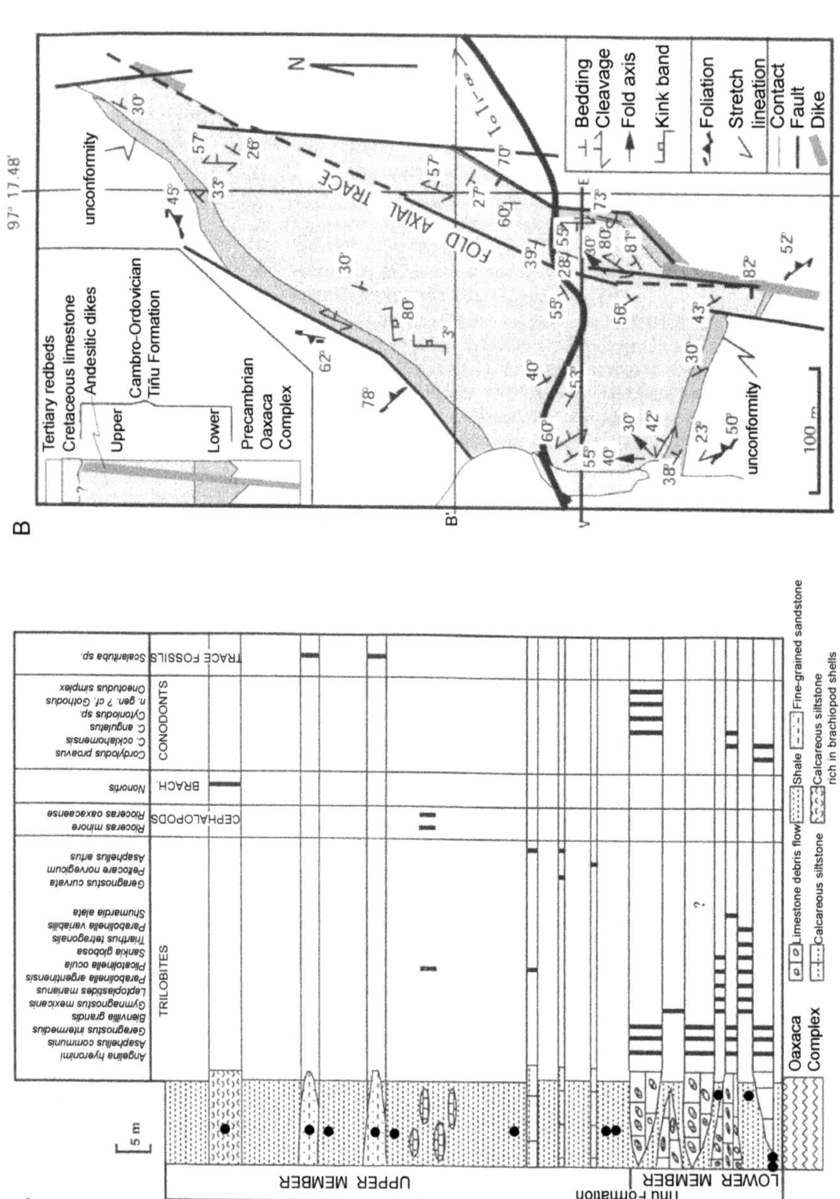

FIG. 2. A. Stratigraphic column and faunal content of the Tiñu Formation in the type section at Rio Salinas (after Robison and Pantoja-Alor, 1968; Shergold, 1975). Sample locations shown by filled circles; letters A–P on the left correspond to samples Tu-A to Tu-P, respectively. B. Geologic map of the Tiñu Formation (after Pantoja-Alor, 1970; Centeno-Garcia and Keppie, 1999).

TABLE 1. Major and Trace Element Analyses of the Tiñu Formation[1]

Sample:	TU-A	TU-B	TU-C	TU-D	TU-E	TU-H	TU-I	TU-J	TU-L	TU-M	TU-N	TU-O	TU-P
Lithology:	Sandstone	Shale	Shale	Shale	Shale	Black shale	Grey shale	Sandstone	Shale	Sandstone	Shale	Sandstone	Sand stone
							wt %						
SiO_2	53.39	61.27	61.86	64.67	65.89	65.20	66.16	8.77	56.04	62.04	58.70	52.51	62.02
TiO_2	0.26	1.29	1.17	1.02	0.91	1.01	1.04	0.12	0.88	0.65	1.01	0.50	0.29
Al_2O_3	3.29	15.08	14.98	14.74	16.01	17.01	17.72	1.62	20.18	12.28	20.19	7.54	7.98
Fe_2O_3	11.82	9.98	8.63	6.86	4.73	2.02	2.73	2.05	6.73	3.76	4.89	2.57	8.96
MnO	0.08	0.05	0.05	0.05	0.03	0.00	0.01	0.18	0.02	0.19	0.01	0.32	0.02
MgO	0.19	0.73	0.91	1.08	0.97	0.93	0.99	0.39	1.14	0.69	1.34	0.26	0.06
CaO	14.79	1.19	0.84	0.76	0.54	0.36	0.23	48.20	0.45	7.07	0.71	17.76	8.02
Na_2O	0.05	0.05	0.02	0.03	0.04	<0.01	0.06	<0.01	0.14	1.07	0.05	0.76	0.05
K_2O	0.74	2.95	3.47	4.08	3.95	4.45	4.47	0.38	4.96	2.45	5.48	1.48	1.15
P_2O_5	10.23	0.35	0.26	0.38	0.24	0.05	0.07	1.71	0.21	0.69	0.25	1.00	6.13
L.O.I.	4.16	8.51	7.12	6.53	6.47	7.68	6.32	36.99	9.24	8.50	7.28	14.80	4.70
Total	98.99	101.45	99.31	100.20	99.77	98.71	99.80	100.41	99.98	99.39	99.91	99.49	99.38
							ppm						
V	54	170	176	174	300	677	288	20	138	85	162	59	82
Cr	17	62	101	60	93	144	93	5	65	24	70	14	48
Co	44	25	31	24	14	<5	6	<5	21	10	12	<5	24
Zr	74	181	153	168	175	204	195	40	135	354	170	600	191
Ba	447	362	399	531	665	634	595	117	676	646	779	492	353
Ni	41	34	206	32	37	14	24	13	20	15	41	10	26
Cu	43	35	43	33	86	55	25	<4	50	10	21	<4	<4
Zn	142	115	72	100	190	86	97	15	75	45	86	112	56
Ga	8	22	21	20	21	23	23	5	27	15	26	9	17
Rb	26	103	132	156	163	188	183	17	194	95	216	52	33
Sr	380	136	96	104	171	61	74	235	203	70	44	121	177
Y	149	26	23	31	31	37	36	32	42	66	39	104	145
Nb	14	16	15	16	16	19	21	8	22	16	23	15	14
Pb	45	13	14	15	15	61	60	86	77	32	33	31	56
Th	16	12	9	10	17	18	22	5	19	11	18	12	18
U	5	4	4	6	7	9	8	3	6	4	7	2	4

[1]Major and trace element analyses were determined by X-Ray Fluorescence at St. Mary's University, Halifax. The precision and accuracy of these determinations is generally better than 5%, with the exception of Nb and Cr which is generally better than 10%.

TABLE 2. Rare-Earth Element Analyses and Y, Zr, Nb, Ba, Hf, Ta, Th Analyses of Representative Samples of the Tiñu Formation[1]

	TU-D	TU-E	TU-I
Y	29.13	26.25	30.67
Zr	210.49	219.25	250.14
Nb	20.39	19.56	18.96
Ba	612.06	799.62	686.37
La	37.26	35.48	54.89
Ce	77.93	63.93	100.81
Pr	9.98	7.58	12.34
Nd	38.74	29.18	45.27
Sm	7.33	5.60	7.05
Eu	1.29	0.94	1.19
Gd	6.48	4.99	5.50
Tb	1.05	0.83	0.89
Dy	6.09	4.87	5.41
Ho	1.16	1.01	1.14
Er	3.31	3.12	3.54
Tm	0.52	0.51	0.56
Yb	3.40	3.53	3.81
Lu	0.50	0.53	0.55
Hf	5.06	5.46	6.08
Ta	1.09	1.13	1.30
Th	13.32	16.50	17.17

[1]REE were analyzed by inductively coupled plasma mass spectrometry (ICP-MS) at Memorial University, Newfoundland, method after Longerich et al., 1990.

0.513144 (DePaolo, 1981, 1988). ε_{Nd} values were calculated based on a Tremadocian depositional age of 485 Ma, assuming a present-day chondritic uniform reservoir of $^{143}Nd/^{144}Nd = 0.512638$ and $^{147}Sm/^{144}Nd = 0.19659$ with a decay rate of λ^{147} Sm of 6.54×10^{-12}/yr. The $^{143}Nd/^{144}Nd$ ratios have a reported error at the 95% confidence level (Jacobsen and Wasserburg, 1980).

Results

Although most samples analyzed are siliciclastic, the high L.O.I. % (loss on ignition) and CaO for some samples (TU-J and O) indicates the high concentration of carbonates, suggesting that these are calcareous siltstones, and are consequently excluded from clastic sediment geochemical plots. Plotted against SiO_2 (Figs. 3A–3F), major and trace element abundances of the remaining clastic samples display two general groupings corresponding to the dominant clastic lithologies of the Tiñu Formation (sandstone and shale). On average, sandstone samples have lower SiO_2, K_2O, MgO, and TiO_2 relative to the shale samples. Taken together, the sandstone and shale display strong positive correlation between SiO_2, K_2O, Al_2O_3, Zr, and TiO_2. Fe_2O_3 vs. SiO_2 shows a negative correlation trend and MgO is relatively constant with increasing SiO_2. The shale generally has a more restricted composition (i.e., SiO_2 between 56 and 72 wt%, TiO_2 between 0.9 and 1.3 wt%, Al_2O_3 between 15 and 21 wt%, MgO between 0.7 and 1.35%). On discrimination diagrams (Fig. 4), the sandstones and shales plot in a variety of fields, suggesting these plots cannot be used as indicators of tectonic environment. The sandstones and shales have similar and wide ranges

TABLE 3. Sm and Nd Contents and Nd Isotopic Composition of Tiñu Formation Rocks[1]

	Nd (ppm)	Sm (ppm)	$^{147}Sm/^{144}Nd$	$^{143}Nd/^{144}Nd$	2σ	$\varepsilon_{Nd(t)}$	$T_{(DM)}$	$\varepsilon_{Nd(0)}$
TU-D	37.8	7.667	0.1226	0.512044	4	–7	1667	–11.6
TU-E	27.18	5.824	0.1295	0.512026	5	–7.8	1837	–11.9
TU-I	39.34	6.745	0.1036	0.511957	3	–7.5	1502	–13.3

[1]Chemical separations and isotopic analyses were determined at the Atlantic Universities Regional Isotopic Facility, Memorial University of Newfoundland. The $^{143}Nd/^{144}Nd$ ratios are measured by thermal ionization mass spectrometry, after chemical separation of Nd from Sm and other REE by ion-exchange chemistry. La Jolla Nd standard gave an average value of 0.511860. ε_{Nd} values are relative to $^{143}Nd/^{144}Nd = 0.512638$ and $^{147}Sm/^{144}Nd = 0.196593$ for present-day CHUR (Jacobsen and Wasserburg, 1980) and λ^{147} Sm = 6.54×10^{-12}/year. $T_{(DM)}$ are calculated using the model of DePaolo (1981, 1988). ε_{Nd} values are calculated for t = 485 Ma (which represents the age of deposition of the Tiñu Formation) and for the present day (t = 0).

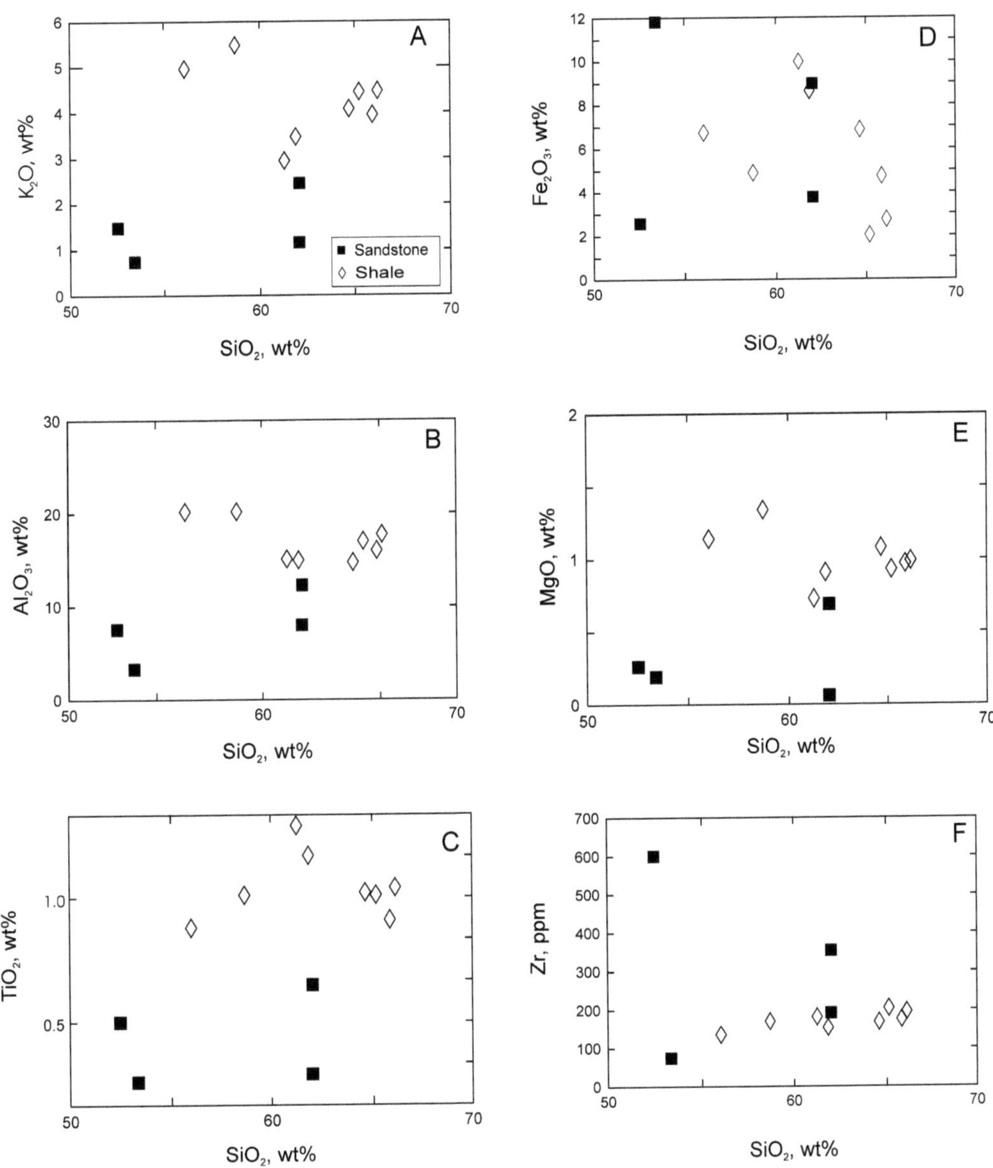

FIG. 3. SiO_2 versus (A) K_2O, (B) Al_2O_3, (C) TiO_2, (D) Fe_2O_3, (E) MgO, and (F) Zr for Tiñu Formation siliciclastic rocks.

in Fe_2O_3 + MgO (ranging from ~4 to 12 wt%), but sandstones are characterized by significantly lower Al_2O_3/SiO_2 and K_2O/Na_2O. One sample, TU-H, with anomalously high K_2O/Na_2O (>400) is excluded from Figure 4B. The distinction between sandstone and shale is also apparent on the Na_2O + CaO–Al_2O_3–K_2O molecular proportion diagram (Fig. 5). The shales plot near the plagioclase-illite/muscovite tie line, suggesting that the relative abundance of these elements is profoundly influenced by alteration of plagioclase. The sandstone samples, on the other hand, have compositions suggesting an important component of mafic minerals, implying derivation from proximal sources.

Trace element ratio plots such as Zr/Nb vs. Ti/Nb, Zr/V vs. Ti/V and Zr/Y vs. Ti/Y (Figs 6A–6C)

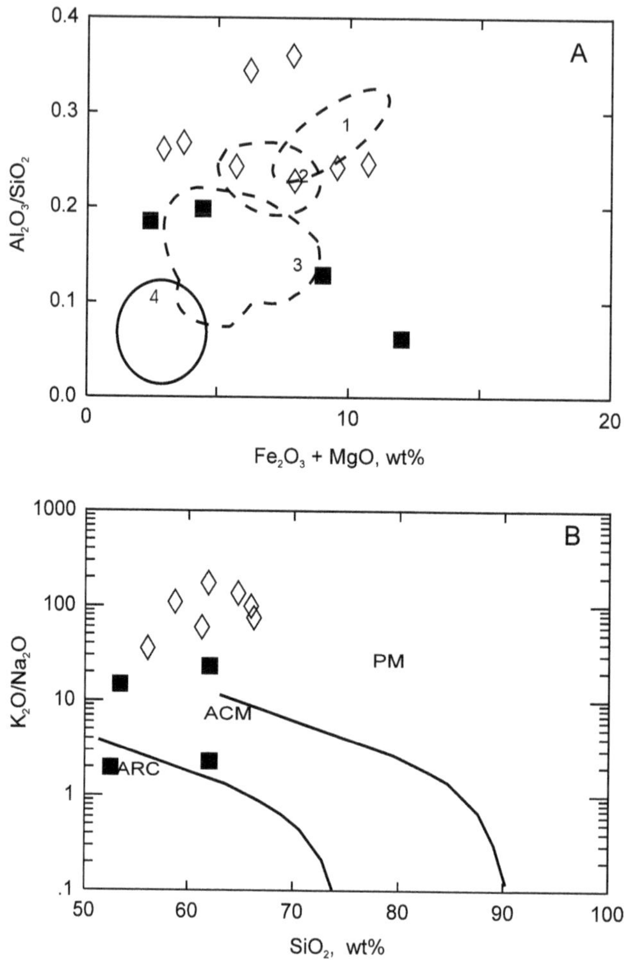

FIG. 4. A. Al_2O_3/ SiO2 vs. MgO + Fe_2O_3 plot (after Bhatia, 1983) for Tiñu Formation siliciclastic rocks. B. K_2O/Na_2O vs. SiO_2 (after Roser and Korsch, 1986) for Tiñu Formation siliciclastic rocks. In (A) 1 = ocean island arc; 2 = continental island arc; 3 = active continental margin; 4 = passive margin. In (B), ACM = active continental margin; PM = passive margin.

also show significant differences between the sandstones and shales. In general, shales show strong positive relationships between these ratios, suggesting that their abundances are controlled by heavy accessory minerals such as zircon, magnetite, and titanite.

Chondrite-normalized patterns for sampled Tiñu Formation shales (Fig. 7) indicate moderate light rare-earth element (LREE) enrichment, (La_N/Sm_N = 3.1 to 4.9), and a relatively flat heavy rare-earth element (HREE) profile (Tb_N/Lu_N ~1.0–1.4). The samples all show a negative Eu anomaly.

The Tiñu samples also show a restricted range in elemental Sm/Nd, which probably reflects the weighted averages of Sm/Nd in the source region. (e.g., Thorogood, 1990). Based on a depositional age of 485 Ma, the three shale samples have negative $\varepsilon_{Nd(t)}$ values, ranging from −7.0 to −7.8 (t = 485 Ma, Table 3; Fig. 8), with T_{DM} ages (after DePaolo, 1981) ranging from 1.50 to 1.83 Ga—i.e., more than 1.0 Ga. older than the depositional age.

Interpretation

To identify the source regions of sedimentary rocks, it is important to consider the effects of weathering, which commonly involve congruent and incongruent dissolution of rock-forming minerals. For example, alkali and alkali earths may be trans-

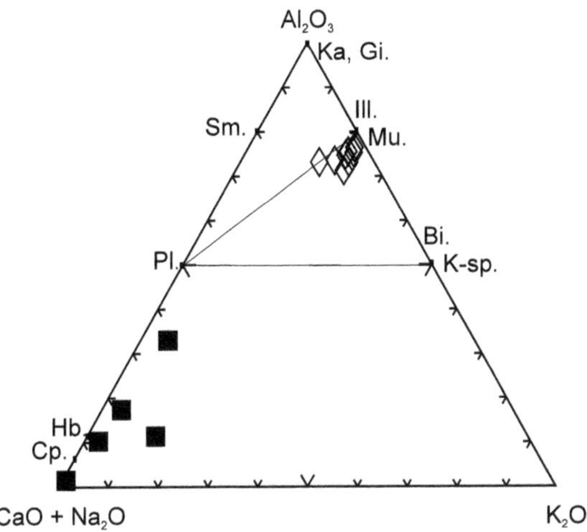

FIG. 5. CaO + Na_2O–Al_2O_3–K_2O molecular proportion diagram (after Nesbitt and Young, 1986) for Tiñu Formation siliciclastic rocks.

ported as dissolved species, and their composition in the sedimentary rocks may not reflect their abundance in source regions. Although some studies have indicated their abundances can be affected by diagenetic and weathering processes (e.g., McDaniel et al., 1994), elements with high charge density, the so-called high-field-strength elements (HFSE; e.g., Ti, P, Y, Zr, and Nb) and rare-earth elements (REE), are generally less soluble than alkalies during weathering, transport, and diagenesis. Instead, these elements are most commonly transported in the solid detritus, and so their abundances in clastic rocks are reliable indicators of provenance (e.g., Holland, 1978; Bhatia and Crook, 1986). Plotting these elements as ratios aids in the reduction of the dilution affect of quartz and the effect of sorting during sediment transport (Murphy, 2002).

Although the Tiñu clastic sediment samples have a wide range in SiO_2, the decreasing abundance of some elements (e.g. Fe_2O_3), increasing abundance in others (e.g., K_2O and Zr), and relatively constant values in yet others (e.g., TiO_2), suggests that major-element trends cannot be simply attributed to the dilution affect of quartz. Overall, the geochemistry of the Tiñu clastic rocks has two important components, felsic and mafic. A continental source is indicated by relationships such as the positive correlation between SiO_2 and Zr (Fig. 3F). The low Ni abundance in most of the shale and sandstone samples together with LREE enrichment (Fig. 8) suggests that the crustal source was moderately differentiated. This signature is typical of mafic to intermediate calc-alkaline source (e.g., Taylor and McLennan, 1985).

Although the samples display evidence of variation in chemical maturity (e.g., decrease in Fe_2O_3 with increasing SiO_2, and lower Hf, Y, Nb, together with a limited range in Zr/Nb and Zr in shale samples with $SiO_2 > 60$ %), in general the lithogeochemistry displays a chemical immaturity, indicating that the sediments are dominated by locally derived components. This lack of chemical maturity suggesting proximal sources is supported by the high Fe_2O_3 and MgO + Fe_2O_3 (up to 10 wt%) and by the lower Al_2O_3/CaO + Na_2O in rocks with high MgO + Fe_2O_3, suggesting that many clastic rocks contain abundant mafic minerals and plagioclase. Inasmuch as these minerals are highly susceptible to weathering, the data indicate that the source rock was proximal. In settings where the clastic sediments are chemically immature, the lithogeochemistry is generally inherited from nearby source rocks, so that standard tectonic discrimination diagrams, such as those in Figure 4, often give spurious results (Murphy, 2000).

Variations in HFSE abundances are consistent with an important mafic contribution to the detritus. Enrichments of titanium relative to zirconium suggest a possible increase in Fe-Ti–bearing phases of mafic rocks, such as titanomagnetite. This is especially evident in Zr/Nb vs. Ti/Nb (Fig. 7A) where the shales show a wide range in Ti/Nb and relatively

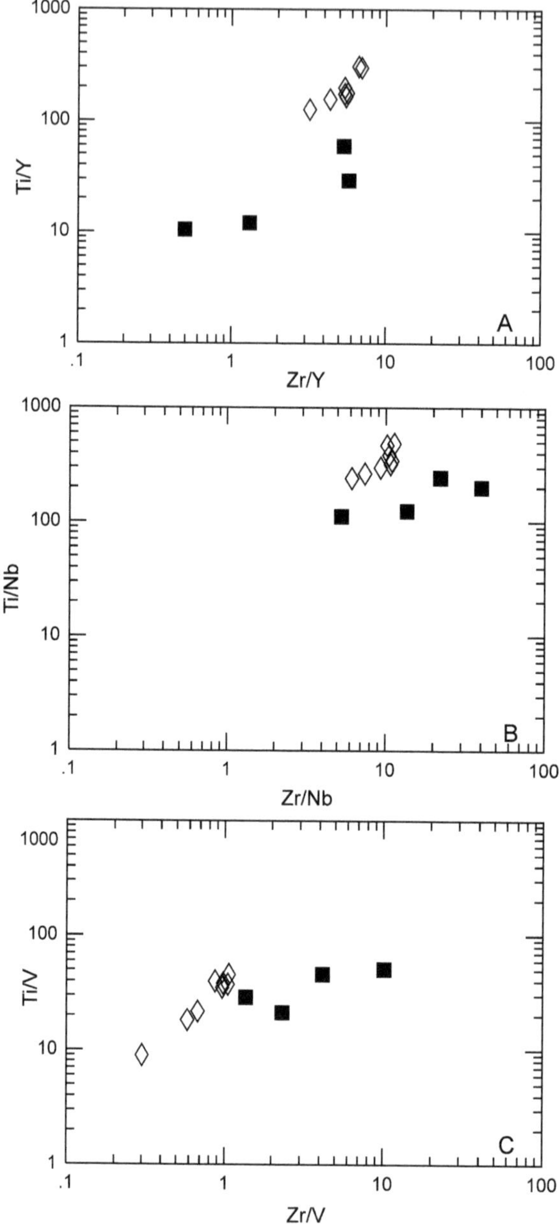

FIG. 6. Interelement ratio plots of (A) Zr/Nb vs. Ti/Nb, (B) Zr/V vs. Ti/V, and (C) Zr/Y vs. Ti/Y for Tiñu Formation siliciclastic rocks.

constant Zr/Nb. Significantly, the Sm-Nd analyses of the most "mafic" shale sample is very similar to the two more felsic samples (Table 3), suggesting that the mafic component is also Mesoproterozoic. Taken together, the geochemical evidence suggests that the source of the Tiñu Formation consists of an older continental basement containing felsic and mafic sources, with no evidence of a significant juvenile input.

This interpretation is supported by Sm-Nd isotopic systematics, which are typical of average continental crust, and T_{DM} ages (Table 3) of 1.5–1.8 Ga

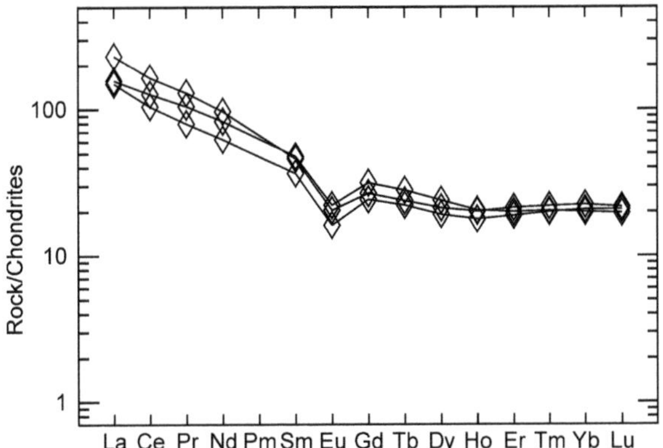

FIG. 7. Chondrite-normalized rare-earth element (REE) for representative samples of the Tiñu Formation siliciclastic rocks.

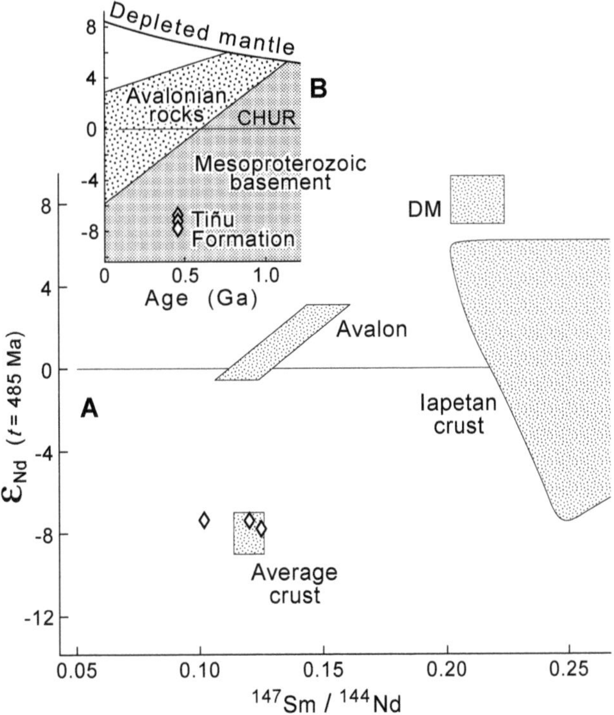

FIG. 8. Sm-Nd isotopic data for representative samples of the Tiñu Formation siliciclastic rocks. (A) ε_{Ndt} versus $^{147}Sm/^{144}Nd$ diagram (t = 485 Ma) comparing Sm-Nd isotopic data for the Tiñu Formation (Table 3) with typical Sm-Nd isotopic compositions of Avalonian crust (e.g., Murphy et al., 1996). The Sm-Nd isotopic characteristics for the average upper crust are bracketed between modern global average river sediment ($^{147}Sm/^{144}Nd$ = 0.114; T_{DM} = 1.52 Ga; Goldstein and Jacobsen, 1988) and the average age of sedimentary mass (Miller et al., 1986). Iapetan crust includes normal and depleted island arc tholeiites, and ophiolitic complexes in Newfoundland and Norway (MacLachlan and Dunning, 1998; Pedersen and Dunning, 1997). (B) ε_{Ndt} vs. time (Ga) diagram (t = 485 Ma) comparing Sm-Nd isotopic data (Table 3) with typical Sm-Nd isotopic compositions of Avalonian and Mesoproterozoic crust (Murphy et al., 1996).

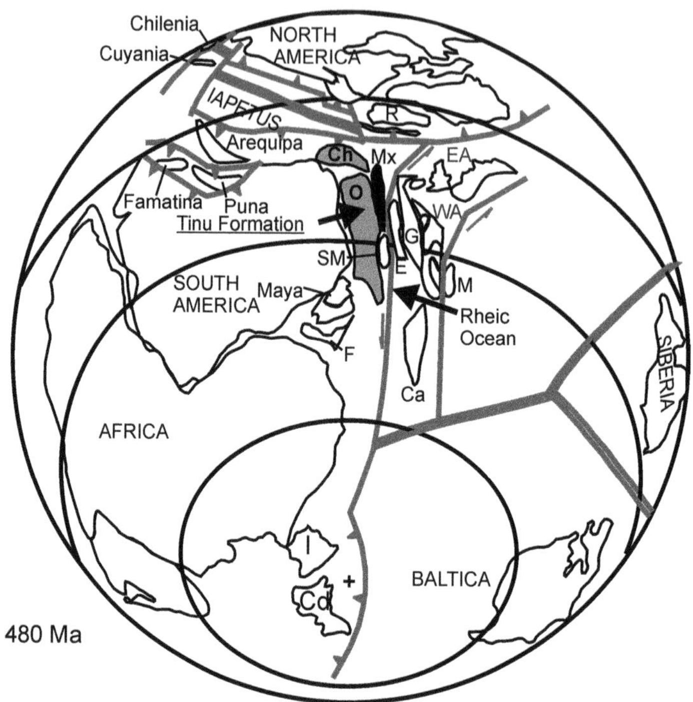

FIG. 9. Palinspastic map of Oaxaquia (O) and related peri-Gondwanan terranes along the Gondwanan margin at ~480 Ma (see Keppie et al., 2003). Abbreviations for terranes: Mx = Mixteca; Ch = Chortis; SM = Sierra Madré; R= Rockall; WA = West Avalonia; EA = East Avalonia; Ca = Carolina; E = Exploits; G = Gander; M = Meguma; F = Florida; I = Iberia; Cd = Cadomia. The rifting of WA, EA, Ca, E, G, and M from this margin in the Early Ordovician resulted in the formation on the Rheic Ocean.

(i.e., more than 1.0 Ga older than the depositional age), which suggest an ancient cratonic source for Tiñu Formation clastic sediments (Fig. 8). Detrital zircon data indicate the dominance of ~992–1155 Ma zircons in these sediments (Gillis et al., 2001). The detrital zircon and Sm-Nd data are typical of Mesoproterozoic or older crust that was recycled during Grenvillian orogenesis. The obvious source for these isotopic characteristics is the underlying Oaxacan Complex.

Conclusion

Despite the importance of the Rheic Ocean to Paleozoic paleogeography and an understanding of the processes that gave rise to Pangea, there are major uncertainties in the identification of its margins, and the mechanisms and timing of its initial rifting and opening. Establishing the provenance of Lower Ordovician clastic successions such as the Tremadocian Tiñu Formation, which was deposited along the Gondwanan margin, provides important constraints on the early evolution of the southern flank of the Rheic Ocean. The geochemical and isotopic data presented herein, combined with the results of Gillis et al. (2001), suggest that the Tiñu Formation had a proximal source, and is consistent with derivation from the underlying ~1 Ga Oaxacan Complex. The lack of evidence for a contribution from a coeval juvenile source suggests that deposition occurred in early stages of ocean development where a proximal source in a continental rift setting would be expected (Keppie and Ramos, 1999). General palinspastic relationships at 480 Ma are illustrated in Figure 9.

Acknowledgments

Murphy acknowledges the continuing support of the National Science and Engineering Research Council (NSERC) Canada. JDK acknowledges funding from PAPIIT project (IN103003). This project is a contribution to IGCP Projects 453 and 497.

REFERENCES

Bhatia, M. R., 1983, Plate tectonics and geochemical composition of sandstones: Journal of Geology, v. 91, p. 611–627.

Bhatia, M. R., and Crook, K. A. W., 1986, Trace element characteristics of greywackes and tectonic setting discrimination of sedimentary basins: Contributions to Mineralogy and Petrology, v. 92, p. 181–193.

Boucot, A. J., Blodgett, R. B., and Stewart, J. H., 1997, European Province Late Silurian brachiopods from the Ciudad Victoria area, Tamaulipas, northeastern Mexico, *in* Klapper, G., Murphy, M. A., and Talent, J. A., eds., Paloeozoic sequence stratigraphy, biostratigraphy, and biogeography: Studies in honor of J. Granville ("Jess") Johnson: Geological Society of America Special Paper 321, p. 273–293.

Centeno-García, E., and Keppie, J. D., 1999, Latest Paleozoic–Early Mesozoic structures in the central Oaxaca terrane of southern Mexico: Deformation near a triple junction: Tectonophysics, v. 301, p. 231–242.

DePaolo, D. J, 1981, Neodymium isotopes in the Colorado front range and crust-mantle evolution in the Proterozoic: Nature, v. 291, p. 193–196.

DePaolo, D. J., 1988, Neodymium isotope geochemistry: An introduction: New York, NY, Springer-Verlag, 187 p.

Dostal, J., Baragar, W. R. A., and Dupuy, C., 1986, Petrogenesis of the Natkusiak continental basalts, Victoria Island, N.W.T.: Canadian Journal Earth Sciences, v. 23 p. 622–632.

Dostal, J., Dupuy, C., and Caby, R., 1994, Geochemistry of the Neoproterozoic Tilesmi belt of Iforas (Mali, Sahara): A crustal section of an oceanic island arc: Precambrian Research, v. 65 p. 55–69.

Dostal, J., Keppie, J. D., .Macdonald, H., and Ortega-Gutiérrez, F., 2004, Sedimentary origin of calcareous intrusions in the ~1 Ga Oaxacan Complex, Southern Mexico: Tectonic implications: International Geology Review, v. 46, p. 528–541.

Elías-Herrera, M., and Ortega-Gutiérrez, F, 2002, Caltepec fault zone: An Early Permian dextral transpressional boundary between the Proterozoic Oaxacan and Paleozoic Acatlán complexes, southern Mexico, and regional implications: Tectonics, v. 21, no. 3 [10.1029/200TC001278].

Gillis, R. J., Gehrels, G. E., Flores de Dios, A., and Ruiz, J., 2001, Paleogeographic implications of detrital zircon ages from the Oaxaca terrane of southern Mexico [abs.]: Geological Society of America, Abstracts with Programs, v. 33, no. 6, A-428.

Goldstein, S. J., and Jacobsen, S. B., 1988, Nd and Sr isotopic systematics of rever water suspended material: Implications for crustal evolution: Earth and Planetary Science Letters, v. 87, p. 249–265.

Holland, H. D., 1978, The chemistry of the atmosphere and oceans: New York, NY, Wiley, 351 p.

Jacobsen, S. B., and Wasserburg, G. J., 1980, Sm-Nd evolution of chondrites: Earth and Planetary Science Letters, v. 50, p. 139–155.

Keppie, J. D., 2004, Terranes of Mexico revisited: A 1.3 billion year odyssey: International Geology Review, v. 46, p. 765–794.

Keppie, J. D., Dostal, J., Cameron, K. L., Solari, L. A., Ortega-Gutiérrez, F., and Lopez, R., 2003, Geochronology and geochemistry of Grenvillian igneous suites in the northern Oaxacan Complex, southern México: Tectonic implications: Precambrian Research, v. 120, p. 365–389.

Keppie, J. D., Dostal, J., Ortega-Gutierrez, F., and Lopez, R., 2001, A Grenvillian arc on the margin of Amazonia: Evidence from the southern Oaxacan Complex, southern Mexico: Precambrian Research, v. 112, No.3-4, p. 165–181.

Keppie, J. D., and Ramos, V.A., 1999, Odyssey of terranes in the Iapetus and Rheic oceans during the Paleozoic, *in* Ramos, V. A., and Keppie, J. D., eds., Laurentia-Gondwana connections before Pangea: Geological Society of America Special Paper 336, p. 267–276.

Keppie, J. D., Sandberg, C. A., Miller, B. V., Sánchez-Zavala, J. L., Nance, R. D., and Poole, F. G., 2004, Implications of latest Pennsylvanian to Middle permian paleontological and U-Pb SHRIMP data from the Tecomate Formation to re-dating tectonothermal events in the Acatlán Complex, southern Mexico: International Geology Review, v. 46, p. 745–753.

Kerr, A., Jenner, G. A. and Fryer, B. J., 1995, Sm-Nd isotope geochemistry of Precambrian to Paleozoic granitoid suites and the deep-crustal structure of the southeast margin of the Newfoundland Appalachians: Canadian Journal of Earth Sciences, v. 32, p. 224–245.

Longerich, H. P., Jenner, G. A., Fryer, B. J., and Jackson, S. E., 1990, Inductively coupled plasma mass spectrometric analysis of geologic samples: A critical evaluation based on case studies: Chemical Geology, v. 83, p. 105–118.

MacLachlan, K., and Dunning, G., 1998, U-Pb ages and tectonomagmatic relationships of early Ordovician low-Ti tholeiites, boninites, and related plutonic rocks in central Newfoundand: Contributions to Mineralogy and Petrology, v. 133, p. 235–258.

McDaniel, D. K., Hemming, S. R., McLennan, S. M., and Hanson, G. N., 1994, Resetting of neodymium isotopes and redistribution of REE's during sedimentary processes: The early Proterozoic Chelmsford Formation, Sudbury Basin, Ontario, Canada: Geochimica et Cosmochimica Acta, v. 54, p. 2015–2050.

McLennan, S. M., Taylor, S. R., McCullouch, M. T., and Maynard, J. B., 1990, Geochemical and isotopic determination of deep sea turbidites: Crustal evolution and plate tectonic associations: Geochimica et Cosmochimica Acta, v. 54, p 2015–2049.

Malone, J. W., Nance, R. D., Keppie, J. D., and Dostal, J., 2002, Deformation history of part of the Acatlán Complex: Late Ordovician–Early Silurian and Early Permian orogenesis in southern Mexico: Journal of South American Earth Sciences, v, 15, p. 511–524.

Miller, R. G., O'Nions, R. K., Hamilton, P. J., and Welin, E., 1986, Crustal residence ages of clastic sediments, orogeny, and crustal evolution: Chemical Geology, v. 57, p. 87–99.

Murphy, J. B. 2000, Tectonic influence on sedimentation along the southern flank of the Late Paleozoic Magdalen Basin in the Canadian Appalachians: Geochemical and isotopic constraints on the Horton Group in the St. Mary's Basin , Nova Scotia: Bulletin of the Geological Society of America, v. 112, p. 997–1011.

Murphy, J. B., 2002, Geochemistry of the Neoproterozoic metasedimentary Gamble Brook Formation, Avalon terrane, Nova Scotia: Evidence for a rifted-arc environment along the West Gondwanan margin of Rodinia: Journal of Geology, v. 110, p 407–419.

Murphy, J. B., Keppie, J. D., Dostal, J., and Cousins, B. L., 1996, Repeated late Neoproterozoic–Silurian lower crustal melting beneath the Antigonish Highlands, Nova Scotia: Nd isotopic evidence and tectonic interpretations, *in* Nance, R. D., and Thompson, M. D., eds., Avalonian and related peri-Gondwanan terranes of the circum–North Atlantic: Geological Society of America Special Paper 304, p. 109–120.

Navarro-Santillán, D., Sour-Tovar, F., and Centeno-Garcia, E., 2002, Lower Mississippian (Osagean) brachiopods from the Santiago Formation, Oaxaca, Mexico: Stratigraphic and tectonic implications: South American Journal of Earth Sciences, v. 15, 2002, p. 327–336.

Nesbitt, H. W., and Young, G. M., 1996, Petrogenesis of sediments in the absence of chemical weathering: Effect of abrasion and sorting on bulk composition and mineralogy: Sedimentology, v. 43, p. 341–358.

Ortega-Gutiérrez, F., Elías-Herrera, M., Reyes-Salas, M., Macías-Romo, C., and Lopez, R., 1999, Late Ordovician–Early Silurian continental collision orogeny in southern Mexico and its bearing on Gondwana-Laurentia connections: Geology, v. 27, p. 719–722.

Ortega-Gutiérrez, F., Ruiz, J., and Centeno-García, E., 1995, Oaxaquia, a Proterozoic microcontinent accreted to North America during the Late Proterozoic: Geology, v. 23, p. 1127–1130.

Ortega-Obregon, C., Keppie, J. D., Solari, L. A., Ortega-Gutiérrez, F., Dostal, J., Lopez, R., Ortega-Rivera, A., and Lee, J. W. K., 2003, Geochronology and geochemistry of the ~917 Ma, calc-alkaline Etla granitoid pluton (Oaxaca, southern Mexico): Evidence of post-Grenvillian subduction along the northern margin of Amazonia: International Geology Review, v. 45, p. 596–610.

Patchett, P. J., and Ruiz, J., 1987, Nd isotopic ages of crust formation and metamorphism in the Precambrian of eastern and southern Mexico: Contributions to Mineralogy and Petrology, v. 96, p. 523–528.

Pedersen, R. B., and Dunning, G. R., 1997, Evolution of arc crust and relations between contrasting sources: U-Pb (age), Nd, and Sr isotopic systematics of the ophiolite terrain of SW Norway: Contributions to Mineralogy and Petrology, v. 128, p. 1–15.

Robison, R., and Pantoja-Alor, J., 1968, Tremadocian trilobites from Nochixtlan region, Oaxaca, Mexico: Journal of Paleontology, v. 42, p. 767–800.

Roser, B. P. and Korsch, R. J., 1986, Determination of tectonic setting of sandstone-mudstone suites using SiO_2 content and K_2O/Na_2O ratio: Journal of Geology, v. 94, p. 635–650.

Shergold, J. H., 1975, Late Cambrian and Early Ordovician trilobites from the Burke River structural belt, western Queensland, Australia: Department of Mineral and Energy, Bureau of Mineral Resources, Geology and Geophysics, Bulletin 153, 221 p.

Solari, L. A., Keppie, J. D., Ortega-Gutiérrez, F., Cameron, K. L., Lopez, R., and Hames, W. E., 2003, 990 Ma and 1,100 Ma Grenvillian tectonothermal events in the northern Oaxacan Complex, southern Mexico: roots of an orogen: Tectonophysics, v. 365, p. 257–282.

Taylor, S. R. and McLennan, S. M. 1985, The continental crust: Its composition and evolution: Oxford, UK, Blackwell Scientific Publications, 312 p.

Thorogood, E. J., 1990, Provenance of the pre-Devonian sediments of England and Wales: Sm-Nd isotopic evidence: Journal of the Geological Society, London, v. 147, p. 591–594.

Phanerozoic Structures in the Grenvillian Northern Oaxacan Complex, Southern Mexico: Result of Thick-Skinned Tectonics

LUIGI A. SOLARI,[1] J. DUNCAN KEPPIE, F. ORTEGA-GUTIÉRREZ,

Instituto de Geología, Universidad Nacional Autónoma de México (UNAM), Ciudad Universitaria, 04510 Del. Coyoacán, México D.F., México

A. ORTEGA-RIVERA,

Centro de Geociencias, Campus Juriquilla, Universidad Nacional Autónoma de México (UNAM), Apdo. Postal 1-742, Centro Querétaro, Qro. 76001, México

W. E. HAMES,

Department of Geology, Auburn University, Auburn, Alabama 36830

J. K. W. LEE

Department of Geology, Queens University, Kingston, Ontario, Canada, K7L 3N6

Abstract

Shear zones and upright folds affecting the northern Oaxacan Complex are documented as follows: (1) SE-directed shearing dated at 479 ± 4 Ma ($^{40}Ar/^{39}Ar$ biotite laser total fusion age); (2) E-directed thrusting dated at 247 ± 3 Ma ($^{40}Ar/^{39}Ar$ biotite plateau age); (3) NW to NNW-trending, steeply inclined folds of mid-Triassic–Jurassic age bracketed between events (2) and (4); and (4) NNW-trending vertical shearing dated at 141 ± 9 Ma ($^{40}Ar/^{39}Ar$ biotite laser total fusion age). Inasmuch as the northern Oaxacan Complex is inferred to have been exposed to conditions well above the ~300°C closure temperature for argon in biotite since it was exhumed at ~710–760 Ma, it is inferred that these ages record thermal re-equilibration associated with hot fluids flowing along active shear zones. These structural events correspond respectively with: (1) the onset of Paleozoic deposition in the Tremadocian, presumably associated with SE-directed, listric normal faulting/shearing; (2) E-vergent thrusting associated with the development of an arc along the length of Mexico; (3) folding in sympathy with the dextral shear during opening of the Gulf of Mexico; and (4) Jurassic–Cretaceous boundary, normal fault readjustments. These data indicate that the northern Oaxacan Complex was involved in thick-skinned tectonics during most Phanerozoic episodes of deformation.

Introduction

THE ROCKS OF MEXICO have been involved in several orogenic events, including the Permo-Carboniferous Alleghanian orogeny and the Cretaceous–Tertiary Laramide Orogeny; however, their inferred effects on the Precambrian basement varies from negligible in thin-skinned models to unknown in thick-skinned models. Thus, De Cserna (1989), and Mitre-Salazar and Roldan-Quintana (1990) interpreted Phanerozoic deformation to be thin-skinned and placed a sole thrust at the top of the Precambrian basement. On the other hand, Ortega-Gutiérrez (1990) in a transect of southern Mexico, showed the Precambrian Oaxacan Complex as wedge-shaped bounded by an east-vergent Laramide thrust and a west-vergent Caledonian thrust. These boundaries have recently been shown to be: (1) an Early Mesozoic(?), east-vergent thrust reactivated by Middle Jurassic dextral shear and late Mesozoic listric normal faulting (Alaniz-Alvarez et al., 1996); and (2) a Permian, dextral flower structure (Elias-Herrera and Ortega-Gutiérrez, 2002), respectively. Nevertheless, Phanerozoic structures within the Precambrian Oaxacan Complex have not been described. Although Centeno-García and Keppie (1999) described Permian–Jurassic structures in the Paleozoic rocks unconformably overlying the Oaxacan Complex, their expression in the underlying Oaxacan Complex was not documented. In this

[1]Corresponding author; email: solari@servidor.unam.mx

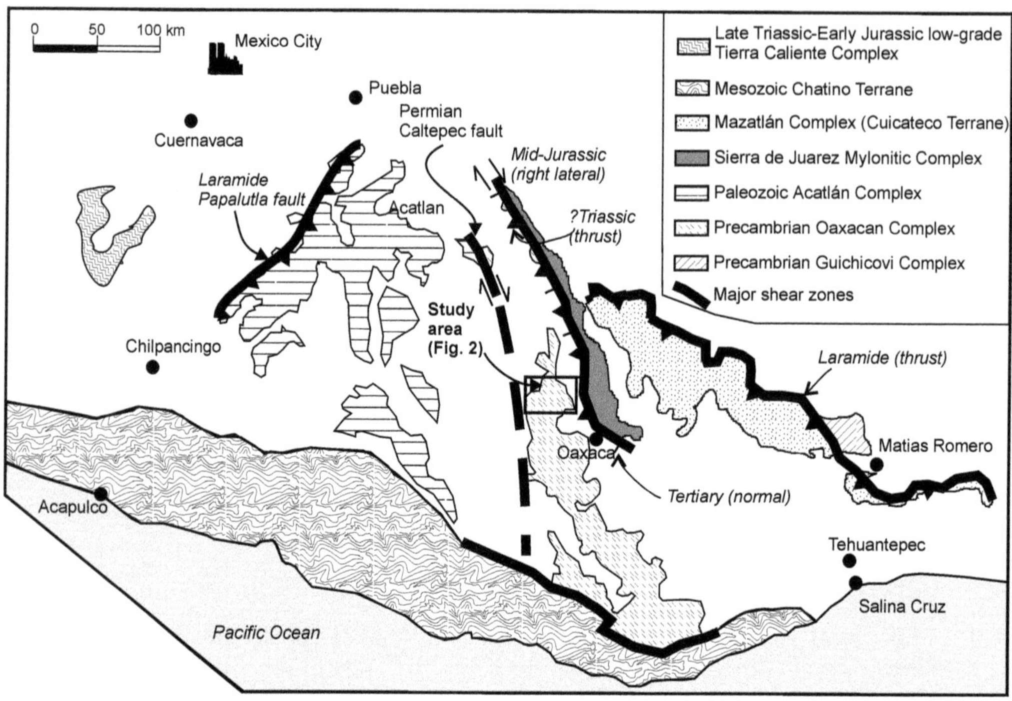

FIG. 1. Terrane subdivision of southern Mexico, modified after Sedlock et al. (1993). Tectonic boundaries between each terrane are also shown with their kinematics.

paper, we present the first documented evidence for Phanerozoic structures within the northern Oaxacan Complex.

Geological Setting

The studied area (Figs. 1 and 2) covers approximately 300 km^2 in the northern Oaxacan Complex, ~25 km northwest of Oaxaca city. Here, the Oaxacan Complex is made up two major tectonic slices separated by a shear zone (Fig. 3C): (1) a lower Huitzo slice is mainly composed of ~1,012 Ma meta-anorthosites and associated ferrodiorites, gabbros, garnet–2 pyroxene mafic gneisses, and garnet-bearing charnockites, which intrude the ≥1,350 Ma El Catrín migmatite; and (2) an upper El Marquez slice made up of ≥1,250 Ma calcareous metasediments, marbles, and quartzofeldspathic granulites, intruded by ~1,150 Ma charnockites, ≥1,130 Ma syenites, ~1,250 Ma gabbros, and undated granitoids (Keppie et al., 2003; Solari et al., 2003).

The northern Oaxacan Complex was affected by two tectonothermal events: the ~1,106 Ma Olmecan migmatitic event and the ~990 Ma granulite facies, Zapotecan orogeny (Keppie et al., 2003; Solari et al., 2003), which was followed by cooling through ~300–350°C by ~856 Ma and may have reached the surface by ~710–760 Ma (Keppie et al., 2004). This is consistent with the fact that all K-Ar biotite ages between the studied area and Ejutla (south of Oaxaca City) are older than ~650 Ma (Fries and Rincon-Orta, 1965), suggesting that this region generally did not subsequently reach temperatures higher than ~300–350°C (estimated closure temperature for argon in biotite: Harrison et al., 1985), i.e., temperatures below the onset of greenschist-facies metamorphism. It is also consistent with: (1) the fact that Tremadocian to Permian fossiliferous sediments (750 m thick) unconformably overlying the northern Oaxacan Complex only underwent subgreenschist–facies metamorphism during the development of Permian–Jurassic bedding-parallel shear zones, and N-S upright folds associated with the development of a slaty cleavage (Centeno-Garcia and Keppie, 1999); and (3) projection of the unconformable base of these Ordovician rocks (~490 Ma) southward across the northern Oaxacan Complex suggests that it does not lie more than a few kilometers above the

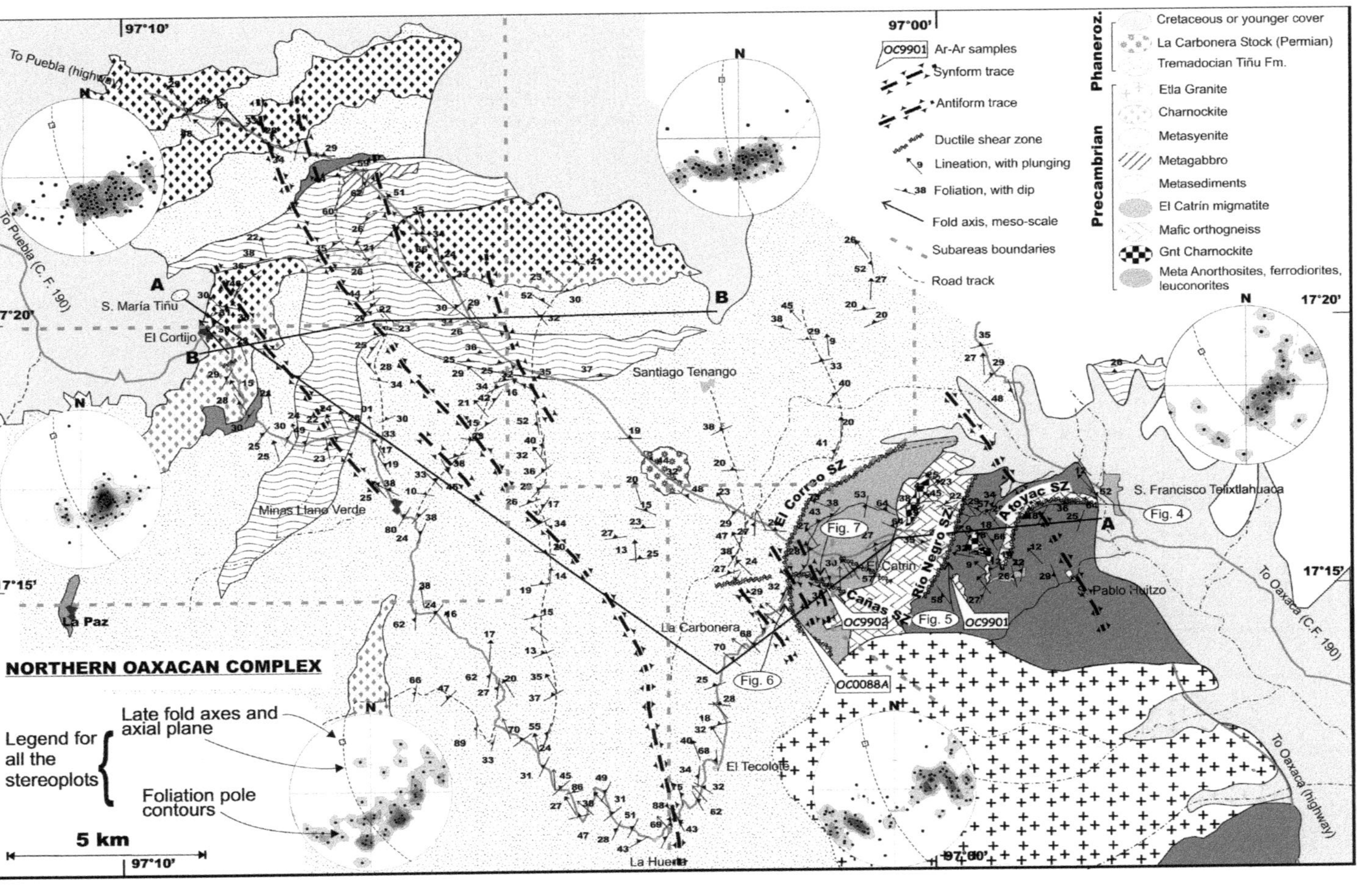

FIG. 2. Geological map of the northern Oaxacan Complex, modified after Solari (2001). Density-contoured stereoplots represent the orientation of late folds in each subarea (subdivision marked by discontinuous grey lines).

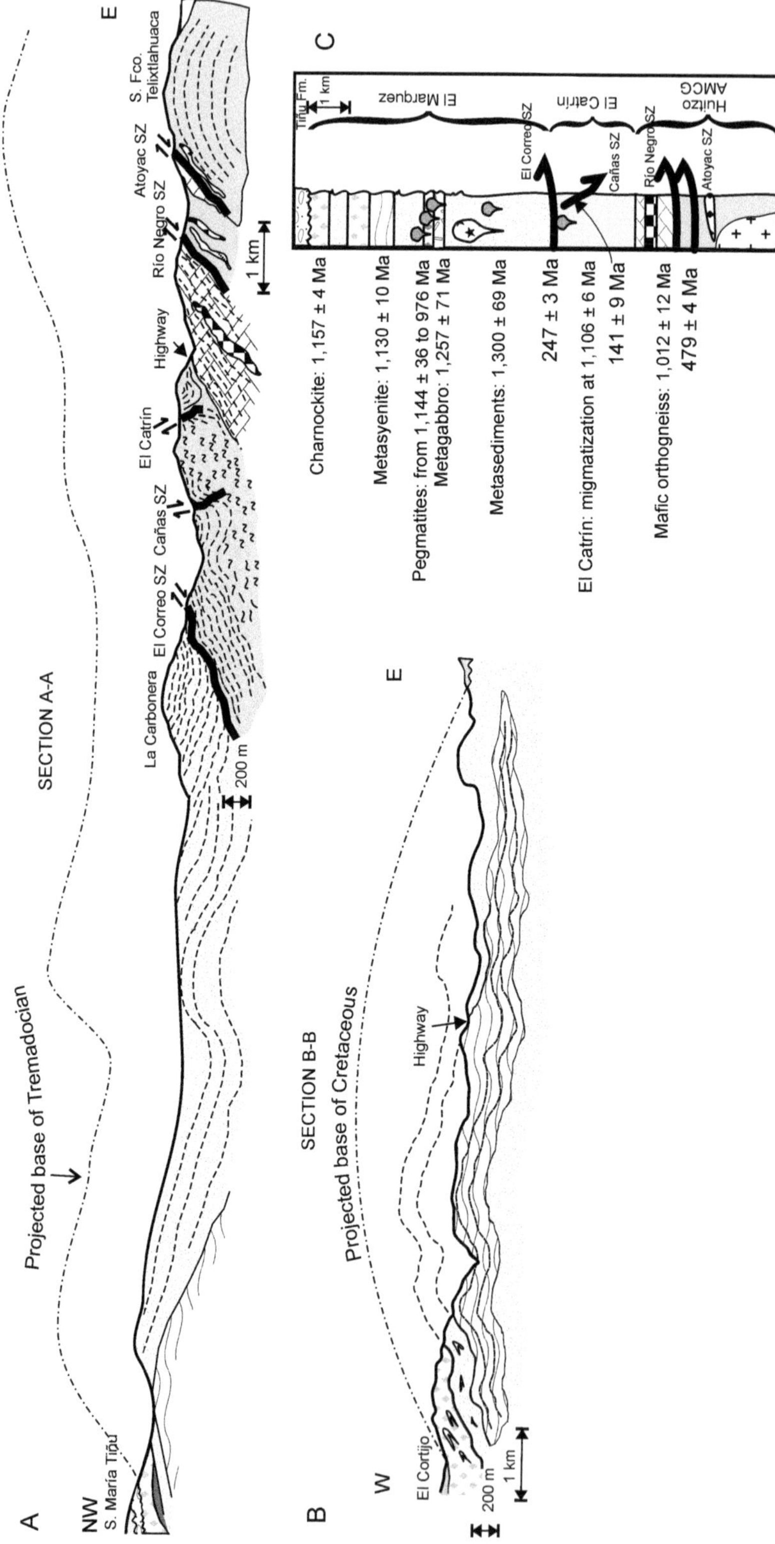

FIG. 3. Geologic section and structural column for the northern Oaxacan Complex. Patterns in the legend are the same as in Figure 2. Sections A–A and B–B show the projection of the base of Tremadocian and Cretaceous sediments across the area. The structural column shows the major shear zones discussed in the text.

present erosion surface (Fig. 3A). Similarly, projecting the sub-Cretaceous (~142 Ma) unconformity across the northern Oaxacan Complex indicates that it lies close to the present erosion surface (Fig. 3B). The purpose of this paper is thus to present structural data and $^{40}Ar/^{39}Ar$ geochronology of greenschist-facies shear zones and folds that deform the high-grade thrust slices in the Oaxacan Complex.

Structure and Age Constraints

The structures we report here are generally greenschist-facies shear zones and associated folds within the Oaxacan basement. The shear zones fall into two groups: parallel and oblique to the dominant foliation. Three major, foliation-parallel, shear zones have been recognized (from structural bottom to top): (1) the Atoyac shear zone occurs within the Lower Huitzo unit and separates the meta-anorthosite unit from a mafic orthogneiss lens; (2) the Rio Negro shear zone occurs within the Huitzo anorthosite unit and marks the boundary between anorthosite and the mafic orthogneiss; and (3) the El Correo shear zone separates the El Catrin migmatite unit from the El Marquez metasedimentary unit (Figs. 2 and 3C). In general, these shear zones formed along boundaries between contrasting lithologies and appear to have reactivated earlier higher-grade shear zones. On the other hand, the Cañas shear zone is a steeply dipping, NNW-trending shear zone that obliquely cuts a NNW-trending upright-steeply inclined fold that is congruent with the major folds shown on the map (Fig. 2).

$^{40}Ar/^{39}Ar$ analyses were performed using both laser fusion analyses of hand-picked biotite crystals (Table 1) at the Massachusetts Institute of Technology using the methodology outlined in Hames and Bowring (1994), and a laser step-heating procedure (Table 2) on small mineral concentrates at Queen's University Geochronology Laboratory as outlined in Clark et al. (1998).

Atoyac shear zone (type locality: Highway 190, 153 km)

This shear zone forms the boundary between the meta-anorthosite (below) and a mafic gneiss lens within the Huitzo unit (Fig. 4). The mafic gneiss displays stretched quartz, feldspar, fine-grained biotite (commonly chloritized), epidote, and sericite. Both biotite and sericite were recrystallized during the shearing, which suggests lower greenschist-facies metamorphism. Sense of shear determined by rare plagioclase sigma porphyroclasts indicates top-to-the-SE.

Rio Negro shear zone (type locality: Toll Highway Oaxaca-México, km 216)

This shear zone separates the top of the meta-anorthosite from the overlying mafic gneiss (Fig. 5). It is composed of planar distributed fine-grained biotite, generally intergrown with uralitic amphibole. Quartz is completely recrystallized to form subgrains with anastomosing boundaries and undulose extinction. In low-strain lenses, remnants of granulitic assemblages characterized by quartz-perthitic feldspar-pyroxenes are almost completely altered to tremolite. Porphyroclasts (<2 mm) composed of perthitic feldspar cores are mantled by decussate albite subgrains (Tullis and Yund, 1985; Pryer, 1993)(Fig. 5I). A top-to-the-south sense of movement is given by: (1) rare, asymmetric σ tails around these porphyroclasts (Fig. 5A inset); (2) S-C plane geometry (Fig. 5H), where the C planes are made up of quartz and plagioclase, and S surfaces are composed of zones rich in biotite and uralitic amphibole; (3) Z-shaped (looking eastwards), E-trending, subhorizontal, close folds ($\leq$10 cm) that locally fold the mylonitic S-planes (Fig. 5C), with a weak axial plane foliation defined by the same biotite + uralitic amphibole association.; and (4) sheath-fold geometries and NW plunging, stretched plagioclase and quartz ($\leq$10:1) at the contact between anorthosites and mylonitized mafic gneisses (Figs. 5B and 5D). The stretching lineation is parallel to elongate lenses of opaque minerals intergrown with apatite and generally surrounded by biotite and uralite coronas.

An attempt to date the uralitic amphibole from this shear zone resulted in an uninterpretable discordant $^{40}Ar/^{39}Ar$ spectrum. On the other hand, $^{40}Ar/^{39}Ar$ laser total fusion analyses of biotite from just below the shear zone yielded a tight cluster of data and three scattered points characterized by minor amounts of extraneous, non-atmospheric argon (Table 1; Fig. 5E). Regression of the clustered data yields an age of 490 ± 20 Ma, with an initial $^{40}Ar/^{36}Ar$ ratio of 213 ± 94 Ma, indicating a mixture of radiogenic and modern atmospheric argon; the scatter is beyond that expected for analytical uncertainties. Thus, the air-corrected clustered data, which yields a mean age of 479 ± 4 Ma, is preferred. This biotite age comes from the same sample that yielded a previously published mean $^{40}Ar/^{39}Ar$ age of 977 ± 12 Ma on hornblende (Solari et al., 2003).

TABLE 1. $^{40}Ar/^{39}Ar$ Laser Total Fusion Analyses of Selected Samples in the Northern Oaxacan Complex[1]

Analysis	$^{39}Ar/^{40}Ar$	$^{36}Ar/^{40}Ar$	$^{38}Ar/^{40}Ar$	$^{37}Ar/^{40}Ar$	K/Ca	K/Cl	Ca/Cl	% $^{40}Ar^*$	Ages
			OC 9901, biotite[2]						
1	3.25E-02 ± 1.75E-04	4.30E-04 ± 2.71E-05	1.13E-04 ± 1.87E-06	0.0024 ± 0.00007	6.9	30.5	4.4	87.3	478.8 ± 2.6
2	2.48E-02 ± 1.70E-04	6.74E-04 ± 1.32E-05	7.80E-05 ± 8.16E-07	0.0019 ± 0.00004	6.6	33.8	5.1	80.1	561.6 ± 3.8
3	3.19E-02 ± 3.08E-04	2.21E-04 ± 1.61E-05	1.07E-04 ± 1.28E-06	0.0060 ± 0.00007	2.8	31.6	11.4	93.5	516.8 ± 5.0
4	3.37E-02 ± 3.28E-04	2.73E-04 ± 1.48E-05	1.02E-04 ± 2.31E-06	0.0030 ± 0.00006	5.9	35.1	5.9	91.9	485.6 ± 4.7
5	3.30E-02 ± 5.02E-04	3.25E-04 ± 2.04E-05	1.13E-04 ± 1.90E-06	0.0033 ± 0.00008	5.2	31	6	90.4	486.8 ± 7.4
6	2.57E-02 ± 2.85E-04	4.13E-04 ± 2.04E-05	9.46E-05 ± 1.69E-06	0.0015 ± 0.00007	9.2	28.9	3.1	87.8	590.7 ± 6.6
7	3.39E-02 ± 3.08E-04	3.44E-04 ± 1.52E-05	1.25E-04 ± 1.91E-06	0.0015 ± 0.00006	11.4	28.8	2.5	89.8	473.6 ± 4.3
8	3.25E-02 ± 3.54E-04	4.23E-04 ± 1.67E-05	1.17E-04 ± 1.77E-06	0.0044 ± 0.00007	3.9	29.7	7.6	87.5	479.3 ± 5.2
9	3.27E-02 ± 4.69E-04	4.83E-04 ± 1.34E-05	1.24E-04 ± 3.70E-06	0.0016 ± 0.00004	10.8	28.1	2.6	85.7	468.8 ± 6.7
10	3.30E-02 ± 2.97E-04	3.58E-04 ± 1.56E-05	1.35E-04 ± 2.15E-06	0.0032 ± 0.00006	5.4	26.1	4.8	89.4	482.5 ± 4.3
			OC 9902, biotite[3]						
1	9.27E-02 ± 2.08E-03	5.72E-04 ± 2.30E-04	8.86E-06 ± 2.57E-06	0.0014 ± 0.0008	34.8	1113.8	32	83.1	174.3 ± 3.9
2	0.111936 ± 4.98E-03	9.44E-04 ± 2.97E-04	1.68E-04 ± 3.61E-05	0.0002 ± 0.0011	381.2	71.1	0.2	72.1	127 ± 5.6
3	0.104591 ± 2.47E-03	8.87E-04 ± 4.35E-05	2.33E-04 ± 1.45E-05	0.0008 ± 0.0002	67.8	47.7	0.7	73.8	138.6 ± 3.3
4	0.123910 ± 5.38E-03	1.06E-03 ± 2.39E-04	4.98E-04 ± 8.87E-05	0.0007 ± 0.0008	87.6	26.5	0.3	68.8	110 ± 4.8
5	0.110986 ± 3.36E-03	7.50E-04 ± 6.82E-05	0 ± 0	0.0006 ± 0.0004	94.6	0	0	77.8	137.8 ± 4.2
6	0.116097 ± 2.17E-03	1.03E-03 ± 7.48E-05	2.24E-04 ± 1.26E-05	0.0007 ± 0.0003	92.5	55.2	0.6	69.5	118.4 ± 2.2
7	0.100850 ± 2.01E-03	7.05E-04 ± 1.29E-04	1.69E-04 ± 1.89E-05	0.0052 ± 0.0005	10.1	63.5	6.3	79.2	153.6 ± 3.1
8	0.106707 ± 4.37E-03	8.98E-04 ± 1.70E-04	2.94E-04 ± 3.58E-05	0.0001 ± 0.0005	1000.1	38.7	0	73.5	135.4 ± 5.5
9	0.091200 ± 3.85E-03	4.33E-04 ± 1.19E-04	3.98E-04 ± 5.70E-05	0.0010 ± 0.0006	46.3	24.4	0.5	87.2	185.5 ± 7.8
10	0.104964 ± 4.10E-03	1.04E-03 ± 1.63E-04	7.88E-06 ± 1.61E-06	0.0002 ± 0.0007	245.4	1416.5	5.8	69.2	129.9 ± 5.1

[1]Mean age calculated as the mean of air-corrected data, with error expressed as the standard error of the mean; ‡ = ages are calculated for each analysis on the basis of analytical precision only; "regression" result is based on the methods of York, 1969. All uncertainties are 2-sigma. The uncertainty in the statistical ages include uncertainties arising from the J-value (0.01132 ± 0.00006), which corresponds to an additional 0.5%. % $^{40}Ar^*$ = percentage of radiogenic ^{40}Ar from total ^{40}Ar.

[2]Mean age = 479 ± 4 Ma; regression = 490 ± 20; initial 40/36 = 213 ± 94; * ages calculated excluding points 2, 3, 6.

[3]Mean age = 141 ± 9 Ma; regression = 236 ± 45; initial 40/36 = 375 ± 315; * ages calculated with all data.

TABLE 2. Step-Heating Analyses of Selected Samples in the Northern Oaxacan Complex[1]

Power	$^{36}Ar/^{40}Ar$	$^{39}Ar/^{40}Ar$	K/Ca	%^{40}Atm	%^{39}Ar	$^{40}Ar^{*}/^{39}K$	Age ± 1SD
1.5	1.012E-03 ± 1.03E-04	5.58E-02 ± 1.08E-03	7.5	29.84	13.15	12.558 ± 0.634	165.45 ± 8
< 2	5.03E-04 ± 4.9E-05	4.39E-02 ± 3.21E-04	9.1	14.86	51.31	19.366 ± 0.363	249.17 ± 4.4
< 3.5>	6.63E-04 ± 1.16E-04	4.35E-02 ± 8.97E-04	3.8	19.56	60.78	18.477 ± 0.912	238.46 ± 11
< 4.5>	5.72E-04 ± 6.6E-05	4.39E-02 ± 3.23E-04	8.2	16.87	77.17	18.942 ± 0.475	244.1 ± 5.7
< 6>	3.65E-04 ± 9.3E-05	4.65E-02 ± 3.73E-04	6.0	10.78	88.56	19.160 ± 0.613	246.7 ± 7.4
< 7	4.14E-04 ± 9.5E-05	4.44E-02 ± 3.64E-04	8	12.23	100	19.757 ± 0.654	253.9 ± 7.8

[1]Biotite OC 0088 A; integrated age = 236.8 ± 3.20 Ma; initial 40/36 = 224.66 ± 145.73 (MSWD = 0.62, isochron between –0.41 and 3.83); correlation age = 254.10 ± 18.26 Ma (37.2% of ^{39}Ar, steps marked by >); plateau age = 247.4 ± 3.4 Ma (86.8% of ^{39}Ar, steps marked by <); J value = .007648 ± .000058.

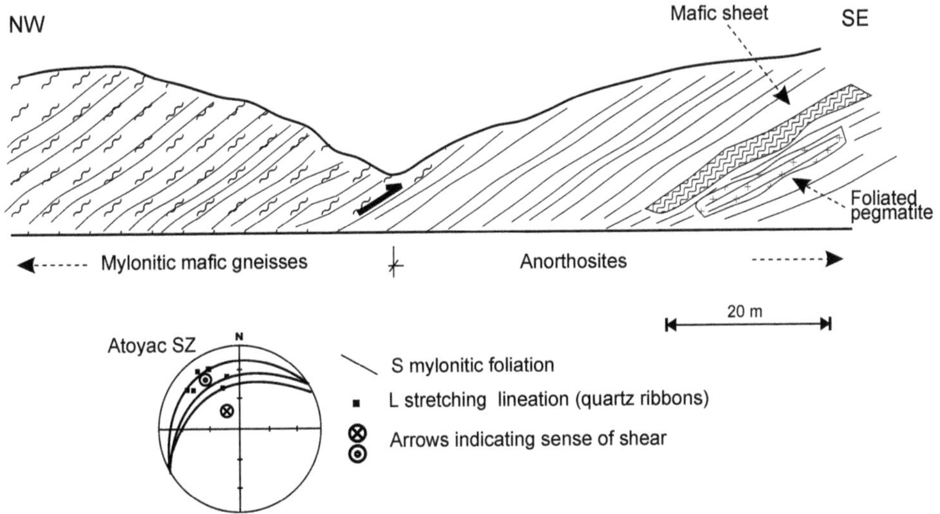

FIG. 4. Schematic profile and kinematic data for the Atoyac shear zone, as measured along Federal Road 190. The location of the outcrop is shown on Figure 2.

The 479 ± 4 Ma age is interpreted as due to resetting adjacent to the shear zone.

The mineral assemblage that recrystallized in the shear planes (uralite + fine-grained biotite), as well as the total absence of granulite-facies minerals characteristic of most of the Oaxacan Complex, suggests that this shear zone developed under greenschist- to lower amphibolite-facies conditions at a temperature estimated at ~300–400°C. However, the stretched plagioclase and quartz of similar aspect ratios are also typical of higher-grade rocks, suggesting that most of the finite strain was inherited with the addition of only a small strain increment during greenschist-facies shearing. This also suggests that the sheath folds may have formed during the ~1 Ga granulite-facies Zapotecan orogeny along a lithologic boundary that was reactivated later under greenschist-facies conditions.

El Correo shear zone (Fig. 6) (Type locality: Highway 190, km 135; reference locality: Toll Highway Oaxaca–Mexico, km 212)

The El Correo shear zone separates the orthogneisses of the Huitzo unit from the overlying metasediments of the El Marquez unit. It is made up of a ~100 m thick zone of phyllonite with S-C fabrics

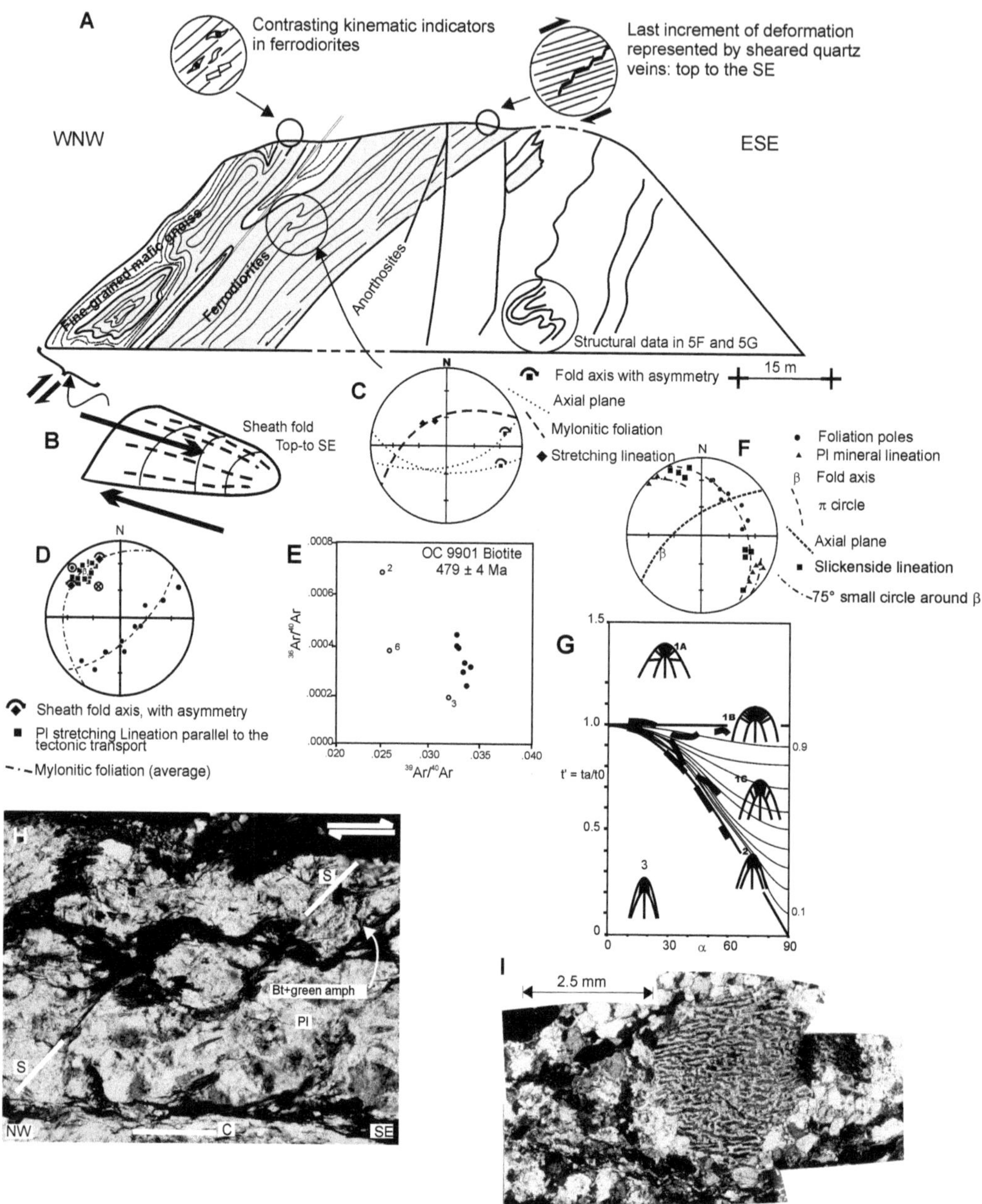

FIG. 5. Rio Negro shear zone. A. Schematic profile of the type outcrop of the Río Negro shear zone, located along the Oaxaca-Mexico highway at km 216. B. 3D view of the main, top-to-the-SE sheath fold recognized in this shear zone. C. Structural data of parasitic folds that affect the mylonitic foliation. D. Structural data and kinematics for the sheath fold depicted in Figure 5B. E. $^{36}Ar/^{40}Ar$ vs. $^{39}Ar/^{40}Ar$ plot for biotite sample OC9901, separated from a gabbroic sheet just below the shear zone. F. Structural data and (G) t' vs. α plot for measurements taken in a NE-trending late fold affecting the main outcrop (position of the fold located in 5A). H. C-S fabric in the mylonite, observed in a thin section cut parallel to the stretching lineation. Long side of the picture corresponds to 2.5 mm. I. Cross-polarized photomicrograph of albite-mantled porphyroclasts in the mylonite.

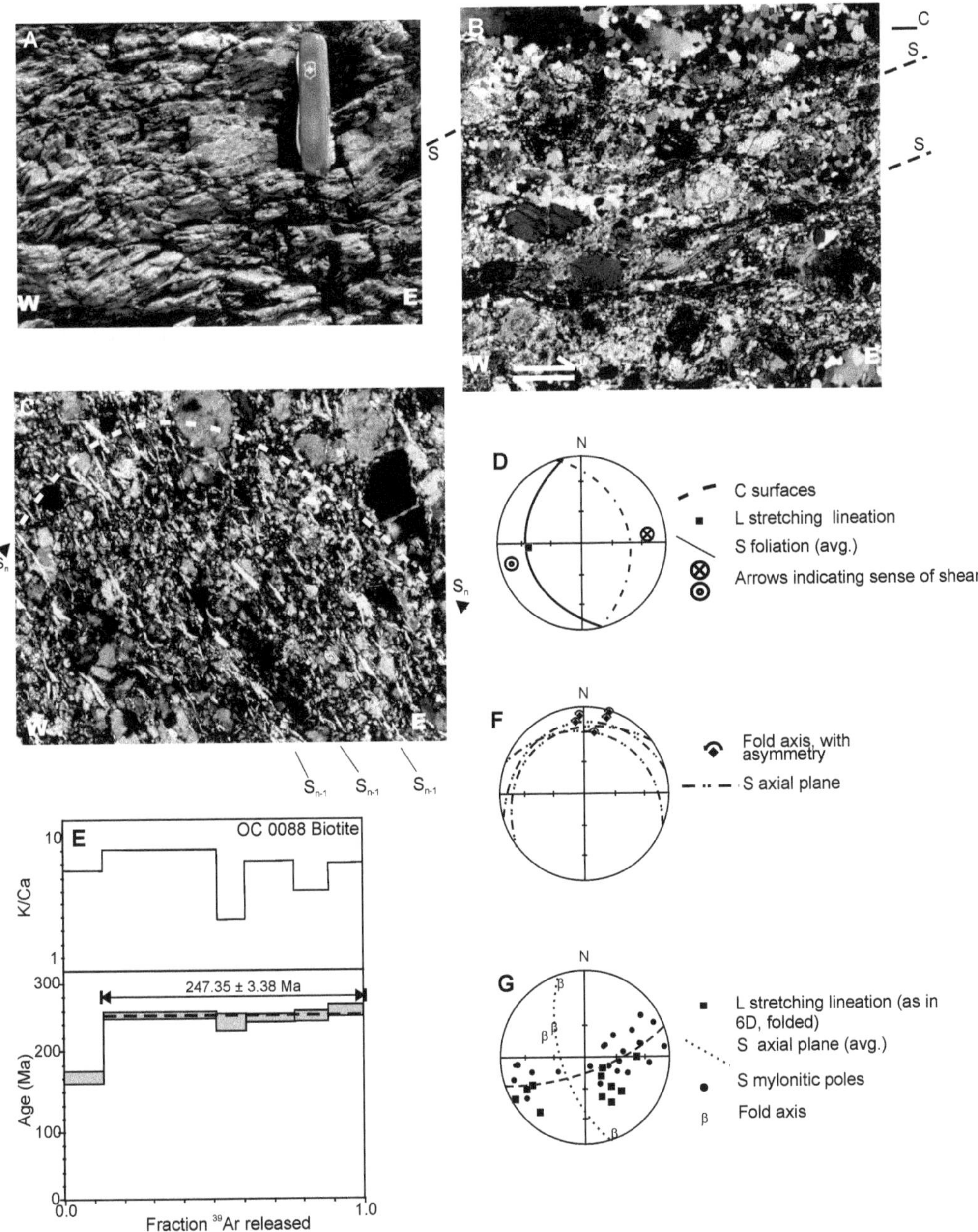

FIG. 6. El Correo shear zone. A. Field picture of phyllonite on Federal Road 190. B. Cross-polarized photomicrograph of CS fabric in oriented phyllonite cut parallel to the stretching lineation; shear sense is top-to-the-E. Long side of the picture equals 2.5 mm. C. Cross-polarized photomicrograph of late, S-shaped close folds in the same outcrop, cut perpendicular to the fold axis; note the fine-grained, white sericite crystals along the axial plane of the fold; scale = 1 mm. D. Stereonet of the same CS fabric illustrated in Figure 6B: shear sense is top-to-the-E. E. $^{39}Ar/^{40}Ar$ Ar spectra for step-heating analysis of biotite OC 0088 separated from the phyllonite, showing a plateau age of 247 ± 3.4 Ma. Stereonet of S-shaped folds measured in this outcrop, with a sense of asymmetry suggesting a westward displacement. G. Stereonet of last phase of folding that affected El Correo shear zone: both mylonitic foliation and stretching lineation are affected by this monoclinal folding.

refolded by two sets of folds. Immediately beneath the shear zone on the Toll Highway Oaxaca–Mexico (km 212), are a series of sheath folds that appear to have formed at high grades of metamorphism, presumably indicating the presence of an older shear zone that was reactivated during phyllonitization. The mylonitic S planes are composed of alternating, fine-grained mica and quartz bands, 1–2 cm thick, in which quartz ribbons have an aspect ratio ≤ 20:1. These S-planes are deflected in C planes at 10–20° to the S planes (Figs. 6A and 6B). A stretched quartz lineation is visible on S planes, and is parallel to aligned mica crystals. In thin sections cut parallel to the lineation and perpendicular to S planes, the C planes are marked by fine-grained (80–100 μm) biotite crystals, and rare, porphyroclastic, inclusion-rich, white mica fish (≤2 mm). The quartz ribbons in the S planes consist of fine-grained subgrains, and are generally accompanied by porphyroclastic plagioclase and perthitic feldspar (Fig. 6B). C-S fabrics and mica fish orientations indicate a top-to-east sense of shear (Fig. 6D). Recrystallization of biotite, the presence of white mica fish, and the formation of subgrains in quartz ribbons indicate a temperature of ~300–350°C (Giletti, 1974; Passchier and Trouw, 1996).

Fine-grained biotite from this shear zone yielded a plateau age of 247 ± 3 Ma from 86.8% of ^{39}Ar (Table 2; Fig. 6E) interpreted as cooling through the closure temperature of 270–290°C estimated using the 80–100 μm grain size. This age is inferred to approximate the time of crystallization and mylonitization along the El Correo shear zone.

On Highway 190, the mylonitic fabrics are deformed by S-shaped, close, gently N-plunging folds (≤5 cm), with gently N-dipping axial planes defined by very fine-grained (up to 30 μm) crystals of white mica, possibly sericite (Fig. 6C). The fold asymmetry suggests that these folds are the product of top-to-the-west shearing (Fig. 6F). These folds and the mylonitic fabrics are deformed by S-shaped, open to closed, NW- to SE-trending, gently to moderately NW- to SE-inclined folds, ≤2 m in amplitude (Fig. 6G). The axial planes of these folds are marked by a fracture cleavage. The fold asymmetry also indicates a top-to-west sense of shear, and because it has the same kinematics, suggests deformation under progressively decreasing temperatures.

Cañas shear zone (type locality: Highway 190, 151 km)

In contrast to the other shear zones, this shear zone is a steeply dipping, NNW-striking shear zone in the middle limb of a monoclinal fold in the main foliation and quartz-feldspar-amphibole banding of the El Catrin migmatitic unit (Fig. 7A). It consists of steeply dipping S-C-C' fabrics and a steeply plunging stretched quartz lineation (Figs. 7B and 7C). The mylonitic fabrics are composed of fine-grained aggregates of green biotite and stretched quartz, with sigma-shaped porphyroclasts of undulose quartz and K-feldspar with a mantle of albite-oligoclase subgrains, which is evidence of grain boundary migration recrystallization (cf. Passchier and Trouw, 1996; e.g., Fig. 7D). All kinematic indicators show relative upward displacement of west side of the zone. Several other shear zones within the northern Oaxacan Complex (Fig. 2) have with similar orientations and kinematics.

Ten Ar laser total fusion analyses were performed on very fine grained, up to 50 μm, irregularly shaped biotite crystals characteristic of dynamic metamorphism and deformation. The results (Fig 7E and Table 1) show a positive correlation of ^{36}Ar/^{40}Ar versus ^{39}Ar/^{40}Ar with a weighted, least-squares linear regression age of 236 ± 45 Ma, interpreted to reflect loss of radiogenic ^{40}Ar accompanied by uptake of atmospheric argon in a surficial weathering environment. Air-corrected data yield a mean age of 141 ± 9 Ma. The small grain size (40–50 μm average) and irregular shapes of the biotite are characterized by small diffusion distances for Ar loss, and inferred lower closure temperatures, estimated at ~250–270°C. The 141 ± 9 Ma age is viewed as a minimum age for the growth of biotite durign mylonitic deformation.

The Cañas shear zone appears to post-date development of a NNW-trending monoclinal fold that is congruent with the late folds described in the next section. This folding is associated with a set of radiating quartz veins preferentially developed in the middle limb (Fig. 7A), roughly parallel to the fold axial plane. Dip isogons (Fig. 7F and 7G) indicate the presence of Class 1A and 1B, and a folded mineral lineation plots near a 20° small circle about the fold axis (Fig. 7H). These data indicate a combination of flexural and tangential longitudinal strain in the development of the fold.

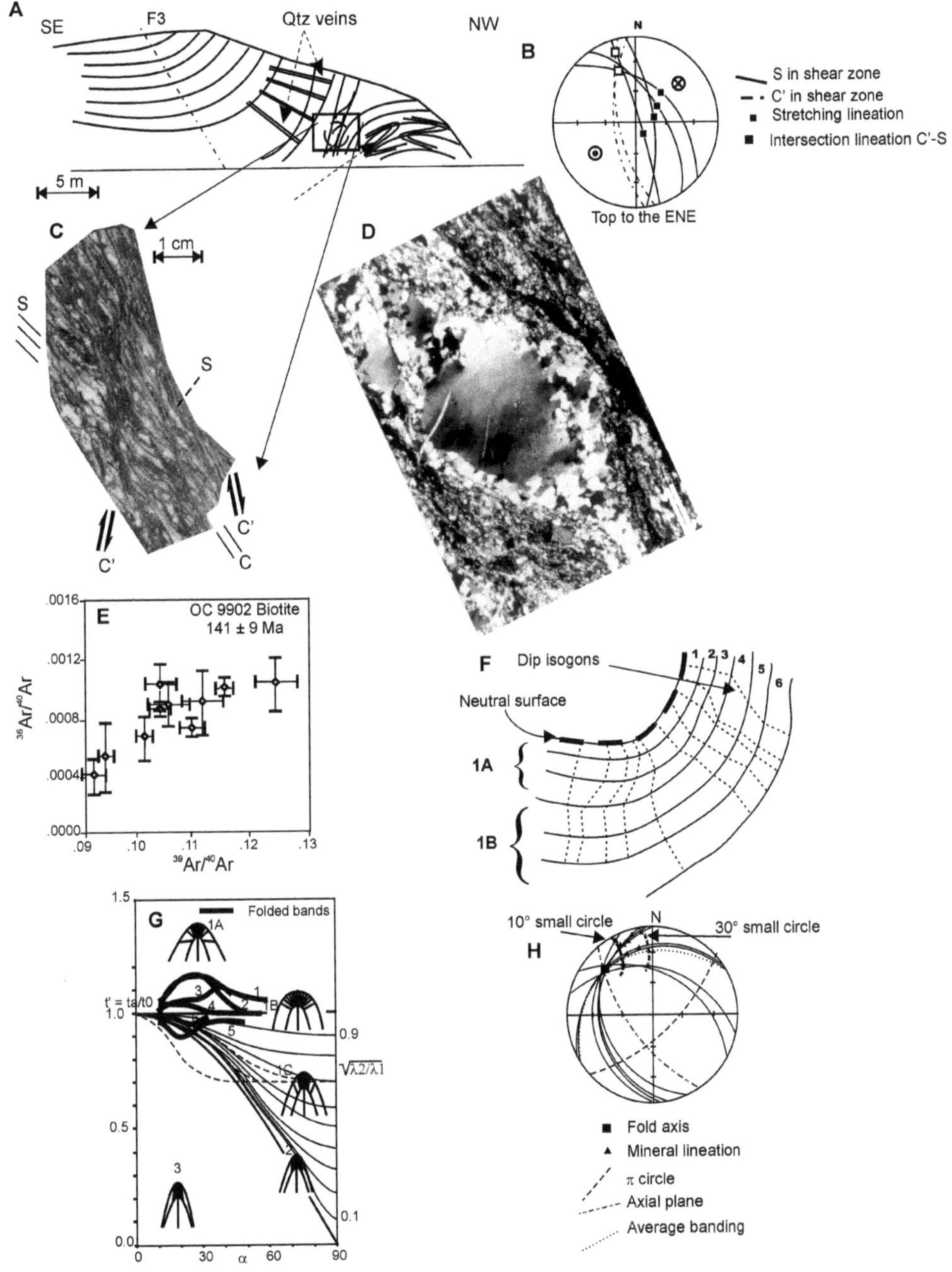

FIG. 7. Cañas shear zone. A. Schematic profile of the type outcrop as measured along Federal Road 190 at km 153. B. Stereonet of SC' fabric observed in mylonite, indicating a top-to-the-ENE shearing. C. Photograph of a polished hand sample, indicating the main fabric elements of this shear zone. D. Cross-polarized photomicrograph of sigma porphyroclasts of quartz, mantled by quartz subgrains affected by grain-boundary migration (Passchier and Trouw, 1996). Shear sense is indicated as top-to-bottom right: long side of the picture equals 1 mm. E. $^{36}Ar/^{40}Ar$ vs. $^{39}Ar/^{40}Ar$ plot for the biotite sample OC9902: weighted mean of 10 analyses yield an age of 141 ± 9 Ma. F. Schematic profile and (G) t' vs. α plot of dip isogons for the main, NW-trending, synformal fold affected by shearing in the subvertical limb. H. Stereonet of measurements taken around the same fold plotted in Figures 7G and 7H: structural data, together with dip-isogons plot, suggest a tangential-longitudinal strain folding mechanism.

Late folds

Two sets of steeply inclined folds were encountered in the northern Oaxacan Complex: (1) a rare NE-trending set; and (2) a dominant NW-trending set. The NE-trending set varies from: (a) open-close, steeply NW-dipping, moderately SW-plunging folds (≤2 m in amplitude) with a poorly developed axial planar fabric of aligned subgrains of plagioclase and oxides in the anorthosite and gabbro; to (b) smaller, close-isoclinal folds with an axial planar fabric defined by fine-grained chlorite replacing biotite in ferrodiorites and migmatites. An inherited mineral lineation in the anorthosite was deformed around these fold hinges and defines a small circle at 75° to the fold axis (Fig. 5F). In mica-rich horizons, a slickenside lineation perpendicular to the fold hinge occurs, suggesting the operation of a flexural slip fold mechanism. Superposed flattening is indicated by the Class 1C-2 fold shape (Fig. 5G). The age of these folds is unconstrained.

The NW-trending fold set consists of open-close, gently to moderately NW-plunging folds that vary in size up to 40 km in wavelength, and are present throughout the area: the best mesoscopic examples occur in the migmatitic El Catrin gneisses. Small, close, Class 1B-1C folds show an axial plane foliation defined by oriented quartz, and rarely, fine-grained biotite, whereas larger open folds do not possess an axial plane foliation. Fine-grained biotite aggregates generally cut across larger biotite crystals along kink axial planes. Axial planar quartz normally displays undulose extinction and deformation lamellae in elongated grains with subgrains surrounded by a thin film of recrystallized, smaller decussate grains. The presence of fine-grained brown biotite and the microstructures present in quartz suggest that the deformation and recrystallization occurred at temperatures less than 300–400°C. These folds are geometrically similar to the last set of folds that deform the El Correo phyllonites. The age of the NW-trending folds is thus bracketed between the age of the El Correo shear zone (~247 ± 3 Ma; cf. Early Triassic = 253 ± 2 Ma to 244 ± 4 Ma; Okulitch, 2003) that they deform, and the Cañas shear zone (~141 ± 9 Ma: cf. Jurassic–Cretaceous boundary = 142 ± 3 Ma), which cuts them.

Summary

These data indicate that several phases of Phanerozoic deformation affected the northern Oaxacan Complex: (1) SE-directed shearing at 479 ± Ma (Rio Negro shear zone); (2) E-directed shearing at 247 ± 3 Ma (El Correo shear zone); and (3) NW to NNW upright to steeply inclined folding that preceded (4) the development of NW to NNW-trending, subvertical shear zones with an up-dip sense of movement at 141 ± 9 Ma (Cañas shear zone). Given the fact that the present surface of the northern Oaxacan Complex appears to have been situated at shallow depths (~1 km and certainly less than the ~10 km required to regionally reset argon closure temperatures of ~300°C in biotite) since reaching the surface at ~710–760 Ma, the biotite ages are interpreted as reflecting thermal re-equilibration due to hot fluid circulation along active shear zones, which raised the local temperature 100–300°C above the surrounding rocks. According to Okulitch (2003), current dating of the time scale boundaries gives the following: (1) Tremadocian = 489 ± 1 Ma to 478 ± 4 Ma, and Arenigian = 478 ± 4 Ma to 467 ± 4 Ma; (2) Early Triassic = 253 ± 2 Ma to 244 ± 4 Ma; and (3) Jurassic–Cretaceous boundary = 142 ± 3 Ma.

Correlation and Tectonic Implications

The Tremadocian–Arenigian shearing recorded in the Rio Negro shear zone immediately postdates or is synchronous with deposition of the oldest sediments on the ~1 Ga Oaxacan Complex. This suggests that the SE-directed shearing may be connected with listric normal faulting associated with the onset of deposition. In this case, the present NW dip of the shear zone is probably due to subsequent folding and tilting of the surface. Palinspastic reconstructions for the Early Ordovician (Fig. 8A) suggest that this event may record the separation of Avalonia from Oaxaquia at the birth of the Rheic Ocean (Keppie et al., 2003).

The Early Triassic E-directed thrusting recorded in the El Correo shear zone preceded the onset of rifting in the Gulf of Mexico, and is synchronous with deformation of the Las Delicias volcanic arc in northern Mexico (Carpenter, 1997). It may have been contemporaneous with the E-directed thrusting of the Oaxacan Complex over the Sierra de Juárez mylonite belt that bounds the Oaxacan Complex along its eastern side (Fig 8B), an event that pre-dated mid-Jurassic dextral shearing at 163 +42/-13 Ma (slightly discordant U-Pb monazite age: Alaniz-Álvarez et al., 1996), and 172 ± 2 Ma ($^{40}Ar/^{39}Ar$ muscovite plateau age: Alaniz-Älvarez et al., 1996). It may be also contemporaneous with the earliest deformation in the Paleozoic sediments that

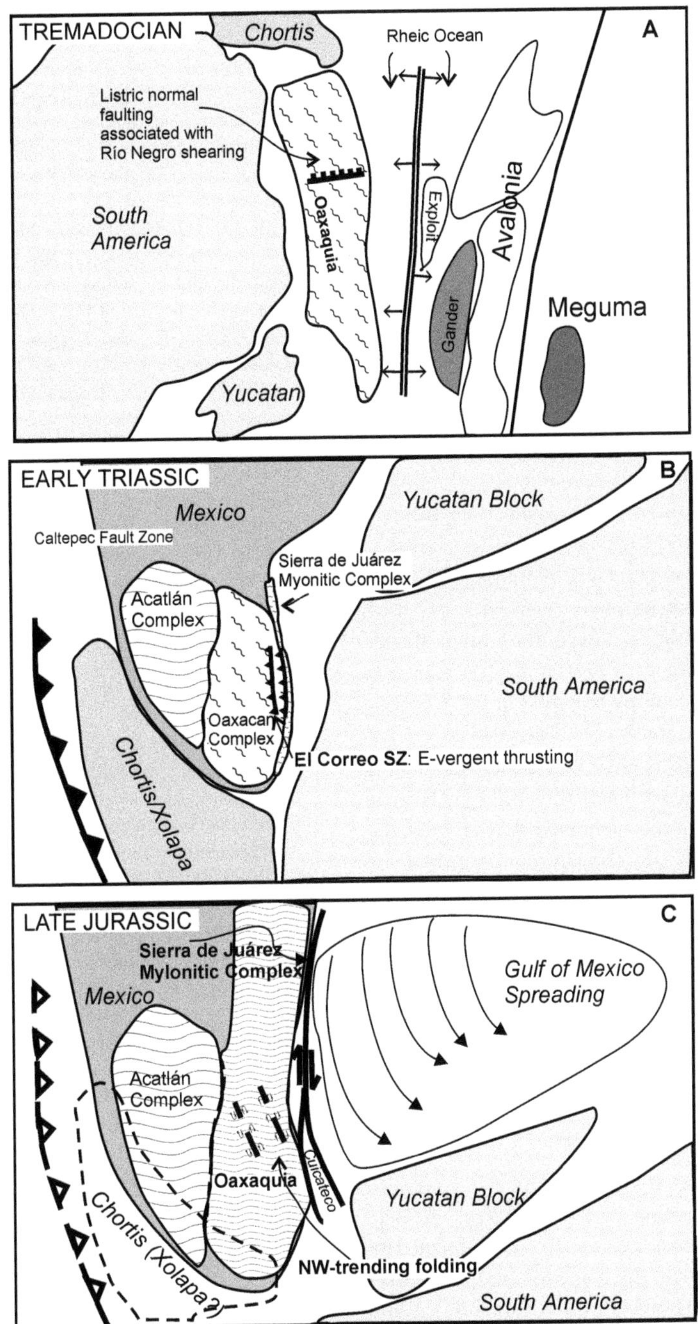

FIG. 8. Palinspastic reconstructions. A. Tremadocian (~490 Ma) for peri-Gondwanan terranes showing rifting in the Rheic Ocean and associated listric normal faulting in the northern Oaxacan Complex (modified after Keppie et al., 2003). B. Early Triassic for southern Mexico, showing E-vergent El Correo thrusting associated with thrusting of the Oaxacan Complex over the Sierra de Juárez (modified after Solari et al., 2001). C. Late Jurassic of southern Mexico and Caribbean region, indicating the opening of the Gulf of Mexico during dextral shear along the Sierra de Juárez Mylonitic Complex, and associated NW-trending folds in the Northern Oaxacan Complex (modified after Centeno-García and Keppie, 1999).

overlie the Oaxacan Complex, bracketed between Late Permian and Early Cretaceous (Centeno-Garcia and Keppie, 1999). This Early Triassic thrusting is synchronous with development of a Permo-Triassic arc along the length of Mexico, suggesting convergence-induced contraction in the arc (Centeno and Keppie, 1999: Solari et al., 2001).

The mid-Triassic–Jurassic, NW-trending folding in the northern Oaxacan Complex was synchronous with opening of the Gulf of Mexico along the N- to NNW-trending Juárez dextral shear zone (Alaniz-Álvarez et al., 1996). The development of NW-trending folds is consistent with such dextral shear (Fig. 8C). The NE-trending folds in the northern Oaxacan Complex are geometrically similar to those affecting the overlying Paleozoic rocks, which have also been constrained to the same time interval (Centeno-Garcia and Keppie, 1999).

The Jurassic–Cretaceous boundary shearing recorded in the Cañas shear zone coincided with terminal stages of the opening of the Gulf of Mexico (Schouten and Klitgord, 1994). It may be related to the listric normal faulting recorded in the Sierra de Juárez by Alaniz-Álvarez et al. (1996).

In conclusion, it is clear that rather than acting as a massive basement block, the northern Oaxacan Complex was internally deformed during Tremadocian extension during Triassic contractional arc deformation, and during opening of the Gulf of Mexico. The late Mesozoic–Early Cretaceous Laramide deformation has yet to be documented; however, it may be hidden in the many brittle structures also present in the northern Oaxacan Complex.

Acknowledgments

Consejo Nacional de Ciencias y Tecnología (CONACyT); Programa de Apoyo a Proyectos de Investigación e Innovación Tecnológica of the Dirección General de Asuntos del Personal Académico (PAPIIT-DGAPA), UNAM; and Instituto de Geología, UNAM are thanked for providing funds to cover different aspects of this work. $^{40}Ar/^{39}Ar$ analytical work at Queen's University was funded by NSERC Discovery and MFA grants to J. K. W. Lee. We also thank Kip Hodges for access to the MIT $^{40}Ar/^{39}Ar$ facility. E. L. Johnson provided a thorough review of the paper.

REFERENCES

Alaniz-Álvarez, S. A., van der Heyden, P., Nieto-Samaniego, A. F., and Ortega-Gutiérrez, F., 1996, Radiometric and kinematic evidence for Middle Jurassic strike-slip faulting in southern Mexico related to the opening of the Gulf of Mexico: Geology, v. 24, p. 443–446.

Carpenter, D. L., 1997, Tectonic history of the metamorphic basement rocks of the Sierra del Carmen, Coahuila, Mexico: Geological Society of America Bulletin, v. 109, p. 1321–1332.

Centeno-García, E., and Keppie, J. D., 1999, Latest Paleozoic–early Mesozoic structures in the central Oaxaca Terrane of southern Mexico: Deformation near a triple junction: Tectonophysics, v. 301, p. 231–242.

Clark, A. H., Archibald, D. A., Lee, A. W., Farrar, E., and Hodgson, C. J., 1998, Laser Probe $^{40}Ar/^{39}Ar$ ages of early- and late-stage alteration assemblages, Rosario porphyry copper-molybdenum deposit, Collahuasi District, I Region, Chile: Economic Geology, v. 93, p. 326–337.

De Cserna, Z., 1989, An outline of the geology of Mexico, *in* Bally, A. W., and Palmer, A. R. eds., The Geology of North America. An overview: Boulder CO, Geological Society of America, Decade of North American Geology, p. 233–264.

Elias,-H., M., and Ortega-Gutiérrez, F., 2002, The Caltepec fault zone: An Early Permian dextral transpressional boundary between the Proterozoic Oaxacan and Palaeozoic Acatlán complexes, southern Mexico, and regional tectonic implications: Tectonics, v. 21, no. 3 [10.1029/2000TC001278].

Fries, C. J., and Rincon-Orta, C., 1965, Nuevas aportaciones y técnicas empleadas en el Laboratorio de Geocronología: Bolletín del Instituto de Geología, UNAM, v. 73, p. 57–133.

Giletti, B. J., 1974, Studies in diffusion I: Argon in phlogopite mica, *in* Hofmann, A. W., ed., Geochemical transport and kinetics: Carnegie Institution of Washington Publication no. 634, p. 107–115.

Hames, W. E., and Bowring, S. A., 1994, An empirical evaluation of the argon diffusion geometry in muscovite: Earth and Planetary Science Letters, v. 124, p. 161–169.

Harrison, T. M., Duncan, I., and McDougall, I., 1985, Diffusion of the ^{40}Ar in biotite: Temperature, pressure, and compositional effects: Geochimica et Cosmochimica Acta, v. 49, p. 2461–2468.

Keppie, J. D. Dostal, J., Cameron, K. L., Solari, L. A., Ortega-Gutiérrez, F., and Lopez, R., 2003, Geochronology and geochemistry of Grenvillian igneous suites in the northern Oaxacan Complex, southern Mexico: Tectonic implications: Precambrian Research, v. 120, p. 365–289.

Keppie, J. D., Nance, R. D., Murphy, J. B., and Dostal, J., 2003, Tethyan, Mediterranean, and Pacific analogues for the Neoproterozoic–Paleozoic birth and development of peri-Gondwanan terranes and their transfer to Laurentia and Laurussia: Tectonophysics, v. 365, p. 195–219.

Keppie, J. D., Solari, L. A., Ortega-Gutiérrez, F., Ortega-Rivera, A., Lee, J. K. W., Lopez, R., and Hames, W. E, 2004, U-Pb and $^{40}Ar/^{39}Ar$ constraints on the cooling history of the northern Oaxacan Complex, southern Mexico: Tectonic implications, *in* Tollo, R. P., Corriveau, L., McLelland, J., and Bartholomew, M. J., eds., Proterozoic tectonic evolution of the Grenville orogen in North America: Geological Society of America Memoir no. 197, in press.

Mitre-Salazar, G. F., and Roldan-Quintana, J., 1990, Ocean-continent transect H-1, La Paz to Saltillo, northwestern and northern Mexico: Boulder, CO, Geological Society of America, scale 1:500,000.

Okulitich, A. V., 2003, Geological time chart, 2003: Geological Survey of Canada, Open File Report no. 3040.

Ortega-Gutiérrez, F., 1990, North American continent-ocean transect program. Transect H3: Acapulco trench to the Gulf of Mexico across southern Mexico: Geological Society of America, DNAG, Continent Ocean transect program.

Passchier, C. W., and Trouw, R. A. J., 1996, Microtectonics: Berlin, Germany, Springer Verlag, 290 p.

Pryer, L. L., 1993, Microstructures in feldspars from a major crustal thrust zone: The Grenville Front, Ontario, Canada: Journal of Structural Geology, v. 15, p. 21–36.

Schouten, H., and Klitgord, K. D., 1994, Mechanistic solutions to the opening of the Gulf of Mexico: Geology, v. 22, p. 507–510.

Sedlock, R. L., Ortega-Gutiérrez, F., and Speed, R. C., eds., 1993, Tectonostratigraphic terranes and tectonic evolution of Mexico: Geological Society of America Special Papers, v. 278, 153 p.

Solari, L. A., 2001, La porción norte del Complejo Oaxaqueño, estado de Oaxaca: Estructuras, geocronología y tectónica: Unpubl. Ph. D. thesis, Universidad Nacional Autónoma de México, Mexico City, 191 p.

Solari, L. A., Dostal, J., Ortega-Gutiérrez, F., and Keppie, J. D., 2001, The 275 Ma arc-related La Carbonera stock in the northern Oaxacan Complex of southern Mexico: U-Pb geochronology and geochemistry: Revista Mexicana de Ciencias Geológicas, v. 18, p. 149–161.

Solari, L. A., Keppie, J. D., Ortega-Gutiérrez, F., Cameron, K. L., Lopez, R., and Hames, W. E., 2003, Grenvillian tectonothermal events in the northern Oaxacan Complex, southern Mexico: Roots of an orogen: Tectonophysics, v. 365, p. 257–282.

Tullis, J., and Yund, R. A., 1985, Dynamic recrystallization of feldspar: A mechanism for ductile shear zone formation: Geology, v. 13, p. 238–241.

York, D., 1969, Least-squares fitting of a straight line with correlated errors: Earth and Planetary Science Letters, v. 5, p. 320–324.

A Late Permian Tectonothermal Event in Grenville Crust of the Southern Maya Terrane: U-Pb Zircon Ages from the Chiapas Massif, Southeastern Mexico

BODO WEBER,[1]

División Ciencias de la Tierra, Centro de Investigación Científica y de Educación Superior de Ensenada (CICESE) B.C., km 107 carr. Tijuana-Ensenada, 22860 Ensenada B.C., Mexico

KENNETH L. CAMERON,

Earth Sciences Division, University of California, Santa Cruz (UCSC), California 95064

MYRIAM OSORIO, AND PETER SCHAAF

Instituto de Geofísica, Universidad Nacional Autónoma de México (UNAM), Ciudad Universitaria, 04510 Coyoacan, D.F., México

Abstract

The Chiapas massif (CM) in southeastern Mexico is the crystalline basement of the southern Maya terrane, which is a crustal block that comprises mainly the Yucatan Peninsula, the Mexican states of Chiapas, Veracruz, and parts of Oaxaca. The CM is composed of igneous rocks and medium- to high-grade metamorphic rocks. Zircon fractions from all samples are discordant, yielding Late Permian lower-intercept ages and >1 Ga upper-intercept ages. The most precise results are from an orthogneiss that yielded intercept ages of 258.4 ± 1.9 Ma and 1046.6 ± 5.6 Ma, and from two augen gneisses which yielded intercepts of 250.9 ± 2.3 Ma and 1017 ± 27 Ma. We interpret the lower intercepts as either igneous crystallization or metamorphic ages and the upper intercepts as the age of inherited components. Results from all other samples are within error of these ages. These results demonstrate that the CM basement contains a Grenvillian component, and that the most important tectonothermal event affecting the CM was of Late Permian age. The results favor a hypothetical model in which the Maya terrane is composed of separated blocks of different geologic histories.

Introduction

PRE-MESOZOIC CRYSTALLINE basement in Mexico is largely covered by younger, mostly volcanic and sedimentary rocks. The scarcity of basement outcrops together with differences in age, metamorphic history, and deformational style of the crystalline rocks are factors that complicate the interpretation of the geology of southern Mexico. Most of southern Mexico, therefore, has been interpreted to be a collage of tectonostratigraphic terranes (e.g., Campa and Coney, 1983; Coney and Campa, 1987; Ortega-Gutiérrez et al., 1990) whose correlation to each other and to North America is uncertain (Fig. 1A). The pre-Mesozoic locations of these land masses, either as peri-Gondwanan blocks between Gondwana and Laurentia or as outboard terranes in the Pacific margin (e.g., Centeno-García and Keppie, 1999; Elías-Herrera and Ortega-Gutiérrez, 2002), are important for the understanding of the late Paleozoic assembly of Pangea.

The crustal blocks with pre-Mesozoic basement in southern Mexico include the Mixteco terrane and its polymetamorphic Acatlan complex of Paleozoic age, the Zapoteco terrane with the Grenville-age Oaxacan complex, and the Maya terrane (Figs. 1A and 1B; e.g., Sedlock et al., 1993) or Maya block (Dengo, 1985). The Maya terrane has been defined as a crustal block that includes the Yucatan Peninsula, parts of the costal plain of the Gulf of Mexico, and southeastern Mexico from the Tehuantepec Isthmus to the border of Guatemala (Sedlock et al., 1993). In the last two decades several studies have been dedicated to the Acatlan complex (e.g., Yañez et al., 1991; Weber et al., 1997; Centeno-García and Keppie, 1999; Ortega-Gutiérrez et al., 1999; Elías-Herrera and Ortega-Gutiérrez, 2002; Meza-Figuera et al., 2003), the Oaxacan complex (e.g., Keppie and Ortega-Gutiérrez, 1995; 1999; Keppie et al., 2001; 2003; Solari et al., 2001; 2003; 2004),

[1]Corresponding author; email: bweber@cicese.mx

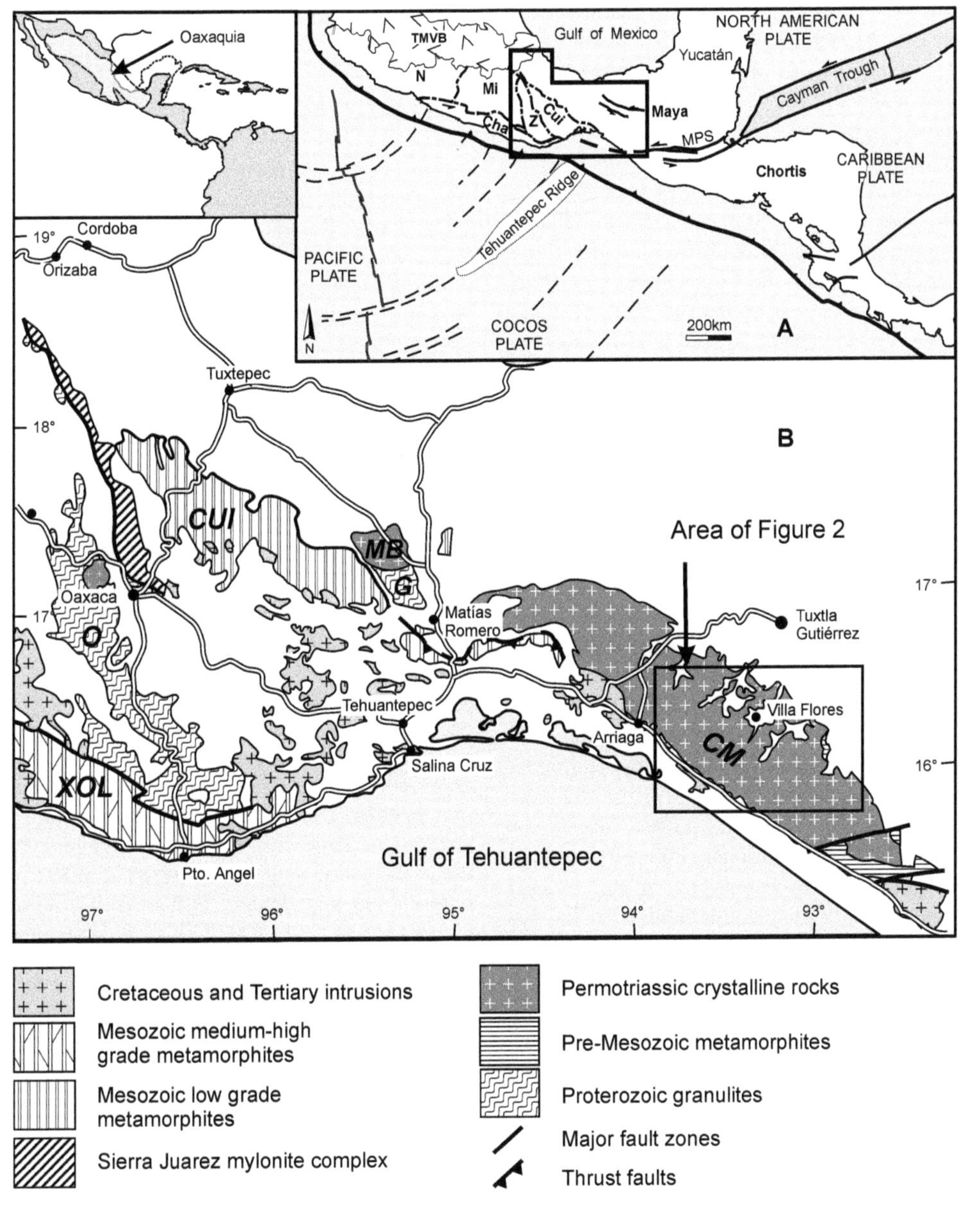

FIG. 1. A. Terrane map of southern Mexico and Central America (modified after Ortega-Gutiérrez et al., 1990). Abbreviations: MPS = Motagua-Polochic fault system; TMVB = Trans-Mexican Volcanic Belt. Terranes: Cha = Chatino; Chortis; Cui = Cuicateco; Maya; Mi = Mixteco; N = Nahuatl; Z = Zapoteco. B. Geologic map of igneous and metamorphic rocks exposed in southern Mexico (modified after Ortega-Gutiérrez et al., 1992). Abbreviations: CM = Chiapas massif; CUI = Cuicateco terrane; G = Guichicovi complex; MB = Mixtequita batholith; O = Oaxacan complex; XOL = Xolapa complex. Inset (upper left) depicts possible extension of Oaxaquia after Ortega-Gutiérrez et al. (1995).

and crystalline basement of the southern Maya terrane, namely the Guichicoi complex (Fig. 1B; Murillo-Muñetón, 1994; Murillo-Muñetón et al., 1994; Weber, 1998; Weber and Köhler, 1999; Weber and Hecht, 2003) and the Chiapas massif (Fig. 1B; Schaaf et al., 2002; Weber et al., 2002). The Chatino "terrane" (Fig. 1A), as suggested by Sedlock et al. (1993), and its metamorphic Xolapa complex (Fig. 1B) are autochthonous and probably represent an exposed magmatic arc of Mesozoic age that was formed within the pre-existing basement complexes of the Zapoteco and Mixteco terranes (Ratschbacher et al., 1991; Herrmann et al., 1994; Ortega-Gutiérrez and Elías-Herrera, 2003), and hence the Chatino is not a terrane in the sense of Howell et al. (1985). Little is known about the Cuicateco terrane (Figs. 1A and 1B), which separates the Maya from the Zapoteco terrane. The Cuicateco terrane has been interpreted as an Early Cretaceous basin inverted during the Laramide orogeny (Barboza-Gudino, 1994).

In contrast to the terrane concept, accumulating evidence suggests that most of Mexico is underlain by a large segment of Precambrian (Grenville) crust, the so-called Oaxaquia microcontinent (Fig. 1 inset; Ortega-Gutiérrez et al., 1995), which extends across several supposed tectonostratigraphic terranes, making it difficult to define these parts of Mexico as "suspect" terranes. The extension of Oaxaquia toward the southeast is still under discussion since Precambrian granulites of the Guichicovi complex (Fig. 1B) were discovered at the southwestern edge of the Maya terrane (Fig. 1B; Murillo-Muñetón, 1994; Murillo-Muñetón et al., 1994; Weber and Köhler, 1999; Weber and Hecht, 2003). These granulites are intruded by Permo-Triassic and Jurassic granitoids of the Mixtequita batholith (Fig. 1B) that have similar ages and isotopic compositions than granitoids from the CM (Damon et al., 1981; Murillo-Muñetón, 1994; Weber, 1998; Schaaf et al., 2002). The CM, which is mainly composed of igneous rocks, constitutes the crystalline basement of the southern Maya terrane (Sedlock et al., 1993) east of the Tehuantepec Isthmus (Fig. 1B). The CM extends over an area of approximately 20,000 km^2 subparallel to the Pacific coast and is bordered by the Motagua-Polochic fault system (MPS, Fig. 1A) near the Guatemalan border. The MPS separates the North American plate from the Caribbean plate (Chortis block).

One of the main objectives of this study is to test whether metamorphic rocks, recently described as the basement of the CM (Schaaf et al., 2002; Weber et al., 2002), form part of Oaxaquia. If not, where do they belong, and what are the implications for the tectonic history of the southern Maya terrane? For this purpose we carried out the first detailed U-Pb zircon dating of metamorphic rocks from the Chiapas massif.

Geologic Setting and Previous Work

Most of the CM comprises Permo-Triassic intrusive rocks (Chiapas batholith; Pantoja-Alor et al., 1971; López Ramos, 1979; Damon et al., 1981; Morán-Zenteno, 1984; Dengo, 1985), interpreted as part of a continental arc that extends NW-SE along the entire length of Mexico, and probably into northwestern South America (e.g., Centeno-García and Keppie, 1999; Torres et al., 1999). Some authors have already noted the presence of orthogneisses and metasedimentary rocks within the CM (Webber and Ojeda-Rivera, 1957; Pantoja-Alor et al., 1971) that have not been interpreted in detail. Geologic mapping recently carried out southwest of Villa Flores, Chiapas (Gross, 2000; Heck, 2000; Möllinger, 2000; Weis, 2000) has shown that igneous rocks in the CM are of heterogeneous composition, ranging from granites to diorites and gabbros. Migmatitic gneisses, augen gneisses, and mylonitic gneisses are widespresd in the CM. These rocks can be regarded as a basement component of the CM, because they are intruded by the Permo-Triassic granitoids, and because they show ductile deformation features (D1) that are not present in the intrusive rocks. However, most of the igneous rocks themselves are also deformed by a subsequent deformation and display E-W–striking foliation planes (Schaaf et al., 2002). This D2 deformation is predominant in the CM, and it is one of the main differences between the granitoids of the CM and other intrusions of the Permo-Triassic continental arc in Mexico (Torres et al., 1999), which do not show that kind of deformation.

Metasedimentary rocks, named the "Sepultura unit", have recently been described in the "El Tablón" riverbed west of Villa Flores (Fig. 2; Weber et al., 2002). The Sepultura unit contains medium- to high-grade pelitic and psammitic clastic metasediments with garnet-, sillimanite-, cordierite-, and Ti biotite–bearing assemblages, as well as wollastonite-garnet-diopside calcsilicates, and olivine-bearing marbles. Our ongoing fieldwork has revealed that the Sepultura unit seems to be a complete sedimentary basin filling, ranging from distal to proximal clastic sediments as well as shelf sediments with marls and limestones. Due to the apparent absence of volcanosedimentary rocks, this basin most likely could be interpreted as a passive

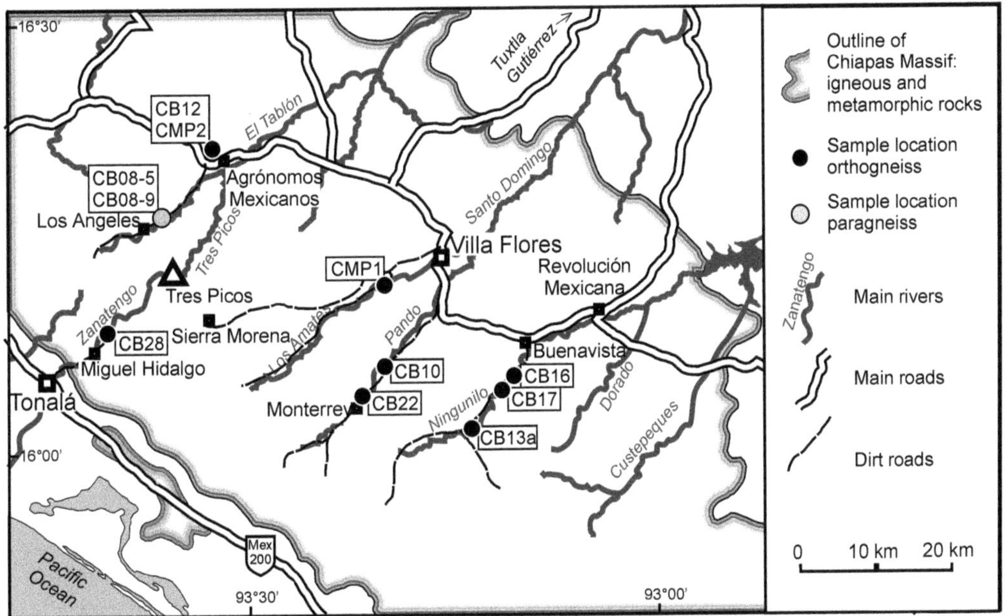

FIG. 2. Geographic sketch map showing sample locations and lithologies.

(continental) margin. High-grade metamorphism and deformation affected these rocks prior to the intrusion of the granitoids and the subsequent low- to medium-grade metamorphic overprint.

Few geochronological data have been previously published from rocks of the CM. Pantoja-Alor et al. (1974) reported Paleozoic to Precambrian ages (390–780 Ma) obtained from zircons using the Pb-α method. Damon et al. (1981) published K-Ar mineral ages ranging from 290 to 170 Ma, and a 10-point Rb-Sr granitoid rock isochron age of 256 ± 10 Ma. Mujica-Mondragón (1987) reported two major magmatic events in the CM, based on a total of 18 K-Ar mineral (biotite, muscovite, feldspar, and hornblende) analyses, 154 to 165 Ma and 222 to 271 Ma, accompanied by three ages between 323 and 512 Ma from micaschist samples. The Permo-Triassic cooling ages have recently been confirmed by Rb-Sr mica whole-rock ages, ranging from 214 ± 11 to 244 ± 12 Ma (Schaaf et al., 2002). The latter study also reported a 7-point Rb-Sr whole-rock (orthogneisses) isochron age of 509 ± 97 Ma, indicating the possible influence of Cambrian re-equilibration of the Rb-Sr isotopic system.

Evidence for the presence of Proterozoic crust is provided from Nd isotopic data of igneous and metamorphic rocks from the CM (Weber, 1998; Schaaf et al., 2002). Felsic granitoids and orthogneisses yielded depleted mantle model ages (T_{DM}) between 1.0 and 1.2 Ga. Some gabbro-amphibolites have older T_{DM} model ages of 1.4 and 1.6 Ga (Schaaf et al., 2002). As it is unlikely that such mafic rocks were formed by mixing of mantle material with older crust, Schaaf et al. (2002) argued that the 1.4 to 1.6 Ga model ages indicate the presence of Proterozoic rocks from Oaxaquia in the CM; however, no granulite-facies rocks like those exposed elsewhere in Oaxaquia (e.g., Ortega-Gutiérrez et al., 1995) have been found in the CM.

Close to the Pacific coast, undeformed plutons intruded the CM (Fig. 1B). These intrusions are distinct from those defining the batholith of the CM and the Permo-Triassic arc. Mineral ages (K-Ar) reported from the undeformed plutons range from 29 to only 2 Ma (Damon and Montesinos, 1978; Mujica-Mondragón, 1987) and they are part of granitoid rock complexes of the Mexican Pacific margin, the intrusion ages of which decrease from Early Cretaceous in the northwest to Miocene in the southeast (Herrmann et al., 1994; Schaaf et al., 1995; Morán-Zenteno et al., 1999).

Sample Overview

In this study we analyzed zircons from 11 samples that are mostly metaigneous rocks, except for

CB08-5 and CB08-9, which are paragneisses from the Sepultura unit. We collected three samples south of Buena Vista (Fig. 2). Two medium-grained, heteroblastic orthogneisses (CB16 and CB17) display augen structures defined by plagioclase crystals in a matrix of recrystallyzed quartz and microcline. The plagioclase "augen" show exsolution textures of microcline (antiperthite, mesoperthite) and recrystallization at their rims. Green biotite is a major constituent in the more mafic sample CB17. Both CB16 and CB17 contain clinopyroxene, accessory titanite reaction rims around ilmenite, abundant apatite, and zircons. The third sample from the Ningunilo River valley (CB13a, Fig. 2) differs from the other two samples by its fine-grained texture, the abundance of untwinned, scarcely perthitic K-feldspar blasts (albite?), and reddish-brown Ti-rich biotite.

We analyzed zircons of two augen gneiss samples from the Pando River valley south of Villa Flores (Fig. 2; CB10, CB22). These samples contain abundant perthitic K-feldspar porphyroclasts that show in some cases deformation-induced recrystallization at their rims, forming microcline and/or flame perthite structures. CB10 contains biotite whereas CB22 contains secondary Fe-rich chlorite, replacing biotite and relict clinopyroxene that is present as a minor constituent.

The orthogneiss sample CB28 is the only one taken from south of the main escarpment of the CM mountain range. It is from the Zanatengo River valley north of the city of Tonalá (Fig. 2); it differs from the other samples by the presence of big (5–10 mm) perthitic K-feldspar porphyroclasts with mesoperthitic exsolutions of plagioclase. This feature defines these feldspars as ternary forms originally formed under high-grade (granulite-facies?) conditions. Both K-feldspar and plagioclase porphyroclasts are deformed and partly recrystallized, forming abundant myrmekite, which is present at the rims of the clasts as well as in the recrystallized quartz-feldspar matrix of the rock. Clinopyroxene is mostly retrogressed to epidote and Fe-rich chlorite. Greenish-brown biotite is also of secondary origin. These petrographic observations lead us to suppose that this sample had previously seen granulite-facies metamorphism, but it was significantly affected by a secondary upper low-grade metamorphism and deformation.

Two samples CMP2 and CB12 are from the same outcrop close to Agrónomos Mexicanos (Fig. 2). These orthogneisses are foliated and banded with alternating biotite-rich and quartz-feldspar–rich layers, and have therefore been interpreted as migmatite (Weis, 2000; Schaaf et al., 2002). CMP1, taken from the Los Amates River valley west of Villa Flores, is a weakly deformed granite from the intrusive complex rather than metamorphic basement component of the CM.

The two paragneiss samples (CB08-5, CB08-9) are from the type locality of the Sepultura unit in the El Tablón River valley south of Agrónomos Mexicanos (Fig. 2; Weber et al., 2002). Sample CB08-5 has an anatectic structure with leucosomes composed of quartz, feldspar, and garnet megablasts, some more than 5 cm in size, and a melanosome mostly composed of reddish-brown biotite. The other sample (CB08-9) is cordierite-bearing, garnet-biotite-migmatitic gneiss with skeletal-shaped garnet porphyroblasts that contain inclusions of biotite, quartz, and cordierite. Biotite is present in two generations. Primary reddish Ti-rich biotite forms layers of melanosome and indicates high-temperature conditions. Secondary green biotite is the result of retrogression within the garnet porphyroblasts. Petrographic observations indicate that the peak metamorphic mineral assemblage was formed by a reaction like biotite + sillimanite + quartz = garnet + cordierite ± K-feldspar + melt. Therefore, metamorphism must have reached high-grade conditions with temperatures greater than 700°C and pressures at ~5–6 kbars (Weber et al., 2002).

Bulk-Rock Geochemistry

Eleven whole-rock powders were commercially analyzed for major and trace elements by X-ray fluorescence spectrometry, and nine of them by inductively coupled plasma mass spectrometry (ICP-MS). The results are listed in Table 1. Figure 3 shows a normative QAPF diagram (A) and a TAS diagram (B) of the rocks analyzed in this study, together with the data from Schaaf et al. (2002) in order to illustrate the wide variation of rock types in the CM. Protoliths of the orthogneisses analyzed in this study can be classified as quartz-monzodiorite (CB17, CB13a), granodiorite (CB12, CB16, CB28, CMP2), quartz-monzonite (CB10), and granite (CB22, CMP1). More mafic rocks are also present in the CM, some of which are even nepheline normative (Fig. 3A).

REE patterns and spider diagrams for the orthogneiss samples (Figs. 4A–4B and 4D–4E) are variable, but some characteristics may be of petrologic

TABLE 1. Chemical Data[1]

Sample:	Orthogneiss									Paragneiss	
	CB10	CB12	CB13a	CB16	CB17	CB22	CB28	CMP1	CMP2	CB08-5	CB08-9
SiO_2	63.8	69.6	64.0	67.7	57.4	70.3	65.9	72.1	68.5	53.8	61.9
TiO_2	0.43	0.53	0.74	0.46	1.01	0.23	0.44	0.22	0.43	1.04	0.80
Al_2O_3	17.2	14.8	16.3	15.6	18.5	14.7	16.1	14.4	15.2	20.0	16.9
Fe_2O_3 (tot)	3.51	3.98	4.72	3.80	6.87	2.15	4.33	1.89	3.33	7.61	6.28
MnO	0.06	0.09	0.09	0.08	0.12	0.06	0.08	0.04	0.08	0.08	0.08
MgO	0.85	1.38	1.81	1.23	2.38	0.78	1.02	0.31	1.24	3.15	2.37
CaO	2.42	3.49	2.44	2.95	5.04	1.54	4.25	1.38	3.60	5.74	4.43
Na_2O	4.14	3.34	4.63	3.81	4.47	3.86	3.92	3.40	3.36	3.7	2.92
K_2O	5.34	2.37	3.17	3.15	2.02	4.34	2.47	5.07	1.91	2.75	2.21
P_2O_5	0.09	0.04	0.32	0.16	0.32	0.10	0.21	0.07	0.11	0.94	0.35
LOI	0.7	0.8	1.6	1.0	1.7	1.5	0.8	0.36	0.7	0.5	1.2
Total	98.54	100.42	99.82	99.94	99.83	99.56	99.52	99.24	98.46	99.31	99.44
Mg#	17.0	22.6	24.5	21.4	22.7	23.3	16.6	12.1	23.8	26.1	24.3
Ba	4568	1048	1290	1140	1130	987	1060	1227	660	419	356
Rb	86	85	42	54	48	74.2	34	108	67	143	125
Sr	479	377	454	424	521	268	480	253	380	360	317
Cs	9	8	0.2	0.3	0.3	0.4	0.2	–	–	6.7	5.1
Ga	20	15	20	20	25	13	16	–	–	28	23
Ta	b.d.	b.d.	0.7	b.d.	0.7	b.d.	b.d.	–	–	1.4	0.9
Nb	10	10	16	9	15	5	3	–	–	25	15
Hf	–	–	9	4	7	3	4	–	–	1	3
Zr	424	245	395	163	302	113	136	191	189	43	136
Y	25	18	31	20	28	7	9	18	15	62	31
Th	14.1	0.8	12.3	0.6	1.1	1.2	0.3	–	–	1.8	0.9
U	0.7	0.7	0.8	0.3	0.5	0.5	0.16	–	–	2.4	1.8
Cr	60	110	–	–	–	–	–	58	214	–	–
Ni	b.d.	b.d.	7	b.d.	75	10	6	12	19	14	9
Co	39	74	7.2	6.6	58	4.1	5.9	–	–	10	10.6
V	31	52	49	44	100	21	48	15	50	133	114
Cu	13	6	22	18	16	b.d.	6	–	–	13	27
Pb	20	16	6	8	6	34	6	–	–	11	14
Zn	39	67	71	69	125	56	56	59	82	122	102
La	121	13.6	75	19.6	30	16.9	17.0	–	–	36.9	18.5
Ce	199	24.3	149	39.4	63.4	29.1	32.3	–	–	93.4	40.8
Pr	16.1	2.4	18.1	5.3	8.3	3.1	3.9	–	–	14.4	5.8
Nd	62.2	11.0	64.6	22	32.7	11.5	17.5	–	–	61.5	23.6
Sm	9	2.5	11.2	4.8	6.6	2.1	3.2	–	–	19.6	5.7
Eu	1.86	0.9	2.4	1.3	2.1	1.02	1.49	–	–	1.8	1.5
Gd	6.1	2.3	10.5	5.0	7.1	1.85	3.0	–	–	22.8	7.0
Tb	0.7	0.4	1.3	0.7	0.9	0.26	0.4	–	–	3.2	1.1
Dy	4.5	2.7	5.9	3.5	4.9	1.35	2.0	–	–	14.1	5.9
Ho	0.78	0.55	1.1	0.8	1.1	0.27	0.38	–	–	2.4	1.1
Er	2.3	1.6	3.2	2.1	3.3	0.83	0.9	–	–	5.9	3.2
Tm	0.3	0.2	0.4	0.3	0.4	0.12	0.12	–	–	0.7	0.4
Yb	1.8	1.6	2.4	1.6	2.7	0.9	0.8	–	–	3.5	2.4
Lu	0.28	0.23	0.41	0.23	0.47	0.10	0.09	–	–	0.57	0.35
La/Nb	12.1	1.4	4.7	2.2	2.0	3.4	5.7	–	–	1.5	1.2

[1]Oxides in weight percents, trace elements in parts per million; b.d. = below detection limit; – = not analyzed.

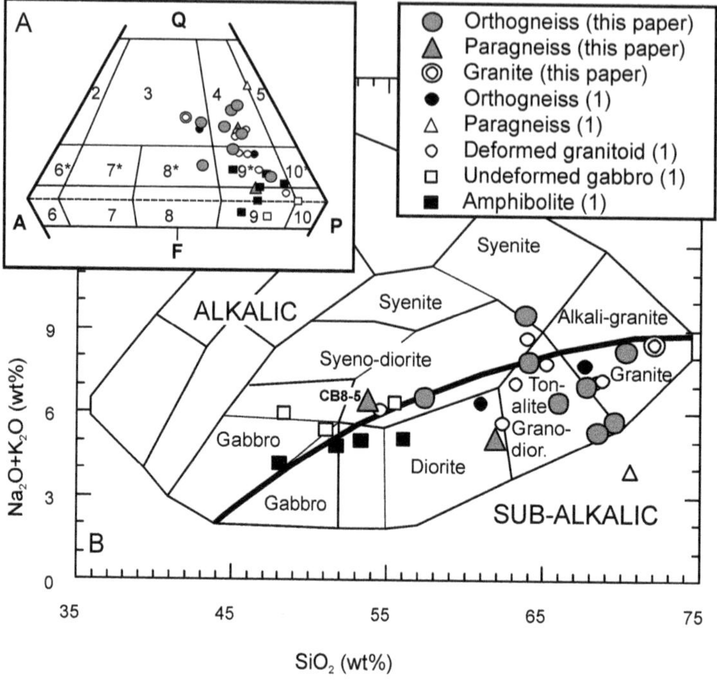

FIG. 3. A. Normative QAPF diagram after Streckeisen (1976). B. Total alkali vs. SiO_2 (wt%) for plutonic and metamorphic rocks of the Chiapas massif after Cox et al. (1979). The dividing line between alkalic and sub-alkalic magma series is from Miyashiro (1978); (1) = data from Schaaf et al. (2002).

significance. A first group of orthogneisses is characterized by strongly fractionated REE patterns with La_N/Yb_N ranging from 13.5 to 48.2, with either positive or negative Eu anomalies, depending on their overall REE contents (Fig. 4A, Table 1). These rocks are fairly depleted in Nb with respect to La (La/Nb from 3.4 to 12.1) and enriched in Ba and K. A second group of orthogneiss samples (CB12, CB16, CB17; Figs. 4B and 4E) has moderately fractionated REE patterns (La_N/Yb_N from 6.1 to 8.8) without significant Eu anomalies, and they are little depleted in HFS elements with respect to La (La/Nb from 1.4 to 2.2). The REE and trace element patterns of these rocks are similar to those of intrusive rocks from the CM (grey fields in Fig. 4; unpubl. data, and from Schaaf et al., 2002) except for a relatively strong Cs depletion in samples CB16 and CB17.

Both metasedimentary samples have less fractionated REE patterns than the orthogneisses (La_N/Yb_N = 5.5 and 7.6, Fig. 4C). The anatectic gneiss CB08-5 has a strong negative Eu anomaly that most probably indicates extraction of feldspar-rich melt that explains its relative depletion in Sr as well. This sample is also characterized by relatively low Th, Hf, and Zr contents with respect to the other paragneiss sample.

Isotopic Analytical Techniques

Zircons were separated by standard procedures using a Wilfley™ table (at UNAM), a Frantz™ isodynamic separator at CICESE and UNAM, heavy liquids at CICESE and UNAM, and handpicking techniques at CICESE, UCSC, and UNAM. Selected zircon fractions (see Table 2) were abraded with pyrite in air abraders at UCSC for 6 hours at 3.0 PSI air pressure. To remove superficial pyrite, the abraded zircons were cleaned with hot 4N HNO_3. Zircon sample dissolution and ion-exchange chemistry were performed following the methods of Krogh (1973) and Mattinson (1987). All zircons, except those from samples CMP1, CMP2, and CB12, were spiked with a $^{205}Pb/^{235}U$ mixed spike solution and analyzed at UCSC. Zircons from samples CMP1, CMP2, and CB12 were spiked with a $^{208}Pb/^{235}U$

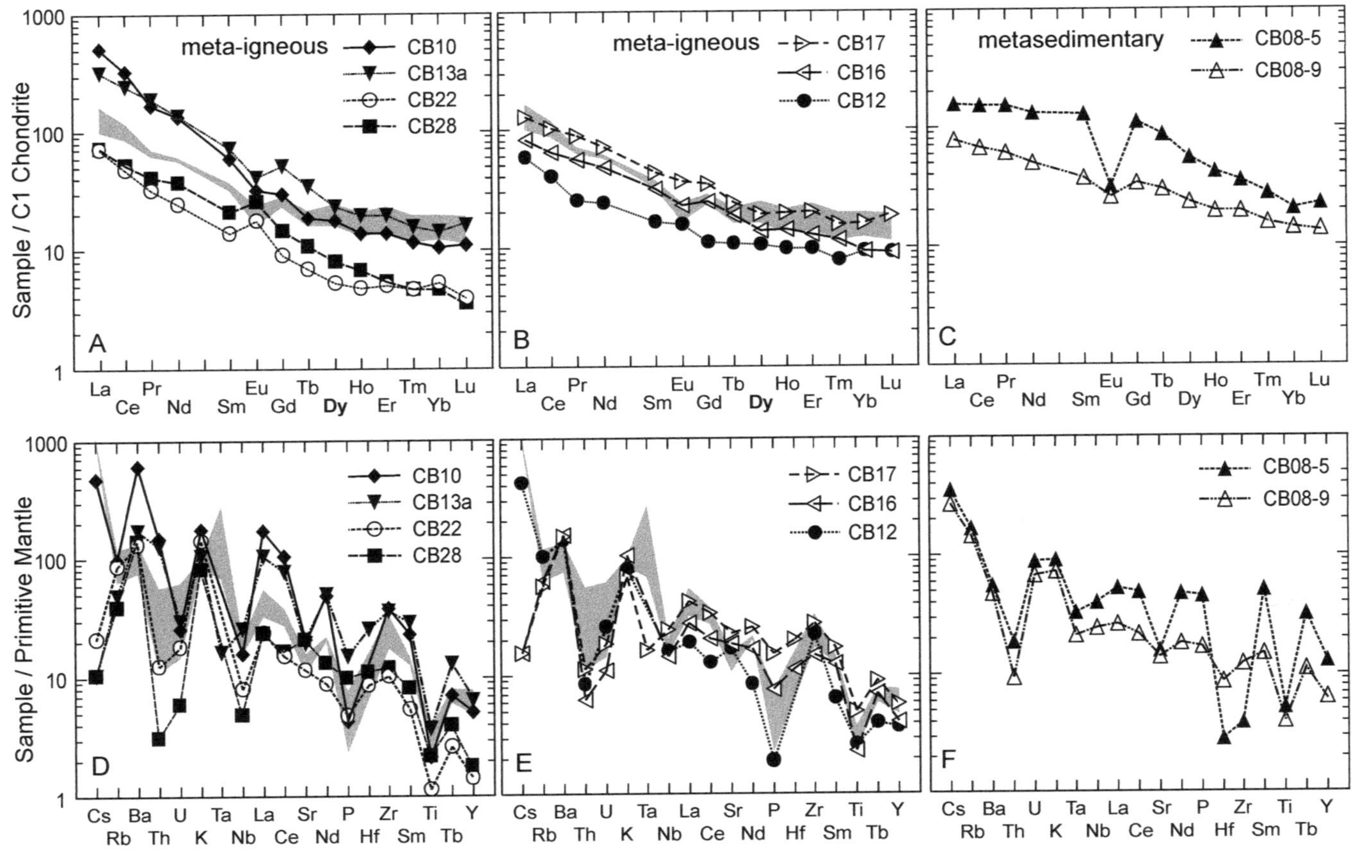

FIG. 4. A–C. REE patterns of metamorphic rocks from the CM. C1 chondrite normalization according to Sun and McDonough (1989). D–F. Primordial mantle-normalized abundances of trace elements in metamorphic rocks from the CM. Normalizing values and element order are according to Wood et al. (1979). Grey field indicates data of igneous rocks from the CM (Schaaf et al., 2002, and our own unpublished data).

mixed spike solution and analyzed at Laboratorio Universitario de Geoquímica Isotópica (LUGIS), UNAM (fractions #41-48), or at LMU München, Germany (fractions #39, 40—for details on the analytical procedure, see Weber and Köhler, 1999). At UCSC the U and Pb isotope analysis was made on the UCSC VG54-WARB with the ^{204}Pb isotope measured on a Daly pulse counter detector, while all other isotopes were measured on Faraday collectors. Samples spiked with ^{205}Pb were analyzed in a dynamic mode with ^{205}Pb alternating between Daly detector and Faraday collector. This enabled an online Faraday/Daly gain calibration for adjusting the measured ^{206}Pb/^{204}Pb ratio. Six Pb isotopic ratios were determined for each zircon fraction. Two-sigma uncertainties on the ^{207}Pb/^{206}Pb ratio and ^{208}Pb/^{206}Pb measured ratios are better than 0.8% and averaged 0.1%. At UCSC, we choose the conservative value of 0.5% as our uncertainty on the U/Pb (^{206}Pb*/^{238}U and ^{207}Pb*/^{235}U) ratio based on replicate analyses of the Geostandard zircon 91500 (Wiedenbeck et al., 1995) and other in-house zircons. Procedural blanks at UCSC are less than 10 pg Pb. At LUGIS, UNAM, the U and Pb isotope analysis was performed with a Finnigan MAT262 TIMS, equipped with eight variable Faraday collectors. ^{204}Pb was analyzed with an ion-counting electron multiplier, whereas all other Pb and U isotopic ratios were determined with the Faraday collectors in static mode. Twenty-five Pb isotopic ratios were determined for each zircon fraction. Procedural blanks during the analyses of CMP1 and CMP2 samples were between 16 and 57 pg Pb.

Geochronology

We analyzed U and Pb isotope ratios and concentrations from a total number of 58 zircon fractions. Most of them were less than 10, 7 less than four zircon grains, and 20 fractions have been air abraded. The results are given in Table 2.

Paragneisses

Both paragneiss samples (CB08-5 and CB08-9) have very low Zr contents (43 and 136 ppm, respectively), and therefore only a few, relatively small zircons could be separated. Most of these are clear, colorless, multifaceted, and rounded (Figs. 5A and 5B) or short prismatic with pointed {211} pyramids. Three fractions (#1 to #3) from sample CB08-5 have slightly discordant isotope ratios (7 to 18% discordant) compared to zircon fractions #4 and #5 from sample CB08-9, the latter with apparent ^{207}Pb*/^{206}Pb* ages of 502–553 Ma. Small (100–150μm), rounded or multifaceted grains of fraction #3 are the only zircons from sample CB08-5, exhibiting a significant inherited component. Larger (>200 μm), rounded and multifaceted grains from sample CB08-9 have a greater amount of inherited zircon material.

Because the two samples are from localities just 300 m apart and from the same sequence in the El Tablón River valley (Fig. 2; Table 2), we put the isotope ratios of zircons from both paragneiss samples together on one Tera-Wasserburg plot (Fig. 6A). Five fractions define a discordia with a lower-intercept age of 241 ± 14 Ma and an upper-intercept at 1024 ± 190 Ma. Isotope ratios of a small (<100 μm) rounded grain fraction (#6) lie off the discordia (this fraction was omitted from the regression calculation), indicating secondary disturbance. It is noteworthy that the short prismatic, {211} bi-pyramidal grains (#2) show little inheritance, and they are therefore most likely of metamorphic origin. Zircons from the metasedimentary rocks show less inheritance than some orthogneiss rocks (see below). The upper-intercept age, however, indicates that the sediments contain inherited zircons of Precambrian (Grenvillian) age. We interpret the lower-intercept age as the time of metamorphic crystallization.

Orthogneisses

Two similar orthogneiss samples with plagioclase "augen", CB16 and CB17, are from two nearby localities, probably from the same unit, and therefore the isotopic results from both samples are shown in one Tera-Wasserburg plot (Fig. 6B). Isotope ratios of five abraded grains (fraction #23) of large (>300 μm), short prismatic to equant, multifaceted zircons from sample CB16 (Fig. 5E) yielded the oldest apparent ^{207}Pb*/^{206}Pb* age of these two samples (710 Ma, Table 2). Smaller rounded or short prismatic zircons (abraded fraction #20 and fraction #21 [not abraded]) are also discordant, and they show an important inherited component. In contrast, small rounded zircons (<100 μm) and big abraded, long, prismatic zircons from the same sample display little inheritance (Fig. 6B). Isotope ratios of those fractions (#22 and #19, respectively) plot close to the lower concordia intercept at 252.9 ± 4.4 Ma. Zircons from sample CB17 have less inherited zircon component than sample CB16. The only fraction with a notable inherited component in the former sample contained rounded multifaceted

TABLE 2. U-Pb Isotopic Data of Zircons from the Chiapas Massif[1]

Sample, type, location					Concentrations			Atomic ratios								Apparent ages, Ma		
No.	Aspect,[2]	Size(μm)	Weight, mg	Comment[3]	U, ppm	Total Pb, ppm	Comm. Pb, pg	$\frac{^{206}Pb}{^{204}Pb}$ [4]	$\frac{^{206*}Pb}{^{238}U}$ [5]	$\pm 2\sigma_m$,[5] %	$\frac{^{207*}Pb}{^{235}U}$ [5]	$\pm 2\sigma_m$, %	$\frac{^{207}Pb}{^{206}Pb}$ [6]	$\pm 2\sigma_m$, (%)	Error corr.	$\frac{^{206}Pb}{^{238}U}$	$\frac{^{207}Pb}{^{235}U}$	$\frac{^{207}Pb}{^{206}Pb}$
Paragneiss anatexite, CB08-5, El Tablón River valley, N 16°16.89', W 93°36.88'																		
1	sp, bipy	~200 *l*	0.148	abr	1110	41	23	31280	0.04008	0.1	0.2879	0.11	0.05209	0.034	0.948	253	257	289
2	bipy	~200 *l*	0.112		1302	45	10	859000	0.03832	0.19	0.2717	0.19	0.05142	0.034	0.984	242	244	260
3	round	100–150	0.134		937	36	97	3585	0.04089	0.1	0.2973	0.08	0.05272	0.048	0.867	258	264	317
Paragneiss anatexite, CB08-9, El Tablón River valley, N 16°16.98', W 93°36.90'																		
4	round	~200	0.086	abr	625	32	26	11060	0.05313	0.16	0.4293	0.17	0.05861	0.043	0.966	334	363	553
5	sp	~200 *l*	0.025		825	38	30	1997	0.04705	0.19	0.3714	0.22	0.05726	0.104	0.879	296	321	502
6	round	<100	0.050		541	22	23	5366	0.04314	0.16	0.3479	0.18	0.05849	0.069	0.919	272	303	548
Augengneiss, CB10, Pando River valley (boulder)																		
7	lp 5:1	~400 *l*	0.165	abr	216	18	29	8915	0.07767	0.19	0.7016	0.20	0.06551	0.052	0.966	482	536	791
8	round	~200	0.121		299	36	51	6170	0.11458	0.14	1.1125	0.14	0.07042	0.038	0.965	699	759	941
9	round	~200	0.059		272	34	59	2290	0.11436	0.13	1.1422	0.15	0.07244	0.066	0.895	698	774	998
10	pris 3:1	~200 *l*	0.107	abr	280	36	11	280600	0.12550	0.09	1.2459	0.10	0.07200	0.038	0.922	762	822	986
11	sp	~200 *l*	0.031	3*	309	49	11	177100	0.14351	0.12	1.4219	0.13	0.07186	0.053	0.917	865	898	982
12	lp 6:1	<250 *l*	0.117		200	9	32	2557	0.04004	0.23	0.2898	0.28	0.05250	0.149	0.841	253	258	307
13	round	~200	0.035	3	296	47	22	8261	0.15316	0.15	1.6483	0.17	0.07805	0.077	0.888	919	989	1148
Orthogneiss, CB13a, Ningunilo River valley, N 16° 02.47', W 93° 13.62'																		
14	round	~350	0.136	abr	516	73	10	1736000	0.13649	0.11	1.3302	0.11	0.07069	0.034	0.953	825	859	948
15	round	~250	0.085	abr	181	23	11	98680	0.11869	0.10	1.1436	0.11	0.06988	0.044	0.912	723	774	925
16	sp 2:1	<200	0.114		267	31	13	73200	0.11403	0.21	1.0795	0.22	0.06866	0.046	0.977	696	743	889
17	flat	~300	0.116		297	36	13	83120	0.11443	0.12	1.0977	0.12	0.06956	0.036	0.954	698	752	916
18	lp 4:1	~400 *l*	0.142		358	31	33	11330	0.08348	0.14	0.7333	0.15	0.06371	0.044	0.955	517	559	732
Orthogneiss, CB16, South of Buenavista, N 16°05.51', W 93°11.01'																		
19	lp 3:1	~300 *l*	0.143	abr	266	12	62	1871	0.04159	0.56	0.2997	0.57	0.05227	0.090	0.988	263	266	297
20	round	~250	0.085	abr	139	8	22	3178	0.05204	0.27	0.4141	0.32	0.05771	0.159	0.864	327	352	519
21	round	<200	0.138		274	14	54	2667	0.05045	0.20	0.3992	0.21	0.05738	0.058	0.960	317	341	506
22	round	<100	0.124		276	12	46	2397	0.04144	0.16	0.2993	0.21	0.05238	0.131	0.775	262	266	302
23	round	~300	0.104	abr	157	11	26	4240	0.06748	0.17	0.5867	0.21	0.06305	0.125	0.802	421	469	710
Orthogneiss, CB17, South of Buenavista, N 16°04.38', W 93°12.35'																		
24	sp 2:1	~200	0.147	abr	218	10	28	4358	0.04085	0.15	0.2934	0.16	0.05210	0.063	0.918	258	261	290
25	lp 5:1	~400 *l*	0.154	abr	228	10	31	4232	0.04073	0.18	0.2931	0.20	0.05220	0.086	0.905	257	261	294
26	lp 5:1	~400 *l*	0.074	abr	264	12	39	1708	0.04094	0.20	0.2971	0.23	0.05263	0.105	0.884	259	264	313
27	round	~150	0.114		255	12	48	2403	0.04384	0.12	0.3270	0.14	0.05410	0.065	0.877	277	287	375
28	lp 5:1	<300 *l*	0.148		331	14	24	8427	0.04060	0.11	0.2900	0.12	0.05180	0.057	0.884	257	259	277
29	sp 2:1	500 *l*	0.076	2* abr	156	7	7	15508	0.04082	0.22	0.3117	0.38	0.05538	0.293	0.636	258	276	428

Augengneiss, CB22, Monterry village, Pando River valley																
30	sp 2:1	~250 *l*	0.105	abr	377	16	35	4094	0.04186	0.17	0.3016	0.18	0.05226	0.076	0.910	264	268	297
31	sp 2:1	~400 *l*	0.117	abr	344	23	40	5184	0.06369	0.12	0.5452	0.14	0.06209	0.058	0.905	398	442	677
32	pris 3:1	~400 *l*	0.066	2 abr	280	12	21	4358	0.04115	0.21	0.2968	0.26	0.05230	0.145	0.825	260	264	299
33	sp <2:1	~300 *l*	0.083		460	24	21	11140	0.05179	0.13	0.4184	0.16	0.05860	0.090	0.822	326	355	552
Orthogneiss, CB28, Zanatengo River valley (south of escarpment)																		
34	round	~400	0.066	2 abr	142	8	17	4552	0.05794	0.31	0.4802	0.36	0.06012	0.191	0.850	363	398	608
35	sp 2:1	~600 *l*	0.087	1 abr	343	14	15	14315	0.04166	4.2	0.2981	4.2	0.05190	0.120	0.999	263	265	281
36	sp 2:1	~350 *l*	0.064	3 abr	324	31	10	570500	0.09517	0.18	0.8962	0.18	0.0683	0.043	0.972	586	650	878
37	pris 3:1	~400 *l*	0.09	abr	366	16	19	9559	0.04276	0.16	0.3110	0.20	0.05274	0.120	0.793	270	275	318
38	sp 2:1	<250 *l*	0.085		308	14	12	45640	0.04770	0.19	0.3654	0.22	0.05555	0.102	0.882	300	316	435
Migmatite, CMP2 and CB12, Villa Flores-Arriaga highway, N 16° 20.35', W 93° 32.63'																		
39	sp	125–200	1.69	CB12[7]	626	26	540	3221	0.04211	0.18	0.3075	0.26	0.05297	0.18	0.73	266	272	327
40	sp	>200	1.91	CB12[7]	740	31	1200	2002	0.04194	0.18	0.3072	0.20	0.05312	0.08	0.91	265	272	334
41	sp, 1:2	>200	0.12	CMP2[8]	513	22	477	392	0.04486	0.20	0.3529	0.58	0.05706	0.52	0.46	283	307	494
42	sp, 1:2	<200	0.75	CMP2[8]	686	45	6147	316	0.05495	0.36	0.4545	0.70	0.05999	0.57	0.59	345	380	603
43	round	200	0.62	CMP2[8]	696	34	1926	687	0.04597	0.21	0.3564	0.46	0.05623	0.39	0.55	290	310	461
44	sp, 1:2	<200	0.81	CMP2[8]	902	46	2666	898	0.04938	0.23	0.4057	0.31	0.05958	0.19	0.78	311	346	588
45	lp	300	0.17	CMP2[8]	511	31	1693	161	0.04244	0.21	0.3134	1.44	0.05355	1.36	0.46	268	277	352
Deformed Granite, CMP1, Las Mercedonas River valley, N 16° 11.27', W 93° 22.35'																		
46	1:3	<200	1.22	8	499	22	2228	736	0.04078	0.27	0.3108	0.4	0.05527	0.28	0.71	258	275	423
47	round	>200	0.16	8	449	21	501	385	0.03973	0.23	0.3123	0.76	0.05702	0.68	0.46	251	276	492
48	1:3	300	1.41	8	712	13	3039	955	0.04403	0.24	0.3425	0.32	0.05642	0.20	0.78	278	299	469
49	lp	>200	0.22	8	567	24	396	791	0.03801	0.33	0.2776	0.77	0.05296	0.66	0.54	240	249	327
50	pris, 1:3	>200	0.17	8	363	21	936	202	0.04305	0.21	0.3405	1.13	0.05736	1.05	0.45	272	298	505
51	sp, 1:2	<200	0.41	8	623	26	642	1055	0.04049	0.19	0.3037	0.27	0.05440	0.19	0.73	256	269	388
52	lp	300	0.24	8	212	10	484	297	0.04103	0.21	0.3205	0.93	0.05665	0.85	0.46	259	282	478
53	lp	>200	0.29	8	1160	26	769	980	0.03413	0.33	0.2544	0.40	0.05406	0.22	0.84	216	230	374
54	lp	>200	0.61	8	632	24	1206	971	0.04624	0.18	0.3654	0.30	0.05732	0.22	0.66	291	316	504
55	pris, 1:3	>200	0.29	8	340	17	693	422	0.04406	0.18	0.3482	0.63	0.05732	0.57	0.46	278	303	504
56	round	<200	0.10	8	756	35	469	437	0.04023	0.19	0.2942	0.80	0.05305	0.73	0.44	254	262	331
57	1:3	300	1.09	8	732	34	2813	840	0.04494	0.21	0.3520	0.36	0.05681	0.28	0.64	283	306	484
58	polished	>200	0.01	8	2498	177	577	149	0.04656	0.25	0.3953	2.14	0.06158	1.99	0.63	293	338	660

[1]Pb* = radiogenic lead.
[2]*sp* = short prismatic; *lp* = long prismatic; *pris* = prismatic, *l* = length .
[3]abr = abraded grains; * = number of grains (if fraction less than 4 grains).
[4]Corrected for fractionation and spike.
[5]Error in 2 sigma mean % (= $2\sigma\sqrt{n}$) .
[6]Corrected for fractionation, spike, blank and common lead (Stacey and Kramers, 1975).
[7]Analyzed at Ludwig-Maximilians-Universität, München, Germany.
[8]Analyzed at LUGIS, Universidad Nacional Autónoma de México.

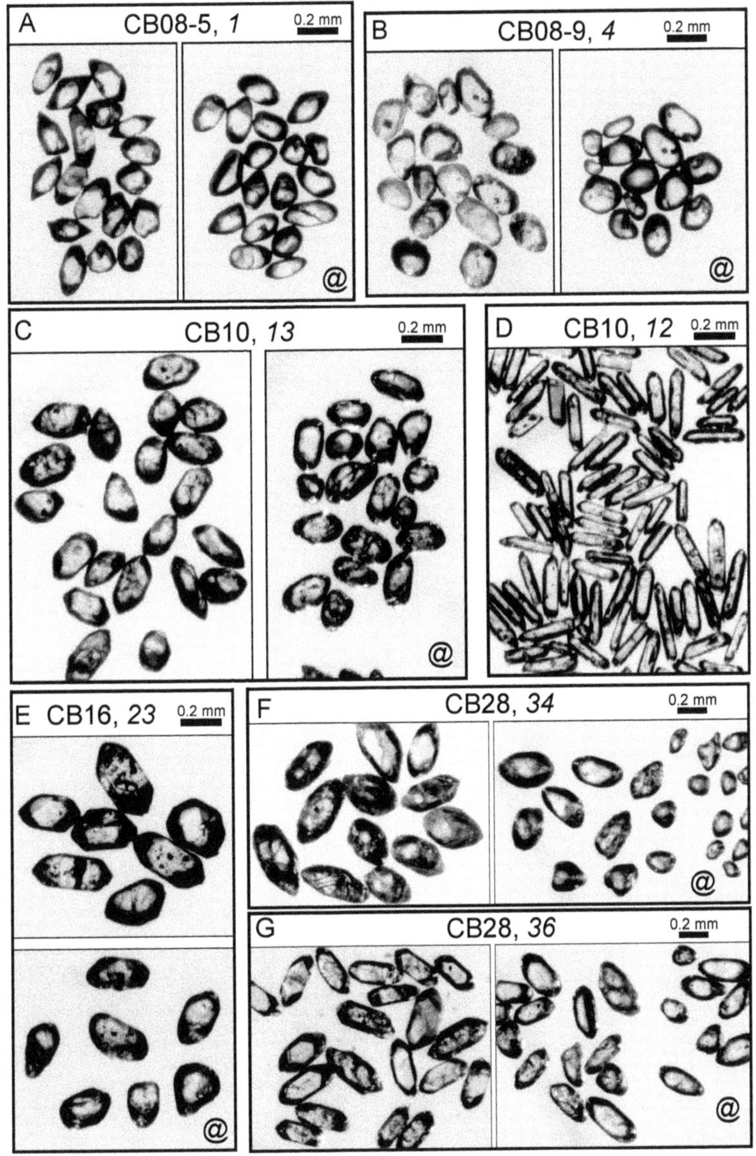

FIG. 5. Transmitted light photographs of zircons extracted from the studied samples. Sample number, *fraction number* (after Table 2); @ = same fraction after abrasion.

grains (fraction #27), whereas much bigger, abraded, long prismatic zircons (fractions #24, #25, #28) plot close to the lower-intercept age of this sample at 250 ± 6 Ma (see Fig. 6B). Fractions #26 and #29 were omitted from the calculation because these fractions seem to display a higher degree of secondary disturbance of the isotopic system.

We regressed the most reliable nine fractions of samples CB16 and CB17, and this discordia defines a lower-intercept age of 250.9 ± 2.3 Ma and an upper-intercept age at 1017 ± 27 Ma. The upper-intercept age can be interpreted as the average age of inherited zircons, and the lower-intercept age as the crystallization age of the main zircon matter. The

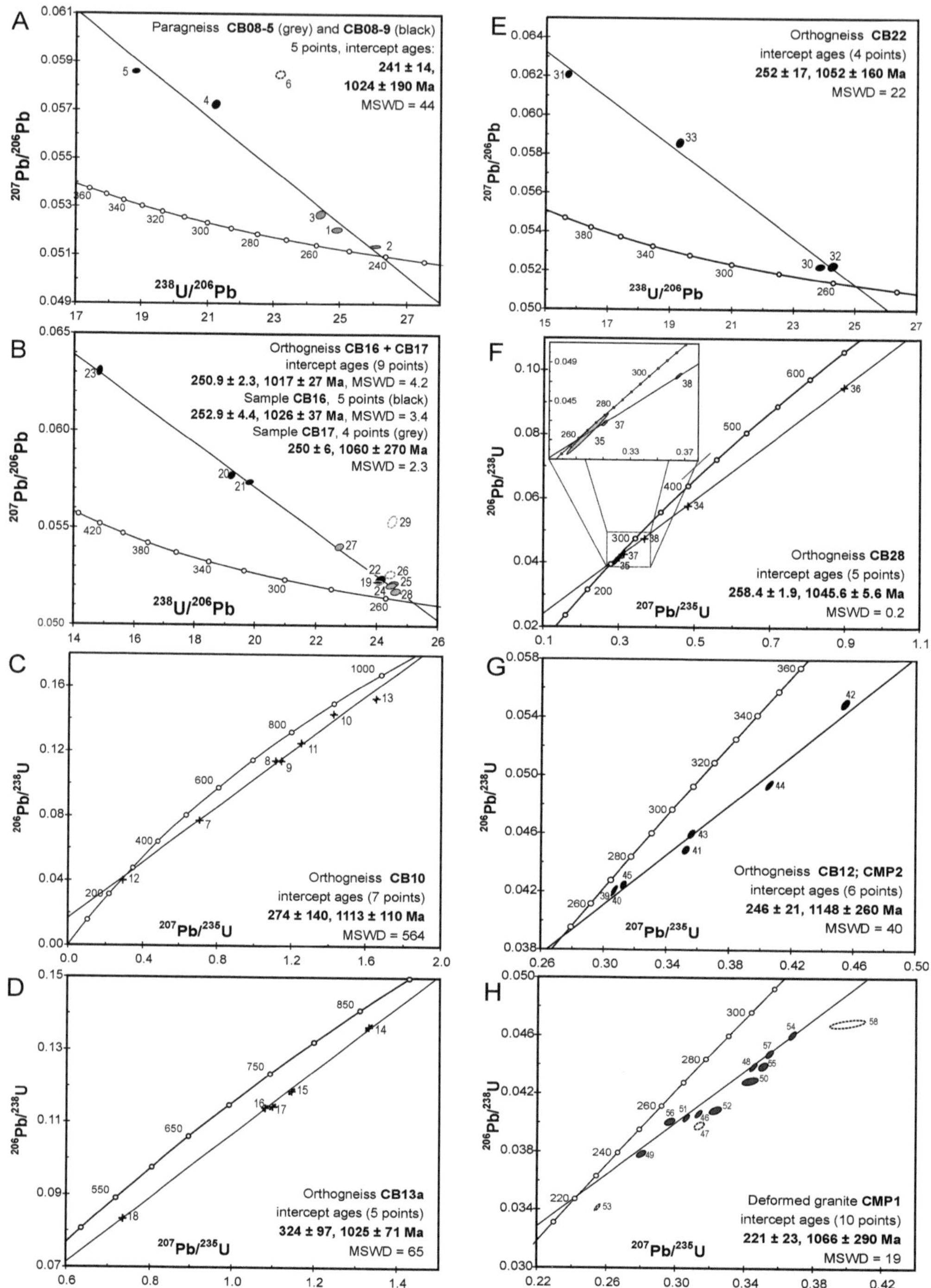

FIG. 6. Tera-Wasserburg (A, B, E) and Wetherhill Concordia (C, D, F, G, H) diagrams with U-Pb results for zircons from the samples analyzed. Ellipses indicate 2σ errors, intercepts calculated with Isoplot/Ex 2.49 (Ludwig, 2001). Fraction numbers are same as those in Table 2.

fact that long prismatic zircons with simple forms ({100}, {101}) are more likely of igneous rather than of metamorphic origin, and that those zircons have isotope ratios close to the lower intercept, indicates that the lower-intercept age most likely reflects igneous crystallization. However, secondary lead loss or the presence of metamorphic zircon overgrowth cannot be discounted.

Another type of augen gneiss with K-feldspar blasts and clasts is represented by three samples: CB13a from the Ningunilo River valley; CB10 and CB22, both from the Pando River valley (Fig. 2). Short prismatic and rounded grains from sample CB10 show large cores that are in some cases visible with the stereomicroscope (fraction #13, Fig. 5C). After abrasion these zircons yield isotope ratios with an apparent $^{207}Pb^{*}/^{206}Pb^{*}$ age of 1148 Ma, the oldest fraction of this study (Table 2, Fig. 6C). Four more, mostly short prismatic and rounded fractions (two of them abraded) yielded $^{207}Pb^{*}/^{206}Pb^{*}$ ages above 940 Ma. Again, the long prismatic, even abraded grains (#7) are the less radiogenic. The smaller grains of fraction #12 (see Fig. 5D) plot below the lower intercept and probably indicate some secondary lead loss. The upper-intercept age, which was calculated from all seven analyzed fractions of sample CB10, is 1113 Ma, with a large error of 110 Ma. Also the lower-intercept ages have large errors, but they are all within the range of the results from other samples.

Mostly short prismatic, rounded, and multifaceted grains from sample CB13a have inherited cores. Isotope ratios of both abraded (#14 and #15) and not abraded (#16 and #17) fractions have $^{207}Pb^{*}/^{206}Pb^{*}$ apparent ages from 948 to 889 Ma (Fig. 6D). These zircons tend to yield results with older apparent ages after air abrasion. One abraded fraction (#14) of five round zircons is the most radiogenic and the least discordant (13%). The long prismatic (4:1) grains (fraction #18) are younger and more discordant with respect to all other fractions of sample CB13a. Isotope ratios of all five fractions analyzed from sample CB13a define a regression line (Fig. 6D) that is not very well constrained because of the strong discordance of all zircon fractions. However, the upper-intercept age at 1025 ± 71 Ma indicates that the cores of these zircons crystallized during the Grenville orogenic event.

Isotope ratios of four fractions from sample CB22 show a distribution similar to the other samples (Fig. 6E) with bigger, short prismatic, and abraded grains being the most radiogenic, and the long prismatic zircons being the least radiogenic. The lower-intercept age using four fractions of that sample is 252 ± 17 Ma and the upper-intercept age is 1052 ± 160 Ma. Compared to samples CB10 and CB13a, the inherited component is less pronounced in sample CB22.

From orthogneiss CB28, which is the only sample that comes from south of the main escarpment of the CM (Fig. 2), we analyzed four fractions with 2–6 grains each and one single grain fraction. These isotope ratios yield the most linear discordia line calculated in this study (MSWD 0.2; Fig. 6F). The lower-intercept age of 258.4 ± 1.9 Ma is significantly older than those of samples CB16 and CB17. Abraded short prismatic grains with {211} and {101} bipyramids (fraction #36, Fig. 5G) yield the most radiogenic isotope ratios, with an apparent $^{207}Pb^{*}/^{206}Pb^{*}$ age of 878 Ma. Although this most radiogenic fraction is 33% discordant, the upper-intercept age at 1046 Ma yielded a comparatively low error of 5.6 Ma (2σ).

Isotopic results of zircons from migmatites CMP2 and CB12, both from the same outcrop (Fig. 2), are shown in Figure 6G. Figure 7 shows cathodoluminescence images of zircons from sample CB12. It is clear that zircons from the CM are fairly complex, with fragments of older inherited grains, magmatic zoning, metamorphic overgrowths, corrosion and continuing growth, etc. Most of the zircon fractions analyzed from these two samples are multigrain fractions and none of them were abraded. The isotope ratios of some fractions scatter above and some below the discordia, but both, upper- and lower-intercept ages (246 ± 21 Ma, 1148 ± 260 Ma) are within the errors of other orthogneisses analyzed.

Deformed granite

The deformed granitoids in the Chiapas massif do not show all the deformation features present in the gneisses; in some cases, intrusive relations indicate that these rocks intruded the gneisses (for more details see Schaaf et al., 2002). Therefore, it seems likely that zircon crystallization was younger in the deformed granite than metamorphism in the gneisses. Isotopic results of zircons from the only sample of deformed granite analyzed in this study are shown in Figure 6H. Isotope ratios of 10 multigrain zircon fractions are moderately scattered, and define a lower-intercept age of 221 ± 23 Ma that is significantly younger than the lower-intercept ages calculated from most of the orthogneisses (Fig. 6). This result possibly is indicative of the time difference between

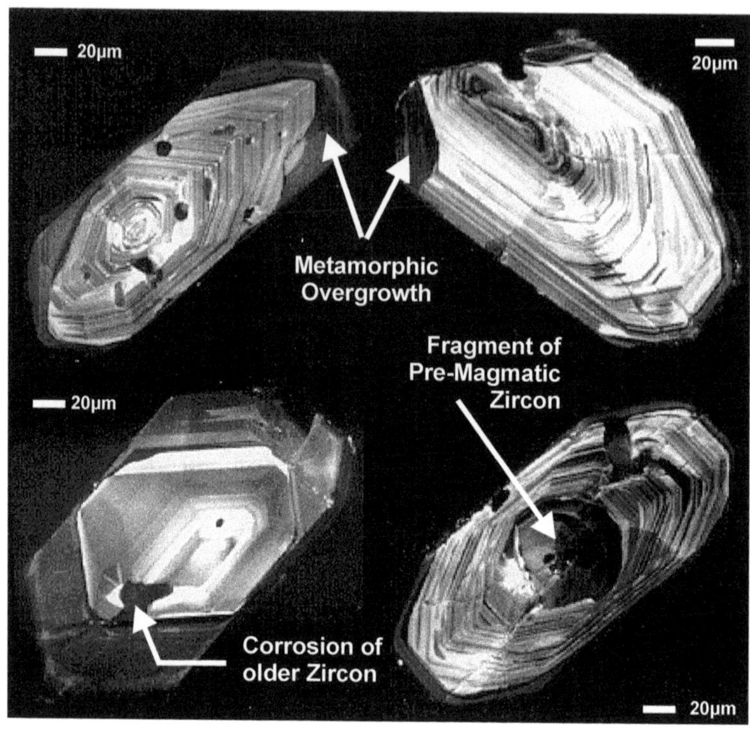

FIG. 7. Cathodoluminescence images of zircons from sample CB12 showing the complexity of zircons from the CM with metamorphic overgrowths surrounding portions exhibiting magmatic zoning, but also with fragments of pre-magmatic zircon cores inside. Corrosion features indicate zircon growth during different phases.

zircon crystallization and metamorphism in the orthogneisses and the subsequent intrusion of the granitoids of the Chiapas batholith. However, some of the zircon fractions analyzed from sample CMP1 indicate secondary lead loss and the results of these fractions were omitted from the regression calculation (fractions #49, #55, #60). Therefore, it remains unclear if the time difference between the lower-intercept ages of the orthogneisses and this deformed granite is of geological significance, or is of analytical and/or secondary origin.

Discussion

In order to summarize the results of our zircon dating, the isotope ratios of 42 zircon fractions from the metamorphic rocks are depicted in one concordia plot (Fig. 8). This is not a valid age calculation, but it illustrates quite well the most important results of this study: (1) most of the analyzed zircon fractions plot close to a lower-intercept age at 252 ± 14 Ma; and (2) the other zircon fractions are strongly discordant and plot along a discordia that intersects the concordia at 1058 ± 29 Ma.

Grenville inheritance

The strongly discordant U-Pb isotope ratios of zircon fractions from all samples point to an upper intercept of Grenvillian age. Although upper-intercept ages of many of the samples have large errors, there is overwhelming evidence that an important amount of the Chiapas massif is recycled >1 Ga crustal material, and hence, the CM has a Grenville-age basement. These results are in good agreement with Nd model ages reported by Schaaf et al. (2002), who discussed the requirement of Proterozoic crust to explain T_{DM} model ages greater than 1.4 Ga in some mafic rocks from the CM.

The exposed pre-batholithic basement rocks in the CM, as known to date, are not comparable with Oaxaquia in terms of metamorphic mineral assemblages. Admittedly, the metamorphic conditions in

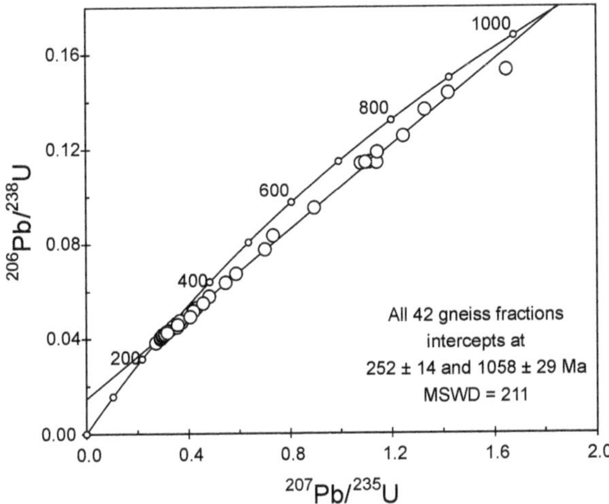

FIG. 8. Concordia plot of all reliable zircon fractions from metamorphic rocks of the CM.

ortho- and paragneisses from the CM reached high-grade P-T conditions, but no orthopyoxene-bearing assemblages typical of granulite facies and of Oaxaquia (e.g., Mora et al., 1986; Weber and Hecht, 2003) could be found in the CM (Weber et al., 2002). In addition, U-Pb isotope ratios of large abraded grains are all discordant and plot far away from the upper intercepts. Their extrapolated >1 Ga ages, therefore, reflect the average age of inherited zircon cores within newly grown zircons, either of magmatic origin in the orthogneisses or metamorphic origin in the paragneisses, and indicate the presence of zircon older than 1 Ga.

It is interesting to note that the two groups of orthogneiss samples that have been distinguished based on their REE and trace-element characteristics apparently differ also in their amount of inherited zircon matter. Zircons from the first set of samples (CB10, CB13a, CB22, CB28) with highly fractionated REE patterns and relatively high Ba and K contents with respect to the second group of samples (CB12, CMP2, CB16, CB17), have generally more radiogenic Pb isotope ratios than zircons from the second group. $^{207}Pb^*/^{206}Pb^*$ ages of 10 out of 21 zircon fractions from the first group are greater than 800 Ma, and 14 greater than 600 Ma (see Table 2). All samples of this group contain reddish-brown Ti-rich biotite, abundant mesoperthite, myrmekite, and ternary feldspar. Only 2 out of 20 zircon fractions from the second group of orthogneisses have yielded $^{207}Pb^*/^{206}Pb^*$ ages greater than 600 Ma. These samples contain greenish low-temperature biotite, but neither titaniferous biotite nor ternary feldspars. These observations indicate that some of the first group of meta-igneous rocks may have been Grenville granulites that were affected by a later medium- to high-grade metamorphic event. This later event erased almost all petrographic evidence of the granulite-facies event of probable Grenville age. The second group of orthogneisses clearly were dominated by the later Permo-Triassic thermal event and the inheritance in these igneous zircons is less pronounced, as it is for the zircons from the granite sample, which also display moderate evidence of Grenville inheritance.

The metasediments from the Sepultura unit, although they have undergone high-grade metamorphism and partial anatexis, and therefore must have been affected by the first deformation event in the CM, do not have as much inherited zircon component compared to the orthogneisses. The metasediments analyzed in this study probably have semipelitic precursors of a distal sedimentary environment with very little zircons from continental sources. Almost all zircon matter in such samples would be of metamorphic origin with very little inheritance and, indeed, the zircons from the paragneisses seem to be metamorphic (see above). However, in order to define the provenance regions of those sediments, it will be an important task to analyze zircons from more proximal sediments of the Sepultura unit.

The Permio-Triassic tectonothermal event

It is quite clear that the most important tectonothermal event documented by the U-Pb zircon data presented here was Permo-Triassic rather than Precambrian in age. The most reliable lower-intercept ages of the orthogneisses are 250.9 ± 2.3 (CB16, CB17), 258.4 ± 1.9 (CB28), 250 ± 16 (CB12, CMP2), and 252 ± 17 (CB22). The slight age difference between the first two values may be significant. It is not ascertainable, however, if this age difference is based on (1) regional differences in timing of the high-grade thermal event, (2) predominance of the metamorphic versus the igneous event in sample CB28 with respect to samples CB16 and CB17, or (3) is simply the result of secondary lead loss. The lower-intercept age of metamorphic zircons from the metasediments of 241 ± 14 Ma is younger than CB28 but within error of all other lower-intercept ages. Metamorphic zircons in the paragneisses most probably grew during the high-grade metamorphic event but again, because the metamorphic zircons are small with high U contents, secondary lead loss cannot be discounted. The lower-intercept age of the granite sample, which is without doubt younger than the gneisses based on field relations, is only slightly younger (221 ± 23 Ma) than—and within error of—most of the analyzed gneisses.

These zircon age results are in perfect agreement with the Rb-Sr isochron age of 254 ± 10 Ma (Damon et al., 1981) and complementary to cooling ages from 214 ± 11 to 244 ± 12 Ma as reported by Schaaf et al. (2002). The difference in timing between two major events: (1) high-grade metamorphism with associated anatexis, D1 deformation, and the major batholith intrusion, and (2) subsequent D2 deformation and its associated low- to medium-grade metamorphism, could not be resolved. It seems possible that both events are the consequence of separate phases of deformation during a single tectonothermal event that lasted roughly from 260 to 230 Ma, with its peak at 260–250 Ma.

Implications for the evolution of the Maya terrane

In a Phanerozoic tectonic evolution model of Mexico, Ortega-Gutiérrez et al. (1994) proposed a relationship between parts of the Maya basement and Paleozoic South America as part of a continental arc concurrent with the collision of the two American continents during the Ouachita orogeny. In more recent reconstructions (Centeno-García and Keppie, 1999; Torres et al., 1999; Solari et al., 2001; Elias-Herrera and Ortega-Gutiérrez, 2002), the Permo-Triassic magmatic arc, which intruded Precambrian basement of Oaxaquia in Mexico and similar basement in northwestern South America, was interpreted to be the result of eastward subduction along the newly formed western margin of Pangea. The CM seems to be of crucial importance in this scenario and we conclude the following.

1. Unlike most other Permo-Triassic intrusive complexes in Mexico, the CM is the only one that exhibits considerable deformation within its plutonic suite (Schaaf et al., 2002) that includes medium- to high-grade metamorphism and ductile deformation of the igneous and sedimentary basement rocks into which plutons of the CM were intruded (Weber et al., 2002).

2. There is a significant difference in the timing of major zircon crystallization between the CM (260–250 Ma) and other intrusions of the supposed arc in southern Mexico (from the Zapoteco and Mixteco terranes) that range from ~270 to ~290 Ma (Yañez et al., 1991; Solari et al., 2001; Elias-Herrera and Ortega-Gutiérrez, 2002). If the latter is true, tectonic processes of Early Permian age in southern Mexico along the Marathon-Ouachita suture (Elias-Herrera and Ortega-Gutiérrez, 2002) are probably not responsible for deformation and metamorphism in the CM, and hence the CM may have a unique history.

3. The metasedimentary sequences of the Sepultura unit show passive-margin characteristics and it is unlikely, therefore, that these sediments formed within the Permo-Triassic arc, which is supposed to be located on an active continental margin. However, the age of sedimentation is very poorly constrained and could be late Proterozoic to Early Permian.

4. A Pan-African or early Paleozoic metamorphic or igneous event as proposed by Schaaf et al. (2002) on the basis of Rb-Sr whole-rock data could not be confirmed by our new U-Pb zircon data. Pan-African ages in Mexico are known from the Yucatan Peninsula (Krogh et al., 1993) and the Mexican state of Coahuila (Lopez et al., 2001), and they are probably remnants of the Pan-African (700–500 Ma) event, which marked the assembly of Gondwana during Neoproterozoic time.

5. Middle Paleozoic (Silurian–Devonian) intrusion ages as known from the Maya Mountains in Belize and from boreholes in Yucatan (e.g., Dengo, 1985, and references therein) are not present in the CM. Such ages are also widespread in the proposed northern South American counterpart of Yucatan,

the Merida Andes (Aleman and Ramos, 2000, and references therein).

6. West of the Isthmus of Tehuantepec, toward the western limit of the Maya terrane, Permian igneous rocks intrude Proterozoic granulites of the Guichicovi complex (Weber, 1998; Weber and Köhler, 1999). This fact indicates that the western part of the Maya terrane is part of Oaxaquia. Although our data support a correlation between the CM and Oaxaquia in terms of crustal precursors, our data do not prove any direct continuation of Oaxaquia from west of the Tehuantepec Isthmus into the CM.

Most of these arguments indicate that the Maya terrane is a composite terrane, and that the CM does not share the same geologic history with the rest of the Maya terrane, particularly Yucatan. Possibly, during the Permian (and earlier), the CM was not in the same position as the Yucatan Peninsula and the western part of the Maya terrane. Therefore, the Late Permian paleogeographic position of the CM in contact with Yucatan within the present day Gulf of Mexico should be dismissed (e.g., Centeno-García and Keppie, 1999). Instead, a Pacific-side position for the CM as a separate crustal block, in close connection to Oaxaquia or other similar rocks in northwestern South America can be considered. In such a hypothesis, the Permian arc affected the CM by intrusion of plutons into the sedimentary sequence of unknown age. Subsequently, the CM collided with the continent, resulting in metamorphism and deformation. A possible suture might be the contact of the CM metamorphic basement with upper Paleozoic sediments of the Santa Rosa Formation east of the CM.

Acknowledgments

This work was supported by UC-MEXUS Consejo National de Ciencia y Tecnología (CONACYT) collaborative grant 2001, CONACYT project D41083-F, Deutsche Forschungsgemeinschaft/ Bundesministerium für Wirtschaftliche Zusammenarbeit und Entwicklung (DFG/BMZ) collaboration project HE2893/3,4-1, CICESE internal project 644111, and UNAM–Instituto de Geofísica internal project B119. Many thanks to Consuelo Macías, Gabriela Solís, and Teodoro Hernández-Treviño for sample preparation at UNAM. We are greatful to Gabriel Rendón-Marquez, Susana Rosas-Montoya (both CICESE), and Ralf Nestler (Uni Freiberg) for their help with preparing the zircon separates. Thanks also go to Victor Perez-Arroyoz (CICESE) for thin section preparation. We are grateful to Fernando Ortega-Gutiérrez and Mariano Elías-Herrera for helpful discussion in the field. Ralf Hiller, Lutz Hecht (both Berlin), and Roberto Molina (Jurriquilla) assisted in the field and contributed with their discussions. We want to thank to Wayne Premo (U.S. Geological Survey) and Ian Fitzsimons (Curtin University) for their constructive and helpful reviews. Last but not least, we thank José Carlos Pisaño-Soto, Pedro Hernández Martínez, and all other staff members from Reserva de la Biosfera La Sepultura (Comision Nacional de Areas Naturales Protegidas) in Tuxtla Gutiérrez, Chiapas, for logistical support in the field.

REFERENCES

Aleman, A., and Ramos, V.A., 2000, Northern Andes, *in* Cordani, U. G., Milani, E. J., Thomaz-Filho, A., and Campos, D. A., eds., Tectonic evolution of South America: Rio de Janeiro, 31st International Geological Congress, p. 453–480.

Barboza-Gudino, J. R., 1994, Regionalgeologische Erkundungen entlang der GEOLIMEX—Traverse in Südmexiko, unter besonderer Berücksichtigung der Sierra de Juarez, Oaxaca: Unpubl. doctoral thesis, Mathematisch-Naturwissenschaftliche Fakultät TU Clausthal, Germany, 105 p.

Campa, M. F., and Coney, P. J., 1983, Tectono-stratigraphic terranes and mineral resource distributions in Mexico: Canadian Journal of Earth Sciences, v. 20, p. 1040–1051.

Centeno-García, E., and Keppie, J. D., 1999, Latest Paleozoic–early Mesozoic structures in the central Oaxaca Terrane of southern Mexico: Deformation near a triple junction: Tectonophysics, v. 301, p. 231–242.

Coney, P. J., and Campa, M. F., 1987, Lithotectonic terrane map of Mexico (west of the 91st meridian): U.S. Geological Survey Miscellaneous Field Studies Map MF-1874-D, scale 1:2,500,000.

Cox, K. G., Bell, J. D., and Pankhurst, R. J., 1979, The interpretation of igneous rocks: New York, NY, Allen and Unwin, 450 p.

Damon, P. E., and Montesinos, E., 1978, Late Cenozoic volcanism and metallogenesis over an active Benioff zone in Chiapas, Mexico: Arizona Geological Society Digest, v. 11, p. 155–168.

Damon, P. E., Shafiqullah, M., and Clark, K., 1981, Age trends of igneous activity in relation to metallogenesis in the southern Cordillera, *in* Dickinson, W., and Payne, W. D., eds., Relations of tectonics to ore deposits in the southern Cordillera: Arizona Geological Society Digest, v. 14, p. 137–153.

Dengo, G., 1985, Mid America: Tectonic setting for the Pacific margin from southern Mexico to northwestern Colombia, *in* Nairn, A. E. M., and Stehli, F. G., eds., The oceanic basins and margins, Vol. 7a: The Pacific Ocean: New York, NY, Plenum Press, p. 123–180.

Elias-Herrera, M., and Ortega-Gutiérrez, F., 2002, Caltepec fault zone: An early Permian destral traspressional boundary between the Proterozoic Oaxacan and Paleozoic Acatlán complexes, southern Mexico, and regional tectonic implications: Tectonics, v. 21, no. 3, p. 1–19.

Gross, A., 2000, Geologische, isotopengeochemische, und geochronologische Untersuchungen an Gesteinen des Chiapas-Massivs, Mexiko: Unpubl. diploma thesis, Universität Freiburg and Universität München, Germany, 106 p.

Heck, M., 2000, Zur Geologie, petrographie und geochemie des Pando-Tales, südwestlich Villa Flores, Chiapas, Mexiko: Unpubl. diploma thesis, Universität Freiburg and Universität München, Germany, 132 p.

Herrmann, U. R., Nelson, B. K., and Ratschbacher, L., 1994, The origin of a terrane: U/Pb zircon geochronology and tectonic evolution of the Xolapa complex, southern Mexico: Tectonics, v. 13, p. 455–474.

Howell, D. G., Jones, D. L., and Schermer, E. R., 1985, Tectonostratigraphic terranes of the Circum-Pacific region, *in* Howell, D. G., ed., Tectonostratigraphic terranes of the Circum-Pacific region, Circum-Pacific Council of Energy and Mineral Resources, Earth Science Series, v. 1, p. 3–30.

Keppie, J. D., Dostal, J., Cameron, K. L., Solari, L. A., Ortega-Gutiérrez, F., and Lopez, R., 2003, Geochronology and geochemistry of Grenvillian igneous suites in the northern Oaxacan Complex, southern Mexico: Tectonic implications: Precambrian Ressearch, v. 120, p. 365–389.

Keppie J. D., Dostal, J., Ortega-Gutiérrez, F., and Lopez, R., 2001, A Grenvillian arc on the margin of Amazonia: Evidence from the southern Oaxacan Complex, southern Mexico: Precambrian Research, v. 112, p. 165–181.

Keppie, J. D., and Ortega-Gutiérrez, F., 1995, Provenance of Mexican Terranes—isotopic constraints: International Geology Review, v. 37, p. 813–824.

Keppie, J. D., and Ortega-Gutiérrez, F., 1999, Middle American Precambrian basement—a missing piece of the reconstructed 1-Ga orogen, *in* Ramos, V. S. and Keppie, J. D., eds., Laurentia-Gondwana connections before Pangea: Geological Society of America, Special Paper 336, p. 199–210.

Krogh, T. E., 1973, A low contamination method for hydrothermal decomposition of zircon and extraction of U and Pb for isotopic age determination: Geochimica et Gosmochimica Acta, v. 37, p. 485–494.

Krogh, T. E., Kamo, S. L., Sharpton, V. L., Martin, L. E., and Hildebrand, A. R., 1993, U-Pb ages of single shocked zircons linking distal K/T ejecta to the Chicxulub crater: Nature, v. 366, p. 731–734.

Lopez, R., Cameron, K. L., and Jones, N. W., 2001, Evidence for Paleoproterozoic, Grenvillian, and Pan-African age crust beneath northeastern Mexico: Precambrian Research, v. 107, p. 195–214.

López Ramos, E., 1979, Geología de México, tomo III: México D.F., Secretaría de Educación Pública, no. 91407, 445 p.

Ludwig, K. R., 2001, ISOPLOT: A plotting and regression program for radiogenic isotope data, Version 2.49: Berkeley, CA, Berkeley Geochronological Center, Special Publication, no. 1a, 55 p.

Mattinson, J. M., 1987, U-Pb ages of zircons: A basic examination of error propagation: Chemical Geology (Isotope Geosciences section), v. 66, p. 151–162.

Meza-Figueroa, D., Ruiz, J., Talavera-Mendoza, O., and Ortega-Gutiérrez, F., 2003, Tectonometamorphic evolution of the Acatlan Complex eclogites (southern Mexico): Canadian Journal of Earth Sciences, v. 40, p. 27–44.

Miyashiro, A., 1978, Nature of alkalic volcanic rock series: Contributions to Mineralogy and Petrology, v. 66, p. 91–104.

Möllinger, S., 2000, Zur geologie, petrologie und geochemie des Las Mercedonas Tales südwestlich von Villa Flores, Chiapas, Mexiko: Unpubl. diploma thesis, Universität Freiburg and Universität München, Germany, 76 p.

Mora, C. I., Valley, J. W., and Ortega-Gutiérrez, F., 1986, The temperature and pressure conditions of Grenville-age granulite-facies metamorphism of the Oaxaca complex, southern Mexico: Revista Instituto de Geología, Universidad Nacional Autónoma de México, v. 6, p. 222–242.

Morán-Zenteno, D., 1984, Geología de la República Mexicana: Mexico, D.F., Facultad de Ingeniería, Universidad Nacional Autónoma de México, Instituto Nacional de Estadística, Geografía e Informática, 88 p.

Morán-Zenteno, D. J.,Tolson, G., Martínez-Serrano, R., Martiny, B., Schaaf, P., Silva-Romo, G., Macías-Romo, C., Alba-Aldave, L., Hernández-Bernal, M. S., and Solís-Pichardo, G., 1999, Tertiary arc-magmatism of the Sierra Madre del Sur, Mexico, and its transition to the volcanic activity of the Trans-Mexican Volcanic Belt: Journal of South American Earth Sciences, v. 12, p. 513–535.

Mujica-Mondragón, R., 1987, Estudio petrogenético de las rocas ígneas y metamóficas en el Macizo de Chiapas: Mexico, D.F, Instituto Mexicano de Petroleo, Proyecto C-2009, unpublished.

Murillo-Muñeton, G., 1994, Petrologic and geochronologic study of Grenville-age granulites and post-granulite plutons from the La Mixtequita area, state of Oaxaca in southern Mexico, and their tectonic significance: Unpubl. M.S. thesis, University of Southern California, Los Angeles, CA, 163 p.

Murillo-Muñetón G., Anderson, J. L., and Tosdal, R. M., 1994, A New Grenville-age granulite terrane in southern Mexico [abs.]: Geological Society of America Abstracts with Programs, v. 26, no. 7, p. A-48.

Ortega-Gutiérrez, F. and Elías-Herrera, M., 2003, Wholesale melting of the southern Mixteco Terrane and origin of the Xolapa complex [abs.]: Geological Society of America, 99th Cordilleran Annual Meeting, paper 27-6.

Ortega-Gutiérrez, F., Elias-Herrera, M., Reyes-Salas, M., Macias-Romo, C, and Lopez, R., 1999, Late Ordovician–Eary Silurian continental collisional orogeny in southern Mexico and its bearing on Gondwana-Laurentia connections: Geology, v. 21, 8, p. 719–722.

Ortega-Gutiérrez, F., Mitre-Salazar, L. M., Roldan-Quintana, J., Aranda-Gómez, J. J., Morán-Zenteno, D., Alaniz-Álvarez, S. A., and Nieto-Samaniego, A. N., 1992, Carta geológica de la República Mexicana. 1:2,000,000: Universidad Nacional Autónoma de México, Instituto de Geología.

Ortega-Gutiérrez, F., Mitre-Salazar, L. M., Roldán-Quintana, J., Sanchez-Rubio, G., and De La Fuente, M., 1990, H-3 Acapulco Trench to the Gulf of Mexico across Southern Mexico: Centennial Continent/Ocean Transect no. 13, Geological Society of America.

Ortega-Gutiérrez, F., Ruiz, J., and Centeno-García, E., 1995, Oaxaquia, a Proterozoic microcontinent accreted to North America during the late Paleozoic: Geology, v. 23, p. 1127–1130.

Ortega-Gutiérrez, F., Sedlock, R. L., and Speed, R. C., 1994, Phanerozoic tectonic evolution of Mexico, *in* Speed, R. C., ed., Phanerozoic evolution of North American continent-ocean transitions: Geological Society of America, DNAG Continental-Ocean Transect Volume, p. 265–306.

Ortega-Gutiérrez, F., Elias-Herrera, M., Reyes-Salas, M., Macias-Romo, C, and Lopez, R., 1999, Late Ordovician-Eary Silurian continental collisional orogeny in southern Mexico and its bearing on Gondwana-Laurentia connections: Geology, v. 21, 8, p. 719–722.

Pantoja-Alor, J., Fries, C., Jr., Rincón-Orta, C., Silver, L. T., and Solorio-Munguia, J., 1974, Contribución a la geocronología del Estado de Chiapas: Boletín Asociación Mexicana de Geólogos Petroleros, v. XXVI, p. 205–223.

Pantoja-Alor, J., Rincón-Orta, C., Fries, C., Jr., Silver, L. T., and Solorio-Munguia, J., 1971, Contribución a la geocronología del Estado de Chiapas: Universidad Nacional Autónoma de México, Instituto de Geología Boletín, v. 100, p. 47–58.

Ratschbacher, L., Riller, U., Meschede, M., Herrman, U., and Frisch, W., 1991, Second look at suspect terranes in Southern Mexico: Geology, v. 19, p. 1233–1236.

Schaaf, P., Morán-Zenteno, D., Hernández-Bernal, M. S., Solís-Pichardo, G., Tolson, G., and Köhler, H., 1995, Paleogene continental margin truncation in southwestern Mexico: Geochronological evidence: Tectonics, v. 14, p. 1339–1350.

Schaaf, P., Weber, B., Weis, P., Gross, A., Ortega-Gutiérrez, F., and Köhler, H., 2002, The Chiapas Massif (Mexico) revised: New geologic and isotopic data for basement characteristics, *in* Miller, H., ed., Contributions to Latin American Geology: Neues Jahrbuch für Geologie und Paläontologie Abhandlung, v. 225, p. 1–23.

Sedlock, R. L., Ortega-Gutiérrez, F., and Speed, R. C., 1993, Tectonostratigraphic teranes and tectonic evolution of Mexico: Geological Society of America Special Paper 278, 153 p.

Solari, L. A., Dostal, J., Ortega-Gutiérrez, F., and Keppie, J. D., 2001, The 275 Ma arc-related La Carbonara stock in the northern Oaxacan Complex of southern Mexico: U-Pb geochronology and geochemistry: Revista Mexicana de Ciencias Geologicas, v. 18, no. 2, p. 149–161.

Solari, L. A., Keppie, J. D., Ortega-Gutiérrez, F., Cameron, K. L., and Lopez, R., 2004, ~990 Ma peak granulitic metamorphism and amalgamation of Oaxaquia, Mexico: U-Pb zircon geochronological and common Pb isotopic data: Revista Mexicana de Ciencias Geologicas, v. 21, no. 2, p. 212–215.

Solari, L. A., Keppie, J. D., Ortega-Gutiérrez, F., Cameron, K. L., Lopez, R., and Hames, W. E., 2003, 990 and 1100 Ma Grenvillian tectonothermal events in the northern Oaxacan Complex, southern Mexico: Roots of an orogen: Tectonophysics, v. 365 , nos. 1-4, p. 257–282.

Stacey, J. S., and Kramers, J. D., 1975, Approximation of terrestrial lead isotope evolution by a two-stage model: Earth and Planetary Science Letters, v. 26, p. 207–221.

Streckeisen, A., 1976, To each plutonic rock its proper name, Earth Science Rewiew, v. 12, p. 1–33.

Sun, S. S., and McDonough, W. F., 1989, Chemical and isotopic systematics of oceanic basalts: Implications for mantle composition and processes, *in* Saunders, A. D., and Norry, M. J., eds., Magmatism in the ocean basins: Geological Society of London Special Publication, v. 42, p. 313–345.

Torres, R., Ruiz, J., Patchett, P. J., and Grajales, J. M., 1999, Permo-Triassic continental arc in eastern Mexico: Tectonic implications for reconstructions of southern North America: Geological Society of America, Special Paper 340, p. 191–196.

Webber, B. N., and Ojeda-Rivera, J., 1957, Investigacion sobre lateritas fosiles en las regiones sureste de Oaxaca y sur de Chiapas: Mexico Instituto Nacional para la Investigación de Recursos Minerales Boletín, v. 37, 66 p.

Weber, B., 1998, Die magmatische und metamorphe Entwicklung eines kontinentalen Krustensegments: Isotopengeochemische und geochronologische Untersuchungen am Mixtequita-Komplex, Südostmexiko: Münchner Geologische Hefte, v. A24, 176 p.

Weber, B., Gruner, B., Hecht, L., Molina-Garza, R., and Köhler, H., 2002, El descubrimiento de basamento metasedimentario en el macizo de Chiapas: la "Unidad La Sepultura": GEOS, v. 22, no. 1, p. 2–11.

Weber, B., and Hecht, L., 2003, Petrology and geochemistry of metaigneous rocks from a Grenvillian basement fragment in the Maya block: The Guichicovi complex, Oaxaca, Southern Mexico: Precambrian Research, v. 124, p. 41–67.

Weber, B., and Köhler, H., 1999, Sm-Nd, Rb-Sr, and U-Pb isotope geochronology of a Grenville terrane in Southern Mexico: Origin and geologic history of the Guichicovi complex: Precambrian Research, v. 96, p. 245–262.

Weber, B., Meschede, M., Ratschbacher, L., and Frisch, W., 1997, Defomation analysis and kinematic history of the Acatlán Compex in the Nuevos Horizontes–San Bernardo region, State of Puebla: Geofísica Internacional, v. 36, no. 2, p. 63–76.

Weis, P., 2000, Geologische und isotopengeochemische Untersuchungen zur magmatischen und metamorphen Entwicklung des Chiapas Massivs, Mexiko: Unpubl. diploma thesis, Universität Freiburg and Universität München, Germany, 127 p.

Wiedenbeck, M., Alle, P., Corfu, F., Griffin, W. L., Meier, M., Oberli, F., von Quadt, A., Roddick, J. C., and Spiegel, W., 1995, Three natural zircon standards for U-Th-Pb, Lu-Hf, trace element and REE analysis: Geostandards Newsletter, v. 19, no. 1, p. 1–23.

Wood, D. A., Joron, J. L., Treuil, M., Norry, M., and Tarney, J., 1979, Elemental and Sr isotope variations in basic lavas from Iceland and the surrounding ocean floor: Contributions to Mineralogy and Petrology, v. 70, p. 319–339.

Yañez, P., Ruiz, J., Patchett, J., Ortega-Gutiérrez, F., and Gehrels, G. E., 1991, Isotopic studies of the Acatlan complex, southern Mexico: Implications for Paleozoic North American Tectonics: Geological Society of America Bull., v. 103, p. 817–828.

Mixteca Terrane

Ordovician and Mesoproterozoic Zircons from the Tecomate Formation and Esperanza Granitoids, Acatlán Complex, Southern Mexico: Local Provenance in the Acatlán and Oaxacan Complexes

J. L. SÁNCHEZ-ZAVALA, F. ORTEGA-GUTIÉRREZ, J. DUNCAN KEPPIE,[1]

Instituto de Geología, Universidad Nacional Autónoma de México, 04510 México D.F., México

G. A. JENNER,

Department of Earth Sciences, Memorial University, St. John's, Newfoundland, Canada A1B 3X5

E. BELOUSOVA,

GEMOC National Key Centre, Department of Earth and Planetary Sciences, Macquarie University, North Ryde 2109, Australia

AND C. MACIÁS-ROMO

Instituto de Geología, Universidad Nacional Autónoma de México, 04510 México DF, México

Abstract

Detrital zircons from the Middle Permian and older Tecomate Formation, formerly the uppermost unit of the Acatlán Complex in southern Mexico, yielded the following U-Pb, LA-ICP-MS ages: (1) a major Cambro-Ordovician population with peaks at 460 and 500 Ma; (2) a limited Neoproterozoic population (729 and 879 Ma); and (3) a major Mesoproterozoic population ranging from ~940 to 1650 Ma with peaks at ~1025 Ma and 1100–1300 Ma. The type locality of the Esperanza blastomylonitic granite yielded concordant U-Pb ages of 471 ± 6 Ma, which, along with similar granitoids in the Acatlán Complex, appears to be the source of the Cambro-Ordovician detrital zircons in the Tecomate Formation. The provenance of the Mesoproterozoic detrital zircons seems to be in the adjacent Oaxacan Complex, which has yielded ages ranging from ~1450 to 917 Ma. The Neoproterozoic detrital zircons may have been recycled from other units in the Acatlán Complex. These data are consistent with Pangea reconstructions that place the Acatlán Complex in its present location relative to the rest of Mexico.

Introduction

THE TECOMATE FORMATION has been considered the uppermost unit in the Paleozoic Acatlán Formation of southern Mexico (Ortega-Gutiérrez et al., 1999) (Figs. 1 and 2). However, it unconformably overlies both the Pelalcingo and Piaxtla groups and was deformed in the Early–Middle Permian during the amalgamation of Pangea (Malone et al., 2002; Keppie et al., 2004c). The Tecomate Formation was previously thought to be of Devonian age based upon the occurrence of cystoids, crinoids, blastoids, and micromolluscs, which were interpreted as pre-Carboniferous in age (Ortega-Gutiérrez, 1993). However, three different limestone horizons in the Tecomate Formation have recently yielded conodonts and fusilinids with ages ranging from latest Pennsylvanian to early Middle Permian (Keppie et al., 2004c). This is consistent with ~320–264 Ma U-Pb SHRIMP ages recorded in granitic pebbles from a deformed conglomerate within the Tecomate Formation (Keppie et al., 2004c). Although these data provide age constraints on the upper part of the Tecomate Formation, no information is available for the lower parts of the unit. This paper presents detrital zircon data for other parts of the type section and a megacrystic granitic locality that may have been a source for K-feldspar clasts in various horizons of the Tecomate Formation.

[1]Corresponding author; email: duncan@servidor.unam.mx

Geological Setting

In its type section the Tecomate Formation occurs in a steeply dipping N-S shear zone (Fig. 2).

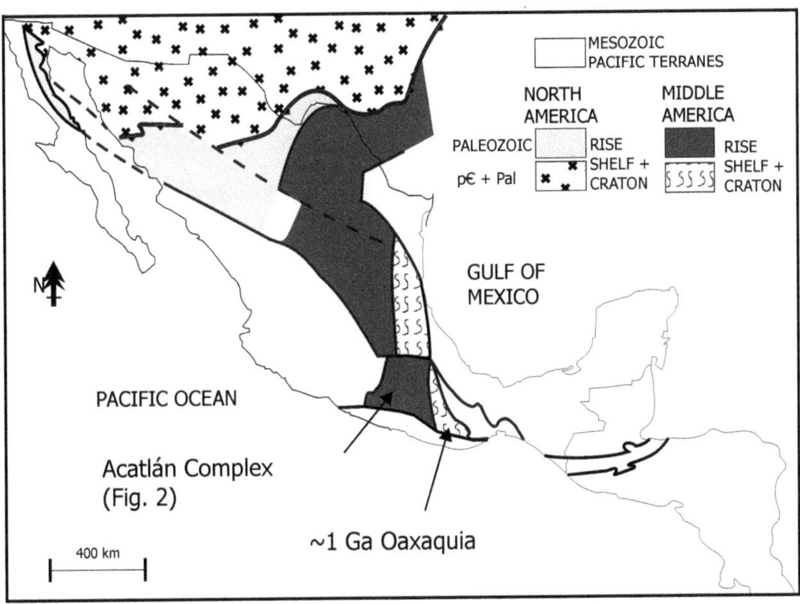

FIG. 1. Sketch map of North and South America showing the location of the Acatlán Complex in southern Mexico.

It consists of low-grade, strongly to mildly deformed pelitic and psammitic rocks, marbles, and conglomerates that disconformably overlie a deformed megacrystic granite that has been correlated with the Esperanza Granitoid (Ortega-Gutiérrez, 1993), and retrograde units of the Xayacatlán Formation, part of the Piaxtla Group (Fig. 3). The Tecomate Formation consists of the following: (1) a basal conglomerate passing upward into interbedded sandstone and slate that grade laterally into conglomerate containing pebbles of granite, metarhyolite, quartzite, phyllite, and gneiss; (2) a volcanosedimentary unit consisting of greywacke, sandstone, conglomerate, slate, and phyllite; (3) interbedded graywacke (containing metagranite, quartzite, mica schist, and phyllite fragments), sandstone, and slate, with impure detrital limestones in the upper half of the unit, one of which contained a conodont with an age close to the Carboniferous–Permian boundary (Keppie et al., 2004c); and (4) capped by a marble horizon that yielded a conodont whose age range straddles the Early-Middle Permian boundary (Keppie et al., 2004c). The type locality of the blastomylonitic, megacrystic Esperanza Granitoid has previously yielded discordant U-Pb zircon data that plot on a chord between 440 ± 14 Ma and 1161 ± 30 Ma, interpreted as the age of intrusion and inheritance, respectively, and a 418 ± 18 Ma concordant U-Pb monazite age (Ortega-Gutiérrez et al., 1999; Fig. 2). The same locality was sampled for this study in an attempt to obtain concordant data. Furthermore, a leucogranite in the western part of the Acatlán Complex has yielded a concordant U-Pb zircon age of 478 ± 5 Ma (Campa-Uranga et al., 2002; Fig. 2). The Tecomate Formation is reported to lie unconformably upon previously deformed and metamorphosed rocks of the Piaxtla and Petlalcingo groups of inferred early Paleozoic age. The Piaxtla Group has yielded Sm-Nd garnet-whole rock ages of 388 ± 44 Ma (Yañez et al., 1991) and a concordant U-Pb zircon age of 346 ± 3 Ma (Keppie et al., 2004a) that are inferred to date a high-grade metamorphic event.

The Tecomate Formation has been deformed by two phases of folding: N-S dextral transpression that produced N-S transcurrent shear zones and S-vergent thrusting, followed by N-S, upright folds (Malone et al., 2002). The dextral shearing occurred synchronously with: (a) intrusion of the 289 ± 1 to 287 ± 2 Ma, earliest Permian Totoltepec pluton (concordant U-Pb zircon ages: Yañez et al., 1991; Malone et al., 2002; Keppie et al., 2004b), i.e., Wolfcampian, Early Permian using the time scale

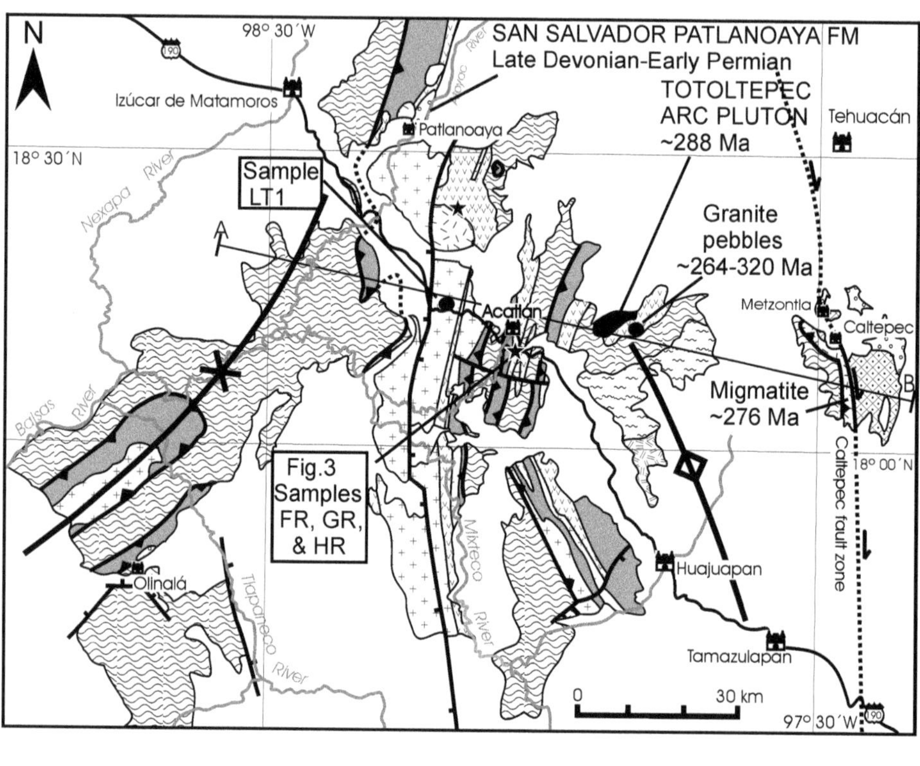

FIG. 2. U-Pb sample localities plotted on a geological map of the Acatlán Complex (modified after Ortega-Gutiérrez et al., 1999).

of Okulitch (2003), and (b) migmatization along the boundary between the Acatlán and Oaxacan complexes that has been dated at 276 ± 1 Ma (concordant U-Pb zircon age from neosome: Eliás-Herrera and Ortega-Gutiérrez, 2002), i.e., Leonardian, Early Permian using the Okulitch time scale (2003). The second set of folds preceded the intrusion of the San Miguel felsic dikes dated at 173 ± 1 Ma (Ruiz-Castellanos, 1979). Using both paleontological and geochronological data, the tectonothermal event affecting the Tecomate Formation appears to have lasted from ~290 Ma to ~240 Ma.

The paleogeographic setting for the Acatlán Complex is contradictory. Böhnel (1999) suggested that it was allochthonous, based on paleomagnetic data indicating that in the Early to Middle Jurassic the Acatlán Complex lay ~25° north of its present location adjacent to eastern Canada, and only reached its present location by the mid-Cretaceous. However, this contradicts: (1) Early Permian paleomagnetic data (Alda-Valdivia et al., 2002); and (2) faunal provinciality data (Vachard et al., 2000), both of which indicate a paleolatitude consistent with its present location in a Pangean reconstruction. The

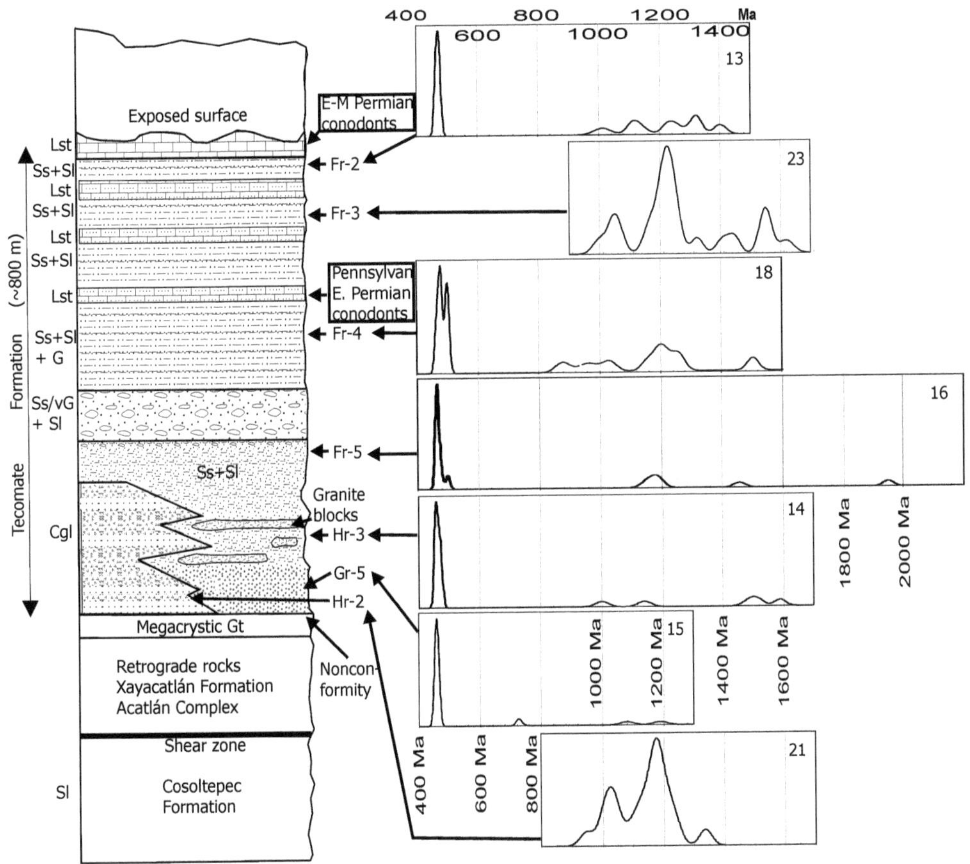

FIG. 3. Reconstructed stratigraphic/structural column of the type section of the Tecomate Formation showing sample locations and histograms of U-Pb zircon ages (number in right hand corner is the total number of zircons). Abbreviations: Cgl = conglomerate; G = graywacke; Lst = limestone; Sl = slate; Ss = sandstone; vG = volcaniclastic graywacke.

present study of detrital zircons adds data that can be applied to this controversy.

Analytical Techniques

Sample preparation

Samples from nine localities were processed for zircons at the Universidad Nacional Autónoma de México using conventional techniques and hand picking. Based on their morphology, the zircons were separated into populations. For example, zircons in sample FR4 (Frijolar, Tecomate Formation) are labeled as FR41-1 signifying sample FR4, population 1, zircon 1.

The zircons were mounted in 2.5 cm-diameter epoxy grain mounts and polished until the zircons were revealed. The zircons were imaged using a backscatter electron (BSE) detector on a CAMECA SX50 electron microprobe at Macquarie University, Australia. Because the BSE detector is a light-sensitive diode, images obtained are a function of the mean atomic number and the cathodoluminescence. The BSE/CL images of the zircons are used to select areas for analysis. All grain mounts, containing samples and standards, were cleaned in 1N nitric acid immediately prior to analysis.

Laser ablation (LA)–inductively coupled plasma (ICP)–mass spectrometry (MS)

U-Pb age determinations were made using LA-ICP-MS at Macquarie University. The laser used in this study was a LUV213 laser ablation system (M = 213 nm) made by New Wave Research/Merchantek. The laser was used with a small diameter (~6 cm)

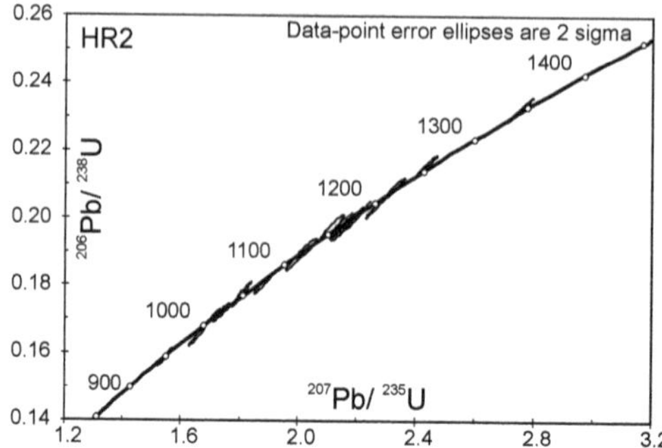

FIG. 4. U-Pb data from sample HR2 of the Tecomate Formation, Acatlán Complex, southern Mexico.

sample cell built at Macquarie. Details of the operating conditions and data acquisiton parameters are provided in the Appendix. The data were processed using GLITTER (van Achterberg et al., 2001). ISOPLOT (Ludwig, 1999) was used to produce the various U-Pb plots and age determinations. The analytical protocol used in this study is that described in detail by Jackson et al. (2004). A synopsis of the salient points of this protocol taken from Jackson et al. are given below.

To correct for fractionation of U and Pb during laser ablation and mass bias of the mass spectrometer, external standardization using a gem quality zircon (GJ-1) was employed. This is a simple and direct technique that involves spot ablations, laser focused above the sample, careful matching of ablation conditions for standard and sample (spot size, laser fluence, integration time) and uses He as the ablation gas, which increases the reproducibility of the U/Pb fractionation. A typical run consists of: 2 analyses of GJ-1; analyses of secondary standards 91500 and MT (Mud Tank); 2 more analyses of GJ-1; analyses of up to 10 unknowns, followed by 2 or more analyses of GJ-1. There is a minimum of 6 analyses of the standard GJ-1 in each run, and analyses of 91500 and MT are made each hour during the run. The mean, unfiltered ^{207}Pb/^{206}Pb age results for the secondary zircon standards 91500 (Wiedenbeck et al., 1995) and MT (Mud Tank) (Black and Gulson, 1978) obtained during this study are: 91500–1069 ± 13 [1.2%]; and MT — 733 ± 12 [1.6%] (95% confidence; see Appendix). The reported ^{207}Pb/^{206}Pb for 91500 is 1065 ± 0.3 and for MT is 732 ± 5.

Low background signals for Pb were achieved by acid cleaning of the sample cell and inserts, sample, and delivery tubes. In addition, the ICP-MS is dedicated to laser ablation analysis only and liquid Ar is used. Typical gas blanks have ^{208}Pb of about 30–40 cps and <10 cps for others elements of interest, except Hg, which has about 300 cps on 202 Hg. Since ^{204}Pb cannot be measured, there are three approaches that can be employed: selective integration of time-resolved signals; graphical corrections (Tera-Wasserburg plots); and mathematical corrections. Each of these is discussed in detail in Jackson et al. (2004); only the first two were used in this study.

Results

Detrital zircons

Sample HR2 (18°09'33", 98°03'0") is a recrystallized quartz-feldspar sandstone consisting of quartz, plagioclase, partially altered K-feldspar, muscovite, and chloritized biotite set in a matrix of chlorite, sericite, and quartz with accessory apatite, zircon, tourmaline, and epidote. This sample yielded 21 concordant zircons (Fig. 4) that range in age from 948 Ma to 1337 Ma (Table 1). The zircons are well rounded, range from equant to elongate, and some show internal zonation. As with the other zircons in this study, they are small. The ages have high-probability peaks at ~1025 Ma and 1175 Ma, with a small peak at ~1350 Ma (Fig. 3).

Sample GR5 (18°07'19", 98°03'05") is a medium-grained, recrystallized sandstone containing mica schist clasts, quartz, K-feldspar, plagioclase, partially alterated mica, and accessory tourmaline, zircon, and apatite. In this sample, 20 zircons were analyzed, with 15 analyses accepted. The zircons are generally small, elongate (<100 μ) and have well-developed internal zonation. One group of zircons (GR5-4) is small, equant grains

TABLE 1. LA-ICCP-MS U-Pb Data and Ages[1]

Analysis	Isotopic ratios $^{207}Pb/^{206}Pb$	±2s	$^{207}Pb/^{235}U$	±2s	$^{206}Pb/^{238}U$	±2s	RHO	Age (Ma) $^{207}Pb/^{206}Pb$	±2 s	$^{207}Pb/^{235}U$	±2 s	$^{206}Pb/^{238}U$	±2 s
FR2 (13/15)													
FR21-2	0.0561	0.0014	0.577	0.017	0.0747	0.0020	0.923	454	53	463	11	**464**	12
FR21-6	0.0561	0.0014	0.592	0.018	0.0765	0.0021	0.892	457	56	472	11	**475**	12
FR22-2	0.0567	0.0013	0.586	0.017	0.0750	0.0020	0.936	479	52	469	11	**466**	12
FR21-5	0.0572	0.0017	0.606	0.021	0.0769	0.0021	0.814	497	66	481	13	**477**	13
FR21-4	0.0593	0.0016	0.624	0.020	0.0764	0.0021	0.868	577	58	493	12	**475**	12
FR21-1	0.0657	0.0020	0.834	0.030	0.0921	0.0027		~~797~~	~~64~~	~~616~~	~~16~~	~~568~~	~~16~~
FR22-1	0.0729	0.0021	1.550	0.050	0.1541	0.0043	0.859	**1012**	57	950	20	924	24
FR21-7	0.0766	0.0016	1.742	0.047	0.1650	0.0045	0.990	**1110**	41	1024	18	985	25
FR21-9	0.0776	0.0021	1.638	0.052	0.1532	0.0043	0.892	**1135**	52	985	20	919	24
FR22-7	0.0784	0.0021	1.380	0.043	0.1277	0.0035		~~1156~~	~~53~~	~~880~~	~~18~~	~~775~~	~~20~~
FR22-5	0.0812	0.0017	2.287	0.062	0.2044	0.0055	0.990	**1226**	41	1208	19	1199	30
FR22-6	0.0825	0.0020	2.305	0.068	0.2026	0.0056	0.934	**1257**	47	1214	21	1189	30
FR21-3	0.0851	0.0018	2.689	0.074	0.2292	0.0062	0.993	**1317**	40	1326	20	1331	33
FR22-4	0.0854	0.0017	2.683	0.070	0.2280	0.0061	0.990	**1324**	38	1324	19	1324	32
FR22-9	0.0889	0.0019	2.979	0.079	0.2432	0.0064	0.994	**1401**	40	1402	20	1403	33
FR3 (23/27)													
FR33-3	0.0724	0.0016	1.560	0.042	0.1562	0.0040	0.970	**998**	43	955	17	936	23
FR33-2	0.0741	0.0015	1.703	0.043	0.1667	0.0042	0.998	**1044**	41	1010	16	994	23
FR32-6	0.0745	0.0017	1.708	0.048	0.1664	0.0044	0.956	**1054**	46	1012	18	992	25
FR32-7	0.0746	0.0017	1.766	0.050	0.1717	0.0046	0.948	**1058**	46	1033	18	1021	25
FR31-4	0.0785	0.0018	2.061	0.058	0.1904	0.0051	0.963	**1160**	44	1136	19	1124	28
FR33-1	0.0789	0.0016	2.121	0.058	0.1951	0.0053	0.994	**1169**	41	1156	19	1149	29
FR31-3	0.0802	0.0016	2.227	0.059	0.2015	0.0054	0.990	**1201**	40	1190	19	1183	29
FR33-5	0.0802	0.0017	1.605	0.043	0.1451	0.0038		~~1203~~	~~41~~	~~972~~	~~17~~	~~874~~	~~21~~
FR31-5	0.0803	0.0019	2.164	0.062	0.1954	0.0053	0.945	**1205**	46	1170	20	1151	29
FR34-1	0.0806	0.0018	2.118	0.058	0.1905	0.0050	0.943	**1213**	44	1155	19	1124	27
FR34-2	0.0809	0.0018	2.201	0.060	0.1973	0.0051	0.942	**1220**	44	1181	19	1161	27
FR34-3	0.0810	0.0017	2.255	0.059	0.2019	0.0052	0.989	**1221**	41	1198	18	1186	28
FR32-1	0.0813	0.0019	2.215	0.064	0.1978	0.0054	0.941	**1228**	46	1186	20	1163	29
FR32-5	0.0813	0.0018	2.297	0.063	0.2048	0.0054	0.966	**1230**	43	1211	19	1201	29
FR35-2	0.0820	0.0019	2.396	0.070	0.2120	0.0059	0.959	**1245**	45	1241	21	1239	31
FR32-4	0.0820	0.0023	2.365	0.076	0.2092	0.0059	0.882	**1246**	53	1232	23	1224	32
FR32-3	0.0822	0.0017	2.303	0.061	0.2033	0.0054	0.990	**1249**	41	1213	19	1193	29
FR35-4	0.0833	0.0017	1.676	0.044	0.1459	0.0039		~~1277~~	~~39~~	~~1000~~	~~17~~	~~878~~	~~22~~
FR31-1	0.0854	0.0017	2.685	0.070	0.2281	0.0060	0.990	**1324**	39	1324	19	1325	32
FR33-4	0.0892	0.0020	2.411	0.068	0.1961	0.0054	0.973	**1409**	42	1246	20	1154	29
FR35-5	0.0910	0.0018	3.207	0.086	0.2555	0.0070	0.990	**1447**	38	1459	21	1467	36
FR32-2	0.0960	0.0020	3.428	0.091	0.2590	0.0068	0.997	**1548**	39	1511	21	1485	35
FR31-2	0.0960	0.0020	3.555	0.094	0.2686	0.0071	0.990	**1548**	39	1540	21	1534	36
FR35-3	0.0964	0.0022	3.146	0.093	0.2369	0.0068	0.974	**1555**	42	1444	23	1371	35
FR35-6	0.0967	0.0022	2.709	0.080	0.2033	0.0059		~~1562~~	~~42~~	~~1331~~	~~22~~	~~1193~~	~~32~~
FR35-1	0.0975	0.0020	2.431	0.065	0.1808	0.0049		~~1577~~	~~37~~	~~1252~~	~~19~~	~~1071~~	~~27~~
FR34-4	0.0998	0.0027	3.306	0.100	0.2403	0.0062	0.856	**1620**	49	1483	23	1389	32

Analysis	Isotopic ratios							Age (Ma)					
	$^{207}Pb/^{206}Pb$	±2s	$^{207}Pb/^{235}U$	±2s	$^{206}Pb/^{238}U$	±2s	RHO	$^{207}Pb/^{206}Pb$	±2 s	$^{207}Pb/^{235}U$	±2 s	$^{206}Pb/^{238}U$	±2 s
FR4 (16/21)													
FR42-8	0.0566	0.0014	0.633	0.019	0.0811	0.0023		~~475~~	~~54~~	~~498~~	~~12~~	~~503~~	~~14~~
FR42-5	0.0568	0.0014	0.603	0.018	0.0770	0.0021	0.917	483	56	479	12	**478**	13
FR42-3	0.0575	0.0015	0.605	0.019	0.0764	0.0022	0.896	509	57	481	12	**475**	13
FR42-6	0.0580	0.0013	0.617	0.017	0.0772	0.0021	0.977	528	48	488	11	**480**	12
FR42-10	0.0583	0.0015	0.651	0.020	0.0810	0.0023	0.896	540	58	509	12	**502**	13
FR42-2	0.0591	0.0014	0.608	0.018	0.0747	0.0021	0.932	570	52	483	11	**465**	12
FR43-11	0.0594	0.0014	0.655	0.019	0.0800	0.0022	0.955	583	50	512	12	**496**	13
FR42-7	0.0600	0.0014	0.672	0.019	0.0813	0.0022	0.944	604	51	522	12	**504**	13
FR43-5	0.0610	0.0014	0.621	0.018	0.0739	0.0020		~~639~~	~~49~~	~~491~~	~~11~~	~~460~~	~~12~~
FR42-4	0.0614	0.0016	0.649	0.020	0.0766	0.0021		~~654~~	~~55~~	~~508~~	~~12~~	~~476~~	~~13~~
FR43-8	0.0625	0.0030	0.647	0.032	0.0752	0.0023		~~690~~	~~100~~	~~507~~	~~19~~	~~467~~	~~14~~
FR43-7	0.0683	0.0015	1.288	0.036	0.1367	0.0037	0.964	**879**	46	840	16	826	21
FR43-10	0.0709	0.0020	1.229	0.041	0.1257	0.0036	0.855	**955**	59	814	19	763	20
FR43-6	0.0733	0.0017	1.365	0.039	0.1350	0.0037	0.961	**1022**	46	874	17	817	21
FR43-9	0.0782	0.0017	1.900	0.052	0.1764	0.0047	0.986	**1151**	43	1081	18	1047	26
FR43-3	0.0797	0.0017	2.190	0.061	0.1994	0.0054	0.987	**1189**	42	1178	19	1172	29
FR43-1	0.0799	0.0016	2.037	0.054	0.1851	0.0050	0.990	**1193**	40	1128	18	1095	27
FR42-9	0.0807	0.0024	0.863	0.029	0.0775	0.0022		**1214**	~~58~~	~~632~~	~~16~~	~~481~~	13
FR43-2	0.0813	0.0017	2.282	0.063	0.2035	0.0056	0.991	**1230**	42	1207	19	1194	30
FR43-4	0.0825	0.0017	2.349	0.062	0.2067	0.0056	0.990	**1257**	39	1227	19	1211	30
FR42-1	0.0934	0.0018	3.319	0.086	0.2578	0.0069	0.990	**1496**	37	1486	20	1478	35
FR5 (16/18)													
FR53-3	0.0555	0.0014	0.578	0.018	0.0756	0.0021		~~432~~	~~56~~	~~463~~	~~11~~	~~470~~	~~12~~
FR51-1	0.0557	0.0013	0.568	0.016	0.0740	0.0019	0.932	440	51	457	10	**460**	11
FR54-5	0.0558	0.0014	0.568	0.016	0.0739	0.0020	0.920	444	54	457	11	**460**	12
FR54-7	0.0561	0.0014	0.596	0.018	0.0770	0.0021	0.906	458	56	474	11	**478**	12
FR53-2	0.0562	0.0013	0.574	0.016	0.0742	0.0019	0.952	458	51	461	10	**462**	12
FR54-8	0.0563	0.0015	0.575	0.017	0.0741	0.0020	0.890	464	57	461	11	**461**	12
FR51-2	0.0565	0.0016	0.579	0.019	0.0743	0.0020	0.848	470	63	464	12	**462**	12
FR52-2	0.0577	0.0013	0.595	0.016	0.0748	0.0020	0.971	518	48	474	10	**465**	12
FR54-4	0.0578	0.0013	0.600	0.017	0.0754	0.0020	0.968	522	50	477	11	**468**	12
FR54-6	0.0591	0.0015	0.610	0.018	0.0750	0.0021	0.917	569	55	484	12	**466**	12
FR51-5R	0.0609	0.0016	0.676	0.020	0.0805	0.0020	0.868	637	57	524	12	**499**	12
FR51-5C	0.0724	0.0019	1.207	0.037	0.1209	0.0034		~~997~~	~~53~~	~~804~~	~~17~~	~~736~~	~~19~~
FR54-3	0.0781	0.0019	1.861	0.054	0.1728	0.0048	0.948	**1150**	48	1067	19	1028	26
FR51-3	0.0785	0.0017	1.960	0.051	0.1810	0.0047	0.998	**1160**	42	1102	17	1073	26
FR52-4	0.0795	0.0016	2.105	0.054	0.1922	0.0051	0.990	**1184**	39	1150	18	1133	28
FR53-1	0.0797	0.0017	2.186	0.057	0.1990	0.0052	0.990	**1189**	42	1176	18	1170	28
FR51-4	0.0915	0.0018	3.051	0.076	0.2418	0.0062	0.990	**1457**	38	1420	19	1396	32
FR52-1	0.1197	0.0023	5.567	0.136	0.3375	0.0086	0.990	**1951**	34	1911	21	1874	42

Table continues

TABLE 1. *Continued*

Analysis	Isotopic ratios $^{207}Pb/^{206}Pb$	±2s	$^{207}Pb/^{235}U$	±2s	$^{206}Pb/^{238}U$	±2s	RHO	Age (Ma) $^{207}Pb/^{206}Pb$	±2 s	$^{207}Pb/^{235}U$	±2 s	$^{206}Pb/^{238}U$	±2 s
GR5 (15/20)													
GR51-3	0.0548	0.0014	0.558	0.016	0.0739	0.0019		~~405~~	~~55~~	~~450~~	~~10~~	~~459~~	~~12~~
GR53-5	0.0566	0.0013	0.576	0.016	0.0738	0.0019	0.953	476	52	462	10	**459**	**12**
GR52-4	0.0567	0.0013	0.584	0.016	0.0748	0.0019	0.952	477	52	467	10	**465**	**12**
GR51-4	0.0567	0.0014	0.575	0.016	0.0736	0.0019	0.931	478	53	461	10	**458**	**11**
GR54-1	0.0569	0.0015	0.577	0.017	0.0736	0.0019	0.892	488	57	463	11	**458**	**11**
GR54-4	0.0573	0.0016	0.592	0.019	0.0749	0.0020	0.849	503	61	472	12	**466**	**12**
GR51-2	0.0574	0.0014	0.582	0.016	0.0735	0.0019	0.916	508	53	466	11	**457**	**11**
GR53-2	0.0577	0.0014	0.586	0.017	0.0736	0.0020	0.927	519	54	468	11	**458**	**12**
GR51-5	0.0578	0.0014	0.583	0.017	0.0732	0.0019	0.925	522	54	467	11	**455**	**12**
GR52-5	0.0581	0.0016	0.584	0.018	0.0729	0.0019	0.862	534	62	467	12	**453**	**12**
GR53-1	0.0586	0.0017	0.600	0.020	0.0743	0.0021	0.844	552	63	477	13	**462**	**12**
GR53-4	0.0590	0.0013	0.588	0.016	0.0723	0.0019	0.967	566	49	469	10	**450**	**12**
GR53-3	0.0591	0.0013	0.603	0.016	0.0741	0.0019	0.996	569	47	479	10	**461**	**12**
GR52-2	0.0631	0.0018	0.664	0.021	0.0764	0.0020		~~710~~	~~59~~	~~517~~	~~13~~	~~474~~	~~12~~
GR54-3	0.0643	0.0024	1.058	0.042	0.1193	0.0034	0.718	752	79	733	20	**727**	**19**
GR54-5	0.0649	0.0041	0.819	0.050	0.0915	0.0031		~~771~~	~~129~~	~~607~~	~~28~~	~~564~~	~~18~~
GR54-2	0.0676	0.0016	0.959	0.027	0.1028	0.0027		~~857~~	~~50~~	~~683~~	~~14~~	~~631~~	~~16~~
GR52-1	0.0754	0.0016	1.420	0.037	0.1366	0.0036	0.990	**1080**	43	897	15	825	20
GR51-1	0.0797	0.0018	1.749	0.046	0.1592	0.0041	0.981	**1189**	43	1027	17	953	23
GR52-3	0.0843	0.0017	0.369	0.009	0.0318	0.0008		~~1299~~	~~39~~	~~319~~	~~7~~	~~202~~	~~5~~
HR2 (21/21)													
HR21A-1	0.0707	0.0016	1.544	0.043	0.1585	0.0043	0.977	**948**	45	948	17	949	24
HR22A-1R	0.0727	0.0015	1.706	0.045	0.1703	0.0045	0.990	**1005**	41	1011	17	1014	25
HR21A-5	0.0729	0.0016	1.653	0.046	0.1646	0.0044	0.971	**1011**	44	991	17	982	24
HR21A-3	0.0733	0.0017	1.737	0.049	0.1718	0.0045	0.936	**1023**	48	1022	18	1022	25
HR21A-7	0.0738	0.0015	1.814	0.047	0.1784	0.0047	0.990	**1035**	39	1051	17	1058	26
HR21A-2	0.0741	0.0015	1.799	0.047	0.1761	0.0047	0.990	**1044**	41	1045	17	1046	26
HR22A-1C	0.0757	0.0016	1.879	0.051	0.1802	0.0048	0.978	**1086**	43	1074	18	1068	26
HR22A-2	0.0770	0.0019	2.028	0.059	0.1910	0.0051	0.929	**1122**	48	1125	20	1127	28
HR22A-4	0.0771	0.0017	1.992	0.055	0.1875	0.0050	0.968	**1123**	44	1113	19	1108	27
HR22A-5	0.0777	0.0018	2.117	0.062	0.1977	0.0055	0.950	**1139**	47	1154	20	1163	30
HR22A-3	0.0786	0.0018	2.151	0.060	0.1986	0.0054	0.967	**1161**	44	1165	19	1168	29
HR21A-6	0.0788	0.0017	2.141	0.059	0.1970	0.0053	0.979	**1167**	43	1162	19	1159	29
HR21B-4	0.0792	0.0019	2.149	0.061	0.1968	0.0053	0.936	**1177**	47	1165	20	1158	28
HR21B-1	0.0792	0.0017	2.188	0.059	0.2003	0.0054	0.993	**1178**	41	1177	19	1177	29
HR21B-3	0.0793	0.0017	2.146	0.057	0.1963	0.0051	0.987	**1179**	41	1164	18	1156	28
HR22B-3	0.0793	0.0019	2.178	0.063	0.1992	0.0054	0.932	**1180**	47	1174	20	1171	29
HR21B-2	0.0794	0.0017	2.212	0.059	0.2019	0.0054	0.997	**1183**	41	1185	19	1186	29
HR22B-1	0.0807	0.0018	2.324	0.064	0.2088	0.0056	0.981	**1215**	42	1220	19	1222	30
HR21A-4	0.0807	0.0016	2.264	0.060	0.2034	0.0054	0.990	**1215**	39	1201	19	1194	29
HR22B-2	0.0819	0.0017	2.432	0.065	0.2155	0.0057	0.992	**1242**	41	1252	19	1258	30
HR22B-4	0.0859	0.0018	2.753	0.074	0.2324	0.0062	0.997	**1337**	40	1343	20	1347	32

Analysis	Isotopic ratios $^{207}Pb/^{206}Pb$	±2s	$^{207}Pb/^{235}U$	±2s	$^{206}Pb/^{238}U$	±2s	RHO	Age (Ma) $^{207}Pb/^{206}Pb$	±2 s	$^{207}Pb/^{235}U$	±2 s	$^{206}Pb/^{238}U$	±2 s
HR3 (14/19)													
HR32-1	0.0553	0.0015	0.564	0.018	0.0740	0.0021		~~422~~	~~61~~	~~454~~	~~12~~	~~461~~	~~12~~
HR31-3	0.0554	0.0013	0.566	0.017	0.0741	0.0020		~~426~~	~~52~~	~~455~~	~~11~~	~~461~~	~~12~~
HR33-3	0.0555	0.0013	0.577	0.017	0.0755	0.0021		~~431~~	~~51~~	~~463~~	~~11~~	~~469~~	~~12~~
HR31-2	0.0580	0.0016	0.582	0.018	0.0729	0.0020	0.872	528	59	466	12	**454**	12
HR31-1	0.0558	0.0013	0.563	0.016	0.0731	0.0020	0.928	445	52	453	11	**455**	12
HR32-2	0.0580	0.0013	0.587	0.016	0.0734	0.0020	0.983	531	48	469	10	**456**	12
HR31-4	0.0564	0.0012	0.570	0.016	0.0734	0.0020	0.978	466	48	458	10	**457**	12
HR32-5	0.0561	0.0012	0.572	0.015	0.0738	0.0020	0.990	458	46	459	10	**459**	12
HR33-7	0.0558	0.0012	0.573	0.016	0.0746	0.0020	0.974	443	47	460	10	**464**	12
HR33-5	0.0569	0.0013	0.594	0.017	0.0757	0.0021	0.966	485	49	473	11	**471**	12
HR33-6	0.0577	0.0013	0.603	0.017	0.0758	0.0020	0.969	519	49	479	11	**471**	12
HR33-8	0.0580	0.0013	0.607	0.017	0.0759	0.0021	0.976	528	49	481	11	**472**	12
HR32-4	0.0679	0.0016	0.850	0.024	0.0908	0.0025		~~866~~	~~48~~	~~625~~	~~13~~	~~560~~	~~15~~
HR32-6	0.0697	0.0015	0.958	0.026	0.0997	0.0027		~~921~~	~~43~~	~~683~~	~~14~~	~~613~~	~~16~~
HR33-2	0.0725	0.0016	1.501	0.041	0.1502	0.0040	0.970	**1000**	44	931	17	902	22
HR32-3	0.0777	0.0017	1.914	0.053	0.1787	0.0048	0.971	**1138**	44	1086	18	1060	26
HR33-1	0.0929	0.0020	3.006	0.080	0.2347	0.0062	0.994	**1486**	40	1409	20	1359	32
HR33-4	0.0943	0.0020	3.258	0.088	0.2506	0.0067	0.990	**1514**	39	1471	21	1441	34
HR33-9	0.0981	0.0020	3.725	0.099	0.2756	0.0075	0.990	**1587**	37	1577	21	1569	38
LT1 (4/5)													
LT1-1	0.0576	0.0016	0.605	0.018	0.0761	0.0020	0.879	516	59	480	12	**473**	**12**
LT1-2	0.0575	0.0014	0.601	0.017	0.0758	0.0020	0.916	511	54	478	11	**471**	**12**
LT1-3	0.0566	0.0013	0.592	0.017	0.0758	0.0020	0.941	476	52	472	11	**471**	**12**
LT1-4	0.0688	0.0019	0.711	0.022	0.0750	0.0020		~~893~~	~~55~~	~~545~~	~~13~~	~~466~~	~~12~~
LT1-5	0.0579	0.0014	0.604	0.017	0.0758	0.0020	0.927	524	54	480	11	**471**	**12**

[1]Bold Typeface = best age used in histograms/probability plots; strikethrough = rejected analysis. Numbers in parentheses after sample = accepted/total analyses.

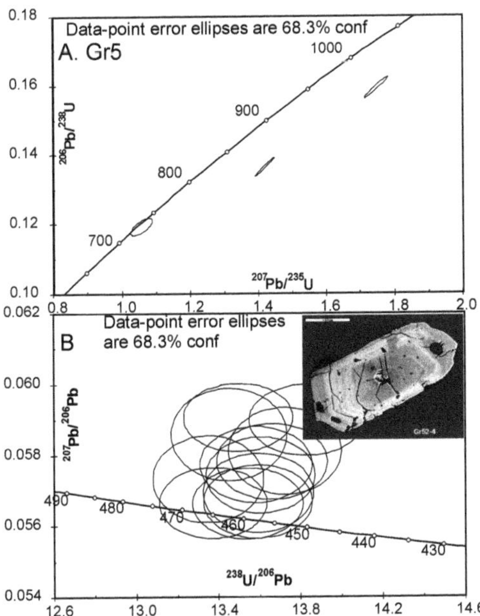

FIG. 5. U-Pb data from sample GR5 of the Tecomate Formation, Acatlán Complex, southern Mexico. White bar in inset (B) is ~50 μm; inset is Gt52-4.

(50–100 μ). The ages range from 459 Ma to 1189 Ma, with high-probability peaks at ~460 Ma and 730 Ma, and scattered analyses between 1000 and 1200 Ma (Fig. 3). Twelve zircons (Fig. 5B) yield a mean $^{206}Pb/^{238}U$ age of 458.4 ± 3.4 Ma (2σ). The single analysis at 729 Ma (GR54-3) is of a concordant zircon (Fig. 5A).

Sample HR3 (18°09'33", 98°02'51") is a medium-grained, recystallized sandstone consisting of grains of quartz, plagioclase, and sericitic K-feldspar set in a matrix of chlorite, quartz and epidote with accessory apatite, zircon, titanite, tourmaline, and amphibole. It contains clasts of metagranite, phyllite, and metavolcanic rocks. This sample yielded 14 ages that range between 454 and 1587 Ma (Table 1). Most of the zircons are equant, euhedral, zoned, and between 50 and 100 μ in size; some elongate zircons occur as well (Fig. 6B). The plot (Fig. 3) shows high-probability peaks at ~460 Ma, 1000 Ma, 1150 Ma, and between 1400 and 1600 Ma. Within the youngest zircons (Fig. 6B) there is evidence for two distinct populations, wherein 4 zircons give a weighted mean $^{206}Pb/^{238}U$ age of 456.1 ± 5.3 Ma (2σ), while 3 others give a mean $^{206}Pb/^{238}U$ age of 471 ± 7 Ma (2σ). A single concordant zircon lies at an intermediate age between these two groups, suggesting that the two populations result from a sampling bias.

Sample FR5 (18°07'20", 98°02'57") is a medium-grained, poorly sorted sandstone consisting of quartz, K-feldspar, plagioclase with accessory zircon, apatite, tourmaline, and rutile. It contains clasts of phyllite and metagranite. In this sample, 18 zircons were analyzed, with 16 analyses accepted. The zircons typically are equant (50–100 μ) with occasional elongate grains. Rounding/abrasion is not well developed. Generally the grains show well-developed internal zonation. Ages range from 460 Ma to 1951 Ma, with a high-probability peak at ~470 Ma, and minor peaks at ~1170, 1460, and 1950 Ma (Figs. 3 and 7A). Nine zircons (Fig. 7B) yielded a mean $^{206}Pb/^{238}U$ age of 464 ± 4 Ma (2σ).

Sample FR4 (18°08'20", 98°02'50") is a recrystallized lithic arenite with clasts of metagranite, mica schist, and phyllite, set in a matrix of quartz and K-feldspar with accessory apatite and zircon. This sample yielded 23 zircons, from which 18 analyses were obtained (Table 1). The zircons in this sample were generally elongate (100–200 μ long, 30–50 μ wide) and internally zoned. Rounding and

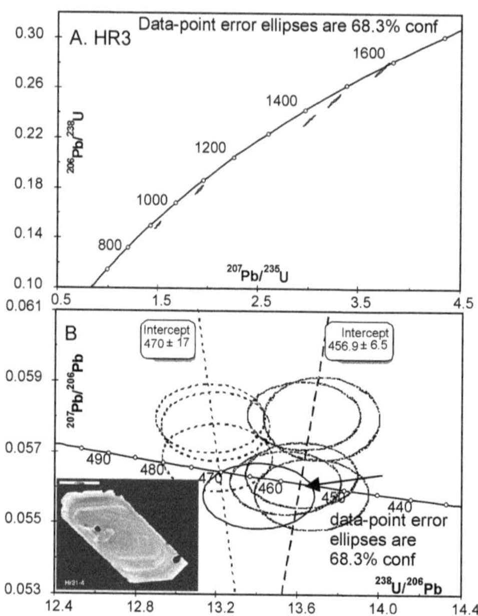

FIG. 6. U-Pb data from sample HR3 of the Tecomate Formation, Acatlán Complex, southern Mexico. White bar in inset (B) is ~50 μm; inset is Hr31-4.

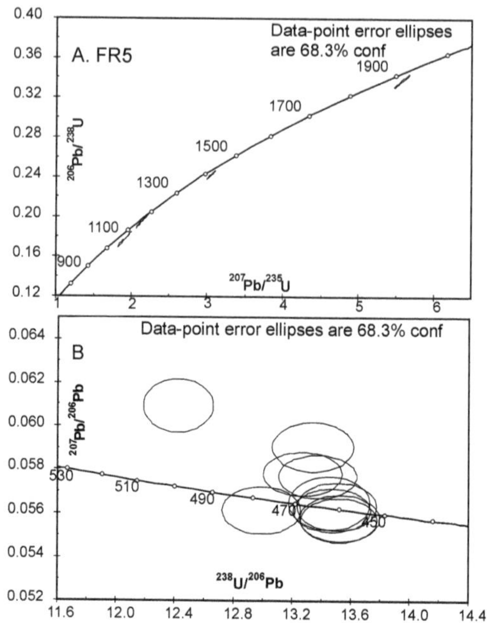

FIG. 7. U-Pb data from sample FR5 of the Tecomate Formation, Acatlán Complex, southern Mexico.

abrasion are not well developed. The ages obtained ranged from 476 Ma to 1496 Ma (Figs. 3 and 8A) with a high-probability peak at ~470–510 Ma, and a continuum of ages between 800 and 1300 Ma. The former zircons (Fig. 8B) can be divided into two groups: (1) 4 zircons with a mean $^{206}Pb/^{238}U$ age of 474 ± 6 Ma (2σ); and (2) 3 zircons with a mean $^{206}Pb/^{238}U$ age of 501 ± 8 Ma (2σ). Again this may reflect a sampling bias.

Sample FR3 (18°06'56", 98°02'53") is a recrystallized sandstone consisting quartz, plagiolcase, and K-feldspar set in a matrix of quartz and sericite, with accessory apatite, zircon, titanite, tourmaline, and epidote. Twenty-seven zircons were picked from this sample and of these 23 yielded reliable analyses (Table 1). The zircon grains are well rounded but range from equant (50–100 μ) to elongate (100–200 μ long; Figs. 3 and 9). Ages range from 998 Ma to 1620 Ma, with high-probability peaks at ~1000 Ma, 1220 Ma, and 1550 Ma.

Sample FR2 (18°06'57", 98°02'50") is a quartz-feldspar sandstone consisting of quartz, K-feldspar, and plagioclase with accessory apatite, zircon, and tourmaline. This sample yielded 15 zircons for which 13 analyses were accepted (Table 1). The zircons in this sample are of two morphological types:

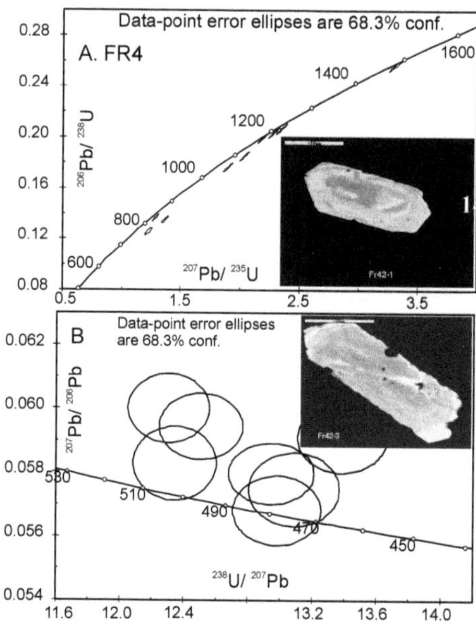

FIG. 8. U-Pb data from sample FR4 of the Tecomate Formation, Acatlán Complex, southern Mexico. White bar in inset (A) is ~50 μm; inset is Fr42-1. White bar in inset (B) is ~50 μm; inset is Fr42-3.

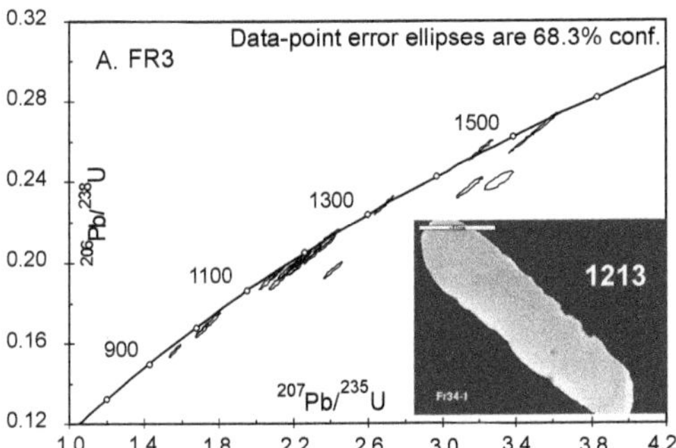

FIG. 9. U-Pb data from sample FR3 of the Tecomate Formation, Acatlán Complex, southern Mexico. White bar in inset is ~50 μm; inset is Fr34-1.

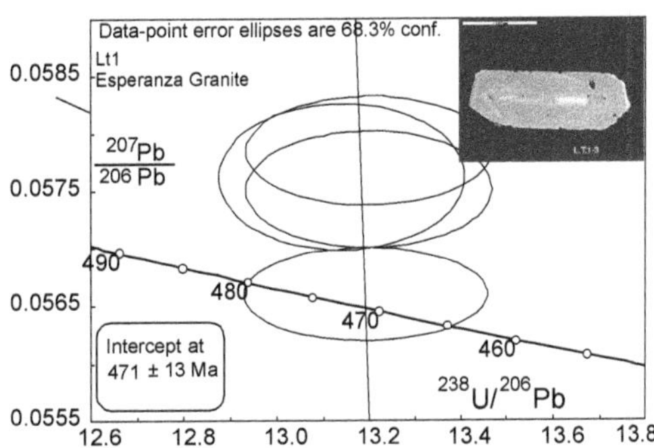

FIG. 11. U-Pb data from the type locality of the megacrystic Esperanza granitoid (sample LT1), Acatlán Complex, southern Mexico. White bar in inset is ~100 μm; inset is L.T.1-3.

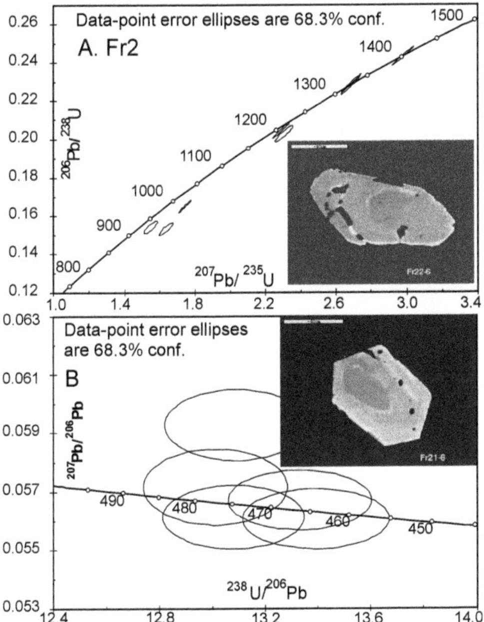

FIG. 10. U-Pb data from sample FR2 of the Tecomate Formation, Acatlán Complex, southern Mexico. White bar in inset (A) is ~50 μm; inset is Fr22-6. White bar in inset (B) is ~50 μm; inset is Fr21-6.

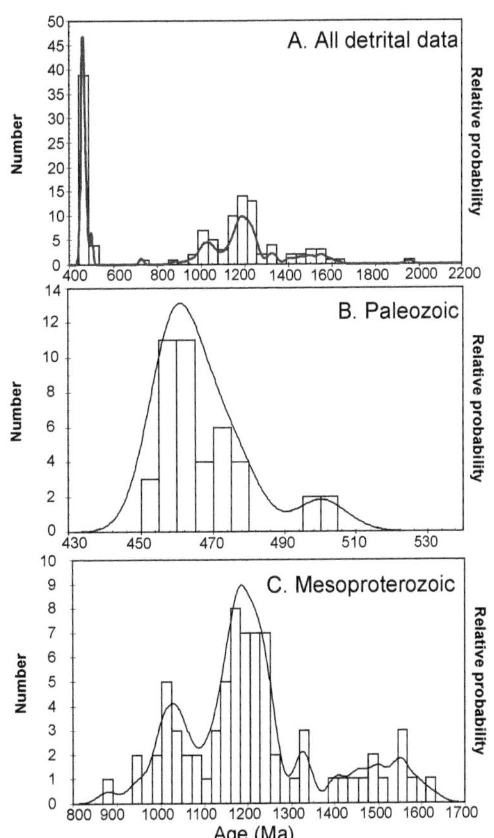

FIG. 12. Histograms of the detrital zircon populations in the Tecomate Formation. A. All the data. B. Paleozoic data. C. Mesoproterozoic data.

equant, 50–100 μ in size; and elongate grains 100–200 μ long and 30–50 μ across. The grains generally show internal zones, and some of the Proterozoic grains are well rounded. Ages obtained range from 464 Ma to 1401 Ma (Figs. 3 and 10), with a high-probability peak at 471 ± 5 Ma (2σ — 5 grains) and a broad spread of ages between 1000 Ma and 1400 Ma (Fig. 3).

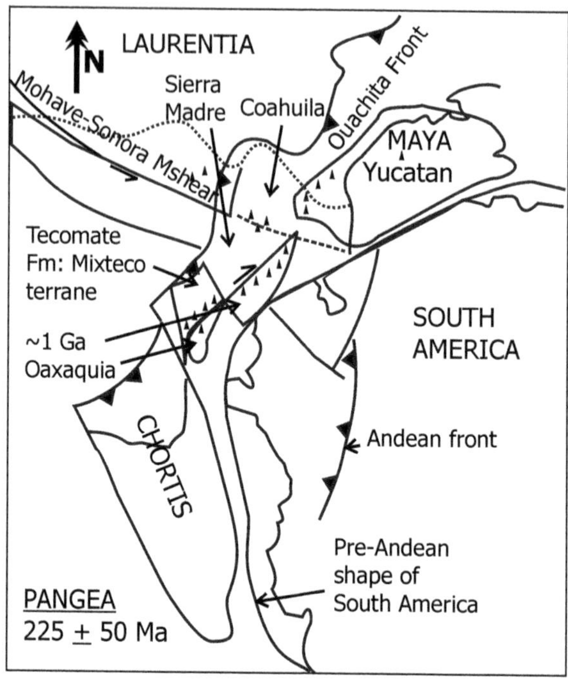

FIG. 13. Potential sources of detrital zircons of the Tecomate Formation in a Pangean reconstruction. Triangles represent location of Permian magmatic arc.

Granitoid zircons

Five, elongate internally zoned zircons were analyzed from the type locality of the Esperanza metagranite (LT1: 98°10'45", 18°13'33"); the unit is pervasively mylonized and metamorphosed at high pressures (Reyes-Salas, 2003). This sample consists of K-feldspar phenoclasts set in a matrix of quartz, K-feldspar, plagioclase, biotite, amphibole, phengite, garnet, epidote, tourmaline, and rutile, the latter minerals suggesting metamorphism at eclogite facies (Ortega-Gutiérrez et al., 1999). The one concordant zircon yielded an age of 471.3 ± 5.9 Ma (2σ) whereas the intercept age of the four nearly concordant points yields an age of 471 ± 13 Ma (Fig. 11).

Discussion

Taking into account all of the zircons in the Tecomate Formation (Fig. 2), 33% are Ordovician (450–480 Ma), 3% are Cambrian (496–504 Ma), 1.7% are Neoproterozoic (727 and 879 Ma), 60.6% are Mesoproterozoic (900–1600 Ma), and 1.7% are Paleoproterozoic (1620 and 1921 Ma). Approximately half of the Mesoproterozoic zircons occur in two samples (FR3 and HR2), which are both dominated by notably rounded and abraded zircons. No Ordovician, Cambrian, or Neoproterozoic zircons were recovered from these samples, probably an artifact of hand-picking. The range of ages in the Mesoproterozoic zircons (Fig. 12C) shows one major high-probability peak at 1100–1300 Ma, a small probability peak at ~1025 Ma, and a few grains between 1400 and 1600 Ma. U-Pb zircons of these ages have been recorded in: (1) granitoid rocks of the Acatlán Complex as upper intercepts of 1161 ± 30 Ma and 1163 ± 30 Ma (Ortega-Gutiérrez et al., 1999; Campa-Uranga et al., 2002); and (2) the adjacent Oaxacan Complex and Neoproterozoic intrusions where ages range from ~917 Ma to 1399 ± 58 Ma (Fig. 13; Keppie et al., 2003; Ortega-Obregón et al., 2003; Solari et al., 2003; Cameron et al., 2004). Thus, although granitoids in the Acatlán Complex may have been the source of some of the Mesoproterozoic zircons in the Tecomate Formation, the span of ages is more consistent with a provenance in the Oaxacan Complex. Further support for this interpretation comes from: (1) the abundant granulite facies grains, such as mesoperthite and rutile-bearing quartz, in the adjacent Patlanoaya Formation of equivalent age to the Tecomate Formation; and (2) pebbles of high-grade gneiss, some of which contain blue quartz in the Tecomate Formation. A similar age range also occurs in other ~1 Ga inliers in Mexico (Cameron et al., 2004; Figs. 1 and 13). The single older detrital zircon (~1950 Ma) is similar to those reported in a Carboniferous conglomerate from northern Mexico (Lopez et al., 2001) and suggests a cratonic source elsewhere either in Amazonia or Laurentia.

The Paleozoic detrital zircons in the Tectomate Formation (Fig. 12B) show a broad major probability peak at 460 Ma and a smaller peak at 500 Ma; however, this may be due to sampling bias. Individual samples (e.g., HR3) show apparently distinct Ordovician populations, but again this is attributable to sampling bias. A local source for these zircons is provided by: (1) the Esperanza Granitoids, which yielded a concordant 471 ± 6 Ma age (Fig. 11); (2) a leucogranite in the western part of the Acatlán Complex that has yielded a concordant U-Pb zircon age of 478 ± 5 Ma (Campa-Uranga et al., 2002); and (3) other correlative, but undated, megacrystic granites in the Acatlán Complex (Fig. 2). In contrast, a conglomerate in the Tecomate Formation in the eastern part of the Acatlán Complex contains a major population of ~290 Ma detrital zircons (Keppie et al., 2004c) derived from the nearby Totoltepec pluton, which forms one of a string of plutons defining a magmatic arc along the backbone of Mexico (Fig. 13). This suggests that the drainage pattern was a significant factor in controlling the sources of detrital zircons. Derivation of the detrital zircons in the Tecomate Formation from sources both in the Acatlán Complex and in the adjacent Oaxacan Complex suggests that, rather than being located adjacent to eastern Canada, the Acatlán Complex lay in its present location relative to the rest of Mexico, which is juxtaposed against southern Laurentia in Pangea reconstructions (Fig. 13).

Acknowledgments

We thank Drs. B. V. Miller and J. C. Bernal for their thorough and constructive reviews. We also acknowledge financial support from PAPIIT to FOG (IN107799), to JDK (IN103003), and NSERC to GAJ. This paper represents a contribution to IGCP projects 453 and 497.

REFERENCES

Alva-Valdivia, L. M., Goguitchaichvli, A., Grajales, M., Flores de Dios, A., Urrutia-Fucugauchi, J., Rosales C., and Morales, J., 2002, Further constraints for Permo-Carboniferous magnetostratigraphy: Case study of the sedimentary sequence from San Salvador-Patlanoaya (Mexico): Compte Rendu Geoscience, v. 334 (2002), p. 1–7.

Black, L. P., and Gulson, B. L., 1978, The age of the Mud Tank carbonatite, Strangways Range, Northern Territory: Bureau of Mineral Resources Journal of Australian Geology and Geophysics, v. 3, p. 227–232.

Böhnel, H., 1999, Paleomagnetic study of Jurassic and Cretaceous rocks from the Mixteca terrane (Mexico): Journal of South American Earth Science, v. 12, p. 545–556.

Campa-Uranga, M. F., Gehrels, G., and Torres de León, R., 2002, Nuevas edades de granitoides metamorfizados del Complejo Acatlán en el Estado de Guerrero: Actas Instituto Nacional de Geoquímica, v. 8, no. 1, p. 248.

Cameron, K. L., Lopez, R., Ortega-Gutiérrez, F., Solari, L., Keppie, J. D., and Schulze, C., 2004, U-Pb geochronology and common Pb isotope compositions of the Grenvillian rocks of eastern and southern Mexico, *in* Tollo, R. P., Corriveau, L., McLelland, J. B., and Bartholemew, G., eds., Proterozoic tectonic evolution of the Grenville Orogen in North America: Geological Society of America Memoir, in press.

Eliás-Herrera, M., and Ortega-Gutiérrez, F., 2002, Caltepec fault zone: An Early Permian dextral transpressional boundary between the Proterozoic Oaxacan and Paleozoic Acatlán complexes, southern Mexico, and regional implications: Tectonics, v. 21, no. 3 [10.1029/200TC001278].

Jackson, S. E., Pearson, N. J., Griffin, W. L., and Belousova, E. A., 2004, The application of laser ablation–inductively coupled plasma–mass spectrometry (LA-ICP-MS) to *in situ* U-Pb zircon geochronology: Chemical Geology, in press.

Keppie, J. D., Dostal, J., Cameron, K. L., Solari, L. A., Ortega-Gutiérrez, F., and Lopez, R., 2003, Geochronology and geochemistry of Grenvillian igneous suites in the northern Oaxacan Complex, southern México: Tectonic implications: Precambrian Research, v. 120, p. 365–389.

Keppie, J. D., Miller, B. V., Nance, R. D., Murphy, J. B., and Dostal, J., 2004a, New U-Pb zircon dates from the Acatlán Complex, Mexico: Implications for the ages of tectonostratigraphic units and orogenic events [abs.]: Geological Society of America Abstracts with Programs, in press.

Keppie, J. D., Nance, R. D., Dostal, J., Ortega-Rivera, A., Miller, B. V., Fox, D., Muise, J., Powell, J. T., Mumma, S. A., and Lee, J. W. K., 2004b, Mid-Jurassic tectonothermal event superposed on a Paleozoic geological record in the Acatlán Complex of southern Mexico:

Hotspot activity during the breakup of Pangea: Gondwana Research, v. 7, no. 1, p. 239–260.

Keppie, J. D., Sandberg, C. A., Miller, B. V., Sánchez-Zavala, J. L., Nance, R. D., and Poole, F. G., 2004c, Implications of latest Pennsylvanian to Middle Permian paleontological and U-Pb SHRIMP data from the Tecomate Formation to re-dating tectonothermal events in the Acatlán Complex, southern Mexico: International Geology Review, in press.

Lopez, R. L., Cameron, K. L., and Jones, N. W., 2001, Evidence for Paleoproterozoic, Grenvillian, and Pan-African age crust beneath northeastern Mexico: Precambrian Research, v. 107, p. 195–214.

Ludwig, K. R., 1999, Isoplot/Ex—a geochronological toolkit for Microsoft Excel: Berkeley, CA, Berkeley Geochronology Center, Special Publication no. 1a, 49 p.

Malone, J. W., Nance, R. D., Keppie, J. D., and Dostal, J., 2002, Deformational history of part of part of the Acatlán Complex: Late Ordovician–Early Silurian and Early Permian orogenesis in southern Mexico: Journal of South American Earth Science, v. 15, p. 511–524.

Okulitch, A. V., 2003, Geological time chart: Geological Survey of Canada, Open File 3040.

Ortega-Gutiérrez, F., 1993, Tectonostratigraphic analysis and significance of the Paleozoic Acatlán Complex of southern México, *in* Ortega-Gutiérrez, F., Centeno-García, E., Morán-Zenteno, D., and Gómez-Caballero, A., eds., Terrane geology of southern Mexico: Universidad Nacional Autónoma de México, Instituto de Geología, First Circum-Atlantic and Circum-Pacific Terrane Conference, Guanajuato, Mexico, Guidebook of Field Trip B, p. 54–60.

Ortega-Gutiérrez, F., Eliás-Herrera, M., Reyes-Salas, M., Macías-Romo, C., and López, R., 1999, Late Ordovician–Early Silurian continental collision orogeny in southern Mexico and its bearing on Gondwana-Laurentia connections: Geology, v. 27, p. 719–722.

Ortega-Obregón, C., Keppie, J. D., Solari, L. A., Ortega-Gutiérrez, F., Dostal, J., Lopez, R., Ortega-Rivera, A., and Lee, J. W. K., 2003, Geochronology and geochemistry of the ~917 Ma, calc-alkaline Etla granitoid pluton (Oaxaca, southern Mexico): Evidence of post-Grenvillian subduction along the northern margin of Amazonia: International Geology Review, v. 45, p. 596–610.

Reyes-Salas, A. M., 2003, Mineralogía y Petrología de los Granitoides Esperanza del Complejo Acatlán, Sur de México: Unpubl. Ph.D. thesis, Universidad Autónoma de Esado de Morelos, 165 p.

Ruiz-Castellanos, M., 1979, Rubidium-strontium geochronology of the Oaxaca and Acatlán metamorphic areas of Southern Mexico: Unpubl. Ph.D. thesis, University of Texas, Dallas, 188 p.

Solari, L. A., Keppie, J. D., Ortega-Gutiérrez, F., Cameron, K. L., Lopez, R., and Hames, W. E., 2003, 990 Ma and 1,100 Ma Grenvillian tectonothermal events in the northern Oaxacan Complex, southern Mexico: Roots of an orogen: Tectonophysics, v. 365, p. 257–282.

Vachard, D., Flores de Dios, A., Pantoja, J., Buitrón, B. E., Arellano, J., and Grajales, M., 2000, Les fusulines du Mexique, une revue biostratigraphique et paléogéographique: Geobios, v. 33, no. 6, p. 655–679.

Van Achterberg, E., Ryan, C. G., Jackson, S. E., and Griffin, W. L., 2001, Date reduction software for LA-ICP-MS: Mineralogical Association of Canada, Short Course Notes 29, p. 239–243.

Wiedenbeck, M., Alle, P., Corfu, F., Griffin, W. L., Meier, M., Oberli, F., von Quadt, A., Roddick, J. C., and Spiegel, W., 1995, Three natural zircon standards for U-Th-Pb, Lu-Hf, trace element, and REE analyses: Geostandard Newsletter, v. 19, p. 1–23.

Yañez, P., Ruiz, J., Patchett, P. J., Ortega-Gutiérrez, F., and Gehrels, G. E., 1991, Isotopic studies of the Acatlán Complex, southern Mexico: Implications for Paleozoic North American tectonics: Geological Society of America Bulletin, v. 103, p. 817–828.

Appendix

LA-ICP-MS Operating Conditions and Data Acquistion Parameters (modified from Jackson et al., 2004), and Figure of the Analytical Data from the Standards[2]

ICP-MS	
Model	Agilent 4500
Forward power	1350W
Gas flows:	
Plasma (Ar)	16L/min
Auxiliary (Ar)	1 L/min
Carrier (He)	0.9-1.2L/min
Make-up (Ar)	0.9-1.2L/min
Shield torch	Used for most analyses
Expansion chamber pressure	~350–360 Pa
Laser	
Model	LUV 213
Wavelength	213 nm
Repetition rate	5 and 10 Hz
Pulse duration (FWHM)	5 ns
Focusing objective	5X, f.l. = 40 mm
Incident pulse energy	ca. 0.1 mJ
Energy density on sample	ca. 8 J/cm^2
Data acquistion parameters	
Data acquisition protocol	Time resolved analysis
Scanning mode	Peak hopping, 1 point per peak
Detector mode	Pulse counting
Isotopes determined	^{206}Pb, ^{207}Pb, ^{208}Pb, ^{232}Th, ^{238}U
Dwell time per isotope	15, 30, 10, 10, 15 ms, respectively
Quadrupole settling time	ca. 2 ms
Time/scan	ca. 89 ms
Data acquistion	180 s (60 s gas blank, up to 120 s ablation
Samples and standards	
Mounts	25 mm diameter polished grain mounts
Standard	Gem zircon "GJ-1", 609 Ma

[2]A = 91500 zircon standard; B = Mud Tank zircon standard.

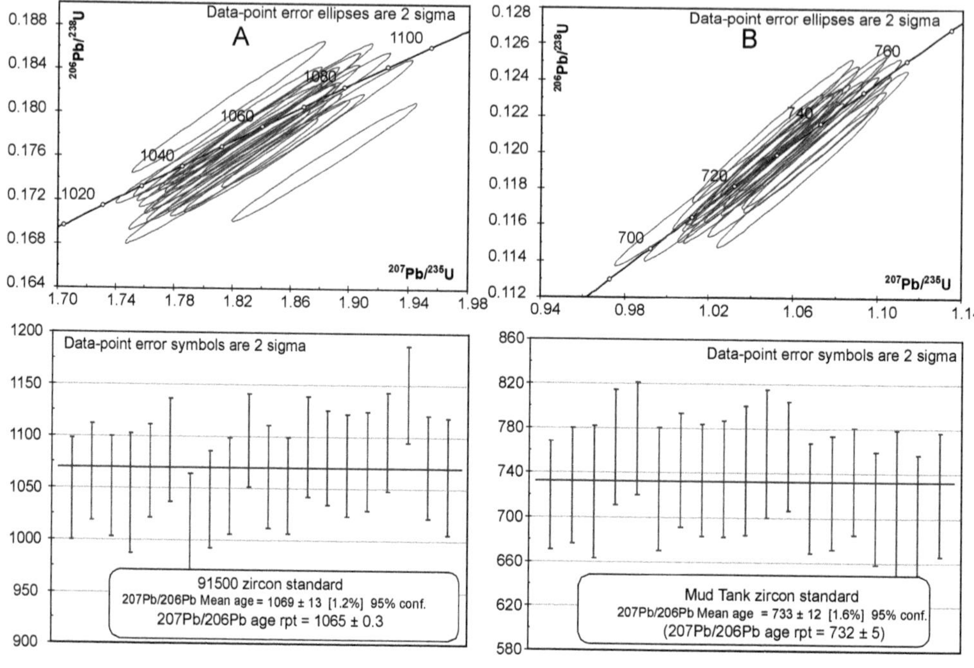

Data-point error ellipses are 2 sigma
A
$^{206}Pb/^{238}U$
$^{207}Pb/^{235}U$
1020
1040
1060
1080
1100
0.188
0.184
0.180
0.176
0.172
0.168
0.164
1.70
1.74
1.78
1.82
1.86
1.90
1.94
1.98
Data-point error ellipses are 2 sigma
B
700
720
740
760
0.128
0.126
0.124
0.122
0.120
0.118
0.116
0.114
0.112
0.94
0.98
1.02
1.06
1.10
1.14
Data-point error symbols are 2 sigma
1200
1150
1100
1050
1000
950
900
91500 zircon standard
207Pb/206Pb Mean age = 1069 ± 13 [1.2%] 95% conf.
207Pb/206Pb age rpt = 1065 ± 0.3
Data-point error symbols are 2 sigma
860
820
780
740
700
660
620
580
Mud Tank zircon standard
207Pb/206Pb Mean age = 733 ± 12 [1.6%] 95% conf.
(207Pb/206Pb age rpt = 732 ± 5)

Implications of Latest Pennsylvanian to Middle Permian Paleontological and U-Pb SHRIMP Data from the Tecomate Formation to Re-dating Tectonothermal Events in the Acatlán Complex, Southern Mexico

J. DUNCAN KEPPIE,[1]

Instituto de Geología, Universidad Nacional Autónoma de México, 04510 México D.F., México

CHARLES A. SANDBERG,

U.S. Geological Survey, Box 25046, MS 939, Federal Center, Denver, Colorado 80225

B. V. MILLER,

Department of Geological Sciences, CB #3315, Mitchell Hall, University of North Carolina at Chapel Hill, Chapel Hill, North Carolina 27599-3315

J. L. SÁNCHEZ-ZAVALA,

Instituto de Geología, Universidad Nacional Autónoma de México, 04510 México D.F., México

R. D. NANCE,

Department of Geological Sciences, Ohio University, Athens, Ohio 45701

AND FORREST G. POOLE

U.S. Geological Survey, Box 25046, MS 939, Federal Center, Denver, Colorado 80225

Abstract

Limestones in the highly deformed Tecomate Formation, uppermost unit of the Acatlán Complex, are latest Pennsylvanian–earliest Middle Permian in age rather than Devonian, the latter based on less diagnostic fossils. Conodont collections from two marble horizons now constrain its age to range from latest Pennsylvanian to latest Early Permian or early Middle Permian. The older collection contains *Gondolella* sp., *Neostreptognathodus* sp., and *Streptognathodus* sp., suggesting an oldest age limit close to the Pennsylvanian–Permian time boundary. The other collection contains *Sweetognathus subsymmetricus*, a short-lived species ranging only from Kungurian (latest Leonardian) to Wordian (earliest Guadelupian: 272 ± 4 to 264 ± 2 Ma). A fusilinid, *Parafusulina* c.f. *P. antimonioensis* Dunbar, in a third Tecomate marble horizon is probably Wordian (early Guadelupian, early Middle Permian). Furthermore, granite pebbles in a Tecomate conglomerate have yielded ~320–264 Ma U-Pb SHRIMP ages probably derived from the ~288 Ma, arc-related Totoltepec pluton. Collectively, these data suggest a correlation with two nearby units: (1) the Missourian–Leonardian carbonate horizons separated by a Wolfcampian(?) conglomerate in the upper part of the less deformed San Salvador Patlanoaya Formation; and (2) the clastic, Westphalian–Leonardian Matzitzi Formation. This requires that deformation in the Tecomate Formation be of Early–Middle Permian age rather than Devonian. These three formations are re-interpreted as periarc deposits with deformation related to oblique subduction. The revised dating of the Tecomate Formation is consistent with new data, which indicates that the unconformity between the Tecomate and the Piaxtla Group is mid-Carboniferous and corresponds to a tectonothermal event.

[1] Corresponding author; email: duncan@servidor.unam.mx

Introduction

THE TECOMATE FORMATION, the uppermost unit of the Acatlán Complex, consists of low-grade, strongly to mildly deformed pelitic and psammitic rocks, marbles, conglomerates, and volcanic rocks that are mainly mafic flows and tuffs with rare felsic units (Ortega-Gutiérrez, 1993). The Tecomate Formation is a pivotal unit for constraining the age of events that occurred in the Acatlán Complex of southern Mexico because it is reported to be bounded above and below by unconformities representing tectonothermal events. Thus, the Tecomate Formation is considered to be unconformably overlain by fossiliferous Late Devonian–Permian rocks of the San Salvador Patlaynoaya Formation (Ortega-Gutiérrez, 1978; Ortega-Gutiérrez et al., 1999; Sánchez-Zavala et al., 1999). On the other hand, the Tecomate Formation is reported to lie unconformably upon previously deformed and metamorphosed rocks of the Piaxtla and Petlalcingo groups of inferred early Paleozoic age: the former has yielded Sm-Nd garnet–whole-rock ages of 388 ± 44 Ma that are inferred to date a high-grade metamorphic event (Yañez et al., 1991).

Previous Interpretations

Previous studies reported cystoids, crinoids, blastoids, and micromolluscs in the Tecomate Formation, which were interpreted as pre-Carboniferous in age (Ortega-Gutiérrez, 1993). The Tecomate is also intruded by the La Noria pluton, which has yielded a U-Pb lower intercept age of 371 ± 34 Ma (Yañez et al., 1991), thereby also suggesting a younger age limit of Devonian to Mississippian for the Tecomate Formation. These data led to the identification of two orogenic events, the Late Ordovician–Early Silurian Acatecan and Devonian Mixteco orogenies (Ortega-Gutiérrez et al., 1999; Sánchez-Zavala et al., 1999). Correlation of these events with those in the Appalachians (Early Silurian Salinian and Devonian Acadian orogenies, respectively) led to the hypothesis that the Acatlán Complex was a fragment of the Cambro-Ordovician Iapetus Ocean crust (Ortega-Gutiérrez et al., 1999).

On the other hand, Keppie and Ramos (1999) proposed that the Acatlán Complex lay adjacent to Amazonia on the margin of the Rheic Ocean throughout the Paleozoic, only reaching its present location during the amalgamation of Pangea. Böhnel (1999) apparently confirmed the allochthonous nature of the Acatlán Complex by reporting paleomagnetic data indicating that in the Early to Middle Jurassic the Acatlán Complex lay ~25°N of its present location, which was reached by the mid-Cretaceous. However, this contradicts (1) Early Permian paleomagnetic data (Alva-Valdivia et al., 2002) and (2) faunal provinciality data (Vachard et al., 2000b) that both indicate a paleolatitude consistent with its present location in the Pangea reconstruction.

Unfortunately, the previously reported fossils from the Tecomate Formation could not be identified to genus or species, and the U-Pb data from the La Noria pluton are discordant. Furthermore, the reported unconformity between the top of the Tecomate Formation and the base of the fossiliferous Upper Devonian–Permian rocks has since been shown to be a fault (Vachard and Flores de Dios, 2002; Hernández-Espriú and Morales-Morales, 2002). In addition, the tectonothermal event affecting the Tecomate Formation was active during: (1) intrusion of the 289 ± 1 to 287 ± 2 Ma, earliest Permian Totoltepec pluton (concordant U-Pb zircon ages: Yañez et al., 1991; Malone et al., 2002; Keppie et al., 2004b)—i.e., Wolfcampian, Early Permian using the time scale of Okulitch (2003); and (2) migmatization along the boundary between the Acatlán and Oaxacan complexes that has been dated at 276 ± 1 Ma (concordant U-Pb zircon age from neosome: Elias-Herrera and Ortega-Gutiérrez, 2002)—i.e., Leonardian, Early Permian using the time scale of Okulitch (2003). Thus the recovery of conodonts and a fusulinid from the Tecomate Formation provides the first reliable age data in the Acatlán Complex and provides critical constraints on the tectonothermal events before and during its deposition.

Revised Dating

Conodonts

A marble sample (#1) collected in the Arroyo el Cualote (Figs. 1 and 2), near the village of Tecomate, about 5 km south of Acatlán de Osorio (Puebla State, Mexico: 18.8° N. Lat., 98.03° W. Long.) yielded fragments of *Gondolella* sp., *Neostreptognathodus* sp., and *Streptognathodus* sp. These conodont genera suggest an age close to the Carboniferous–Permian boundary, and most likely latest Pennsylvanian. This is consistent with the discovery of *Streptognathodus bellus* in the nearby San Salvador Patlanoaya Formation, a conodont that

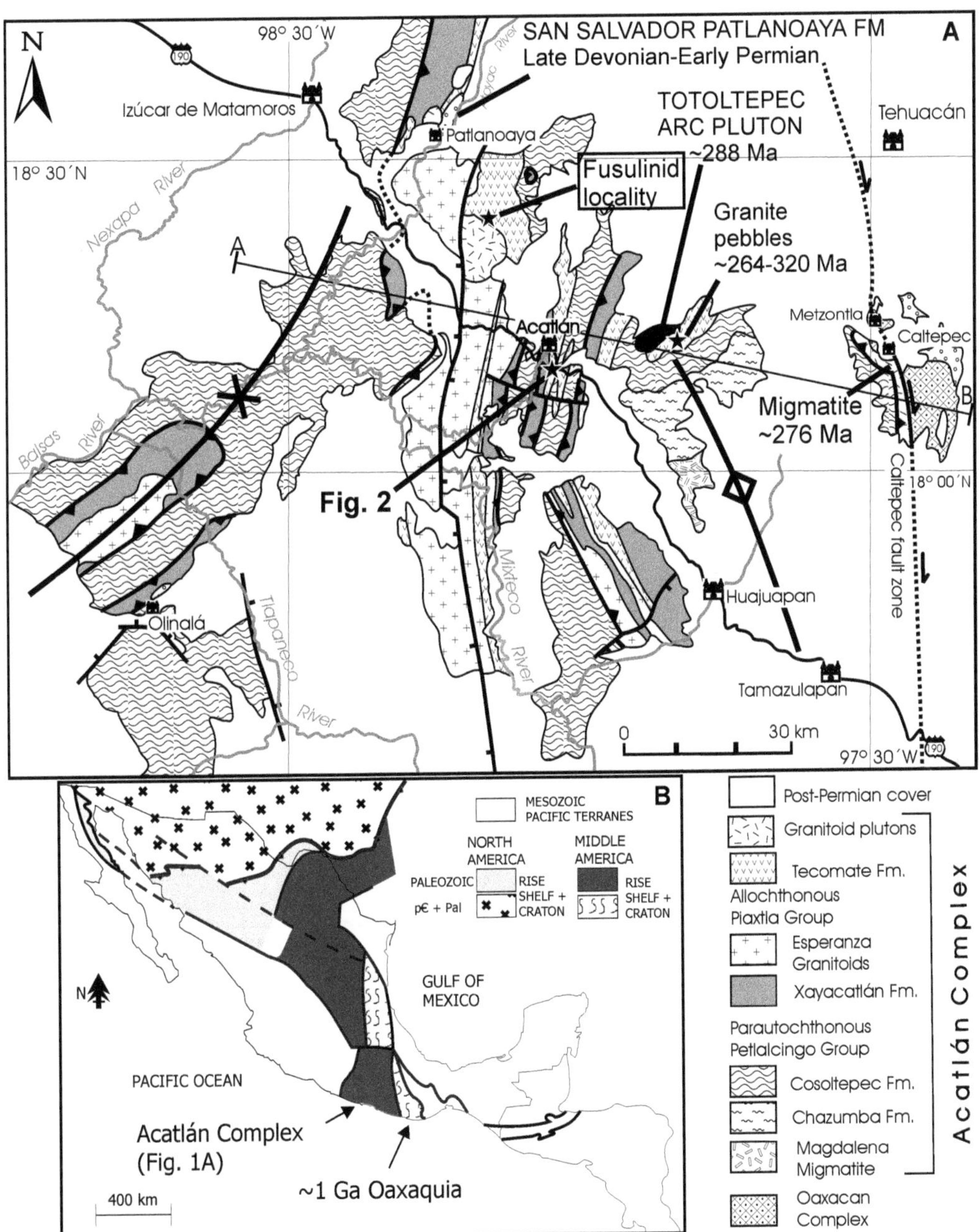

FIG. 1. A. Geological map of the Acatlán Complex showing fossil localities and U-Pb sample location reported in this paper (modified after Ortega-Gutiérrez et al., 1999). B. Index map showing location of Figure 1A (modified after Keppie et al., 2003).

is latest Virgilian, latest Pennsylvanian in age and has faunal affinities with the southern Urals (Cardiroit et al., 2002).

A second marble sample (#2) was collected by Ricardo Vega-Granillo from a different horizon in the Tecomate Formation about 2 km ENE of sample

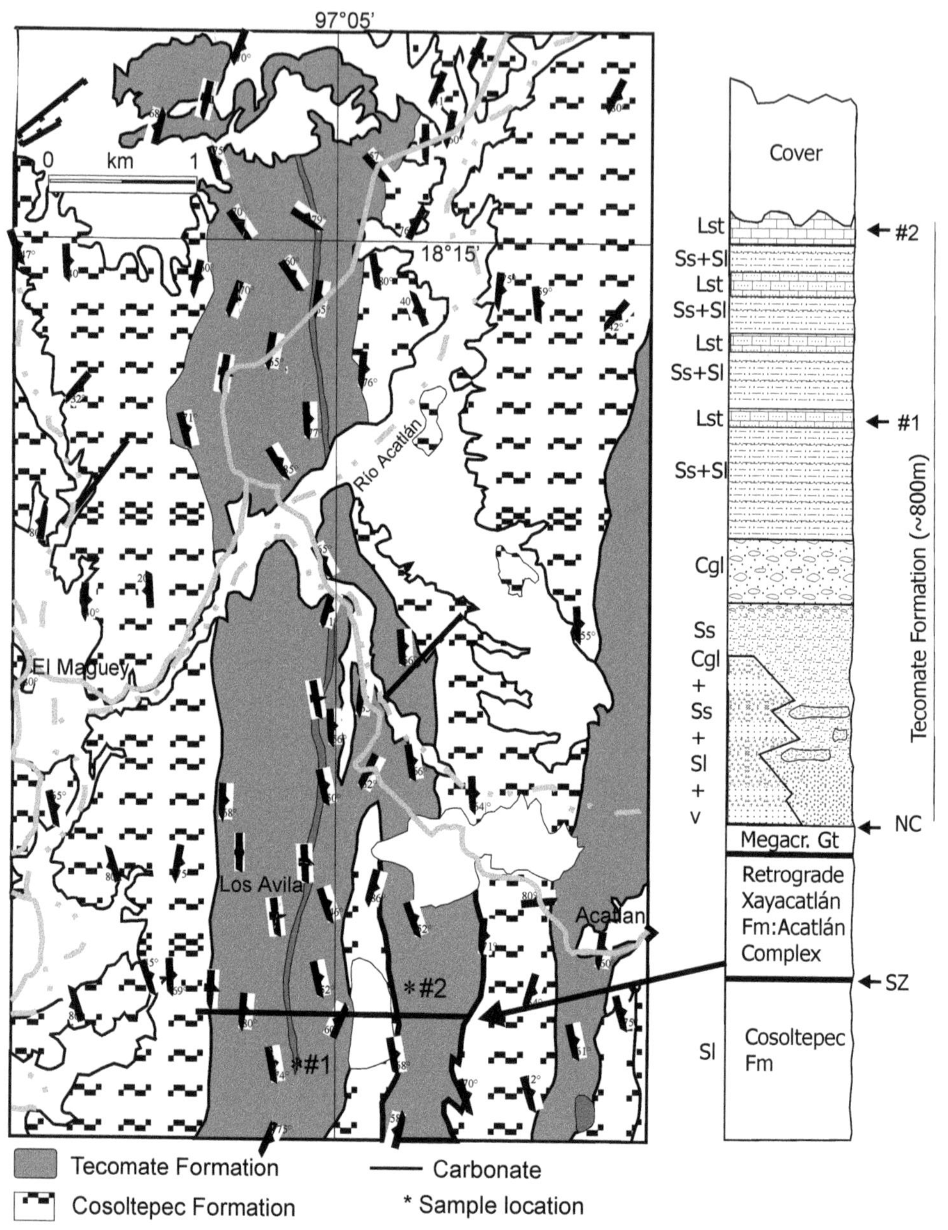

FIG. 2. Detailed geological map of the area near Acatlán, and reconstructed stratigraphic/structural section showing the fossil localities (#1 and #2) in the Tecomate Formation referred to in the text. Abbreviations: Cgl = congomerate; Lst = limestone; Megacr Gt = megacrystic granite; NC = nonconformity; Sl = slate; Ss = sandstone; SZ = shear zone; v = volcanic rocks.

#1 (Figs. 1 and 2; 18°08'30" N. Lat., 98°03' W. Long.). This second sample yielded a nearly complete juvenile specimen of *Sweetognathus subsymmetricus* Wang, Ritter, and Clark (Fig. 3). According to Mei et al. (2002), this species ranges in age only from upper Kungurian (latest Leonardian, Early Permian) to Wordian (early Guadelupian, Middle Permian). This conodont is a tropical species that

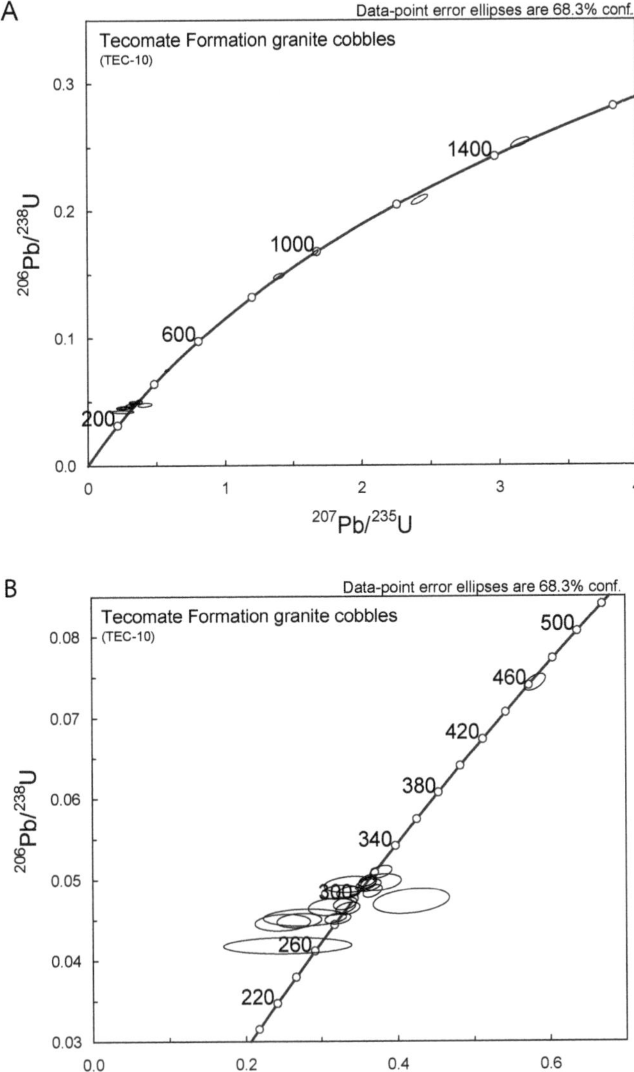

FIG. 3. Concordia diagrams of SHRIMP-RG U-Pb data for zircons separated from granitoid cobbles in the Tecomate Formation.

has been found previously in South China and possibly in Iran, but not in North America (Mei and Henderson, 2001; Mei et al., 2002).

Fusulinid

A third marble sample was collected near the village of Los Hornos de Zaragoza by Ricardo Vega-Granillo (18°24'N. Lat., 98°11'30" W. Long.) where it lies stratigraphically above a conglomerate that rests nonconformably upon the Los Hornos granitic pluton. This sample contained *Parafusulina* c.f. *antimonioensis* Dunbar, and the large size of the specimen strongly suggests a Wordian (early Guadelupian) age (Calvin H. Stevens, pers. comm. to Poole, 2002).

Geochronology

Analytical methods. Twenty-seven cobbles, each roughly 10 cm in diameter, were removed from a deformed conglomerate outcrop of the Tecomate

TABLE 1. SHRIMP-RG Data for Zircons Separated from Granite Cobbles in the Tecomate Formation[1]

Grain	U, ppm	Th, ppm	206Pb*, ppm	206Pb*/238U	207Pb*/235U	206Pb/238U, Ma
TEC10-1	267	97	11.1	0.0483(48)	0.3350(107)	304.3(3.0)
TEC10-2	236	76	10.3	0.0509(51)	0.3776(94)	319.9(3.2)
TEC10-3	159	104	6.31	0.0463 (51)	0.3340(107)	291.8(3.3)
TEC10-4	67	58	2.6	0.0446 (67)	0.2480(243)	281.2(4.0)
TEC10-5	81	39	3.19	0.0452(68)	0.2770(360)	284.9(4.2)
TEC10-6	414	96	26.4	0.0741(67)	0.5799(93)	460.9(4.0)
TEC10-7	203	145	7.88	0.0451 (45)	0.3210(112)	284.2(2.9)
TEC10-8	430	265	18.3	0.0495(46)	0.3618(69)	311.7(2.8)
TEC10-9	429	287	17.8	0.0485(53)	0.3666(84)	305.1(3.4)
TEC10-10	158	72	34.3	0.2521(247)	3.1580(442)	1450(13)
TEC10-11	84	47	3.36	0.0465(70)	0.3130(207)	293.0(4.4)
TEC10-12	146	26	26.1	0.2076(270)	2.4290(413)	1216(15)
TEC10-13	68	32	2.89	0.0495(69)	0.3760(184)	311.6(4.3)
TEC10-14	297	226	12.7	0.0498(48)	0.3599(76)	313.1(2.9)
TEC10-15	147	105	5.92	0.0469(52)	0.3300(102)	295.7(3.2)
TEC10-16	150	51	19.1	0.1481(148)	1.4020(252)	890.2(8.3)
TEC10-17	702	52	100	0.1661(146)	1.6690(167)	990.3(8.0)
TEC10-18	132	44	5.59	0.0492(54)	0.3610(112)	309.6(3.5)
TEC10-19	147	54	6.27	0.0493(69)	0.3370(222)	310.2(4.4)
TEC10-20	94	47	3.66	0.0448(58)	0.2700(194)	282.6(3.7)
TEC10-21	91	39	3.4	0.0417 (67)	0.2540(559)	263.6(4.2)
TEC10-22	139	55	5.79	0.0483(58)	0.3240(162)	304.3(3.4)
TEC10-23	59	26	2.35	0.0472(104)	0.4170(329)	297.1(6.5)

[1]Errors (terms in parentheses) are 1-sigma; Pb* indicates radiogenic portion. Error in standard calibration was 0.34% (not included in above errors but required when comparing data from different mounts). Common Pb is corrected using measured ^{204}Pb.

Formation at 18°11.977/97°48.988, halfway between Totoltepec de Guerrero and Chichihualtepec (Fig. 1, Table 1). The cobbles were crushed and zircons were separated using standard techniques (e.g., Ratajeski et al., 2001). It was impossible to remove completely the matrix from around some cobbles, and thus small adhering pieces of matrix material may have contributed some detrital zircon grains to the main population of zircon phenocrysts from the cobbles. Zircon grains were carefully hand-picked in an effort to select clear, well-faceted, acicular grains without inclusions or obvious inherited cores. These grains are commonly interpreted as igneous phenocrysts. A few rounded, frosted, purple-pink grains were also selected to get an idea of the age of grains most likely to be detrital. Selected zircon grains were mounted in epoxy, polished, imaged in cathodoluminescence and photographed under reflected light, then gold-coated prior to analysis.

Uranium-lead isotopic analyses were conducted on the SHRIMP-RG (sensitive high-resolution ion microprobe–reverse geometry) ion microprobe at Stanford University following techniques described

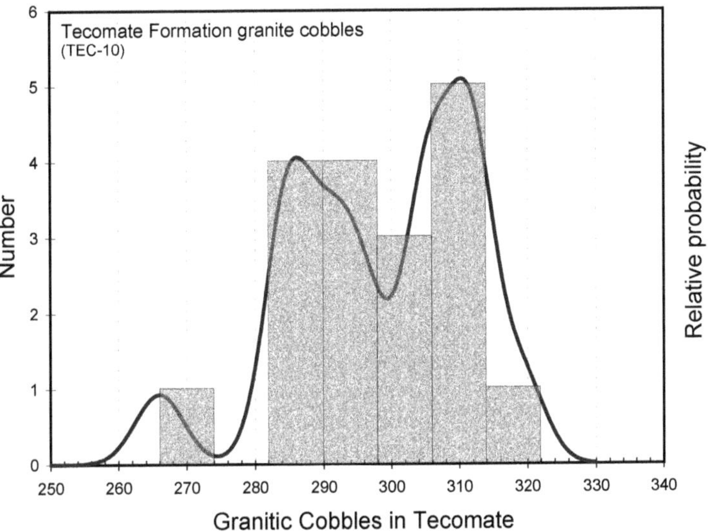

FIG. 4. Histogram of U-Pb data from the conglomerate horizon in the Tecomate Formation.

in DeGraaff-Surpless et al. (2002). Uranium and Thorium concentrations were measured relative to the CZ3 zircon standard and isotopic ratios and concentrations were measured relative to the R33 zircon standard from a quartz diorite of the Braintree complex, Vermont (J. Wooden, unpubl. data). Data were reduced with the SQUID v.1.02 program of Ludwig (2001) and plotted with Isoplot Ex, v. 3.00 (Ludwig, 2001). Errors in Table 1 are quoted at 1-sigma and ellipses on concordia diagrams are 68.3% confidence level. See footnote to Table 1 for additional analytical details.

Results. The ages of zircons separated from granite cobbles in the Tecomate Formation show a distinct cluster in $^{206}Pb/^{238}U$ ages between 320 and 264 Ma, and five other analyses are older, up to 1450 Ma (Table 1; Fig. 3). There is an apparent bimodality to the cluster of young ages (Fig. 4). However, this may be an artifact of the discordance, which is perhaps due to Pb loss. Both the $^{207}Pb/^{235}U$ ages and the $^{207}Pb/^{206}Pb$ ages are too imprecise to evaluate fully the significance of discordance or the nature of true concordance in these data. The cluster of analyses between 320 and 264 Ma, therefore, suggests that the youngest zircon grains, and thus the maximum age for depositional of the Tecomate Formation, are Late Carboniferous or Early Permian. These ages are similar to those from the ~289–287 ± 2 Ma age Totoltepec pluton (Yañez et al., 1991; Keppie et al., 2004b), and the granite pebbles and the Totoltepec pluton are petrographically identical.

Five other analyzed zircon grains yield ages of 461 Ma, and 1450 to 890 Ma (Table 1; Fig. 3). The group of four analyses from 1450 to 890 Ma consisted of rounded, frosted, pitted, and purple-pink grains that are suspected to be detrital, and were likely derived from the matrix material. The grain that yielded an age of 461 Ma was acicular and prismatic and may have been detrital, or may have been from a cobble of Esperanza granitoid.

Revised Interpretations

The syntectonic nature of the 288 ± 1 Ma earliest Permian Totoltepec pluton and the congruence of the structures within the pluton and the adjacent Tecomate Formation (Malone et al., 2002) closely defines the age of part of the younger tectonothermal event as Wolfcampian (Early Permian), rather than Devonian as was previously inferred. This negates evidence for the Devonian Mixtecan orogeny. The deformation involved south-vergent thrusting and dextral shear in N-S vertical shear zones that generally took place under greenschist-facies conditions (Malone et al., 2002). However, amphibolite-facies rocks and syntectonic migmatites dated at 276 ± 1 Ma (late Early Permian) have been recorded in the positive flower structure (Caltepec fault zone) that separates the Acatlán and ~1 Ga Oaxacan complexes (Elias-Herrera and Ortega-Gutiérrez, 2002). Although this tectonothermal event is synchronous with the terminal stages of Pangea amalgamation, it is also contemporaneous with the Permian magmatic arc that has been traced along the length of Mexico (Torres et al., 1999) of which the Totoltepec pluton forms a part (Keppie et al., 2004b). In a Pangean reconstruction (Keppie et al., 2003), the Acatlán Complex lies west of the suture zone between South and North America, and so the Early–Middle Permian deformation is more likely related to oblique subduction of a Pacific plate beneath the Acatlán Complex. In this context, the Tecomate, Matzitzi, and upper Patlanoaya formations may be interpreted as periarc deposits, with the alluvial fan and deltaic deposits of the Matzitzi Formation (Hernández-Láscares, 2000) passing laterally into shallow marine units in the Tecomate and San Salvador Patlanoaya formations.

The age of the tectonothermal event and the unconformity separating the Tecomate Formation and the Esperanza granitoid (part of the Piaxtla Group) in the Tecomate type section (Fig. 2) was previously assumed to be Late Ordovician–Silurian (Ortega-Gutiérrez et al., 1999). Recent U-Pb dating of the Esperanza granitoid in its type section has yielded a concordant age of 472 ± 6 Ma (Sanchez-Zavala et al., 2004), and the eclogite-facies metamorphism of the Piaxtla Group has been dated at 346 ± 3 Ma (U-Pb concordant analysis: Keppie et al., 2004a)—i.e., early Osagean on the time scale of Okulitch (2003). The Osagean–Late Pennsylvanian hiatus between the Tecomate Formation and the Piaxtla Group appears to correspond to that in the San Salvador Patlanoaya Formation (Osagean–Desmoinesian/Missourian: Vachard et al., 2000a). Thus this event is synchronous with the initial stages of Pangea amalgamation. This is consistent with the appearance of Laurentian fauna in the Mississippian–Pennsylvanian rocks of Oaxaquia (Stewart et al., 1999; Navarro-Santillán et al., 2002). The kinematics of this event in southern Mexico remains to be better documented (Malone et al., 2002).

Acknowledgments

We would like to acknowledge a Papiit grant IN103003 to JDK that facilitated the field work, and NSF grants EAR 0308105 to RDN and EAR 0308437 to BVM. This paper represents a contribution to IGCP Project 453: Modern and Ancient orogens. We thank Ricardo Vega-Granillo (University of Sonora, Hermosillo) for providing two marble samples; Calvin H. Stevens, San José State University, for identifying the fusulinid; and J. Utting for a constructive review.

REFERENCES

Alva-Valdivia, L. M., Goguitchaichvli, A., Grajales, M., Flores de Dios, A., Urrutia-Fucagauchi, J., Rosales, C., and Morales, J., 2002, Further constraints for Permo-Carboniferous magnetostratigraphy: Case study of the sedimentary sequence from San Salvador Patlanoaya (Mexico): Compte Rendu Geoscience, v. 334, p. 1–7.

Böhnel, H., 1999, Paleomagnetic study of Jurassic and Cretaceous rocks from the Mixteca terrane (Mexico): Journal of South American Earth Sciences, v. 12, p. 545–556.

Caridroit, M., Lamerandt, A., Dégardin, J-M., Flores de Dios, A., and Vachard, D., 2002, Discovery of radiolaria and conodonts in the Carboniferous–Permian of San Salvador Patlanoaya (Puebla, Mexico): biostratigraphic implications: Compte Rendu Palevol, v. 1, p. 205–211.

DeGraaff-Surpless, K., Graham, S. A., Wooden, J. L., and McWilliams, M. O., 2002, Detrital zircon provenance analysis of the Great Valley Group, California: Evolution of an arc-forearc system: Geological Society of America Bulletin, v. 114, p. 1564–1580.

Elias-Herrera, M., and Ortega-Gutiérrez, F., 2002, Caltepec fault zone: An Early Permian dextral transpressional boundary between the Proterozoic Oaxacan and Paleozoic Acatlán complexes, southern Mexico, and regional tectonic implications: Tectonics, v. 21, no. 3, p. 4-1– 4-19.

Hernández-Espriú, J. A., and Morales-Morales, F., 2002, Geologia metamorfica del complejo Acatlán y su cobertura Paleozoica, del area de San Miguel Las Minas-Patlanoaya-Ahuatlan, Estado de Puebla: Unpubl. B.Sc. thesis, Universidad Nacional Autónoma de México, 102 p.

Hernández-Láscares, D., 2000, Contribución al conocimiento de la estratigrafía de la Formación Matzitzi área: Los Reyes Metzontla-Santiago Coatepec, extremo suroriental del estado de Puebla. Unpubl. M.Sc. thesis, Instituto de Geologia, Universidad Nacional Autonoma de Mexico, 117 p.

Keppie, J. D., Miller, B. V., Nance, R. D., Murphy, J. B., and Dostal, J., 2004a, New U-Pb zircon dates from the Acatlán Complex, Mexico: Implications for the ages of tectonostratigraphic units and orogenic events [abs.]: Geological Society of America Abstracts with Programs, in press.

Keppie, J. D., Nance, R. D., Powell, J. T., Mumma, S. A., Dostal, J., Fox, D., Muise, J., Ortega-Rivera, A., Miller, B. V., and Lee, J. W. K., 2004b, Mid-Jurassic tectonothermal event superposed on a Paleozoic geological record in the Acatlán Complex of southern Mexico: Hotspot activity during the breakup of Pangea: Gondwana Research, v. 7, no. 1, p. 239–260.

Keppie, J. D., and Ramos, V. A., 1999, Odyssey of terranes in the Iapetus and Rheic oceans during the Paleozoic, *in* Ramos, V. A., and Keppie, J. D., eds., Laurentia-Gondwana connections before Pangea: Boulder, CO, Geological Society of America Special Paper 336, p. 267–276.

Keppie, J. D., Solari, L. A., Ortega-Gutiérrez, F., Elias-Herrera, M., and Nance, R. D., 2003, Paleozoic and Precambrian rocks of southern Mexico—Acatlán and Oaxacan complexes, *in* Geologic transects across Cordilleran Mexico, Guidebook for the field trips of the 99th Geological Society of America Cordilleran Section Annual Meeting, Puerto Vallarta, Mexico: Universidad Nacional Autonoma de Mexico, Insituto de Geologia, Publicación Especial 1, Field Trip 12, p. 281–314.

Ludwig, K. R., 2001, Isoplot/Ex, version 2.49, a geochronological toolkit for Microsoft Excel: Berkeley Geochronological Center, Special Publication 1a.

Malone, J. W., Nance, R. D., Keppie, J. D., and Dostal, J., 2002, Deformational history of part of the Acatlán Complex: Late Ordovician–Early Silurian and Early Permian orogenesis in southern Mexico: Journal of South American Earth Sciences, v. 15, p. 511–524.

Mei, Shilong, and Henderson, C. M., 2001, Evolution of Permian conodont provincialism and its significance in global correlation and paleoclimate implication: Palaeogeography, Palaeoclimatology, Palaeoecology, v. 170, p. 237–260.

Mei, Shilong, Henderson, C. M., and Wardlaw, B. R., 2002, Evolution and distribution of the conodonts *Sweetognathus* and *Iranognathus* and related genera during the Permian, and their implications for climate change: Palaeogeography, Palaeoclimatology, Palaeoecology, v. 180, p. 57–91.

Navarro-Santillán, D., Sour-Tovar, F., and Centeno-Garcia, E., 2002, Lower Mississippian (Osagean) brachiopods from the Santiago Formation, Oaxaca, Mexico: Stratigraphic and tectonic implications: Journal of South American Earth Sciences, v. 15, p. 327–336.

Okulitch, A. V., 2003, Geological time chart: Geological Survey of Canada, Open File 3040.

Ortega-Gutiérrez, F., 1978, Estratigrafía del Complejo Acatlán en la Mixteca Baja, Estados de Puebla y Oaxaca: Universidad Nacional Autónoma de México, Instituto de Geologia, Revista, v. 2, p. 112–131.

______, 1993, Tectonostratigraphic analysis and significance of the Paleozoic Acatlán Complex of southern México, *in* Ortega-Gutiérrez, F., Centeno-García, E., Morán-Zereno, D., and Gómez-Caballero, A., eds., Terrane geology of southern Mexico: Universidad Nacional Autónoma de México, Instituto de Geología, First Circum-Atlantic Terrane Conference, Guanajuato, Mexico, Guidebook of Field Trip B, p. 54–60.

Ortega-Gutiérrez, F., Elías-Herrera, M., Reyes-Salas, M., Macías-Romo, C., and López, R., 1999, Late Ordovician-Early Silurian continental collisional orogeny in southern Mexico and its bearing on Gondwana-Laurentia connections: Geology, v. 27, p. 719–722.

Ratajeski, K., Glazner, A. F., and Miller, B. V., 2001, Geology and geochemistry of mafic to felsic plutonic rocks in the Cretaceous intrusive suite of Yosemite Valley, California: Geological Society of America Bulletin, v. 113, p. 1486–1502.

Sánchez-Zavala, J. L., Centeno-García, E., and Ortega-Gutiérrez, F., 1999, Review of Paleozoic stratigraphy of Mexico and its role in Gondwana-Laurentia connections, *in* Ramos, V. A., and Keppie, J. D., eds., Laurentia-Gondwana connections before Pangea: Boulder, CO, Geological Society of America Special Paper 336, p. 211–226.

Sánchez-Zavala, J. L., Ortega-Gutiérrez, F., Keppie, J. D., Jenner, G., and Belousava, E., 2004, Cambro-Ordovician and Mesoproterozoic U-Pb zircon detrital ages from the Tecomate Formation, Acatlán Complex, southern Mexico: Local provenance in the Acatlán and Oaxacan complexes: International Geology Review, in press.

Stewart, J. H., Blodgett, R. B., Boucot, A. J., Carter, J. L., and López, R., 1999, Exotic Paleozoic strata of Gondwanan provenance near Ciudad Victoria, Tamaulipas, México, *in* Ramos, V. A., and Keppie, J. D., eds., Laurentia-Gondwana connections before Pangea: Boulder, CO, Geological Society of America Special Paper 336, p. 227–252.

Torres, R., Ruíz, J., Patchett, P. J., and Grajales-Nishimura, J. M., 1999, Permo-Triassic continental arc in eastern Mexico; tectonic implications for reconstructions of southern North America, *in* Bartolini, C., Wilson, J. L., and Lawton, T. F., eds., Mesozoic sedimentary and tectonic history of north-central Mexico: Boulder, CO, Geological Society of America Special Paper 340, p. 191–196.

Vachard, D., and Flores de Dios, A., 2002, Discovery of latest Devonian/earliest Mississippian microfossils in San Salvador Patlanoaya (Puebla, Mexico): Biogeographic and geodynamic consequences: Compte Rendu Geoscience, v. 334, p. 1095–1101.

Vachard, D., Flores de Dios, A., Buitrón, B.E., and Grajales, M., 2000a, Biostratigraphie par fusulines des calcaires Carbonifères et Permiens de San Salvador Patlanoaya (Puebla, Mexique): Geobios, v. 33, No. 1, pp. 5–33.

Vachard, D., Flores de Dios, A., Pantoja, J., Buitrón, B. E., Arellano, J., and Grajales, M., 2000b, Les fusulines du Mexique, une revue biostratigraphique et paléogéographique: Geobios, v. 33, no. 6, p. 655–679.

Yañez, P., Ruiz, J., Patchett, P. J., Ortega-Gutiérrez, F., Gehrels, G. E., 1991, Isotopic studies of the Acatlán Complex, southern Mexico: Implications for Paleozoic North American tectonics: Geological Society of America Bulletin, v. 103, p. 817–828.

Detrital Zircon Data from the Eastern Mixteca Terrane, Southern Mexico: Evidence for an Ordovician–Mississippian Continental Rise and a Permo-Triassic Clastic Wedge Adjacent to Oaxaquia

J. DUNCAN KEPPIE,[1]

Instituto de Geología, Universidad Nacional Autónoma de México, 04510 México D.F., México

R. D. NANCE,

Department of Geological Sciences, Ohio University, Athens, Ohio 45701

JAVIER FERNÁNDEZ-SUÁREZ,

Departamento de Petrología y Geoquímica, Universidad Complutense, 28040 Madrid, Spain

CRAIG D. STOREY,

Department of Earth Sciences, The Open University, Walton Hall, Milton Keynes, MK7 6AA, United Kingdom

TERESA E. JEFFRIES,

Natural History Museum, Cromwell Road, London, SW7 5BD, United Kingdom

AND J. BRENDAN MURPHY

Department of Earth Sciences, St. Francis Xavier University, Antigonish, Nova Scotia, Canada, B2G 2W5

Abstract

The eastern part of the Mixteca terrane of southern Mexico is underlain by the Petlalcingo Group (part of the Acatlán Complex), and has been interpreted as either a Lower Paleozoic passive margin, or a trench/forearc sequence deposited in either the Iapetus or Rheic oceans. The group, from bottom to top, consists of: (1) the Magdalena Migmatite protolith (metapsammites, metapelites, calsilicates, and marbles), which grades up into (2) the meta-psammitic Chazumba Formation; overthrust by (3) the Cosoltepec Formation (phyllites and quartzites with minor mafic meta-volcanic horizons). The group is unconformably overlain by the Pennsylvanian–Middle Permian Tecomate Formation, which is overthrust by the ~288 Ma Totoltepec pluton and unconformably overlain by Middle Jurassic rocks. In contrast to previous inferences that the protoliths of the units (1) to (3) were early Paleozoic in age, detrital zircon LA-ICPMS ages combined with published data constrain depositional ages as follows: (i) Magdalena Migmatite protolith: post-303 Ma–pre-171 Ma (Permian–Early Jurassic); (ii) Chazumba Formation: post-239 Ma–pre-174 Ma (Middle Triassic–Early Jurassic); and (iii) Cosoltepec Formation: post-455 Ma–pre-310 Ma (uppermost Ordovician–Mississippian). Given the different ages and depositional environments of the Cosoltepec Formation versus the Chazumba Formation and Magdalena protolith, we recommend redefining the Chazumba and Magdalena as lithodemes grouped in the Petlalcingo Suite and excluding the Cosoltepec Formation. Detrital zircons in all three units show a population peak at ~850–1200 Ma, suggesting derivation from the adjacent ~1 Ga Oaxacan Complex. A ~470–640 Ma peak is limited to the Cosoltepec Formation whose source may be found in ~470 Ma plutons in the Acatlan Complex, beneath the Yucatan Peninsula, and in the Brasiliano orogens of South America. The inferred turbiditic protolith of the Chazumba Formation and Magdalena protolith suggests that it represents a clastic wedge deposited in front of S-verging Permo-Triassic thrusts on the western margin of Pangea. The mainly oceanic affinity of the basalts in the Cosoltepec Formation suggests deposition of sedimentary protoliths in a continental rise fringing Oaxaquia. These data are more consistent with deposition of the Cosoltepec Formation in the Rheic Ocean than in the Iapetus Ocean.

[1]Corresponding author; email: duncan@servidor.unam.mx

Introduction

THE ACATLÁN COMPLEX forms the basement of the Mixteca terrane of southern Mexico (Fig. 1A). The complex is juxtaposed on its eastern side against the ~1 Ga Oaxacan Complex of the Oaxaquia terrane along a Permian dextral flower structure, where syntectonic migmatites have yielded an age of 276 ± 1 Ma (Elías-Herrera and Ortega-Gutiérrez, 2002). The ~1 Ga rocks of southern and central Mexico belong to the Middle American microcontinent, which was fringed by Paleozoic passive margin and oceanic sequences, one of which is the Acatlán Complex (Fig. 1B). Ortega-Gutiérrez et al. (1999) claimed that the Acatlán Complex consists of two major thrust sequences: (1) a lower Petlalcingo Group made up of the Magdalena Migmatite, Chazumba and Cosoltepec formations of inferred early Paleozoic age (Fig. 2), and (2) an upper Piaxtla Group. Both groups are unconformably overlain by the Pennsylvanian–Lower Permian Tecomate Formation: ~305–270 Ma (Keppie et al., 2004c) (Fig. 2), and the upper Fammenian–Middle Permian Patlanoaya Formation (~370–260 Ma: Upper Devonian) is reported to rest unconformably on the Cosoltepec Formation in the northern Mixteca terrane (Vachard and Flores de Dios, 2002). The Petlalcingo Group is interpreted to be either a passive margin sequence (Ramirez-Espinoza, 2001) or a trench and forearc deposit (Ortega-Gutiérrez et al., 1999), and the complex as a whole is inferred to be the vestige of either the Iapetus Ocean (Ortega-Gutiérrez et al., 1999) or the Gondwanan margin of the Rheic Ocean (Keppie and Ramos, 1999; Keppie, 2004).

Constraining the protolith ages of the sedimentary successions such as the Petlalcingo Group is a key to resolving this controversy. In the absence of fossils in the Petlalcingo Group, this paper presents detrital zircon ages from the Magdalena Migmatite paleosome, a Chazumba Formation metapsammite, and a semipelite of the Cosoltepec Formation, which provide an older limit on the time of deposition of the units. An upper limit is provided by previously published Jurassic ages for igneous and metamorphic events (Keppie et al., 2004b). These ages, together with geochemical data, provide the basis for deducing the depositional environment of the units, and permit evaluation of their paleogeography.

Geological Setting

Petlalcingo Group

The Petlalcingo Group consists of the Magdalena Migmatite, and the Chazumba and Cosoltepec formations (Figs. 1 and 2), the most complete section of which is exposed in an antiform in the eastern part of the Mixteca terrane. The Chazumba Formation consists of a thick, polydeformed sequence of metapsammites and metapelites that were metamorphosed under amphibolite-facies conditions during the Jurassic and contain several tectonic lenses of Jurassic, mafic-ultramafic rocks (Keppie et al., 2004b). The formation appears to grade structurally downwards into a similar lithological unit that additionally includes calcsilicate and marble lenses and bands. This unit was pervasively migmatized and repeatedly deformed during the Jurassic (~175–170 Ma; Keppie et al., 2004b) producing a mapable unit called the Magdalena Migmatite (Keppie et al., 2004b).

The Cosoltepec Formation structurally overlies the Chazumba Formation and comprises extensive phyllites and quartzites and minor mafic metavolcanic units (Figs. 1B and 1C). These rocks have been penetratively deformed three times, generally at greenschist facies; however, a mafic unit at the base of the formation was metamorphosed in the amphibolite facies. This latter unit has yielded $^{40}Ar/^{39}Ar$ plateau ages of 218 ± 11 Ma (hornblende) and 224 ± 2 Ma (muscovite; Keppie et al., 2004b). A minimum depositional age for the Cosoltepec Formation is provided by the fault-modified unconformity with the overlying Pennsylvanian–Middle Permian Tecomate Formation (Figs. 1B and 1C; Keppie et al., 2004c).

Piaxtla Group

The Petlalcingo Group is structurally juxtaposed against locally eclogitic mafic and ultramafic rocks, high-grade metasedimentary units, granitoid rocks, and migmatites of the Piaxtla Group that are inferred to have been thrust over the Petlalcingo Group (Ortega-Gutiérrez et al., 1999; Fig. 1). A mafic eclogite from the northern part of the Mixteca terrane has yielded a concordant U-Pb zircon age of 346 ± 3 Ma that is inferred to date the eclogite-facies metamorphism followed by migmatization that has yielded SHRIMP ages of ~350–330 Ma (Keppie et al., 2004a). The rocks in sheared contact with the Piaxtla Group are generally of lowgrade, suggesting that the Piaxtla Group had cooled before the two groups were

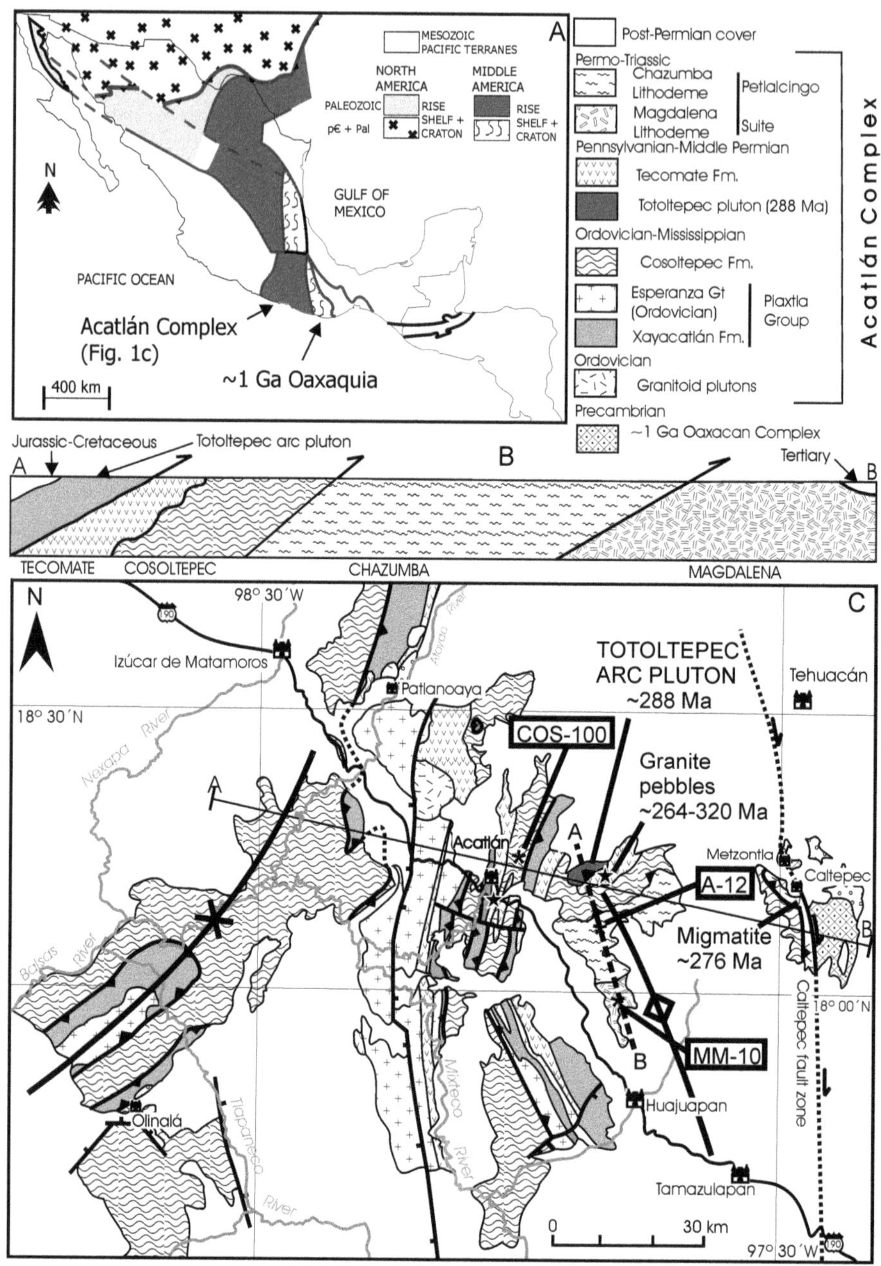

FIG. 1. A. Map showing the location of the Mixteca terrane (modified after Keppie et al., 2003b). B. Cross-section A–B located on map below. C. Geological map of the Mixteca terrane (modified after Ortega-Gutiérrez et al., 1999).

tectonically juxtaposed. The Piaxtla Group may comprise several thrust slices and is inferred to represent obducted oceanic and/or continental lithosphere (Ortega-Gutiérrez et al., 1999). As with the Petlalcingo Group, a minimum depositional age for the Piaxtla Group is provided by the unconformably overlying Pennsylvanian–Middle Permian Tecomate Formation (Ortega-Gutiérrez et al., 1999).

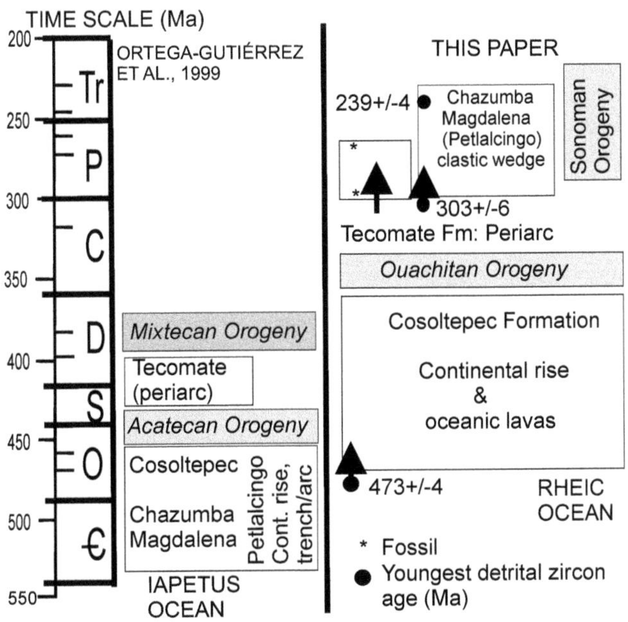

FIG. 2. Stratigraphic columns for the eastern part of the Mixteca terrane: Petlalcingo Group, Tecomate Formation, and Totoltepec pluton; Acatlán Complex after Ortega-Gutiérrez et al. (1999) and using data in this paper and Keppie et al. (2004c). Abbreviations: € = Cambrian, O = Ordovician, S = Silurian, D = Devonian, C = Carboniferous, P = Permian, Tr = Triassic.

Tecomate Formation

The Tecomate Formation consists of conglomerate, sandstone, slate, within-plate mafic and felsic volcanic rocks, and limestones that contain latest Pennsylvanian–Middle Permian conodonts (Keppie et al., 2004c). Zircons from granite pebbles in a conglomerate horizon have yielded SHRIMP ages of ca. 264–320 Ma and were likely sourced in the Totoltepec pluton (287 ± 2 Ma, Yañez et al, 1991; 289 ± 1 Ma, Keppie et al., 2004b), which overthrusts the Tecomate and Cosoltepec formations (Figs. 1B and 1C). The Tecomate Formation has been penetratively deformed by two sets of structures under greenschist-facies conditions: (1) isoclinal folding associated with N-S dextral shearing and south-vergent thrusting during which the syntectonic Totoltepec pluton was emplaced; and (2) NW-through N- to NE-trending upright open folding with a axial planar crenulation cleavage (Malone et al., 2002). K-Ar data on muscovite from the Tecomate Formation has yielded an age of 288 ± 14 Ma (Weber et al., 1997). In the antiform in the eastern part of the Mixteca terrane, the Acatlán Complex in the antiform is unconformably overlain Middle Jurassic rocks in the north, and Eocene–Oligocene rocks in the south (Keppie et al., 2004b).

Orogenic events

Taking all these data together, several tectonothermal events appear to have affected the Petlalcingo Group, including: (1) a Jurassic event localized in the southeastern part of the Mixteco terrane (Keppie et al., 2004b); (2) a Permian event that started in the Early Permian synchronous with intrusion of the ~288 Ma Totoltepec pluton and deforms rocks as young as lower Middle Permian that are unconformably overlain by Middle Jurassic rocks (Keppie et al., 2004c); and (3) an event yet to be dated that produced the first phase of deformation in the Cosoltepec Formation—this may have been synchronous with polyphase deformation under eclogite-facies metamorphic conditions in the Piaxtla Group, which yielded an age of 346 ± 3 Ma, i.e. Mississippian (Keppie et al., 2004a).

Analytical Techniques

Zircons were separated from three samples of clastic rocks, one from each of the Magdalena Migmatite protolith, the Chazumba Formation, and the Cosoltepec Formation. Details of the widely used separation methodology can be found in Fernández-Suárez et al. (2002) and Jeffries et al. (2003). Separated zircons were examined optically under a binocular microscope, and representative good-quality grains were picked and mounted in epoxy resin. The mounts were then ground down so that ~50% of the grains were exposed, and then polished to high quality. Grains were then imaged by cathodoluminescence (CL) in a JEOL 5900LV scanning electron microscope (SEM) at the Natural History Museum, London (NHM). Finally, the mounts were cleaned thoroughly by immersion in an ultrasonic bath containing a dilute HNO_3 acid, and dried before being introduced to the laser ablation chamber. Analytical instrumentation, analytical protocol and methodology, data reduction, age alculation and common Pb correction followed those described in Fernández-Suárez et al. (2002) and Jeffries et al. (2003).

In this study, nominal laser beam diameter was 30 μm for >75% of the analyses, but where the area to be analyzed was deemed to be large enough a 45μm beam was used to ensure the analysis was collected with the optimal signal strength that the analyte volume allowed.

Data were collected in discrete runs of 20 analyses, comprising 12 unknowns bracketed before and after by 4 analyses of the standard zircon 91500 (Wiedenbeck et al., 1995). Concordia age calculations, and concordia and frequency histograms/probability density distribution plots were performed using Isoplot v.3.00 (Ludwig, 2003).

Results

Two hundred eighty-eight (288) analyses, nearly all representing one analysis per grain, were performed on zircons from samples MM-10 (Magdalena Migmatite paleosome: 96 analyses), A-12 (Chazumba metapsammite: 96 analyses), and COS-100 (Cosoltepec semipelite: 96 analyses). Of those, 15 were rejected (4 in MM-10, 3 in A-12, and 8 in COS-100) based on the presence of features such as discordance >20%; high common Pb detected in the U-Pb, Th-Pb, and Pb-Pb isotope ratio plots; and/or elemental U-Pb fractionation or inconsistent behavior of U-Pb and Th-Pb ratios in the course of ablation (see Fernández-Suárez et al., 2002; Jeffries et al., 2003). Figure 3 shows concordia plots and combined binned frequency and probability density distribution plots for the three samples, with the data presented in Table 1: 2σ errors are quoted throughout. Where the analyses overlap concordia, we assign a U-Pb Concordia age (*sensu* Ludwig, 2003) as the best age estimate (see bold type in Table 1). Where analyses are normally discordant (i.e., they plot below concordia), we assign the $^{207}Pb/^{206}Pb$ age inasmuch as we are confident that any discordance is not a result of excess common Pb in the analysis or analytically induced problems such as laser-induced elemental fractionation (see Fernández-Suárez et al., 2002 and Jeffries et al., 2003 for details). Consequently, these ages will approximate the "correct" age, assuming a zero-age Pb-loss event, and there is a small danger that a non zero-age thermal event could result in these ages representing minimum ages. However, the amount of discordance within these zircons is minor (see Fig. 3 and Table 1), and therefore this phenomenon is unlikely to affect any of the conclusions reached regarding this dataset.

Magdalena Migmatite

The sample of the paleosome of the Magdalena Migmatite (MM-10) was collected south of Magdalena (17°59.26', 97°48.42': Fig. 1B) and consists of quartz, plagioclase (An_{37-26}), biotite, hornblende, and accessory zircon, titanite, apatite, rutile, tourmaline, and opaques: chlorite and muscovite are common alteration products. Most of the zircons yielded concordant to nearly concordant data ranging from ~850 to 1250 Ma with one concordant point at ~1575 Ma (Fig. 3, Table 1). The youngest detrital zircon yielded a U-Pb Concordia age of 303 ± 6 Ma. And another grain yielded a Concordant age of 521 ± 8 Ma.

Chazumba Formation

The metapsammite sample of the Chazumba Formation (A-12) was collected at the village of Tultitlan (18°04.65', 97°02.93') and consists of quartz, plagioclase (An_{35-20}), biotite, muscovite, garnet, and accessory zircon, tourmaline, and opaques. Most of the zircons are concordant to nearly concordant, with ages ranging from ~920 to 1150 Ma (Fig. 3, Table 1). The youngest grain has a Concordia age of 239 ± 4 Ma.

TABLE 1. U-Pb LA-ICPMS Data from samples MM-10, A-12, and COS-100

Sample/ analysis	Isotopic ratios (2σ errors) $^{206}Pb/^{238}U$	Pct. error	$^{207}Pb/^{235}U$	Pct. error	$^{207}Pb/^{206}Pb$	Pct. error	Age, Ma $^{206}Pb/^{238}U$	(2σ)	$^{207}Pb/^{235}U$	(2σ)	$^{207}Pb/^{206}Pb$	(2σ)	Best age estimate, Ma	
							MM-10							
au11a05	0.1955	1.30	2.1988	2.92	0.0815	2.56	1151	15	1181	34	1234	36	1234	36
au11a06	0.0483	2.32	0.3456	2.78	0.0518	4.28	304	7	301	8	278	8	**303**	**6**
au11a07	0.1431	2.08	1.4144	2.18	0.0717	3.04	862	18	895	20	976	21	976	21
au11a08	0.1415	1.88	1.3169	5.12	0.0675	5.92	853	16	853	44	852	44	**853**	**15**
au11a09	0.2022	1.02	2.2317	2.14	0.0801	2.54	1187	12	1191	25	1198	26	**1188**	**10**
au11a10	0.1838	2.00	1.9862	2.36	0.0784	2.86	1088	22	1111	26	1156	27	**1102**	**13**
au11a11	0.1811	2.62	1.9301	3.58	0.0773	3.12	1073	28	1092	39	1128	40	**1084**	**22**
au11a12	0.2141	1.40	2.5144	1.46	0.0852	1.44	1251	18	1276	19	1319	19	1319	19
au11a13	0.1487	1.72	1.4219	4.46	0.0693	3.78	894	15	898	40	909	41	**894**	**14**
au11a14	0.1474	1.38	1.4045	3.36	0.0691	2.84	886	12	891	30	902	30	**886**	**11**
au11a15	0.1853	1.04	2.0219	1.96	0.0791	2.12	1096	11	1123	22	1175	23	1175	23
au11a16	0.1917	2.02	2.1471	4.10	0.0812	3.22	1131	23	1164	48	1227	50	1227	50
au11b06	0.1724	1.22	1.8072	1.88	0.0760	1.86	1025	13	1048	20	1096	21	1096	21
au11b09	0.2061	1.48	2.3503	2.66	0.0827	3.22	1208	18	1228	33	1261	34	**1216**	**13**
au11b10	0.1817	2.82	1.9477	2.64	0.0777	4.28	1076	30	1098	29	1139	30	**1093**	**16**
au11b11	0.1476	0.98	1.4023	1.84	0.0689	1.72	887	9	890	16	895	16	**888**	**8**
au11b13	0.1801	1.02	1.9571	2.52	0.0788	2.58	1068	11	1101	28	1167	29	1167	29
au11b14	0.1916	1.78	2.2176	2.46	0.0839	3.14	1130	20	1187	29	1291	32	1291	32
au11b15	0.1770	1.12	1.8959	1.10	0.0777	1.68	1051	12	1080	12	1139	13	1139	13
au11b16	0.1808	1.18	1.9702	2.48	0.0790	2.76	1071	13	1105	27	1173	29	1173	29
au12a05	0.1526	1.80	1.5184	2.60	0.0721	3.02	916	16	938	24	990	26	990	26
au12a06	0.1792	1.20	1.9294	1.84	0.0781	1.96	1063	13	1091	20	1149	21	1149	21
au12a07	0.2159	1.76	2.5188	1.66	0.0846	0.64	1260	22	1277	21	1306	22	1306	22
au12a08	0.1956	1.66	2.1297	2.34	0.0790	2.90	1152	19	1158	27	1171	27	**1155**	**12**
au12a09	0.1877	4.08	2.0493	2.36	0.0792	3.80	1109	45	1132	27	1176	28	**1132**	**16**
au12a10	0.1868	1.58	1.9826	1.50	0.0770	1.98	1104	17	1110	17	1120	17	**1108**	**9**
au12a11	0.1956	1.14	2.1210	2.06	0.0786	2.38	1152	13	1156	24	1163	24	**1153**	**9**
au12a12	0.1832	1.56	2.0010	2.36	0.0792	2.30	1084	17	1116	26	1178	28	1178	28
au12a13	0.1875	2.34	2.0015	2.12	0.0774	1.56	1108	26	1116	24	1132	24	**1118**	**14**
au12a14	0.1902	3.66	2.0907	2.36	0.0797	3.88	1122	41	1146	27	1190	28	**1144**	**16**
au12a15	0.1512	2.08	1.5077	2.08	0.0723	1.56	908	19	934	19	994	21	994	21
au12a16	0.1588	2.84	1.6378	4.20	0.0748	3.88	950	27	985	41	1063	45	1063	45
au12b05	0.1843	2.68	1.9864	2.58	0.0782	1.62	1090	29	1111	29	1151	30	1151	30
au12b06	0.1834	1.08	1.9229	1.48	0.0760	1.44	1086	12	1089	16	1096	16	**1088**	**9**
au12b07	0.1844	2.10	2.0044	1.94	0.0788	1.32	1091	23	1117	22	1168	23	1168	23
au12b08	0.2011	2.14	2.2302	1.92	0.0804	3.34	1181	25	1191	23	1208	23	**1189**	**13**
au12b09	0.1602	1.08	1.5958	1.28	0.0722	1.00	958	10	969	12	992	13	992	13
au12b10	0.2069	1.32	2.3392	2.64	0.0820	1.94	1212	16	1224	32	1245	33	**1213**	**15**
au12b11	0.0490	0.94	0.3533	3.30	0.0523	3.52	308	3	307	10	298	10	**308**	**3**
au12b12	0.1899	2.30	1.9789	4.24	0.0756	5.48	1121	26	1108	47	1084	46	**1116**	**21**
au12b13	0.1321	1.22	1.3093	1.16	0.0719	0.62	800	10	850	10	982	11	982	11
au12b14	0.1803	1.42	1.9533	2.16	0.0786	2.24	1069	15	1100	24	1161	25	1161	25
au12b15	0.1908	1.00	2.0652	1.52	0.0785	1.06	1126	11	1137	17	1159	18	1159	18
au12b16	0.2316	1.60	2.8960	2.80	0.0907	2.32	1343	21	1381	39	1440	40	1440	40
au13a05	0.1976	1.52	2.1983	1.64	0.0807	1.94	1162	18	1181	19	1214	20	1214	20
au13a06	0.1897	1.92	2.0508	2.94	0.0784	2.46	1120	21	1133	33	1157	34	**1126**	**18**
au13a07	0.1854	1.38	1.9937	1.22	0.0780	1.50	1096	15	1113	14	1146	14	1146	14
au13a08	0.1873	1.28	2.0032	2.10	0.0775	2.06	1107	14	1117	23	1135	24	**1111**	**11**
au13a09	0.2108	1.08	2.3357	2.12	0.0804	2.22	1233	13	1223	26	1206	26	**1229**	**10**
au13a10	0.1856	2.36	1.9589	1.92	0.0765	1.60	1097	26	1102	21	1109	21	**1102**	**12**
au13a11	0.1914	2.38	2.0061	4.60	0.0760	5.04	1129	27	1118	51	1096	50	**1125**	**20**
au13a12	0.1924	0.94	2.1208	1.44	0.0799	1.66	1134	11	1156	17	1195	17	1195	17

Table continues

TABLE 1. (*Continued*)

Sample/ analysis	Isotopic ratios (2σ errors) ^{206}Pb/ ^{238}U	Pct. error	^{207}Pb/ ^{235}U	Pct. error	^{207}Pb/ ^{206}Pb	Pct. error	Age, Ma ^{206}Pb/ ^{238}U	(2σ)	^{207}Pb/ ^{235}U	(2σ)	^{207}Pb/ ^{206}Pb	(2σ)	Best age estimate, Ma	
							MM-10 (*continued*)							
au13a13	0.1920	2.66	2.1097	3.02	0.0797	1.60	1132	30	1152	35	1189	36	1189	36
au13a14	0.0836	2.40	0.6729	2.18	0.0584	2.80	518	12	522	11	543	12	**521**	**8**
au13a15	0.2765	2.56	3.7296	2.34	0.0978	1.50	1574	40	1578	37	1583	37	**1579**	**17**
au13a16	0.1729	1.28	1.8574	1.96	0.0779	1.48	1028	13	1066	21	1144	22	1144	22
au13b05	0.2033	1.66	2.2292	2.24	0.0795	2.56	1193	20	1190	27	1185	27	1191	13
au13b06	0.1865	0.74	2.0125	1.04	0.0783	0.98	1102	8	1120	12	1153	12	1153	12
au13b07	0.1831	0.90	1.9686	1.30	0.0780	1.14	1084	10	1105	14	1146	15	1146	15
au13b08	0.1838	2.36	1.9682	2.44	0.0776	2.20	1088	26	1105	27	1138	28	1103	16
au13b09	0.1947	2.40	2.0375	3.28	0.0759	3.14	1147	28	1128	37	1092	36	1135	20
au13b10	0.1921	1.84	2.1578	2.56	0.0815	1.56	1133	21	1168	30	1233	32	1233	32
au13b11	0.1944	1.64	2.0869	1.72	0.0778	1.40	1145	19	1145	20	1143	20	**1145**	**12**
au13b12	0.1922	1.76	2.1143	2.48	0.0798	2.02	1133	20	1153	29	1191	30	1191	30
au13b13	0.2590	0.98	3.5396	1.36	0.0991	1.14	1485	15	1536	21	1607	22	1607	22
au13b14	0.1744	1.74	1.8355	2.06	0.0763	1.68	1036	18	1058	22	1103	23	1103	23
au13b15	0.1740	1.28	1.8247	1.24	0.0761	1.34	1034	13	1054	13	1097	14	1097	14
au13b16	0.1836	1.50	1.9408	2.20	0.0766	3.12	1087	16	1095	24	1112	24	**1091**	**12**
							A12							
au18a05	0.1437	2.48	1.3984	4.80	0.0706	4.28	866	21	888	43	945	40	**870**	**19**
au18a06	0.1554	1.06	1.6125	1.46	0.0753	0.72	931	10	975	14	1075	8	1075	8
au18a08	0.1540	1.02	1.5166	1.88	0.0714	1.76	923	9	937	18	969	17	969	17
au18a09	0.1573	1.28	1.6055	2.46	0.0740	2.92	942	12	972	24	1041	30	1041	30
au18a10	0.1623	1.60	1.6155	1.68	0.0722	1.86	970	16	976	16	991	18	**974**	**10**
au18a11	0.1956	1.16	2.0937	1.84	0.0776	2.10	1152	13	1147	21	1137	24	**1149**	**9**
au18a12	0.2017	2.24	2.1546	2.00	0.0775	0.66	1184	27	1167	23	1133	7	1133	7
au18a13	0.1679	1.52	1.6452	2.02	0.0711	1.36	1001	15	988	20	959	13	959	13
au18a14	0.1494	1.00	1.4708	1.94	0.0714	1.68	898	9	918	18	969	16	969	16
au18a15	0.1744	1.66	1.8429	2.20	0.0766	1.56	1036	17	1061	23	1112	17	1112	17
au18a16	0.1658	1.36	1.6448	1.86	0.0719	2.04	989	13	988	18	984	20	**988**	**10**
au18b05	0.1575	2.18	1.5344	4.66	0.0706	5.14	943	21	944	44	946	49	**943**	**16**
au18b06	0.1526	2.96	1.5253	3.48	0.0725	3.90	916	27	941	33	999	39	**931**	**18**
au18b07	0.1828	1.24	1.9914	1.98	0.0790	2.16	1082	13	1113	22	1172	25	1172	25
au18b08	0.1463	2.32	1.4357	3.60	0.0712	2.78	880	20	904	33	962	27	962	27
au18b09	0.1777	1.94	1.8694	3.74	0.0763	4.00	1054	20	1070	40	1102	44	**1060**	**16**
au18b10	0.1545	1.62	1.5702	1.34	0.0737	1.02	926	15	959	13	1033	11	1033	11
au18b11	0.1585	2.00	1.6255	1.94	0.0744	0.90	948	19	980	19	1052	9	1052	9
au18b12	0.1812	1.44	1.9051	1.90	0.0762	1.96	1074	15	1083	21	1101	22	**1079**	**11**
au18b13	0.1538	0.82	1.5243	1.22	0.0719	1.06	922	8	940	11	982	10	982	10
au18b14	0.1558	1.40	1.5350	2.12	0.0714	2.12	933	13	945	20	970	21	**938**	**10**
au18b15	0.1354	2.36	1.3384	2.52	0.0717	0.96	819	19	863	22	977	9	977	9
au18b16	0.1516	0.96	1.5102	1.82	0.0722	1.38	910	9	935	17	992	14	992	14
au18c05	0.1563	1.88	1.5432	2.98	0.0716	2.92	936	18	948	28	974	28	**941**	**14**
au18c06	0.1601	1.38	1.5617	1.66	0.0707	2.10	957	13	955	16	950	20	**956**	**8**
au18c07	0.1582	1.02	1.4932	2.26	0.0684	2.40	947	10	928	21	882	21	882	21
au18c08	0.1533	2.02	1.4822	1.56	0.0701	2.56	919	19	923	14	932	24	**922**	**8**
au18c09	0.1635	1.74	1.6680	1.84	0.0740	2.14	976	17	996	18	1041	22	1041	22
au18c10	0.1560	2.34	1.5829	3.38	0.0736	3.30	935	22	964	33	1030	34	1030	34
au18c11	0.1786	1.62	1.8958	2.06	0.0770	1.26	1059	17	1080	22	1120	14	1120	14
au18c12	0.1577	1.34	1.5562	2.22	0.0716	2.10	944	13	953	21	974	20	**947**	**11**
au18c13	0.1562	1.08	1.5600	1.48	0.0724	1.88	936	10	954	14	998	19	998	19
au18c14	0.1545	2.36	1.5268	2.76	0.0717	2.54	926	22	941	26	976	25	**936**	**16**
au18c15	0.1837	0.98	2.0096	1.72	0.0793	1.74	1087	11	1119	19	1180	21	1180	21
au18c16	0.1766	1.54	1.8701	2.20	0.0768	1.74	1048	16	1071	24	1116	19	1116	19

Table continues

TABLE 1. (*Continued*)

Sample/ analysis	Isotopic ratios (2σ errors) $^{206}Pb/^{238}U$	Pct. error	$^{207}Pb/^{235}U$	Pct. error	$^{207}Pb/^{206}Pb$	Pct. error	Age, Ma $^{206}Pb/^{238}U$	(2σ)	$^{207}Pb/^{235}U$	(2σ)	$^{207}Pb/^{206}Pb$	(2σ)	Best age estimate, Ma	
							A12 (*continued*)							
au19a05	0.1906	1.20	2.0713	1.50	0.0788	1.24	1125	13	1139	17	1167	14	1167	14
au19a06	0.1610	1.30	1.6133	2.54	0.0727	2.04	962	13	975	25	1004	20	1004	20
au19a07	0.1925	1.58	2.0870	3.00	0.0786	2.64	1135	18	1145	34	1163	31	**1138**	**15**
au19a08	0.1816	1.60	1.9229	1.58	0.0768	1.08	1076	17	1089	17	1116	12	1116	12
au19a09	0.1856	1.16	1.9722	2.22	0.0771	2.28	1097	13	1106	25	1123	26	**1100**	**10**
au19a10	0.1576	1.10	1.5449	2.22	0.0711	2.54	943	10	948	21	959	24	**945**	**8**
au19a11	0.1617	1.44	1.6174	1.68	0.0726	1.58	966	14	977	16	1001	16	**974**	**10**
au19a12	0.1552	1.20	1.5298	2.48	0.0715	2.88	930	11	942	23	971	28	**934**	**9**
au19a13	0.1391	2.48	1.3531	2.54	0.0706	1.04	840	21	869	22	944	10	944	10
au19a14	0.1632	1.32	1.6269	1.86	0.0723	1.54	975	13	981	18	994	15	**978**	**11**
au19a15	0.1813	1.46	1.9410	3.24	0.0776	2.22	1074	16	1095	35	1138	25	1138	25
au19a16	0.1615	3.38	1.6593	5.22	0.0745	5.34	965	33	993	52	1055	56	**977**	**25**
au19b05	0.1818	1.10	1.9239	1.50	0.0768	1.50	1077	12	1089	16	1115	17	1115	17
au19b06	0.1632	2.90	1.6347	3.92	0.0726	2.38	975	28	984	39	1004	24	**980**	**24**
au19b07	0.1684	3.46	1.6569	6.72	0.0714	5.38	1003	35	992	67	967	52	**1001**	**32**
au19b08	0.1610	1.16	1.6448	2.68	0.0741	2.30	962	11	988	26	1043	24	1043	24
au19b09	0.1623	1.74	1.6401	2.06	0.0733	2.22	970	17	986	20	1021	23	**980**	**12**
au19b10	0.1709	1.24	1.7162	1.26	0.0728	0.98	1017	13	1015	13	1009	10	**1015**	**8**
au19b11	0.1946	1.14	2.1276	1.32	0.0793	1.10	1146	13	1158	15	1179	13	1179	13
au19b12	0.1575	1.08	1.5681	1.30	0.0722	1.58	943	10	958	12	991	16	991	16
au19b13	0.1595	1.78	1.5632	1.92	0.0711	1.20	954	17	956	18	959	12	**956**	**12**
au19b14	0.1575	1.58	1.5977	2.82	0.0736	2.80	943	15	969	27	1029	29	1029	29
au19b15	0.1835	2.32	1.9820	2.90	0.0783	2.06	1086	25	1109	32	1155	24	1155	24
au19b16	0.1570	1.94	1.6286	2.86	0.0752	3.22	940	18	981	28	1074	35	1074	35
au19c05	0.1678	1.98	1.7591	3.90	0.0760	3.74	1000	20	1031	40	1096	41	1096	41
au19c06	0.1574	0.76	1.5812	1.30	0.0729	1.36	942	7	963	13	1010	14	1010	14
au19c07	0.1594	1.00	1.5998	1.76	0.0728	1.54	953	10	970	17	1008	16	1008	16
au19c08	0.1539	0.74	1.5250	1.78	0.0719	1.52	923	7	940	17	982	15	982	15
au19c09	0.1546	1.72	1.5663	3.98	0.0735	3.78	927	16	957	38	1027	39	1027	39
au19c10	0.1546	2.24	1.4563	1.54	0.0683	2.50	927	21	912	14	878	22	**915**	**9**
au19c13	0.1630	2.52	1.6428	3.04	0.0731	3.08	973	25	987	30	1016	31	**982**	**17**
au19c14	0.1704	0.98	1.7271	2.26	0.0735	2.10	1014	10	1019	23	1027	22	**1015**	**9**
au19c15	0.0378	1.76	0.2605	6.26	0.0500	6.82	239	4	235	15	196	13	**239**	**4**
au19c16	0.1579	1.58	1.5666	2.16	0.0720	2.50	945	15	957	21	985	25	**951**	**10**
							COS-100							
au02a05	0.0832	1.58	0.6462	3.50	0.0563	3.68	515	8	506	18	465	17	**513**	**7**
au02a06	0.0915	1.60	0.7367	2.12	0.0584	1.80	564	9	560	12	544	10	**563**	**8**
au02a07	0.1669	1.58	1.7406	2.12	0.0756	2.22	995	16	1024	22	1085	24	1085	24
au02a08	0.0915	1.98	0.7491	2.86	0.0594	2.52	564	11	568	16	581	15	**566**	**10**
au02a09	0.0785	2.46	0.6345	2.52	0.0586	1.92	487	12	499	13	554	11	554	11
au02a10	0.1918	1.56	2.0850	1.58	0.0788	1.84	1131	18	1144	18	1168	21	**1141**	**10**
au02a11	0.2126	0.84	2.5192	1.50	0.0859	1.52	1243	10	1278	19	1336	20	1336	20
au02a12	0.1922	3.36	2.0551	3.24	0.0775	1.90	1133	38	1134	37	1135	22	**1134**	**21**
au02a13	0.0879	2.02	0.7378	1.58	0.0608	1.92	543	11	561	9	634	12	634	12
au02a14	0.0914	1.66	0.7585	1.90	0.0602	2.96	564	9	573	11	609	18	**569**	**7**
au02a15	0.0974	1.94	0.8165	1.40	0.0608	1.80	599	12	606	8	631	11	**606**	**6**
au02a16	0.0937	1.66	0.7718	2.86	0.0597	1.78	577	10	581	17	593	11	**577**	**9**
au03a05	0.0989	1.14	0.8254	1.74	0.0605	1.56	608	7	611	11	622	10	**609**	**6**
au03a06	0.1315	1.64	1.2361	0.80	0.0682	1.46	796	13	817	7	873	13	873	13
au03a08	0.0868	1.72	0.7292	3.04	0.0609	3.66	537	9	556	17	636	23	636	23
au03a09	0.1020	1.24	0.8459	2.50	0.0602	2.66	626	8	622	16	609	16	**625**	**7**

Table continues

TABLE 1. (*Continued*)

Sample/ analysis	Isotopic ratios (2σ errors) $^{206}Pb/^{238}U$	Pct. error	$^{207}Pb/^{235}U$	Pct. error	$^{207}Pb/^{206}Pb$	Pct. error	Age, Ma $^{206}Pb/^{238}U$	(2σ)	$^{207}Pb/^{235}U$	(2σ)	$^{207}Pb/^{206}Pb$	(2σ)	Best age estimate, Ma	
							COS-100 (*continued*)							
au03a10	0.1333	1.08	1.2841	1.96	0.0699	1.06	807	9	839	16	924	10	924	10
au03a11	0.0850	0.96	0.6731	3.30	0.0574	3.10	526	5	523	17	507	16	**526**	**5**
au03a12	0.0746	1.04	0.5889	1.28	0.0573	1.14	464	5	470	6	502	6	502	6
au03a13	0.1026	0.94	0.8531	1.50	0.0603	1.44	630	6	626	9	613	9	**629**	**5**
au03a14	0.1741	1.50	1.7751	2.12	0.0739	2.32	1035	16	1036	22	1040	24	**1036**	**11**
au03a16	0.1281	3.42	1.1796	1.94	0.0668	2.62	777	27	791	15	831	22	**793**	**10**
au03b05	0.2029	2.32	2.2772	1.40	0.0814	2.76	1191	28	1205	17	1230	34	**1204**	**10**
au03b07	0.0761	0.98	0.5952	1.26	0.0567	1.34	473	5	474	6	479	6	**473**	**4**
au03b08	0.1607	1.92	1.6495	2.56	0.0744	0.72	961	18	989	25	1053	8	1053	8
au03b10	0.2782	1.10	4.0681	1.16	0.1061	1.08	1582	17	1648	19	1733	19	1733	19
au03b11	0.2047	0.48	2.2888	0.84	0.0811	0.82	1201	6	1209	10	1223	10	1223	10
au03b12	0.1513	3.72	1.6345	3.60	0.0783	2.34	908	34	984	35	1155	27	1155	27
au03b13	0.1046	1.14	0.8920	1.92	0.0618	1.40	641	7	647	12	669	9	**642**	**7**
au03b14	0.0823	1.38	0.6554	2.20	0.0578	2.00	510	7	512	11	521	10	**510**	**6**
au03b15	0.1836	1.02	1.9537	0.88	0.0771	1.08	1087	11	1100	10	1125	12	1125	12
au03b16	0.0864	1.60	0.6995	2.96	0.0587	2.36	534	9	538	16	557	13	**534**	**8**
au04a05	0.0834	2.04	0.6830	2.78	0.0594	2.48	516	11	529	15	581	14	581	14
au04a06	0.1682	1.40	1.7148	0.80	0.0739	1.30	1002	14	1014	8	1039	14	1039	14
au04a07	0.0904	0.98	0.7548	1.84	0.0605	1.98	558	5	571	11	623	12	623	12
au04a08	0.2820	0.94	3.9943	0.84	0.1027	0.72	1601	15	1633	14	1673	12	1673	12
au04a09	0.0842	0.92	0.6694	1.26	0.0577	1.18	521	5	520	7	517	6	**521**	**4**
au04a10	0.0972	1.14	0.7945	2.30	0.0593	2.68	598	7	594	14	578	15	**597**	**6**
au04a11	0.0817	0.86	0.6484	1.74	0.0576	1.36	506	4	507	9	513	7	**506**	**4**
au04a12	0.1850	1.76	1.8590	2.40	0.0729	3.40	1094	19	1067	26	1010	34	1010	34
au04a13	0.1371	2.06	1.3380	3.74	0.0708	2.94	828	17	862	32	950	28	950	28
au04a14	0.2079	1.70	2.3279	1.82	0.0812	1.32	1218	21	1221	22	1226	16	**1221**	**13**
au04a16	0.2027	0.94	2.2733	1.04	0.0813	0.98	1190	11	1204	13	1229	12	1229	12
au05a05	0.0815	1.24	0.6597	1.04	0.0587	1.18	505	6	514	5	557	7	557	7
au05a06	0.3195	0.74	5.0701	0.90	0.1151	0.70	1787	13	1831	16	1881	13	1881	13
au05a07	0.1909	1.08	2.0856	1.72	0.0792	1.64	1126	12	1144	20	1178	19	1178	19
au05a08	0.0821	1.30	0.6670	2.38	0.0589	2.08	509	7	519	12	565	12	565	12
au05a10	0.0732	0.90	0.5633	1.88	0.0558	1.72	455	4	454	9	444	8	**455**	**4**
au05a11	0.1568	1.40	1.5494	2.12	0.0716	1.20	939	13	950	20	976	12	976	12
au05a12	0.0866	1.38	0.6967	2.54	0.0583	2.46	535	7	537	14	542	13	**536**	**7**
au05a13	0.3823	1.36	7.0795	1.20	0.1343	0.98	2087	28	2121	25	2154	21	2154	21
au05a15	0.0819	0.98	0.6524	1.44	0.0578	1.50	507	5	510	7	520	8	**508**	**4**
au05a16	0.1571	1.18	1.5737	1.86	0.0726	1.42	941	11	960	18	1004	14	1004	14
au05b05	0.1270	0.90	1.1504	2.12	0.0657	1.94	771	7	777	16	796	15	**771**	**6**
au05b06	0.2413	2.02	3.4451	1.80	0.1036	0.78	1393	28	1515	27	1689	13	1689	13
au05b07	0.0918	1.56	0.7814	3.18	0.0617	3.60	566	9	586	19	664	24	664	24
au05b08	0.3587	1.22	6.0778	1.34	0.1229	0.68	1976	24	1987	27	1998	14	**1992**	**10**
au05b09	0.0931	0.82	0.7586	1.28	0.0591	1.48	574	5	573	7	571	8	**574**	**4**
au05b10	0.0733	0.68	0.5921	2.30	0.0586	2.36	456	3	472	11	551	13	551	13
au05b11	0.0851	1.80	0.6896	2.10	0.0588	2.34	526	9	533	11	558	13	**530**	**7**
au05b12	0.3319	1.34	5.6099	0.90	0.1226	0.62	1848	25	1918	17	1994	12	1994	12
au05b13	0.1324	1.54	1.2365	1.96	0.0677	1.08	802	12	817	16	859	9	859	9
au05b15	0.4747	1.06	12.2978	1.70	0.1878	1.08	2504	27	2627	45	2723	29	2723	29
au05b16	0.0872	1.32	0.7016	1.24	0.0584	1.24	539	7	540	7	543	7	**540**	**5**

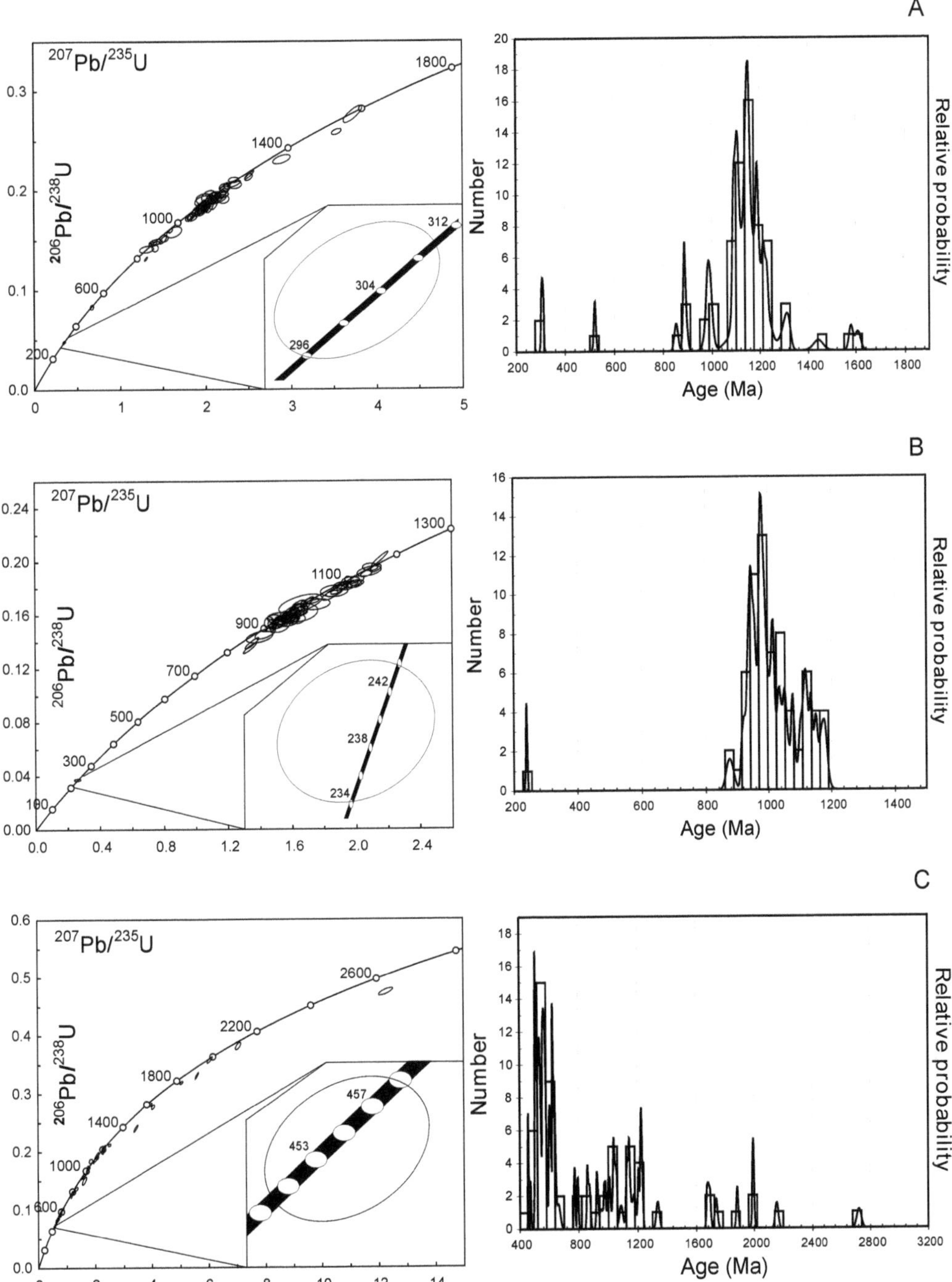

FIG. 3. U-Pb detrital zircon data plotted on concordia diagrams from samples of Magdalena Lithogene (A), Chazumba Lithogene (B), and Cosoltepec Formation (C). Error ellipses are shown with 2σ errors on the Concordia diagrams and input values for the histograms and probability density plots are 2σ age errors from the "best age estimate" discussed in the text and included in Table 1.

Cosoltepec Formation

The low-grade semipelitic sample of the Cosoltepec Formation (COS-100) was collected on the Acatlán-Totoltepec road near La Huerta (18°15.244', 98°00.132') and consists of quartz, biotite, phengite, chlorite, and opaques. Analyzed zircons are generally concordant to nearly concordant with two main populations at ~455–630 Ma and ~770–1200 Ma (Fig. 3, Table 1). There are several other discordant points with $^{207}Pb/^{206}Pb$ ages ranging from ~1700 to 2700 Ma. The youngest zircon gives a concordant age of 455 ± 4 Ma.

Depositional Ages of Units

Magdalena Migmatite protolith

The depositional age of the Magdalena Migmatite protolith is constrained between the youngest detrital zircon (303 ± 6 Ma) and the age of migmatization (171 ± 1 Ma; Keppie et al., 2004b)—that is, between the Permo-Carboniferous boundary and the Middle Jurassic (Gradstein et al., 2004). However, because migmatization took place at depths of 19 ± 2 km (Keppie et al., 2004b), the younger time limit is probably somewhat older to allow time for sedimentary+tectonic burial (Fig. 2).

Chazumba Formation

The youngest detrital zircon has a concordant age of 239 ± 4 Ma (Middle Triassic; Gradstein et al., 2004) and provides an older limit on the time of formation deposition. Emplacement of the Tultitlan mafic lens into the Chazumba Formation at 174 ± 1 Ma provides a younger limit near the base of the Middle Jurassic (Keppie et al., 2004b). Inasmuch as intrusion was synchronous with the growth of garnet in the country rocks at a minimum depth of ~15 km (Keppie et al., 2004b), the younger constraint on deposition is also somewhat older to allow time for sedimentary + tectonic burial. The 224 ± 2 Ma $^{40}Ar/^{39}Ar$ plateau age on metamorphic muscovite from the adjacent Cosoltepec Formation (Keppie et al., 2004b) probably provides a tighter constraint of the time of deposition of the Chazumba Formation—that is, post-Anisian–pre-Norian (Gradstein et al., 2004). This younger time constraint also applies to the Magdalena Migmatite protolith. But because zircons from only one sample of each of the Magdalena protolith and Chazumba Formation have been analyzed, deposition of these two units may have extended throughout the Permian and into the Early Triassic (Fig. 2). Clearly more sampling is required; however, the polydeformed nature of the two units makes it difficult to determine stratigraphic top and bottom.

Cosoltepec Formation

Deposition of the Cosoltepec Formation is constrained between the youngest concordant detrital zircon age, which has a $^{206}Pb/^{238}U$ age of 455 ± 4 Ma (Mid-Caradoc, Upper Ordovician; Gradstein et al., 2004), and the Pennsylvanian unconformity beneath the Tecomate Formation: ~305 Ma (Keppie et al., 2004c), although an older constraint is provided by the upper Fammenian unconformity beneath the Patlanoaya Formation: 370 Ma (Vachard and Flores de Dios, 2002). However, the youngest detrital zircons in the Pennsylvanian–Middle Permian Tecomate Formation (Sanchez-Zavala et al., 2004) are ~480–450 Ma (Early–Late Ordovician; Gradstein et al., 2004), suggesting caution should be exercised in assigning a more precise age to the Cosoltepec Formation (Fig. 2).

Redefinition of units

The significantly different ages of the Cosoltepec Formation versus the Chazumba Formation and Magdalena protolith suggests that the Cosoltepec be removed from the Petlalcingo Group. Furthermore, the complex structure and composite metasedimentary and meta-igneous nature of the Chazumba and Magdalena units makes it impossible to determine the stratigraphy, suggesting that they be designated lithodemes grouped in the Petlalcingo Suite. Until such time as further work is carried out on the Cosoltepec Formation, its name and formational status are retained.

Provenance

Magdalena and Chazumba lithodemes

The ~239 Ma zircon in the Chazumba Lithodeme may have been derived from the Permo-Triassic arc that extended throughout eastern and central Mexico (Fig. 4; Centeno-Garcia and Silva-Romo, 1997; Torres et al., 1999; Dickinson and Lawton, 2001). The ~303–308 Ma zircon in the protolith of the Magdalena Lithodeme is similar to those from granitic pebbles in the Tecomate Formation, which were inferred to have been derived from the Totoltepec pluton that is also part of the Permo-Triassic arc (Figs. 1C and 4; Keppie et al., 2004b). A source for the ~920–1250 Ma detrital zircons in the Magdalena and Chazumba lithodeme samples

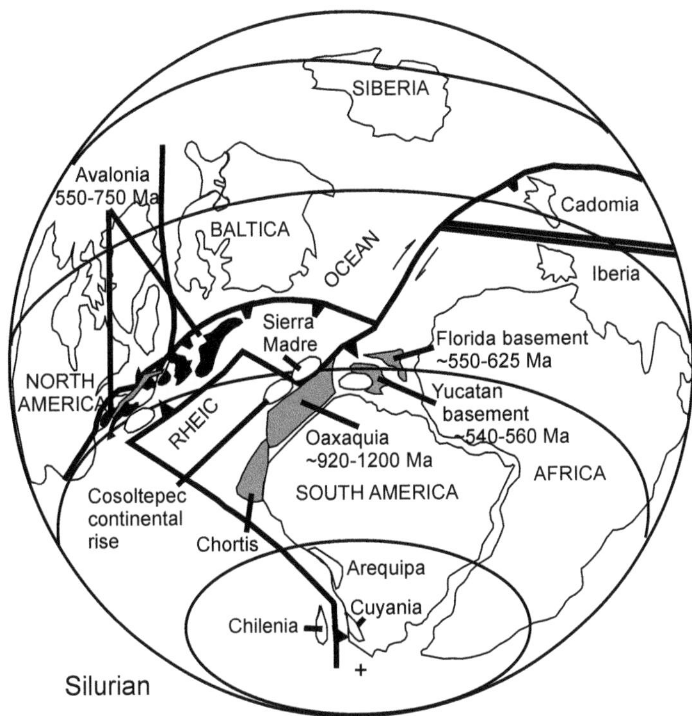

FIG. 4. Silurian reconstructions showing the location of the Cosoltepec Formation as a continental rise deposit adjacent to Oaxaquia (modified after Keppie, 2004).

may by found either in the adjacent Oaxacan Complex (Figs. 1C and 4; Keppie et al., 2001, 2003a; Solari et al., 2003; Ortega-Obregon et al., 2003) or recycled from various units of the Acatlán Complex (e.g., Cosoltepec and Tecomate formations; this paper and Sanchez-Zavala et al., 2004). Possible provenances for the ~850-920 Ma detrital zircons include the basement of Avalonia (e.g., Murphy et al., 2004) and the Goiás magmatic arc of eastern Amazonian (Pimental et al., 2000)(Fig. 4).

Cosoltepec Formation

The ~473 Ma detrital zircon in the Cosoltepec Formation may have come directly from plutons of this age in the Acatlán Complex (such as the Esperanza Granitoids), which have yielded a concordant age of 471 ± 6 Ma (U-Pb zircon, Sánchez-Zavala et al., 2004), and a leucogranite in the western part of the complex that has yielded a concordant age of 478 ± 5 Ma (U-Pb zircon; Campa-Uranga et al., 2002). On the other hand, the Neoproterozoic–Cambrian detrital zircons could have come from either the basement beneath the Yucatan Peninsula (Figs. 1C and 5; Krogh et al., 1993a, 1993b), or the Brasiliano orogens of South America (e.g., Pimental et al., 2000). Most of the older zircons probably had a source in the Oaxacan Complex.

Tectonic and Stratigraphic Implications

The Permian–Late Triassic depositional age of the Magdalena and Chazumba lithodemes overlaps the time of deposition of the Tecomate Formation and the intrusion of the syntectonic Totoltepec pluton (Keppie et al., 2004c). Kinematic studies indicate that this deformation involved S-vergent thrusting associated with dextral, N-S vertical shear zones (Malone et al., 1999; Elías-Herrera and Ortega-Gutiérrez, 2002). Such overthusting would have led to depression of the lithosphere and deposition of a clastic wedge (Magdalena-Chazumba lithodemes) in front of the advancing thrust sheets. In particular, overthrusting of the Totoltepec pluton and Tecomate Formation would have provided a source for the ~304 Ma detrital zircon in the protolith of the Magdalena Lithodeme. The absence

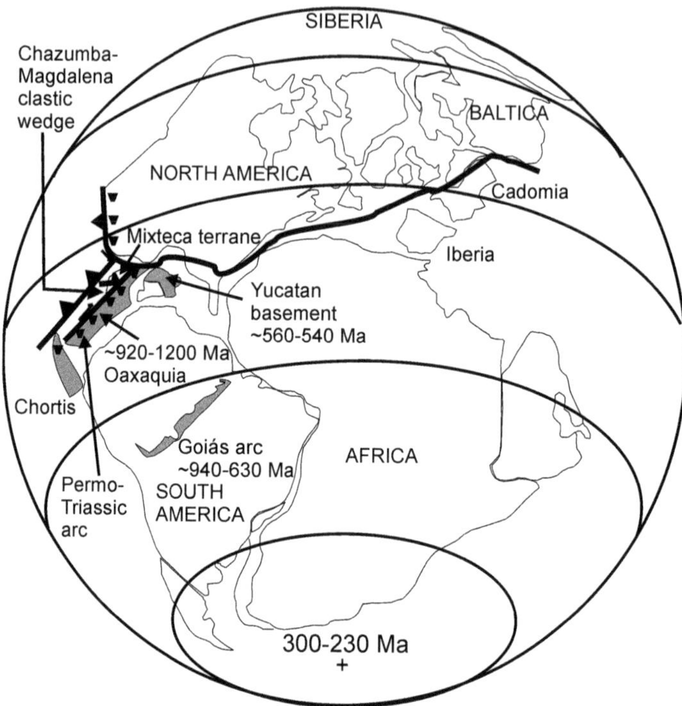

FIG. 5. 300–230 Ma reconstruction showing the location of the Petlalcingo Suite (Chazumba and Magdalena lithogenes) as a clastic wedge (modified after Keppie, 2004).

of 460–630 Ma detrital zircons in the Magdalena and Chazumba samples may have a number of explanations: (1) that the Cosoltepec Formation was also was being depressed beneath the overlying allochthons and so was not exposed to erosion; (2) that Cosoltepec detritus may be present elsewhere in the Chazumba and Magdalena lithodemes either exposed or buried beneath Mesozoic and Teriary units; or (3) that units containing Cosoltepec detritus have been eroded away. Clearly further sampling is in order. However, progressive southward advance of the thrust front eventually placed the Cosoltepec Formation above the Chazumba Lithodeme in a manner consistent with a clastic wedge setting. Support for the deposition of the Chazumba and Magdalena lithodemes in a clastic wedge is also provided by the turbiditic nature of the metasediments in these two units. The synchroneity of their deposition with arc magmatism farther east and with dextral N-S shear zones suggests oblique subduction of the paleo-Pacific plate beneath the western Mexican part of Pangea at this time (Fig. 4A). Such oblique subduction has also been postulated for the Middle Permian–earliest Triassic Sonoma orogeny in the U.S. Cordillera (Saleeby and Busby-Spera, 1992).

The Latest Ordovician—Middle Devonian depositional age of the Cosoltepec Formation, its distal turbiditic nature with continental-derived detritus (this paper), and the presence within it of interleaved oceanic tholeiitic basalts (authors' unpubl. data) suggests that it represents the part of a continental rise deposited on oceanic lithosphere. Its age range is more consistent with deposition in the Rheic Ocean (Ordovician–Carboniferous; Fig. 4B) than in the Iapetus Ocean (latest Precambrian–Early Paleozoic), which had closed by latest Ordovician.

Acknowledgments

We would like to acknowledge a Papiit grant IN103003 to JDK that facilitated the field work; an NSF grant (EAR 0308105) and an Ohio University 1804 Award to RDN; and an NSERC Discovery grant to JBM. We also thank Miguel Morales for assistance with drawing the figures. This paper represents a contribution to IGCP Project 453 (Modern and Ancient Orogens) and IGCP Project 498 (The Rheic Ocean).

REFERENCES

Campa-Uranga, M. F., Gehrels, G., and Torres de Leon, R., 2002, Nuevas edades de granitoides metamorfizados del complejo Acatlán en el Estado de Guerrero: Actas Instituto Nacional de Geoquimica, v. 8, no. 1, p. 248.

Centeno-García, E., and Silva-Romo, G., 1997, Petrogenesis and tectonic evolution of central Mexico during Triassic-Jurassic time: Revista Mexicana de Ciencias Geologicos, v. 14, p. 244–260.

DeGraaff-Surpless, K., Graham, S. A., Wooden, J. L., and McWilliams, M. O., 2002, Detrital zircon provenance analysis of the Great Valley Group, California: Evolution of an arc-forearc system: Geological Society of America Bulletin, v. 114, p. 1564–1580.

Dickinson, W. R., and Lawton, T. F., 2001, Carboniferous to Cretaceous assembly and fragmentation of México: Geological Society of America Bulletin, v. 113, p. 1142–1160.

Elías-Herrera, M., and Ortega-Gutiérrez, F., 2002, Caltepec fault zone: An Early Permian dextral transpressional boundary between the Proterozoic Oaxacan and Paleozoic Acatlán complexes, southern Mexico, and regional implications: Tectonics, v. 21, no. 3 [10.1029/200TC001278].

Fernández-Suárez, J., Gutiérrez-Alonso, G., and Jeffries, T. E., 2002, The importance of along-margin terrane transport in northern Gondwana: Insights from detrital zircon parentage in Neoproterozoic rocks from Iberia and Brittany: Earth and Planetary Science Letters, v. 204, p. 75–88.

Gradstein, F. M., Ogg, J. G., Smith, A. G., Bleeker, W., and Lourens, L. J., 2004, A new geologic time scale, with special reference to Precambrian and Neogene: Episodes, v. 27, p. 83–100.

Jeffries, T., Fernández-Suárez, J., Corfu, F., and Gutiérrez Alonso, G., 2003, Advances in U-Pb geochronology using a frequency quintupled Nd:YAG based laser ablation system (lambda = 213 nm) and quadrupole based ICP-MS: Journal of Analytical Atomic Spectrometry v. 18, p. 847–855.

Keppie, J. D., 2004, Terranes of Mexico revisited: A 1.3 billion year odyssey: International Geology Review, v. 46, p. 765–794.

Keppie, J. D., Dostal, J., Cameron, K. L., Solari, L. A., Ortega-Gutiérrez, F., and Lopez, R., 2003a, Geochronology and geochemistry of Grenvillian igneous suites in the northern Oaxacan Complex, southern México: Tectonic implications: Precambrian Research, v. 120, p. 365–389.

Keppie, J. D., Dostal, J., Ortega-Gutierrez, F., and Lopez, R., 2001, A Grenvillian arc on the margin of Amazonia: Evidence from the southern Oaxacan Complex, southern Mexico: Precambrian Research, v. 112, nos. 3–4, p. 165–181.

Keppie, J. D., Miller, B. V., Nance, R. D., Murphy, J. B., and Dostal, J., 2004a, New U-Pb zircon dates from the Acatlan Complex, Mexico: Implications for the ages of tectonostratigraphic units and orogenic events [abs.], *in* Geological Society of America, Abstracts with Programs, v. 36, no. 2, p. 104.

Keppie, J. D., Nance, R. D., Powell, J. T., Mumma, S. A., Dostal, J., Fox, D., Muise, J., Ortega-Rivera, A., Miller, B. V., and Lee, J. W. K., 2004b, Mid-Jurassic tectonothermal event superposed on a Paleozoic geological record in the Acatlán Complex of southern Mexico: Hotspot activity during the breakup of Pangea: Gondwana Research, v. 7, p. 239–260.

Keppie, J. D., and Ramos, V. S., 1999, Odyssey of terranes in the Iapetus and Rheic Oceans during the Paleozoic, *in* Ramos, V. S., and Keppie, J. D., eds. Laurentia–Gondwana connections before Pangea: Boulder, CO, Geological Society of America Special Paper 336, p. 267–276.

Keppie, J. D., Sandberg, C. A., Miller, B. V., Sánchez-Zavala, J. L., Nance, R. D., and Poole, F. G., 2004c, Implications of latest Pennsylvanian to Middle Permian paleontological and U-Pb SHRIMP data from the Tecomate Formation to re-dating tectonothermal events in the Acatlán Complex, southern Mexico: International Geology Review, v. 46, p. 745–754.

Keppie, J. D., Solari, L. A., Ortega-Gutiérrez, F., Elías-Herrera, M., and Nance, R. D., 2003b, Paleozoic and Precambrian rocks of southern Mexico—Acatlán and Oaxacan complexes, *in* Geologic transects across Cordilleran Mexico, Guidebook for the field trip of the 99th Geological Society of America Cordilleran Section Annual Meeting, Puerto Vallarta, Jalisco, Mexcio, April 4–10, 2003: Mexico. DF, Universidad Nacional Autonoma de Mexico, Instituto de Geologia, Publicacion Especial 1, Field Trip 12, p. 281–314.

Krogh, T. E., Kamo, S. L., and Bohor, B. F., 1993a, Fingerprinting the K/T impact site and determining the time of impact by U(Pb dating of single shocked zircons from distal ejecta: Earth and Planetary Science Letters, v. 119, p. 425–429.

Krogh, T. E., Kamo, S. L., Sharpton, B., Marin, L., and Hildebrand, A. R., 1993b, U-Pb ages of single shocked zircons linking distal K/T ejecta to the Chicxulub crater: Nature, v. 366, p. 232–236.

Ludwig, K. R., 2001, SQUID 1.00. A user's manual: Berkeley, CA , Berkeley Geochronology Center Special Publication No. 2, 17 p.

Ludwig, K. R., 2003, Isoplot 3.00: Berkeley, CA, Berkeley Geochronology Center Special Publication No. 4, 70 p.

Malone, J. W., Nance, R. D., Keppie, J. D., and Dostal, J., 2002, Deformational history of part of the Acatlán Complex: Late Ordovician–Early Silurian and Early

Permian orogenesis in southern Mexico: Journal of South American Earth Sciences, v. 15, p. 511–524.

Murphy, J. B., Dostal, J., Nance, R. D., and Keppie, J. D., 2004, Grenville-aged juvenile crust development in the peri-Rodinian ocean, *in* Tollo, R. P., Corriveau, L., McLelland, J. B., and Bartholemew, G., eds., Proterozoic tectonic evolution of the Grenville Orogen in North America: Geological Society of America Memoir 197, p. 135–144.

Ortega-Gutiérrez, F., Elías-Herrera, M., Reyes-Salas, M., Macias-Romo, C., and López, R., 1999, Late Ordovician–Early Silurian continental collision orogeny in southern Mexico and its bearing on Gondwana-Laurentia connections: Geology, v. 27, p. 719–722.

Ortega-Obregon, C., Keppie, J. D., Solari, L. A., Ortega-Gutiérrez, F., Dostal, J., Lopez, R., Ortega-Rivera, A., and Lee, J. W. K., 2003, Geochronology and geochemistry of the ~917 Ma, calc-alkaline Etla granitoid pluton (Oaxaca, southern Mexico): Evidence of post-Grenvillian subduction along the northern margin of Amazonia: International Geology Review, v. 45, p. 596–610.

Pimentel, M. M., Fuck, R. A., Jost, H., Ferreira Filho, C. F., and de Araújo, S. M., 2000, The basement of the Brasilia fold belt and the Goiás magmatic arc, *in* Cordani, U. G., Thomaz Filho, A., and Campos, D. A., eds., Tectonic evolution of South America, *in* 31st International Geological Congress, Rio de Janeiro, Brasil, p. 195–230.

Ramirez-Espinosa, J., 2001, Tectono-magmatic evolution of the Paleozoic Acatlán Complex in southern Mexico, and its correlation with the Appalachian system: Unpubl. Ph.D. thesis, University of Arizona, 170 p.

Saleeby, J. B., and Busby-Spera, C., 1992, Early Mesozoic tectonic evolution of the western U.S. Cordillera, *in* Burchfiel, B. C., Lipman, P. W., and Zoback, M. L., eds., The Cordilleran Orogen: Counterminous U.S.: Boulder, CO, Geological Society of America, The Geology of North America, v. G-3, p. 107–168.

Sánchez-Zavala, J. L., Ortega-Gutiérrez, F., Keppie, J. D., Jenner, G. A., Belousova, E., and Maciás-Romo, C., 2004, Ordovician and Mesoproterozoic zircons from the Tecomate Formation and Esperanza granitoids, Acatlán Complex, southern Mexico: Local provenance in the Acatlán and Oaxacan complexes: International Geology Review, v. 46, p. 1005–1021.

Solari, L. A., Keppie, J. D., Ortega-Gutiérrez, F., Cameron, K. L., Lopez, R., and Hames, W. E., 2003, 990 Ma and 1,100 Ma Grenvillian tectonothermal events in the northern Oaxacan Complex, southern Mexico: Roots of an orogen: Tectonophysics, v. 365, p. 257–282.

Torres, R., Ruíz, J., Patchett, P. J., and Grajales-Nishimura, J. M., 1999, Permo-Triassic continental arc in eastern Mexico: Tectonic implications for reconstructions of southern North America, *in* Bartolini, C., Wilson, J. L., and Lawton, T. F., eds., Mesozoic sedimentary and tectonic history of north-central Mexico: Geological Society of America, Special Paper 340, p. 191–196.

Vachard, D., and Flores de Dios, A., 2002, Discovery of latest Devonian/earliest Mississippian microfossils in San Salvador Patlanoaya (Puebla, Mexico): Biogeographic and geodynamic consequences: Compte Rendu Geoscience, v. 334 , p. 1095–1101.

Weber, B., Meschede, M., Ratschbacher, L., and Frisch, W., 1997, Structure and kinematic history of the Acatlán Complex in Nuevos Horizontes–San Bernardo region, Puebla: Geofisica Internacional, v. 36, no. 2, p. 63–76.

Wiedenbeck, M., Allé, P., Corfu, F., Griffin, W. L., Meier, M., Orbeli, F., von Quadt, A., Roddick, J. C., and Spiegel, W., 1995, Three natural zircon standards for U-Th-Pb, Lu-Hf, trace element, and REE analyses: Geostandards Newsletter, v. 19, p. 1–23.

Yañez, P., Ruiz, J., Patchett, P. J., Ortega-Gutiérrez, F., and Gehrels, G., 1991, Isotopic studies of the Acatlán Complex, southern Mexico: Implications for Paleozoic North American tectonics: Geological Society of America Bulletin, v. 103, p. 817-828.

North American Craton

Geochronology and Geochemistry of the Francisco Gneiss: Triassic Continental Rift Tholeiites on the Mexican Margin of Pangea Metamorphosed and Exhumed in a Tertiary Core Complex

J. DUNCAN KEPPIE,[1]

Instituto de Geología, Universidad Nacional Autónoma de México, 04510 México D.F., México

J. DOSTAL,

Department of Geology, Saint Mary's University, Halifax, Nova Scotia B3H 3C3, Canada

B. V. MILLER,

Department of Geological Sciences, CB #3315, Mitchell Hall, University of North Carolina at Chapel Hill, Chapel Hill, North Carolina 27599-3315

A. ORTEGA-RIVERA,

Centro de Geociencias, Campus Juriquilla, Universidad Nacional Autónoma de México, Apdo. Postal 1-742, Centro Querétaro, Qro. 76001, Mexico

JAIME ROLDÁN-QUINTANA,

Instituto de Geología, Universidad Nacional Autonoma de Mexico, Estación Regional del Noroeste, Apartado Postal 1039, Hermosillo, Sonora 83000, México

AND J. W. K. LEE

Department of Geology, Queens University, Kingston, Ontario, Canada, K7L 3NG

Abstract

Migmatized amphibolite-facies gneisses and amphibolites of the Francisco Gneiss exposed in the northern part of the Guerrero composite arc terrane have been interpreted as either Precambrian basement, a Triassic metamorphic complex, or a distinct terrane. Field observations suggest exposure in a metamorphic core complex of a protolith composed of interleaved bimodal igneous and sedimentary rocks. Geochemical data indicate that the amphibolites are within-plate, continental tholeiites. Recalculated T_{DM} ages for the rhyolites are consistent with partial melting of the Grenvillian basement of North America projected beneath the area. U-Pb isotopic analyses of zircon from two felsic rocks yielded concordant ages between 216 and 197 Ma due to a combination of inheritance and Pb-loss: the best estimate of protolith age is ~206 Ma—i.e., Norian, Late Triassic. Concordant U-Pb titanite ages range from 112 to 98 Ma, whereas nearly concordant U-Pb xenotime ranges from 91 to 51 Ma. These are inferred to result from partial-complete resetting during the high-grade metamorphic event. $^{40}Ar/^{39}Ar$ analyses from the gneisses yielded plateau ages of 16.5 ± 1 Ma (muscovite) and 13 ± 1 Ma (biotite), which date cooling through ~370°C and ~300°C, respectively (early Middle Miocene). Biotite from the granitic sheet yielded a plateau age of 13 ± 2 Ma. These data are interpreted in terms of Miocene exhumation in a core complex of high-grade metamorphic rocks developed either over a slab window or as a result of tectonic burial during the Laramide orogeny. The extrusion of Upper Triassic, continental rift tholeiites is consistent with emplacement in a back-arc environment.

[1]Corresponding author; email: duncan@servidor.unam.mx

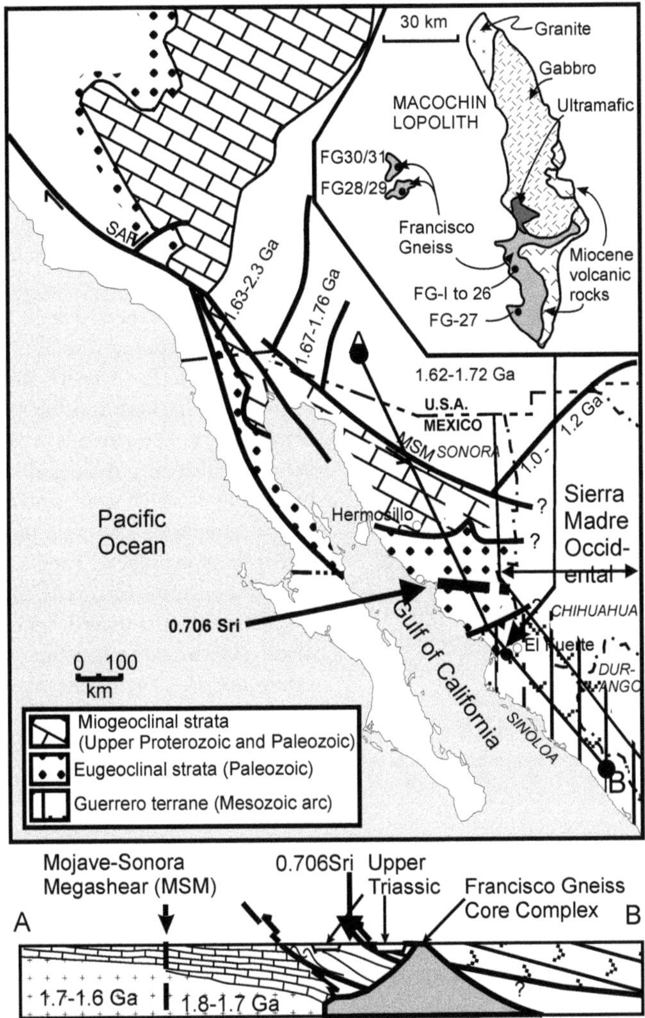

FIG. 1. Geological map and cross-section of southwestern North America showing the location of the Francisco Gneiss (modified after Valencia-Moreno et al., 2001). Inset is a geological map of the Francisco Gneiss and adjacent areas (modified after Mullan, 1978). Abbreviations: MSM = Mojave-Sonora megashear; SAF = San Andreas fault.

Introduction

METAMORPHIC CORE COMPLEXES are a common feature of the Cordillera of western North America where they appear to be associated with the extensional Basin and Range province (Dickinson, 2002). However, even though the Basin and Range province extends as least as far south as the Trans-Mexican Volcanic Belt, relatively few have been documented in northern Mexico, and none south of the latitude of Hermosillo (~29°N; Nourse et al., 1994). Of these, none have been studied using a combination of geochronological techniques in order to record their cooling history, which may then be used to infer their tectonic setting. Their apparent absence south of 29°N may be due to a lack of recognition that the high-grade rocks represent lower-plate rocks rather than basement. We suggest that this is the case for the Francisco Gneiss that is exposed ~300 km north of the mouth of the Gulf of California (Fig. 1).

The amphibolite-facies Francisco Gneiss crops out in the Sierra de Francisco at the southern end of the Sierra de Sonobari, which projects through the

Sonora-Sinaloa coastal plain on the eastern side of the Gulf of California (Fig. 1). This unit, along with the Paleozoic Rio Fuerte Group and the Mesozoic El Zapote Group, were originally combined into the Sonobari Complex (de Cserna and Kent, 1961); however, the contact between the Francisco Gneiss and the Paleozoic rocks is not exposed. This has led to two different interpretations of the Francisco Gneiss: (1) Precambrian basement (de Cserna and Kent, 1961; Mullan, 1978); and (2) a Triassic metamorphic complex based upon an unpublished U-Pb concordant age of ~220 Ma (Anderson and Schmidt, 1983, data repository): it is unclear if this age represents a metamorphic or a protolith age. Campa and Coney (1983) isolated it as a separate terrane (Sonobari terrane) until its relationships were better known, whereas Sedlock et al., (1993) placed it in the Tahué terrane (part of the Guerrero terrane of Campa and Coney, 1978). On the other hand, Ortega-Gutiérrez et al. (1995) inferred that the gneisses represented a retrograde part of the ~1 Ga granulite-facies basement of eastern Mexico. Given these uncertainties, we undertook this geological, geochemical, and geochronological study to provide a more complete database within which to evaluate the tectonic significance of the Francisco Gneiss.

Geological Setting

The Francisco Gneiss occurs in three inliers that crop out on the western side of the extensive Tertiary magmatic arc of the Sierra Madre Occidental (Fig. 1). These inliers are situated just within the northern edge of the Guererro terrane, a greenschist facies, Mesozoic island arc complex that was thrust over the Laurentian Precambrian craton and bordering Paleozoic miogeoclinal and eugeoclinal strata during the Late Cretaceous–Eocene Laramide Orogeny (Centeno-Garcia et al., 1993). Using Sr and Nd isotopes in Upper Cretaceous–Early Tertiary plutons, Valencia-Moreno et al. (2001) have proposed that the southern edge of the Laurentian basement lies ~100 km to the north of the inliers (Fig. 1). Quaternary deposits surround the western two inliers and also flank the western margin of the Sierra de Francisco. Tectonically juxtaposed along the contact between the Francisco Gneiss and Tertiary Fuerte Formation is the tholeiitic, layered Macochin lopolith consisting of cumulus pargasitic hornblendite overlain by plagioclase-hornblende gabbro and cut by gabbroic dikes (Mullan and Bussell, 1977). It was intruded at temperatures and pressures of 950-800°C and 5 kbar, and was affected by amphibolite- and greenschist-facies metamorphism, which produced the following assemblages: cummingtonite–high Mg hornblende, and epidote-albite-tremolite ± biotite-titanite under conditions estimated to have been 5 kbar/600–700°C and 5kbar/350°C, respectively (Mullan and Bussell, 1977). Associated deformation produced a hornblende lineation, a marginal penetrative foliation, shear zones, and folds in the dikes and southern contact (Mullan and Bussell, 1977; Mullan, 1978: Fig. 1). The eastern flank of the Sierra de Francisco is in contact with 600–700 m thick ignimbrites (Fuerte Formation) overlain by 1–2 km thick, fluvial sandstone and conglomerate (Maune Formation), both of which were gently folded before eruption of ignimbrites of the Cerros Formation (Mullan, 1978). In many places, such as 10 km northeast of El Fuerte near the Miguel Hidalgo Dam, the Maune Formation is juxtaposed against Paleozoic rocks along a low-angle fault, perhaps related to Miocene extension. The Fuerte and Maune formations both contain metamorphic and granitic pebbles (Mullan, 1978). Regional correlations suggest that the Fuerte and Maune formations are Miocene (Ortega-Gutiérrez et al., 1992).

The main outcrop of the Francisco Gneiss in the Sierra de Francisco consists mainly of banded felsic and mafic gneisses with some pelitic gneiss. Migmatization is pervasive along the western margin of this inlier and in the two western inliers. In the lowest-grade part of the complex, field relationships suggest that the protolith of the rocks was either an interbedded sequence of mafic and felsic flows and sedimentary rocks, or a series of dikes or sills intruding sedimentary rocks.

The structural and metamorphic history includes the following: (1) isoclinal folds, F_1, with an axial planar foliation consisting of hornblende, biotite, and feldspar; (2 and 3) two sets of isoclinal folds, F_2 and F_3, with a crenulation cleavage that developed during migmatization: coaxial subhorizontal fold axes are parallel to the L_1 hornblende lineation and a stretched quartz lineation; (4) open to tight, recumbent folds, F_4; (5) NNE- and ENE-trending upright folds; (6) intrusion of garnetiferous pegmatites and granitic sheets; and (7) gentle folds and a fracture cleavage in the pegmatites and granitic sheets parallel to the S_1 foliation in the gneisses.

Petrography

Francisco gneisses are predominantly a bimodal suite of amphibolites and biotite gneisses, although there are transitional rocks. Some amphibolites contain biotite and grade into amphibole-biotite gneisses whereas the biotite gneisses have variable proportions of amphibole.

The amphibolites are typically homogeneous, although some are migmatitic or layered. All of them are granoblastic, and medium to coarse grained. They are composed of amphibole and plagioclase (andesine or andesine-labradorite). Titanite, quartz, and Fe-Ti oxides occur in minor to accessory amounts. Apatite is a common accessory. The dominant amphibole is typically about 1–2 mm in size and has distinct green pleochroic colors. Biotite is absent in most amphibolites but in the rest it occurs in variable amounts, ranging from trace to minor amounts. Relict plutonic textures are absent.

The biotite gneisses are light to dark green, medium grained, and equigranular. The principal mineral constituents are plagioclase (oligoclase-andesine), K-feldspar, quartz, and biotite. Amphibole, epidote, and titaniferous magnetite occur in minor amounts, and apatite is common accessory. Modal proportions are variable, ranging from the rocks with only a minor amount of biotite through the samples with major amounts of biotite to biotite-amphibole gneisses.

Geochemistry

Twenty-four samples of amphibolite and gneisses were collected for chemical analyses. The samples were analyzed by X-ray fluorescence spectrometer for major and several trace (Rb, Sr, Ba, Zr, Nb, Y, Zn, V, Cr and Ni) elements in the Nova Scotia Regional Geochemical Centre at Saint Mary's University, Halifax. Additional trace elements (rare earth elements, Hf, Zr, Nb, Ta and Th) in nine samples were determined by inductively coupled plasma—mass spectrometer (ICP-MS) in the Geochemical Laboratory of the Ontario Geological Survey in Sudbury. Precision and accuracy of the X-ray data are reported by Dostal et al. (1986) and for ICP-MS data by Ayer and Davis (1997). Briefly, analytical errors for trace elements are generally < 5 rel. %. The analyses of representative samples are given in Table 1.

The use of chemical data of volcanic rocks metamorphosed under high-grade conditions for tectonomagmatic models assumes that the elements under consideration remained essentially immobile during secondary processes. There are several lines of evidence to suggest that most major elements, such as Al, Mg and Si, as well as many trace elements including high-field-strength elements (Zr, Hf, Nb, Ta, Th, and Ti), rare-earth elements (REE), and the transition elements (Cr, Ni, and V) were not redistributed in the analyzed rocks. In particular, their consistent trends and similarities to modern volcanic rocks suggest that they retained the original distribution.

According to geochemistry and petrography, three groups of rocks are present: amphibolites, leucocratic gneisses, and biotite-amphibolite gneisses. The biotite-amphibolite gneisses are migmatized and heterogeneous, and it appears that their protoliths represent a mixture of protoliths of the amphibolites and leucocratic gneisses: such an observation is most consistent with the inference that these rocks were originally sediments of volcaniclastic origin. In view of this, the biotite-amphibolite gneisses were not considered for petrogenetic and tectonomagmatic evaluations.

Amphibolites

The Zr/TiO_2 vs SiO_2 and Zr/TiO_2 vs Nb/Y diagrams (Figs. 2 and 3) confirm the distinctly bimodal nature of the Francisco Gneiss suite, and suggest that the amphibolites were originally subalkaline basalts. The amphibolites have SiO_2 contents ranging between 47 and 54 wt% (LOI-free), and plot along a tholeiitic trend, showing an increase of TiO_2 with FeO_{tot}/MgO ratio (Fig. 4). Also, most of them plot within the tholeiitic field on the AFM diagram (Fig. 5). They differ from island-arc tholeiites by having high Ti/V (~35)(Fig. 6) and Zr/Y (~3-4) ratios. Furthermore the Cr-Ti relationships of these rocks (Fig. 7) indicate that they are within-plate tholeiites.

Their chondrite-normalized REE patterns are flat, with $(La/Yb)_n$ and $(La/Sm)_n$ ranging from 1 to 1.6 and from 0.7 to 1.2, respectively (Fig. 8), and have the $(La)_n$ ~ 20–25. The mantle-normalized trace element patterns are also flat from Lu to Th and are accompanied by a negative Nb-Ta anomaly, similar to those of some primitive continental tholeiites (Fig. 9). Recalculating the Nd data in a mafic sample from the Francisco Gneiss presented by Valencia-Moreno et al. (2001) for ~210 Ma yields an initial ε_{Nd} value of +4.19 and a T_{DM} model age of 1238.4 Ma, values that are consistent with

TABLE 1. Major and Trace Element Compositions of Rocks from the Francisco Gneiss[1]

	Mafic rocks									Felsic rocks						
Sample	FG-1	FG-3	FG-4	FG-5	FG-6	FG-7	FG-9	FG-12	FG-13	FG-16	FG-17	FG-19	FG-21	FG-25	FG-26	FG-27
SiO_2	48.55	48.54	47.92	47.34	47.55	70.65	47.18	47.53	74.62	74.06	72.54	72.42	47.62	73.24	47.03	72.23
TiO_2	1.63	1.47	1.52	1.76	1.67	0.36	2.06	1.75	0.17	0.25	0.24	0.28	1.78	0.27	2.08	0.33
A_2O_3	16.11	16.95	16.17	14.60	15.44	15.16	14.55	14.75	13.40	14.09	13.73	14.23	14.64	13.95	14.42	14.63
Fe_2O_3	12.22	10.97	11.93	13.43	12.77	3.68	13.77	13.02	1.61	2.42	2.25	2.20	12.92	2.55	13.87	2.99
MnO	0.28	0.29	0.22	0.25	0.23	0.03	0.24	0.22	0.03	0.04	0.03	0.04	0.18	0.03	0.27	0.04
MgO	6.58	6.15	5.81	6.26	6.06	1.07	5.77	6.55	0.39	0.71	0.58	0.66	6.18	0.71	6.62	1.45
CaO	8.95	8.20	10.54	10.03	10.23	3.29	10.30	10.62	1.94	2.55	2.13	3.43	10.41	1.98	10.58	1.63
Na_2O	2.93	4.01	3.19	2.96	3.15	4.63	2.58	2.45	4.07	4.02	3.59	3.84	2.67	4.10	2.71	3.79
K_2O	0.85	0.58	0.80	0.95	1.01	1.29	0.96	0.84	2.48	1.96	3.57	1.57	0.95	3.01	0.96	2.44
P_2O_5	0.22	0.21	0.20	0.22	0.21	0.10	0.28	0.22	0.05	0.07	0.07	0.08	0.23	0.07	0.23	0.09
LOI	1.03	1.68	0.79	0.78	1.14	0.46	0.88	0.84	0.53	0.40	0.30	0.39	1.11	0.45	0.60	1.10
Total	99.35	99.05	99.09	98.58	99.46	100.72	98.57	98.79	99.29	100.57	99.03	99.13	98.69	100.36	99.37	100.72
Mg#	0.52	0.53	0.49	0.48	0.48	0.37	0.45	0.50	0.32	0.37	0.34	0.37	0.49	0.36	0.49	0.49
Cr	200	192	203	211	198	25	197	216	26	26	23	26	206	19	182	20
Ni	76	60	51	63	63	11	50	67	6	62	6		55		68	
Co	52	54	46	49	50	8	50	51					51	5	56	7
V	278	250	255	296	280	59	336	295	32	43	43	47	298	46	332	55
Zn	131	111	106	87	88	25	91	64	27	28	26	22	66	19	94	35
Rb	16	13	8	15	14	47	15	13	62	47	75	47	17	55	14	57
Ba	76	161	27	0	0	879			1472	1276	1716	793	0	1611	0	1953
Sr	213	271	299	222	242	276	220	196	236	229	206	214	210	193	242	158
Ga	18	18	17	19	18	15	18	17	9	11	12	11	17	12	18	14
Ta	0.20		0.20			0.37	0.00	0.18	0.97				0.18	0.37	0.00	0.42
Nb	2.9	3.0	2.9	5.0	4.0	7.9	6.0	2.7	10.2	7.0	7.0	9.0	2.7	7.2	2.1	8.0
Hf	2.10		2.00			1.20	0.00	2.40	1.20				2.50	1.10	2.40	1.30
Zr	72	98	72	107	96	44	122	88	34	111	110	111	89	35	84	43
Y	30	27	30	28	26	15	31	33	14	18	21	17	34	25	38	15
Th	0.80		0.98			10.15		0.56	14.41	10.00	10.00	8.00	0.57	15.77	0.25	10.42
La	8.14		7.90			29.70		6.50	33.16	21.00	26.00	12.00	7.48	30.44	6.46	25.72
Ce	21.25		19.92			56.01		17.26	59.59				19.62	59.99	18.94	51.62
Pr	3.13		2.91			5.96		2.66	6.16				2.98	6.59	3.20	5.60
Nd	15.34		14.52			21.08		14.05	20.56				15.34	23.35	16.73	20.11
Sm	4.35		4.08			3.73		4.25	3.37				4.58	4.40	5.18	3.57
Eu	1.55		1.34			0.83		1.50	0.64				1.65	0.88	1.76	0.74
Gd	5.32		4.89			3.05		5.39	2.72				5.69	3.90	6.63	3.06
Tb	0.89		0.85			0.48		0.94	0.40				0.95	0.65	1.14	0.47
Dy	5.68		5.46			2.71		6.02	2.36				6.12	4.11	7.08	2.77
Ho	1.20		1.17			0.55		1.30	0.47				1.31	0.90	1.52	0.56
Er	3.51		3.52			1.57		3.80	1.42				3.77	2.72	4.41	1.60
Tm	0.50		0.50			0.22		0.54	0.22				0.55	0.40	0.62	0.22
Yb	3.27		3.29			1.32		3.49	1.57				3.53	2.55	3.99	1.38
Lu	0.49		0.50			0.18		0.53	0.25				0.54	0.36	0.59	0.20

[1]Mg# = Mg/Mg + Fetot; major elements in wt%, trace elements in ppm.

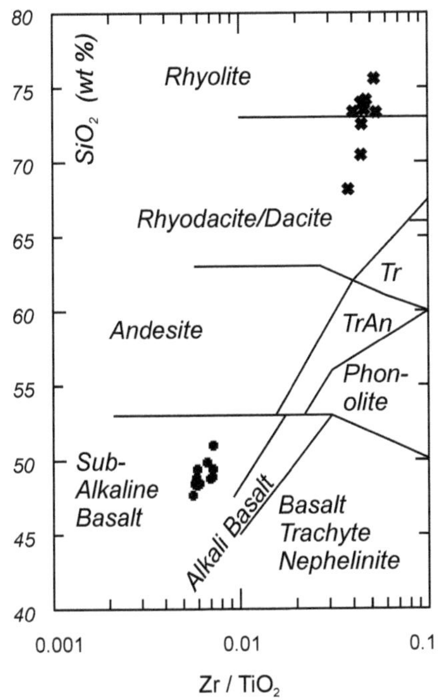

FIG. 2. Zr/TiO_2 versus SiO_2 (wt%) diagram of Winchester and Floyd (1977) for the meta-igneous rocks of the Francisco Gneiss complex (+ = mafic rocks; x = felsic rocks). Abbreviations: TrAn = trachyandesite; Tr = trachyte.

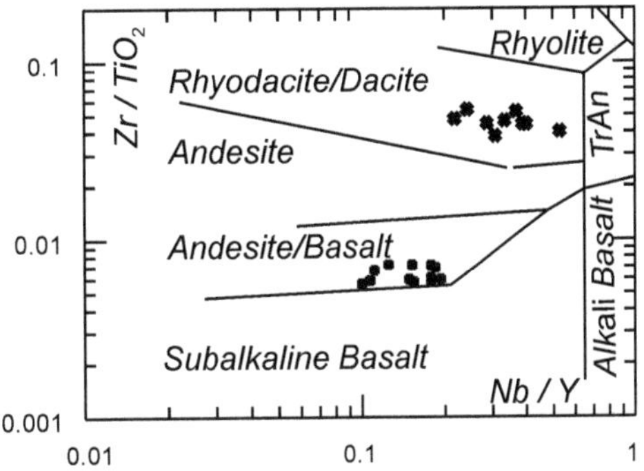

FIG. 3. Zr/TiO_2 versus Nb/Y diagram of Winchester and Floyd (1977) for the meta-igneous rocks of the Francisco Gneiss complex (+ = mafic rocks; x = felsic rocks). Abbreviations: TrAn = trachyandesite.

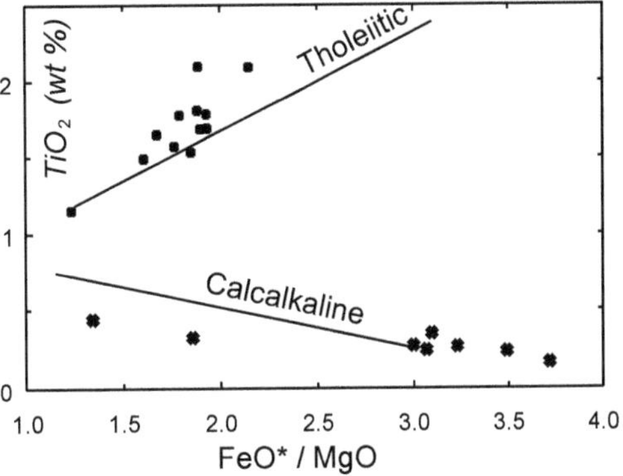

FIG. 4. FeO*/MgO versus TiO_2 (wt%) diagrams for the meta-igneous rocks of the Francisco Gneiss complex (+ = mafic rocks; x = felsic rocks). Vectors for tholeiitic and calc-alkaline trends are after Miyashiro (1974).

projecting the Grenvillian basement in the North American craton beneath the Sonobari area (Fig. 1).

Leucocratic gneisses

Felsic gneisses have mostly granitic/rhyolitic composition with SiO_2 ranging between 70 to 75 wt% and with $Al_2O_3 > 13$ wt%. Their contents of ($Na_2O + K_2O$) range from 5.5 to 8 wt%, whereas $Na_2O > K_2O$. Compositionally they resemble calc-alkaline rhyolites (Fig. 4). Their REE patterns are enriched in light REE but have flat heavy REE (Fig. 8) with $(La/Yb)_n$ between 7.5 and 14. The patterns display small negative Eu anomalies. The significant differences in the shapes of the REE patterns between mafic and felsic rocks cannot be accounted for by fractional crystallization. The mantle-normalized incompatible trace element patterns of the felsic rocks (Fig. 9), marked by negative Nb, Ta, and Ti anomalies, resemble modern rhyolites, including those associated with continental rifting and arc environments. Recalculation of the Nd data from a felsic sample analyzed by Valencia-Moreno et al. (2001) for ~210 Ma gives and initial ε_{Nd} value of –2.0 and a T_{DM} model age of 923 Ma, suggesting that the felsic magma was derived from a calc-alkaline Grenvillian basement beneath the Sonobari area.

Summary

The chemical composition of the amphibolites and leucocratic gneisses is compatible with a volcanic origin. The amphibolites resemble rift-related tholeiites with rather primitive distribution of incompatible trace elements. The presence of Ta-Nb depletion in these rocks, however, suggests that they were derived from a source previously fluxed by subduction fluids. The leucocratic gneisses compositionally resemble rhyolites erupted in continental rift or suprasubduction zone environments. If, as

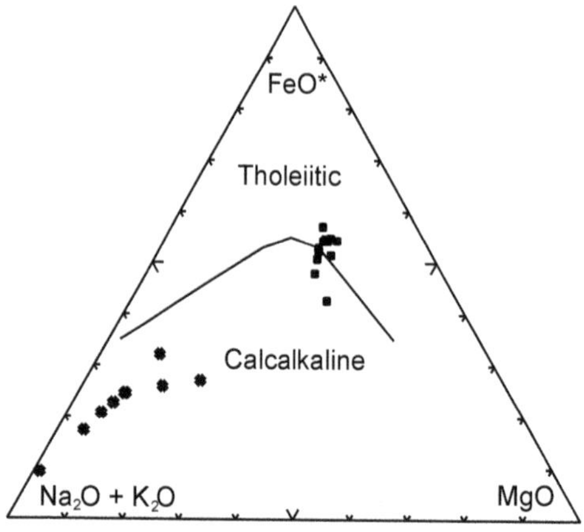

FIG. 5. AFM diagram for meta-igneous rocks of the Francisco Gneiss complex. (+ = mafic rocks; x = felsic rocks) showing tholeiitic and calc-alkaline fields. FeO* = Fe+FeO_{tot}.

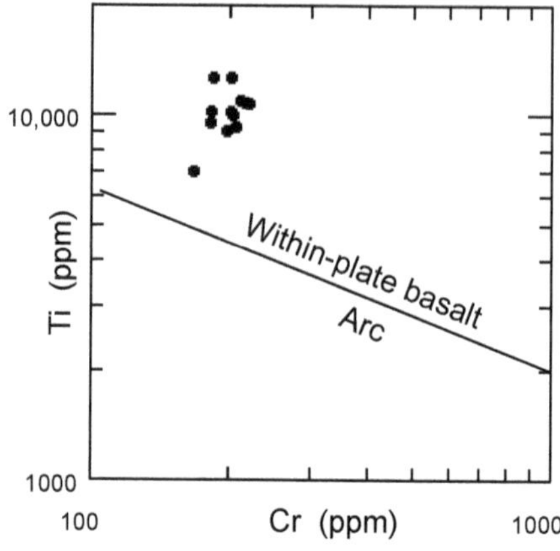

FIG. 7. Ti (ppm) versus Cr (ppm) plot of Pearce (1975) for the mafic meta-igneous rocks of the Francisco Gneiss showing the compositional fields for island-arc tholeiites and within-plate, ocean-floor basalts.

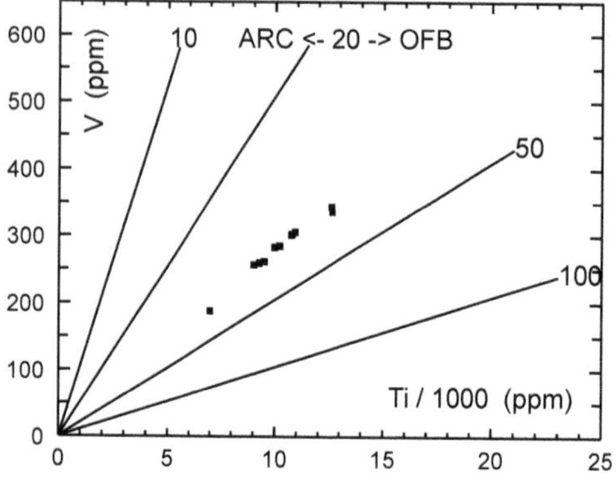

FIG. 6. V (ppm) vs. Ti/1000 (ppm) plot of Shervais (1982) for the metaigneous mafic rocks of the Francisco Gneiss.

seems likely, the rhyolites are interbedded with the tholeiitic mafic rocks and volcaniclastic rocks, then a continental rift environment is more likely. In this case, upwelling of the mafic magmas is inferred to have induced melting of the crust, which produced the felsic rocks, and both were extruded together.

The tholeiitic Macochin lopolith could be comagmatic with either: (1) the bimodal continental rift tholeiites of the Francisco Gneiss; or (2) the Tertiary volcanic rocks. Unfortunately neither geochemistry nor geochronology is available for the Macochin lopolith and the adjacent Tertiary volcanic rocks, which remain topics for future research.

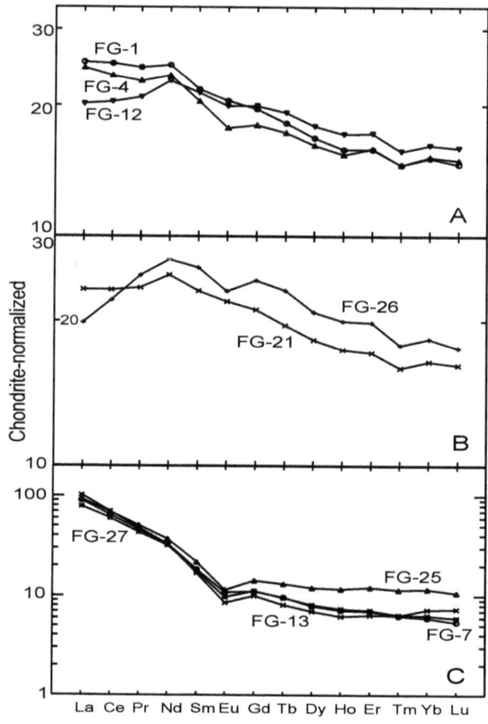

FIG. 8. Chondrite-normalized rare-earth element abundances in the meta-igneous rocks of the Francisco Gneiss complex. A and B. Mafic rocks C. Felsic rocks. Normalizing values are after Sun and McDonough (1989).

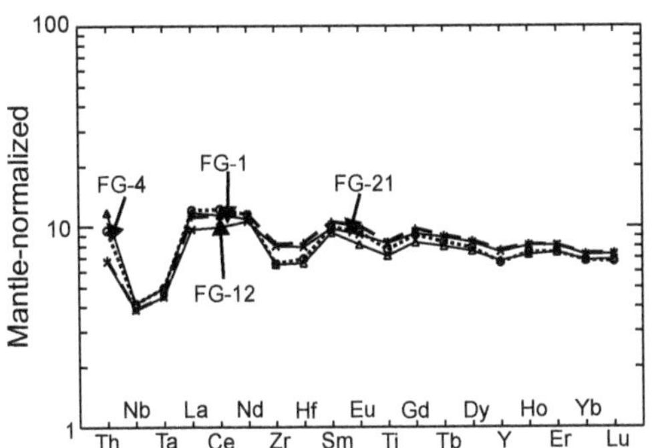

FIG. 9. Mantle-normalized incompatible trace element abundances in the mafic meta-igneous rocks of the Francisco Gneiss complex. Normalizing values after Sun and McDonough (1989).

U-Pb Geochronology

Analytical methods

Two samples were collected for U-Pb analysis: a felsic orthogneiss (FG-19) and a coarse migmatitic felsic gneiss (FG-27)(Fig. 1, Table 1). FG-19 is composed of plagioclase (oligoclase-andesine), K-feldspar, quartz, and biotite, with minor amounts of amphibole, epidote, and titaniferous magnetite, and accessory apatite. FG-27 is similar except for the addition of leucosomes of quartz and feldspar. The samples were crushed and zircons were separated, processed, and analyzed in the Radiogenic Isotope Geochemistry Laboratory at University of North Carolina, Chapel Hill using the techniques and equipment described in Ratajeski et al. (2001). Zircon, xenotime, and titanite grains were carefully handpicked in an effort to select clear, well-faceted, inclusion- and core-free grains. These grains are commonly the least likely to contain inherited components. Most zircon analyses were conducted on single grains or a single fragment broken from a grain; the smallest zircons required fractions of 2–5 grains (Table 1).

Results

The array of zircon analyses on concordia diagrams, and thus the age interpretation, is complicated by Pb-loss and probably also by inheritance of only slightly older zircon (Fig. 10A). Sample FG-19 yielded six concordant or nearly concordant analyses of high precision (a result of relatively high radiogenic Pb and low common Pb contents; Table 2); two less precise analyses are significantly discordant. Five analyses scatter slightly at about 206 Ma and one concordant analysis is about 216 Ma (Fig. 10A). Our best interpretation of these data is that the analyses at about 206 Ma closely approximate the crystallization age of the gneissic protolith. The individual analyses are precise enough to rule out simple Pb-loss from 216 Ma, but instead the data cluster around 206 Ma, which is best explained by small degrees of Pb-loss along with inheritance of only slightly older age. Three titanite analyses scatter along concordia from about 112 to about 98 Ma (Fig. 10A, inset). This array most likely represents cooling through the titanite closure temperature sometime between 206 Ma and 112 Ma, followed by partial resetting after 98 Ma, possibly even as young as the ~15 Ma reheating event indicated by the $^{40}Ar/^{39}Ar$ data (see below).

The data array from sample FG-27 also shows the effects of slight degrees of Pb-loss and inheritance (Fig. 10B), but several of the analyses are of much lower precision due to lower Pb contents and higher blanks relative to FG-19 (Table 2). The data array does not adequately fit a discordia line, but the general trend of concordant and nearly concordant analyses all point to about 205 Ma, which we suggest is best interpreted as the protolith crystallization age of the migmatitic gneiss. Five very high precision fractions of xenotime scatter from about 91 Ma down to about 51 Ma (Fig. 10B, inset) and do not form any readily interpretable discordant array. Similar to titanite from FG-19, we interpret this array to be the result of partial resetting of xenotime sometime after about 51 Ma, the youngest xenotime analysis.

In summary, U-Pb zircon data from both samples suggest protolith crystallization ages of about 206 Ma, but because of complex Pb loss and young inheritance, rigorous data reduction and error propagation are not possible. Titanite and xenotime data are consistent with partial resetting of U-Pb systematics in these minerals at <98 Ma and <51 Ma, respectively; resetting by the same ~15 Ma event that reset the $^{40}Ar/^{39}Ar$ system is possible.

$^{40}Ar/^{39}Ar$ Geochronology

Analytical methods

Biotite and muscovite grains were separated from three samples, a mafic orthogneiss (FG-20), a migmatitic felsic gneiss (FG-27), and a late garnetiferous granitic sheet (FG-18)(Fig. 1). These minerals were pre-treated and concentrated by standard techniques and later selected by handpicking under a binocular microscope from fractions that ranged in

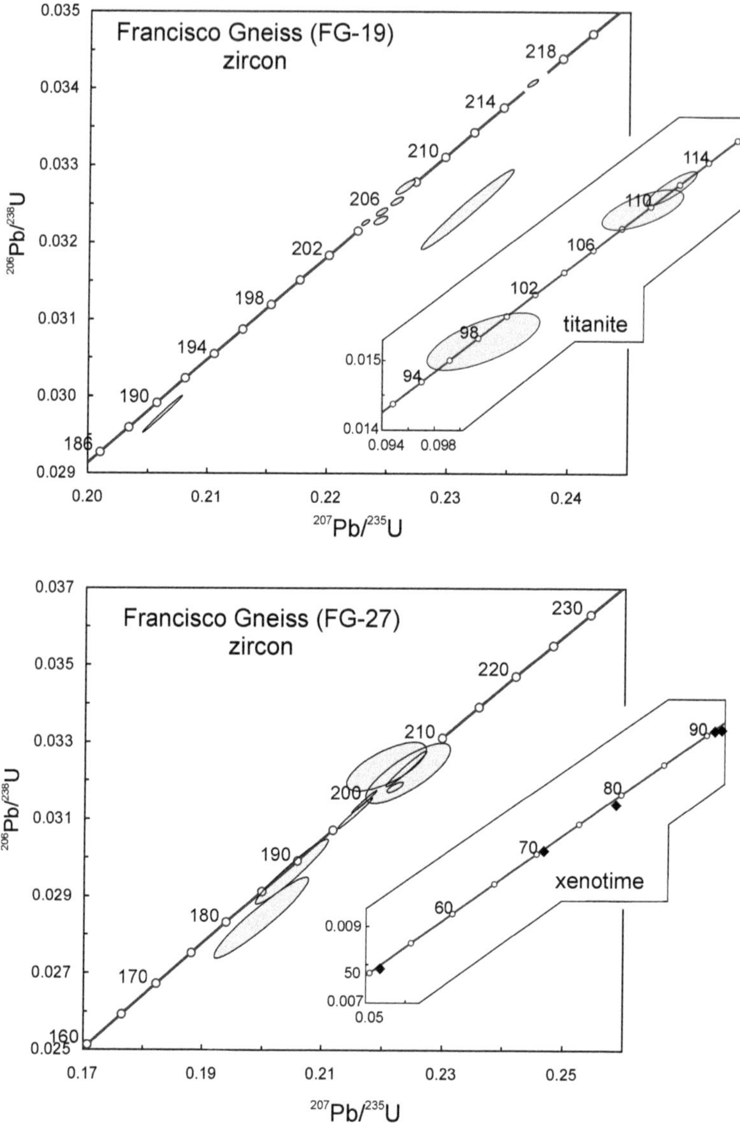

FIG. 10. U-Pb isotopic data from the Francisco Gneiss plotted on concordia diagrams. A. Sample FG-19. B. Sample FG-27. Error ellipses are 2σ.

size from 40 to 60 mesh at the mineral separation laboratory at UNICIT–Universidad Nacional Autónoma de México, Campus-Juriquilla, Querétaro, Qro. Mineral separates were loaded into Al foil packets and irradiated together with Hb3gr (1072 Ma) as a neutron-fluence monitor at the McMaster Nuclear Reactor (Hamilton, Ontario). $^{40}Ar/^{39}Ar$ analyses were performed by standard laser step-heating techniques described in detail by Clark et al. (1998) at the Geochronology Research Laboratory of Queen's University, Kingston, Ontario, Canada. The data are given in Table 3 and plotted in Figure 11. All data have been corrected for blanks, mass discrimination, and neutron-induced interferences. For the purposes of this paper, a plateau age is obtained when the apparent ages of at least

TABLE 2. U-Pb Isotopic Data for Minerals of the Francisco Gneiss

Analysis no., fraction (number of grains)	Weight (mg)[1]	Total[1] U (ng)	Total[2] Pb (pg)	Total[2] Com.Pb, (pg)	U (ppm)	Pb ppm	Atomic ratios: $^{206}Pb^2/^{204}Pb$	$^{206}Pb^3/^{208}Pb$	$^{206}Pb^3/^{238}U$	% Error4	$^{207}Pb^3/^{235}U$	% Error4	$^{207}Pb^3/^{206}Pb$	% Error[4]	Ages, Ma: $^{206}Pb/^{238}U$	$^{207}Pb/^{235}U$	$^{207}Pb/^{206}Pb$	ρ^5
	2.000								FG-19									
1. Large stubby prism (1)	0.0067	2.83	102.9	1.50	423	15	3926	4.048	0.03253	0.137	0.22584	0.183	0.05035	0.120	206	207	211	0.75
2. Medium prisms (2)	0.0052	3.01	109.3	1.38	576	21	4518	3.972	0.03239	0.126	0.22458	0.176	0.05029	0.121	205	206	208	0.72
3. Short thin acicular prisms (5)	0.0055	2.12	76.74	1.15	389	14	3814	3.981	0.03228	0.138	0.22449	0.209	0.05044	0.153	205	206	215	0.68
4. Small thin prisms (2)	0.0040	1.79	63.2	0.95	448	16	3957	4.874	0.03271	0.232	0.22650	0.281	0.05022	0.153	207	207	205	0.84
5. Large equant multi-faceted (1)	0.0235	3.18	108	1.17	135	4.6	5605	5.695	0.03225	0.098	0.22321	0.123	0.05019	0.073	205	205	204	0.80
6. Large stubby prism (1)	0.0061	1.59	51.5	0.62	260	8.4	4905	4.645	0.02976	0.660	0.20638	0.664	0.05030	0.073	189	191	209	0.99
7. Long acicular prism (1)	0.0066	1.04	39.3	4.08	158	5.9	544	3.805	0.03242	1.316	0.23165	1.345	0.05182	0.269	206	212	277	0.98
8. Large broken prismatic tip (1)	0.0350	7.33	253	1.02	209	7.2	15600	7.878	0.03407	0.120	0.23698	0.149	0.05045	0.088	216	216	216	0.81
9. Titanite, medium lensoid	0.0654	0.58	26.0	18.1	8.92	0.4	49.7	4.043	0.01527	2.208	0.10178	3.453	0.04835	2.525	98	98	116	0.68
10).Titanite, largest lensoid	0.2963	2.15	101	64.3	7.27	0.3	54.8	3.734	0.01716	1.375	0.11380	2.205	0.04810	1.639	110	109	104	0.67
11. Titanite, smallest lensoid	0.0604	1.27	44.2	15.7	21.0	0.7	107.7	1.837	0.01747	1.141	0.11600	1.337	0.04816	0.666	112	111	107	0.87
									FG-27									
1. Large stubby prism (1)	0.0029	0.73	39.4	17.5	249	14	104	5.407	0.03235	1.618	0.22065	2.439	0.04947	1.734	205	202	170	0.70
2. Large equant multi-faceted (1)	0.0082	1.90	62.1	4.30	231	7.6	900	7.407	0.03143	0.763	0.21704	0.773	0.05009	0.121	199	199	199	0.99
3. Long acicular prism (1)	0.0036	0.85	27.4	5.04	234	7.6	335	5.848	0.02962	2.317	0.20510	2.425	0.05023	0.680	188	189	205	0.96
4. Large stubby prism (1)	0.0081	1.36	46.1	3.50	168	5.7	790	5.075	0.03113	1.061	0.21556	1.077	0.05023	0.178	198	198	205	0.99
5. Long acicular prism (1)	0.0057	1.54	52.6	4.17	272	9.3	778	5.767	0.03229	1.185	0.22393	1.218	0.05030	0.267	205	205	209	0.98
6. Thin prism (1)	0.0029	0.67	26.2	6.31	236	9.2	236	5.554	0.03218	1.990	0.22420	2.552	0.05053	1.553	204	205	220	0.79
7. Medium prisms (2)	0.0038	0.75	24.8	1.42	197	6.6	1080	6.047	0.03181	0.373	0.22207	0.489	0.05063	0.304	202	204	224	0.78
8. Short acicular prism (1)	0.0004	0.50	17.2	5.15	1307	45	193	5.427	0.02842	3.025	0.20007	3.199	0.05106	1.013	181	185	244	0.95
9. Xenotime1 (1)	0.0066	17.3	146	5.82	2623	22	1511	5.972	0.00790	0.328	0.05178	0.346	0.04756	0.106	51	51	77	0.95
10. Xenotime2 (1)	0.0071	170	2409	41.9	23979	339	3685	11.296	0.01424	0.050	0.09409	0.100	0.04793	0.083	91	91	96	0.56
11. Xenotime3 (1)	0.0058	71.1	808	19.7	12258	139	2529	7.869	0.01100	0.061	0.07226	0.269	0.04765	0.248	71	71	82	0.44
1.)Xxenotime4 (1)	0.0037	41.4	617	91.7	11203	167	373	3.998	0.01225	0.085	0.08134	0.618	0.04816	0.578	78	79	107	0.51
13. Xenotime5 (1)	0.0041	30.2	507	46.2	7375	124	614	3.940	0.01420	0.103	0.09472	0.791	0.04836	0.741	91	92	117	0.53

[1]Weight estimated from measured grain dimensions and assuming densities: zircon = 4.67g/cm3, xenotime = 5.0g/cm3, titanite = 3.48g/cm3 ~20% uncertainty affects only U and Pb concentrations.
[2]Corrected for fractionation (0.15 ± 0.1%/amu - Daly) and spike.
3Corrected fractionation, blank, and initial common Pb.
[4]Errors quoted at 2σ.
[5] $^{207}Pb/^{235}U$-$^{206}Pb/^{238}U$ correlation coefficient of Ludwig, 1989.

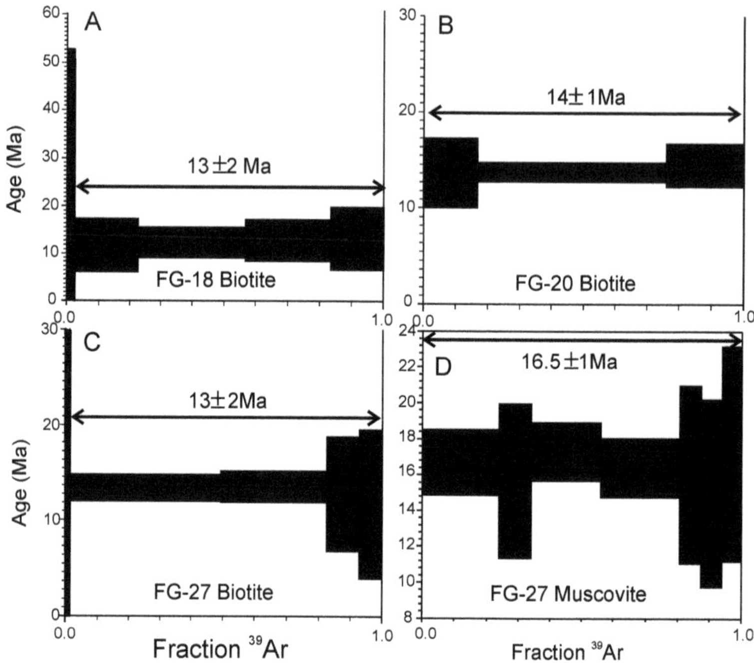

FIG. 11. $^{40}Ar/^{39}Ar$ data plotted on incremental release spectra. A. Biotite from sample FG-18. B. Biotite from sample FG-20. C. Biotite from sample FG-27. D. Biotite from sample FG-27.

three consecutive steps, comprising a minimum of 55% of the $^{39}Ar_k$ released, agree within 2σ error with the integrated age of the plateau segment. Errors shown in Table 2 and on the age spectrum represent the analytical precision at ± 2σ.

Results

Biotites from the three gneiss samples yielded plateau ages of 13 ± 2 Ma (FG-18), 14 ± 1 Ma (FG-20), and 13 ± 1 Ma (FG-27) (Figs. 11A and 11B), whereas muscovite from the latter sample yielded a nearly concordant spectrum with an integrated age of 16.5 ± 1 Ma (Fig. 11D). Biotite from the granitic sheet yielded concordant data with a plateau age of 13 ± 2 Ma (Fig. 11C). Closure temperatures using the actual grain size and the experimental data from Harrison et al. (1985) and Hames and Bowring (1994) were calculated for our samples, and yielded the following values; biotite at ~300°C (FG-18, 27) and ~360°C (FG-20), and in muscovite at ~369°C (FG-27). Thus, these $^{40}Ar/^{39}Ar$ ages are considered to record cooling through these closure temperatures for ^{40}Ar diffusion.

Discussion

Triassic

The presence of continental tholeiites in the Sierra de Francisco suggests that the feather-edge of the Laurentian craton at depth extends at least this far south (c.f. Fig. 1). The ~206 Ma age of these tholeiites corresponds to a Norian (Late Triassic) age (Gradstein et al., 2004). These rocks are the same age as the Barranca Group that occur in a graben just to the north of the Francisco Gneiss (Stewart and Roldán-Quintana, 1991: Fig. 1, cross-section), suggesting that both developed in a rifting environment. Rocks of similar age have been recorded elsewhere in the Guererro terrane (Fig. 12A): (1) in the Norian Zacatecas Formation in the northeastern Guerrero terrane, pillow basalts of MORB or primitive oceanic arc affinity are interlayered with shale and sandstone of recycled orogen provenance with model T_{DM} ages of 1.28–1.57 Ga (Centeno-García and Silva-Romo, 1997); and (2) in the Ladinian–Carnian (Middle–Upper Triassic) Arteaga and Placeres complexes in the southwestern Guerrero

TABLE 3. $^{40}Ar/^{39}Ar$ Isotopic Analyses of Minerals from the Francisco Gneiss[1]

Step	Laser power (Watts)	Isotope ratios: $^{40}Ar/^{39}Ar$	$^{38}Ar/^{39}Ar$	$^{37}Ar/^{39}Ar$	$^{36}Ar/^{39}Ar$	Ca/K	Cl/K	$\%^{40}Ar$ atm	f ^{39}Ar	$^{40}Ar^*/^{39}ArK$	Age
AOR-L215: FG-27 Ms 40/60											
1**	1	1.989 ± 0.009	0.014 ± 0.064	0.001 ± 0.316	0.003 ± 0.135	0.005	–0.000	10.71	24.19	1.223 ± 0.134	16.68 ± 1.82
2**	2	2.655 ± 0.013	0.015 ± 0.114	0.002 ± 0.596	0.002 ± 0.618	0.005	–0.000	–0.01	9.71	1.146 ± 0.316	15.63 ± 4.30
3**	3	2.711 ± 0.006	0.014 ± 0.072	0.001 ± 0.479	0.006 ± 0.070	0.002	–0.000	36.98	21.95	1.266 ± 0.120	17.27 ± 1.63
4**	4	1.444 ± 0.006	0.013 ± 0.068	0.001 ± 0.507	0.001 ± 0.349	0.002	–0.000	14.40	24.95	1.202 ± 0.121	16.39 ± 1.64
5**	5	1.355 ± 0.010	0.015 ± 0.134	0.002 ± 0.526	0.002 ± 0.590	0.005	–0.000	8.56	6.87	1.173 ± 0.364	16.01± 4.95
6**	6	1.364 ± 0.009	0.014 ± 0.130	0.002 ± 0.526	0.002 ± 0.566	0.005	–0.000	14.57	6.16	1.097 ± 0.383	14.96 ± 5.20
7**	7	1.333 ± 0.013	0.014 ± 0.150	0.002 ± 0.691	0.002 ± 0.763	0.004	–0.000	–0.01	6.18	1.258 ± 0.440	17.16 ± 5.97
			J = 0.007594 ± 0.000068					Isotope correlation date = 16.01 ± 6.68			
			Volume ^{39}ArK = 138.16					Initial $^{40}Ar/^{36}Ar$ ratio = 331.24 ± 423.54			
			Integrated date = 16.51±1.02					MSWD = 0.11			
			Plateau date = 16.51±1.02					$\%^{39}ArK$ for CA = 100.00			
			$\%^{39}ArK$ for PA = 100.00								
AOR-L316: FG-27 Bt 40/60											
1	0.50	16.795±0.018	0.055 ± 0.260	0.019 ± 0.580	0.066 ± 0.172	0.111	0.005	100.26	1.29	–0.076 ± 3.519	–1.04 ± 48.19
2**	2.00	1.417 ± 0.006	0.027 ± 0.024	0.001 ± 0.225	0.002 ± 0.208	0.010	0.003	28.90	48.37	0.981 ± 0.103	13.38 ± 1.40
3**	4.00	1.135 ± 0.007	0.027 ± 0.040	0.002 ± 0.231	0.001 ± 0.526	0.013	0.003	10.16	33.21	0.992 ± 0.120	13.52 ± 1.63
4**	6.00	1.096 ± 0.010	0.028 ± 0.084	0.004 ± 0.325	0.002 ± 0.894	0.025	0.003	11.14	10.07	0.936 ± 0.441	12.77 ± 5.99
5**	7.00	1.133 ± 0.010	0.028 ± 0.108	0.007 ± 0.254	0.002 ± 0.825	0.050	0.003	20.60	7.05	0.858 ± 0.573	11.70 ± 7.79
			J = 0.007587 ± 0.000070					Isotope correlation date = 13.68 ± 2.50			
			Volume ^{39}ArK = 133.24					Initial $^{40}Ar/^{36}Ar$ ratio = 260.72 ± 175.64			
			Integrated date = 13.06±1.35					MSWD = 0.34			
			Plateau date = 13.25±1.21					$\%^{39}ArK$ for CA = 100.00			
			$\%^{39}ArK$ for PA = 98.71								
AOR-L298: FG-18 Bt 60/80											
1	0.50	5.689 ± 0.019	0.061 ± 0.199	0.028 ± 0.383	0.032 ± 0.319	0.113	0.007	87.31	2.60	0.714 ± 3.145	9.77 ± 42.91
2**	1.50	1.752 ± 0.012	0.044 ± 0.064	0.005 ± 0.333	0.005 ± 0.288	0.027	0.006	49.37	20.06	0.859 ± 0.407	11.75 ± 5.55
3**	3.00	1.318 ± 0.010	0.044± 0.039	0.003 ± 0.288	0.002 ± 0.349	0.021	0.007	29.28	34.08	0.904 ± 0.238	12.36 ± 3.24
4**	4.00	1.233 ± 0.012	0.044 ± 0.048	0.004 ± 0.328	0.002 ± 0.489	0.017	0.007	21.22	26.38	0.942 ± 0.322	12.88 ± 4.38
5**	7.00	1.187 ± 0.013	0.046 ± 0.066	0.006 ± 0.314	0.003 ± 0.617	0.036	0.007	14.17	16.88	0.986 ± 0.481	13.47 ± 6.55
			J = 0.007604±0.000068					Isotope correlation date = 14.06 ± 5.08			
			Volume ^{39}ArK = 62.88					Initial $^{40}Ar/^{36}Ar$ ratio = 205.02 ± 375.93			
			Integrated date = 12.49±2.50					MSWD = 0.59			
			Plateau date = 12.57±2.30					$\%^{39}ArK$ for CA = 100.00			
			$\%^{39}ArK$ for PA = 97.40								

AOR-L300: FG-20 Bt 40/60											
1	0.50	7.424 ± 0.024	0.061 ± 0.306	0.031 ± 0.526	0.041 ± 0.363	0.116	0.007	88.78	0.95	0.842 ± 4.681	11.50 ± 63.74
2**	1.50	1.498 ± 0.009	0.038 ± 0.049	0.003 ± 0.314	0.003 ± 0.312	0.017	0.005	31.64	16.00	0.996 ± 0.265	13.60 ± 3.60
3**	5.00	1.158 ± 0.006	0.036 ± 0.022	0.005 ± 0.066	0.001 ± 0.350	0.039	0.005	10.86	59.25	1.006 ± 0.075	13.74 ± 1.03
4**	7.00	1.147 ± 0.009	0.036 ± 0.044	0.007 ± 0.104	0.001 ± 0.583	0.051	0.005	4.99	23.79	1.062 ± 0.166	14.49 ± 2.26

J = 0.007597 ± 0.000068
Volume ^{39}ArK = 118.87
Integrated date = 13.87±1.17
Plateau date = 13.90±1.01
%^{39}ArK for PA = 99.05

Isotope correlation date = 14.80 ± 3.02
Initial ^{40}Ar/^{36}Ar ratio = 122.83 ± 510.32
MSWD = 13.50
%^{39}ArK for CA = 100.00

[1]Isotope production ratios: (^{40}Ar/^{39}Ar)K = 0.0302; (^{37}Ar/^{39}Ar)Ca = 1416.4306; (^{36}Ar/^{39}Ar)Ca = 0.3952; Ca/K = 1.83 x (^{37}ArCa/^{39}ArK).
** = steps used in calculating plateau age.

terrane, N-MORB pillow basalts are interbedded with terrigenous sediments with model T_{DM} ages of 1.3–1.4 Ga (Centeno-Garcia et al., 1993, 2003). The sedimentary rocks have been correlated with the continental-rise turbidites of the Carnian La Ballena Formation in the adjacent Sierra Madre terrane, which also has a 1.28 Ga T_{DM} model age (Centeno-García and Silva-Romo, 1997). These T_{DM} model ages are similar to those of the adjacent ~1.0–1.4 Ga complexes of Oaxaquia that runs along the backbone of Mexico, and probably was the source of the sediments.

The combination of continental tholeiites, MORB basalts and continental rise sediments suggests a rifted continental-ocean margin. On the other hand, an active Late Triassic margin is indicated by: (1) a rifted continental magmatic arc extending from northern Sonora across Arizona and into California (Saleeby and Busby-Spera, 1992); and (2) Late Triassic oceanic-arc rocks in the Vizcaíno Norte and Sur terranes of Baja California and structurally underlying Late Triassic blueschists of the Western Baja terrane that were accreted to North America by latest Jurassic–earliest Cretaceous (Sedlock et al., 1993). The offset in these two arcs may be due to changes in dip of the subduction zone: shallower under the United States than under Baja California. However, the rifted Late Triassic margin in mainland Mexico appears to replace an Early Permian–Middle Triassic arc that extends along the Pacific margin of Pangea from southern Laurentia along the backbone of Mexico nearly to the southern coast of Mexico (Torres et al., 1999). The Middle–Late Triassic switch from an active to a rifted margin in mainland Mexico appears to be approximately synchronous with a change in the absolute plate motion from WNW to ENE and the initiation of rifting within the magmatic arc (Saleeby and Busby-Spera, 1992). Given this scenario, it is possible that rifting of the arc in the United States became a backarc rift in mainland Mexico, in which case the continental-rift tholeiites and rise would represent the North American margin of the backarc basin (Fig. 12A). In the Middle–Late Jurassic, subduction along the eastern margin of this backarc basin is indicated by the development of the Nazas volcanic arc in northeastern Mexico, whence it was offset to the Colombian Central Cordillera (Dickinson and Lawton, 2001).

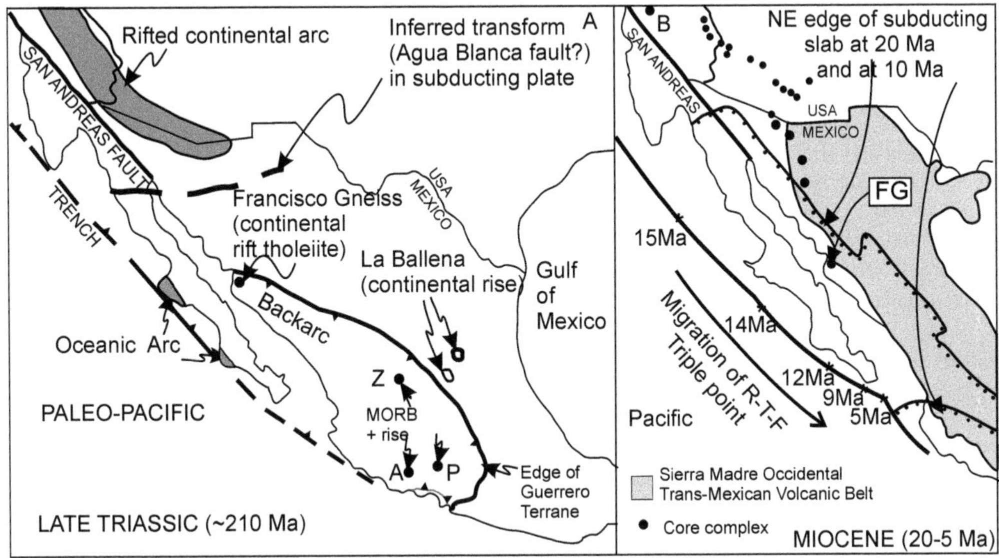

FIG. 12. Plate tectonic settings for the Francisco Gneiss in a southwestern North American context during (A) the Late Triassic (~210 Ma) and (B) the Miocene (20–5 Ma). The reconstructions show a closed Gulf of California. Abbreviations: A = Arteaga; FG = Francisco Gneiss; P = Placeres, Z = Zacatecas.

Late Cretaceous–Miocene

The main Mesozoic–Tertiary tectonic events affecting northern Mexico are the Late Cretaceous and Eocene Laramide orogeny (~90–50 Ma) that thrust the Guerrero terrane over the North American continental margin during greenschist-facies metamorphism (Keppie, 2004), and the Oligocene–Miocene extension (~30-12 Ma) that produced upper amphibolite–facies metamorphic core complexes (Dickinson, 2002). Thus, either or both of these events could have caused the resetting recorded in the U-Pb system, but temperatures must have been relatively high because closure temperatures for U and Pb isotopes in titanite and xenotime have been estimated at >660°C (Frost et al., 2000) and >720°C (Aleinikoff and Grauch, 1990), respectively. Inasmuch as peak metamorphic temperatures have been estimated at 600–700°C (Mullan and Bussell, 1977), xenotime would have been only partially reset. Two scenarios are envisaged: (1) burial of the Triassic protolith as a result of Laramide tectonic loading; and (2) a rise of the geoisotherms as a result of Oligocene–Miocene extension. Laramide tectonic loading of the lithosphere would be followed by a slow rise in the isotherms over ~25–40 m.y. (McKenzie, 1978), possibly resulting in migmatization. In this case, the youngest xenotime age (~51 Ma) could represent the final stage of re-equilibration of the isotherms after initial onset of the Laramide loading 40 m.y. earlier. However, this model does not explain their domal exhumation, which is recorded by cooling of the Francisco Gneiss through ~350°C and ~300°C at ~16 Ma and 13 ± 1 Ma, respectively (early Middle Miocene: Gradstein et al., 2004). Their exhumation is more consistent with core complex development during Miocene extension (Stewart et al., 1998). Extending this hypothesis, it is equally possible that the temperature rise could also result from the Miocene extension, in which case the youngest xenotime age (~51 Ma) would represent a mixed protolith-metamorphic age. The age data are also explicable in terms of a combination of Laramide tectonic loading followed by Miocene extension. Existing data cannot resolve these different scenarios.

By Miocene times, the Francisco Gneiss was cool enough to allow intrusion of coherent granitic sheets at 13 ± 2 Ma. These ages are synchronous with deposition of the Miocene ignimbrites and continental sediments along the eastern margin of the Sierra de Francisco, and suggest that the granitic sheets may have been feeders for the ignimbrites. Rogers et al.

(2002) have suggested that widespread eruption of ignimbrites immediately preceded the development of a slab window. The difference in metamorphic grade between the Francisco Gneiss and the flanking sediments suggests that the contact is a listric normal fault and that the structure represents a core complex. This is consistent with the recumbent attitude of the F_{1-4} folds in the gneisses. The late upright folds in the Francisco Gneiss may be correlated with those in the overlying Miocene rocks.

The discovery of a core complex in the Francisco Gneiss extends their distribution southward from northern Sonora into northern Sinaloa. The development of core complexes is generally attributed to extensional deformation (Dickinson, 2002). Collision of the East Pacific Rise with the trench led to the development of a slab window. Severinghaus and Atwater (1990) have calculated the location of the slab window through time, taking into account the subduction rate, the age of the plate at the trench, and slab duration, and show that the southern edge of the slab window swept along the mainland coast of Baja California between 20 and 10 Ma and would have passed the Sierra de Francisco area at ~15 Ma (Fig. 12B). This coincides with the development of the Francisco Gneiss core complex; thus rise of the asthenosphere in the slab window could have produced the migmatization and the Macochin lopolith if it was intruded in the Miocene.

Acknowledgments

We would like to acknowledge a Papiit grant IN103003 to JDK that facilitated the fieldwork, and a NSERC Discovery grant to JD. We would also like to thank Miguel Morales for assistance with drawing the figures. This paper represents a contribution to IGCP Project 453 (Modern and Ancient Orogens) and IGCP Project 498 (The Rheic Ocean).

REFERENCES

Aleinikoff, J. N., and Grauch, R. I., 1990, U-Pb geochronologic constraints on the origin of a unique monazite-xenotime gneiss, Hudson Highlands, New York. American Journal of Science, v. 290, p. 522–546.

Anderson, T. H., and Schmidt, V. A., 1983, The evolution of Middle America and the Gulf of Mexico–Caribbean Sea region during Mesozoic time: Geological Society of America Bulletin, v. 94, p. 941–966.

Ayer, J. A., and Davis, D. W., 1997, Neoarchean evolution of differing convergent margin assemblages in the Wabigoon Subprovince: Geochemical and geochronological evidence from the Lake of the Woods greenstone belt, Superior province, northwestern Ontario: Precambrian Research, v. 81, p. 155–178.

Campa, M. F., and Coney, P. J., 1983, Tectono-stratigraphic terranes and mineral resource distributions in México: Canadian Journal of Earth Science, v. 20, p. 1040–1051.

Centeno-García, E., Corona-Chávez, P., Talavera-Mendoza, O., and Iriondo, A., 2003, Geology and evolution of the western Guerrero terrane—a transect from Puerto Vallarta to Zihuatanejo, México, *in* Guidebook for field trips of the 99th Annual Meeting of the Cordilleran Section of the Geological Society of America, Publ. Esp. 1, Instituto de Geologia, Universidad Nacional Autónoma de México, p. 201–228.

Centeno-García, E., Ruiz, K., Coney, P. J., Patchett, P. J., and Ortega-Gutiérrez, F., 1993, Guerrero terrane of México: Its role in the Southern Cordillera from new geochemical data: Geology, v. 21, p. 419–422.

Centeno-García, E., and Silva-Romo, G., 1997, Petrogénesis and tectonic evolution of central México during Triassic–Jurassic time: Revista Mexicana de Ciencias Geológicas, v. 14, p. 244–260.

Clark, A. H., Archibald, D. A, Lee, A. W., Farrar, E., and Hodgson, C. J., 1998, Laser Probe $^{40}Ar/^{39}Ar$ ages of early- and Late-stage alteration assemblages, Rosario porphyry copper-molybdenum deposit, Collahuasi District, I Region, Chile: Economic Geology, v. 93, p. 326–337.

de Cserna, Z., and Kent, B. H., 1961, Mapa geológico de reconocimineto y secciónes estrcutruales de la región de San Blas y El Fuerte, Estados de Sinaloa y Sonora: México, DF, Instituto de Geologia, Universidad Nacional Autónoma de México, Cartas Geología y Mineras 4, scale 1:1,000,000.

Dickinson, W. R., 2002, The Basin and Range Province as a composite extensional domain: International Geology Review, v. 44, p. 1–38.

Dickinson, W. R., and Lawton, T. F., 2001, Carboniferous to Cretaceous assembly and fragmentation of México: Geological Society of America Bulletin, v. 113, p. 1142–1160.

Dostal, J., Baragar, W. R. A., and Dupuy, C., 1986, Petrogenesis of the Natkusiak continental basalts, Victoria Island, Northwest Territories, Canada: Canadian Journal of Earth Sciences, V. 23, p. 622–632.

Frost, B. R., Chamberlain, K. R., and Schumacher, J. C., 2000, Sphene (titanite): Phase relations and role as a geochronometer: Chemical Geology, v. 172, p. 131–148.

Gradstein, F. M., Ogg, J. G., Smith, A. G., Bleeker, W., and Lourens, L. J., 2004, A new geologic time scale, with special reference to Precambrian and Neogene: Episodes, v. 27, p. 83–100.

Hames, W. E., and Bowring, S. A., 1994, An empirical evaluation of the argon diffusion geometry in muscovite: Earth and Planetary Science Letters, v. 124, p. 161–167.

Harrison, T. M., Duncan, I., and McDougall, I., 1985, Diffusion of ^{40}Ar in biotite: Temperature, pressure, and compositional effects: Geochimica et Cosmochimica Acta, v. 49, p. 2461–2468.

Keppie, J. D., 2004, Terranes of Mexico revisited: A 1.3 billion year odyssey. International Geology Review, v. 46, p.765–794.

Ludwig, K. R., 1989, Pb dat: A computer program for processing raw Pb-U-Th isotope data: U. S. Geological Survey, Open-File Report no. 88-667.

McKenzie, D. P., 1978, Some remarks on the development of sedimentary basins: Earth and Planetary Science Letters, v. 40, p. 25–32.

Miyashiro, A., 1974, Volcanic rock series in island arcs and active continental margins: American Journal of Science, v. 274, p. 321–355.

Mullan, H. S., 1978, Evolution of part of the Nevadan orogen in northwestern Mexico: Geological Society of America Bulletin, v. 89, p. 1175–1188.

Mullan, H. S., and Bussell, M. A., 1977, The basic rocks series in batholithic associations: Geological Magazine, v. 114, p. 265–280.

Nourse, J. A., Anderson, T. A., and Silver, L. T., 1994, Tertiary metamorphic core complexes in Sonora, northwestern Mexico: Tectonics, v. 13, p. 1161–1182.

Ortega-Gutiérrez, F., Mitre-Salazar, L. M., Roldan-Quintana, J., Aranda-Gomez, J., Moran-Zenteo, D. J., Alaniz-Alvarez, S., and Nieto-Samaniego, A., 1992, Carta Geologica de la Republica Mexicana, scale 1:2,000,000: Mexico, DF, Instituto de Geologia, Universidad Nacional Autónoma de México.

Ortega-Gutiérrez, F., Ruiz, J., and Centeno-García, E., 1995, Oaxaquia, a Proterozoic microcontinent accreted to North America during the late Paleozoic: Geology, v. 23, p. 1127–1130.

Pearce, J. A., 1975, Basalt geochemistry used to investigate past tectonic settings on Cyprus. Tectonophysics, v. 25, p. 41–67.

Ratajeski, K., Glazner, A. F., and Miller, B. V., 2001, Geology and geochemistry of mafic to felsic plutonic rocks in the Cretaceous intrusive suite of Yosemite Valley, California: Geological Society of America Bulletin, v. 113, p. 1486–1502.

Rogers, R. D., Kárason, H., and van der Hilst, R. D., 2002, Epeirogenic uplift above a detached slab in northern Central America: Geology, v. 30, p. 1031–1034.

Saleeby, J. B., and Busby-Spera, C., 1992, Early Mesozoic tectonic evolution of the western U.S. Cordillera, *in* Burchfield, B. C. et al., eds., The Cordilleran orogen: Conterminous U.S.: Boulder, CO, Geological Society of America, Geology of North America, v. G3, p. 107–168.

Sedlock, R. L., Ortega-Gutiérrez, F., and Speed, R. C., 1993, Tectonostratigraphic terranes and tectonic evolution of México: Geological Society of America, Special Paper 278, 153 p.

Severinghaus, J., and Atwater, T., 1990, Cenozoic geometry and thermal condition of the subducting slabs beneath western North America, *in* Wernicke, B., ed., Basin and Range extensional tectonics near the latitude of Las Vegas, Nevada: Geological Society of America, Memoir 176, p. 1–22.

Shervais, J. W., 1982, Ti-V plots and the petrogenesis of modern and ophiolitic lavas: Earth and Planetary Science Letters, v. 59, p. 101–118.

Stewart, J. H., Anderson, R. E., Aranda Gómes J. J., Roldán Quintana, J., and 16 others, 1998, Map showing Cenozoic tilt domains and associated structural features, western North America: Plate I: Geological Society of America, Special Paper 323.

Stewart, J. H., and Roldán-Quintana, J., 1991, Upper Triassic Barranca Group: Nonmarine and shallow-marine rift-basin deposits of northwestern Sonora, México, *in* Jacques, C., and Pérez, F., eds., Studies in Sonoran geology: Geological Society of America, Special Paper 254, p. 19–36.

Sun, S. S., and McDonough, W. F., 1989, Chemical and isotopic systematics of oceanic basalts: Implications for mantle composition and processes, *in* Saunders, A. D., and Norry, M. J., eds., Magmatism in the ocean basins: Geological Society, London, Special Publication 42, p. 313–345.

Torres, R., Ruiz, J., Patchett, P. J., and Grajales, J. M., 1999, A Permo-Triassic continental arc in eastern México: Tectonic implications for reconstructions of southern North America, *in* Bartolini, C., et al., eds., Mesozoic sedimentary and tectonic history of north-central México: Geological Society of America Special Paper 340, p. 191–196.

Valencia-Moreno, M., Ruiz, J., Barton, M. D., Patchett, P. J., Zürcher, L., Hodkinson, D. G., and Roldán-Quintana, J., 2001, A chemical and isotopic study of the Laramide granitic belt of northwestern Mexico: Identification of the southern edge of the North American Precambrian basement: Geological Society of America Bulletin, v. 113, p. 1409–1422.

Winchester, J. A., and Floyd, P. A., 1977, Geochemical discrimination of different magma series and their differentiation products using immobile elements: Chemical Geology, v. 20, p. 325–343.

Sierra Madre Terrane

Deformational History of the Granjeno Schist, Ciudad Victoria, Mexico: Constraints on the Closure of the Rheic Ocean?

DAVID S. DOWE, R. DAMIAN NANCE,[1]

Department of Geological Sciences, 316 Clippinger Laboratories, Ohio University, Athens, Ohio 45701

J. DUNCAN KEPPIE,

Instituto de Geología, Universidad Nacional Autónoma de México, 04510 México D.F., Mexico

K. L. CAMERON,

Department of Earth Sciences, University of California, Santa Cruz, California 95064

A. ORTEGA-RIVERA,

Centro de Geociencias, Campus Juriquilla, Universidad Nacional Autónoma de México, Apartado Postal 1-742, Centro Querétero, Qro. 76001, Mexico

F. ORTEGA-GUTIÉRREZ,

Instituto de Geología, Universidad Nacional Autónoma de México, 04510 México D.F., Mexico

AND J. W. K. LEE

Department of Geology, Queens University, Kingston, Ontario, Canada, K7L 3NG

Abstract

Exposed in the core of a NNW-trending frontal anticline of the Laramide fold-thrust belt of northeastern Mexico, the Granjeno Schist comprises a polydeformed assemblage of Paleozoic metasedimentary and metavolcaniclastic rocks and serpentinized mafic-ultramafic units. The earliest deformation (D_1) predates emplacement of a leucogranite at 351 ± 54 Ma and may record obduction of this oceanic unit. Subsequent deformations (D_{2a}–D_{2c}) record tectonic juxtapositioning of the Granjeno Schist against the ~1 Ga Novillo Gneiss by NNW-directed dextral shear under conditions of decreasing temperature. Cooling ages of 313 ± 13 Ma and 300 ± 4 Ma are considered to date the onset of dextral motion, which continued into the Permian. These events are linked to the Late Paleozoic closing of the Rheic Ocean.

Introduction

NORTHWEST OF CIUDAD VICTORIA in the northern Mexican state of Tamaulipas (Fig. 1), the Huizachal-Peregrina anticlinorium exposes metamorphosed Paleozoic (Granjeno Schist) and ~1 Ga Precambrian (Novillo Gneiss) rocks potentially correlative with those of the Appalachian-Ouachita and Grenville orogens, respectively (Ortega-Gutiérrez, 1978; Ramírez-Ramírez, 1992). Although such a correlation would have profound implications for late Precambrian–Paleozoic continental reconstructions, the nature of the linkages is uncertain because critical aspects of the geology of the anticlinorium are unknown.

The ~1 Ga rocks of the Novillo Gneiss may also be linked to those of the ~1 Ga Oaxacan Complex in southern Mexico, which are thought to have been sutured to Laurentia following closure of either the Iapetus (Ortega-Gutiérrez et al., 1999) or Rheic (Keppie and Ramos, 1999) oceans. But whereas the Iapetus Ocean spanned the Cambrian–Ordovician, the Rheic Ocean lasted from Ordovician to Permo-Carboniferous time. This age difference provides a test for the two rival hypotheses, a topic addressed in this paper.

Like the gneisses of the Oaxacan Complex, the Novillo Gneiss is unconformably overlain by unmetamorphosed lower Paleozoic (Silurian) sedimentary strata that contain fauna of Gondwanan provenance (Stewart et al., 1999), suggesting that both represent portions of the Gondwanan margin of

[1]Corresponding author; e-mail: nance@ohio.edu

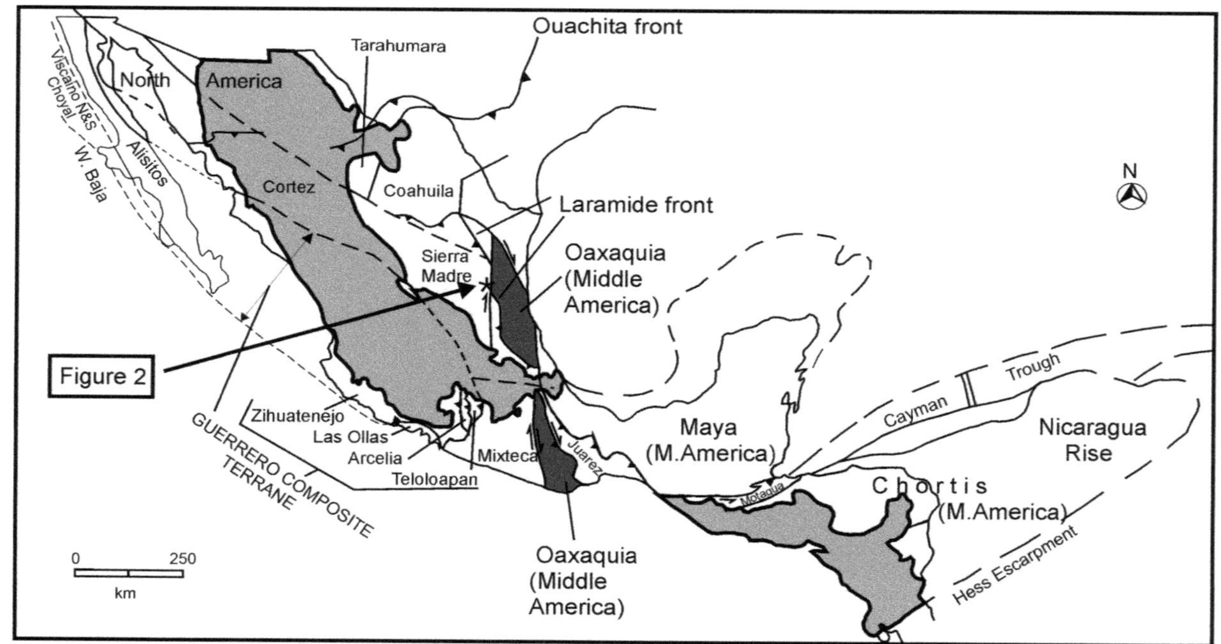

FIG. 1. Generalized tectonic map of Mexico showing the distribution of terranes and their inferred boundaries (after Keppie, 2004). Oaxaquia (dark shading) identifies that part of Mexico underlain by ~1.0 Ga basement. Light shaded areas show distribution of Cenozoic volcanic rocks. The Granjeno Schist in the Huizachal-Peregrina Anticlinorium (asterisk) lies in the eastern Sierra Madre (arrow) and is bound to the east by the ~1 Ga Novillo Gneiss (part of Oaxaquia). The Acatlán Complex forms the basement of the Mixteca terrane adjacent to Oaxaquia to the south.

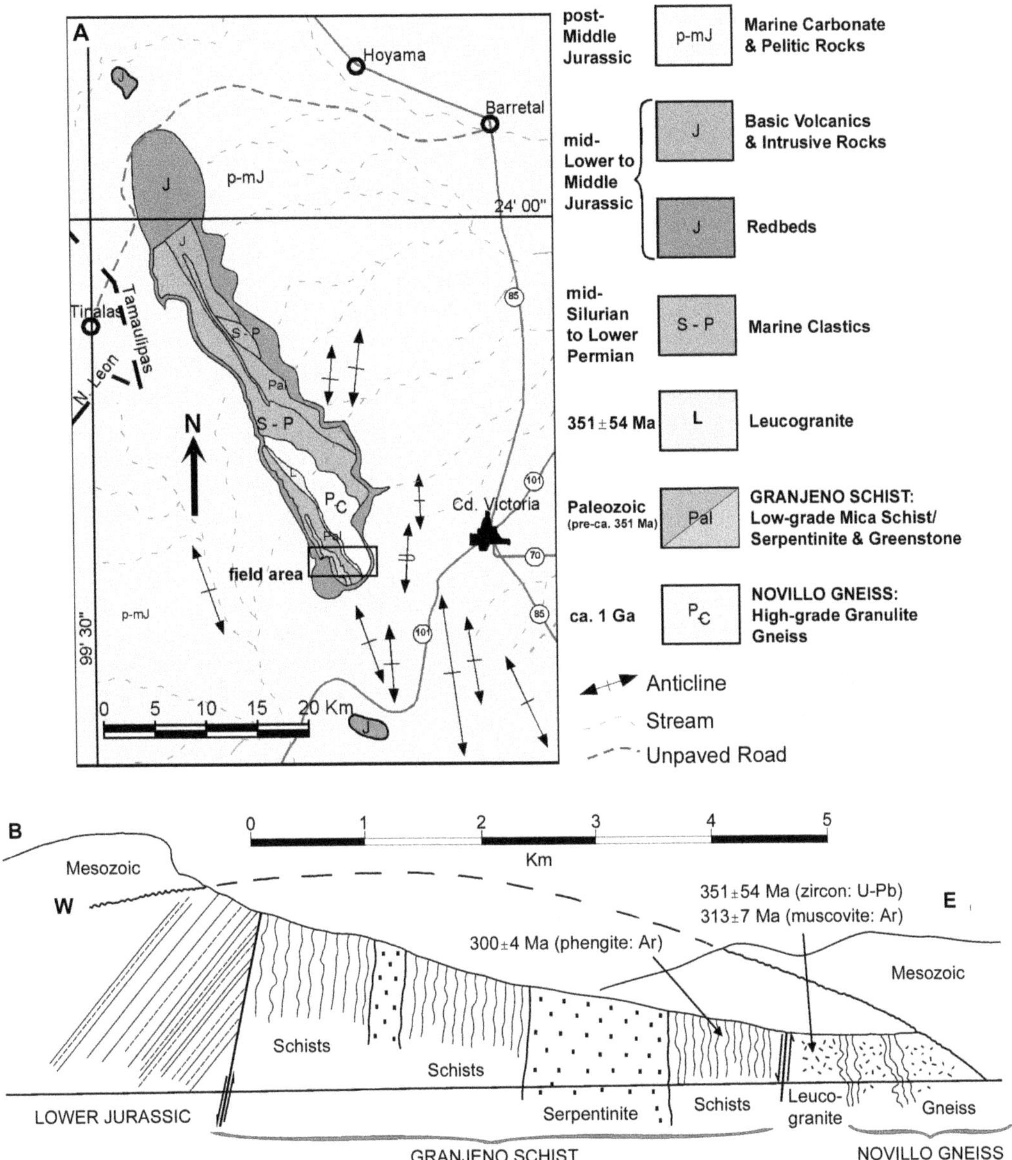

FIG. 2. A. Generalized geologic map of the Huizachal-Peregrina anticlinorium (after Ramírez-Ramírez, 1992) showing location of field area. B. E-W cross-section through center of field area (modified after Carrillo-Bravo, 1961).

these oceans (Yañez et al., 1991; Ramírez-Ramírez, 1992; Ortega-Gutiérrez et al., 1999). By contrast, lower Paleozoic strata of Laurentian affinity unconformably overlie Grenville basement rocks of the Appalachian-Ouachita orogen.

Geological Setting

The Huizachal-Peregrina anticlinorium is a basement-involved structure in the front ranges of the Laramide fold-thrust belt of the Sierra Madre

Oriental (Fig. 2A). Precambrian and Paleozoic rocks that crop out in the core of the anticlinorium constitute the largest exposure of such rocks in northeastern Mexico. Campa and Coney (1983) placed all of these rocks in the Sierra Madre terrane (Guachichil terrane of Sedlock et al., 1993). However, the contrast between the ophiolitic-oceanic nature of the Granjeno Schist and the miogeoclinal-continental strata overlying the Novillo Gneiss led Keppie (2004) to exclude the Novillo Gneiss from the Sierra Madre terrane. Instead, correlation of the Novillo and Oaxacan complexes support their inclusion in the Oaxaquia terrane (Ortega-Gutiérrez et al., 1995; Keppie, 2004). Juxtaposition of these two terranes has been interpreted as either: (1) pre-Wenlock (Fries et al., 1962; Cserna et al., 1977; Cserna and Ortega-Gutiérrez, 1978), based upon the identification of Granjeno Schist fragments in Wenlockian conglomerates unconformably overlying the Novillo Gneiss; or (2) post-metamorphic (Sedlock et al., 1993), based upon the absence of reset ~350–250 Ma ages in the Novillo Gneiss that are common in the Granjeno Schist.

Novillo Gneiss and Paleozoic Cover

The Novillo Gneiss consists of a variety of high-grade rocks including gabbro-anorthosite, granite, amphibolite, metaquartzite, talc-silicate units, and marble (Ortega-Gutiérrez, 1978; Ramírez-Ramírez, 1992). U-Pb geochronology (Cameron et al., 2004) reveals two groups of metaigneous protoliths, an older assemblage emplaced between ~1235 Ma and ~1115 Ma, and a younger AMCG (anorthosite-mangerite-charnockite-granite) suite emplaced between ~1035 Ma and ~1010 Ma. Emplacement of a post-tectonic anorthositic pegmatite at 978 ± 13 Ma followed granulite-facies metamorphism dated at 990 ± 5 Ma. K-Ar mineral ages recorded in the Novillo Gneiss range from 928 ± 28 Ma in phlogopite (Denison et al., 1971), through 919 ± 18 Ma and 880 ± 17 Ma in hornblende (Denison et al., 1971), to 744 ± 25 Ma in biotite (Fries et al., 1962). The Novillo Gneiss is in fault contact with the Granjeno Schist, along which is intruded an elongate, foliated plagioclase leucogranite ("plagiogranite" of Ramírez-Ramírez, 1992; Fig. 2B) reportedly containing xenoliths of both the Novillo Gneiss (Ortega-Gutiérrez, 1978) and Granjeno Schist (Gursky, 1996). This leucogranite has yielded Rb-Sr errorchrons with Precambrian ages of 774 ± 256 Ma and 570 ± 181 Ma (Garrison et al., 1980).

The Novillo Gneiss is unconformably overlain by an unmetamorphosed Paleozoic succession, the base of which comprises Middle Silurian (early to mid-Wenlock; 430–424 Ma; Okulitch, 2002) shallow-marine strata containing fauna of Gondwanan affinity (Boucot et al., 1997; Stewart et al., 1999). Pebbles of inferred Granjeno provenance have been reported in the Silurian conglomerates (Fries et al., 1962; Cserna et al., 1977; Cserna and Ortega-Gutiérrez, 1978); however, this was not confirmed by Stewart et al. (1999). The Silurian rocks are unconformably overlain by Lower Mississippian (early Osagean; 351–342 Ma; Okulitch, 2002) sandstones and shales, the fauna of which suggest shallow-marine deposition proximal to the North American craton. A flow-banded rhyolite at the top of this sequence has yielded a mid-Mississippian lower intercept U-Pb zircon age of 334 ± 34 Ma (Stewart et al., 1999). The Mississippian rocks are unconformably overlain by Lower to Middle Pennsylvanian (314–306 Ma; Okulitch, 2002) strata dominated by turbiditic lime grainstones that are, in turn, overlain by extensive Lower Permian (Wolfcampian and Leonardian) volcanogenic flysch deposited in a forearc basin (Gursky and Michalzik, 1989). All of these rocks are deformed by symmetric-asymmetric, open-close, NNW-trending folds, a few of which are associated with NE-vergent thrusts (Carrillo-Bravo, 1961; Gursky, 1996; Stewart et al., 1999).

Granjeno Schist

The Granjeno Schist consists of rocks of both sedimentary and igneous protolith (Carillo-Bravo, 1961; Ortega-Gutiérrez, 1978; Ramírez-Ramírez, 1974, 1992). Metasedimentary lithologies are dominated by pelitic schist, but also include psammitic, silicic (metachert), graphitic, and volcaniclastic schist, whereas meta-igneous rocks include metabasite, metagabbro, and serpentinite. Pelitic lithologies are essentially phyllites that occur rhythmically interlayered on a centimeter scale with fine-grained psammitic schist. Quartz veins intruded parallel to the bedding provide a record of the various phases of deformation. Small (0.2–0.4 mm), subangular to subrounded, homogeneously distributed quartz and albite porphyroblasts are common (up to 40%) in these pelitic interlayers. Mineral assemblages are dominated by quartz with phengite, chlorite (after biotite), minor albite and graphite, and rare pre- to syntectonic garnet. Interbedded graphitic schists contain equal amounts of quartz and graphite, whereas fine-grained volcaniclastic schists show rhythmic color alternations in light blue-green and dark grey.

Stratigraphically interlayered with the metasedimentary rocks are metabasites, including massive metabasalts that rarely exceed a few meters in thickness, and a single mafic volcaniclastic unit that is greenish in color, aphanitic, and up to 30 meters in thickness. Mineral assemblages in both metabasites are dominated by varying amounts of actinolite, chlorite, albite, clinozoisite, epidote, calcite, and quartz.

A large metaperidotite-metagabbro body (~0.5 × 10 km) and several smaller lenses are tectonically enclosed within the metasedimentary and meta-igneous rocks of the Granjeno Schist. The ultramafic body is a strongly serpentinized, foliated, lizardite-chrysotile rock with rare pyroxenite (cpx-opx) and calcium-rich metasomatic borders, whereas the metagabbro is massive and occasionally preserves relict cumulate textures. Metagabbro mineral assemblages, like those of the metabasites, are dominated by actinolite, albite, chlorite and epidote group minerals. Brown microcrystalline silica-rich layers in proximity to the serpentinite mass may represent metacherts.

An early Paleozoic age for the unfossiliferous Granjeno Schist has been assumed on the basis of the oldest whole-rock Rb-Sr age of 452 ± 45 Ma on the schists (Cserna and Ortega-Gutiérrez, 1978, recalculated for revised decay constant by Sedlock et al., 1993). Other Rb-Sr whole-rock and whole rock–muscovite isochron ages on the metasediments range from 373 ± 37 Ma to 320 ± 12 Ma (Cserna and Ortega-Gutiérrez, 1978 recalculated; Garrison et al., 1980). Muscovite K/Ar age data range from 318 ± 10 Ma to 257 ± 8 Ma (Fries et al., 1962; Cserna et al., 1977), and suggest an episode of late Paleozoic greenschist-facies metamorphism (Ramírez-Ramírez, 1992).

Overstep Sequence

Folds in the Paleozoic rocks are truncated by the unconformity below the Mesozoic sequence, the oldest unit of which (Huizachal Formation), originally thought to extend into the Late Triassic on the basis of Rhaetic plant remains (Carrillo-Bravo, 1961), is now considered to be no older than the middle Early Jurassic (Stewart et al., 1999). This Mesozoic sequence consists of Middle Jurassic conglomerates and redbeds, Middle Jurassic redbeds and limestones, and Upper Jurassic to Cretaceous carbonates and evaporites (Stewart et al., 1999) that together represent a rift-drift succession associated with the opening of the Gulf of Mexico (Sedlock et al., 1993). The sequence was deformed during the latest Cretaceous to mid-Eocene Laramide orogeny, which produced a broad open anticline in the sub-Jurassic unconformity and E- to NE-vergent thin-skinned thrusting and tighter overturned folds in the overlying rocks that appear to be coaxial with those in the Paleozoic rocks (Carrillo-Bravo, 1961). The Laramide Orogeny occurred in response to east-vergent subduction along the Pacific margin (Sedlock et al., 1993).

Deformational History

To clarify the deformational history of the Granjeno Schist, its structural geometry was examined in detail in well-exposed sections along the Cañón de Novillo and its tributaries at the southern end of the Huizachal-Peregrina anticlinorium (Fig. 2A). At least four sets of fold structures can be recognized in outcrop. The earliest fold set (F_1) is assigned to a deformational event designated D_1. The remaining folds (F_{2a}–F_{2c}), however, are considered to record phases of a single progressive deformation event designated D_2. Figure 3 shows equal area stereographic projections of the minor structures associated with each of the four fold phases.

D_1 structures

The first recognizable fabric consists of a bedding-parallel foliation (S_1), which is axial planar to mesoscopic isoclinal folds (F_1) that lie in a steeply dipping, NNW-SSE trending plane and plunge moderately to steeply northwest or southeast (Fig. 3A). Bedding-parallel quartz veins folded by F_1 provide the best record of this phase of deformation in outcrop (Fig. 4A), but microstructural analysis shows that F_1 closures are also preserved as rootless isoclinal hinges in the composite matrix foliation. They are also preserved in albite porphyroblasts as folded trails of graphite and white mica inclusions aligned at an angle to the external foliation (Fig. 5), indicating that the porphyroblasts grew late to post-kinematically with respect to D_1. Garnet appears to have grown at the same time. Any lineation that may have been associated with D_1 has been entirely overprinted by later deformation.

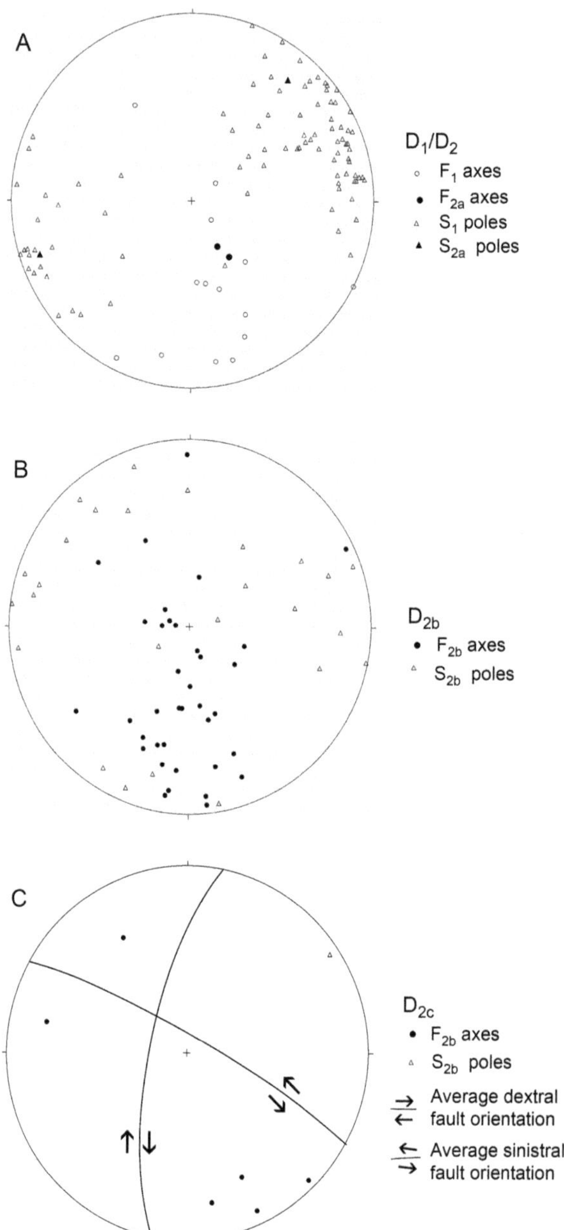

FIG. 3. Equal-area stereographic projections of fold axes and poles to foliation within the Granjeno Schist. A. D_1/D_{2a} structures. B. D_{2b} structures. C. D_{2c} structures.

D_{2a} structures

The second deformation also produced a bedding-subparallel foliation (S_{2a}), which is axial planar to mesoscopic tight to isoclinal folds (F_{2a}) that plunge predominantly southeast (Fig. 3A). These folds, which are coplanar but non-coaxial with those of F_1 (Fig. 4B), are most evident where S_{2a} is at high angles to the bedding-parallel S_1 foliation in F_{2a} closures, particularly where F_1 closures occur near F_{2a} hinges (inset Fig. 4). On the limbs of F_{2a} folds,

FIG. 4. Field photographs of D_1/D_{2a} structures. A. Rootless F_1 isoclinal closures within composite S_1/S_{2a} foliation in pelitic schist. B. Tight F_{2a} fold closure in bedding-parallel S_1. Inset shows F_1 isocline refolded about F_{2a} in graphitic schist. Pencil (left) is for scale.

however, S_{2a} and S_1 are mutually parallel and indistinguishable, such that the foliation, which is defined by stretched and flattened quartz grains and the alignment of phengite, chlorite (after biotite), and graphite in pelitic lithologies and the alignment of actinolite and chlorite in metabasites, is a composite (S_1/S_{2a}) fabric throughout much of the Granjeno Schist. Given that the orientations of the F_1 and F_2 fold axes are quite similar and their styles of deformation as seen in outcrop are difficult to distinguish unless F_1 hinges are found near F_{2a} closures, separation of the two phases is largely incomplete. Stereonets showing D_1–D_{2a} fold data (Fig. 3) consequently show only two definitive F_2 axes, although others are likely present among the plotted axes. The style of deformation related to D_{2a} (isoclinal folding and penetrative axial planar fabric) is nearly identical to that of D_1, yet the two episodes must be separated in time by the intrusion of the leucogranite, given that the latter contains xenoliths of both Novillo Gneiss and Granjeno Schist (Fries and Rincon-Orta, 1965) and is affected by D_{2a}. In general, the composite S_1/S_{2a} strikes NW-SE and dips steeply to the southwest or northeast. In thin section, S_{2a} is preserved as concordant inclusions in the rims of metavolcaniclastic albite porphyroblasts that are otherwise enveloped by the S_1/S_{2a} foliation, suggesting that rim growth occurred early during F_{2a}. Where present, biotite replaced by chlorite defines the composite S_1/S_{2a} foliation, which wraps around rare garnet porphyroblasts, suggesting that the garnet grew at the same time as the albite porphyroblasts. Like those of F_1, F_{2a} axes plunge southeast at moderate angles (Fig. 3A), but where both occur together, F_1 axes lie at about 30 degrees to F_{2a} in the S_1/S_{2a} plane, producing shallowly SE-plunging F_1 axes on the west limbs of F_{2a} folds, and steeply SE- to NE-plunging F_1 axes on the east limb of F_{2a} folds. F_{2a} is not accompanied by a recognizable lineation, suggesting either over-

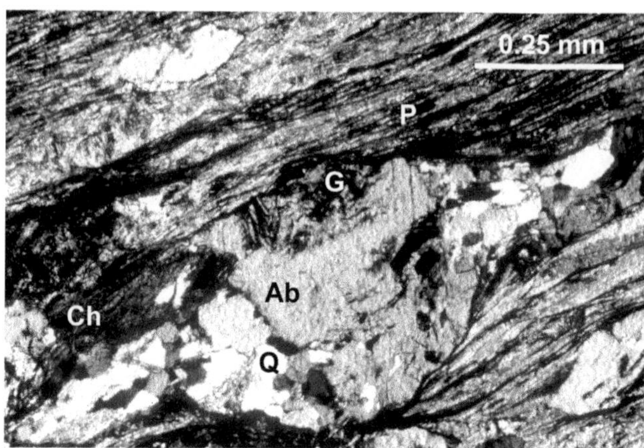

FIG. 5. Photomicrograph (in crossed-polars) of pelitic schist showing albite (Ab) porphyroblast containing folded inclusions of graphite (G) defining S_1. Abbreviations: P = phengite; Ch = chlorite; Q = quartz. 0.25 mm bar is for scale.

printing during D_{2b} or the lack of a significant simple shear component during F_{2a} deformation.

D_{2b} *structures*

The third fabric-forming event produced a prominent crenulation cleavage (S_{2b}), which is axial planar to mesoscopic and microscopic folds (F_{2b}) that are broadly coaxial with F_{2a} and predominantly plunge south at gentle to steep angles (Fig. 3B). These structures refold earlier fabrics and are particularly prominent in pelitic lithologies (Fig. 6A). In thin section, the crenulation cleavage is defined by the alignment of phengite and chlorite, either as a result of folding (zonal crenulation cleavage), or through the development of discontinuities along the cleavage planes (discrete crenulation cleavage). F_{2b} shortening is estimated at 30% in the metapelites, based on unfolding the folded layers (e.g., Davis and Reynolds, 1996), but only about 10% in the metapsammitic and metaigneous lithologies. S_{2b} dips SE or west to NW at moderate to steep angles and intensifies toward the tectonic boundaries between the Granjeno Schist and the serpentinite-gabbro body to the west, and the leucogranite to the east, the latter being a steeply dipping, NNW-trending ductile shear zone containing dextral kinematic indicators (Fig. 7A). The presence of S_{2a}-C fabrics in the schists within this ductile shear zone suggests that D_{2a-2b} records progressive stages of tectonic juxtapositioning of the Granjeno Schist and the Novillo Gneiss. Similar dextral ductile shear zones developed within the serpentinite (Fig. 7B) suggest that the final tectonic juxtapositioning of the Granjeno Schist and the serpentinite-gabbro body also occurred at this time. Associated with S_{2b} is a lineation (L_{2b}) produced either by the crenulation of S_1/S_{2a} foliation surfaces or by the intersection of these surfaces with S_{2b}. L_{2b} parallels the F_{2b} axes and plunges SW to SE at shallow to steep angles.

The orientation of D_{2b} structures is highly variable, largely because the S_{2b} crenulation cleavage is oblique to the S_1/S_{2a} composite foliation, which defines the geometry of subsequent F_{2c} folds. This has the effect of deforming S_{2b} into cone shapes. Unfolding of S_{2b} about F_{2c} structures produces ambiguous results, but F_{2b} axes restored in this fashion plunge at shallow angles to the south and SW or to the north and NE. In the southwest corner of the field area, which is unaffected by D_{2c} and so may preserve the original orientation of D_{2b} structures, F_{2b}/L_{2b} axes plunge gently south and S_{2b} is shallow dipping.

D_{2c} *structures*

The fourth set of structures are steeply inclined, moderately SE-plunging, close class 1C-2 folds (F_{2c}) that lack an associated cleavage (Fig. 3C). These structures fold the S_{2b} crenulation cleavage and the L_{2b} lineation, such that L_{2b} plunges gently NNW on the northeast limbs of F_{2c} folds and moderately SSW on southwest limbs of F_{2c} (Fig. 6B). F_{2c} axes are slightly oblique to those of F_{2b} and plunge steeply NNW or SSE. When plotted stereographically, deformation of L_{2b} about F_{2c} axes produces a broadly small-circle distribution (at 30–40° to F_{2c}), suggesting a component of flexural flow in the

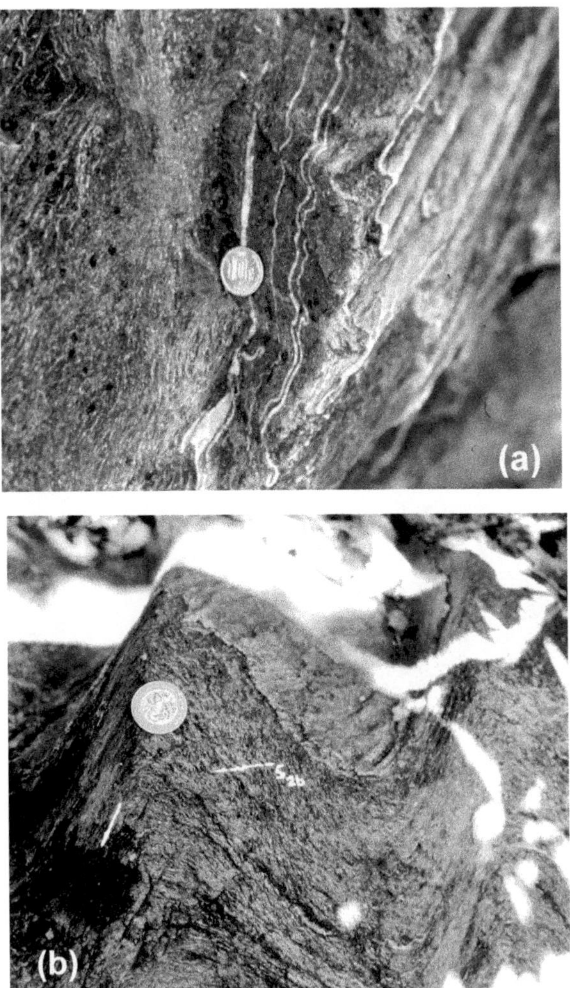

FIG. 6. Field photographs of D_{2b} and D_{2c} structures in pelitic schist. A. Small-scale F_{2b} closures and S_{2b} crenulation cleavage. B. S_{2b} crenulation cleavage (outlined in white) folded about upright F_{2c} fold.

development of F_{2c} folds. This, together with the shape of the fold profile, suggests a fold mechanism involving buckling followed by homogeneous flattening. The clockwise sense of rotation of F_{2b} relative to F_{2c} is consistent with continued dextral shear along the boundary between the Granjeno Schist and Novillo Gneiss, suggesting that F_{2a}–F_{2c} may represent phases of a single progressive deformation, albeit at decreasing metamorphic temperatures. In most areas of the schist, F_{2c} folds are developed in association with steeply dipping brittle faults, whereas D_{2a} and the early stages of D_{2b} developed under greenschist-facies metamorphism.

Plotted stereographically, the dextral and sinistral faults, and F_{2c} axes, show a geometric relationship to the bounding NNW-trending dextral shear zones, suggesting that D_{2c} records post-metamorphic deformation associated with final tectonic juxtapositioning in the late Paleozoic. The F_{2c} folds are geometrically similar to those recorded in the Silurian–Lower Permian rocks resting unconformably on the Novillo Gneiss, which include NNW-trending, open-tight, upright to asymmetric folds, in places associated with thrusts that involve the basement gneisses (Stewart et al., 1999). These folds appear to have been truncated by the unconformity

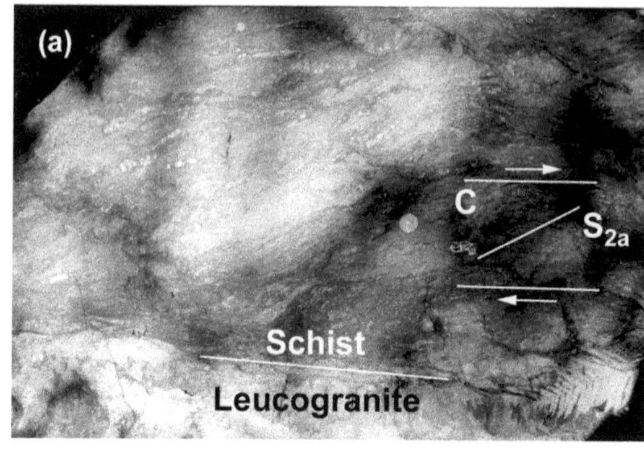

FIG. 7. Field photographs of dextral C-S_{2a} fabrics. A. In ductile shear zone separating Granjeno Schist and leucogranite. B. In shear zone within serpentinite body.

at the base of the Mesozoic sequence (Carrillo-Bravo, 1961; Ramírez-Ramírez, 1992), bracketing their age, and that of D_{2c}, between Early Permian and Early Jurassic.

Metamorphic History

Stable mineral assemblages of the Granjeno Schist are those of the lower greenschist facies (chlorite zone), characterized by quartz-albite-phengite-chlorite-graphite ± calcite in the metasedimentary protoliths, albite-chlorite-actinolite-clinozoisite-epidote-quartz-calcite ± titanite in the metabasites, and chrysotile-lizardite in the serpentinites. Such assemblages suggest metamorphic P-T conditions of about 350–400°C at 300–600 MPa (e.g., Bucher and Frey, 1994). Within serpentinites and metabasites, evidence of an earlier, higher-pressure metamorphism may be represented by the reports of pumpellyite and a green lavender-colored actinolite (Ramírez-Ramírez, 1992) that might be glaucophane. Similarly, the rare preservation of relict biotite and almandine garnet in the metasedimentary schists may indicate the existence of an earlier upper greenschist (garnet zone) event in at least this component of the Granjeno Schist. The garnets are enveloped by the composite $S_{1\text{-}2a}$ foliation and, together with the folded s-trails in albite, suggest that this metamorphism occurred essentially between D_1 and D_{2a}. The concordant penetration of these trails into the rims of albite porphyroblasts, and the fact that the composite foliation wraps about them, suggest that the retrograde growth of these relatively inclusion-free rims also occurred synkinematically with respect to D_{2a}. Retrogression to the chlorite zone is likely to have coincided with D_{2a} inasmuch as the composite S_1/S_{2a} foliation is additionally defined by retrograde chlorite (after biotite) and phengite.

D_{2b} likely occurred under chlorite-zone to sub-greenschist-facies conditions. The S_1/S_{2a} phengite-

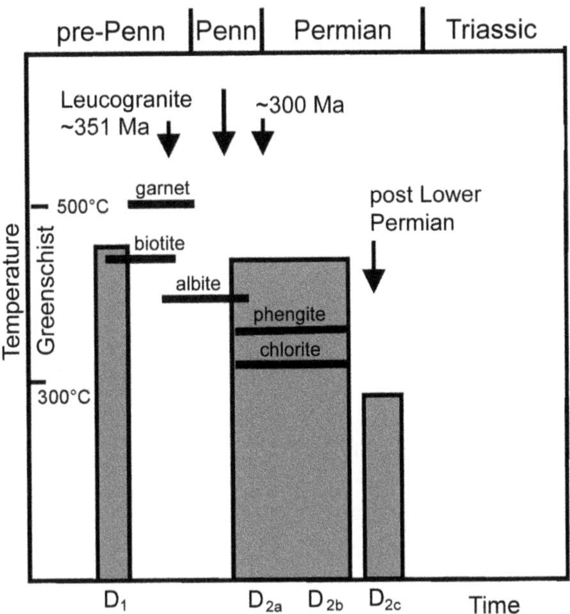

FIG. 8. Schematic temperature-time diagram for the Granjeno Schist, showing approximate temperatures within the greenschist facies represented by metamorphic mineral parageneses (solid bars), and the timing of their appearance with respect to deformational phases (shaded columns) and geochronological/stratigraphic data (arrows). Abbreviation: Penn = Pennsylvanian.

chlorite fabric appears to have remained stable during F_{2b} crenulation to produce the S_{2b} cleavage, and new growth of phengite-chlorite accompanied dextral movement within the ductile shear zone bordering the Granjeno Schist. This suggests that the early stages of F_{2b} development occurred under similar metamorphic conditions to those of F_{2a}. Thus, the 313 ± 7 Ma and 300 ± 4 Ma cooling ages recorded, respectively, in muscovite and phengite (see below) probably closely post-date the development of S_{2a} and early S_{2b} fabrics. D_{2b}, however, likely extended into the Permian under subgreenschist-facies conditions. Hence, D_{2a} and D_{2b} are considered contemporary with a single period of metamorphic retrogression. D_{2c} is not associated with new mineral growth, but occurred under metamorphic conditions no higher than those of D_{2b}. The approximate temperatures represented by the various metamorphic mineral parageneses and the timing of their appearance with respect to the deformational phases are schematically summarized in Figure 8.

Geochronology

U-Pb zircon geochronology

Zircons separated from the leucogranite emplaced along the fault separating the Granjeno Schist from the Novillo Gneiss were analyzed using the methodology described by Lopez et al. (2001). Analyses vary from single, irregular-prismatic, abraded zircons to small populations of prismatic grains (Table 1). All the data are discordant and fall on a chord that intersects concordia at 990 ± 46 Ma and 351 ± 54 Ma (Fig. 9), which are interpreted to approximate the age of inheritance and the time of intrusion, respectively. Although poorly constrained, the latter is identical within error to the 334 ± 39 Ma lower intercept age determined for the El Aserradero Rhyolite (Stewart et al., 1999), which occurs in the Paleozoic sequence on top of the Novillo Gneiss between the Lower Mississippian Vicente Guerrero Formation and the Middle–Upper Pennsylvanian Del Monte Formation (~335–314 Ma; Okulitch, 2002). Hence, it is possible that the leucogranite may have fed the rhyolite. The inheritance age is also poorly constrained but falls within the range of those reported from the Novillo Gneiss (Cameron et al., 2004).

$^{40}Ar/^{39}Ar$ laser step-heating analyses

A muscovite separate from the leucogranite and a phengite separate from a quartz keratophyre unit in the Granjeno Schist were analyzed by $^{40}Ar/^{39}Ar$ laser step-heating thermochronology using the methodology reported in Keppie et al. (2004b). Amphibole separates from a metagabbro gave highly discordant results and the data are not reported.

Muscovite that defines the foliation in the leucogranite yielded an Ar-loss profile with an age of 138 ± 10 Ma in the lowest temperature step, rising to a plateau of 313 ± 7 Ma in the four highest temperature steps (Fig. 10A; Table 2). Given the range in muscovite grain size, the plateau age is inferred to date cooling through ~350°C, based on the diffusion equations of Dodson (1973).

The phengite separate from the quartz keratophyre yielded a plateau with an age of 300 ± 4 Ma (Fig. 10B; Table 2), which is likewise interpreted to date cooling through ~350°C (Dodson, 1973). This date falls on the Carboniferous–Permian boundary (Okulitch, 2002).

Discussion

A younger age limit on the first deformation of the Granjeno Schist may be provided by the reports of Granjeno pebbles in Middle Silurian rocks overlying the Novillo Gneiss (Fries et al., 1962; Cserna et al., 1977; Cserna and Ortega-Gutiérrez, 1978). However, Stewart et al. (1999) failed to confirm this observation despite detailed petrographic study. In view of this, the significance of the oldest recalculated whole-rock Rb-Sr age of 452 ± 45 Ma reported for the schists (Cserna and Ortega-Gutiérrez, 1978) is uncertain. The lithologic association of pelitic sediments, mafic volcanic rocks, serpentinites, and gabbros, are typical of oceanic settings, in particular, those of accretionary prisms. The high initial $^{87}Sr/^{86}Sr$ ratio (0.7120 ± 0.0027) of the pelitic sediments suggests derivation from a cratonic source (Garrison et al., 1980). Hence, subduction and obduction close to a continental margin are likely to have been involved in its early deformational history. This is supported by the emplacement of the serpentinite-gabbro body into the Granjeno Schist during D_1. Whether such processes are recorded in the earliest (D_1) phase of deformation is uncertain,

TABLE 1. U-Pb Geochronologic Data

Fraction	Description1	Weight, mg	U, ppm	Total Pb, ppm	Com. Pb, pg	$^{206}Pb/^{204}Pb$ Measured	$^{207}Pb^*/^{206}Pb^*$ Atomic ratio	pct. err.	$^{208}Pb^*/^{206}Pb^*$ Atomic ratio	$^{206}Pb^*/^{238}U$ Atomic ratio	pct. err.	$^{207}Pb^*/^{235}U$ Atomic ratio	pct. err.	$^{206}Pb^*/^{238}U$ Age (Ma)	$^{207}Pb^*/^{235}U$ Age (Ma)	$^{207}Pb^*/^{206}Pb^*$ Age (Ma)	%Dis.	Tot. Pb on fil. (pg)	Pb/Pb age		Rho 6/8-7/5
N28 Post-Grenville granite																					
A	abr., 1:2 prisms (3)	49	192	23	57	1170	0.067058	0.11	0.09602	0.11289	0.19	1.0438	0.22	690	726	840 ± 2	17.9	1110	840	5%	0.86
C	abr., 1:2 prisms (6)	18	244	22	14	1797	0.06404	0.18	0.0698	0.09260	0.38	0.8177	0.42	571	607	743 ± 4	23.2	401	743	3%	0.91
D1	1:2 prisms (3)	33	128	12.6	11	2317	0.06617	0.22	0.1031	0.09736	0.39	0.8883	0.45	599	645	812 ± 5	26.2	416	812	3%	0.88
D2	1:2 prisms (3)	46	115	12.9	26	1379	0.06618	0.23	0.1091	0.10835	0.31	0.9886	0.39	663	698	812 ± 5	18.3	593	812	4%	0.80
E1	irreg., abr. (1)	113	88	10.8	74	918	0.06774	0.18	0.1719	0.10943	0.18	1.0221	0.26	669	715	861 ± 4	22.2	1220	861	6%	0.72
E2	irreg., abr. (1)	39	71	10.8	11	2181	0.07081	0.36	0.2046	0.13929	0.58	1.3599	0.68	841	872	952 ± 7	11.7	421	952	3%	0.85
E3	irreg., abr. (3)	48	130	19.9	16	3617	0.07072	0.08	0.1405	0.14619	0.22	1.4254	0.23	880	900	949 ± 2	7.3	955	949	2%	0.94
F2	1:2 prisms (1)	22	194	20.5	11	2230	0.06274	0.40	0.3539	0.08654	0.43	0.7486	0.59	535	567	699 ± 9	23.5	451	699	2%	0.73
G	1:2 prisms (6)	23	322	26.4	16	2449	0.06418	0.13	0.0065	0.08377	0.32	0.7413	0.35	519	563	748 ± 3	30.6	607	748	3%	0.93

[1]Pb* = radiogenic lead; abr. = abraded, prsm. = prismatic; irreg. = irregular; 1:2 = aspect ratio; com. = common; f. = fraction; fil. = filament; Tot. = total.

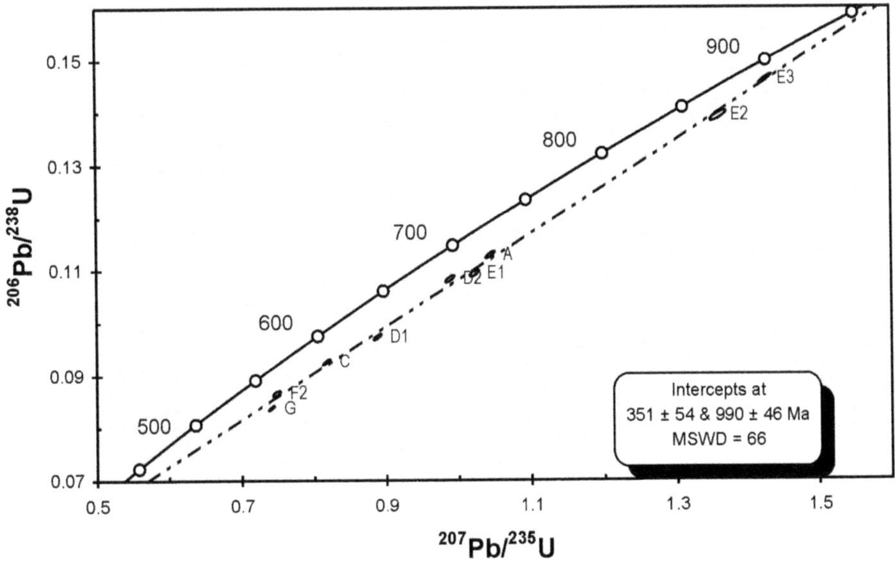

FIG. 9. U-Pb concordia diagram for zircons from the leucogranite.

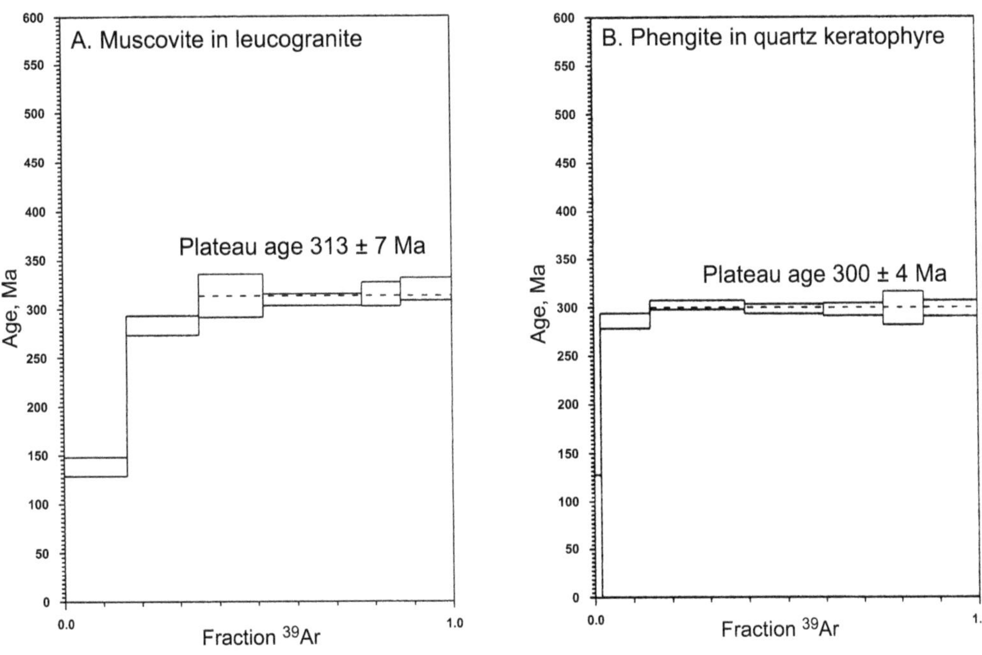

FIG. 10. $^{40}Ar/^{39}Ar$ apparent age spectra of (A) muscovite in leucogranite, and (B) phengite in quartz keratophyre.

although the absence of mineral stretching lineations suggests that S_1 developed largely under conditions of simple flattening.

Subsequent Carboniferous and Permian deformation are indicated by several factors. The 351 ± 54 Ma leucogranite appears to have been intruded

TABLE 2. $^{40}Ar/^{39}Ar$ Data for Muscovite in Leucogranite and Phengite in Quartz Keratophyre[1]

Power	$^{36}Ar/^{40}Ar$	$^{39}Ar/^{40}Ar$	r	Ca/K	%^{40}Atm	%^{39}Ar	$^{40}Ar^{*}/^{39}K$	Age, Ma	^{40}Ar	^{39}Ar	^{38}Ar	^{37}Ar	^{36}Ar	Blank Ar	Atmos 40/36
								Muscovite							
1.00	0.000269 ± 0.000225	0.087642 ± 0.000784	0.006	0.016	7.92	16.37	10.504 ± 0.764	138.40 ± 9.69	8.847 ± 0.061	0.782 ± 0.004	0.019 ± 0.003	0.003 ± 0.002	0.004 ± 0.000	0.028	288.024
2.00	0.000046 ± 0.000110	0.044048 ± 0.000836	0.007	0.005	1.35	18.62	22.395 ± 0.853	283.25 ± 9.98	19.970 ± 0.357	0.890 ± 0.005	0.016 ± 0.003	0.002 ± 0.002	0.003 ± 0.000	0.035	288.024
<3.00	0.000027 ± 0.000210	0.039648 ± 0.001804	0.004	0.004	0.79	16.73	25.022 ± 1.937	313.74 ± 22.29	19.970 ± 0.721	0.807 ± 0.022	0.014 ± 0.005	0.003 ± 0.004	0.002 ± 0.000	0.070	288.024
<4.00	0.000022 ± 0.000068	0.040324 ± 0.000196	0.001	0.008	0.65	25.13	24.638 ± 0.516	309.32 ± 5.95	29.420 ± 0.048	1.198 ± 0.005	0.021 ± 0.003	0.002 ± 0.002	0.002 ± 0.000	0.028	288.024
<6.00	0.000027 ± 0.000141	0.039498 ± 0.000302	0.001	0.018	0.78	10.03	25.119 ± 1.074	314.85 ± 12.35	12.001 ± 0.028	0.483 ± 0.003	0.014 ± 0.003	0.003 ± 0.002	0.002 ± 0.000	0.028	288.024
<7.00	0.000024 ± 0.000130	0.038850 ± 0.000293	0.0001	0.008	0.70	13.13	25.561 ± 1.004	319.93 ± 11.52	15.961 ± 0.041	0.629 ± 0.004	0.014 ± 0.003	0.002 ± 0.002	0.002 ± 0.000	0.028	288.024
								Phengite							
0.50	0.002162 ± 0.001544	0.086636±0.002736	0.009	0.077	63.71	1.67	41.70 ± 5.270	56.81 ± 70.67	1.291 ± 0.018	0.119 ± 0.003	0.008 ± 0.002	0.004 ± 0.002	0.005 ± 0.000	0.032	288.024
2.00>	0.000181 ± 0.000088	0.042261 ± 0.000386	0.011	0.014	5.33	12.94	22.400 ± 0.654	286.07 ± 7.72	19.948 ± 0.136	0.855 ± 0.005	0.016 ± 0.002	0.004 ± 0.002	0.006 ± 0.000	0.033	288.024
<3.00>	0.000063 ± 0.000052	0.041241 ± 0.000284	0.005	0.007	1.86	24.45	23.796 ± 0.406	302.48 ± 4.75	38.586 ± 0.202	1.607 ± 0.007	0.025 ± 0.003	0.004 ± 0.002	0.005 ± 0.000	0.032	288.024
<4.00>	0.000051 ± 0.000051	0.041999 ± 0.000334	0.006	0.005	1.52	20.77	23.448 ± 0.406	298.41 ± 4.76	32.188 ± 0.212	1.367 ± 0.006	0.021 ± 0.002	0.004 ± 0.002	0.004 ± 0.000	0.032	288.024
<5.00>	0.000040 ± 0.000071	0.042213 ± 0.000457	0.005	0.006	1.18	15.66	23.409 ± 0.561	297.95 ± 6.58	24.158 ± 0.230	1.032 ± 0.005	0.018 ± 0.002	0.004 ± 0.002	0.003 ± 0.000	0.032	288.024
<6.00>	0.000046 ± 0.000191	0.041932 ± 0.000939	0.005	0.000	1.36	10.52	23.524 ± 1.445	299.30 ± 16.94	16.382 ± 0.324	0.703 ± 0.007	0.013 ± 0.004	0.004 ± 0.003	0.003 ± 0.000	0.062	288.024
<7.00>	0.000035 ± 0.000090	0.042130 ± 0.000411	0.003	0.005	1.03	13.97	23.491 ± 0.673	298.91 ± 7.89	21.599 ± 0.175	0.922 ± 0.005	0.018 ± 0.002	0.003 ± 0.002	0.003 ± 0.000	0.032	288.024

[1]For leucogranite: mass = 1.0 mg; volume 39K = 4.71 × 1E-10^3 NTP; integrated age = 280.27 ± 5.60 Ma; for initial 40/36, linear regression has positive slope; correlation age = 288.66 ± 50.26 Ma (98.3% of ^{39}Ar, steps marked by >); plateau age = 313.46 ± 7.34 Ma (65.0% of ^{39}Ar, steps marked by <); J = .007591 ± 0.000068. For phengite: mass = 1.0 mg; volume 39K = 6.49 × 1E-10^3 NTP; integrated age = 294.11 ± 3.65 Ma; initial 40/36 = 1046.06 ± 3489.73 (MSWD = 2.44, isochron between 0.29 and 2.41); correlation age = 288.66 ± 50.26 Ma (98.3% of ^{39}Ar, steps marked by >); plateau age = 299.68 ± 3.77 Ma (85.4% of ^{39}Ar, steps marked by <); J = .007671 ± 0.000052.

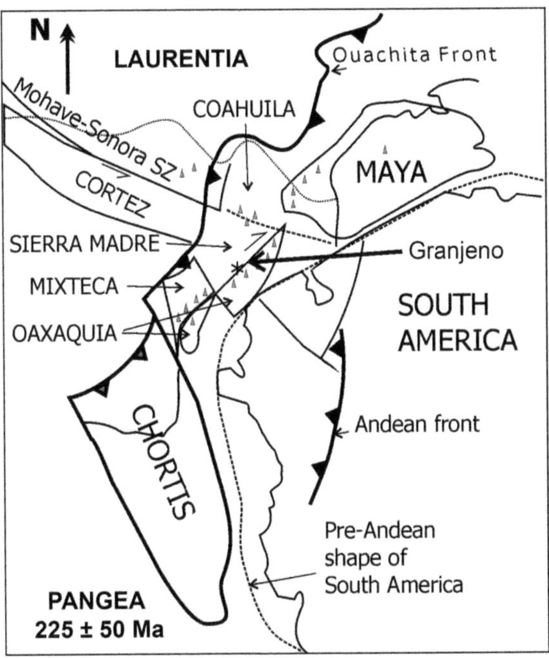

Fig. 11. Pangea reconstruction showing approximate location of Granjeno Schist (asterisk), Actalán Complex (Mixteca terrane), and Oaxaquia with respect to Permian arc (small triangles) in Permo-Triassic reconstruction of North, Middle, and South America (modified from Keppie, 2004).

along the fault between the Granjeno Schist and Novillo Gneiss, which is inferred to have been active during intrusion. The muscovite cooling age of 313 ± 7 Ma in the leucogranite coincides with the hiatus in the late Paleozoic stratigraphy between early Osagean and Lower–Middle Pennsylvanian. This suggests that dextral movements along the NNW-trending fault started in the mid-Carboniferous. That deformation continued into the Early Permian is suggested by: (1) the 300 ± 4 Ma phengite plateau age (Carboniferous–Permian boundary; Okulitch, 2002); (2) thrusting and folding that affects Lower Permian rocks (Stewart et al., 1999); and (3) F_{2c} structures that appear to have formed in the same dextral shear regime. The orientation of F_{2a}–F_{2c} axes suggests a strong component of transpression and progressive clockwise rotation consistent with dextral motion. The combination of dextral shearing along sub-vertical faults and folding/thrusting adjacent to such faults suggests that the regional geometry is that of a positive flower structure.

The unconformably overlying Lower Jurassic redbeds are folded but unmetamorphosed, their deformation reflecting the latest Cretaceous to mid-Eocene Laramide orogeny. Low-temperature increments of the $^{40}Ar/^{39}Ar$ muscovite spectra that yield Mesozoic ages likely record argon loss concomitant with the development of the Laramide fold-thrust belt, of which the Huizachal-Peregrina anticlinorium is part.

Conclusions

The Late Carboniferous to post-Early Permian dextral shear deformation (D_2) in the Granjeno Schist, which records its progressive juxtaposition against the Novillo Gneiss, encompasses the interval of dextral ductile shear deformation in the Acatlán Complex of southern Mexico, which records its juxtaposition against the ~1 Ga Oaxacan Complex in the Early Permian (Weber et al., 1997; Elías-Herrera and Ortega-Gutiérrez, 2002; Malone et al., 2002; Keppie et al., 2004b). This lends support to the correlation of the Novillo Gneiss and the Oaxaca Complex (Ortega-Gutiérrez et al., 1995), the boundary between Paleozoic metasedimentary rocks and ~1 Ga basement being apparently marked by a gravity gradient from positive to negative values, respectively (de la Fuente Duch et al., 1991).

In the Acatlán Complex, deformation involved: (1) tight to isoclinal folding associated with the development of a composite foliation that records S-vergent thrusting and dextral transpression under lower greenschist facies conditions, followed by (2) upright, N-S folding and the development of a crenulation cleavage (Malone et al., 2002). Both phases of deformation are of Early–Middle Permian age, and may be correlated with F_{2a-2b} in the Granjeno Schist. The Early–Middle Permian deformation is synchronous with arc magmatism on the western margin of Pangea that extends along the backbone of Mexico (Torres et al., 1999). In this context, the Granjeno Schist/Novillo Gneiss lie in the forearc region between the inferred trench and the arc (Fig. 11). That Pangea had begun to amalgamate in the Carboniferous is shown by the Laurentian affinity of Mississippian fauna in the Paleozoic strata overlying the Novillo Gneiss and the Oaxacan Complex (Stewart et al., 1999; Navarro-Santillan et al., 2002). Amalgamation of Pangea is estimated by Dickinson and Lawton (2001) to have ended at ~281 Ma, i.e., in the mid-Early Permian.

On the other hand, the age of the D_1 structures in the Granjeno Schist is uncertain because the reports of Granjeno pebbles in pre-Middle Silurian conglomerate overlying the Novillo Gneiss were not confirmed by Stewart et al. (1999). D_1 eclogite-facies structures in the Acatlán Complex have recently been dated as Early Mississippian (Middleton et al., 2004; Nance et al., 2004; Keppie et al., 2004a), and suggest the presence of a subduction zone off the western side (present coordinates) of Oaxaquia. It is therefore possible that D_1 structures in the Granjeno Schist developed during obduction associated with a Permo-Carboniferous arc. The age of this subduction/obduction event is contemporaneous with the last stages of closure of the Rheic Ocean, rather than the closure of the Iapetus Ocean in Late Ordovician–Early Silurian time.

Acknowledgments

This project was made possible by Program for North American Mobility in Higher Education grants from the Fund for the Improvement of Post-secondary Education (to RDN) and the Secretaria de Educatión Pública de México (to JDK), and an Ohio University 1804 Award to RDN. RDN and JDK also acknowledge the support of NSF grant EAR 0308105 and PAPIIT grant IN103003, respectively. This paper is a contribution to IGCP Project 497: The Rheic Ocean. This paper has benefited from the constructive comments of Luigi Solari and Brendan Murphy.

REFERENCES

Boucot, A. J., Blodgett, R. B., and Stewart, J. H., 1997, European Province Late Silurian brachiopods from the Ciudad Victoria area, Tamaulipas, northeastern Mexico, *in* Klapper, G., Murphy, M. A., and Talent, J. A., eds., Paleozoic sequence stratigraphy, biostratigraphy, and biogeography: Studies in honor of J. Granville ("Jess") Johnson: Geological Society of America Special Paper 321, p. 273–293.

Bucher, K., and Frey, M., 1994, Petrogenesis of metamorphic rocks. Berlin, Germany, Springer-Verlag, 318 p.

Cameron, K. L., Lopez, R., Ortega-Gutiérrez, F., Solari, L. A., Keppie, J. D., and Schulze, C., 2004, U-Pb constraints and Pb isotopic compositions of leached feldspars: Constraints on the origin and evolution of Grenvillian rocks from eastern and southern Mexico, *in* Tollo, R. P., Corriveau, L., McLelland, J., and Bartholomew, M. J., eds., Proterozoic tectonic evolution of the Grenville Orogen in North America: Geological Society of America Memoir 197, p. 755–770.

Campa, M. F., and Coney, P. J., 1983, Tectono-stratigraphic terranes and mineral resource distributions in México: Canadian Journal of Earth Sciences, v. 20, p. 1040–1051.

Carillo-Bravo, J., 1961, Geología del anticlinoria Huizachal-Peregrina al NW de Ciudad Victoria, Tamaulipas: Asociación Mexicana de Geólogos Petroleros Boletín, v. 13, p. 1–98.

Cserna, Z. de, and Ortega-Gutiérrez, F., 1978, Reinterpretación tectónica del Equisto Granjeno de ciudad Victoria, Tamaulipas; Contestación: Universidad Nacional Autónoma de México, Instituto de Geología, Revista, v. 2, p. 212–215.

Cserna, Z. de, Graf, J. L., Jr. and Ortega-Gutiérrez, F., 1977, Aloctono del Paleozoico inferior en la region de Ciudad Victoria, Estado de Tamaulipas: Universidad Nacional Autónoma de México, Instituto de Geología, Revista, v. 1, p. 33–43.

Davis, G. H., and Reynolds, S. J., 1996, Structural geology of rocks and regions, 2nd ed. New York, John Wiley, 776 p.

de la Fuente Duch, M. F., Aiken, C. L. V., and Manuel Mena, J., 1991, Cartas gravimetricas de la Republica Mexicana: Universidad Nacional Autónoma de México, Instituto de Geofisica, scale 1:3,000,000.

Denison, R. E., Burke, W. H., Jr., Hetherington, E. A., Jr., and Otto, J. B., 1971, Basement rock framework of parts of Texas, southern New Mexico, and northern Mexico, *in* Seewald, K., and Sundeen, D., eds., The geologic framework of the Chihuahua tectonic belt: West Texas Geological Society Publication 71-59, p. 3–14.

Dickinson, W. R., and Lawton, T. F., 2001, Carboniferous to Cretaceous assembly and fragmentation of México: Geological Society of America Bulletin, v. 113, p. 1142–1160.

Dodson, M. H., 1973, Closure temperature in cooling geochronological and petrological systems: Contributions to Mineralogy and Petrology, v. 40, p. 259–274.

Elías-Herrera, M., and Ortega-Gutiérrez, F., 2002, Caltepec fault zone: An Early Permian dextral transpressional boundary between the Proterozoic Oaxacan and Paleozoic Acatlán complexes, southern Mexico, and regional implications: Tectonics, v. 21, p. 1–18.

Fries, C., and Rincon-Orta, C., 1965, Nuevas aportaciones geocronológicas y tecnicas empleadas en al laboratorio de geocronología: Universidad Nacional Autónoma de México, Instituto de Geología, Boletín, v. 73, p. 57–134.

Fries, C., Jr., Schmitter, E., Damon, P. E., Livingston, D. E., and Erikson, R., 1962, Edad de las rocas metamorficas en los Canones de la Peregrina y de Caballeros, parte centro-occidental de Tamaulipas: Universidad Nacional Autónoma de México, Instituto de Geologia, Boletín, v. 64, p. 55–59.

Garrison, J. R., Ramírez-Ramírez, C., and Long, L. E., 1980, Rb/Sr isotopic study of the ages and provenance of Precambrian granulite and Paleozoic greenschist near Ciudad Victoria, Mexico, *in* The origin of the Gulf of Mexico and the early opening of the central North Atlantic Ocean: Baton Rouge, LA, Louisiana State University, p. 37–49.

Gursky, H.-J., 1996, Paleozoic stratigraphy of the Peregrina Canyon area, Sierra Madre Oriental: Zentralblatt für Geologie und Paläontologie, Teil I, Heft 7/8, p. 973–989.

Gursky, H.-J., and Milchalzik, D. 1989, Lower Permian turbidites in the northern Sierra Madre Oriental, México: Zentralblatt für Geologie und Paläontologie, Teil I, Heft 5/6, p. 821–838.

Keppie, J. D. 2004, Terranes of Mexico revisited: A 1.3 billion year odyssey: International Geology Review, v. 46, no. 9, p. 765–794.

Keppie, J. D., and Ramos, V. S., 1999, Odyssey of terranes in the Iapetus and Rheic Oceans during the Paleozoic, *in* Keppie, J. D. and Ramos, V. A., eds., Laurentia-Gondwana Connections before Pangea: Geological Society of America Special Paper 336, p. 267–276.

Keppie, J. D., Miller, B. V., Nance, R. D., Murphy, J. B., and Dostal, J., 2004a, New U-Pb zircon dates from the Acatlán Complex, Mexico: Implications for the ages of tectonostratigraphic units and orogenic events [abs.]: Geological Society of America, Abstracts with Programs, v. 36, no. 2, p. 104.

Keppie, J. D., Sandberg, C. A., Miller, B. V., Sánchez-Zavala, J. L., Nance, R. D., and Poole, F. G., 2004b, Implications of latest Pennsylvanian to Middle Permian paleontological and U-Pb SHRIMP data from the Tecomate Formation to re-dating tectonothermal events in the Acatlán Complex, southern Mexico: International Geology Review, v. 46, no. 8, p. 745–753.

Lopez, R. L., Cameron, K. L., and Jones, N. W., 2001, Evidence for Paleoproterozoic, Grenvillian, and Pan-Africa age crust beneath northeastern Mexico: Precambrian Research, v. 107, p. 195–214.

Malone, J. W., Nance, R. D., Keppie, J. D., and Dostal, J., 2002, Deformational history of part of the Acatlán Complex: Late Ordovician–Early Silurian and Early Permian orogenesis in southern Mexico: Journal of South American Earth Sciences, v. 15, p. 511–524.

Middleton, M. C., Keppie, J. D., Murphy, J. B., and Nance, R. D., 2004, Deformational history of the Piaxtla Group within the Acatlán Complex [abs.]: Geological Society of America, Abstracts with Programs, v. 36, no. 2, p. 156.

Nance, R. D., Keppie, J. D., and Miller, B. V., 2004, Record of the closure of the Rheic Ocean in the Acatlán Complex of Southern Mexico [abs.]: Geological Society of America, Abstracts with Programs, v. 36, no. 2, p. 156.

Navarro-Santillán, D., Sour-Tovar, F., and Centeno-Garcia, E., 2002, Lower Mississippian (Osagean) brachiopods from the Santiago Formation, Oaxaca, Mexico: Stratigraphic and tectonic implications: Journal of South American Earth Science, v. 15, p. 327–336.

Okulitch, A. V., 2002, Geological time scale 2002: Geological Survey of Canada, Open File 3040, National Earth Science Series, Geological Atlas—Revision.

Ortega-Gutiérrez, F., 1978, El Gneiss Novillo y rocas metamorficas asociadas en los canones del Novillo y la Peregrina, area Ciudad Victoria, Tamaulipas: Universidad Nacional Autónoma de México, Instituto de Geología, Revista, v. 2, p. 19–30.

Ortega-Gutiérrez, F., Elías-Herrera, M., Reyes-Salas, M., Macias-Romo, C., and Lopez, R., 1999, Late Ordovician–Early Silurian continental collisional orogeny in southern Mexico and its bearing on Gondwana-Laurentia connections: Geology, v.27, p. 719–722.

Ortega-Gutiérrez, F., Ruíz, J., and Centeno-Garcia, E., 1995, Oaxaquia, a Proterozoic microcontinent accreted to North America during the late Paleozoic: Geology, v. 23, p. 1127–1130.

Ramírez-Ramírez, C., 1974, Reconocimiento geologico de las zonas metamorficas al poniente de Ciudad Victoria, Tamaulipas: Mexico, D.F., Universidad Nacional Autónoma de México, Facultad de Ingenieria, 78 p.

Ramírez-Ramírez, C., 1992, Pre-Mesozoic geology of Huizachal-Peregrina Anticlinorium, Ciudad Victoria, Tamaulipas, and adjacent parts of eastern Mexico:

Unpubl. Ph.D. thesis, University of Texas, Austin, 317 p.

Sedlock, R. L., Ortega-Gutiérrez, F., and Speed, R. C., 1993, Tectonostratigraphic terranes and tectonic evolution of Mexico: Geological Society of America Special Paper 278, 146 p.

Stewart, J. H., Blodgett, R. B., Boucot, A. J., Carter, J. L., and Lopez, R., 1999, Exotic Paleozoic strata of Gondwanan provenance near Ciudad Victoria, Tamaulipas, Mexico, *in* Ramos, V. A., and Keppie, J. D., eds., Laurentia-Gondwana connections before Pangea: Geological Society of America Special Paper 336, p. 227–252.

Torres, R., Ruíz, J., Patchett, P. J., and Grajales-Nishimura, J. M., 1999, Permo-Triassic continental arc in eastern Mexico; tectonic implications for reconstructions of southern North America, *in* Bartolini, C., Wilson, J. L., and Lawton, T. F., eds., Mesozoic sedimentary and tectonic history of north-central Mexico: Geological Society of America Special Paper 340, p. 191–196.

Weber, B., Meschede, M., Ratschbacher, L., and Frisch, W., 1997, Structure and kinematic history of the Acatlán Complex in Nuevos Horizontes–San Bernardo region, Puebla: Geofisica Internacional, v. 36, p. 63–76.

Yañez, P., Ruiz, J., Patchett, P. J., Ortega-Gutiérrez, F., and Gehrels, G., 1991, Isotopic studies of the Acatlán Complex, southern Mexico: Implications for Paleozoic North American tectonics: Geological Society of America Bulletin, v. 103, p. 817–828.

The Tuzancoa Formation: Evidence of an Early Permian Submarine Continental Arc in East-Central Mexico

L. ROSALES-LAGARDE, E. CENTENO-GARCÍA,[1]

Instituto de Geología, Universidad Nacional Autónoma de México, Ciudad Universitaria, Delegación Coyoacán, México D.F. 04510, México

J. DOSTAL,

Department of Geology, Saint Mary's University, 923 Robie Street, Halifax, Nova Scotia, B3H 3C3 Canada

F. SOUR-TOVAR,

Museo de Paleontología, Facultad de Ciencias, Universidad Nacional Autónoma de México, Ciudad Universitaria, Delegación Coyoacán, México D.F. 04510, México

H. OCHOA-CAMARILLO,

Procesos Analíticos Informáticos, Bernabuco 792, Col. Lindavista, C.P. 07300 México, D.F.

AND S. QUIROZ-BARROSO

Museo de Paleontología, Facultad de Ciencias, Universidad Nacional Autónoma de México, Ciudad Universitaria, Delegación Coyoacán, México D.F. 04510, México

Abstract

Paleozoic rocks exposed in northeast Hidalgo State, Mexico, have traditionally been assigned to the Guacamaya Formation, a orogenic flysch assemblage associated with the collision of South and North America during the formation of Pangea. However, major differences exist in the stratigraphy of these rocks and those of the Guacamaya Formation at its type section. The rocks of Hidalgo are here redefined as the Tuzancoa Formation, and comprise interbedded submarine andesitic-basaltic lava flows, siliciclastic turbidites, volcaniclastic turbidites, calcareous debris-flows, and lenses of conglomerate. The Tuzancoa Formation contains abundant Permian (Wolfcampian–Leonardian) fossils.

The lavas and volcaniclastics are andesitic to basaltic in composition, and are related to subduction. Their REE patterns are similar to those of recent island arc magmas, with slight enrichment in LREE, flat HREE, and no Eu anomaly. The $\varepsilon Nd_{(280)}$ of +4.38 in a representative sample is also similar to island arcs and suggests little crustal contamination. However, field evidence indicates that the arc was built upon Precambrian continental crust. Thus, we infer that the "primitive" arc geochemical signatures reflect emplacement on thin crust or under extensional conditions. The Tuzancoa Formation is interpreted to have been deposited in an intra-arc basin, and is probably related to rocks of the Las Delicias Formation in Coahuila and to the Guacamaya Formation in Tamaulipas. All are thought to have formed within a continental arc that extended throughout east-central Mexico in the Early Permian.

Introduction

PALEOZOIC ROCKS EXPOSED in the Tianguistengo-Calnali region of the Sierra Madre Oriental in eastern Mexico (Fig. 1) were first described by Carrillo-Bravo (1965) as the southern extension of the Guacamaya Formation (Carrillo-Bravo, 1961), a Lower Permian sedimentary succession exposed northwest of Ciudad Victoria, Tamaulipas State. In the type section, the Guacamaya Formation comprises rhythmic marine successions made up of dark shale interbedded with sandstone and conglomerate. The formation has been considered in several paleogeographic reconstructions to be part of the collisional belt between Laurentia and Gondwana (orogenic flysch), formed during the Carboniferous–

[1]Corresponding author; email: centeno@servidor.unam.mx

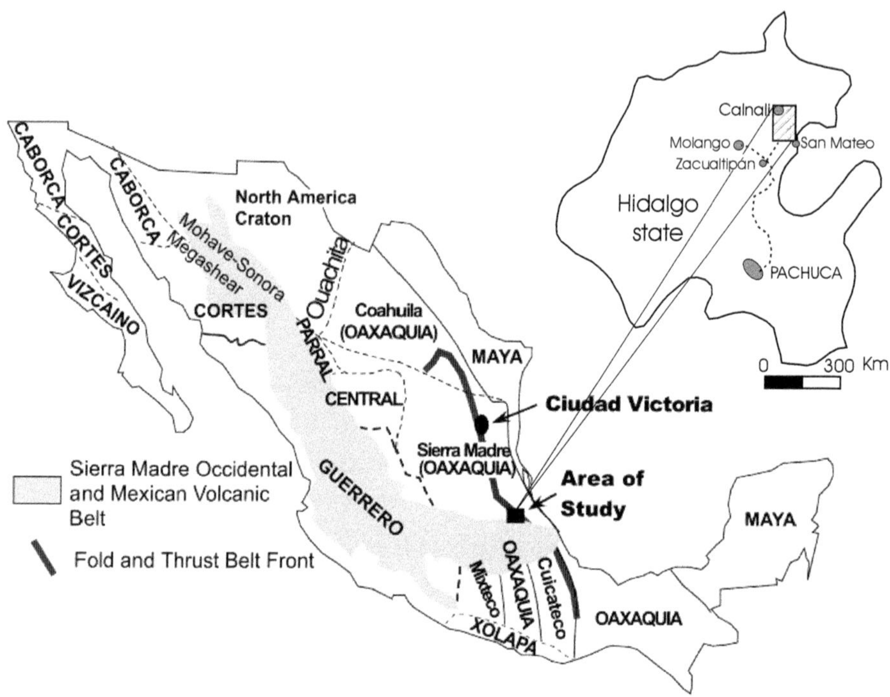

FIG. 1. Location map of the Tuzancoa Formation, geographic references used in the text, and main tectonostratigraphic terranes, modified from Centeno-García, 2003.

Permian assembly of Pangea (Pindell and Dewey, 1982; Pindell, 1985; Ross and Scotese, 1988). This possible collisional belt was extended by those authors from Ciudad Victoria to northwest Hidalgo State, where the studied rocks are exposed. However, detailed regional mapping of the succession in Hidalgo State has shown it to contain abundant volcanic and volcaniclastic rocks representative of a continental arc rather than an orogenic flysch (Rosales-Lagarde et al., 1997). Similar to our conclusions, Gursky and Michalzik (1989) interpreted the Guacamaya Formation as a forearc deposit.

In this paper, we document major differences between the Guacamaya Formation in Hidalgo State and the type section in Tamaulipas State, and rename the Hidalgo succession the Tuzancoa Formation (Rosales-Lagarde, 2002). We describe the stratigraphy, sedimentology, and depositional environment of the Tuzancoa Formation and report the elemental and isotopic composition of its volcanic and volcaniclastic rocks, and their tectonic significance. The results are intended to contribute to our understanding of the tectonic evolution of eastern North America and the assembly of Pangea during Late Paleozoic time.

Geological Setting

Outcrops of the Tuzancoa Formation are located in the east-central part of Hidalgo State, Mexico, at 20°45'–20°49.5' N. Lat., and 98°35'–98°31' W. Long. This region was described as the Huayacocotla anticlinorium (Carrillo-Bravo, 1965), and has most recently been reinterpreted as the frontal part of the Laramide fold-thrust belt (Suter, 1984, 1987, 1990; Sedlock et al., 1993; Ochoa-Camarillo, 1996). The area forms part of the Sierra Madre terrane of Campa and Coney (1983), and the central part of the Oaxaquia micro-continent of Ortega-Gutiérrez et al. (1995) (Fig. 1).

The oldest rocks in the area are Precambrian ortho- and paragneisses (the Huiznopala Gneiss), with magmatic ages between 1200 and 1150 Ma, and Grenvillian metamorphic ages around 1000 Ma (Ortega-Gutiérrez et al., 1997; Lawlor et al., 1999). These gneisses have characteristics common to those exposed to the north (the Novillo Gneiss in Ciudad Victoria) and to the south in Oaxaca (the Oaxaca Complex) (Patchett and Ruiz, 1987; Ruiz et al., 1988; Ortega-Gutiérrez et al., 1995; Lawlor et al., 1999). The only Paleozoic rocks exposed in the area are those of the Permian Tuzancoa Formation. To the east, however, granitic intrusions of Permian–Triassic age occur (Torres-Vargas et al., 1999), but their relationship to the Tuzancoa Formation is uncertain. The stratigraphy continues with a continental to shallow marine succession of Early Jurassic age (Huayacocotla Formation), followed by continental redbeds of probable Jurassic age (Cahuasas Formation) (Ochoa-Camarillo et al., 1999) and a thick marine limestone succession that ranges in age from Oxfordian to Cenomanian (Tepexic, Santiago, Chipoco, Tamán, Pimienta, Tamaulipas, Otates, Agua Nueva, San Felipe, and Méndez formations) (Carrillo-Bravo, 1965; Ochoa-Camarillo, 1996; Ochoa-Camarillo et al., 1999). All pre-Cenozoic units are folded and thrusted into large nappes with several kilometers of displacement (Suter, 1984, 1987, 1990; Ochoa-Camarillo, 1996). Basaltic flows and other volcanic rocks of Cenozoic age cap the area's hilltops.

Facies and Depositional Environments

A detailed stratigraphic section was measured along the Tlacolula River, between the towns of Otlamalacatla and Chapula (Fig. 2). This river exposes one of the most complete successions of the Tuzancoa Formation, and is considered to be the type section. Five different lithofacies were recognized in the type section (Fig. 3): Facies 1, lava flows and breccias; Facies 2, siliciclastic turbidites; Facies 3, volcaniclastic turbidites; Facies 4, calcareous debris flows; and Facies 5, channel-filled conglomerate. These lithofacies are repeated throughout the section to form a rhythmic succession, except for Facies 2, which is found only at the lowermost part of the stratigraphy. The section is disrupted by thrust faults and folds, so the original thickness is unknown. The approximate thickness of sections between thrust planes is ~700 m.

Facies 1

This volcanic facies consists mostly of andesitic to basaltic lava flows, with a few flows of dacitic composition at the top of the succession. Flows are mainly aphanitic but some are porphyritic with large plagioclase phenocrysts. Flow textures are variable: some are massive, others show pseudo-pillowed structures with radial fracturing, and yet others are capped by brecciated crusts formed by fast cooling by contact with water. They are submarine lava flows, interbedded with the other facies. This facies also contains layers of massive volcanogenic shale that may represent air-fall tuffs. The facies varies in thickness from a few meters to more than 70 m, and occurs at several stratigraphic levels (Figs. 2 and 3).

Facies 2

This facies comprises interbedded shale, sandstone, and conglomeratic sandstone. Shale has laminar bedding and sandstone beds have erosional bases with intraclasts of shale. Colors vary from black to light brown. Sandstone and conglomeratic sandstone predominantly consist of quartz grains, with less abundant plagioclase, detrital mica, chert, and volcanic lithics. The sandstones are fine- to coarse-grained quartz arenite to sublitharenite. Clast sizes in conglomeratic sandstones vary from granules to pebbles. Primary structures suggest deposition by turbidity currents. Facies 2 is approximately 120 m thick, and occurs only at the base of the succession (Figs. 2 and 3). The facies does not contain lava flows, but volcanic clasts occur in the sandstone. The high percentage of quartz and mica grains, however, suggests that a Precambrian basement source was supplying sediments to the basin.

Facies 3

The most common facies in the Tuzancoa Formation is volcaniclastic (Figs. 2 and 3), and comprises interbedded shale, siltstone, and sandstone that range in color from black to olive green. Sandstone is mostly fine-grained with abundant lithic (volcanic) fragments (50% or higher), feldspar and quartz grains. This facies again shows bedding features and primary structures typical of turbidites. Sandstone beds are parallel to lenticular in shape; they have erosional bases, and most of them contain shale intraclasts of varying size. The shale and siltstone locally contain thin beds of black limestone or calcareous shale. In lower parts of the succession, the turbidites comprise massive sandstone with large shale intraclasts (up to 2 m) (Bouma, 1962; facies T_A and T_E), interbedded with thin rhythmically bedded shale and sandstone (Bouma facies T_A). Sedimentary folding, slumps, and sandstone beds with cross and convolute bedding (Bouma facies T_C and T_D) are more abundant toward the top of the succession. One slump comprises mega-breccia with large

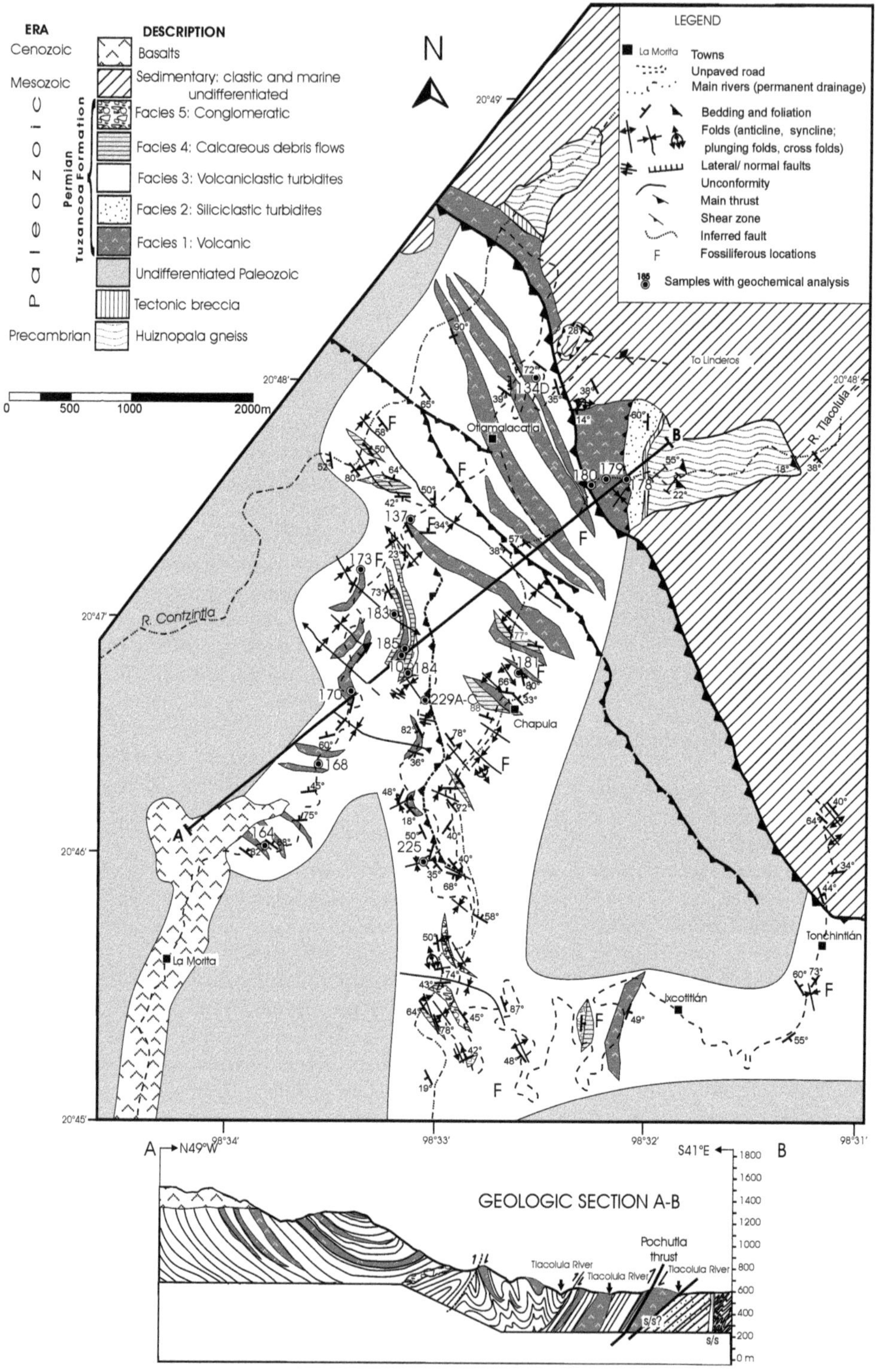

FIG. 2. Geologic map and cross section of the Tuzancoa Formation.

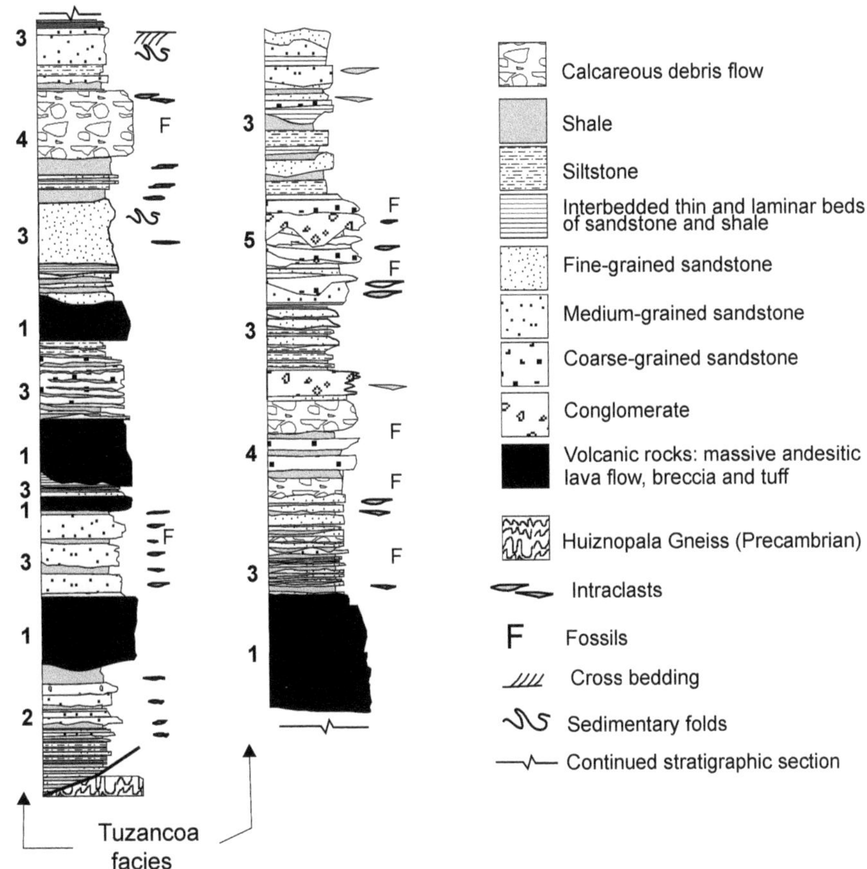

FIG. 3. Simplified stratigraphic column and facies distribution of the Tuzancoa Formation. Numbers at the left of the column indicate the facies number.

blocks of volcanic sandstone (8 m or more) within a siltstone matrix. Facies 3 contains all the other facies, and varies in thickness from a few meters to 400 m (Figs. 2 and 3). It contains scarce fossils, most of which occur at the base of the succession (Fig. 3).

Facies 4

This facies is characterized by calcareous debris flows that contain abundant crinoid stems, fusulinids, and other fossil fragments. Lithologies vary from wackestone and calcareous shale, to calcarenite and calcareous breccia (rudstone). The debris flows have lenticular bedding and erosional bases, and contain intraclasts of shale and sandstone. Facies 4 ranges in thickness from 50 cm to 8 m; it is commonly associated with sedimentary folds in facies 3, and is found at various levels throughout the succession (Figs. 2 and 3).

Facies 5

This facies consists of lenticular bodies of conglomerate and coarse-grained sandstone. Conglomeratic beds have erosional bases and flat tops, and show normal graded bedding. They are poorly sorted, matrix supported, and contain both exoclasts and intraclasts. The exoclasts are subangular to subrounded and vary in size from granules to cobbles. They include crystalline limestone, fossiliferous limestone, volcanic fragments (andesite to rhyolite), granite, green (rhyolitic) tuff, white quartz, and gneiss. Intraclasts are mostly volcanic sandstone, shale, and limestone with fusulinids. Conglomerate is interpreted as channel-fill deposits. The thickness of this facies, which occurs in the middle and upper part of the succession, varies from 1 to 8 m (Figs. 2 and 3).

The facies association of the Tuzancoa Formation suggests that it was deposited in a submarine environment, while the sedimentary characteristics of the turbidites indicates a submarine fan and/or distal deltaic setting. There are no chert layers or black shales that would suggest deep water. On the contrary, the presence of calcareous shale and black limestone indicates a relatively shallow deposition, above the carbonate compensation depth. The abundance of lava flows (Facies 1) suggests that the depositional basin was located close to a volcanic source, while the thickness of the lava flows and the occurrence of calcareous debris flows indicate that the volcanic edifices were of considerable size. These edifices are also likely to have supplied the volcanic clasts in the turbiditic and conglomeratic facies. Submarine canyons that cut the volcaniclastic sequence were filled by conglomerate. The amount of quartz and detrital mica, and the occurrence of gneiss and granite clasts in conglomerate beds at various stratigraphic levels suggest that the Precambrian basement was continuously exposed during magmatism. Basement-derived sediments were more abundant during the initial stages of deposition (Facies 2), whereas the presence of sedimentary rock fragments (chert and limestone) suggests that older sedimentary units were also being eroded. The abundance of volcanic fragments, however, indicates that the proportion of volcanic rocks being eroded was larger than that of the basement rocks.

Fossils and Age of the Tuzancoa Formation

The Tuzancoa Formation contains abundant fossils (Carrillo-Bravo, 1965; Pérez-Ramos, 1978; Buitrón et al., 1987; Arellano-Gil, et al., 1998; Sour-Tovar et al., in press), which are more common and better preserved in the calcareous debris flows than they are in the volcaniclastic facies. They include crinoid stems, fusulinids, algae, foraminifera, and bivalve fragments. Fossils in the volcanic sandstone and shale include brachiopods, pelecypods, ammonites, bryozoans, and trilobites. The species reported are listed in Table 1. Most are Wolfcampian to Leonardian in age (Asselian to Artinskian, 290–260 Ma; Okulitch, 1999), but some taxa have ranges that go down to the Virgilian (Late Pennsylvanian) and others have ranges that go up to the Late Permian. Fossil plants of Leonardian age have been reported in Calnali, north of the studied area (Silva-Pineda, 1987), and others that may be Permian in age occur in continental strata at San Mateo, a few kilometers south of the studied area (Weber, 1997). However, their relationship with the marine facies of the Tuzancoa Formation is unknown. Overall, the association of marine fossils suggests shallow water and, although redeposition by gravity flows (turbidites and debris flows) is possible, deep-marine fossils have not been observed at any level of the succession. Given the presence of well preserved fossil plants at Calnali, and assuming the continental facies found south of the studied area are likewise part of the Tuzancoa Formation, a distal deltaic setting, rather than a submarine fan, is the favored depositional environment of the succession.

With regard to the paleogeographic affinity of the fauna, most of the brachiopod and trilobite species have been reported in west Texas localities, which form part of the Mid-continent Province of North America (Sour-Tovar et al., 2005). The fusulinid species show strong affinities with those from localities of the same age in Chiapas in southern Mexico (Arellano-Gil et al., 1998).

Contact Relationships and Structure of the Tuzancoa Formation

The Tuzancoa Formation is overlain by Jurassic marine sandstones and shales, and it is in tectonic contact with the Precambrian Huiznopala Gneiss and Jurassic–Cretaceous limestones at its base. This basal contact is a complex structure (Fig. 4). The siliciclastic Facies 2, which forms the base of the Paleozoic succession in the Tlacolula River, is in strike-slip fault contact with the Precambrian Huiznopala Gneiss (Fig. 2), forming a thick tectonic breccia. The same tectonic relationship is observed in the Contzintla River to the north, where the volcanic Facies 1 is separated from the gneiss by a thick breccia associated with lateral faulting. Upper Jurassic and Cretaceous rocks unconformably overlie the strike-slip zone (Fig. 4), and are in turn overthrust by rocks of the Tuzancoa Formation as part of the Sierra Madre fold-and-thrust belt.

Original contact relationships between the Huiznopala Gneiss and the Tuzancoa Formation are not exposed, but are assumed to be depositional, based on (1) the regional distribution of the Tuzancoa Formation with respect to Precambrian outcrops, which are found to the east and west, and (2) the nature of the Paleozoic succession and its depositional environment.

TABLE 1. Fossils Reported from the Tuzancoa Formation

Taxa	Author	Taxa	Author
Microforaminifera		Brachiopods	
Diplosphaerina/Eotuberitina	Arellano-Gil et al., 1998	*Krotovia* sp.	Sour-Tovar et al., 2005
Earlandia sp.		*Dasysaria* sp.	
Endothyra sp.		*Derbyoides* cf. *D. dunbari* cf.	
Tetrataxis sp.		*Neospirifer* cf. *N. amphigyus*	
Climacammina sp.		*Neospirifer* sp.	
Globivalvulina sp.		*Spiriferellina tricosa*	
Calcitornella sp.		*Holosia* sp.	
Arenovidalina sp.			
Nodosaria sp.		Trilobites	
Geinitzina sp.		*Anisopyge whitei*	Sour-Tovar et al., 2005
Fusulinids			
Skinnerella sp.	Arellano-Gil et al., 1998	Crinoids	
Staffella sp.		*Cyclocaudex plenus*	Arellano-Gil et al., 1998
Schwagerina sp.	Perez-Ramos, 1978	*Cyclocaudex jucundus*	
Triticites sp.		*Preptopremmum rugosum*	
Tetrataxis sp.		*Cyclocrista cheneyi*	
Paraschwagerina roveloi	Carrillo Bravo, 1965		
Skinnerella figueroai		Algae	
S. cf. *gruperaensis*		*Tubiphytes* sp.	Arellano-Gil et al., 1998
Chalaroschwagerina aff. *Ch. chiapasensis*		*Epimastopora* sp.	
Pseudoschwagerina sp.			
Monodiexodina sp.		Plants	
Schubertella sp.		*Equisetum* sp.	Silva-Pineda, 1987
Triticites aff. *T. victoriaensis*		Pecopteris arrescens	
Schwagerina cf. S. Thompsoni		*Pecopteris* sp.	
Paraschwagerina sp.		*Taeniopteris* sp.	
Triticites sp.		*Gamgamopteris* sp.	
Paraschwagerina sp.		*Neuropteris* sp.	
Pseudofusulina sp.		cf. *Odontopteris* sp.	
Schwagerina sp.		*Walchia* sp.	

The contact between the Tuzancoa Formation and overlying Jurassic marine rocks has not been mapped in detail outside the study area, but within the map area the two successions are concordant. The Tuzancoa Formation is overlain discordantly by Cenozoic basalts, probably belonging to the Mexican Volcanic Belt.

Internally, the Tuzancoa Formation shows tight to open folding with some chevron folds and internal thrust faults. Bedding, thrust faults, and fold axis orientation are NW-SE and are parallel to major thrust faults of the Late Cretaceous fold-thrust belt of the Sierra Madre Oriental. Except for strike-slip faults located near the contact with the Huiznopala Gneiss, older structures were not detected.

Geochemistry of Volcanic and Volcaniclastic Rocks

Twenty-eight representative samples of volcanic and volcaniclastic rocks were collected at various stratigraphic levels of the Tuzancoa Formation for

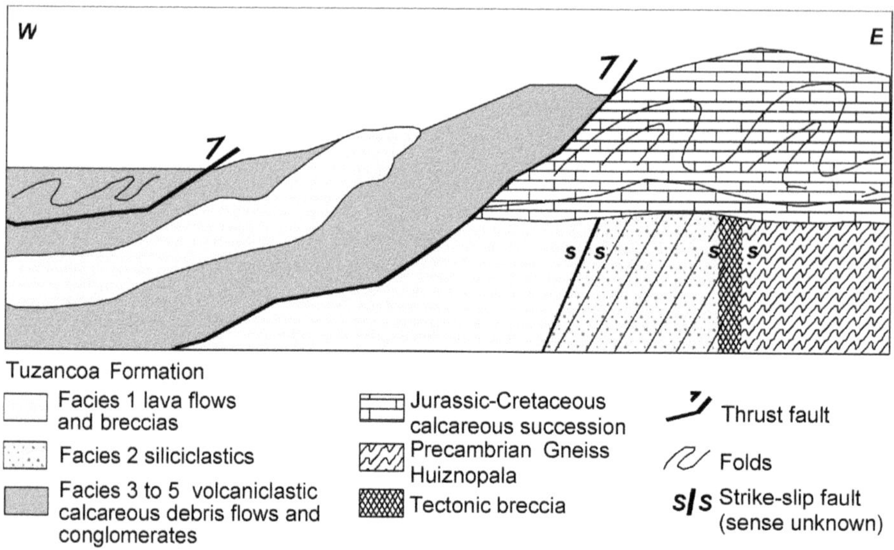

FIG. 4. Sketch showing complex contact relationships at the base of the Tuzancoa Formation.

petrographic, major and trace element, and isotopic analyses. Sixteen of these samples were collected from massive volcanic rocks of Facies 1, and 12 samples from volcaniclastic rocks of Facies 3. Major and selected trace (Rb, Sr, Ba, Zr, Nb, Y, Cr, Ni, Co, Sc, V, Cu, Pb, and Zn) elements were analyzed by X-ray fluorescence (XRF) at the Geochemical Centre of the Department of Geology, Saint Mary's University, Halifax (Table 2). Additional trace elements (REE, Hf, Ta, Nb, U and Th) were analyzed in 21 samples by inductively coupled plasma-mass spectrometry (ICP-MS) using Na_2O_2–sintering (16 samples) and acid digestion (5 samples) techniques at the Department of Earth Sciences of the Memorial University of Newfoundland (Table 3). The precision for the trace elements is between 2% and 8 % of the values cited (Jenner et al., 1990; Longerich et al., 1990).

One representative sample was selected for Sm-Nd isotopic analysis (Table 4) at the Department of Geosciences of the University of Arizona in Tucson. The whole-rock Sm-Nd isotopic analysis was performed on a VG-354 mass spectrometer following procedures described in Patchett and Ruiz (1987). Blanks were Nd < 300 pg. Duplicates are less than ± 0.2 epsilon-Nd units.

The Tuzancoa volcanic and volcaniclastic rocks are mafic to intermediate in composition with SiO_2 ranging from 52% to 65%. Most are andesites and basaltic andesites (Fig. 5A), with no obvious compositional difference between volcanic and volcaniclastic rocks. All are calc-alkaline (Fig. 5B), with Al_2O_3 between 15.5% and 19.4%, and Mg# (= 100 × Mg/Mg + Fe_{tot}) between 57.4 and 38.5. Major-element compositions are typical of calc-alkaline andesites (Table 2).

Chondrite-normalized REE patterns of both volcanic and volcaniclastic rocks are slightly enriched in light REE (LREE) and display flat heavy REE (HREE), with $(La/Yb)_N$ between 1.98 and 7.65, and $(La/Sm)_N$ between 0.8 and 1.0 (Figs. 6A and 6B). They do not show europium anomalies. Most samples show similar REE contents with the exception of two samples (178 and 180) from the lowest stratigraphic levels, which have more fractionated LREE (Fig. 6A).

MORB-normalized trace element profiles (Figs. 7A and 7B) show enrichment in incompatible and depletion in compatible elements compared to N-MORB and primitive oceanic island arc basalts. They are particularly enriched in low ionic potential elements (Sr, K, Rb, Ba, and Th), but have a negative Nb-Ta anomaly (Nb/Ta ≅ 20). These patterns are characteristic of magmas derived from an enriched subcontinental lithosphere modified by subduction processes. This is consistent with the position of these rocks on tectonic discrimination diagrams such as those of Gill (1981), Pearce (1983), and

TABLE 2. Whole-Rock Major (wt%) and Trace Elements (ppm) Analyses of Tuzancoa Samples (measured by XRF)

Sample	Lat.	Long.	Rock type	SiO_2	TiO_2	Al_2O_3	Fe_2O_3	MnO	MgO	CaO	Na_2O	K_2O	P_2O_5	LOI	Rb	Sr	Ba	Zr	Nb	Y	Cr	Ni	Co	Sc	V	Cu	Pb	Zn	Ta	Hf	Th	U
10	20°46.95'	98°33.18'	Lava flow	58.3	0.9	16.8	8.2	0.1	4.8	7.3	3.3	0.2	0.1	4.2	7	259	59	61	2.5	12	54	27	28	13	224	23	5	98	0.1	1.6	0.9	1
137	20°47.4'	98°33.1'	Epiclastic	56.8	0.8	18.5	8.7	0.1	5	5	4.4	0.5	0.1	4.3	12	338	123	60	2.3	10	50	25	25	7	207	22	3	97	0.1	1.5	0.8	1
152	20°43.77'	98°33.75'	Lava flow	60.3	0.8	18.5	5.9	0.1	3.5	4	6.4	0.3	0.1	3.2	7	615	165	55	1.4	9	68	34	18	14	171	26	2	82	0.1	1.3	0.9	0
152Ai	20°43.77'	98°33.75'	Lava flow	62.1	0.7	16.9	6.6	0.1	3.9	4.6	4.9	0.1	0.1	3.7	2	373	101	41	1.9	8	56	30	17	7	137	18	2	72	0.3	1.9	1.3	0.4
157	20°43.64'	98°33.82'	Lava flow	62.1	0.7	17.4	3.6	0.1	1.2	10.9	4	0	0.1	3.1	2	43	0	0	0	0	24	16	9	0	112	14	5	58	0	0	1	2.1
161	20°43.6'	98°33.93'	Lava flow	62	0.8	17.7	5.8	0.1	3.3	3.3	6.7	0.2	0.1	2.8	3	236	132	55	2.3	10	50	30	49	12	156	31	7	82	0.3	1.2	0.7	1
164	20°46.16'	98°33.78'	Epiclastic	60.9	0.8	17.2	6.6	0.1	4.3	4.8	4.6	0.5	0.1	3.5	9	498	159	57	1.5	10	53	27	37	12	197	24	2	83	0.1	1.5	0.8	0
168	20°46.4'	98°33.6'	Pyroclastic	63	0.7	16.5	6	0.1	3.9	4.5	4.7	0.4	0.1	3.8	10	246	461	0	0	0	35	23	23	10	148	21	10	79	0	0	3.1	3.1
170	20°46.26'	98°33.44'	Lava flow	60.5	0.6	17.3	6.4	0.1	3.8	6.2	4.5	0.5	0.1	3.6	16	300	453	0	0	0	43	24	22	4	144	24	6	82	0	0	2.1	2.1
178	20°47.6'	98°32.2'	Lava flow	57.5	0.8	17.4	7.7	0.1	4.1	8.5	2.8	0.9	0.1	3.8	32	559	294	91	5.5	17	32	10	26	8	189	36	13	76	0	2.2	3.8	1
179	20°47.58'	98°32.28'	Lava flow	57.2	0.8	17.6	8.3	0.1	5.3	6.3	2.3	2	0.1	4.1	75	470	685	0	0	0	36	9	34	7	226	20	13	78	0	0	5.2	2.1
180	20°47.57'	98°32.32'	Lava flow	55.1	0.8	19.4	8.8	0.1	4.9	3.1	5.9	1.7	0.2	3.2	55	436	867	133	8.4	19	18	4	27	7	171	26	6	94	0.5	3	5.8	1
181	20°46.8'	98°32.63'	Pyroclastic	61.6	0.9	16.4	8.4	0.1	3.1	5	3.6	0.7	0.1	5.5	22	220	152	80	3.2	25	14	29	19	8	97	29	4	84	0.2	1.8	1.6	1.8
183-1	20°47.22'	98°33.24'	Epiclastic	58.5	1.1	16.7	10.3	0.1	6	2.8	4.2	0.1	0.1	4.7	7	400	331	0	0	0	81	25	30	15	215	16	2	114	0	0	2.1	1
183-2	20°47.22'	98°33.24'	Epiclastic	59.5	1.2	15.5	10.5	0.1	6.2	2.5	4.4	0.1	0.1	5.2	6	402	320	0	0	0	84	23	30	17	238	15	2	115	0	0	3.2	1.1
184	20°46.75'	98°33.2'	Epiclastic	56.4	0.7	18	5.8	0.4	3.9	8.6	5.2	0.9	0.1	7.3	21	425	247	0	0	0	24	24	19	9	158	24	2	78	0	0	1.1	1.1
185-A	20°46.95'	98°33.27'	Epiclastic	58.1	0.9	17	8.1	0.1	4.4	7.8	3.2	0.3	0.1	4.2	11	254	82	67	2.5	12	55	27	26	3	215	19	6	93	0	1.7	0.9	1
185-B	20°46.95'	98°33.27'	Epiclastic	58	1	16.6	8.6	0.1	4.3	7.4	3.5	0.3	0.1	4.5	12	342	398	0	0	0	56	23	40	12	208	23	4	101	0	0	2.1	1
192	20°54.4'	98°37.3'	Lava flow	57.7	0.8	17.7	7.5	0.1	4.7	6.4	4.4	0.4	0.1	4.1	10	443	135	55	2.2	10	45	27	29	5	186	35	3	90	0.1	1.3	0.7	1
201	20°48.5'	98°35.5'	Epiclastic	60.6	0.8	17.4	7.5	0.1	4.2	4	4.3	1	0.1	3.5	18	412	610	82	2.7	15	40	18	28	7	210	22	7	104	0.1	2.1	1.5	1
206	20°50.73'	98°33.82'	Lava flow	57.5	0.9	18.3	8.6	0.1	4.2	3.1	7	0.1	0.1	3.6	4	222	64	76	2.7	16	26	18	19	16	176	64	5	88	0	2.2	1.2	2.1
207	20°50.8'	98°33.7'	Lava flow	55.7	0.9	18.6	8.4	0.2	4.7	6	5.1	0.1	0.3	4.2	7	556	73	70	2.4	17	22	11	34	2	216	24	4	95	0	1.9	1	0
207B	20°50.8'	98°33.7'	Lava flow	55.9	0.9	18.4	8.3	0.1	4.6	6.7	4.8	0.1	0.2	4.1	6	605	51	0	0	0	18	12	22	8	200	24	5	93	0	0	1	0
210Bi	20°50.32'	98°34.1'	Lava flow	57.9	0.6	19.3	7	0.1	4.8	4.8	4.4	1.1	0.1	4.7	22	539	433	40	1.4	10	34	22	15	9	140	19	3	86	0.1	1.9	1.2	0.4
212	20°49.48'	98°34.57'	Lava flow	67.2	0.5	15.6	4.2	0.1	2.1	4.4	4.3	1.5	0.1	2.9	27	460	488	72	2.2	8	30	19	45	7	93	17	5	64	0.4	1.8	1.9	2.1
225	20°47.4'	98°33.1'	Lava flow	61.9	0.7	17.1	6.2	0.1	3.9	3.6	5.2	1.1	0.1	3.2	18	234	396	98	2.7	14	55	27	37	13	145	10	5	102	0.2	2.5	2.1	3.1
229C	20°46.68'	98°33.1'	Epiclastic	55.3	0.9	18	9.1	0.2	5.9	7	3.2	0.4	0.1	4.7	13	312	84	64	2.1	11	68	28	32	8	234	25	4	113	0	1.7	0.8	2.1
229Bi	20°46.68'	98°33.1'	Epiclastic	58.8	0.8	16.7	7.8	0.1	4.3	7.4	3.6	0.3	0.1	4.1	6	280	67	49	1.8	12	48	25	23	9	171	21	3	93	0.1	1.7	1.1	0.4

TABLE 3. Whole-Rock Rare-Earth-Element Analyses (ppm) of Selected Tuzancoa Samples (obtained by ICP-MS)

Sample	La	Ce	Pr	Nd	Sm	Eu	Tb	Dy	Ho	Er	Tm	Yb	Lu
10	5.56	12.4	1.7	7.84	2.2	0.7	0.3	2.1	0.4	1.2	0.2	1.1	0.2
137	5.09	11.5	1.6	7.34	1.9	0.7	0.3	1.8	0.4	1.1	0.2	1	0.2
152	6.21	12.7	1.8	8.1	2	0.7	0.3	1.7	0.3	0.9	0.1	0.8	0.1
152A	6.04	12.9	1.8	7.63	1.9	0.7	0.3	1.6	0.3	0.8	0.1	0.7	0.1
161	4.64	11.7	1.7	7.95	2.1	0.8	0.3	1.9	0.4	1	0.1	0.8	0.1
164	5.38	11.3	1.6	7.19	2	0.7	0.3	1.7	0.3	0.9	0.1	0.9	0.1
178	11.9	25.6	3.3	13.1	3.2	0.9	0.5	2.9	0.6	1.8	0.3	1.8	0.3
180	18.2	36	4.7	16.9	3.8	1.2	0.5	3.2	0.6	1.8	0.3	1.8	0.3
181	7.76	18.4	2.7	12.6	3.5	1.2	0.6	4.1	0.9	2.6	0.4	2.6	0.4
185-A	5.71	13	1.8	8.1	2.2	0.8	0.3	2.1	0.4	1.2	0.2	1.2	0.2
192	4.94	11.4	1.6	6.85	1.9	0.7	0.3	1.7	0.3	1	0.1	0.9	0.1
201	8.17	16.3	2.3	10	2.5	0.8	0.4	2.2	0.5	1.3	0.2	1.1	0.2
206	8.28	18.3	2.4	10.5	2.7	0.9	0.4	2.8	0.6	1.7	0.2	1.6	0.3
207	6.53	15.1	2.1	9.84	2.6	0.9	0.4	2.9	0.6	1.7	0.3	1.7	0.2
210B	5.2	11.3	1.6	6.37	1.7	0.7	0.3	1.8	0.4	1.1	0.2	1	0.2
212	7.79	16.5	2.1	8.88	2	0.7	0.2	1.4	0.3	0.8	0.1	0.7	0.1
225	7.84	17.5	2.4	10.5	2.7	0.8	0.4	2.5	0.5	1.4	0.2	1.4	0.2
229C	5.06	11.7	1.6	7.45	2	0.7	0.3	2	0.4	1.2	0.2	1.2	0.2
229B	5.98	13.2	1.9	7.71	2.1	0.8	0.3	2.1	0.4	1.1	0.2	1.1	0.2

TABLE 4. Whole-Rock Nd Isotopic Composition for a Tuzancoa Formation Sample[1]

Sample	Rock type	Sm	Nd	$^{147}Sm/^{144}Nd$	$^{143}Nd/^{144}Nd$	T_{DM}	$\varepsilon Nd_{(0)}$	$\varepsilon Nd_{(280)}$
Mo-10	lava flow	2.076	8.256	0.152011	0.51249	672	2.79	4.38

[1]Nd and Sm concentrations (ppm) were analyzed by isotopic dilution (depleted mantle age T_{DM} calculated using the evolution curve of DePaolo, 1981).

Wood (1980), in which both volcanic and volcaniclastic rocks plot in the orogenic andesites-arc field (Figs. 8A, 8B, and 8C). The similarity in composition between volcanic and volcaniclastic rocks suggests that the volcaniclastic rocks lack contributions of other sedimentary components.

Overall, the geochemical composition of the volcanic and volcaniclastic rocks of the Tuzancoa Formation suggests that they originated in an orogenic environment (arc or backarc). They are compositionally similar to present oceanic volcanic island arcs, with neither a significant contribution of old continental crust nor a high degree of crystal fractionation. However, geological evidence indicates that the Tuzancoa arc was build on Precambrian continental crust.

The single sample analyzed for Sm/Nd isotopes has a $\varepsilon Nd_{(280)}$ value of +4.38, which is comparable with other subduction-related magmas (e.g., Wilson, 1989), and yielded a model age (T_{DM}) of 672 Ma.

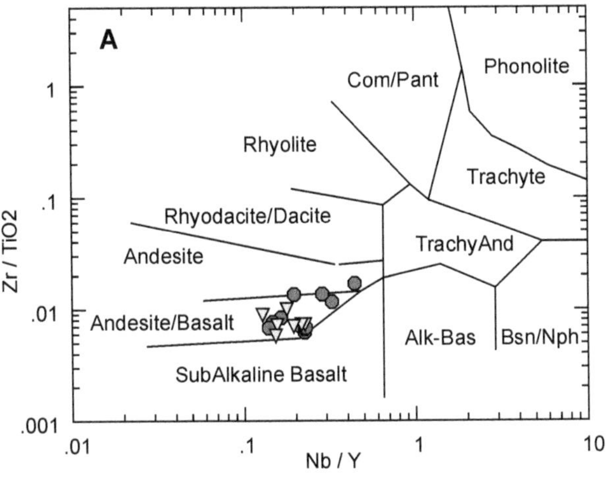

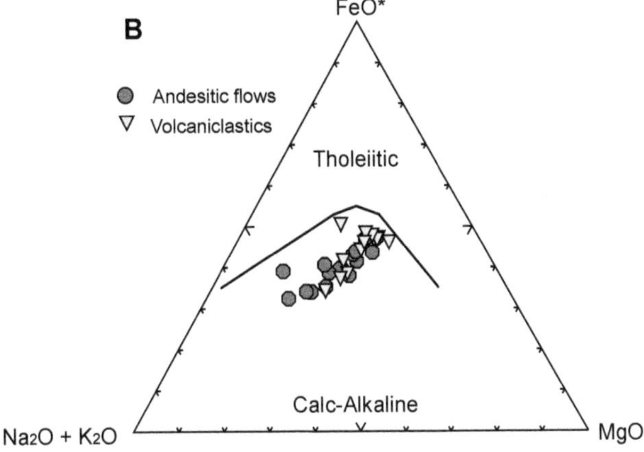

FIG. 5. Discrimination diagrams (A) Zr/TiO_2 vs. Nb/Y (Winchester and Floyd, 1977), and (B) $(Na_2O + K_2O)$–FeO_{tot}–MgO (AFM) diagram (Irvine and Baragar, 1971). All samples are andesites to basalts and belong to the calc-alkaline series.

This suggests some contribution of old crustal material to the magma generation (subducted sediments and/or the crust underneath the arc).

Regional Correlation with Other Permian Successions

Carboniferous–Permian sedimentary rocks are abundant in Mexico (Sánchez-Zavala et al., 1999). However, they are exposed in relatively small isolated areas, making correlation difficult. Upper Paleozoic rocks in southern Mexico are mostly sedimentary and do not contain volcanic rocks like those found in the Tuzancoa Formation. Only two successions with volcanic rocks occur in southern Mexico (Fig. 9A). The first is the Matzitzi Formation, which contains an ignimbrite flow (Centeno-García et al., 1997). The second is the Juchatengo Complex, which includes a deformed marine succession comprising basaltic pillow lavas and siliciclastic sediments (Grajales-Nishimura et al., 1999). Neither of these units shows any stratigraphic or volcanic compositional similarity to the Tuzancoa Formation, nor is there any apparent correlation between the Tuzancoa Formation and contemporaneous units in Chihuahua and Sonora in northwestern Mexico, which mostly comprise shallow marine limestones and siliciclastics, and lack evidence of magmatism.

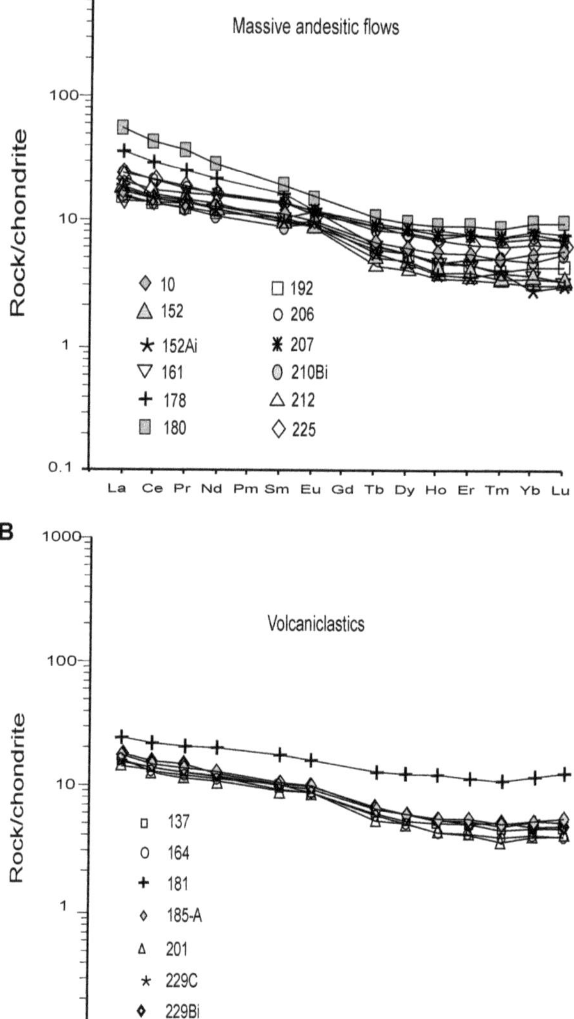

FIG. 6. Chondrite-normalized rare-earth-element data from (A) volcanic and (B) volcaniclastic rocks of the Tuzancoa Formation. Both have similar REE concentrations (normalizing values from Nakamura, 1974).

The only Upper Paleozoic unit in Mexico that is reported to contain basaltic and andesitic rocks is the one in Delicias, Coahuila (Fig. 9A). This unit forms a marine succession of debris flows and turbidites that contains large blocks (up to tens of meters long) of limestone, volcanic, and volcaniclastic rocks, and some conglomerate (McKee et al, 1988, 1999; Lopez, 1997; Lopez et al., 2001). Their ages range from Late Mississippian to Late Permian (Guadalupian). The volcanic rocks range in composition from rhyolite to andesite, and rare basalt (Lopez, 1997), compositionally indicative of a continental arc setting (McKee et al., 1999). The conglomerates contain metamorphic clasts that include Precambrian gneisses, granite, quartz, schist, and limestone, suggesting a mixed provenance of volcanic arc and uplifted basement (Lopez, 1997, Lopez et al., 2001).

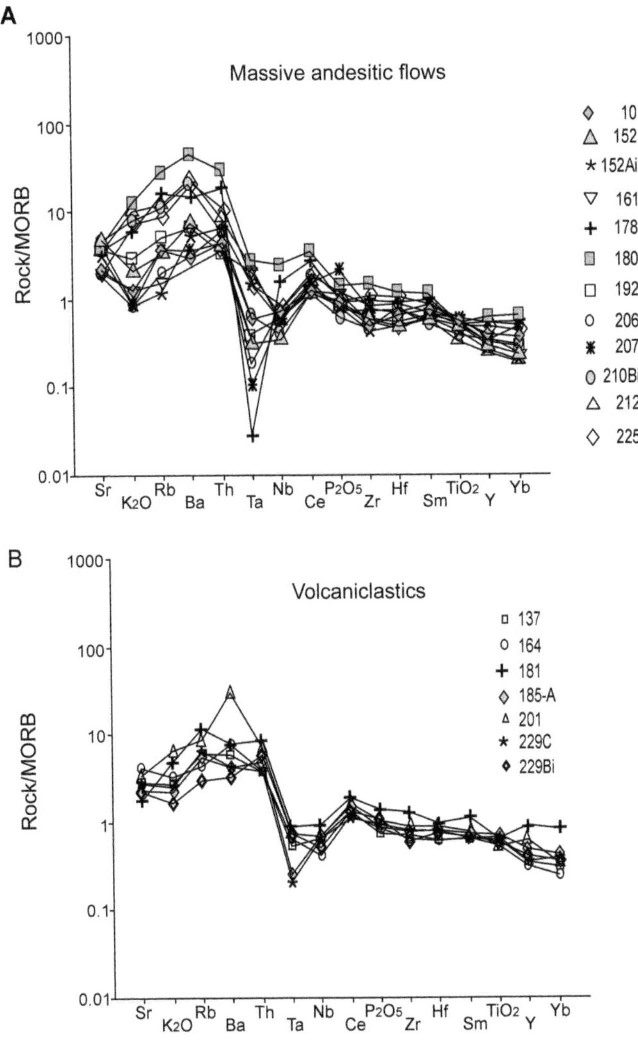

FIG. 7. Selected trace and other elements for (A) volcanic and (B) volcaniclastic rocks from the Tuzancoa Formation normalized to mid-ocean ridge basalts (MORB) (normalizing values from Pearce, 1983)

Although volcanic rocks have not been found in the Guacamaya Formation (Fig. 9A), turbidites of this formation in Tamaulipas contain abundant volcanic clasts (volcanic sandstone; Gursky and Michalzik, 1989; Centeno-Garcia et al., 1997). The Guacamaya Formation therefore may have been deposited in a basin receiving sediments from the Tuzancoa arc (backarc or forearc basin). Gursky and Michalzik (1989) suggested a forearc setting for the Guacamaya Formation.

Although most ages from a granitoid belt of arc affinity that extends along eastern Mexico from Delicias, Coahuila to Oaxaca (260–232 Ma; Torres-Vargas et al., 1999) are younger than the Tuzancoa volcanism, contemporary U-Pb zircon ages (287 and 289 Ma) have been reported from the Totoltepec pluton in the Acatlán Complex (Keppie et al., 2004), suggesting that a genetic relation between the intrusive and extrusive magmatism for this time might exist.

REE patterns of volcanic rocks of intermediate composition ([SiO_2] = 57–60%) from the Tuzancoa Formation are very similar to those shown by volcanic blocks of intermediate composition in the Las

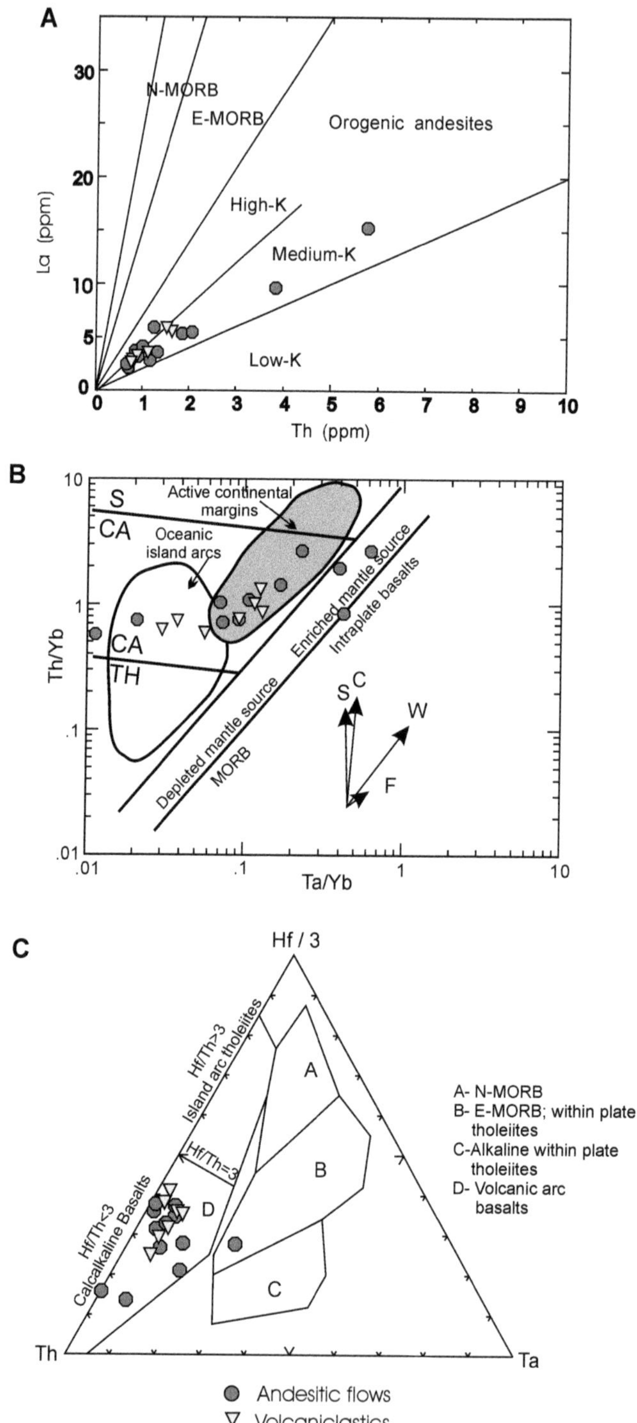

FIG. 8. Tectonic discrimination diagrams for volcanic and volcaniclastic rocks of the Tuzancoa Formation. A. Th-La (Gill, 1981). B. Ta/Yb-Th/Yb (Pearce, 1983) where vectors shown indicate the influence of subduction components (S), crustal contamination (C), within-plate enrichment (W), and fractional crystallization (F); TH = tholeiitic, CA = calc-alkaline, and S = shoshonitic fields are also shown. C. Hf/3-Th-Ta (Wood, 1980). Samples plot mostly in volcanic arc fields.

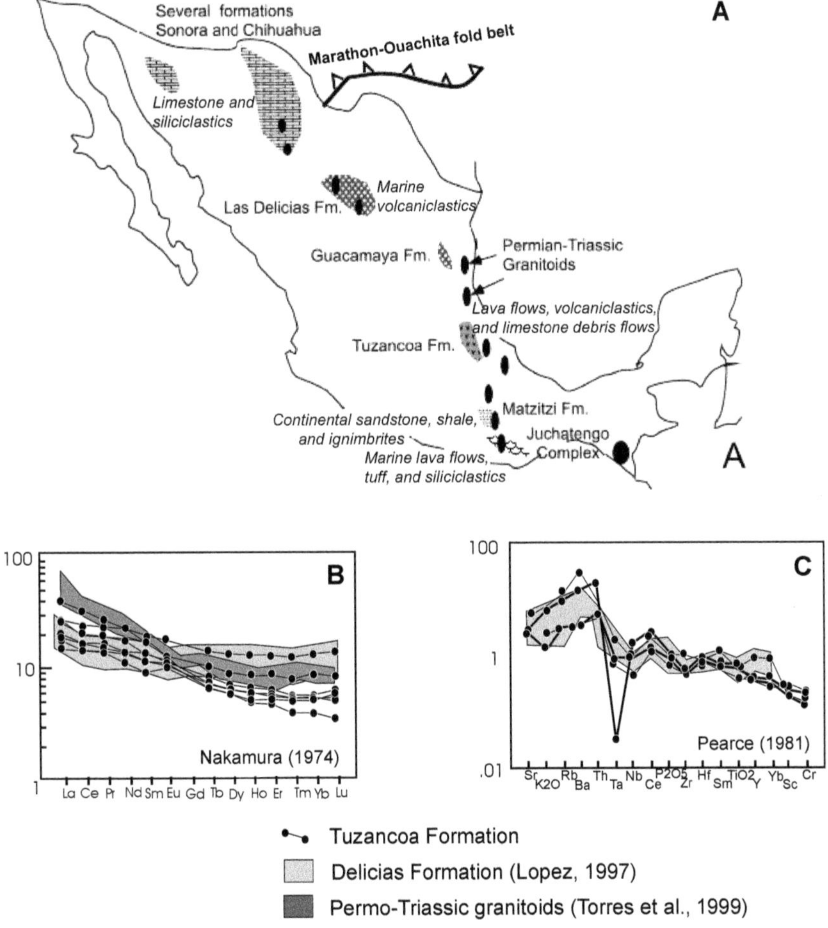

FIG. 9. A. Geographic location of Las Delicias, Guacamaya, and Tuzancoa formations and other Paleozoic locations mentioned in the text. B. Comparison of whole-rock chondrite-normalized REE. C. MORB-normalized trace element data from the Tuzancoa Formation with volcaniclastic rocks of Las Delicias Formation (light-shaded area; Lopez, 1997) and the granitoid belt of eastern Mexico (dark-shaded area; Torres-Vargas et al., 1999); normalizing values after Nakamura (1974) and Pearce (1983), respectively.

Delicias Formation (Lopez, 1997, Lopez et al., 2001) (Fig. 9B). By contrast, REE patterns reported from the Permo-Triassic granitoid belt are more fractionated, with an enrichment of LREE (Torres-Vargas et al., 1999), as shown by Figure 9B. These data suggest that the successions at Hidalgo and Coahuila probably formed part of the same Carboniferous–Permian continental arc. MORB-normalized trace and other elements from the Las Delicias basic and intermediate volcanic clasts (Lopez, 1997) are also similar to those from the volcanic and volcaniclastic rocks of the Tuzancoa Formation (Fig. 9C).

Paleogeographic Reconstruction and Tectonic Setting

Our data suggests that Upper Paleozoic rocks of the Tuzancoa Formation were deposited in a marine basin associated with a volcanic arc (Fig. 10). That this basin lay close to the volcanic edifices is indicated by abundant lava flows interbedded with the volcaniclastic rocks. Facies associations suggest that the main transport mechanism of the volcaniclastic rocks was that of turbiditic and debris flows. Calcareous sediments, initially deposited on narrow, shallow-marine platforms surrounding the volcanic

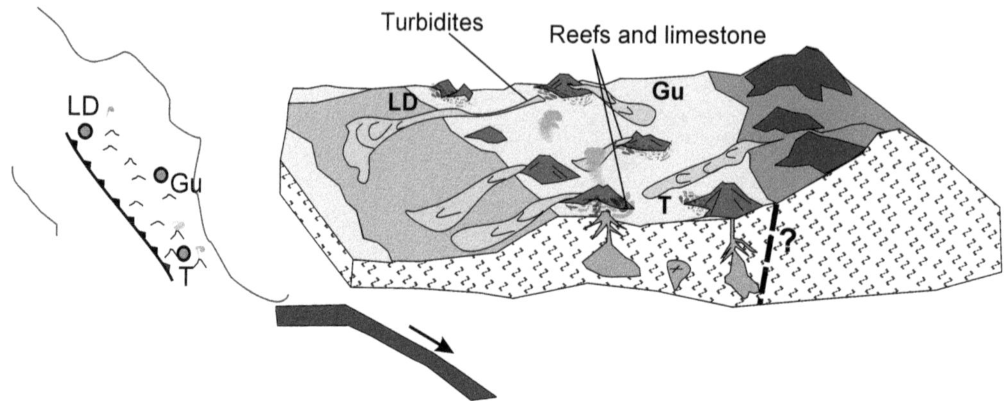

FIG. 10. Proposed paleoenvironmental and paleogeographic model for the Tuzancoa Formation (T), and possible regional distribution of other arc related facies (LD = Las Delicias Formation, Gu = Guacamaya Formation).

edifices, were remobilized and redeposited in deeper areas of the basin (Fig. 10).

Arellano et al. (1998) suggested that the Tuzancoa Formation was deposited on the continental slope. However, lithofacies and fossils indicative of deep-marine environments are absent. Based on the composition of the volcanic rocks and the characteristics of the volcaniclastic rocks, we interpret the Tuzancoa Formation to be a marine intra-arc basin deposit. Facies associations of the Las Delicias and Guacamaya formations are more suggestive of back-arc or forearc sedimentation than those of the Tuzancoa Formation. The Tuzancoa basin received sediments from volcanoes and, to a minor extent, from uplifted Precambrian basement and older sedimentary rocks. Given relatively primitive oceanic arc-like geochemical signatures of the volcanic rocks, submarine conditions of the arc, and the contribution of sediments derived from older rocks, we suggest that the arc developed in an extensional setting that allowed magmas to reach the surface rapidly without large amounts of crust being assimilated. The volcanoes probably developed in a marine basin with uplifted blocks of basement (Fig. 10).

Lopez (1997) suggested that rocks of Las Delicias Formation were deposited in a forearc basin, and placed the subduction zone between North America and Oaxaquia (Sierra Madre and Coahuila terranes). However, the North American affinity of the fossil associations in the Tuzancoa Formation and other contemporaneous units suggests that Oaxaquia was already attached to North America by Carboniferous to Early Permian time (Stewart et al., 1999; Navarro-Santillán et al., 2002; Sour-Tovar et al., in press).

Other authors, like Dickinson and Lawton (2001) have inferred continuous subduction beneath western Oaxaquia from the Carboniferous to the Triassic. Our data supports the existence of a subduction zone during the Early Permian. Although accretionary complexes associated with the Permian arc have not been found, we favor a model with the subduction zone along the Pacific margin of Oaxaquia, as shown in Figure 10.

Acknowledgments

The authors are grateful to the Universidad Nacional Autónoma de México (DGAPA-UNAM) for funding through projects PAPIIT IN101095 and PAPIIT IN116599 and to the GSA Foundation, grant 5873-96, for geochemical analyses. We also thank J. Patchett and C. Isachsen (University of Arizona) for the Nd analysis, Pablo Peñaflor E. (Institute of Geology, UNAM) for his help in the preparation of samples, and Rufino Lozano (Instituto de Geología, UNAM) for XRF analyses. We are also grateful to D. Slauenwhite of Saint Mary's University, Halifax, Nova Scotia; and to Pam King, Mike Turbrett, and Lakmali Hewa of Memorial University, Saint John's, Newfoundland for their support during XRF and ICP-MS analyses, respectively. This project was one of the requirements for fulfillment of a Master's Degree in Science (Geology) by Laura Rosales-Lagarde (Posgrado en Ciencias de la Tierra, Instituto de Geología, UNAM). Reviewers Damian Nance and Martin Valencia provided valuable comments and suggestions. This is a Contribution to IGCP project 453: Modern and Ancient Orogens.

REFERENCES

Arellano-Gil, J., Vachard, D., Yussim, S., and Flores de Dios-González, L., 1998, Aspectos estratigráficos, estructurales y paleogeográficos del Pérmico inferior al Jurásico inferior en Pemuxco, Estado de Hidalgo, México: Revista Mexicana de Ciencias Geológicas, v. 15, p. 9–13.

Bouma, 1962, Sedimentology of some flysch deposits: A graphic approach to facies interpretation: Amsterdam, Netherlands, Elsevier, 168 p.

Buitrón, B. E., Patiño-Ruíz, J., and Moreno-Cano, A., 1987, Crinoides del Paleozoico tardío (Pensilvánico) de Calnali, Hidalgo: Revista de la Sociedad Mexicana de Paleontología, v. 1, p. 125–136.

Campa, M. F., and Coney, P. J., 1983, Tectono-stratigraphic terranes and mineral resource distributions in Mexico: Canadian Journal of Earth Sciences, v. 20, p. 1040–1051.

Carrillo-Bravo, J., 1961, Geología del Anticlinorio Huizachal-Peregrina al NW de Ciudad Victoria, Tamaulipas: Boletín de la Asociación Mexicana de Geólogos Petroleros, v. 13, p. 1–98.

Carrillo-Bravo, J., 1965, Estudio Geológico de una parte del Anticlinorio Huayacocotla: Boletín de la Asociación Mexicana de Geólogos Petroleros, v. 17, nos. 5–6, p. 73–92.

Centeno-García, E., 2003, Terrane configuration of central-western Mexico and the Mesozoic evolution of the continental margin, *in* 99th Annual Cordilleran Section, Geological Society of America: Puerto Vallarta, Jalisco, Mexico, paper no. 27-3.

Centeno-García, E., Sánchez-Zavala, J. L., Patchett J., Isachsen, C., Sour-Tovar, F., and Ortega-Gutierrez, F., 1997, Stratigraphy, Nd isotopes, sediment provenance, and paleogeography of Paleozoic sequences in southern Mexico [abs.], *in* Terrane Dynamics–97, International Conference on Terrane Geology: Christchurch, New Zealand, Abstracts, p. 42–45.

DePaolo, D. J., 1981, Neodymium isotopes in the Colorado Front range and crust-mantle evolution in the Proterozoic: Nature, v. 291, p. 193–196.

Dickinson, W., and Lawton, T., 2001, Carboniferous to Cretaceous assembly and fragmentation of Mexico: Geological Society of America Bulletin, v. 113, no. 9, p. 1142–1160.

Gill, J. B., 1981, Orogenic andesites and plate tectonics: Berlin, Germany, Springer-Verlag, 358 p.

Grajales-Nishimura, J. M., Centeno-García, E., Keppie, J. D., and Dostal, J., 1999, Geochemistry of Paleozoic basalts form the Juchatengo complex of southern Mexico: Tectonic implications: Journal of South American Earth Sciences, v. 12, p. 537–544.

Gursky, H. J., and Michalzik, D., 1989, Lower Permian turbidites in the northern Sierra Madre Oriental, Mexico: Zentralblatt für Geologie und Paläontologie Teil I, v. 5/6, p. 821–838.

Irvine, T. N., and Baragar, W. R. A., 1971, A guide to the chemical classification of the common volcanic rocks: Canadian Journal of Earth Sciences, v. 8, p. 523–548.

Jenner, G. A., Longerich, H. P., Jackson S. E., and Fryer, B.J., 1990, ICP-MS—a powerful tool for high-precision trace–element analysis in Earth sciences: Evidence from analysis of selected U.S.G.S. reference samples: Chemical Geology, v. 83, p. 133–148.

Keppie, J. D., Nance, R. D., Dostal, J., Ortega-Rivera, A., Brent, V. M., Fox, D., Muise, J., Powell, J. T., Mumma, S. A., and Lee, J. W. K., 2004, Mid-Jurassic tectonothermal event superposed on a Paleozoic geological record in the Acatlán Complex of southern Mexico: Hotspot activity during the breakup of Pangea: Gondwana Research, v. 7, p. 239–260.

Lawlor, P. J., Ortega-Gutiérrez, F., Cameron, K. L., Ochoa-Camarillo, H., Lopez, R., Sampson, D. E., 1999, U-Pb geochronology, geochemistry, and provenance of the Grenvillian Huiznopala Gneiss of Eastern Mexico: Precambrian Research, v. 94, p. 73–99.

Longerich, H. P., Jenner, G. A., Fryer B. J., and Jackson, S. E., 1990, Inductively coupled plasma-mass spectrometric analysis of geological samples: A critical evaluation based on case studies: Chemical Geology, v. 83, p. 105–118.

Lopez, R., 1997, The Pre-Jurassic geotectonic evolution of the Coahuila terrane, northwestern Mexico: Grenville basement, a late Paleozoic arc, Triassic plutonism, and the events south of the Ouachita suture: Unpubl. Ph.D. thesis, University of California, Santa Cruz, 82 p.

Lopez, R., Cameron, K. L., and Jones, N. W., 2001, Evidence for Paleoproterozoic, Grenvillian and Pan-African age Gondwana crust beneath northeastern Mexico: Precambrian Research, v. 107, p. 195–214.

McKee, J. W., Jones, N. W., and Anderson, T. H., 1988, Las Delicias Basin: A record of late Paleozoic arc volcanism in northeastern Mexico: Geology, v.16, p. 37–40.

McKee, J. W., Jones, N. W., and Anderson, T. H., 1999, Late Paleozoic and early Mesozoic history of the Las Delicias terrane, Coahuila, Mexico, *in* Bartolini, C., Wilson, J. L., and Lawton, T. F., eds., Mesozoic sedimentary and tectonic history of north-central Mexico: Geological Society of America, Special Paper 340, p. 161–188.

Nakamura, N., 1974, Determination of REE, Ba, Fe, Mg, Na, and K in carbonaceous and ordinary chondrites: Geochimica et Cosmochimica Acta, v. 38, p. 757–775.

Navarro-Santillán, D., Sour-Tovar, F., and Centeno-García, E., 2002, Lower Mississippian (Osagean) brachiopods from the Santiago Formation, Oaxaca,

Mexico: Stratigraphy and tectonic implications: Journal of South American Earth Sciences, v. 15, p. 327–336.

Ochoa-Camarillo, H. R., 1996, Geología del anticlinorio de Huayacocotla en la región de Molango, Estado de Hidalgo: Unpubl. M.Sc. thesis, Universidad Nacional Autónoma de México, 91 p.

Ochoa-Camarillo, H., Buitrón-Sánchez, B., and Silva-Pineda, A., 1999, Red beds of the Huayacocotla Anticlinorium, state of Hidalgo, east-central Mexico, *in* Bartolini, C., Wilson, J. L., and Lawton, T. F., eds., Mesozoic sedimentary and tectonic history of north-central Mexico: Geological Society of America Special Paper 340, p. 59–68.

Okulitch, A. V., 1999, Geological time scale 1999: Geological Survey of Canada, Open File 3040, National Earth Science Series, Geological Atlas-Revision.

Ortega-Gutiérrez, F., Lawlor, P., Cameron, K. L., and Ochoa-Camarillo, H., 1997, New studies of the Grenvillean Huiznopala Gneiss, Molango area, State of Hidalgo, Mexico—preliminary results, *in* Instituto de Investigaciones en Ciencias de la Tierra de la Universidad Autónoma del Estado de Hidalgo e Instituto de Geología de la Universidad Nacional Autónoma de México, eds., II Convención sobre la Evolución Geológica de México y Recursos Asociados: Pachuca, Mexico, Libro-guía de las excursiones geológicas, Excursión 1, p. 19–25.

Ortega-Gutiérrez, F., Ruiz, J., and Centeno-García, E., 1995, Oaxaquia, a Proterozoic microcontinent accreted to North America during the late Paleozoic: Geology, v. 23, p. 1127–1130.

Patchett, P. J., and Ruiz, J., 1987, Nd isotopic ages of crust formation and metamorphism in the Precambrian of eastern and southern Mexico: Contributions to Mineralogy and Petrology, v. 96, p. 523–528.

Pearce, J. A., 1983, Role of the sub-continental lithosphere in magma genesis at active continental margins, *in* Hawkesworth, C. J., and Norry, M. J., eds., Continental basalts and mantle xenoliths: Nantwich, UK, Shiva, p. 230–249.

Pérez-Ramos, O., 1978, Estudio Bioestratigráfico del Paleozoico Superior del Anticlinorio de Huayacocotla en la Sierra Madre Oriental: Boletín de la Sociedad Geológica Mexicana, v. XXXIX, p. 126–135.

Pindell, J. L., 1985, Alleghenian reconstructions and subsequent evolution of the Gulf of Mexico, Bahamas, and proto-Caribbean: Tectonics, v. 4, p. 1–40.

Pindell, J., and Dewey, J. F., 1982, Permo-Triassic reconstruction of western Pangea and the evolution of the Gulf of Mexico/Caribbean region: Tectonics, v. 1, p. 179–211.

Rosales-Lagarde, L., 2002, Estratigrafía y geoquímica de la secuencia volcanosedimentaria paleozoica del noreste del Estado de Hidalgo, México: Unpubl. M.Sc. thesis, Universidad Nacional Autónoma de México, 89 p.

Rosales-Lagarde, L., Centeno-García, E., Ochoa-Camarillo, H., and Sour-Tovar, F., 1997, Permian volcanism in eastern Mexico—preliminary report, *in* Instituto de Investigaciones en Ciencias de la Tierra de la Universidad Autónoma del Estado de Hidalgo e Instituto de Geología de la Universidad Nacional Autónoma de México, eds., II Convención sobre la Evolución Geológica de México y Recursos Asociados: Pachuca, Mexico, Libro-guía de las excursiones geológicas, Excursión 1, p. 27–32.

Ross, M. I., and Scotese, C. R., 1988, A hierarchical tectonic model of the Gulf of Mexico and Caribbean region: Tectonophysics, v. 155, p. 139–168.

Ruíz, J., Patchett, P. J., and Ortega-Gutiérrez, F., 1988, Proterozoic and Phanerozoic basement terranes of Mexico from Nd isotopic studies: Geological Society of America Bulletin, v. 100, p. 274–281.

Sánchez-Zavala, J.L., Centeno-García, E., and Ortega-Gutiérrez, F., 1999, Review of Paleozoic stratigraphy of México and its role in the Gondwana-Laurentia connections, *in* Ramos, V. A., and Keppie, J. D., eds., Laurentia-Gondwana connections before Pangea: Geological Society of America, Special Paper 336, p. 1–16.

Sedlock, R. L., Ortega-Gutierrez, F., and Speed, R. C., 1993, Tectonostratigraphic terranes and tectonic evolution of Mexico: Geological Society of America, Special Paper 278, 153 p.

Silva-Pineda, 1987, Algunos elementos paleoflorísticos del Pérmico de la región de Calnali, Estado de Hidalgo: Revista de la Sociedad Mexicana de Paleontología, v. 1, p. 313–327.

Sour-Tovar, F., Pérez-Huerta, A., Quiroz-Barroso, S. A., and Centeno-García, E., 2005, Braquiópodos y trilobites del Pérmico del noroeste del estado de Hidalgo, México: Revista Mexicana de Ciencias Geológicas, v. 22, p. 24–35.

Stewart, J. H., Blodgett, R. B., Boucot A. J., Carter, J. L., and Lopez, R., 1999, Exotic Paleozoic strata of Gondwanan provenance near Ciudad Victoria, Tamaulipas, Mexico, *in* Ramos, V. A., and Keppie, J. D., eds., Laurentia-Gondwana connections before Pangea: Geological Society of America, Special Paper 336, p. 227–252.

Suter, M., 1984, Cordilleran deformation along the eastern edge of the Valles-San Luis Potosi carbonate platform, Sierra Madre Oriental fold-thrust belt in east-central Mexico: Geological Society of America Bulletin, v. 95, p. 1387–1397.

Suter, M., 1987, Structural traverse across the Sierra Madre Oriental fold-thrust belt in east-central Mexico: Geological Society of America Bulletin, v. 98, p. 249–264.

Suter, M., 1990, Hoja Tamazunchale 14-Q-e(5), Geología de la Hoja Tamazunchale, Estados de Hidalgo, Querétaro y San Luis Potosí, Carta Geológica de México, Serie de 1:100,000: México, DF, Universidad

Nacional Autónoma de México, Instituto de Geología, 55 p.

Torres-Vargas, R., Ruíz, J., Patchett, P. J., and Grajales-Nishimura, J. M., 1999, Permo-Triassic continental arc in eastern Mexico: Tectonic implications for reconstructions of southern North America, *in* Bartolini, C., Wilson, J. L., and Lawton, T. F., eds., Mesozoic sedimentary and tectonic history of north-central Mexico: Geological Society of America Special Paper 340, p. 191–196.

Weber, R., 1997, How old is the Triassic flora of Sonora and Tamaulipas and news on Leonardian floras in Puebla and Hidalgo, Mexico: Revista Mexicana de Ciencias Geológicas, v. 14, p. 225–243.

Wilson, M., 1989, Igneous petrogenesis: London, UK, Unwin Hyman Ltd, 465 p.

Winchester, J. A., and Floyd, P. A., 1977, Geochemical discrimination of different magma series and their differentiation products using immobile elements: Chemical Geology, v. 20, p. 325–343.

Wood, D. A., 1980, The application of a Th-Hf-Ta diagram to problems of tectonomagmatic classification and to establishing the nature of crustal contamination of basaltic lavas of the British Tertiary volcanic province: Earth and Planetary Science Letters, v. 50, p. 11–30.

Chuacús
Terrane

Orogeny in Time and Space in the Mesoamerican Region and Appalachian-Variscan Connections

FERNANDO ORTEGA-GUTIÉRREZ,[1] LUIGI A. SOLARI, JESÚS SOLÉ,
Instituto de Geología, Universidad Nacional Autónoma de México (UNAM), 04510 México D.F., México

UWE MARTENS,
Universidad de San Carlos de Guatemala—Centro Universitario del Norte, Cobán, Guatemala

ARTURO GÓMEZ-TUENA,
Centro de Geociencias, Universidad Nacional Autónoma de México (UNAM), Juriquilla, Querétaro, México

SERGIO MORÁN-ICAL,
Universidad de San Carlos de Guatemala—Centro Universitario del Norte, Cobán, Guatemala

MARGARITA REYES-SALAS, AND CARLOS ORTEGA-OBREGÓN
Instituto de Geología, Universidad Nacional Autónoma de México (UNAM), 04510 México D.F., México

Abstract

This paper describes the first discovery of eclogite-facies rocks in the Paleozoic Chuacús basement complex of north-central Guatemala. In this area, the complex comprises a thick, polydeformed sequence of high-Al metapelite, amphibolite, and quartzofeldspathic banded gneisses and schists characterized by garnet, phengite, and kyanite. Detailed petrographic, electronprobe microanalyses, and a late Carboniferous U-Pb zircon apparent age indicate that this deeply rooted orogenic terrane may be related to the Alleghenian suturing between Gondwana and Laurentia. Eclogite-facies metamorphism is established by assemblages with omphacite-garnet-rutile ± phengite ± zoisite in mafic rocks, which are consistent with garnet-kyanite-zoisite-rutile-quartz-phengite ± staurolite ± chloritoid assemblages in pelitic rocks, and amphibole-calcite/dolomite/aragonite?-rutile-quartz-zoisite ± clinochlore ± diopside in marbles. Moreover, various textural and mineralogical features (such as radial cracks in garnet and kyanite around quartz inclusions; palisade-like coronas of a silica mineral around quartz in some carbonates; lamellar inclusions of a titaniferous phase in garnet, zoisite, and phengite; and plagioclase or white mica in some omphacite; as well as the relatively high Na_2O content of garnet [up to 0.12 wt%]), suggest relict ultrahigh-pressure metamorphism (UHPM). These conditions predated high-temperature–high-pressure hydration and decompression melting that occurred between 18 and 23 kbar and 700–770°C. This decompressional melting event of eclogitic rocks is dated as late Carboniferous by U-Pb on discordant zircons from a leucocratic neosome, and may be associated with the initial closure of Pangea. K-Ar ages of ~70–75 Ma on micas and amphibole, stable at 14 kbar and 597°C, are interpreted to record the Cretaceous obduction of Caribbean ophiolites and arc assemblages onto the Chuacús complex and the southern edge of the Maya block, along the paleo-Motagua fault zone.

Introduction

METAMORPHIC ROCKS of the Chuacús series in the Central Cordillera of Guatemala, located between the Motagua and Polochic faults, constitute one of the most enigmatic and critically located basement units of all Mesoamerica, because these rocks are emplaced at the present tectonic boundary between the North America and the Caribbean plates, and they have been difficult to correlate with adjacent basement terranes of southern Mexico and nuclear Central America. The rocks exposed in the Sierra de Chuacús have been named the Chuacús series (McBirney, 1963, p. 187) or Chuacús Group (Kesler

[1]Corresponding author; email: fortega@servidor.unam.mx

et al., 1970). However, the sequence in the studied area (Fig. 1) is not composed of mappable layered units following a stratigraphic order, but constitutes a high-grade metamorphic complex with a poorly understood polyphase orogenic history; thus we prefer to use the name Chuacús complex, as it is more in accord with current international stratigraphic practice (Hedberg, 1976).

Although several geologic and geochronologic studies have been published on the sequence (McBirney, 1963; Gomberg et al., 1968; Kesler et al., 1970; Kesler, 1971; Clemons et al., 1974; Newcomb, 1978; Roper, 1978), petrologic aspects have been treated only by McBirney (1963) and in an unpublished thesis by van den Boom (1972), who reported assemblages with garnet, rutile, phengite, or barroisite, but did not recognize eclogite-facies rocks. Reconnaissance studies across the Sierra de Chuacús and extensive sampling in the Pachalúm, Palibatz, and El Chol areas (Fig. 1) reveal for the first time that most mafic, pelitic, quartzofeldspathic, and calcareous units were in fact metamorphosed to eclogite-facies conditions. These rocks show high-pressure mineralogy with omphacite, garnet, rutile, kyanite, zoisite, and probably aragonite. High-pressure metamorphism supports the existence of a deep-seated regional tectonothermal event that may require the profound revision of the entire evolution of the Maya-Chortís boundary from Precambrian to Cenozoic times. Considering the preliminary nature of this work, the main contributions of the paper are: (1) a detailed petrographic description and metamorphic petrology of representative lithologies, including the recognition that the P-T path may extend to ultrahigh-pressure conditions; (2) the U-Pb and K-Ar dating of key units; and (3) on the basis of these data, the proposal of two alternative tectonic settings for the origin of the Chuacús complex.

Pre-Cretaceous Geologic Setting

The basement geology of the southern margin of the Maya block located between the Polochic and Motagua fault zones (Fig. 1) is complicated and remains essentially unresolved (Weyl, 1980; Donnelly et al., 1990). Identified components include low to high-grade pre-Mesozoic metamorphic rocks (Chuacús Group of Kesler et al., 1970), the Paleozoic Rabinal and younger granitoids, as well as volcanic and sedimentary rocks of probable late Paleozoic age, most of which have uncertain depositional, intrusive, or faulted contacts. Previous isotopic ages have identified a pre-Mesozoic history as old as the Middle Proterozoic and as young as latest Paleozoic (Permo-Triassic), but most of these data represent reset, mixed, or cooling ages that reflect the latest intrusive or tectonic event and inherited components (i.e., Gomberg et al., 1968). The typical structural history of the area includes at least three main superposed phases of penetrative deformation and metamorphism in Paleozoic, Cretaceous, and Cenozoic times, in the latter case associated with active tectonics along the Motagua-Polochic transform system (Donelly et al., 1990, and references therein). The Paleozoic sedimentary rocks of the Cuchumatanes area (Anderson et al., 1973), exposed just north of the Polochic fault in the Maya block (see Fig. 1), are faulted against metamorphic rocks composed of granitic-dioritic gneisses and lower-grade layered sequences that may not be continuous with the Chuacús complex exposed south of the fault. Although upper Paleozoic and possibly lower Paleozoic strata of the Maya block appear to overlie the Rabinal granite just south of the Polochic fault (Bohnenberger, 1966), we found that the southern limit of this granite with the Chuacús complex is not a gradual contact but, as originally proposed by McBirney (1963), is tectonic and revealed by intense blastomylonitic retrogression and deformation of the high-grade metamorphic sequence, as well as mylonitic shear zones within the granitoid. Thus, the common assumption that high-grade metamorphic rocks (Chuacús complex) between the Motagua and Polochic faults are part of the Maya block (e.g., Kesler et al., 1970; Weyl, 1980; Donelly et al., 1990) may not be valid.

The Chuacús complex was considered a Lower Carboniferous metamorphic unit unconformably covered by sedimentary marine rocks of the Santa Rosa Group assigned to the Permian (van den Boom, 1972; Pindell, 1985). However, according to Clemons et al. (1974), the base of the Santa Rosa Group in the area mapped by McBirney (1963) is also obscured by shearing and metamorphism throughout a "transition zone" above the Chuacús complex. North of Rabinal, the Santa Rosa Group unconformably overlies "granitized" Chuacús rocks (Rabinal granite) dated at 345 ± 20 Ma by Gomberg et al. (1968), whereas at Sacapulas, 80 km west of Cobán, Bohnenberger (1966) described unconformable relationships between the Chuacús "Group" and the Sacapulas Formation, a sedimentary unit transitional with the Tactic Formation of late Paleozoic

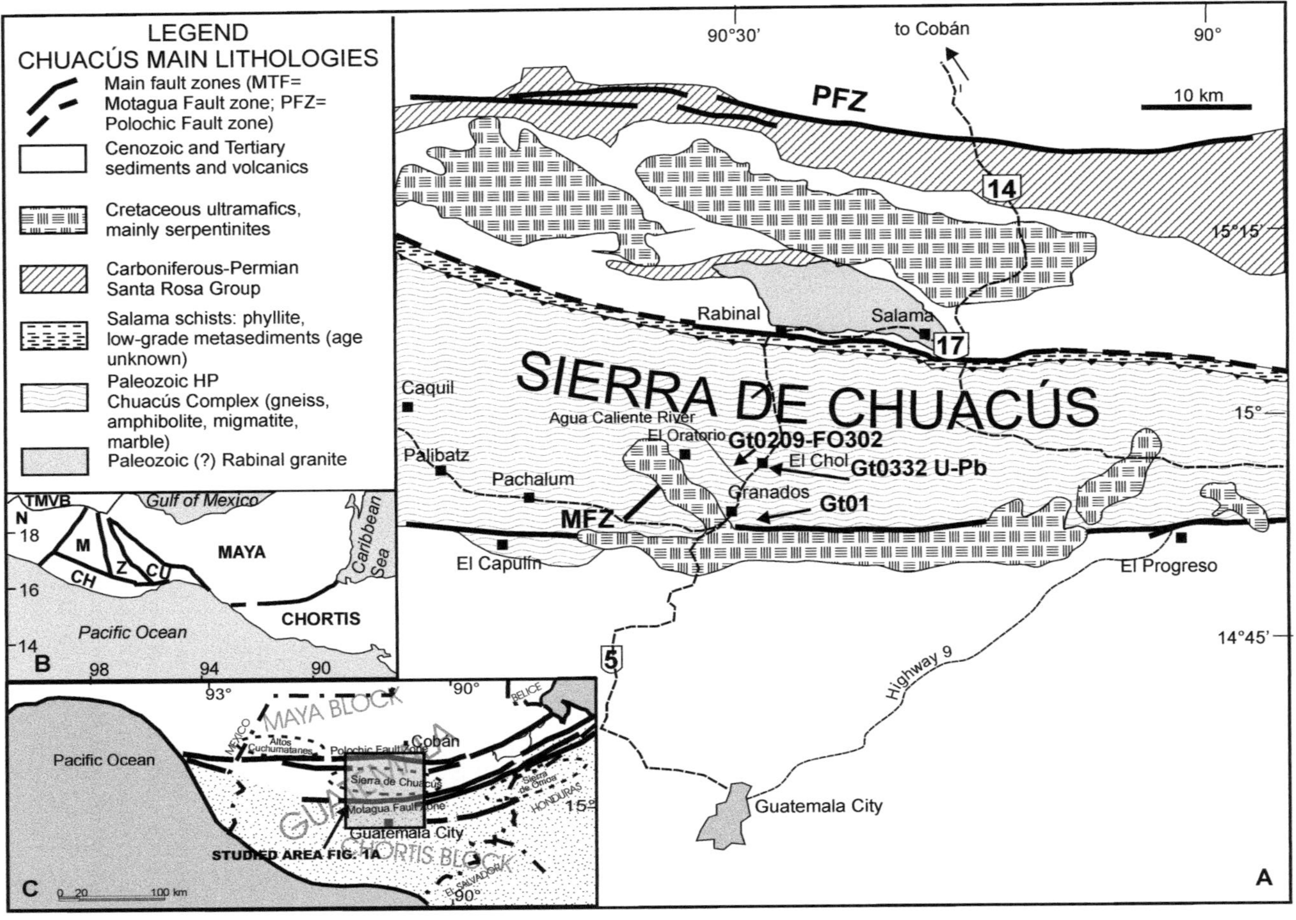

FIG. 1. Location map. A. Geologic sketch map of north central Guatemala showing the major locations referred to in the text. Modified from Weyl, 1980. B. Terrane subdivision of southern Mexico and Central America. C. Main neotectonic features characterizing the studied area.

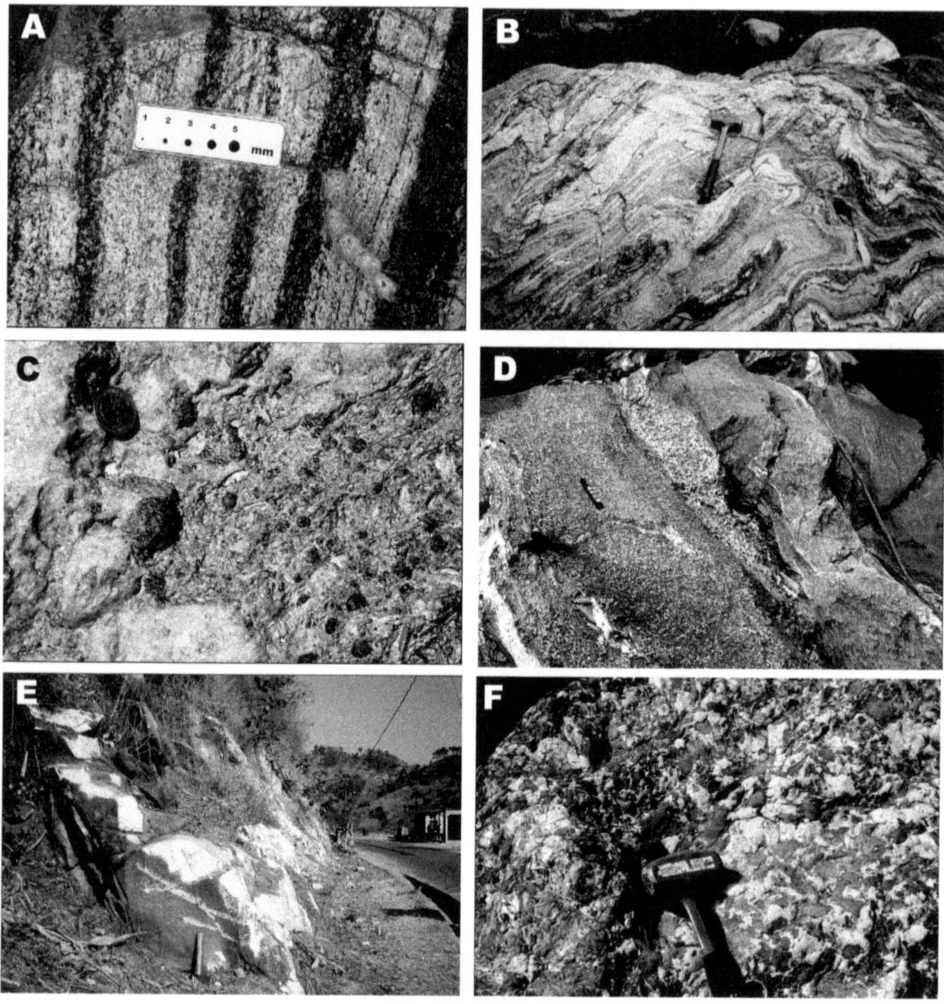

FIG. 2. A. Banded facies of Chuacús complex defined by alternating albite-quartz–rich felsic layers and eclogitic amphibolites. Barranca Agua Caliente. B. Migmatitic facies of the Chuacús complex, showing intense polyphase deformation, El Chol. C. Garnet-kyanite metapelite. Garnets are pyrope-rich (up to 41 mol%) and preserved abundant inclusions of chloritoid, staurolite, zoisite, rutile, and white mica. The coin is about 2 cm across. Palibatz area. D. Gabbroic orthogneisses of Chuacús complex characterized by the paucity of garnet and absence of omphacite. Palibatz area. E. White dolomite marble at El Oratorio, Palibatz area. F. Albite-phengite-quartz pegmatite, San Antonio Granados area.

age, and both forming part of the Santa Rosa Group. In the Cuchumatanes uplift, the contact between the Carboniferous–Permian Santa Rosa Group and the underlying "Undifferentiated Metamorphic Rocks" (Anderson et al., 1973, Fig. 2) is shown as an unconformity. It also is important to note that Donnelly et al. (1990) reported an Ar-Ar unpublished hornblende age of 238 Ma in the Chuacús complex, and Late Triassic ages (whole rock–mineral Rb-Sr isochron, and Ar-Ar on muscovite) for the Matanzas granite which, according to McBirney (1973), intruded the Chuacús complex in the studied area.

South of Sierra Chuacús beneath allochthonous ultramafic bodies of the Cretaceous El Tambor terrane (proto-Caribbean oceanic crust), along the Motagua fault, and cropping out to the south in the Chortís continental block, is a high-grade metamorphic unit known as the Las Ovejas complex formed by garnet-staurolite-sillimanite schists and gneisses, amphibolite, migmatite, and marble

(Schwartz, 1972). This complex is continuous with the Omoa complex of northwestern Honduras, where orthogneisses (Bañaderos complex) were dated imprecisely (Rb-Sr whole-rock "errorchron") by Horne et al. (1976) as early Paleozoic to Precambrian (460–980 Ma), and intruded by late Paleozoic granites (302 ± 12 Ma, Rb-Sr isochron). More recently, 1 Ga granites (concordant U-Pb ages) that intruded chlorite-biotite-garnet schists have been reported from the Sula Valley of northwestern Honduras (Manton, 1996). Unconformably overlying that basement, low-grade metamorphic rocks comprising the San Diego Phyllite (Giunta et al., 2002) are intruded by the 180 Ma San Isidro granite batholith (Horne et al., 1976), and unconformably covered by Lower Cretaceous sedimentary rocks. The most easterly part of the Chortís block in Honduras and Nicaragua includes low-grade crystalline sequences (Cacaguapa schist and Palacaguina Formation) unconformably covered by, or faulted against, fossiliferous Middle–Lower Jurassic shallow-marine and continental rocks of the Agua Fría Formation (Gordon, 1992, 1993). No sedimentary rocks of Paleozoic age have been found in the Chortís block; nonetheless they are abundant in the Maya block.

Petrography of the Chuacús Complex in the Studied Area

The Chuacús complex in the Sierra de Chuacús of north-central Guatemala constitutes a thick sequence of dominantly metasedimentary rocks that together with some mafic and felsic orthogneisses underwent metamorphism and deformation at high pressures and temperatures. Abundant garnet and potassic micas with common rutile, kyanite, staurolite, and chloritoid crystallized from Al-rich sediments. Accompanying rocks include calcic and magnesian marbles, quartzite, and graphitic schist that may indicate a protolith of mature sediments typical of passive margins. The sequence forms a banded (Fig. 2A), multiply folded, and locally migmatitic complex (Fig. 2B) of metapelitic, quartzofeldspathic (Fig. 2C), and amphibolitic gneisses (Fig. 2D), and schists, most of which are rich in garnet, mica, albite, and rutile, together with some white to dark impure marble layers (Fig. 2E). Thin layers of serpentinite in the Chuacús complex were reported by van den Boom (1972) but were not found during our study. Pegmatitic bodies (Fig. 2F) composed of phengite-quartz-albite ± tourmaline ± garnet were emplaced parallel to the predominantly vertical, E-W–trending foliation. The eclogitic phases identified in thin section of mafic rocks (see Fig. 1 for location of samples) consist of green omphacite, garnet, phengite, zoisite/epidote/allanite, and rutile, that may be replaced by pargasitic and Na-Ca amphibole, aegirine-augite, albite, biotite, quartz, and titanite. Minor jadeite and probable lawsonite in late veins occur in some mafic rocks associated with the metacarbonates. Radial fractures in garnet and kyanite, as well as coronas of garnet around green omphacite and apatite, and laminae of rutile/ilmenite included in the main eclogitic phases, are important textural features that suggest UHPM. Coronitic textures of titanite around rutile, biotite around phengite, new garnet around hornblende and biotite, fine- and coarse-grained symplectites, as well as abundance of polycrystalline pseudomorphs after omphacite, garnet, and phengite, formed later during decompression. Altogether, 36 different minerals have been identified (Table 1).

Metapelite and quartzite

The most characteristic lithology of the Chuacús complex in the studied area is micaschist, most of which contains large garnet grains (up to 4 cm, Fig. 2C) and kyanite (up to 7 cm). The complete assemblage of representative metapelites sampled at Caquil River, Palibatz area, consists of quartz–garnet–white mica–kyanite–staurolite–chloritoid–biotite–zoisite–tourmaline–rutile–zircon ± apatite ± monazite. Kyanite is riddled with inclusions of rutile and quartz, and contains single megacrystic garnets that contain inclusions of rutile/ilmenite, staurolite, phengite, chloritoid, quartz, and zoisite. Some kyanite megacrysts also enclose biotite, monazite, dravite, and zoisite, as well as abundant ilmenite and phengite-clinochlore aggregates. Radial cracks and kink-bands in kyanite associated with inclusions of quartz are conspicuous (Fig. 3A). Garnet megacrysts contain abundant inclusions of rutile, phengite, and ore, with minor amounts of biotite, kyanite, staurolite, chloritoid, a bright green mica, zoisite/clinozoisite, apatite, carbonate, and zircon. Rutile in several shapes and colors occurs as inclusions in garnet, kyanite, phengite, torumaline, and locally in monazite. All phengites show a low to moderate 2V, and besides rutile also may contain inclusions of apatite, tourmaline, kyanite, and zircon.

Quartzite in layers from centimeters up to 20 m thick is common and contains the assemblage quartz ±

TABLE 1. Representative Mineral Associations in Lithologies of the Chuacús Complex, Guatemala[1]

Sample:	Omp	Grt	Rt	Phe	Zo	Qtz	Jd	Lw	Ky	Ar	Cc	Ttn	Tr	Di	Dol	Pr	Prg	Am	Ep	Pl	Bt	St	Tur	Chl	Cld	Tlc	Aln	Kf	Ilm	Ore	Mnz	Ap	Zrn
Ky-megaycryst		X	X	X	X	X																	X	X		X			X				
Grt-megacryst			X	X	X	X			X													X	X						X				
GT0339		X	X	X		X													X				X	X	X					X			X
FO302	X	X	X	X	X	X						X					X	X	X	X	X						X	X	X			X	X
2602-A			X		X	X				X	X	X	X	X		X	X	X						X								X	
FO3603			X	X		X					X																						X
FO3703			X		X	X	X	X			X	X	X							X													
FO3803		X	X	X		X			X				X								X		X									X	
FO3903			X	X		X			X			X									X		X	X					X			X	X
FO4003				X		X															X		X										
FO4303	X	X	X	X		X						X						X	X	X	X								X			X	
FO4403				X		X																								X			
FO4503		X		X		X						X							X	X	X		X				X	X				X	X
FO4603		X		X		X						X						X	X	X	X						X	X				X	X
FO4703				X	X	X					X	X								X												X	X
FO4803	X	X			X	X						X						X	X											X			X
FO4903				X		X														X	X												X
FO5003		X			X		X					X						X		X													
FO5103		X		X		X													X											X		X	X
FO5203				X		X																											
FO5303		X	X	X		X												X	X	X	X			X					X		X	X	X
FO5403	X				X	X						X						X															
FO5503															X																		
FO5603	X																	X		X										X			
FO5703						X					X					X				X													X
FO5803	X	X	X			X					X	X						X	X	X	X									X			
FO6003	X				X	X				X	X	X				X		X			X												
FO6103		X	X	X		X						X				X		X	X	X	X									X		X	X
FO6203	X	X	X	X	X	X						X						X	X	X	X											X	

FO6403						X		X	X	X	X								X	
FO6503				X		X		X		X	X	X							X	X
FO6603		X	X		X	X		X	X	X	X	X		X				X	X	X
FO6703				X		X				X	X	X		X					X	
FO6803		X	X	X				X	X	X	X	X	X			X	X		X	X
FO6803-A				X		X		X		X	X	X					X			
FO6903		X	X	X	X	X		X		X	X	X	X	X					X	X
FO702			X	X			X	X	X	X	X	X	X						X	
FO7103			X	X	X	X	X	X			X						X		X	
FO7203				X		X	X										X			
FO7403				X	X		X	X	X								X			
FO7503				X	X		X	X	X											
FO7603		X		X		X		X		X	X	X								
FO7603-A				X		X		X		X	X	X		X					X	
FO8303		X		X		X				X	X			X	X					
FO8403		X			X	X	X	X	X	X	X	X	X	X					X	
FO8503		X	X	X	X	X		X	X								X		X	
FO8703						X	X		X			X	X				X			
FO8903-A				X			X				X	X	X				X	X	X	X
FO9103		X		X	X	X		X	X	X	X	X								X
FO9303				X	X	X	X	X	X											
FO9403	X	X	X	X	X	X	X		X	X	X	X	X				X			
FO9503				X		X		X	X	X	X	X	X			X			X	X
FO9803		X		X					X	X	X	X								
FO9903		X	X	X		X	X	X	X		X	X	X				X			
FO10003		X	X			X		X	X	X	X	X	X				X		X	
FO10103						X														
FO10203		X	X		X	X		X	X	X	X						X		X	
FO10903	X	X	X	X	X	X		X	X	X	X	X	X			X			X	X
FO11403		X	X	X		X						X								X

[1]Mineral abbreviations are from Kretz , 1983, with the following changes: Phe = phengite; Amp = amphibole; Sf = sulfide.

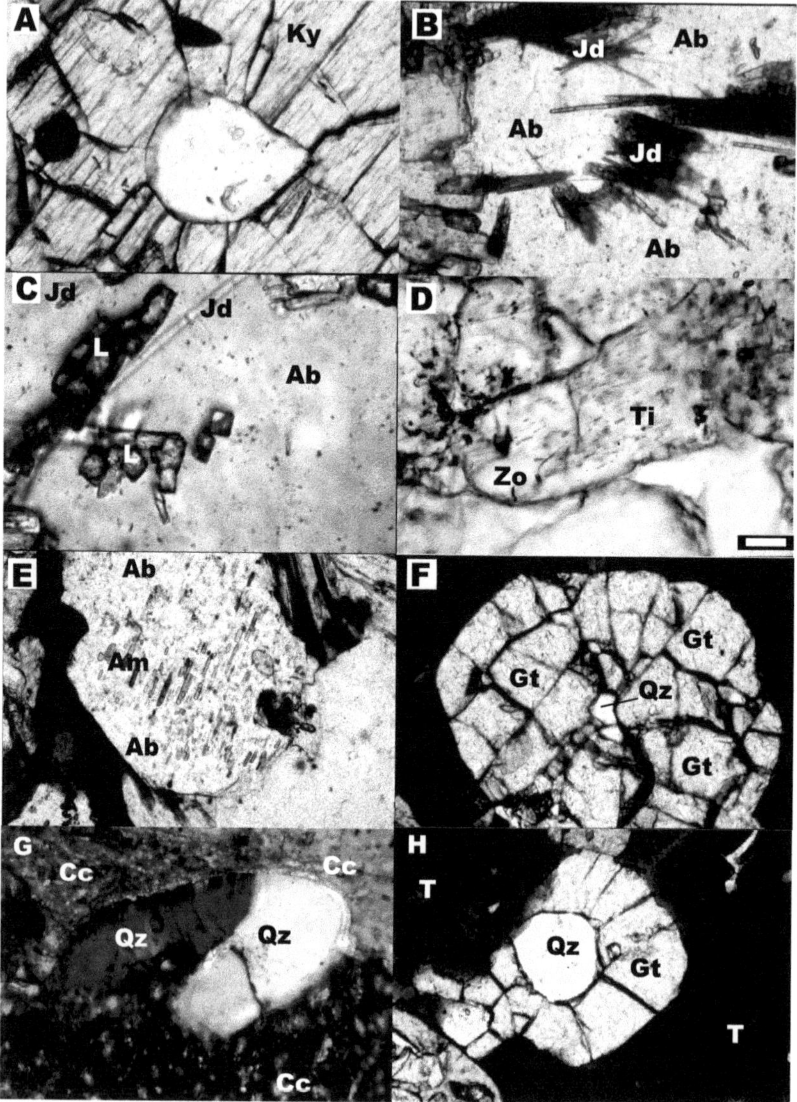

FIG. 3. A. Radial cracks in kyanite (Ky) around quartz (Qz), Palibatz area. Plane polarized light, field of view is ~ 560 µm. B. Jadeitic pyroxene (Jd) set in a groundmass of metasomatic albite vein (Ab). Plane-polarized light, field of view ~560 µm, village of El Capulín. C. Lawsonite(?) laths (L) associated with jadeitic pyroxene (Jd) inside an albite metasomatic vein (Ab). Plane polarized light, field of view ~220 µm, village of El Capulín. D. Lamellar inclusions of TiO_2 phase (Ti) within zoisite crystal (Zo). Plane polarized light, field of view is about 220 µm. Impure marbles north of Rabinal, Chuacús complex. E. "Exsolved" oriented crystals of Na-amphibole (Am) within albite neoblast (Ab), developed possibly from a highly sodic omphacite within albite-mica gneisses of the Chuacús complex. Plane polarized light, field of view about 560 µm. F. Imperfect radial cracking of garnet (Gt) around quartz inclusions (Qz), possibly due to the coesite-quartz phase transition in eclogitic amphibolite, Barranca Agua Caliente. Plane polarized light, field of view is about 560 µm. G. Palisade-like structure around quartz (Qz) within carbonates (Cc) of impure marbles. El Chol-Rabinal area, northern Chuacús complex. Crossed polarized light, field of view about 220 µm. H. Radial cracking in garnet (Gt) around quartz (Qz) enclosed in taramite (T). Radial cracks never invade the host. Plane polarized light, field of view is about 560 µm. Eclogitic rock at Barranca Agua Caliente.

garnet ± amphibole ± phengite ± tourmaline. Accessory minerals are rutile, biotite, and zircon. This last mineral occurs as rounded, small grains (probably detrital) and larger ones (up to 165 μm) with squared sections. Rutile occurs in colorless, euhedral tiny prisms, and as large, stubby, yellow-black grains. Amphibole has a strong pleochroism in shades of deep grey-blue and green, indicating a highly sodic composition. Quartzite grades into quartzose gneisses with a similar asssemblage as the micaschists.

Quartzofeldspathic gneisses

Coarse grained quartzofeldspathic gneisses are characterized by abundant poikiloblastic sodic plagioclase, white mica, and biotite. They commonly grade into other lithologies including quartzite, amphibolite, and metapelite. The composition of these rocks and polyphase metamorphism produced a complex mineralogy that may include the high-pressure minerals clinopyroxene, rutile, garnet, phengite, and zoisite as inclusions in poikilitic albite, associated with biotite, white mica, amphibole, garnet, and accessory minerals such as apatite, Fe-Ti oxides, zircon, and titanite. Potassium feldspar as microcline occurs rarely. Albite porphyroblasts commonly enclose small but abundant prismatic greenish-blue amphiboles oriented in the crystal lattice of the plagioclase (Fig. 3E), which, if reintegrated into a single phase, would probably yield highly sodic omphacite. This last mineral was rarely preserved within albite. Phengite is characterized by its small 2V (even uniaxial), high relief, and marked pleochroism in yellow, green, and brown shades. Symplectitic coronas of biotite-albite ± titanite ± rutile around white mica record the breakdown of high-pressure phengites in some gneisses. High-temperature biotite outside symplectites is distinctive because of its black to yellow intense pleochroism and skeletal habit, or as very thin plates (down to 400 μm wide and 4 μm thick). Amphibole is also strongly colored in shades of brown and green. Carbonate commonly coexists in some of these gneisses with other high-pressure phases (rutile, garnet) within albite porphyroblasts. A late generation of garnet, probably formed at lower temperature than previous ones, may rim greenish biotite. Although these albitic gneisses could be of volcanic or intrusive origin, the high-Al content (abundant potassic micas), rounded nature of accessory zircon, and gradational contacts with surrounding clearly metasedimentary units, suggest also a dominant sedimentary protolith.

Migmatitic gneisses

These rocks are both banded and folded with evidence of neosomatic mobilization clearly expressed by folded felsic bands and concordant and discordant albite-quartz-phengite ± zosite ± garnet pegmatitic bodies. The banded facies consists of mafic layers with omphacite-amphibole-garnet-biotite-albite-rutile/titanite-epidote/zoisite/allanite-apatite, whereas the thin leucrocratic bands (neosome) are formed of quartz-phengite-albite-epidote/allanite-apatite. Omphacite, rutile, garnet, and zoisite occur in textural equilibrium. Common symplectites of amphibole-albite-quartz, probably formed after omphacite. Biotite occurs as thin plates with random orientation in the quartz-albite matrix. Rutile in the mafic bands is abundant and occurs in four types: (1) large anhedral grains and tiny needles as inclusions in garnet, amphibole, phengite, epidote, zoisite, clinopyroxene, and rarely within apatite and zircon; (2) large individual anhedral to subhedral grains in the matrix showing red to yellow color and commonly replaced by ilmenite, or mantled by titanite; (3) colorless to yellow gem-quality crystals in equilibrium with the albite-quartz matrix; and (4) large inclusions within garnets, with very dark to purple-brown pleochroism. The dark purple rutile and the euhedral needles tend to lie toward the centers of garnet, probably indicating that they were the first phases that crystallized at the onset of the eclogite-facies event, whereas euhedral rutile (type c) may have precipitated from the melt.

Marbles and calcsilicates

White marble occurs at El Capulín and El Oratorio villages. Those at El Capulín consist of quartz–carbonate–rutile–white mica–sulfide–zircon. Quartz commonly forms hexagonal unstrained plates. Mica shows a small 2V and a random orientation within a granoblastic mosaic of carbonates (calcite/aragonite?, and probably dolomite). Rutile occurs as individual inclusions in white mica and as clusters within the carbonate matrix. Carbonate is granoblastic with complex sutured contacts; some grains show a biaxial character, suggesting metastable aragonite. Euhedral sulfide is a characteristic minor phase, whereas zircon occurs as oval and subhedral grains within the carbonate groundmass. Marbles at El Oratorio are only composed of dolomite (determined by X-ray diffraction), and very small quantities of white mica, quartz, and zircon with abundant solid and fluid inclusions.

Interbedded amphibolite and calcsilicate bands are associated with the El Capulín marbles that preserved two superposed tectonic fabrics, both with associated distinctive high-pressure mineralogies. S_1 forms a nematoblastic foliation defined by the assemblage zoisite-tremolite-titanite/rutile, superposed at right angles by a crosscuting foliation S_2, defined by a plexus of veins showing the assemblage jadeitic clinopyroxene–albite–lawsonite?–carbonate, which partly replaced the S_1 foliation. Jadeitic clinopyroxene occurs as large altered crystals, and as tiny elongate prisms with an average extinction angle of 38 ± 2° (Fig. 3B), whereas lawsonite was observed tentatively in one of the veins as small, sharp rectangular plates (Fig. 3C).

Calc-silicate rocks and marble that crop out near the northern edge of the Chuacús complex between the villages of El Chol and Rabinal contain porphyroblasts of dark amphibole up to 3 cm long concentrated in tightly folded bands. In thin section, the rock exhibits a medium-grained matrix composed of Mg-calcite and calcite, with zoisite, quartz, white mica, and clinochlore as common accessory minerals; pargasite and tremolite form large porphyroblasts whose dark color is produced by abundant carbonaceous? and Fe-oxide inclusions. Diopside is rare, whereas some rutile is present in the matrix coexisting with carbonates and quartz, and more rarely as lamellar (5–20 μm long and 0.1–1 μm wide) inclusions in zoisite (Fig. 3D). Quartz exhibits an equigranular polygonal and interlobate texture, but in some cases it displays distinctive coronas of palisade-like fibrous quartz (Fig. 3G) that are similar to structures commonly interpreted to develop from the coesite-quartz inversion in some UHP metamorphic rocks (Chopin and Sobolev, 1995; Gilottti and Krogh, 2002). Symplectitic arrangements of two intergrown carbonates suggest dolomite exsolution from former magnesian calcite, whereas the apparent biaxial nature of some carbonate (2V estimated up to 20°) may indicate deformed calcite or more likely metastable aragonite.

High-pressure metamorphism in these rocks is indicated by: (1) microprobe EDS analyses that reveal high-silica paragonite; (2) the abundance of lamellar titanium oxide inclusions in zoisite (Fig. 3D), probably formed by exsolution from high pressure Ti-zoisite (cf. Tropper et al., 2002); (3) the stable assemblage quartz-rutile-carbonate (aragonite?); and (4) highly pleochroic pink titanite, some of which has inclusions of corundum (EDS analysis), possibly indicating a former aluminous high-pressure titanite (cf. Franz and Spear, 1985).

Eclogitic amphibolites

Garnet amphibolites contain both relict and preserved eclogite-facies minerals, including abundant green omphacite. The effects of high-temperature decompression and probably partial melting on the eclogitic assemblages produced a stable mineralogy of 14 to 16 coexisting phases (see Table 1). The primary eclogitic mineral associations consisted of omphacite-garnet-rutile-quartz (coesite?), with phengite, zoisite, and probably aragonite as the main accessory minerals, whereas pargasitic and sodic-calcic amphiboles, biotite, calcite, clinopyroxene, new generations of garnet, rutile, titanite, and phengite developed later in equilibrium with sodic plagioclase. Rare primary omphacite may be distinguished by elongate inclusions of ilmenite, white mica, biotite/phlogopite, or plagioclase, which probably formed by exsolution from a slightly potassic clinopyroxene. Porphyroblasts composed of symplectitic albite and Na-amphibole wrapped around by elongate clinopyroxene are considered retrograde pseudomorphs after a jadeitic omphacite porphyroclast. Garnet may exhibit radial cracking around large inclusions of quartz (Fig. 3A) or quartz-orthoclase polycrystals, and contains abundant rutile inclusions of two types: large subhedral grains, and numerous needles and dusty (0.25–0.5μm) grains arranged in globular zones. Phengite is scarce but found outside or inside amphibole less commonly, with lamellar (100 × 2 μm) inclusions of barite. A colorless, probably high-Al titanite, in places coexisting with rutile inside garnet, may be of high-pressure origin. Apatite, monazite, zircon, sulfide, and Fe-Ti oxides are minor accessory phases; monazite occurs both as grains or forming veins that traverse an entire thin section. Epidote group minerals are euhedral, commonly zoned with strong brown, yellow, or purple-brown cores of allanite mantled by purplish gray zoisite, and at the outermost rims by colorless to yellowish epidote. Zoisite is commonly rimmed by epidote, although both polymorphs may co-exist in textural equilibrium, and some epidotes contain abundant inclusions of rutile. Amphibole is colored with different patterns of pleochroism in brown, brownish-green, and blue-green, all with distinctive gray shades. Large rutile inclusions or numerous small rutile grains probably formed by exsolution within the amphiboles. Rare primary phengite may be distinguished by its uniaxial to small 2V angle, and by submicroscopic lamellar inclusions parallel to cleavage formed by barite, or elongate opaque blebs that under strong magnification are tentatively identified as ilmenite; the latter possibly formed by exsolution from a Ti-rich phengite. Biotite, as inclusions in hornblende and in symplectites with albite and epidote, is a product of phengite breakdown, inasmuch as this last mineral remained partially to completely rimmed by brown biotite. However, both micas may occur together in apparent textural equilibrium. Symplectites include albite-hornblende after omphacite replaced by neocrystallized high-Na-amphibole. Other types of symplectites identified in a single sample include: (1) hornblende-quartz after garnet or omphacite; (2) epidote-quartz after garnet; (3) epidote-biotite after garnet-phengite; (4) albite-hornblende after clinopyroxene; (5) biotite-orthoclase after phengite; (6) epidote-hornblende after clinopyroxene; and (7) epidote-albite after hornblende. Chlorite, hematite, stilpnomelane, and actinolite formed by low-temperature retrogression.

Mineral Chemistry

Minerals from a representative garnet amphibolite (FO302) were analyzed at the National University of Mexico, Laboratory of Petrology (LUP) using a JEOL JXA 8900R microprobe. Complementary qualitative analyses from other rocks and minerals were made with EDS techniques using a JEOL JSM 35C electronic microscope with a Tracor Noran analyzer. WDS Chemical analyses are shown in Table 2.

Although potassium feldspar is scarce in the Chuacús complex, it may occur in the quartzofeldspathic gneisses, and with quartz as inclusions in garnet. This last feldspar is zoned, with sodium and barium enriched in the core, and iron in the rim. Plagioclase forming the groundmass in eclogitic mafic gneisses shows a uniform composition of $Ab_{94}An_{5.5}Or_{0.5}$ and is very homogeneous (albite-oligoclase) in all other units throughout the Chuacús complex, whereas garnet in these mafic rocks is essentially a grossular-almandine solid solution with the average formula $Al_{58}Gr_{28}Pyr_{10}Spes_4$ (Fig. 4A). Selected garnets were analyzed with the intent of measuring Na, in order to assess its possible very high-pressure origin. These analyses were based on the use of an appropriate low-Na standard (sanidine) and applying longer acquisition times (30–60 sec). The obtained results (60 point analyses) reveal up to 0.12 Na_2O wt% and an average of 0.07 Na_2O wt% (Table 2), reinforcing the idea that part of the Chuacús metamorphic history once involved very high pressures. A backscattered electron map of a garnet, in accordance with its high temperature of equilibration, revealed very faint compositional zoning, with manganese and calcium slightly more concentrated in the central zone and magnesium toward the periphery, whereas iron remained constant. The pyrope low content (average of 2.35 ± 0.17 MgO wt%) was probably dictated by the original high iron composition of the rock and not by the P-T conditions, inasmuch as garnets in adjacent metapelites are rich in pyrope. A garnet megacryst handpicked from a garnetiferous metapelite yielded (EDS analysis) the formula Alm_{54}-Py_{41}-$Gross_4$-$Spess_1$, consistent with a high temperature and pressure of formation. Omphacitic clinopyroxene shows a moderate Mg# of 46 and low contents of Al (6–8 wt%), with high Na and Fe that plot very near the aegirine field (Fig. 4B). Jadeite + acmite molecules make up over 45% of the clinopyroxene, indicating its high-pressure origin.

All analyzed amphiboles (Table 2) are of the Na-Ca and Ca groups plotting in the fields (Fig. 4C) of taramites and pargasites, with a mean Mg# of 44.6, a high content of potassium (average of 0.95 K_2O wt%), and without Cr or Mn. The moderate content of TiO_2 (up to 1.1 wt%), reflected in their green-brown pleochroic shades, supports a relative high temperature of formation.

Although most grains of phengite seem to be of retrograde origin, with measured Si of 3.24–3.29 apfu, high iron (5–6 wt% total FeO), and enriched in titanium (> 1 wt% TiO_2), one crystal (Table 2) yielded an Si content of 3.39 a.p.f.u., with lower Fe and Ti and higher Mg than the former. All phengites are low in paragonite (< 1 wt % Na_2O).

Epidote contains between 22 and 25 mol% pistacite, with very small amounts of piedmontite (0.3–0.7 mol%). EDS analyses of epidotes indicate that their different color zones are due to relative abundance of REE. Grey to brown and purplish anhedral allanitic cores contain considerable La and Ce (>10 oxide wt%), whereas colorless euhedral epidote overgrowing the anhedral core shows none or small amounts of those elements. Rutile and kyanite show distinctive contents of iron oxide.

TABLE 2. Representative Microprobe Analyses of Minerals from a Basic Eclogite FO302 and a Garnet Metapelitic Megacryst[1]

Mineral	Garnet			Biotite			Amphibole			Plagioclase			Clinopyroxene			Phengite	Titanite	Garnet megacryst (meta-pelites)
Sample	49	50	51	36	37	39	22	44	49	23	25	11	1	2	12	18	9	2
SiO_2	38.140	37.939	38.078	37.530	37.450	37.410	41.017	41.676	41.464	65.190	65.410	64.900	52.843	52.956	52.660	50.489	30.20	41.17
TiO_2	0.106	0.092	0.053	1.980	1.910	1.910	1.010	0.836	0.972	–	–	–	0.195	0.203	0.224	0.12	36.34	–
Al_2O_3	20.918	20.930	21.042	15.500	15.320	15.390	13.715	12.698	13.405	21.490	21.470	21.390	7.356	7.767	6.976	26.605	2.03	24.15
Cr_2O_3	0.007	0.010	0.001	0.010	–	0.010	0.014	0.001	0.012	–	–	–	0.010	0.017	0.012	–	–	–
FeO	26.772	26.006	26.489	20.730	20.980	20.900	20.634	20.712	20.432	0.040	0.190	0.080	12.374	11.936	12.633	4.475	0.901[3]	22.78
MnO	1.405	2.385	3.099	0.210	0.180	0.220	0.244	0.207	0.273	–	–	–	0.082	0.106	0.120	–	0.17	0.53
MgO	2.586	2.322	2.206	9.060	9.050	8.870	6.651	6.818	6.697	–	–	0.010	6.175	5.988	6.331	2.962	0.11	9.59
CaO	10.144	9.883	8.840	0.020	0.020	0.030	8.778	8.316	9.028	1.460	1.380	1.500	12.977	12.660	13.437	0.01	27.10	1.39
Na_2O	0.118	0.111	0.100	0.210	0.160	0.120	4.240	3.876	3.836	10.940	10.730	10.880	6.597	6.728	6.234	0.21	0.17	–
K_2O	–	–	–	10.580	10.790	10.840	1.999	1.766	1.561	0.070	0.110	0.100	–	0.017	0.158	11.27	0.88	–
P_2O_5	0.085	0.098	0.077	–	–	–	–	–	–	–	–	–	–	–	–	–	–	–
F																	0.67	–
Totals	100.281	99.776	99.985	95.830	95.880	95.710	98.302	96.906	97.680	99.190	99.290	98.860	98.609	98.378	98.785	96.137	97.65	99.61
Oxygens	12	12	12	11	11	11	23	23	23	8	8	8	6	6	6	11	5.00	12
Si	3.009	3.013	3.024	2.878	2.878	2.880	6.234	6.407	6.336	2.887	2.891	2.873	1.957	1.963	1.953	3.387	1.001	3.08
Ti	0.006	0.005	0.003	0.114	0.111	0.110	0.115	0.097	0.112	–	–	–	0.005	0.006	0.006	0.006	0.906	–
Al	1.945	1.960	1.970	1.401	1.388	1.397	2.458	2.301	2.415	1.122	1.119	1.134	0.321	0.339	0.305	2.102	0.079	2.13
Cr	–	0.001	–	0.001	–	0.001	0.002	–	0.001	–	–	–	–	–	–	–	–	–
Fe[3]	0.043	0.019	–	–	–	–	0.404	0.333	0.217	0.002	0.007	0.003	0.227	0.207	0.232	–	0.022	–
Fe[2]	1.723	1.708	1.760	1.330	1.349	1.346	2.219	2.330	2.394	–	–	–	0.157	0.163	0.160	0.251	–	1.43
Mn	0.094	0.160	0.208	0.013	0.012	0.014	0.031	0.027	0.035	–	–	–	0.003	0.003	0.004	–	0.005	0.03
Mg	0.304	0.275	0.261	1.035	1.036	1.017	1.507	1.562	1.525	–	–	–	0.341	0.331	0.350	0.296	0.005	1.07
Ca	0.857	0.841	0.752	0.002	0.002	0.002	1.430	1.370	1.478	0.069	0.066	0.072	0.515	0.503	0.534	–	0.963	0.11
Na	0.018	0.017	0.015	0.031	0.024	0.018	1.250	1.155	1.137	0.939	0.919	0.949	0.474	0.484	0.448	0.028	0.011	–
K	–	–	–	1.036	1.059	1.066	0.388	0.347	0.305	0.004	0.006	0.006	–	0.001	0.007	0.995	0.037	–
F				–	–	–	–	–	–	–	–	–	–	–	–	–	0.070	–
Sum	8.000	8.000	7.995	7.841	7.858	7.852	16.037	15.930	15.956	5.023	5.008	5.037	4	4	4	7.065	3.030	7.85

[1]Natural minerals and synthetic phases were used as standards, and an accelerating voltage of 15–20 KeV, a beam current of 15 nA, and variable counting times (5–50 sec) according to the element and mineral under analysis.
[2]analysis performed by EDS
[3]Fe^{3+}

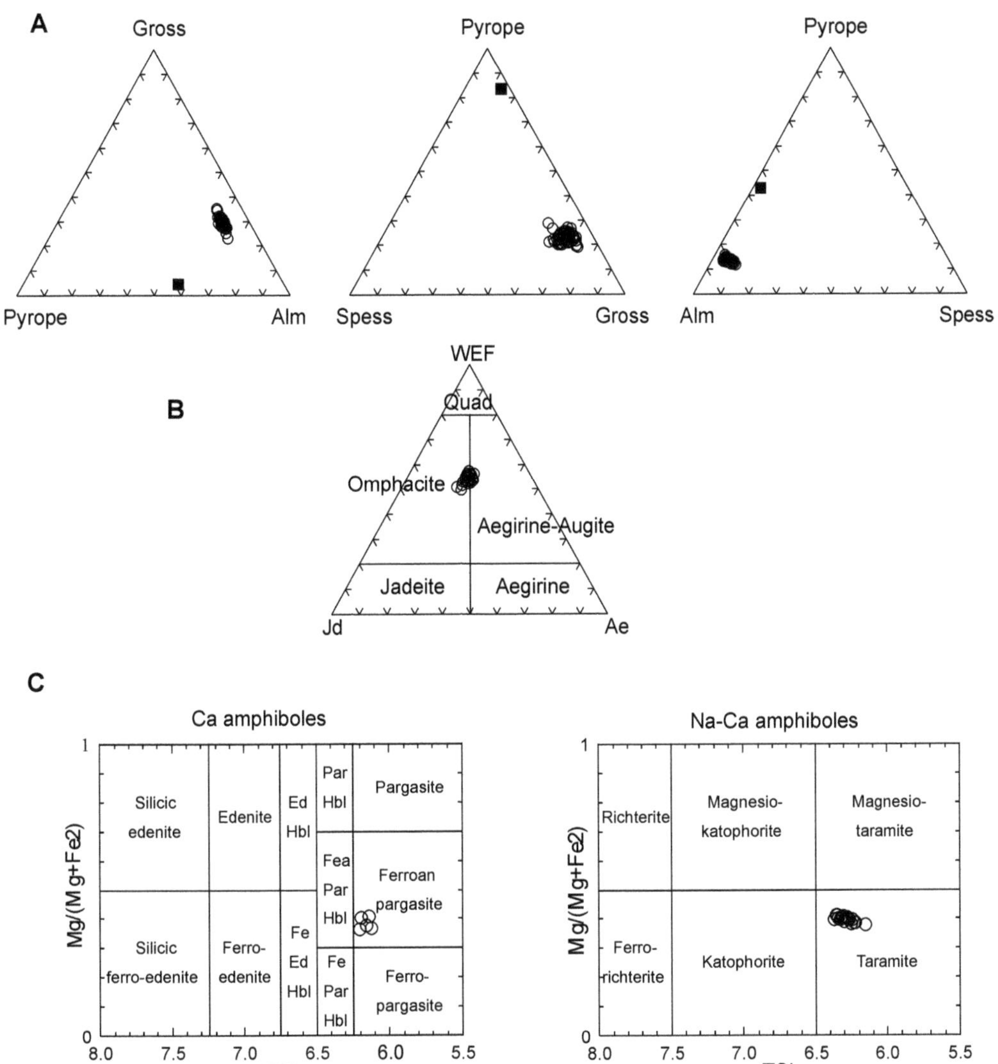

FIG. 4. A. Classification diagram of garnets from eclogitic (circles) and a metapelite (square) of Chuacús rocks. Note the high pyrope content of the metapelitic garnet and the high grossular-low pyrope of the eclogitic types. B. Classification diagram for clinopyroxenes (after Morimoto, 1988). Chuacús clinopyroxenes straddle the omphacite-aegirine boundary. C. Classification diagrams for Chuacús amphiboles. All fall in the taramite and ferroan pargasite fields (after Leake et al., 1997), implying high temperatures and pressures of formation.

Metamorphic Evolution of the Chuacús Complex

The prevalent metamorphic subdivision of the studied area (van den Boom, 1972) includes three zones (chlorite, biotite, and garnet) that record prograde Barrovian regional metamorphism of the Chuacús complex. These zones imply stability in the greenschist to lower amphibolite facies for the complex (see also Kesler, 1971). However, our study demonstrates that the Chuacús complex underwent eclogite facies metamorphism, and that it probably recorded high-pressure partial melting during decompression. Lower-grade retrogression only developed along regional shear zones where most high P-T phases disappeared completely and were replaced by chlorite, green biotite, hematite, stilpnomelane, and actinolite. Our data indicate a

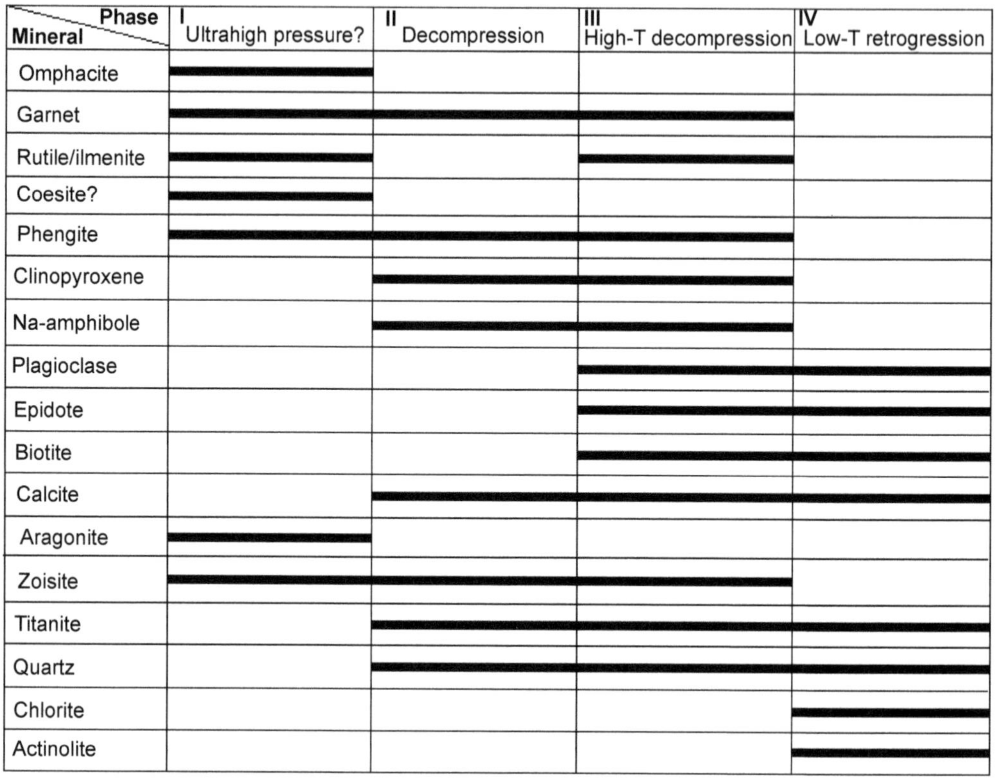

Phase / Mineral	I Ultrahigh pressure?	II Decompression	III High-T decompression	IV Low-T retrogression
Omphacite	▬▬▬			
Garnet	▬▬▬	▬▬▬	▬▬▬	
Rutile/ilmenite	▬▬▬		▬▬▬	
Coesite?	▬▬▬			
Phengite	▬▬▬	▬▬▬	▬▬▬	
Clinopyroxene		▬▬▬	▬▬▬	
Na-amphibole		▬▬▬	▬▬▬	
Plagioclase			▬▬▬	▬▬▬
Epidote			▬▬▬	▬▬▬
Biotite			▬▬▬	▬▬▬
Calcite		▬▬▬	▬▬▬	▬▬▬
Aragonite	▬▬▬			
Zoisite	▬▬▬	▬▬▬	▬▬▬	
Titanite		▬▬▬	▬▬▬	▬▬▬
Quartz		▬▬▬	▬▬▬	▬▬▬
Chlorite				▬▬▬
Actinolite				▬▬▬

FIG. 5. Mineral stability during the 4-phase protracted metamorphic event that affected the Chuacús complex, starting probably at very high pressures (Phase I), following an isothermal high-T decompressional path (phases II and III), and shallowing during Phase IV. A possible younger high-pressure metamorphic event is not shown.

protracted four-phase metamorphic history of the Chuacús complex (Fig. 5), probably culminating at ultrahigh pressure (phase I), followed by a fast, hot decompression (phases II and III), and terminating with a locally registered low-grade event associated with intense shearing (phase IV). The breakdown of original plagioclase-free eclogite-facies minerals occurred at high temperatures and pressures, as indicated by new parageneses containing clinopyroxene, rutile, garnet, biotite, and taramitic amphibole in equilibrium with albite.

The rare presence of jadeitic clinopyroxene and probably lawsonite along veins filled with albite that replaced a zoisite-tremolite-titanite/rutile tectonite fabric, is puzzling and difficult to explain in the postulated frame of a high-T decompression path proposed below for the Chuacús complex. The event could be much younger (Cretaceous?) and was clearly of lower grade, requiring the introduction of sodium and CO_2 in order to form the albite-carbonate veins from zoisite-tremolite-chlorite rocks.

Eclogitization or amphibolitization?

Although textures such as apparent coronas of garnet around hornblende and plagioclase, or crystallization of rutile and clinopyroxene in textural equilibrium with albite in some of the mafic rocks could be taken as the progressive eclogitization of amphibolites, the following petrographic criteria indicate that eclogite-facies metamorphic rocks underwent recrystallization in the eclogite-amphibolite transitional facies: (1) coarse, anhedral rutile with coronas of titanite is present inside and outside amphibole; (2) symplectites of hornblende-quartz ± albite after omphacite or garnet, amphibole-albite ± clinopyroxene after omphacite, and of biotite-albite after phengite occur commonly; (3) inclusions of garnet, some of which show radial cracks, within amphibole is the dominant texture over the reverse relation; (4) presence of rutile lamellae in garnet or of ilmenite in white mica and some omphacite; (5) absence of rutile-ilmenite coronas around titanite, or omphacite coronas around amphibole; (6) radial cracks within garnet do not cut across host amphibole or groundmass plagioclase (Fig. 3H). On these bases, it is concluded that the present plagioclase-rich rocks of the Chuacús complex represent an arrested process of amphibolitization of former high or very high pressure, plagioclase-free eclogitic rocks.

Ultrahigh-pressure metamorphism?

Several textural observations and mineralogical data suggest that UHPM once affected some or all of the Chuacús rocks. Radial cracking (Fig. 3F) is evident around inclusions of quartz within garnet or kyanite as described above. However, this feature has some important differences from the dense and commonly curved radial cracking in confirmed UHPM terrains (i.e., the Dora Maira massif; Chopin, 2003), inasmuch as cracks in Chuacús garnets are more widely separated, straight, fewer per grain, and shorter than those in most UHPM rocks. This sort of cracking could be due to the thermal expansion of low quartz with a positive volume change of up to 7–8 % (Wendt et al., 1993) or to the coesite-quartz phase transition (10 % volume increase) with cracks partially annealed by the high-temperature decompressional event. In Chuacús rocks, quartz inclusions in garnet and kyanite are typically very large single crystals and lack palisade structure, although this last structure appears to be present around a few quartz grains within carbonates of some of the marbles, as already mentioned. Although the multiple lamellar inclusions of rutile in garnet described above may be due to exsolution from former very high pressure garnet (cf. Liou, et al. 1998, p. 71), it instead could represent a restite mineral after digestion of a Ti-rich phase (cf. Vogel, 1967), such as ilmenite, biotite, Ti-augite, titanite, or hornblende, and the total replacement of that phase by the growing garnet. In addition, the presence of multiple inclusions of white mica, biotite/phlogopite, and plagioclase in clinopyroxene, and abundant ilmenite rods in some phengites as well as the existence of the stable assemblage carbonate (aragonite?)–rutile–quartz in marbles, also support the possibility that the Chuacús complex underwent UHPM prior to its present high-temperature–high-pressure "retrobaric" state. The Chuacús complex shows specific petrological similarities to UHPM terrains and differs somewhat from comparable features in HP terrains. Table 3 offers a compilation of textural, mineralogical, and geologic features reported in the literature related to UHP and HP metamorphic terrains compared with those found in the Chuacús complex. Radial cracking, Na-rich garnets, and lamellar inclusions (exsolution?) or palisade structure described above for the Chuacús minerals, may eventually characterize it as a new and one of the first UHPM areas of the Americas affecting continental crust, albeit almost completely overprinted by metamorphic recrystallization and partial melting along a high-T decompressional path.

In conclusion, the overall convergence of the above observations indicates UHPM. Unfortunately, however, these textural and mineralogical criteria remain inconclusive in the absence of actual coesite relicts or definite geobarometric calculations (e.g. Schreyer, 1995; Liu et al., 2001; Klemd, 2003).

High-pressure decompressional melting of Chuacús eclogites?

An albite-quartz ± white mica matrix present in many of the mafic rocks contains euhedral epidote/zoisite, rutile, garnet, clinopyroxene, amphibole, and biotite that commonly define intersertal or poikilitic textures typical of igneous rocks. These textures are interpreted to have formed in response to decompressional partial melting of hydrated eclogitic rocks. During this process, a melt of albite-quartz-mica ± zoisite/epidote ± rutile ± garnet formed, leaving a residue of recrystallized mafic phases. P-T conditions calculated for the event (see below) are permissive of anatexis of eclogitic rocks if water pressure is high (Rapp et al., 1991; Poli and Schmidt, 1997). White mica–albite rich pegmatites and abundant hydrous minerals such as amphibole, epidote, biotite, and phengite, as well as the coronitic alteration of coarse rutile to titanite within the groundmass or inside amphiboles, indicate that magmatic or high-temperature supercritical fluids were rich in water, sodium, calcium, potassium, and silica. This albite-quartz-mica-epidote-zoisite groundmass of Chuacús eclogitic rocks is similar in mineralogy and composition to leuco-tonalitic segregations ascribed to decompression partial melting of eclogites in the Müncheberg Masssif of Germany (Franz and Smelik, 1995). Under these conditions, UHP metamorphic phases and textures, if they ever existed, would have been easily recrystallized and

TABLE 3. Textural and Mineralogical Characteristics of HP and UHP Metamorphic Terranes Compared to Those Found in the Chuacús Complex

Feature	UHPM	HPM	Chuacús complex	Min. P GPa, 700–800 °C	Main reference
Na-rich garnet	Common	Absent	Present		Chopin and Sobolev, 1995
K-feldspar exsol. lamellae in Cpx	Present	Absent	Not detected	4.0	Becker and Altherr, 1992
Phengite exsol. lamellae in Cpx	Present	Absent	Apparent	4.0	Schmädicke and Müller, 2000
Quartz exsol. lamellae in Cpx	Present	Absent	Possible	3.0–4.0	Smith, 1984
Olig. exsol. lamellae in met. Cpx	Present	Absent	Prersent		Schmädicke and Müller, 2000
Rt ± Ap ± Cpx exsol. lamellae in Gt	Common	Present	Present (Rt)		Ye et al., 2000
Garnet/Ilm exsol. lamellae in Cpx	Rare	Absent	Apparent (Ilm)		Zhang et al., 2002
Radial cracks in garnet around Qz	Ubiquitous	Absent	Present	2.8	Multiple references
Radial cracks in kyanite around Qz	Present	Not reported	Present		Sabau, 2000
Radial cracks in Cpx on phengite	Rare	Absent	Not detected		Schmidt et al., 2000
Radial cracks in Grt around calcite	Present	Absent	Not detected		Zhang et al., 2003
Fibrous quartz (palisade structure)	Common	Absent	Probable		Gilotti and Krogh-R., 2002
High (Al-F) titanite	Common	Rare	Probable		Carswell et al., 1996
High-Si phengite (> 3.5 Si apfu)	Ubiquitous	Common	Present		Carswell, et al., 1997
Coesite/Qtz lamellae in titanite	Present	Absent	Not detected	6.0	Ogasawara et al., 2002
Coesite	Ubiquitous	Absent	Not detected	2.8	Chopin, 1984
Diamond	Rare	Absent	Possible	3.5	Sobolev and Shatsky, 1990
Garnet peridotite	Common	Rare	Not detected		Medaris, 1999
Magnesite-diopside	Present	Absent	Not detected	2.9–4.5	Smith, 1988
Magnesite-aragonite	Rare	Absent	Not detected	6–6.5	Zhang et al., 2003
K-omphacite	Present	Absent	Not detected		Sobolev and Shatsky, 1990
High-Ti garnet	Common	Present	Present		Zhang et al., 2002
Continental protolith	Common	Common	Present		Coleman and Wang, 1995
Ophiolites	Rare	Common	Absent		Schereyer, 1995
Arg-Rt-SiO_2 for XCO_2< 0.05	Present	Absent	Probable		Ye et al., 2002
Clinoenstatite lamellae in diopside	Present	Absent	Not detected		Bozhilov et al., 1999
Opx precipitates in garnet	Rare	Absent	Not detected	< 6.0	Van Roermund et al., 2001
$FeTiO_3$ rods in olivine	Present	Present	Not detected		Dobrzhinetskaya et al., 1996
α-PbO_2–type TiO_2 phases	Present	Absent	Not detected	4.0–5.0	Hwang et al., 2000
Lisetite	Present	Abesent	Not dtected		Smith, 1988
Nyböite	Present	Absent	Not detected	2.0–2.5	Smith, 1988
Ellenbergerite	Present	Absent	Absent	2.8	Chopin, 1984
Pyrophanite-rutile-hematite	Present	Absent	Not detected		Smith, 1988

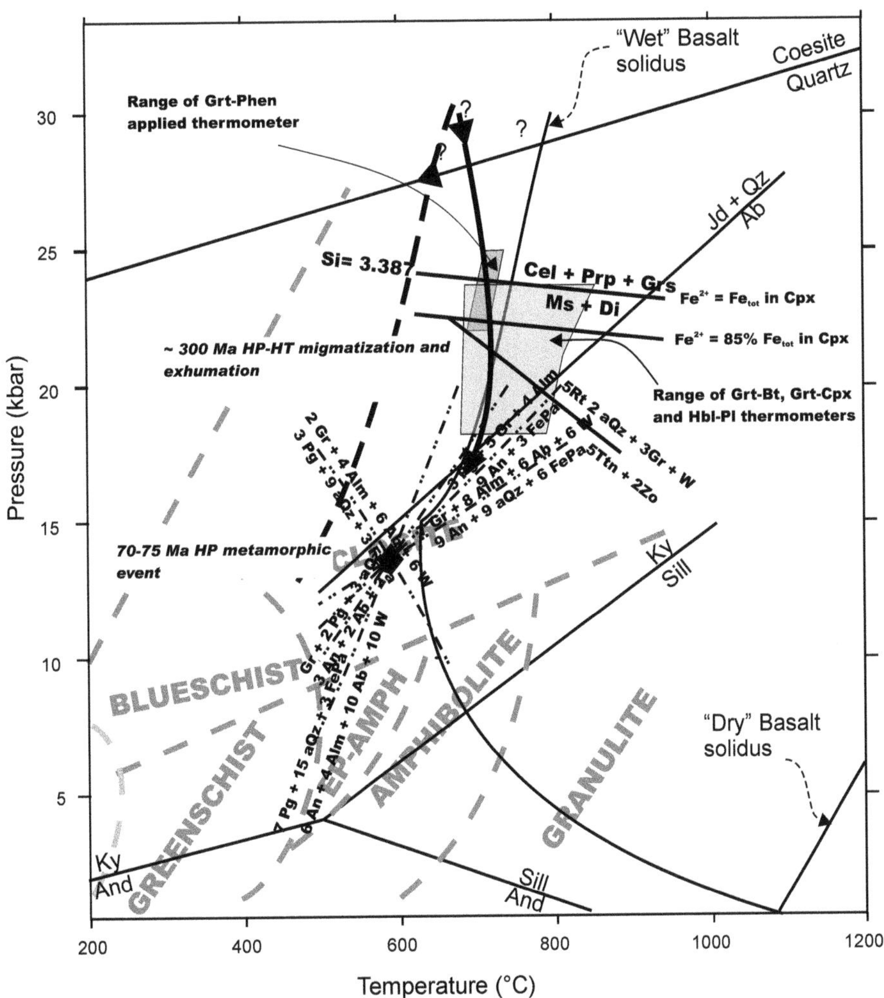

FIG. 6. Proposed P-T diagram for studied sample FO302. Clockwise P-T path culminating in HP-HT decompression during the Late Carboniferous is built on the mineral reactions discussed in the text. The black diamond may record the Late Cretaceous HP event. See text for further details.

erased. In fact many rocks of the Alps that underwent UHPM conditions constitute inconspicuous gneisses and schists, or even igneous rocks with no textural or mineralogical traces of high-pressure metamorphism (Schreyer, 1999).

Geothermobarometry

We carried out microprobe analyses on several samples collected in the Sierra de Chuacús, in order to acquire P and T data. Representative analyses of sample FO302, a basic eclogite sampled at Barranca Agua Caliente (Fig. 1C), are reported in Table 2. Thermometers used on this rock were garnet-biotite (calibrations of Ferry and Spear, 1978; Hodges and Spear, 1982; Dasgupta et al., 1991; Bhattacharya et al., 1992), hornblende-plagioclase (Holland and Blundy, 1994), garnet-clinopyroxene (calibration of Pattison and Newton, 1989), and garnet-phengite (calibration of Green and Hellman, 1982). These pairs yielded temperatures of 685–832°C, 730–750°C, 687°C and 738 ± 20°C respectively, assuming a pressure of 20 kbar. The temperatures decrease 3–5% if the assumed pressure is 15 kbar, and increase the same amount if a pressure of 25 kbar is assumed. The garnet-phengite-omphacite barometer of Waters and Martin (1993) was also applied using the highest Ca-garnet, highest Jd-clinopyroxene, and highest Si-phengite, as suggested by Carlswell et al. (2000). The highest pressure obtained with this barometer, using all the iron as Fe^{2+} in clinopyroxene as measured (analysis # 12 in Table 2), is 23.2 kbar at a temperature estimated to be 738 ± 20°C (Fig. 6). The inferred prograde path of the P-T curve (dashed) to its peak conditions is based on the presence of chloritoid and staurolite as inclusions in high-pyrope garnet and kyanite of metapelites, and on textural evidence for the possible former presence of coesite. The decompressional part of the curve (solid) is the result of geothermobarometry applied to high-temperature reequilibrated assemblages, which is consistent with textural and mineralogical evidence of apparent high-pressure partial melting mafic lithologies in the presence of water.

We also performed calculations using the software TWQ (Berman, 1991), on the equilibrium assemblage garnet (lowest Fe/Mg)-phengite (lowest Si)-taramite/Fe-pargasitic hornblende-titanite-zoisite-albite, which has the advantage of yielding both T and P. The computation used an internally consistent thermodynamic dataset, allowing compositional disequilibria between the involved minerals to be detected. Our result for this, probably Cretaceous, metamorphic event (see below), obtained by performing the calculation in the presence H_2O, indicates equilibrium at 13.6 kbar and 592°C (diamond in Fig. 6), which is difficult to reconcile with the jadeite-lawsonite apparent stability in some of these rocks. Intermediate-pressure conditions were estimated using the equation 5 rutile + 3 grossular + 2 quartz + H_2O = 5 titanite + 2 zoisite, with activity coefficients for titanite and zoisite as suggested in Carlswell et al. (2000). This equation is plotted in Figure 6, considering an a_{H2O} of 1.0.

Geochronology

In an attempt to constrain the age of metamorphism, we sampled a leucocratic migmatite from El Chol for zircon separation, as well as white micas and amphibole for K-Ar dating. Muscovite and fuchsite were separated from a foliated megapegmatite (100 × 30 m, sample Gt01 in Fig. 1 and Tables 4 and 5), consisting of white mica (phengite and fuchsite), quartz, and albite exposed near the village of San Antonio Granados. Phengite was also separated from retrograde phengite-bearing gneisses at the locality of Barranca Agua Caliente (sample Gt0209 in Fig. 1 and Table 5), which is the same place where we separated taramitic hornblende from the mafic eclogite FO302 used for thermobarometry.

U-Pb geochronology

About 15 kg of a leucocratic, folded migmatite were collected at El Chol (Fig. 1C). Zircons were separated using standard crushing, magnetic, and heavy liquid techniques. Only diamagnetic grains at 2.0 A were selected for analysis, handpicked under binocular microscope, and abraded, following the technique of Krogh (1982). A total of three single zircons, as well as two small populations, were finally chosen, spiked with $^{205}Pb/^{235}U$ mixed spike and digested in a HF-HNO_3 mixture, using Parrish (1987) microcapsules, in a Parr steel vessel for 5 days at 220°C. After evaporation and 24 h in 3.1N HCl on hot plate, Pb and U were separated by ion exchange, performed in 150 µl capacity Teflon microcolumns filled with BioRad anion-exchange resin. U and Pb were loaded separately on previously outgassed Re filament and isotopic ratios measured in a Finnigan MAT 262 mass spectrometer at Laboratorio Universitario de Geoquímica Isotópica (LUGIS), UNAM. SRM 982 and 983 Pb reference standards, and U500 were used to calibrate the spectrometer gains, as well as to correct for isotopic mass fractionation. Faraday/SEM gain yields were performed at the beginning of each analysis. Common Pb blanks for these analyses ranged between 32 and 72 pg. U-Pb data reduction and error calculations were performed with the program PbDat (Ludwig, 1991), whereas the isochron plot is from the software Isoplot (Ludwig, 1999).

The leucosomatic migmatite contains the assemblage Qz-Ab-Ep/Cz-Phe-Ap, but evidence that this rock underwent high-pressure metamorphism is given by: (1) the small 2V angle of phengite (20–30°); (2) aggregates of Ep/Cz-Qz interpreted as possible pseudomorphs after omphacite; and (3) the presence of adjacent omphacite-garnet-rutile–bearing melanosomatic bands. Only euhedral zircons were chosen for analysis. Zircon # 2 was 350 µm long, stubby with a 2:1 ratio, and amber yellow in color. Zircon # 4 was a pale yellow elongated prism, 4:1 ratio, with well-developed pyramidal terminations. Zircon # 5 was a 400 µm stubby, multifaceted, yellow crystal. Population # 3 was composed of seven bypyramidal grains, slightly pink in color, whereas population # 6 was composed of 20 thin prismatic grains, colorless and < 200 µm in length.

TABLE 4. U-Pb Geochronology of El Chol Migmatite (Sample Gt0332), Chuacús Complex, Guatemala[1]

Fraction[2]	Weight mg	U ppm	Total Pb ppm	Com. Pb pg	$^{206}Pb/^{204}Pb$ Observed ratios[3]	$^{207}Pb/^{206}Pb$ Observed ratios[3]	$^{208}Pb/^{206}Pb$ Observed ratios[3]	$^{206}Pb/^{238}U$ Atomic ratios[4]	$^{207}Pb/^{235}U$ Atomic ratios[4]	$^{207}Pb/^{206}Pb$ Atomic ratios[4]	$^{206}Pb/^{238}U$ Age (Ma)[5]	$^{207}Pb/^{235}U$ Age (Ma)[5]	$^{207}Pb/^{206}Pb$ Age (Ma)[5]
#2, sng, stby 2:1, colorless, multifac, abr	0.034	250.018	27.82	32	923	0.07662	0.10215	0.108627	1.03339	0.068996	665	721	899 ± 11
#3, 8 xls, short byp, abr	0.158	173.239	15.451	94	804	0.07686	0.09498	0.086869	0.759545	0.063414	537	574	722 ± 18
#4, byp, prsm, 3:1, yellow amber, abr	0.048	469.643	24.619	31	1218	0.06139	0.06676	0.053617	0.410487	0.055526	337	349	433 ± 11
#5, single, stby, yellow, abr	0.086	208.882	26.825	69	1419	0.07774	0.10571	0.124986	1.22233	0.07093	759	811	955 ± 6
#6, 21 grains, abr, long prsm	0.048	377.649	52.28	43	1387	0.07393	0.11848	0.134242	1.29563	0.069999	812	844	928 ± 8

[1]Zircon sample dissolution and ion exchange chemistry modified after Krogh, 1973, and Mattinson, 1987 in Parrish (1987) -type microcapsules.

[2]All diamagnetic fractions at 2.0 Amp. Xls = crystal number; sng = single crystal; stby = stubby grains; prsm = prismatic; byp = bypiramidal; abr = abraded grains; multifac = multifaceted

[3]Observed isomtopic ratios are corrected for mass fractionation of 0.12% for ^{205}Pb spiked fraction. Two sigma uncertainties on the $^{207}Pb/^{206}Pb$ and $^{208}Pb/^{206}Pb$ ratios are < 0.4%, generally better than 0.2%; uncertainties in the $^{206}Pb/^{204}Pb$ ratio vary from 0.4% to 1.4%.

[4]Decay constants used: $^{238}U = 1.55125 \times 10^{-10}\ a^{-1}$; $^{235}U = 9.48485 \times 10^{-10}\ a^{-1}$; $^{238}U/^{235}U = 137.88$. Uncertainities on the U/Pb ratio is 0.5%

[5] $^{207}Pb/^{206}Pb$ age uncertainties are ±2 sigma calculated from the data reduction program PBDAT of K. Ludwig, 1991. Total processing Pb blank amount ranged between 32 pg and 72 pg. Initial Pb composition are from isotopic analyses of feldspar separates. Isotopic data were measured on a Finnigan MAT 262 mass Spectrometer with SEM Ion Counting at LUGIS, UNAM, Mexico City.

TABLE 5. K-Ar Analytical Results for Selected Samples of the Chuacús Complex, Guatemala

Sample	Mineral	Grain size[1]	%K	$^{40}Ar^{*2}$	$\%^{40}Ar^{*3}$	Age, Ma[4]
Gt01	Muscovite	> 1cm core	8.54	11.50	92.7	75.9 ± 1.2
Gt01	Muscovite	> 1cm border	8.51	10.91	96.2	72.4 ± 1.1
Gt01	Muscovite	500–1000 μm	8.06	10.50	98.4	73.9 ± 1.0
Gt01	Muscovite	150–300 μm	7.61	8.175	93.1	60.9 ± 1.6
Gt01	Fuchsite	500–1000 μm	8.67	10.7	99.9	69.5 ± 1.1
Gt01	Fuchsite	200–500 μm	8.53	9.28	97.9	61.7 ± 1.0
Gt0209	Phengite	500–1000 μm	8.61	10.5	87.3	69.0 ± 1.0
Gt0209	Phengite	300–500 μm	8.53	10.9	87.8	72.3 ± 1.2
F0302	Amphibole	300–500 μm	1.13	1.45	91.2	73.0 ± 1.2

[1]Grain size = physical diameter of analyzed crystals.
[2] $^{40}Ar^*$ = radiogenic argon in 1×10^{-10} moles/g.
[3] $\%^{40}Ar^*$ = percentage of radiogenic Ar from total Ar
[4]Age = error is stated as 1σ.

Analytical results are presented in Table 4 and the concordia plot is shown in Figure 7. A chord through all the analyses yields a lower intercept of 302 ± 52 Ma, whereas using only the single-grain analyses, it yields a much better constrained intercept of 302 ± 4.6 Ma, with a low MSWD of 0.78. We interpret these apparent ages as indicating the time of migmatization in the El Chol sequence (image in Fig. 2B). Because this dated leucosome is twice folded together with HP melanosomatic bands, it would also constitute a minimum age for a first HP event in the rocks exposed at El Chol. The upper intercept may be interpreted in terms of material inherited from some Mesoproterozoic source for the Chuacús rocks.

K-Ar geochronology

K-Ar analyses were performed at LUGIS, UNAM, measuring K by X-ray fluorescence spectrometry following the method of Solé and Enrique (2001). Argon was measured by isotope dilution with a VG1200 mass spectrometer. Calibration was done with LP-6 Bt and B4M Ms standards. The constants recommended by Steiger and Jäger (1977) were used throughout. Results shown in Table 5 indicate a restricted range of dates between ~ 61 to ~ 76 Ma for pegmatitic fuchsite and muscovite, and 69–72 Ma for the retrograde phengite in HP gneiss. The taramitic amphibole from sample FO302 yields an age of 73 Ma. These ages may be interpreted as recording a Late Cretaceous burial and reheating event that affected Paleozoic continental crust either during collision of the southern Chuacús complex against the western part of the Greater Antillean arc, or during collision between the Chortís and Maya blocks.

Tectonic Implications

The Chuacús complex is currently considered a Paleozoic crystalline basement forming the southernmost continental edge of the Maya block (Yucatán platform), an interpretation that is partly supported by our new U-Pb zircon age, petrologic data, and field structural observations discussed in this paper. However, the fully documented presence of Cretaceous eclogitic rocks (McBirney et al., 1967; Harlow, 1994, Sisson et al., 2003), including some with possible UHPM (Tsujimori et al., 2003) associated with the collision of the Caribbean plate against the southern margin of the Maya block, and the abundance of Late Cretaceous K-Ar ages of crystalline rocks throughout the Polochic-Motagua zone, requires a brief discussion of these two possible Paleozoic or Cretaceous tectonic scenarios where the Chuacús complex could has been formed. A full development of this central issue is beyond the scope of the present paper, and a more complete

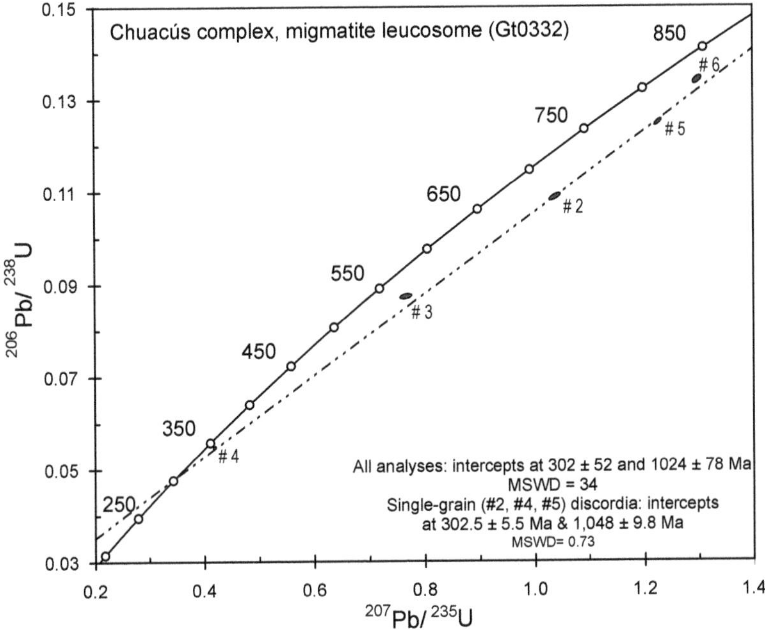

FIG. 7. U-Pb concordia plot for the selected sample at El Chol, Sierra de Chuacús, Guatemala.

discussion must await further detailed geochronological, structural, geochemical and, if possible, paleomagnetic studies.

The Appalachian (Alleghenian-Ouachitan) Scenario

The Chuacús complex has been traditionally linked to the Maya microcontinental block within a group of other ancient blocks of Mesoamerica such as Chortís and Oaxaquia, which together constitute the pre-Mesozoic basement bridge between North and South America. However, as documented above, the Chuacús complex is in fact separated from the Maya block by large faults and is petrologically unique compared to any of the known pre-Mesozoic metamorphic complexes of southern Mexico and the Maya block (Oaxacan, Acatlán, Cuchumatanes, and Chiapas massif), and is also different from those in the Chortís block (Omoa, Las Ovejas, Palacaguina, and Cacaguapa). Because of its dominant sedimentary protoliths and scarcity of originally igneous bodies, the Chuacús complex probably constituted a continental margin sedimentary prism trapped in past collisions between two continental masses.

The Late Carboniferous age obtained for the Chuacús main metamorphism indicates that it could have formed in the deepest root so far exposed of the Alleghenian suture between the Laurentia and Gondwana supercontinents, which collided in the Late Paleozoic to form Pangea. Whether the Chortís block and the Yucatán platform formed the leading edges of those continents is difficult to tell because long range displacements of Mesozoic through Tertiary times may be associated with their present contact at the Motagua-Polochic fault system (Anderson and Schmidt, 1983; Deaton and Burkart, 1984). It should be noted in this context that, while the Chortís block apparently lacks sedimentary rocks of Paleozoic age associated with the postulated orogeny, thick sedimentary packages are present in the Maya block, represented by the distinctly metamorphosed and strongly deformed Mississippian–Lower Pennsylvanian lower Santa Rosa Group of the Altos Cuchumatanes, north central Guatemala, and Chicomuselo, Chiapas areas. The full documentation of an angular unconformity between these rocks and the unmetamorphosed upper Santa Rosa Group in the Chiapas area (Hernández-García, 1973), in that area supports the idea that the southern Maya block underwent strong tectonic disturbance during the early Alleghenian–Ouachitan orogeny. However, because the Chuacús complex conforms a fault-bounded tectonostratigraphic terrane with unknown displacements and apparently unconstrained by a pre-Mesozoic sedimentary cover, no definite correlations with those Late Paleozoic units can be established without further studies.

Moreover, it should be noted that most Pangean paleogeographic models show the Chortís block attached to southwestern Mexico in the Cordilleran margin or outboard in the paleo-Pacific ocean, precluding any possible collisional interactions with the Maya terrane, which is generally placed in the hinterland of the Ouachita-Marathon orogen attached to northern South America (Pindell, 1985; 1993; Stanek and Voigt, 1994; Centeno-García and Keppie, 1998). Thus, although U-Pb zircon dating and metamorphic styles of the Chuacús complex are compatible with Late Paleozoic collision during the Alleghenian-Ouachitan accretion of Pangea, the specific paleogeographic setting of the Chuacús complex in relation to the sutured blocks cannot be fully reconstructed with the present data.

The Caribbean scenario

In a Caribbean scenario, the southern edge of the Yucatán platform (Maya block) developed as a passive margin from the Late Jurassic to Late Cretaceous, as South America moved southward during the break-up of Pangea (Ross and Scotese, 1988). This margin would be represented by the predominantly sedimentary rocks that formed the Chuacús complex. At the same time, the Caribbean plate originated in the Pacific, and its frontal Antillean arc moved and finally overthrust the Bahamas continental platform by the Eocene, but involving first an earlier collision of the western arc segment during ~84–72 Ma (Campanian) against the southern margin of the Maya block (Pindell and Dewey, 1982; Ross and Scotese, 1988, Pindell, 1993). This event in the Guatemalan foreland region of the collisional front was marked by the inception and full development of the Sepur basin during the Late Cretaceous (Mann, 1999), accompanied by thrusting of large ophiolitic massifs over crystalline basement (Chuacús complex) and sedimentary rocks of the Maya block (Rosenfeld, 1993). In this case the intervening El Tambor oceanic basin, together with the Antillean arc, would be the only overthrust blocks onto the continental margin of the Yucatán platform, but in other similar models (e.g., Dickinson and Lawton, 2001) the Chortís block also is shown as being involved in the collision.

In our view, however, the high to very high pressures shown by most rocks of the Chuacús complex would imply protracted continental (A-type) subduction followed by collision, crustal thickening, and rapid exhumation and the minimum burial of about 85 km (~23 kbar) documented for the Chuacús complex, could hardly be built up by a relatively thin ocean crust-arc system obducted onto the Chuacús complex, particularly if the Chortís block did not form part of the upper plate because of its much later arrival (Pindell and Barret, 1990; Riller et al., 1992; Schaaf et al., 1995). In fact, the Chuacús complex lacks pervasive tectonic fabrics and metamorphic recrystallization related to this Antillean Late Cretaceous orogeny, as fresh high-grade gneisses of the Chuacús complex in the studied area rest steeply underneath the subhorizontal, low-grade mafic-ultramafic Cretaceous nappes. It should be mentioned that although Cretaceous eclogites also have been found north of the Motagua fault zone (Sisson et al., 2003) and thus in contact with the Chuacús complex, all of them formed at low temperatures (max. 550°C), and occur associated with jadeitites (Harlow, 1994) or blueschists. Therefore, these eclogites certainly formed well below the temperatures documented for the Chuacús complex eclogitic rocks. The rare presence of jadeite-lawsonite-albite low-temperature associations in veins cutting previous high-pressure foliated fabrics of the Chuacús complex may represent a Cretaceous high-pressure burial event superposed on an older Paleozoic eclogitic orogenic terrane. The former K-Ar ages presented in this paper would fit this scenario, involving a collision between the southern Chuacús complex against the western part of the Greater Antillean arc, as it moved eastward in the front of the Caribbean oceanic plate. Accordingly, the overall data presented above allow us to conclude that the pre-Mesozoic orogenic scenario better suits our new and former petrologic, field, and age data to explain the origin of the Chuacús complex.

Conclusions

We have documented the first finding of eclogitic rocks of mafic, pelitic, quartzofeldspathic, and calcareous composition in the Chuacús complex of Guatemala. High temperature (~ 770–600°C) decompression and partial melting of these deeply buried rocks (>23 kbar) were dated by U-Pb in zircons as probably Late Carboniferous. Furthermore, and although the evidence is not yet conclusive, several textural and mineralogical observations converge to the former occurrence of UHPM, opening the opportunity for further petrologic research in the area, and the eventual finding of diagnostic minerals, possibly including coesite and microdiamond.

Unfortunately, the full tectonic significance of this finding can only be attained when the precise structural history and age of the metamorphic event are resolved. A Carboniferous age would be more consistent with current stratigraphic data and deep-seated orogenic interactions between Laurentia and Gondwana in an Appalachian-Ouachitan-Marathon frame, but Cretaceous continental collisions (e.g., Chortís-Maya blocks) during Caribbean tectonic evolution, or possibly older Grenvillian or Pan-African scenarios, cannot be precluded by the present data alone.

Acknowledgments

Funds for fieldwork and analyses were provided by PAPIIT-UNAM grants IN116999, IN107999, and IN100002 to John Duncan Keppie, Fernando Ortega, Luigi Solari, and Jesús Solé, respectively. CONACyT project # 32475-T to J. Solé funded the Ar instrumentation. We acknowledge the continuous support and comments of John Duncan Keppie, as well as help during X-ray diffraction determinations of carbonates by Teresa Pi, and sample preparation and microprobing by Consuelo Macías and Carlos Linares. Gabriela Solís, Juán Julio Morales, and Peter Schaaf are thanked for maintaining ultrapure laboratories and mass spectrometric facilities of LUGIS, UNAM, and to Hugo Delgado Granados for the use of the Laboratorio Universitario de Petrología (LUP). $^{205}Pb/^{235}U$ mixed spike was kindly provided by Dr. R. Romer, Potsdam, Germany. Last but not least, we thank Thomas Anderson and Brent Miller for their most careful and thoughtful reviews, which improved the manuscript substantially.

REFERENCES

Anderson, T. H., Burkart, B., Clemons, R. E., Bohneberger, O. H., and Blount, D. N., 1973, Geology of the western Altos Cuchumatanes, Northwestern Guatemala: Geological Society of America Bulletin, v. 84, p. 805–826.

Anderson, T. H., and Schmidt, V. A., 1983, The evolution of Middle America and the Gulf of Mexico–Caribbean region during Mesozoic time: Bulletin of the Geological Society of America, v. 94, p. 941–966.

Becker, H., and Altherr, R., 1992, Evidence from ultrahigh pressure marbles for recycling of sediments in the upper mantle: Nature, v. 358, p. 745–748.

Berman, R. G., 1991, Thermobarometry using multi-equilibrium calculations: A new technique, with petrological applications: Canadian Mineralogist, v. 29, p. 833–855.

Bhattacharya, A. Mohanty, L. Maji, A., Sen, S. K., and Raith, M., 1992, Non-ideal mixing in the phlogopite-annite binary; constraints from experimental data on Mg-Fe partitioning and a reformulation of the biotite-garnet geothermometer: Contributions to Mineralogy and Petrology, v. 111, p. 87–93.

Bohnenberger, O. H., 1966, Nomenclatura de las Capas Santa Rosa en Guatemala: Publicaciones Geológicas del ICAITI (Guatemala), no. 1, p. 47–51.

Bozhilov, K. N., Green, H. W., and Dobrzhinetskaya, L., 1999, Clinoenstatite in Alpe Arami peridotite: Additional evidence of very high pressure: Science, v. 284, p. 128–132.

Carswell, D. A., O'Brien, P. J., Wilson, R. N., and Zhai, M., 1997, Thermobarometry of phengite-bearing eclogites in the Dabie Mountains of central China: Journal of Metamorphic Geology, v. 15, p. 239–252.

Carswell, D. A., Wilson, R. N., and Zhai, M. G., 1996, Ultrahigh pressure aluminous titanites in carbonate-bearing eclogites at Shuanghe in Dabieshan, central China: Mineralogical Magazine, v. 60, p. 361–471.

______, 2000, Metamorphic evolution, mineral chemistry, and thermobarometry of schists and orthogneiss hosting ultra-high pressure eclogites in the Dabie Shan of central China: Lithos, v. 52, p. 121–155.

Centeno-García, E., and Keppie, J. D., 1998, Latest Paleozoic–early Mesozoic structures in the central Oaxaca terrane of southern Mexico: Deformation near a triple junction: Tectonophysics, v. 301, p. 231–242.

Chopin, C., 1984, Coesite and pure pyrope in high-grade blueschists of the western Alps: A first record and some consequences: Contributions to Mineralogy and Petrology, v. 86, p. 107–118.

______, 2003, Ultrahigh-pressure metamorphism: Tracing continental crust into the mantle: Earth and Planetary Science Letters, v. 212, p. 1–14.

Chopin, C., and Sobolev, N. V., 1995, Principal mineralogical indicators of of UHP in crustal rocks, *in* Coleman, R. G., and Wang, X., eds., Ultrahigh pressure metamorphism: Cambridge, UK, Cambridge University Press, p. 96–132.

Clemons, R. E, Anderson, T. H., Bohnenberger, O. H., and Burkart, B., 1974, Stratigraphic nomenclature of recognized Paleozoic and Mesozoic of western Guatemala: American Association of Petroleum Geologists Bulletin, v. 58, p. 313–320.

Coleman, R. G., and Wang, X., 1995, Overview of the geology and tectonics of UHPM, *in* Coleman, R. G., and Wang, X., eds., Ultrahigh pressure metamorphism: Cambridge, UK, Cambridge University Press, 528 p.

Dasgupta, S., Sengupta, P., Guha, D., and Fukuoka, M., 1991, A refined garnet-biotite Fe-Mg exchange geothermometer and its application in amphibolites and granulites: Contributions to Mineralogy and Petrology, v. 109, p. 130–137.

Deaton, B. C., and Burkart, K., 1984, Time of sinistral slip along the Polochic fault of Guatemala: Tectonophysics, v. 102, p. 297–313.

Dickinson, W. R., and Lawton, T. F., 2001, Carboniferous to Cretaceous assembly and fragmentation of Mexico: Geological Society of America Bulletin, v. 113, 1142–1160.

Dobrzhinetskaya, L., Green, H. W., and Wang, S., 1996, Alpe Arami: A peridotite massif from depths of more than 300 kilometers: Science, v. 271, p. 1841–1846.

Donnelly, T. W., Horne, G. S., Finch, R. C., and López-Ramos, E., 1990, Northern Central America: The Maya and Chortís blocks, *in* Dengo, G., and Case, J. E., eds., The Caribbean region: Boulder, CO, Geological Society of America, The Geology of North America, v. H, p. 37–76.

Ferry, J. M., and Spear, F. S., 1978, Experimental calibration of the partitioning of Fe and Mg between biotite and garnet: Contributions to Mineralogy and Petrology, v. 66, p. 113–117.

Franz, G., and Spear, F. S., 1985, Aluminous titanite (sphene) from the eclogite zone, south-central Tauern window, Austria: Chemical Geology, v. 50, p. 33–46.

Franz, G., and Smelik, E. A., 1995, Zoisite-clinozoisite bearing pegmatites and their importance for decompressional melting in eclogites: European Journal of Mineralogy, v. 7, p. 1421–1436.

Gilottti, J. A., and Krogh Ravna, E. J., 2002, First evidence for ultra-high pressure metamorphism in the North-East Greenland Caledonides: Geology, v. 30, p. 551–554.

Giunta, G., Beccaluva, L., Coltorti, M., Mota, B., Padoa, F., Siena, F., Dengo, C., Harlow, G. E., and Rosenfeld, J., 2002, The Motagua suture zone in Guatemala: Ofioliti, v. 27, p. 1–42.

Gomberg, D. M., Banks, P. O., and McBirney, A. R., 1968, Guatemala: Preliminary zircon ages from central cordillera: Science, v. 162, p. 121–122.

Gordon, M. B., 1992, Northern Central America (The Chortís block), *in* Westerman, G. E. G., ed., Jurassic of the Circum-Pacific: Cambridge, UK, Cambridge University Press, p. 107–113.

______, 1993, Revised Jurassic and Early Cretaceous (Pre-Yojoa Group) stratigraphy of the Chortis Block: Paleogeographic and tectonic implications, *in* Pindell, J. L., and Perkins, R. F., eds., Mesozoic and early Cenozoic development of the Gulf of Mexico and Caribbean region: Gulf Coast Section, Society of Economic Paleontologists and Mineralogists, 13th Annual Research Conference, p. 143–154.

Green, T. H., and Hellman, P. L., 1982, Fe-Mg partitioning between coexisting garnet and phengite at high pressure, and comments on a garnet-phengite geothermometer: Lithos, v. 15, p. 253–256.

Harlow, G. E., 1994, Jadeitites, albitites and related rocks from the Motagua Fault Zone, Guatemala: Journal of Metamorphic Geology, v. 12, p. 49–68.

Hedberg, H. D., ed., 1976, International stratigraphic guide: New York, NY, John Wiley, 200 p.

Hernández-García, R., 1973, Paleogeografía del paleozoico de Chiapas, México: Boletín de la Asociación Mexicana de Geólogos Petroleros, v. 25, no. 1-3, p. 77–134.

Hodges, K. V., and Spear, F. S., 1982, Geothermometry, geobarometry and the Al_2SiO_5 triple point at Mt. Moosilauke, New Hampshire: American Mineralogist, v. 67, p. 1118–1134

Holland, T. J. B., and Blundy, J. D., 1994, Non-ideal interactions in calcic amphiboles and their bearing on amphibole-plagioclase thermometry: Contributions to Mineralogy and Petrology, v. 116, p. 433–447.

Horne, G. S., Clark, G. S., and Pushkar, P., 1976, Pre-Cretaceous rocks of northwestern Honduras: Basement terrane in Sierra de Omoa: American Association of Petroleum Geologists Bulletin, v. 60, p. 566–583.

Hwang, S.-L., Shen, P., Chu, H.-T., and Yui, T.-F., 2000, Nanometer-size alfa-PbO_2-type TiO_2 in garnet: A thermobarometer for ultrahigh-pressure metamorphism: Science, v. 288, p 321–324.

Kesler, S. E., 1971, Nature of ancestral orogenic zone in nuclear Central America: American Association of Petroleum Geologists Bulletin, v. 55, p. 2116–129.

Kesler, S. E., Josey, W. L., and Collins, E. M., 1970, Basement rocks of western nuclear Central America: The western Chuacús Group, Guatemala: Geological Society of America Bulletin, v. 81, p. 3307–3322.

Klemd, R., 2003, Ultra-high pressure metamorphism in eclogite from the western Tianshan high-pressure (Xinjiang, western China)—comment: American Mineralogist, v. 88, p. 1153–1156.

Kretz, R., 1983, Symbols for rock forming minerals: American Mineralogist, v. 68, p. 277–279.

Krogh, T. E., 1973, A low-contamination method for hydrothermal decomposition of zircon and extraction of U and Pb for isotopic age determinations: Geochimica et Cosmochimica Acta, v. 37, p. 485–494.

______, 1982, Improved accuracy of U-Pb zircon ages by the creation of more concordant systems using an air abrasion technique: Geochimica et Cosmochimica Acta, v. 46, p. 637–649.

Leake, E. B., and 21 others, 1997, Nomenclature of amphiboles: Report of the Subcommittee on Amphiboles of the International Mineralogical Association, Commission on New Minerals and Mineral Names: The Canadian Mineralogist, v. 35, p. 219–246.

Liou, J. G., Zhang, R. Y., Ernst, W. G., Rumble, D., and Maruyama, S., 1998, High pressure minerals from deeply subducted metamorphic rocks, *in* Hemley, R. J., ed.., Ultra-high pressure mineralogy: Physics and chemistry of the Earth's deep interior: Reviews in Mineralogy, v. 37, p. 33–96.

Liu, J., Ye, K., Maruyama, S., Cong, B., and Fan, H., 2001, Mineral inclusions in zircon from gneisses in the ultra-high-pressure zone of the Dabie Mountains, China: Journal of Geology, v. 109, p. 523–535.

Ludwig, K. R., 1991, PbDat: A computer program for processing Pb-U-Th isotope data, version 1 24: Reston, VA, U.S. Geological Survey.

______, 1999, Isoplot/Ex, ver. 2.49, A geochronological toolkit for Microsoft Excel, 1a: Berkeley, CA, Berkeley Geochronology Center.

Mann, P., 1999, Caribbean sedimentary basins: Classification and tectonic setting from Jurassic to present, *in* Mann, P., ed., Caribbean basins: Amsterdam, Netherlands, Elsevier Science B.V., Sedimentary Basins of the World, v. 4, p. 3–31.

Manton, W. I., 1996, The Grenville of Honduras [abs.], *in* Geological Society of America, Abstracts with Programs, v. 26. no. 7, p. A-493.

Mattinson, J. M., 1987, U-Pb ages of zircons: A basic examination of error propagation: Chemical Geology, v. 66, p. 151–162.

McBirney, A. R., 1963, Geology of a part of the central Guatemalan cordillera: California University Publications in Geological Sciences, v. 38, p. 177–242.

McBirney, A. R., Aoki, K. J., and Bass, M., 1967, Eclogites and jadeite from the Motagua fault zone, Guatemala: American Mineralogist, v. 52, p. 908–918.

Medaris, L. G., Jr., 1999, Garnet peridotite in Eurasian HP and UHP terranes: A diversity of origins and thermal histories: International Geology Review, v. 41, p. 799–815.

Morimoto, N., 1988, Nomenclature of pyroxenes: American Mineralogist, v. 73, p. 1123–1133.

Newcomb, W. E., Retrograde cataclastic gneiss north of the Motagua fault zone, Guatemala: Geologie en Mijnbouw, v. 57, 271–276.

Ogasawara, Y., Fukasawa, K., and Maruyama, S., 2002, Coesite exsolution from supersilicic titanite in UHP marble from the Kokchetav Massif, northern Kazakhstan: American Mineralogist, v. 87, p. 454–461.

Parrish, R. R., 1987, An improved micro-capsule for zircon dissolution in U-Pb geochronology: Chemical Geology, v. 66, p. 99–102.

Pattison, D. R. M., and Newton, R. C., 1989, Reversed experimental calibration of the garnet-clinopyroxene Fe-Mg exchange thermometer: Contributions to Mineralogy and Petrology, v. 101, p. 87–103.

Pindell, J. L., 1985, Alleghenian reconstruction and the subsequent evolution of the Gulf of Mexico, Bahamas and proto-Caribbean Sea: Tectonics, v. 3, p. 133–156.

______, 1993, Mesozoic–Cenozoic paleogeographic evolution of northern South America: American Association of Petroleum Geologists Bulletin, v. 77, p. 340

Pindell, J. L., and Barret, S. F., 1990, Geological evolution of the Caribbean region; a plate-tectonic perspective, *in* Dengo, G., and Case, J. E., eds., The Caribbean region: Boulder, CO, Geological Society of America, The Geology of North America, v. II, p. 405–432.

Pindell, J. L., and Dewey, J. F., 1982, Permo-Triassic reconstruction of western Pangea and the evolution of the Gulf of Mexico/Caribbean region: Tectonics, v. 1, p. 179–211.

Poli, S., and Schmidt, M. W., 1997, The high-pressure stability of hydrous phases in orogenic belts: An experimental approach on eclogite-forming processes: Tectonophysics, v. 273, p. 169–184.

Rapp, R. P., Watson, E. B., and Miller, C. F., 1991, Partial melting of amphibolite/eclogite and the origin of Archean trondhjemites and tonalites: Precambrian Research, v. 51, p. 1–25.

Riller, U., Ratsbacher, L., and Frisch, W., 1992, Left-lateral transtension along the Tierra Colorada deformation zone: Journal of South American Earth Sciences, v. 5, p. 237–249.

Roper, P. J., 1978, Stratigraphy of the Chuacús Group on the south side of the Sierra las Minas Range, Guatemala: Geologie en Mijnbouw, v. 57, p. 309–313.

Rosenfeld, J. H., 1993, Sedimentary rocks of the Santa Cruz ophiolite, Guatemala—a proto-Caribbean history, *in* Pindell, J. L., and Perkins, R. F., eds., Mesozoic and early Cenozoic development of the Gulf of Mexico and Caribbean region: Gulf Coast Section, Society of Economic Paleontologists and Mineralogists, 13th Annual Research Conference, p. 173–180.

Ross, M. I., and Scotese, C. R., 1988, A hierarchical tectonic model of the Gulf of Mexico and Caribbean region: Tectonophysics, v. 155, p. 139–168.

Sabau, G., 2000, A possible UHP-eclogite in the Leota Mts. (South Carpathians) and its history from high-pressure melting to retrograde inclusion in a melange: Lithos, v. 52, p. 253–276.

Schaaf, P., Morán-Zenteno, D., Hernández-Bernal, M. S., Solis-P, G., Tolson, G., and Kohler, H., 1995, Paleogene continental margin truncation in southwestern Mexico: Geochronological evidence: Tectonics, v. 14, p. 1339–1350.

Schmadicke, E., and Muller, W. F., 2000, Unusual exsolution phenomena in omphacite and partial replacement of phengite by phlogopite + kyanite in an eclogite from the Erzgebirge: Contributions to Mineralogy and Petrology, v. 139, p. 629–642.

Schmidt, R., Franz, L., Oberhansli, R., and Dong, S., 2000, High-Si phengite, mineral chemistry, and P-T evolution of ultra-high-pressure eclogites and calc-silicates from the Dabie Shan, eastern China: Geological Journal, v. 35, p. 185–207.

Schreyer, W., 1995, Ultradeep metamorphic rocks: The retrospective viewpoint: Journal of Geophysical Research, v. 100, p. 8353–8366.

______, 1999, Experimental aspects of UHP metamorphism: Granite systems: International Geology Review, v. 41, p. 701–710.

Schwartz, D. P., 1972, Petrology and structural geology along the Motagua fault zone, Guatemala [abs.]: Transactions of the Caribbean Geological Conference, Guadalupe, v. 6, p. 299.

Sisson, V. B., Harlow, G. E., Sorensen, S., S., Brueckner, H. K., Saham, E., Hemming, S. R., and Lallemmant, H. G., Lawsonite eclogite and other high-pressure assemblages in the southern Motagua fault zone, Guatemala: Implications for Chortís collision and subduction zone [abs.]: Geological Society of America, Abstracts with Programs, v. 35, no. 6, p. 639.

Smith, D. C., 1984, Coesite in clinopyroxene in the Caledonides, and its implications for geodynamics: Nature, v. 310, p. 641–644.

______, 1988, A review of the peculiar mineralogy of the "Norwegian coesite-eclogite province," with crystal-chemical, petrological, geochemical and geodynamical notes and an extensive bibliography, in Smith, D. C., ed., Eclogites and eclogite-facies rocks: Amsterdam, Netherlands, Elsevier, p. 1–206.

Sobolev, N. V., and Shatsky, V. S., 1990, Diamond inclusions in garnet from metamorphic rocks: Nature, v. 343, p. 742–746.

Solé, J., and Enrique, P., 2001, X-ray fluorescence analysis for the determination of potassium in small quantities of silicate minerals for K-Ar dating: Analytica Chemica Acta, v. 440, p. 199–205.

Stanek, K .P., and Voigt, S., 1994, Model of Meso-Cenozoic evolution of northwestern Caribbean: Zentralblatt für Geologie und Palaeontologie, Teil I, 1993, p. 499–511.

Steiger, R. H., and Jager, E., 1977, Subcomission on geochronology: Earth and Planetary Science Letters, v. 36, p. 359–362.

Tropper, P., Manning, C. E., and Essene, E. J., 2002, The substitution of Al and F in titanite at high pressure and temperature: Experimental constraints on phase relations and solid solution properties: Journal of Petrology, v. 43, p. 1787–1814.

Tsujimori, T., Liou, J. G., Coleman, R. G., Rohtert, W., and Clearly, J. G., 2003, Eclogitization of a cold subducting slab: Prograde evolution of lawsonite-eclogites from the Motagua fault zone, Guatemala [abs.]: Geological Society of America, Abstracts with Programs, v. 35, no. 6, p. 639.

van den Boom, G., 1972, Petrofazielle Gleidrung des metamorphem Grundgebirges in der Sierra de Chuacús, Guatemala: Beihefte Geologisches Jahrbuch, v. 122, p. 5–49.

Van Roermund, H. L. M., Drury, M. R., Barnhoorn, A., and de Ronde, A., 2001, Relict majoritic garnet microstructures from ultra-deep orogenic peridotites in western Norway: Journal of Petrology, v. 42, p. 117–130.

Vogel, D. E., 1967, Petrology of an eclogite and pyrigarnite-bearing polymetamorphic rock complex at Cabo Ortegal, NW Spain: Leidse Geologische Mededelingen, v. 40, p. 121–213.

Waters, D. J., and Martin, H. N., 1993, Geobarometry of phengite-bearing eclogites [abs.]: Terra Abstracts, v. 5, p. 410-411 [updated calibration of 1996 at http://www.earth.ox.ac.uk/~davewa/ecbar.html].

Wendt, A. S., D'Arco, P., Goff, B., and Oberhansli, R., 1993, Radial cracks around alfa-quartz inclusions in almandine: constrains on the metamorphic history of the Oman mountains: Earth and Planetary Science Letters, v. 114, p. 449–461.

Weyl, R. 1980, Geology of Central America: Berlin, Germany, Gebruder Borntraeger, 371 p.

Ye, K., Cong, B., and Ye, D., 2000, The possible subduction of continental material to depths greater than 200 km: Nature, v. 407, p. 734–736.

Ye, K., Liu, J., Cong, B., Ye, D., Xu, P., Omori, S., and Maruyama, S., 2002, Ultrahigh-pressure (UHP) low-Al titanites from carbonate-bearing rocks in Dabieshan-Sulu UHP terrane, eastern China: American Mineralogist, v. 87, p. 875–881.

Zhang, L., Ellis, D. J., and Jiang, W., 2002, Ultra-high pressure metamorphism in western Tianshan, China: Part I. Evidence from inclusions in coesite pseudomorphs in garnet and from quartz exsolution lamellae in omphacite in eclogites: American Mineralogist, v. 87, p. 853–860.

Zhang, L., Ellis, D. J., Williams, S., and Jiang, W., 2003, Ultrahigh-pressure metamorphism in eclogites from the western Tianshan high-pressure belt (Xinjiang, western China)—reply: American Mineralogist, v. 88, p. 1157–1160.

Chortis Terrane

Tectonic Implications of Alternative Cenozoic Reconstructions for Southern Mexico and the Chortis Block

J. DUNCAN KEPPIE[1] AND DANTE J. MORÁN-ZENTENO

Instituto de Geología, Universidad Nacional Autónoma de México, 04510 México D.F., México

Abstract

Most current Eocene reconstructions juxtapose the Chortis block of northern Central America against southern Mexico, and invoke ~1100 km Cenozoic sinistral displacement on the Acapulco-Motagua-Cayman fault zone, the inferred northern margin of the Caribbean plate. Such a hypothesis is incompatible with the presence of undeformed Upper Cretaceous–Recent sediments that cross the projected trace of the Motagua fault zone in the Gulf of Tehuantepec, minimal offset of the Permian Chiapas batholith, and the absence in Honduras of several major features in southern Mexico. These problems may be overcome if the Chortis block is back-rotated anticlockwise about a pole near Santiago, Chile, i.e. ~1100 km along the Cayman transform faults during the Cenozoic. Such a reconstruction when combined with reconstructions of features in the Pacific Ocean, suggests that Middle Miocene collision of the Tehuantepec aseismic ridge with the Acapulco Trench led to: (1) asymmetric flattening of the subduction zone; (2) an anticlockwise rotation of the Mexican magmatic arc to its present location by the Middle Miocene; (3) the development of a volcanic arc gap in southeastern Mexico, in which the late Middle Miocene Chiapas fold-and-thrust belt developed: as the Tehuantepec Ridge swept westward, arc volcanism was re-established in the gap. Eocene collision of the Chumbia Seamount Ridge (inferred mirror image of the Moonless Mountains–unnamed seamount ridge between the Molokai and Clarion fracture zones) with the Acapulco Trench followed by its ESE migration during the Oligocene led to: (a) flattening of the subducting slab inducing subduction erosion and exhumation of the southern Mexican margin; (b) anticlockwise rotation of the volcanic arc; and (c) sinistral strike-slip faulting in the Sierra Madre del Sur. This contrasts with the region north of the projected Molokai fracture zone where the dip of the subduction zone appears to have steepened, producing extension. Eocene(–Late Cretaceous) subduction along the southern coast of Mexico explains the remnants of a Late Cretaceous arc in the Gulf of Tehuantepec and neighboring Guatemala.

Introduction

PLATE TECTONICS provides a powerful tool in constructing palinspastic maps by backward modeling, especially where seafloor magnetic lineations and transform faults are available. Such data have enabled the history of plate motions in the Pacific Ocean west off Middle America to be elegantly unraveled (e.g., Mammerickx and Klitgord, 1982; Meschede et al., 2002). However, the Pacific plates (Farallon and its daughters: Guadalupe, Rivera, and Cocos) are separated from the North American and Caribbean plates by a trench, which requires that the plate motions of the North American and Caribbean plates be determined independently. Using hotspots and plate circuit data, Engebretson et al. (1985), DeMets et al. (1990), and Atwater and Stock (1999) reconstructed the relative plate motions between the North American and Pacific plates during the Cenozoic and Mesozoic. In the case of the Caribbean plate, seafloor magnetic lineations are limited to the Cayman Trough located along its northwestern border with the North American plate (Fig. 1). The Cayman Trough contains a NNW-trending, spreading ridge truncated at both ends by ENE-trending transform faults across which ~1100 km of sinistral offset between the North American and Caribbean plates has been calculated (Rosencrantz and Sclater, 1986; Ladd et al., 1990). The magnetic lineations indicate that the Cayman Trough has been spreading at 1.5–2.5 cm/yr since before the latest Middle Eocene, ~44 Ma (Rosencrantz and Sclater, 1986; Rosencrantz et al., 1988; Fig. 1). Combining these data with the orientation of transform faults bounding the Cayman Trough allowed Pindell et al. (1988) and Ross and Scotese (1988) to identify a

[1]Corresponding author; email: duncan@servidor.unam.mx

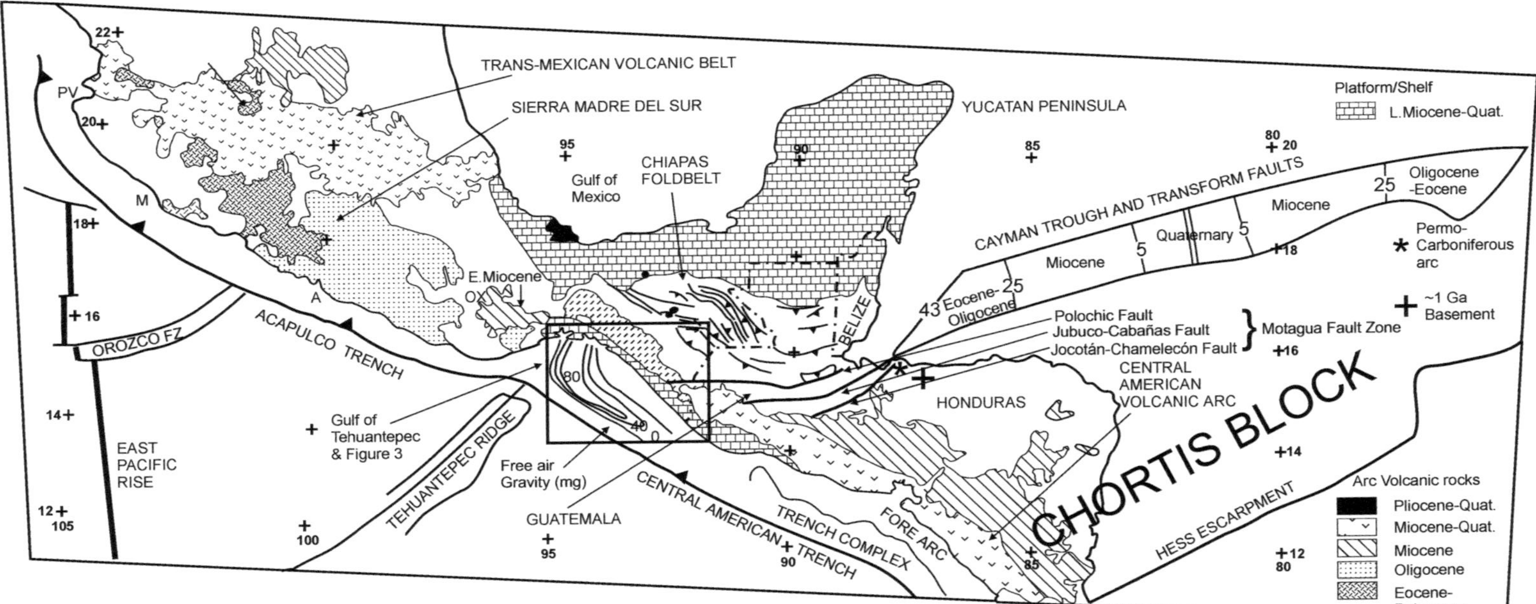

FIG. 1. Geological map of southern Mexico and northern Central America (modified after Muehlberger, 1992; Ferrari et al., 1999; and Moran-Zenteno et al., 1999). Abbreviations: A = Acapulco; M = Manzanillo; OX = Oaxaca; PV = Puerto Vallarta; +95 = geographic coordinates; numbers in Cayman Trough = age of oceanic lithosphere.

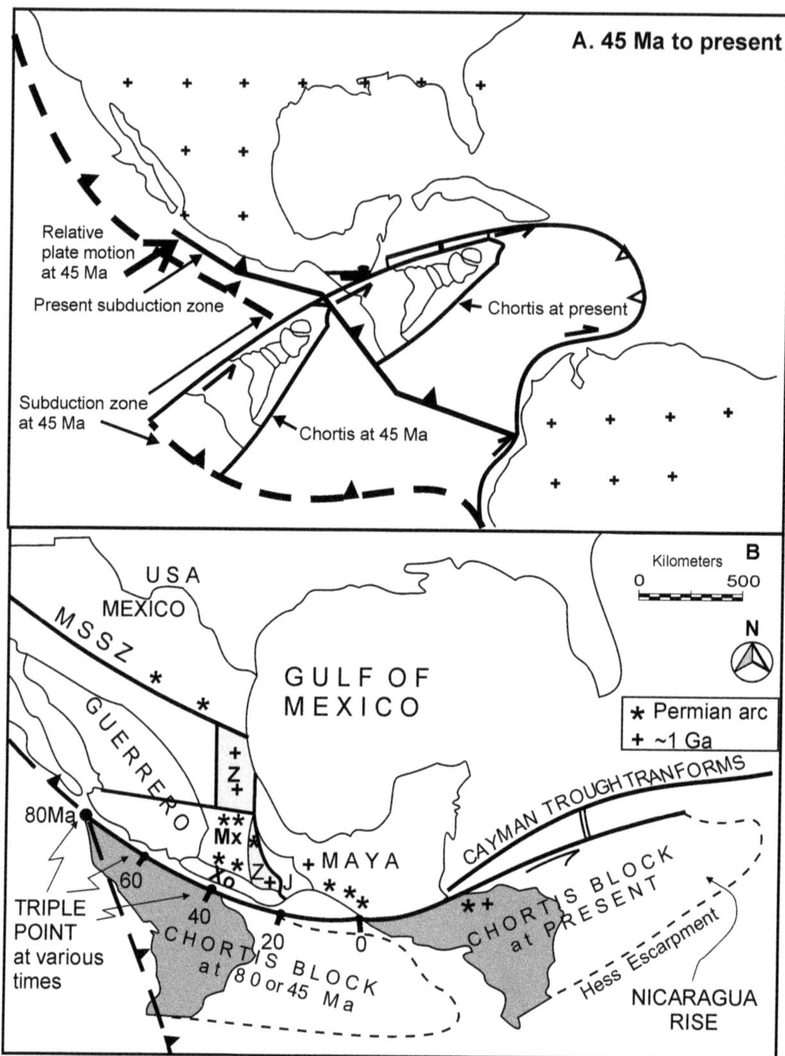

FIG. 2. Cenozoic reconstructions of the Chortis block: (A) using the rotation pole near Santiago (Chile) determined by Ross and Scotese (1988) and Pindell et al. (1988), and ~1100 km sinistral offset on the Cayman transform faults (Rosencranz et al., 1988); (B) assuming a connection between the Cayman transform faults and the Acapulco Trench through the Motagua fault zone, and juxtaposing the Chortis block against southern Mexico (modified after Pindell et al., 1988; Ross and Scotese, 1988; Schaaf et al., 1995; Meschede and Frisch, 2002; Guinta et al., 2002). Abbreviations: J = Juarez terrane; MSSZ = Mohave-Sonora shear zone; Mx = Mixteco terrane; TMVB = Trans-Mexican Volcanic Belt axis; Xo = Xolapa terrane, Z = Zapotecan terrane.

pole of rotation to be located near Santiago, Chile. Using these constraints places the Chortis block ~1100 km WSW of its present position in the Early Eocene (Fig. 2A).

On the other hand, most authors have connected the Cayman transforms (convex northward) with the E-trending Motagua fault zone and the WNW-trending Acapulco Trench (convex southward; Fig. 1), and infer that it represents the Caribbean–North American plate boundary (e.g., Pindell et al., 1988; Ross and Scotese, 1988; Meschede and Frisch, 2002). Applying ~1100 km displacement along this boundary places the Chortis block off southern Mexico in the Eocene (Fig. 2B). On the other hand, some authors believe that the sinistral displacement on the Motagua fault zone is only ~170 km, and place the Chortis block south of the Maya block in the Eocene (e.g., Donnelly et al., 1990; Guinta et al., 2002). We first examine the Motagua fault zone–Acapulco Trench correlation and the inference that the Motagua fault zone represents the transform boundary between the North American and Caribbean plates.

Can the Motagua Fault Zone Be Connected with the Acapulco Trench, and Is It the Transform Fault between the North American and Caribbean Plates?

Most workers have assumed that the Motagua fault zone joins the Acapulco Trench at a T-T-F triple point and that the Motagua fault zone is the sinistral, transform boundary between the North American and Caribbean plates (e.g., Anderson and Schmidt, 1983; Pindell et al., 1988; Ross and Scotese, 1988; Guzmán-Speciale et al., 1989; Velez-Scholvink, 1990; Vazquez-Meneses et al., 1992; Schaaf et al., 1995; Barrier et al., 1998; Meschede and Frisch, 2002). In this context, the southeastward migration of the Tertiary magmatism in the Sierra Madre del Sur has been interpreted in terms of the passage of the T-T-T triple junction that accompanied the southeasward displacement of the Chortis block as part of the Caribbean plate (Herrmann et al., 1994; Schaaf et al., 1995). However, such an interpretation is at odds with several important observations.

1. There is no geological or geophysical evidence of an E-W fault zone in the Pacific coastal plain and continental shelf along the projected trace of the Motagua fault zone. On the contrary, an undeformed Upper Cretaceous–Recent sedimentary and volcanic basin oversteps the fault trace (Fig. 3; Sanchez-Barreda, 1981): this is consistent with the flat-lying sediments predicted by gravity data (Couch and Woodcock, 1981). In the Pacific continental shelf, seismic reflection data show that the Polochic fault is overstepped by Upper Cretaceous–Quaternary sediments. However, there is a change of thickness of the Upper Cretaceous–Eocene sediments along the projected trace of the Polochic fault (Fig. 3; Sanchez-Barreda, 1981). Other seismic reflection data reveal that these continental shelf sediments are cut by WNW-trending normal faults (Sanchez-Barreda, 1981; Barrier et al., 1998). Furthermore, gravity and magnetic contours run parallel to the coast with no offset across the projected trace of the Motagua fault zone: the observed bend in the contours is probably related to the Tehuantepec Ridge (Fig. 1; de la Fuente-Duch et al., 1991; Hernandez-Quintero, 2002).

2. The Motagua fault zone consists of a series of parallel faults (from south to north): Jocotán-Chamelecon, Jubuco-Cabañas, and Polochic (Figs. 1 and 3). The Jocotán-Chamelecón fault along the southern side of the Motagua fault zone appears to have only Cretaceous dip-slip movement (Donnelly et al., 1968). ~20 km Cenozoic, sinistral offset of alluvial fans and feeder rivers across the Jubuco fault in the east decreases to zero towards the west (Muller, 1979; Johnson, 1984). The Jocotán-Chamelecón and Jubuco-Cabañas faults are overstepped by the Upper Miocene–Quaternary volcanic rocks in western Guatemala (Donnelly et al., 1990). Estimates of Cenozoic displacement on the Cabañas fault vary from very limited to ~20 km sinistral offset (McBirney, 1963; McBirney et al., 1967; Lawrence, 1975). Estimates of Cenozoic displacement on the Polochic fault in Guatemala and adjacent Mexico have varied from zero (Bonis, 1967) through a few kilometers (Anderson et al., 1985) to ~130 km sinistral offset (Burkart et al., 1987): the latter based mainly upon offset map boundaries, which give apparent rather than true displacements. On the other hand, in Mexico, the Polochic fault cuts some Tertiary–Quaternary plutons (Burkart et al., 1987), and the Permian Chiapas batholith, the axis of which shows little or no offset, and the coastal plain sediments are unaffected by the fault (Muehlberger, 1992). The Middle America Trench shows no offset along the projected trace of the Polochic fault (Fig. 3): the main bend in the Middle America Trench occurs at its intersection with the Tehuantepec Ridge.

3. Earthquake fault plane solutions on shallow earthquakes (<70 km depth) show a gradual change from east to west: sinistral, transcurrent motions along the Motagua fault zone in central-eastern Guatemala and adjacent Caribbean, through normal movements in western Guatemala, and adjacent Mexico and Pacific continental shelf, to reverse motions and down-dip extension along the subduction zone (Guzmán-Speziale et al., 1989; Barrier et al., 1998). Thus, these observations are consistent with field observations along the various faults.

4. Structural data in southern Mexico between the Gulf of Tehuantepec and Puerto Vallarta indicate a complex stress regime during the Cenozoic:

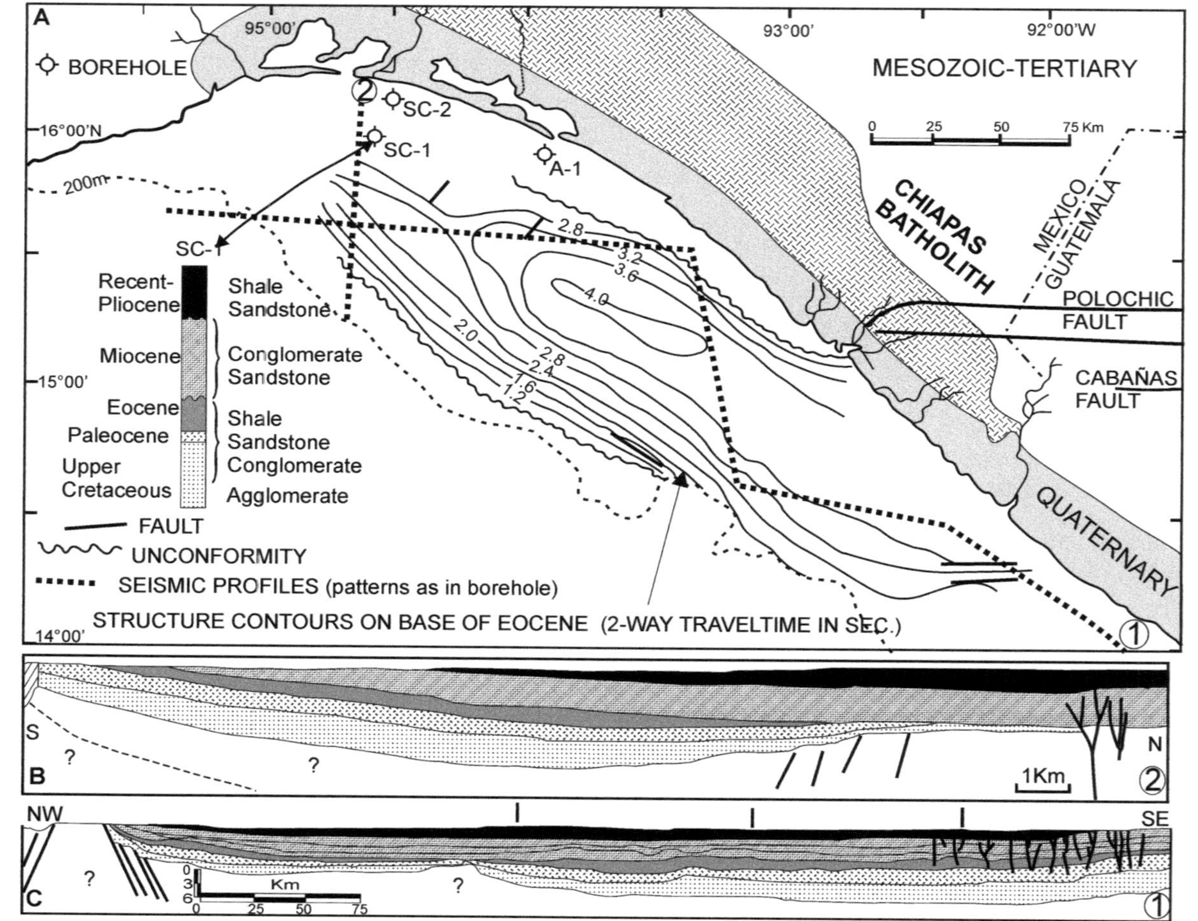

FIG. 3. A. Map of the Gulf of Tehuantepec and neighboring coastal area showing structure contours (two-way travel time in seconds) on the base of the Eocene and log of drill core (after Sánchez-Barreda, 1981). Note the absence of fault displacement across the projected trace of the Motagua fault zone; B and C. Reflection seismic sections located on (A) interpreted by Sanchez-Barreda (1981).

(1) sinistral movement on NW-trending faults in a NNE-compressional/WNW extensional regime during the Late Eocene; (2) dextral movement on NW-trending faults during the Early Oligocene in an ENE-extensional/WNW-compressional regime in the north-central Sierra Madre del Sur (Alaniz-Álvarez et al., 2002); (3) sinistral movement on E-trending faults in the Late Oligocene; (4) normal faulting on E-W graben-bounding faults in the Early–Middle Miocene (Morán-Zenteno et al., 1999); and (5) Recent earthquake fault plane solutions of shallow earthquakes show only normal and inverse kinematics, although rivers indicate a cumulative sinistral offset across five faults of ~11 km (Tolson-Jones, 1998). Earthquake data are inferred by Ego and Ansan (2002) to be the result of the slightly oblique convergence between the Cocos-Rivera and North American plates, which is partitioned between a trench-normal component and a sinistral-extensional component along the Trans-Mexican Volcanic Belt and the accretionary wedge. The oblique *extensional* episodes are not in accord with the sinistral, but *convergent* deformation required on the Motagua fault zone–Acapulco Trench if it forms the northern boundary of the Caribbean plate using the rotation pole near Santiago (Chile).

5. The eastward migration of the Cocos–North America–Caribbean triple point as it trails behind the Chortis block would lead to an eastward elongation of the southern Mexican magmatic arc, but Late Cretaceous arc magmatism has been reported from Chiapas north of the Motagua fault zone (68 ± 3 Ma: K-Ar age on K-feldspar; Burkart et al., 1987) and from the Gulf of Tehuantepec (Turonian–Santonian, basalt, dacite, and tonalitic agglomerate; Sanchez-Barreda, 1981; Fig. 3). These data imply the existence of a trench to the south of the Gulf of Tehuantepec and western Guatemala during the Late Cretaceous, a conclusion incompatible with placing the Chortis block adjacent to southern Mexico.

6. No counterparts have been found in the Chortis block for two features occurring in the Xolapa Complex of southern Mexico (Fig. 1): (a) the southward-thickening Jurassic–Cretaceous sedimentary rocks (Ortega-Gutiérrez and Elias-Herrera, 2003); and (b) the southward increase in Early Cretaceous (~132 Ma; Herrmann et al., 1994) high-temperature/low-pressure metamorphism (Ortega-Gutiérrez and Elias-Herrera, 2003). Instead unmetamorphosed Jurassic and Cretaceous rocks (El Plan Formation and Honduras and Valle de Angeles groups; Horne et al., 1990) occur in the north-central Chortis block.

7. None of the N-S terrane boundaries recognized in southern Mexico (Guerrero-Mixteco-Zapoteco-Juarez-Maya; Campa and Coney, 1983; Sedlock et al., 1993; Keppie, 2004) that are truncated by the coast of Mexico have been recorded in the Chortis block (Donnelly et al., 1990; Figs. 1 and 2). On the other hand, correlatives of the Precambrian granulite-facies Oaxacan Complex of southern Mexico (Keppie et al., 2003) have been reported in the Chortis block as ~1 Ga, amphibolite-facies gneisses (Manton, 1996; Fig. 1). Also a Permo-Carboniferous tectonomagmatic event has been recorded in the basement rocks of northern Honduras (305 ± 12 Ma: Rb-Sr isochron on orthogneisses: Horne et al., 1990), and may be correlated with a similar event in the Acatlán Complex of southern México (Fig. 1).

Thus, there would appear to be very little data to support either tracing the Motagua fault zone into the Acapulco Trench or locating the Chortis block adjacent to southern Mexico in the Early Cenozoic. Recognition of this problem led some workers to suggest other solutions. For example, Malfait and Dinkleman (1972) and Plafker (1976) suggested that the western corner of the Caribbean plate was pinned by compression between the Cocos and North American plates; however, shallow earthquake data indicate oblique extension, not compression. Vasquez-Meneses et al. (1992) proposed a wide transform zone, the northern margin of which passes through the northern margin of the Gulf of Tehuantepec, whereas Guzmán-Speciale and Meneses-Rocha (2000) inferred that the Chiapas fold-and-thrust belt represents a transfer zone between the Motagua fault zone and some unspecified fault to the north that is overstepped by Upper Miocene strata of the coastal plain. However, both of these proposals are still subject to many of the problems outlined above.

Thus an alternative solution is proposed: that the transform boundary between the North American and Caribbean plates identified in the Cayman Trough extends west-southwestward beneath the Pacific Volcanic Chain of Guatemala. Although its surface trace is largely obscured by the voluminous arc volcanic rocks, the positive Bouguer anomaly in the continental shelf is abruptly terminated just west of longitude 90°W and a positive Bouguer anomaly may be traced across Guatemala to join the western end of the Cayman tranform fault (de la

Fuente-Duch et al., 1991). Application of a purely transform motion on the Cayman transform faults throughout the Cenozoic as required by the plate tectonic paradigm results in the arrangement shown in Figure 2A. Also note that the transform offset in the trench and arc would be progressively eliminated, and the lithosphere west of the transform would be subducted, leaving very little record of the ~1100 km offset except to the north of the Central American Volcanic Arc. In such a scenario, the southern margin of Mexico would have bordered on the Pacific Ocean throughout the Cenozoic, and this necessitates a re-examination of the onshore Cenozoic record in the light of Pacific tectonics.

Flat-Slab Subduction: Possible Andean Analogues for Southern Mexico

An examination of the western margin of South and Middle America reveals that there are three principal areas where flat-slab subduction is presently active, each connected with the overriding of one or two ridges: Pampean flat-slab related to the Juan Fernandez ridge, Peruvian flat-slab related to the Nazca and Inca ridges, and by analogy, the Mexican flat-slab related to the Tehuantepec ridge (Fig. 4A). All of the South American ridges are aseismic oceanic island chains produced above a hotspot (Gutscher et al., 2000). Of the still extant flat-slab segments, the Mexican and Pampean ones are most similar because the subduction zone flattens asymmetrically (i.e., dip direction of the subduction zone is oblique to the trench). In southern Mexico, the present dip of the subduction zone varies from ~30°N near its western end through ~25°N to subhorizontal as one approaches the Tehuantepec Ridge, reverting east of the Tehuantepec Ridge to ~30°N. As a consequence the distance between the trench and the arc along the Trans-Mexican Volcanic Belt varies from ~180 km in the west to ~400 km in the east (Pardo and Suarez, 1995): this is similar to that seen in the Pampean segment. In both the Mexican and Pampean regions, the directions of convergence are only slightly oblique to the aseismic ridges (Fig. 4A), which would imply slow westward and southward motions of the ridge-trench intersection, respectively.

On the other hand, the Inca Ridge/plateau has been completely subducted beneath the Andean margin, and its former existence is inferred to be the mirror image of the Marquesas Plateau, presently located in the Pacific Ocean (Gutscher et al., 1999). A similar seamount chain, called the Moonless Mountains, is present between the Molokai and Murray fracture zones in the Pacific Ocean off Mexico and the United States, and it appears to continue into an unmamed seamount chain between the Molokai and Clarion fracture zones (Atwater, 1989; Muehlberger, 1992). Its inferred subducted mirror image is here called the Chumbia Seamount Ridge (named after an extinct language of southern Mexico). In order to evaluate the geological consequences of subducting an aseismic ridge, we first summarize selected South American examples, and then examine the Cenozoic record of southern Mexico for potential analogues.

In the central Andes, where the Benioff zone has a normal dip of 30–35° the magmatic arc is relatively narrow (~50–100 km wide), with the magmatic front occurring where the Benioff zone reaches a depth of ~80–100 km. The nature of the volcanism changes from calc-alkaline to alkaline with distance from the trench, and crustal contamination increases with time as the crust is thickened (Kay et al., 1991). However, where the dip of the Benioff zone decreases, the volcanic zone widens to >200 km, and with time it migrates inland from ~180 to 700 km from the trench, eventually dying out to produce a gap in the magmatic arc (Ramos et al., 2002). The extinction of the arc has been related to retreat of the asthenospheric wedge as it is replaced by juxtaposition of the subducting and overriding lithosphere (Ramos et al., 2002).

Collision of the Nazca and Juan Fernandez ridges with the trench has also resulted in extensive subduction erosion of the forearc, which has been uplifted ~1 km followed by extensional subsidence (Jaillard et al., 2000; Ramos, 2000). Progressive coupling between the two plates appears to have caused a migration of the deformation front from 275 km to 360 km from the trench through the arc into the foreland basin (Ramos et al., 2002). This was associated with the migration of alluvial fans. The geometry of the Andes across the Pampean flat-slab at 31°S indicates a triangle zone with opposing thrust vergence: eastward in the west and westward in the east (Fig. 4C).

Cenozoic Features of Southern Mexico

Volcanic arcs and gaps

Following accretion of the Guerrero island arc complexes during the Laramide orogeny, a continental volcanic arc was established on the mainland

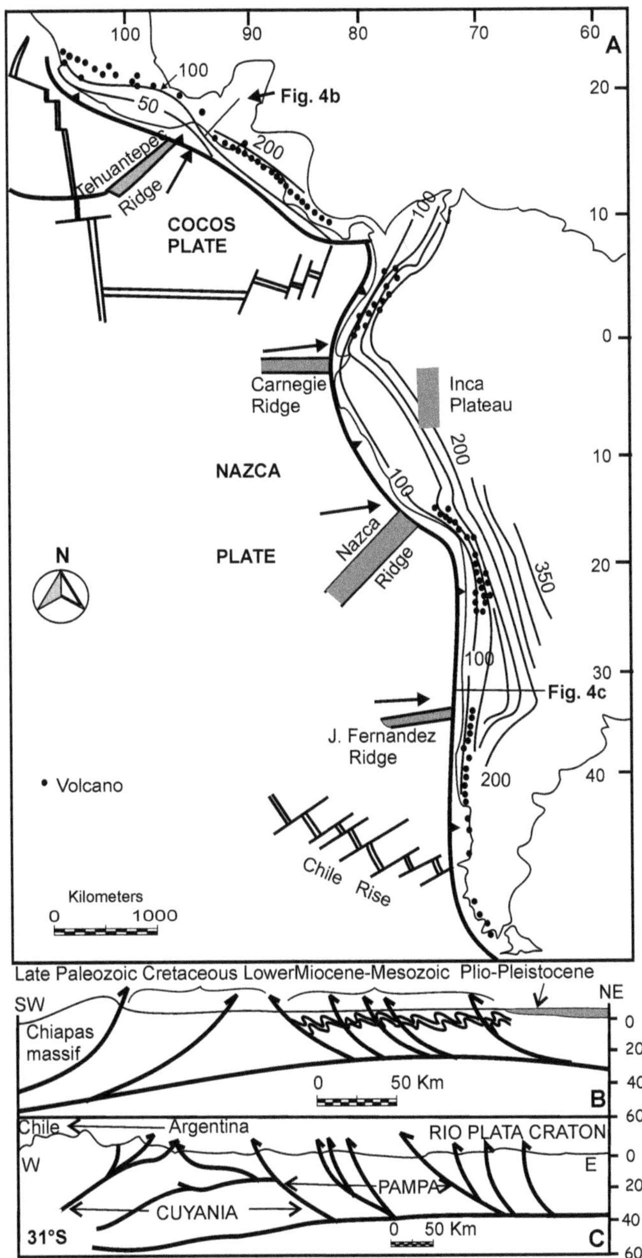

FIG. 4. A. Map of Middle and South America showing the association of mid-ocean and aseismic ridges with areas of flat-slab subduction and volcanic arc gaps (modified after Ramos and Aleman, 2000). B. Cross-section through Central America. C. Cross-section through South America.

of southwestern Mexico, which, after crossing an apparent gap in Chiapas where volcanism is relatively rare, continues in the Central American Volcanic Arc (Fig. 1). This volcanic arc has been related to subduction of the Farallon plate and its daughters (e.g., Carr and Stoiber, 1990; Ferrari et al., 1999; Morán-Zenteno et al., 1999). However, this arc has migrated through time (Figs. 1 and 5):

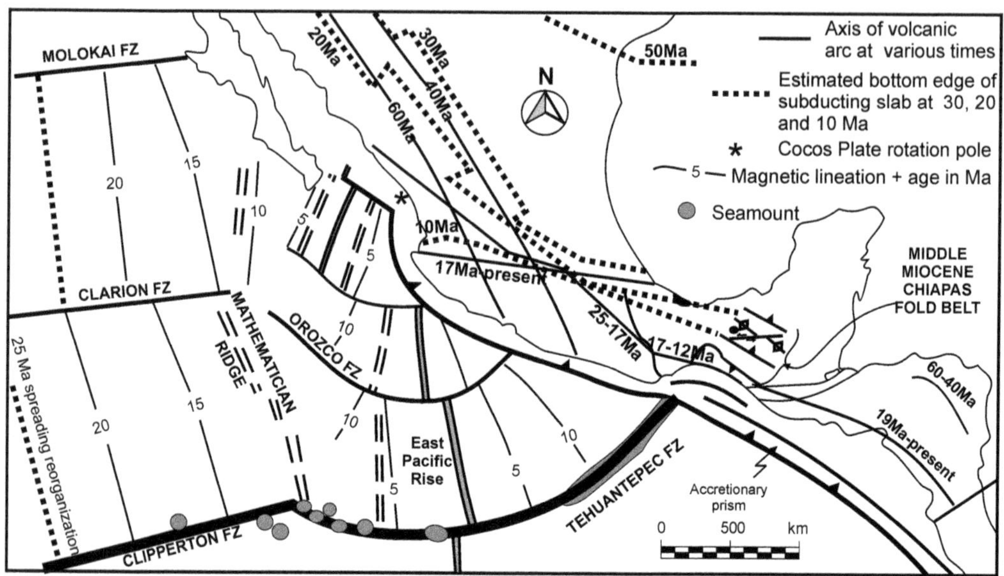

FIG. 5. Map of Middle America and adjacent Pacific Ocean at the present time showing axes of volcanic arc at various times (after Ferrari et al., 1999), estimated bottom edge of the subducted slab at 30, 20, and 10 Ma (after Severinghaus and Atwater, 1989), the East Pacific Rise, various fracture zones abandoned mid-ocean ridges and sea floor anomalies (after Mammerikx and Klitgord, 1982).

(1) in the Oligocene, the arc rotated ~30° anticlockwise, with the southeastern end moving eastward from a position west of Acapulco in the Paleocene and Eocene through Acapulco-Oaxaca in the Oligocene to east of Oaxaca in the Early Miocene (Morán-Zenteno et al., 1999); and (2) in the Middle Miocene, after an hiatus of ~10 m.y., it moved to its present position along the Trans-Mexican Volcanic Belt, migrating slightly southward in the late Neogene and Quaternary (Ferrari et al., 1999).

Also the width and arc-trench gap appear to have changed through time: subduction erosion along the southern coast of Mexico during the Oligocene (Morán-Zenteno et al., 1996) makes estimates of the arc-trench gap unreliable. Thus, in the latest Cretaceous (Maastrichtian: ~70–65 Ma), the arc was ~50–100 km wide and lay ~150–200 km from the trench: arc-backarc remnants of this arc in drill core near Salina Cruz (Fig. 3: Late Cretaceous porphyritic basalt, dacite, and tonalitic agglomerate; Sanchez-Barreda, 1981), and adjacent to the Polochic fault (Fig. 1: 68 ± 3 Ma: K-Ar on microcline perthite; Burkart et al., 1987). During the Paleocene–Eocene (~65–35 Ma) the arc widened to ~300 km wide, widening further to ~350 km in the Oligocene (~35–23 Ma; Fig. 1). In the Middle Miocene (~16–11 Ma), the width of the arc appears to have decreased to a uniform ~200 km, despite the fact that the arc-trench gap increases from ~130 km in the west to ~350 km at the eastern end of the Trans-Mexican Volcanic Belt (Fig. 1; Ferrari et al., 1999). Seismic tomography (Rogers et al., 2002), and the lack of subduction-related earthquakes farther north of the Trans-Mexican Volcanic Belt (Pardo and Súarez, 1995), suggest that the bottom edge of the subducted slab lies close to the present northern edge of the Trans-Mexican Volcanic Belt (Fig. 6), which would limit the northern extent and width of arc volcanism. Using relationships between subduction rate, plate age at the trench, and slab duration, Severinghaus and Atwater (1990) calculated that the bottom edge of the subducted slab at ~10 Ma also lay along the Trans-Mexican Volcanic Belt, and at ~20 and ~30 Ma it also lay along the eastern half of the Trans-Mexican Volcanic Belt (Fig. 5).

Such a location for the bottom edge of the slab beneath the Trans-Mexican Volcanic Belt in the Middle Miocene–Recent may provide an explanation for the contemporaneous eruption of calc-alkaline rocks derived from the mantle wedge, adakites produced by melting the subducting slab (Gómez-

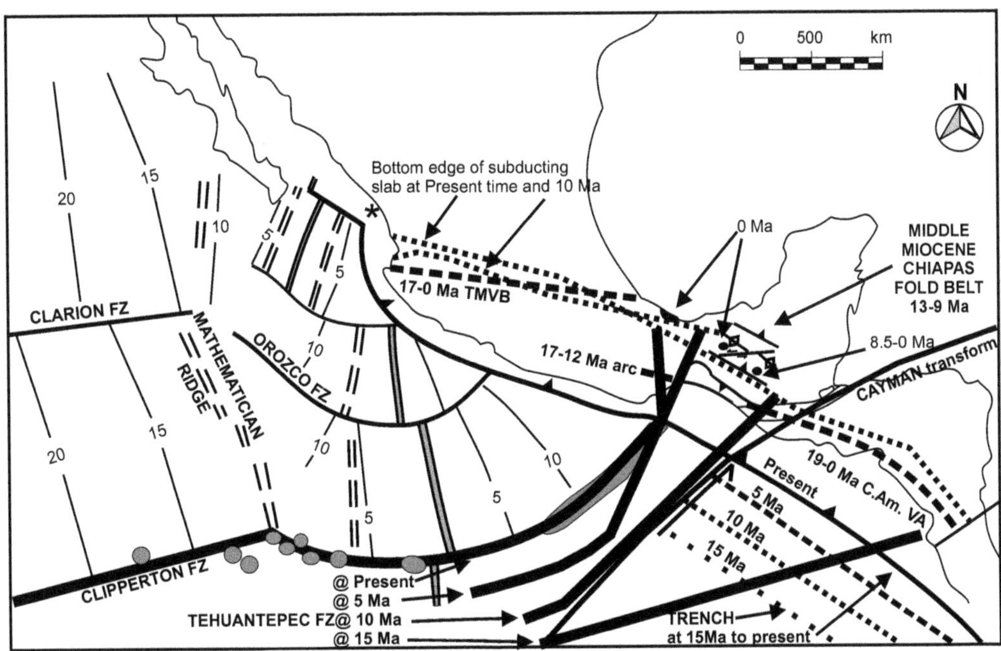

FIG. 6. Reconstruction of the location of the trench off the Chortis block and the position of the Tehuantepec Ridge at 15 Ma, 10 Ma, 5 Ma, and present times.

Tuena et al., 2002; Martinez-Serrano et al., 2004), and within-plate oceanic-island basalts from the sub-slab asthenosphere and/or enriched mantle wedge. Maury et al. (2003) have suggested that adakites can be produced by slab melting during the transition from steep to flat subduction. This would obviate the need for invoking either a plume (Marquez et al., 1999), or a continental rift (Verma, 2002) to explain the oceanic-island volcanism. Seismic tomography suggests that a fragment of the subducted Farallon plate is present north of the Trans-Mexican Volcanic Belt at a depth of ~400–650 km, and a low S-wave velocity layer between this and the presently subducting slab corresponds to a slab window (van der Lee and Nolet, 1997; Rogers et al., 2002). Ferrari (2004) has proposed that the Farallon fragment was detached in the Late Miocene (11.6–6 Ma). However, the estimated southern edge of the slab window at ~30 Ma roughly coincides with the eastern Trans-Mexican Volcanic Belt, suggesting that the detachment took place during or before earliest Oligocene (Fig. 5). Rogers et al. (2002) have suggested that extrusion of widespread Middle Miocene ignimbrites immediately preceded slab detachment, which might provide a possible analog for the Lower Oligocene ignimbrites in the Sierra Madre Occidental of Mexico (McDowell and Clabaugh, 1979). The Central American Volcanic Arc appears to have migrated toward the trench and remained approximately parallel to the trench throughout the Cenozoic (Carr and Stoiber, 1990).

The Trans-Mexican and the Central American volcanic arcs are separated in Chiapas by an Eocene–Late Miocene volcanic arc gap (Fig. 1). Cenozoic magmatism in the Chiapas gap is limited to a few isolated volcanoes: the Pliocene and Quaternary (~8.5–0 Ma): 7–0 Ma, alkaline, Los Tuxtlas volcanic field, 1.1–0 Ma, alkaline, El Chichon, 8.5–0 Ma, calc-alkaline–alkaline, Chiapanecan volcanoes; and 0.04-0 Ma, calc-alkaline, Tacana volcano (Mora et al., 2004a; Garcia-Paloma et al., 2004). These volcanoes lie near the northern edge of the subducting Cocos slab (Fig. 6; Rogers et al., 2002).

Chiapas and the Gulf of Tehuantepec

In Mexico, reflection seismic data in the Gulf of Tehuantepec show two depositional hiatii: (1) Upper Cretaceous–Eocene sediments are unconformably overlain by Lower Miocene volcanoclastic conglomerate; and (2) Lower Miocene sediments are unconformably overlain by Upper Miocene–Recent sediments (Fig. 3; Sanchez-Bareda, 1981). The earlier exhumation event may also be recorded in the southernmost Oaxacan Complex near Puerto Angel, which is bracketed between a latest Eocene, ~36 Ma K-feldspar cooling age in the Oaxacan Complex (Schulze et al., 2004) and unconformably overlying late Lower Miocene rocks dated at 17.5 Ma (Morán-Zenteno et al., 1999). The earlier hiatus is also present to the west between the Acapulco Trench and the southern coast of Mexico where Middle Miocene–Recent sediments lie unconformably upon ~32 Ma, earliest Oligocene, calc-alkaline, dioritic basement (Morán-Zenteno et al., 1996). The Oligocene hiatus is synchronous with subduction erosion of the southwestern coast of Mexico during the Late Oligocene–Middle Miocene (Moran-Zenteno et al., 1996).

The second hiatus is also recorded in the Chiapas foldbelt, where the following sequence has been recorded: (1) an almost continuous sequence with minor discontinuities from latest Cretaceous to early Middle Miocene (12-15 ± 1 Ma: K-Ar ages on biotite; Ferrusquía-Villafranca, 1996); (2) deformation of these rocks by NW-WNW–trending folds associated with SW- and NE-vergent thrusts, E-trending dextral faults, N-trending sinistral faults, and NE-trending normal faults (De la Rosa et al., 1989), all attributable to NE-SW to NNE-SSW compression: this is synchronous with deformation in the southeastern Gulf of Mexico that produced sinistral positive flower structure, whose development is tightly bracketed between in the late Middle Miocene (Pacheco-Gutiérrez, 2002); (3) these structures are unconformably overlain by Upper Miocene–Quaternary sediments and volcanic rocks (8.5–0 Ma; Mora et al., 2004b; Garcia-Paloma et al., 2004); and (4) episodic normal faulting on the easterly dipping N-S Isthmus fault zone has taken place since the Late Miocene (Barrier et al., 1998). The overall geometry is similar to the Andes at 31°S, and indicates a triangle zone across which thrust vergence changes (Fig. 4B). The Chiapas foldbelt coincides with a gap in the Neogene magmatic arc between the eastern end of the Trans-Mexican Volcanic Belt and the Central American Volcanic Arc (Fig. 1).

Thus, in summary, there appear to be depositional hiatii in southern Mexico: (1) an Oligocene hiatus represented by exhumation and subduction erosion; and (2) a latest Middle Miocene hiatus expressed as exhumation and development of a fold-and-thrust belt.

Tehuantepec-Clipperton Ridge and associated seamounts in the Pacific Ocean

Off southern Mexico, the Acapulco trench represents the region where the Cocos Plate containing the East Pacific Rise and a number of fracture zones and ridges is presently subducting (Fig. 5). Farther north between the R-T-F triple point (East Pacific Rise–Acapulco Trench–Gulf of California transform-ridge) and the F-F-T triple point (Mendocino–San Andreas–Juan de Fuca) subduction has ceased, and the Farallon plate has been completely subducted.

The Tehuantepec Ridge lies within the Cocos plate and approximately coincides with the Tehuantepec fracture zone, a transform offset of the East Pacific Rise (Muehlberger, 1992). A line of seamounts occurs along the Tehuantepec Ridge and is intersected by the Middle American Trench at Anomaly 5A-B (= ~12.5–15 Ma; Mammerickx and Klitgord, 1982), which coincides with a sharp bend in the ridge. Near the trench, the oceanic crustal depth across the Tehuantepec fracture zone reflects the ~22 Ma difference in age of the oceanic crust (older to the southeast), a difference that amounts to ~700 meters at the Middle America trench (Sclater et al., 1971). Local seamounts along the Tehuantepec Ridge reach elevations >1000–2000 meters above the surrounding seafloor (Addicott, 1984; Mammerickx, 1989). West of the East Pacific Rise, the Tehuantepec fracture zone continues into the Clipperton facture zone, which is coincident with seamounts that can be traced back at least to Anomaly 20 (= ~45 Ma): some of the Clipperton seamounts lie south of the fracture zone (Addicott, 1984; Mammerickx, 1989). The subducted portion of the Tehuantepec Ridge may be a mirror image of the Clipperton fracture zone. Its present location beneath Mexico and the Pacific continental shelf is probably indicated by: (1) the N-S line of intermediate-depth earthquakes located near the top of the subducting slab (Ponce et al., 1992; Pardo and Suarez, 1995); (2) a gravity low (de la Fuente et al., 1991); (3) a low S-wave velocity zone crossing the Isthmus of Tehuantepec that elsewhere is characteristic of extension (Goes and van der Lee, 2002); and (4) a change in the dip of the subduction zone from shallow to steep (Pardo and Suarez, 1995). The northern limit of the subducted Tehuantepec Ridge probably coincides with the northern edge of the subducted slab, which, based on seismic tomography (Rogers et al., 2002), lies ~50 km south of the Gulf of Mexico (Fig. 6). This coincides with the northern edge of Quaternary volcanoes in the Trans-Mexican Volcanic Belt where deep earthquakes associated with the subduction terminate (Fig. 6; Pardo and Súarez, 1995).

Moonless Mountains seamounts and their unnamed southeasterly continuation

In the Pacific Ocean between the Clarion and Molokai fracture zones at longitude 120–127°W, an unnamed NW-trending line of seamounts locally reaches 2000–2500 meters above the general seafloor (Addicott, 1984; Mammerickx, 1989). North of the Molokai fracture zone, this seamount chain appears to continue northwestward into the Moonless Mountains seamount ridge near the Murray fracture zone (Addicott, 1984; Mammerickx, 1989). These seamounts occur in oceanic crust ranging in age from ~25 to 50 Ma (Atwater, 1989; Muehlberger, 1992), and they are parallel to the Hawaiian Ridge, suggesting that the Moonless-unnamed seamounts formed in a similar way—i.e., above a plume. Inasmuch as the oldest seamounts occur in ~50 Ma oceanic lithosphere, it is inferred that they were formed by a plume located on the East Pacific Rise at that time. By analogy with the Marquesas Plateau seamounts and their mirror image Inca Plateau (Gutscher et al., 1999), we assume that there was a mirror image to the Moonless Mountains/unnamed ridge, which would have been oriented NE-SW and located in the Farallon Plate: it is here named the Chumbia seamount ridge. The ~25–50 Ma part of the Farallon lithosphere has already been subducted beneath the western margin of Mexico and/or United States.

Comparison of Andean and Mexican flat-slab segments

Some of the Cenozoic features recorded in southern Mexico are similar to those in the Pampean flat-slab segment (Figs. 1 and 4A): (1) the gradual change of dip of the subduction zone—in Mexico, the arc presently lies ~20° anticlockwise of the trench; (2) an anticlockwise swing in the arc occurs in both regions—in Mexico, this rotation appears to have taken place in the Oligocene and in the Miocene; (3) the presence of a volcanic arc gap—in Mexico, there is a Neogene gap in the Chiapas region; (4) a fold-and-thrust belt in the magmatic arc gap; and (5) exhumation and subduction erosion in the forearc region. However, several features appear to be unique to Mexico: (a) the width of the Trans-Mexican Volcanic Belt only reaches ~150 km, probably due to the termination of the slab near the present northern margin of the Trans-Mexican Volcanic Belt (Figs. 1, 5, and 6); (b) arc magmatism returned to the Chiapas gap during the Pliocene and Quaternary (Figs. 1 and 5); and (c) deformation appears to have been synchronous and limited to the Middle Miocene in the Chiapas foldbelt, whereas it was diachronous in the Pampean region (Ramos et al., 2002). In order to explore the reasons for these observations, we now present some reconstructions.

Reconstructions

Our Cenozoic reconstructions use the plate motions deduced by previous workers (Mammerickx and Klitgord, 1982; Engebretson et al., 1985; Atwater, 1989; Meschede et al., 2002). These studies have shown that main spreading reorganizations occurred at ~25 Ma (Late Oligocene) and at ~12.5 Ma (Middle Miocene) of the East Pacific Rise, and two jumps of the Cocos Ridge at 19.5 Ma and 14.7 Ma (early–Late Eocene and Middle Miocene, respectively. An ENE convergence rate across the Acapulco Trench decreased from a maximum of ~170 mm/yr in the Middle Eocene to a presently NNE-directed rate of convergence that varies from ~23 mm/yr in the west to ~64 mm/yr in the east, with a subsidiary peak of ~100 mm/yr at ~17 Ma (Schaaf et al., 1995; Ego and Ansan, 2002). Furthermore, for the Pacific plate, an increase in the rate of motion occurred at ~12.5 Ma and a change of direction took place at ~8 Ma from WNW to NNW (Atwater and Stock, 1998).

Acapulco Trench

Although the relative motions between North American and Farallon/Cocos plates during the Cenozoic are relatively well constrained (Mammerickx and Klitgord, 1982; Engebretson et al., 1985; Atwater, 1989; Atwater and Stock, 1998; Meschede et al., 2002), subduction erosion along the south coast of Mexico makes it difficult to locate the Acapulco Trench throughout the Cenozoic. However, its location in the Eocene (Fig. 6) may be estimated using the following assumptions: (1) the back side of the latest Cretaceous–Middle Eocene magmatic arc is to be found in the ~68 Ma plutons just north of the Polochic fault in western Guatemala (Burkart et al., 1987), and in the drill core from the Gulf of Tehuantepec (Sanchez-Barreda, 1981); and (2) the back side of the magmatic arc is located ~300 km from the trench if the dip of the Benioff zone was normal

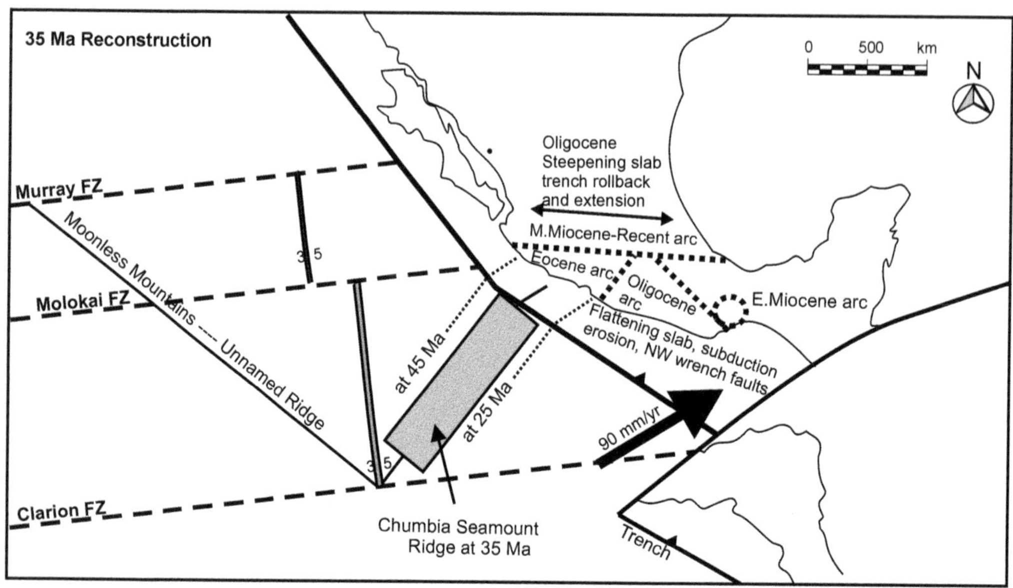

FIG. 7. Reconstruction for 35 Ma showing the locations of the mid-ocean ridge, the Molokai-Shirley and Clarion fracture zones, and the Moonless–unnamed seamount ridge and its inferred mirror image, the Chumbia seamount provinces.

for western America (~30–35°; Ramos and Aleman, 2000). This would place the Acapulco Trench ~250 km south of its present location east of Acapulco during the Eocene and latest Cretaceous. We assume that during the Oligocene, subduction erosion caused migration of the trench to its present location.

Middle America Trench

The position of the Middle America Trench off the Chortis block through the Cenozoic is extrapolated by assuming a uniform rate of displacement on the Cayman Transform faults throughout the total sinistral offset of ~1100 km (Rosencranz et al., 1988).

15 Ma to Present reconstruction and the Tehuantepec Ridge (Fig. 6)

Using the relative plate outlined above, we estimate that at ~15 Ma, the Clipperton-Tehuantepec Ridge was located ~700 km south of its present position (Fig. 6). Between ~15 Ma and the present, northward motion of this ridge was accompanied by anticlockwise rotation of the Tehuantepec Ridge on the Cocos plate about a pole located near the mouth of the Gulf of California (Fig. 6), culminating in a N-S orientation of the ridge across the Isthmus of Tehuantepec. Prior to ~10 Ma, the Tehuantepec Ridge was subducting beneath the Chortis block; however, the asymmetric shallowing of the slab to the west of it caused anticlockwise rotation of the eastern end of the Trans-Mexican Volcanic Arc (Ferrari et al., 1999) in a manner similar to that associated with the Juan Fernandez Ridge off Chile (Fig. 4A). At ~10 Ma, the Tehuantepec Ridge was located parallel to and just west of the projected Cayman transform fault and impinged upon the Acapulco Trench in Guatemala (Fig. 6). We infer that collision of the Tehuantepec Ridge with the Acapulco Trench caused an abrupt flattening of the subducting slab, producing a gap in the magmatic arc and the development of the Chiapas foldbelt. Between ~10 and 5 Ma, the Tehuantepec Ridge rotated anticlockwise across Chiapas and the Tehuantepec Ridge–Acapulco Trench intersection migrated westward across the Gulf of Tehuantepec. This westward migration probably explains the lack of across-strike diachronism in the deformation of the Chiapas foldbelt. Between ~5 Ma and the present, although the Tehuantepec Ridge continued to rotate anticlockwise, its point of intersection with the trench appears to have remained essentially stationary (Fig. 6). As the subducting slab steepened behind the rotating ridge, it led to the reestablish-

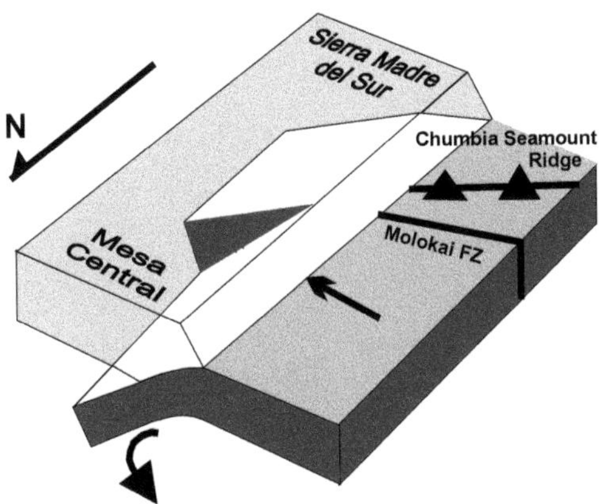

FIG. 8. Block diagram for Mexico during the Oligocene showing the subducting slab steepening beneath the Mesa Central and shallowing beneath the Sierra Madre del Sur, the latter a consequence of subducting the inferred Chumbia seamount ridge. Note that the Molokai fracture zone roughly coincides with the change in dip of the Benioff zone.

ment of the magmatic arc in the gap at 8.5 Ma in Chiapas, followed by its reappearance in the Tuxtlas volcanoes at 7 Ma (Mora et al., 2004b; Fig. 6). Furthermore, it induced normal faulting in the overriding slab, in which the east side of the N-S Isthmus fault zone was downthrown (Barrier et al., 1998).

35 Ma reconstruction and the Chumbia seamount ridge (Fig. 7)

A reconstruction at ~35 Ma (near the Eocene–Oligocene time boundary) places the portion of the Guadalupe plate (daughter of the Farallon plate) between the Clarion and Molokai fractures zones off southern Mexico (Fig. 7). Assuming that the Chumbia seamount ridge was the mirror image of the Moonless Mountains–unnamed seamount ridge, the Chumbia ridge at ~35 Ma would be entering the trench midway between Puerto Vallarta and Acapulco. Using the convergence directions and rate of convergence deduced by Engebretson et al. (1985), the Chumbia-trench intersection probably migrated from near Manzanillo at ~45 Ma to Acapulco at ~25 Ma (Fig. 7). This eastward migration is supported by the migration of the volcanic arc, suggesting a genetic connection—i.e., that subduction of the Chumbia ridge caused flattening of the slab and rotation of the arc. However, subduction erosion also took place during the Oligocene, suggesting that subduction of seamounts caused subduction erosion at the bottom of the overriding slab and exhumation at the top. Furthermore, intermittent coupling between the flattening subducting slab and the overriding slab may have caused thick-skinned deformation that was taken up by the strain partitioning and strike-slip deformation observed in southern Mexico (Alaniz-Álvarez et al., 2002). The reconstructed location of the Molokai fracture zone is roughly coincident with the change in inferred dip of the subduction zone from shallowing in the south to steepening to the north (Figs. 7 and 8; Nieto-Samaniego et al., 1999).

45 Ma reconstruction

The ~45 Ma reconstruction depicts a trench along the southern coast of Mexico. Its position is based upon the assumption that the rear of the arc during the Late Cretaceous and Paleocene is preserved in the Gulf of Tehuantepec (borehole SC-1; Fig. 3), and in the ~68 Ma arc plutons in western Guatemala north of the Motagua fault zone (Fig. 9). Assuming a normal 30° dip for the subduction zone places it ~250 km south of its present position and suggests a WNW trend for the trench swinging northwest from Puerto Vallarta. Such a location is consistent with collision of the northern part of the Chortis block with the Yucatan block in southern Mexico, which may have been responsible for the Late Cretaceous–Early Eocene obduction of ophiolitic lenses onto the southern margin of Mexico (Guinta et al., 2001).

Discussion

Our Cenozoic reconstructions suggest that collision of aseismic ridges with the Acapulco Trench caused migration of arc volcanism and deformation in southern Mexico. The Miocene collision of the Tehuantepec Ridge caused flattening of the subducting slab, anticlockwise rotation of the volcanic arc to its present location in the Trans-Mexican Volcanic Belt, produced a volcanic arc gap in Chiapas, the Middle Miocene Chiapas foldbelt, and exhumation in the Gulf of Tehuantepec. Earlier collision of the inferred Chumbia Ridge with the trench also caused flattening of the subducting slab, anticlockwise rotation of the arc during the Oligocene, and transpressional strike-slip deformation.

In the Andes, other factors besides collision of aseismic ridges with the trench appear to have been synchronous with deformational episodes—e.g., changes in convergence rate and direction (Jaillard et al., 2000; Ramos and Aleman, 2000). In the

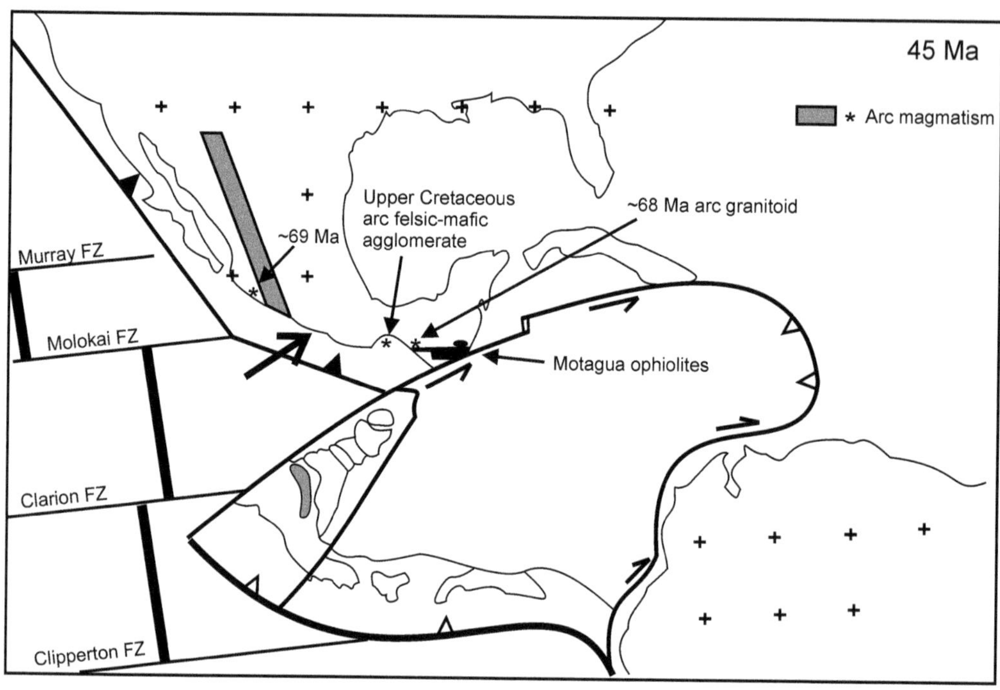

FIG. 9. 45 Ma reconstruction showing the locations of the mid-ocean ridge, various fracture zones, the volcanic arc, the Acapulco and Middle America trenches, and the Chortis block.

Pacific Ocean adjacent to Mexico, plate reorganizations took place at: (1) ~25 Ma when the Farallon plate broke into several smaller plates, including the Guadalupe and Cocos plates (Mammerickx and Klitgord, 1982; Meschede et al., 2002); (2) at 19.5 Ma the Cocos Ridge jumped, accompanied by a change in the spreading direction from NW-SE to NNW-SSE (Meschede et al., 2002); (3) at 14.7 Ma the Cocos Ridge jumped again, leading to a N-S spreading direction that is still active (Meschede et al., 2002); (4) at 12.5–11 Ma the Pacific-Cocos ridge started spreading on a much closer pole of rotation, producing a marked curvature in the Tehuantepec transform aseimic ridge compared to its earlier E-W trend (Fig. 1) (Mammerickx and Klitgord, 1982); and (5) at 6.5–3.5 Ma the present East Pacific Rise became the only active ridge between the Cocos and Pacific plates (Mammerickx and Klitgord, 1982). An ENE-directed rate of convergence across the Acapulco Trench decreased from a maximum of ~170 mm/yr in the Middle Eocene to a presently NNE-directed rate of convergence that varies from ~23 mm/yr in the west to ~64 mm/yr in the east, with a subsidiary peak of ~100 mm/yr at ~17 Ma (Schaaf et al., 1995; Ego and Ansan, 2002). Furthermore, there was an increase in the rate of motion in the Pacific Plate at ~12.5 Ma and a change of direction of motion at ~8 Ma from WNW to NNW (Atwater and Stock, 1998). Because the absolute motion of the North American plate is relatively constant throughout the Cenozoic (Engebretson et al., 1985), these changes of motion of the Pacific plates would seem to be the main factors to consider in the tectonics of southern Mexico. Of these, the Late Oligocene exhumation and subduction erosion and the development of the Chiapas foldbelt coincide with reorganizations of the East Pacific spreading ridge. Thus these reorganizations may also have contributed to events in the North American plate.

Acknowledgments

We acknowledge constructive reviews by Drs. J. B. Murphy and O. Campos-Enriquez. We are grateful for funds provided by a CONACyT grant (0255P-T9506), and a PAPIIT grant (IN116999) to JDK, and we thank C. Ortega-Obregon, M. Morales-Gómez, and L. A. Solari for help in preparation of the figures. This paper represents a contribution to IGCP project 453.

REFERENCES

Addicott, W. O., 1984, Geodynamic map of the Circum-Pacific region: Northeast quadrant: Scale 1:10,000,000. Tulsa, OK, American Association of Petroleum Geologists.

Alaniz-Álvarez, S. A., Nieto-Samaniego, A. F., Morán-Zenteno, D. J., and Alba-Aldave, L., 2002, Rhyolitic volcanism in extension zone associated with strike-slip tectonics in the Taxco region, southern Mexico: Journal of Volcanology and Geothermal Research, v. 2483, p. 1–14.

Anderson, T. H., Erdlac, R. J., and Sandstrom, M. A., 1985, Late-Cretaceous allochthons and post-Cretaceous strike-slip displacement along the Cuico-Chixoy-Polochic Fault, Guatemala: Tectonics, v. 4, p. 353–374.

Anderson, T. H., and Schmidt, V. A., 1983, The evolution of Middle America and the Gulf of Mexico–Caribbean Sea region during the Mesozoic: Geological Society of America Bulletin, v. 94, p. 941–966.

Atwater, T., 1989, Plate tectonic history of the northeast Pacific and western North America, *in* Winterer, E. L., Hussong, D. M., and Decker, R. W., eds., The eastern Pacific Ocean and Hawaii: Boulder, CO, Geological Society of America, Geology of North America, v. N, p. 21–72.

Atwater, T., and Stock, J., 1998, Pacific-North America plate tectonics of the Neogene southwestern United States: An update: International Geology Review, v. 40, p. 375–402.

Barrier, E., Velasquillo, L., Chavez, M., and Gaulon, R., 1998, Neotectonic evolution of the Isthmus of Tehuantepec (southeastern Mexico): Tectonophysics, v. 287, p. 77–96.

Bonis, S. B., 1967, Geologic reconnaissance of the Alta Verapaz fold belt, Guatemala: Unpubl. Ph.D. thesis, Louisiana State University, Baton Rouge, 146 p.

Burkart, B., Deaton, B. C., Dengo, C., and Moreno, G., 1987, Tectonic wedges and offset Laramide structures along the Polochic fault of Guatemala and Chiapas, Mexico: Reaffirmation of large Neogene displacement: Tectonics, v. 6, p. 411–422.

Campa, M. F., and Coney, P. J., 1983, Tectono-stratigraphic terranes and mineral resource distributions in Mexico: Canadian Journal of Earth Sciences, v. 20, p. 1040–1051.

Carr, M. J., and Stoiber, R. E., 1990, Volcanism, *in* Dengo, G., and Case, J. E. eds., The Caribbean region: Boulder, CO, Geological Society of America, Geology of North America, v. H, p. 375–392.

Couch, R., and Woodcock, S., 1981, Gravity and structure of the continental margins of southwestern Mexico and northwestern Guatemala: Journal of Geophysical Research, v. 86, p. 1829–1840.

de la Fuente-Duch, M., Mena-J., M., and Aiken, C. L. V., 1991, Cartas gravimetricas de la Republica Mexicana: 1. Carta de Anomalia de Bouguer: Instituto de Geofisica, Universidad Nacional Autonoma de Mexico, scale 1:3,000,000.

De la Rosa-Z., J. L., Eboli-M, P. A., Yamasaki,-M., F., and Ballinas-G., R., 1989, Geologia del Estado de Chiapas: Mexico, DF, Departamento de Geologia, Unidad de Estudios de Ingenieria Civil, Subdireccion de Construccion, 192 p.

DeMets, C., Gordon, R. G., Argus, D. F., and Stein, S., 1990, Current plate motions: Geophysical Journal International, v. 101, p. 425–478.

Donnelly, T. W., Crane, D., and Burkart, B., 1968, Geologic history of the landward extension of the Bartlett Trough—preliminary notes: Transactions 4th Caribbean Geological Conference (Trinidad), p. 225–228.

Donnelly, T. W., Horne, G. S., Finch, R. C., and López-Ramos, E., 1990, Northern Central America; the Maya and Chortis blocks, *in* Dengo, G., and Case, J. E. eds., The Caribbean region: Boulder, CO, Geological Society of America, The geology of North America, v. H, p. 37–76.

Ego, F., and Ansan, V., 2002, Why is the Central Trans-Mexican Volcanic Belt (102°–99°W) in transtensive deformation?: Tectonophysics, v. 359, p. 189–208.

Engebretson, A. C., Cox, A., and Gordon, R. G., 1985, Relative motions between oceanic and continental plates in the Pacific Basin: Geological Society of America Special Paper 206, 59 p.

Ferrari, L., 2004, Slab detachment control on mafic volcanic pulse and mantle heterogeneity in central Mexico: Geology, v. 32, p. 77–80.

Ferrari, L., López-Martínez, M., Aguirre-Díaz, G., and Carrasco-Nuñez, G., 1999, Space-time patterns of Cenozoic arc volcanism in central Mexico: From the Sierra Madre Occidental to the Mexican Volcanic Belt: Geology, v. 27, p. 303–306.

Ferrusquia-Villafranca, I., 1996. Contribución al conocimiento geológica de Chiaps-el área Ixtapa-Soyaló: Instituto de Geologia, Universidad Nacional Autónoma de México, Boletin 109, 130 p.

Garcia-Paloma, A., Macias, J. L., and Espindola, J. M., 2004, Strike-slip faults and K-alkaline volcanism at El Chichon volcano, southeastern Mexico: Journal of Volcanology and Geothermal Research, v. 136, p. 247–268.

Goes, S., and van der Lee, S., 2002, Thermal structure of the North American uppermost mantle inferred from seismic tomography: Journal of Geophysical Research, v. 107, p. ETG 2, [DOI 10.1029/2 2000JB000049].

Gómez-Tuena, A., LaGatta, A. B., Langmuir, C. H., Goldstein, S. L., Ortega-Gutiérrez, F., and Carrasco-Núñez, G., 2002, Temporal control of subduction magmatism in the eastern Trans-Mexican Volcanic Belt: Mantle sources, slab contributions, and crustal contamination: Geochemistry, Geophysics, Geosystems, v. 4, No. 8, p. 8912 [doi: 10.1029/2003GC000524].

Guinta, G., Beccaluva, L., Coltorti, M., Sienna, F., Mortellaro, D., and Cutrupia, D., 2002, The peri-Caribbean ophiolites: Sructure, tectono-magmatic significance,and geodynamic implications: Caribbean Journal of Earth Sciences, v. 36, p. 1–20.

Gutscher, M.-A., Olivet, J.-L., Aslanian, D., Eissen, J.-P., and Maury, R., 1999, The "lost Inca Plateau": Cause of flat subduction beneath Peru?: Earth and Planetary Science Letters, v. 171, p. 335–341.

Gutscher, M. A., Spakman, W., Bijwaard, H., and Engdahl, E. R., 2000, Geodynamics of flat subduction: Seismicity and tomographic constraints from the Andean margin: Tectonics, v. 19, p. 814–833.

Guzmán-Speciale, M., and Meneses-Rocha, J. J., 2000, The North America–Caribbean plate boundary west of the Motagua-Polochic fault system: A fault jog in southeastern Mexico: Journal of South American Earth Sciences, v. 13, p. 459–468.

Guzmán-Speziale, M., Pennington, W. D., and Matumoto, T., 1989, The triple junction of the North Anerica, Cocos, and Caribbean plates: Seismicity and tectonics: Tectonics, v. 8, p. 981–997.

Hernandez-Quintero, J. E., 2002, Interpreacion global de anomalias magneticas corticales satelitales sobre la Republica Mexicana: Unpubl. M.Sc. thesis, Instituto de Geofisica, Universidad Nacional Autonoma de Mexico, 135 p.

Herrmann, U. R., Nelson, B. K., and Ratschbacher, L., 1994, The origin of a terrane: U/Pb geochronology and tectonic evolution of the Xolapa complex (southern Mexico): Tectonics, v. 13, p. 455–474.

Horne, G. S., Finch, R. C., and Donnelly, T. W., 1990, The Chortis Block, *in* Dengo, G., and Case, J. E., eds., The Caribbean region: Boulder, CO, Geological Society of America, Geology of North America, v. H, p. 55–76.

Jaillard, E., Hérail, G., Monfret, T., Diaz-Martinez, E., Baby, P., Levanu, A., and Dumont, J. F., 2000, Tectonic evolution of the Andes of Ecuador, Peru, Bolivia, and northernmost Chile, *in* Cordani, U. G., Thomaz Filho, A., and Campos, D. A., eds., Tectonic evolution of South America: 31st International Geolological Congress, Rio de Janeiro, Brasil, p. 481–559.

Johnson, K. R., 1984, Geology of the Gualán and southern Sierra de las Minas quadrangles, Guatemala: Unpubl. Ph.D. thesis, State University of New York at Binghampton, 300 p.

Kay, S. M., Mpodozis, C., Ramos, V. A., and Munizaga, F., 1991, Magma source variations for mid to late Tertiary volcanic rocks erupted over shallowing subduction zone and through a thickening crust in the Main Andean Cordillera (28–33°S, *in* Harmon, R. S., and Rapela, C., eds., Andean magmatism and its tectonic settings: Geological Society of America Special Paper 265, p. 113–137.

Keppie, J. D., 2004, Terranes of Mexico revisited: A 1.3 billion year odyssey: International Geology Review, v. 46, p. 765–794.

Keppie, J. D., Dostal, J., Cameron, K. L., Solari, L. A., Ortega-Gutiérrez, F., and Lopez, R., 2003, Geochronology and geochemistry of Grenvillian igneous suites in the northern Oaxacan Complex, southern México: Tectonic implications: Precambrian Research, v. 120, p. 365–389.

Ladd, J. W., Holcombe, T. L., Westbrooke, G. K., and Edgar, N. T., 1990, Caribbean marine geology: Active margins of the plate boundary, *in* Dengo, G., and Case, J. E. eds., The Caribbean region: Boulder, CO, Geological Society of America, Geology of North America, v. H, p. 261–290.

Lawrence, D. P., 1975, Petrology and structural geology of the Sanarate–El Progresso area, Guatemala: Unpubl. Ph.D. thesis, State University of New York at Binghampton, 255 p.

Malfait, B. T., and Dinkleman, M. G., 1972, Circum-Caribbean tectonic and igneous activity and evolution of the Caribbean plate: Geological Society of America Bulletin, v. 83, p. 251–272.

Mammerickx, J., 1989, Bathymetry of the North Pacific Ocean, *in* Winterer, E. L., Hussong, D. M., and Decker, R. W., eds., The Eastern Pacific Ocean and Hawaii: Boulder, CO, Geological Society of America, Geology of North America, v. N, plates 1B and 1C.

Mammerickx, J., and Klitgord, K. D., 1982, Northern East Pacific Rise: Evolution from 25 m.y. to the present: Journal Geophysical Research, v. 87, No.B8, p. 6751–6759.

Manton, W. L., 1996, The Grenville in Honduras [abs.]: Geological Society of America Abstracts with Programs, v. 28, no. 7, p. A493.

Marquez, A., Oyarzun, R., Doblas, M., and Verma, S. P., 1999, Alkalic (OIB type) and calc-alkalic volcanism in the Mexican volcanic belt: A case for plume-related magmatism and propogating rifting at an active margin?: Geology, v. 27, p. 51–54.

Martinez-Serrano, R.G., Schaaf, P., Solis-Pichardo, G., Hernández-Bernal, Ma. del Sol, Hernádez-Treviño, T., and Morales-Contreras, J. J., 2004, Sr, Nd, and Pb isotope and geochemical data from the Quaternary volcanism of Nevado de Toluca volcano—adakitic magmatism in central Mexico?, *in* Aguirre-Diaz, G. J., Macías-Vásquez, J. L., and Siebe, C., eds. Proceedings of the Geological Society of America Penrose Conference: Neogene–Quaternary continental margin volcanism, Universidad Nacional Autonoma de Mexico, Instituto de Geologia Publicacíon Especial 2, p. 52.

Maury, R., Gutscher, M. A., Bourdon, E., and Peacock, S., 2003, Slab melting during the transition from steep to flat subduction: Evidence from numerical models of forearc thermal structure: Geophysical Research Abstracts, v. 5, p. 05922.

McBirney, A. R., 1963, Geology of a part of the central Guatemalan Cordillera: University of California publications in Geological Sciences, v. 38, p. 177–242.

McBirney, A. R., Aoki, K. I., and Bass, M., 1967, Elcogite and jadeite from the Motagua fault zone, Guatemala: American Mineralogist, v. 52, p. 908–918.

McDowell, F. W., and Clabaugh, S., 1979, Ignimbrites of the Sierra Madre Occidental and their relation to the tectonic history of western Mexico: Geological Society of America Special Paper 180, p. 113–124.

Meschede, M., Barckhausen, U., and Worm, H-U., 2002, The evolution of the Cocos-Nazca spreading center and overprinting hotspot traces, *in* Jackson, T. A., ed., Caribbean geology into the third millennium: Transactions of the 15th Caribbean Geological Conference, University of West Indies Press, Kingston, Jamaica, p. 63–72.

Meschede, M., and Frisch, W., 2002, The evolution of the Caribbean Plate and its relation to global plate motion vectors: Geometric constraints for an Inter-American origin, *in* Jackson, T. A., ed., Caribbean geology into the third millennium: Transactions of the 15th Caribbean Geological Conference, University of West Indies Press, Kingston, Jamaica, p. 1–14.

Mora, J. C., Macias, J. L., Garcia-Paloma, A., Arce, J. L., Espindola, J. M., Manetti, P., Vaselli, O., and Sanchez, J. M., 2004a, Petrology and geochemistry of the Tacana Volcanic Complex, Mexico-Guatemala: evidence for the last 40,000 y of activity. Geofisica Internacional, v. 43, p. 331–359.

Mora, J. C., Macias, J. L., Godines, L., Espindola, J. M., Garcia-Paloma, A., and Sánchez, G. S., 2004b, Miocene to Recent volcanism in southern Mexico, *in* Aguirre-Diaz, G. J., Macías-Vásquez, J. L., and Siebe, C., eds., Proceedings Geological Society of America Penrose Conference: Neogene–Quaternary continental margin volcanism: Mexico, DF, Universidad Nacional Autonoma de Mexico, Instituto de Geologia Publicacíon Especial 2, p. 56.

Morán-Zenteno, D. J., Corona-Chávez, P., and Tolson, G., 1996, Uplift and subduction erosion in southwestern Mexico since the Oligocene: pluton geobarometry constraints: Earth and Planetary Science Letters, v. 141, p. 51–65.

Morán-Zenteno, D. J., Tolson, G., Martinez-Serrano, R. G., Martiny, B., Schaaf, P., Silva-Romo, G., Machias-Romo, C., Alba-Aldave, L., Hernández-Bernal, M. S., and Solis-Pichardo, G. N., 1999, Tertiary arc-magmatism of the Sierra Madre de Sur, Mexico, and its transition to the volcanic activity of the Trans-Mexican Volcanic Belt: Journal of South America Earth Sciences, v. 12, p. 513–535.

Muehlberger, W. R., 1992, Tectonic map of North America: Scale 1:5,000,000: Tulsa, OK, American Association of Petroleum Geologists.

Muller, P. D., 1979, Geology of the Los Amates quadrangles and vicinity, Guatemala: Unpubl. Ph.D. thesis, State University of New York at Binghampton, 326 p.

Nieto-Samaniego, A. F., Ferrari, L., Alaniz-Álvarez, S. A., Labarthe-Hernández, G., and Rosas-Elguera, J., 1999, Variation of Cenozoic extension and volcanism across the southern Sierra Madre Occidental volcanic province, Mexico: Geological Society of America Bulletin, v. 111, p. 347–363.

Ortega-Gutiérrez, F., and Elias-Herrera, M., 2003, Wholesale melting of the southern Mixteco terrane and origin of the Xolapa Complex [abs.]: Geological Society of America Abstracts with Program, v. 35, No. 4, p. 66.

Pacheco-Gutiérrez, A. C., 2002, Deformacion transpresiva Miocenica y el desarrollo de sistemas de fracturas en la porcion nororiental de la Sonda de Campeche: Unpubl. M.Sc. thesis, Universidad Nacional Autonoma de Mexico, 98 p.

Pardo, M., and Suarez, G., 1995, Shape of the subducted Rivera and Cocos Plates in southern Mexico: Seismic and tectonic implications: Journal of Geophysical Research, v. 6, No. B7, p. 12,357–12,373.

Pindell, J. L., Cande, S. C., Pitman, W. C., III, Rowley, D. B., Dewey, J. F., Labrecque, J., and Haxby, W., 1988, A plate kinematic framework for models of the Caribbean evolution: Tectonophysics, v. 155, p. 121–138.

Plafker, G., 1976, Tectonic aspects of the Guatemala earthquake of 4 February 1976: Science, v. 193, p. 1201–1208.

Ponce, L. R., Gaulon, R., Súarez, G., and Lomas, E., 1992, Geometry and the state of stress of the downgoing Cocos plate in the Isthmus of Tehuantepec, Mexico: Geophysical Research Letters, v. 19, p. 773–776.

Ramos, V. A., 2000, The South Central Andes, *in* Cordani, U. G., Thomaz Filho, A., and Campos, D. A., eds., Tectonic evolution of South America, *in* Proceedings 31st International Geological Congress, Rio de Janeiro, Brazil, p. 561–604.

Ramos, V. A., and Aleman, A., 2000, Tectonic evolution of the Andes, *in* Cordani, U. G., Thomaz Filho, A., and Campos, D. A., eds., Tectonic evolution of South America, *in* 31st International Geological Congress, Rio de Janeiro, Brazil, p. 635–688.

Ramos, V. A., Cristallini, E. O., and Pérez, D. J., 2002, The Pampean flat-slab of the Central Andes: Journal of South American Earth Sciences, v. 15, p. 59–78.

Rogers, R. D., Kárason, H., and van der Hilst, R. D., 2002, Epeirogenic uplift above a detached slab in northern Central America: Geology, v. 30, p. 1031–1034.

Rosencrantz, E., Ross, M. I., and Sclater, J. G., 1988, The age and spreading history of the Cayman Trough as determined from depth, heat flow, and magnetic anomalies: Journal of Geophysical Research, v. 93, p. 2141–2157.

Rosencrantz, E., and Sclater, J. G., 1986, Depth and age in the Cayman Trough: Earth and Planetary Science Letters, v. 79, p. 133–144.

Ross, M. I., and Scotese, C. E., 1988, A hierarchical model of the Gulf of Mexico and Caribbean region: Tectonophysics, v. 155, p. 139–168.

Sánchez-Barreda, L. A., 1981, Geologic evolution of the continental margin of the Gulf of Tehuantepec in southern Mexico: Unpubl. Ph.D. thesis, University of Texas, Austin, Texas, 192 p.

Schaaf, P., Morán-Zenteno, D., and Hernández-Bernal, M. del., 1995, Paleogene continental margin truncation in southwestern Mexico: Geochronological evidence: Tectonics, v. 14, p. 1339–1350.

Schulze, C. H., Keppie, J. D., Ortega-Rivera, A., Ortega-Gutiérrez, F., and Lee, J. K. W., 2004, Mid-Tertiary cooling ages in the Precambrian Oaxacan Complex of southern Mexico: Indication of exhumation and inland arc migration: Revista Mexicana de Ciencias Geologicas, in press.

Sclater, J. G., Anderson, R. N., and Bell, M. L., 1971, The elevation of ridges and the evolution of the central eastern Pacific: Journal of Geophysical Research, v. 76, p. 7888–7915.

Sedlock, R. L., Ortega-Gutiérrez, F., and Speed, R. C., 1993, Tectonostratigraphic terranes and tectonic evolution of Mexico: Geological Society of America Special Paper 278, 153 p.

Severinghaus, J., and Atwater, T., 1990, Cenozoic geometry and thermal condition of the subducting slab beneath western North America, *in* Wernicke, B., ed., Basin and Range extensional tectonics near the latitude of Las Vegas, Nevada: Geological Society of America Memoir 176, p. 1–22.

Tolson-Jones, G., 1998, Deformación, exhumación y Neotectónica de la margen continental de Oaxaca: Datos estructurales, petrológicos y geotermobarométricos: Unpubl. Ph.D. thesis, Universidad Nacional Autonoma de Mexico, 98 p.

van der Lee, S., and Nolet, G., 1997, Seismic image of the subducted trailing fragments of the Farallon plate: Nature, v. 386, p. 266–269.

Vazquez-Meneses, M. E., Villaseñor-Rojas, P. E., Sanchez-Quiñones, R., and Islas-Carrion, M. A., 1992, Neotectonica del sureste de Mexico: Revista del Instituto Mexicano del Petroleo, v. 24, p. 12–37.

Velez-Scholvink, D., 1990, Modelo transcurrente en la evolucion tectonico-sedimentaria de Mexico: Boletin de la Asociacion Mexicana de Geologos Petroleros, v. 11, No. 2, p. 1– 25.

Verma, S. P., 2002, Absence of Cocos plate subduction-related mafic volcanism in southern Mexico: A unique case on Earth?: Geology, v. 30, p. 1095–1098.

POTENTIAL CORRELATIVES

Cadomian and Early Paleozoic Magmatic Events: Insights from Zircon Morphology and Geochemical Signatures of I- and S-Type Granitoids (Saxo-Thuringia/Bohemian Massif/Central Europe)

MICHAEL GEHMLICH[1] AND KERSTIN DROST

Staatliche Naturhistorische Sammlungen Dresden, Forschungsmuseum für GeoBioWissenschaften, Königsbrücker Landstrasse 159, D-01109 Dresden, Germany

Abstract

This paper presents studies of zircon morphologies and geochemical signatures of Neoproterozoic and Early Paleozoic metagranitoids and metarhyolites of the Saxo-Thuringian Zone (Central Europe). The Saxo-Thuringian Zone was derived from the northern periphery of Gondwana and became part of the Variscan belt of Central Europe during Paleozoic orogenic processes.

Five tectonomagmatic events are preserved in the Saxo-Thuringian Neoproterozoic and Early Paleozoic igneous rocks. These different geotectonic scenarios are each characterized by distinctive geochemical signatures and zircon morphologies. Samples of felsic lithologies with age ranges of ~580–560 Ma, ~500 Ma, and ~380–360 Ma show whole-rock compositions of dry I-type with a tendency to A-type granites and zircon crystals with strong dominance of {100} prisms and {101} pyramids. The ~580–560 Ma and ~500 Ma periods of predominantly juvenile magmatism both were followed by tectonothermal events producing S-type granitoids at ~550–530 Ma and 495–480 Ma. Associated zircon populations are dominated by {110} prisms and {211} pyramids. Inasmuch as the errors of individual geochronological analyses may overlap two different geotectonic episodes, the relationship between geochemical features and zircon morphologies of the Saxo-Thuringian Neoproterozoic and Early Paleozoic igneous rocks can aid in assigning the dated sample to the proper stage of formation.

Introduction

MAGMATIC ZIRCON tends to develop distinct crystal morphologies depending on the composition of the granitic magma (e.g., Pupin, 1980; Speer, 1980; Vavra, 1990; Benisek and Finger, 1993). The idea of using zircon morphology to distinguish granites from different tectonic settings and different crystallization temperatures was introduced by Pupin (1980). Vavra (1990) and Benisek and Finger (1993) showed that the morphology, especially the dominant prism, strongly depends on the supersaturation of $ZrSiO_4$ in the melt and the concentration of several trace elements, such as U. According to Vavra (1994), zircon crystals with a dominance of {100} over {110} prisms were formed in relatively dry I-type granites. For the zircon pyramids, Pupin (1980) discussed a connection with the chemical composition, especially the (Na + K/Al) ratio of the magma. Based on Pupin (1980), Rottura et al. (1989) distinguished between alkaline, subalkaline, calkalkaline, and anatectic granitoids. Furthermore, Vavra (1990) showed that the {211} pyramid is typical for zircon grains with inherited cores.

Inasmuch as the chemistry changes during magma crystallization and the distribution of trace elements is heterogeneous, granites can be characterized by an assemblage of zircon grains of different morphologies. Although zircons in many granitoids show a relatively narrow range in morphologies (Gehmlich, 2003), studies that relate zircon morphology to tectonic setting are rare (Schermaier et al., 1992; Brätz, 2000; Gehmlich, 2003).

In the paper we describe zircon morphologies of granitoids with distinct formation ages and different formation conditions. The implications of the crystal morphology will be discussed in combination with the lithogeochemical signatures.

[1]Corresponding author: E-mail: rgundmg@gmx.de

Geological Setting

The Saxo-Thuringian zone (for short, Saxo-Thuringia) represents the northwestern part of the Bohemian Massif (Central Europe) (Fig. 1). Saxo-Thuringia and the entire Bohemian Massif are derived from the Gondwana margin. According to Linnemann et al. (2000), the separation of the massif from Gondwana took place in Late Ordovician time when it became part of the Armorican terrane assemblage. This assemblage collided with the northern continents during the Variscan orogeny in Late Devonian to Carboniferous time (Franke, 2000).

U-Pb and Pb-Pb (zircon) geochronological studies of Saxo-Thuringian rocks (e.g., Kröner et al., 1995; Linnemann et al., 2000; Tichomirowa et al., 2001; Tichomirowa, 2002; Gehmlich, 2003) yield Cadomian (Late Neoproterozoic) and Cambro-Ordovican ages for magmatic rocks, and indicate that Saxo-Thuringia was derived from Gondwana. Several tectonic events resulting in episodes of felsic magmatism occurred between ~580 Ma and ~480 Ma (~580–560 Ma, 550–530 Ma, and 500–480 Ma, Fig. 2), and reflect volcanic arc development, Late Neoproterozoic–Early Cambrian subduction, and Cambro-Ordovician rifting (Linnemann et al., 2000). For Silurian to Middle Devonian time, no magmatic activity is known in Saxo-Thuringia. Linnemann et al. (1999) interpreted Silurian time as the drift stage of Saxo-Thuringia from Gondwana. The first magmatic event related to the Late Paleozoic (Variscan) collision is characterized by felsic rocks with ages of ~380–360 Ma (Gehmlich, 2003). Most uncertainties in the age determinations are in the range of 5 to 10 Ma (e.g., Linnemann et al., 2000; Tichomirowa et al., 2001; Gehmlich, 2003).

Methods

To compare the zircon morphology and the geochemical signatures, several meta-igneous rocks were investigated. The geochemical data and the formation ages of these rocks used in this study were published in several papers (Table 1). Our chemical data (Gehmlich, 2003) were produced by the Mineralogical and Petrological Institute of the University of Köln and by ACTLABS Activation Laboratories Ltd., Ancaster, Ontario, Canada.

In each sample, the zircon morphologies of ~100–300 zircon grains were investigated by scanning electron microscopy, and the morphology was entered in the Pupin scheme. It is possible that zircon grains cannot be uniquely assigned to a specific type. For instance, in Figure 3, a zircon is shown with a dominant {211} pyramid on the left part and dominant {101} pyramid on the right part. In other cases, slight differences in morphology are apparent if the zircon is revolved around the c-axis. Therefore, the data are presented only as a field in the Pupin scheme. Such a field is characterized by >50% zircon grains of the whole zircon population of a sample, and is therefore typical for the sample.

Samples and their Geochemical Characterization

The investigated rocks are metagranitoids and metarhyolitoids. Most of the samples have been taken from weakly metamorphosed areas (Schwarzburg anticline, Ziegenrück-Teuschnitz syncline, Berga anticline, Vogtland anticline, Elbe zone, Lausitz anticline) (Fig. 1). Some metamorphic rocks are from the Saxonian Zwischengebirge of Wildenfels and Frankenberg as well as from the Erzgebirge. A further sample is from the neighboring Tepla-Barrandian unit (Fig. 1).

The oldest sedimentary rocks in Saxo-Thuringia are Neoproterozoic siliciclastic units deposited in a backarc basin of the Cadomian island arc. They are dated by ash layers with an age of ~580–560 Ma (U-Pb, zircon; Buschmann et al., 2001). Within these units (and in younger sediments) detrital zircon grains and island arc–derived granitoid pebbles of the same age occur (Linnemann et al., 2000; Tichomirowa, 2002). The samples are characterized by high values of SiO_2, LREE, relatively high Ga/Al ratios, and low values of MgO and FeO_{total} (Figs. 4A–4C). Using the discrimination diagram of Whalen et al. (1987), these samples tend to be A-type granites.

The second sample group consists of ~550–530 Ma granites and granodiorites, which intrude sedimentary units of the Cadomian backarc basin and represent the final stage of the Cadomian orogeny. The geochemical signatures of these Late Cadomian metagranitoids show lower Ga/Al ratios than the ~580–560 Ma pebbles, mentioned above (Fig. 4A). The metagranitoids are characterized by A/CNK >1.1 and high Cr and Ni contents. They are interpreted as anatectic S-type granitoids (Kröner et al., 1995; Tichomirowa, 2002; Gehmlich, 2003). Tichomirowa (2002) emphasized the geochemical similarity of the metagranitoids and the surrounding

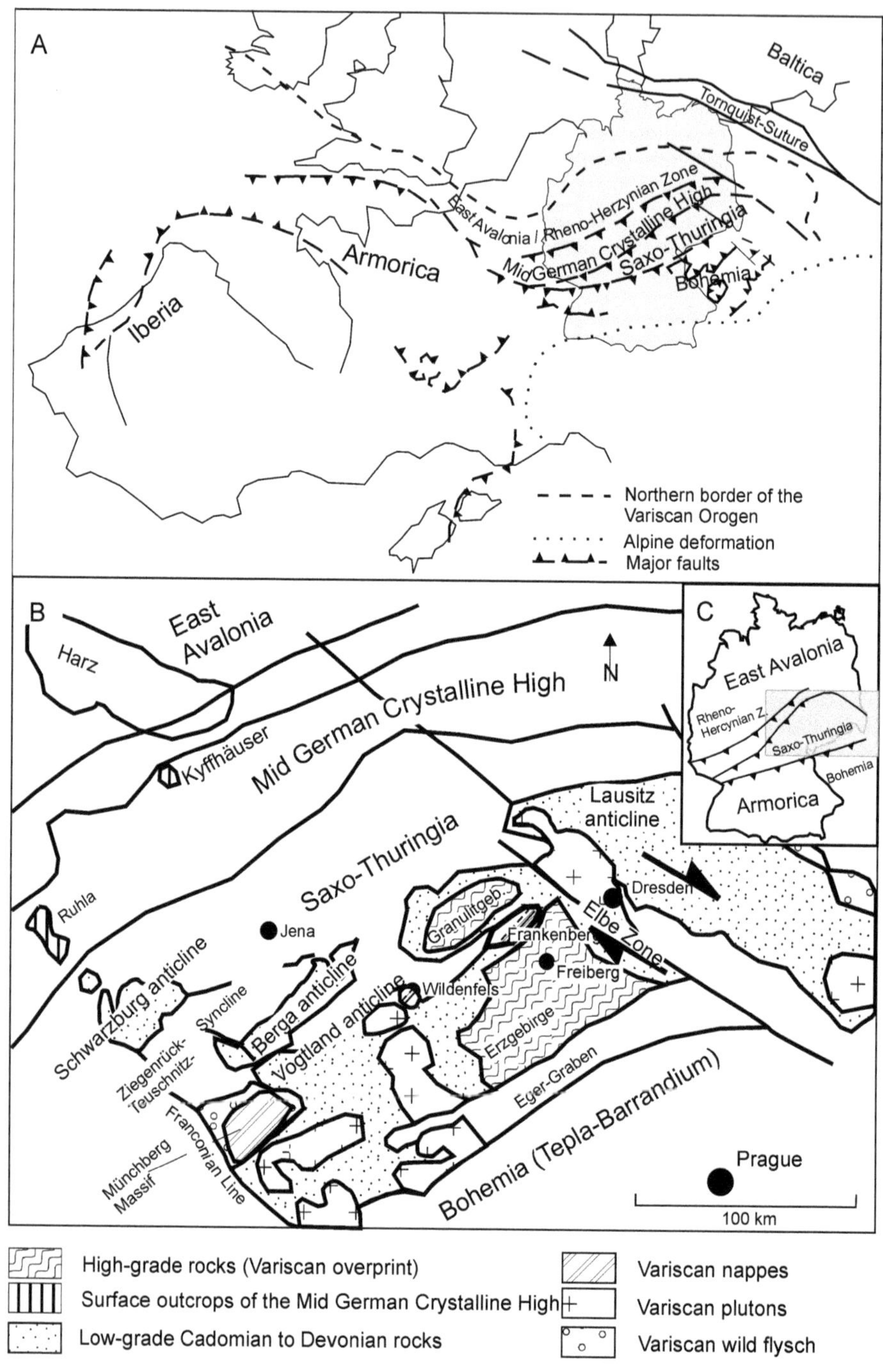

FIG. 1. Geological map of Saxo-Thuringia within the Variscan orogenic belt and the major tectonometamorphic units. A. Europe. B. Saxo-Thuringia. C. Germany.

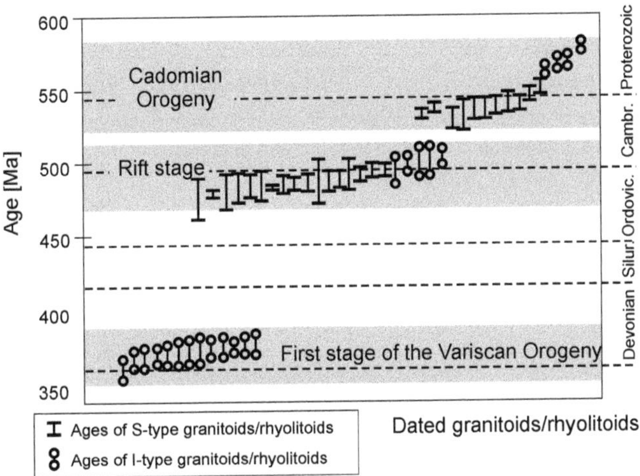

FIG. 2. Cadomian and Early Paleozoic zircon ages of granitoids/rhyolitoids from Saxo-Thuringia. Furthermore, the subdivision of these rocks in I-type (with a tendency toward A-type) and S-type rocks is evident. For references see Table 1.

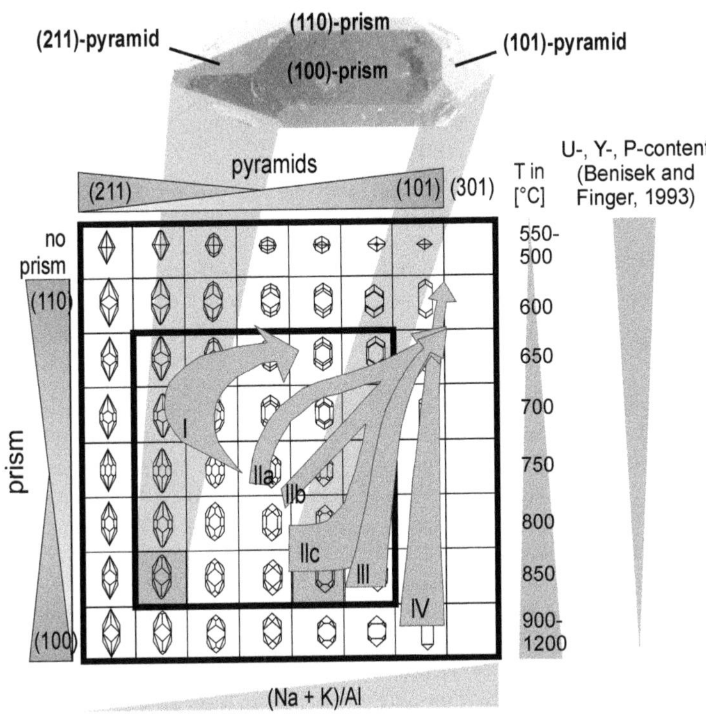

FIG. 3. Pupin scheme showing the zircon morphology depending on (Na + K)/Al, and formation temperature (Pupin, 1980) or on the content of U, Y, and P (Benisek and Finger, 1993) with different trends: Legend: I = Al-rich anatectic granitoids; IIa,b,c = calc-alkaline granitoids; III = K-rich subalkaline granitoids; IV = alkaline granitoids (Rottura et al., 1989). A zircon with very different peaks cannot be classified by the Pupin scheme.

TABLE 1. Used Samples, Their Ages, and References with Comments

Age	References	Sample (age in Ma)	Comments
~ 580 Ma ~560 Ma	Linnemann et al., 2000 and unpubl. chem. data	WeG1 (568 ± 4), WeG2 (559 ± 3), Clanz1 (577 ± 3)	
	Gehmlich, 2003	SSA3 (570 ± 4)	
~550 Ma ~530 Ma	Tichomirowa, 2002	DDR16 (538 ± 3), Kin1 (534 ± 4), Kit1b (545 ± 6), Seid1 (532 ± 6)	Anatexites S-type granites
	Kröner et al., 1995	DDR13 (555 ± 7), DDR16 (550 ± 7)	Anatexites S-type granites
	Gehmlich, 2003	Gl-Gt (541 ± 7), Laub1 (533 ± 10), Dohna1 (537 ± 7), Laas1 (531 ± 7), Klotz1 (541 ± 9)	S-type granites
~500 Ma ~ 480 Ma	Gehmlich, 2003	Bla-Rhy1 (487 ± 6), Böhl1* (495 ± 9), Frank3 (484 ± 8), Frank7* (501 ± 10), P-Bt (479 ± 2), S-Gn (486 ± 4), T-Gt (485 ± 6), Tek1a* (502 ± 10), Tek1b* (503 ± 5), Wild2b (486 ± 4), Wild2d (485 ± 4), Wild2a (496 ± 3)	S-type granites/ rhyolites * I-type granites/ rhyolites with tendency to A-type
	Drost et al., 2004	OKR* (499 ± 4)	
	Kröner and Willner, 1998	DDR17 (481.7 ± 1.1)	S-type granite
	Mohnicke et al., 2001	CosGra (479 ± 20)	S-type granite
	Tichomirowa, 2002	Rumb (486 ± 5), Non1 (492 ± 14), KE445 (495 ± 5), Mem1 (483 ± 6)	S-type granites
	Mingram and Rötzler, 1999	EG94-3 (480 ± 12), EG95-2 (485 ± 12), EG95-5 (492 ± 4)	Anatexites S-type granites
~380 Ma ~ 360 Ma	Gehmlich, 2003	Chl-Gn[1] (377 ± 7), Chl-Gn2[1] (369 ± 5), Hart2[1] (375 ± 7), Hart3[1] (374 ± 6), Hbg1[2] (373 ± 9), Hbg2[2] (372 ± 8), Hbg3[2] (360 ± 7), Neum1[2] (367 ± 7), Neum2[2] (366 ± 6), Neum3[2] (370 ± 7), Pol[1] (371 ± 8), Post1[1] (375 ± 4), Olz1[1] (376 ± 6)	[1]Rhyolitoids with tendency to A-type composition; [2] I-type granitoids

siliciclastic rocks, deposited ~580–570 Ma, and concluded that these Cadomian (meta-)greywackes are the protolith that melted to yield the granitoids.

After a period of relative magmatic quiescence Cambro-Ordovician (~500–480 Ma) felsic magmatism occurred. According to Kemnitz et al. (2002) these igneous rocks reflect rifting and separation of Saxo-Thuringia from Gondwana. Most samples of this suite are rhyolites and subvolcanic equivalents. Some metamorphosed rocks are interpreted as intrusive rocks. Other samples include a granitic dike within the Cadomian basement and granitic pebbles of Lower Carboniferous conglomerates.

The samples of the Cambro-Ordovician range can by subdivided into three groups according to chemical signatures. (1) The dike and the pebbles are characterized by high values of SiO_2; low values of CaO, MgO, and FeO_{total}; high Ga/Al ratios; and a strong negative Eu-anomaly (Figs. 4A, 4B, and 4D). These samples are I-type granites/rhyolites with a tendency toward an A-type composition. (2) The two other groups are both S-type granites/rhyolites and can be distinguished by the SiO_2 content (Tichomirowa, 2002; Gehmlich, 2003). The samples with low SiO_2 (< 73 wt%) have relatively high values of MgO, FeO_{total}, and LREE (Figs. 4B and 4E) and are very similar to the ~550–530 Ma S-type rocks (see above, Figs. 4B and 4C). (3) The third group has high SiO_2 (> 73 wt%); very low LREE, Zr, Nb, Ba, Sr, MgO, FeO_{total}, and TiO_2; and a strong negative Eu-anomaly (Mingram and Rötzler, 1999) (Figs. 4B and 4E). In contrast to the I-type granitoids, both groups of S-type rocks show lower Ga/Al ratios (Fig. 4A). The two S-type groups differ in the contents of

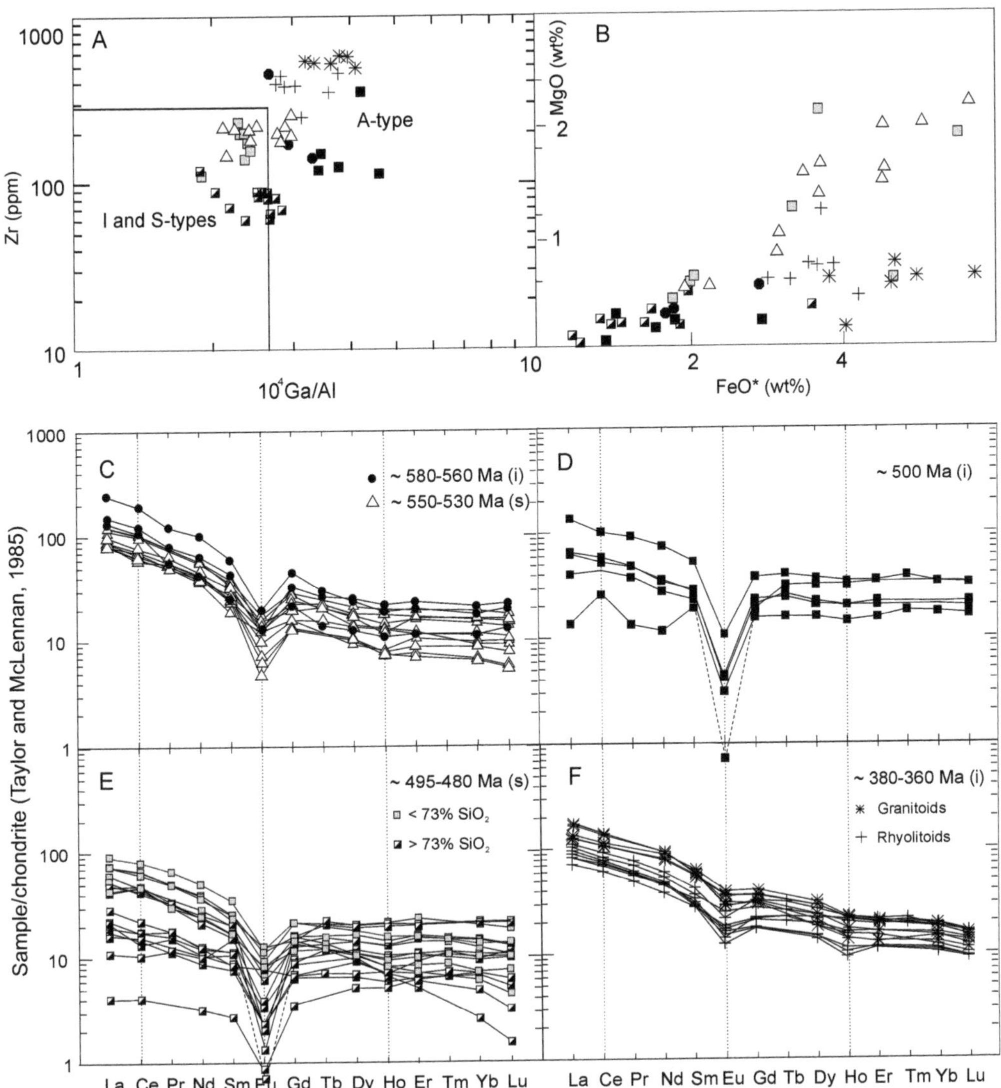

FIG. 4. Chemical characterization for granitoids/rhyolitoids of the different age groups. A. Discrimination diagram for rocks of A-type compositions (Whalen et al., 1987). B. Harker diagram for mafic components MgO-FeO_{total}. C–F. Chondrite-normalized REE diagrams (normalization values from Taylor and McLennan, 1985). C. Neoproterozoic and Late Cadomian. D. Upper Cambrian. E. Tremadocian. F. Upper Devonian. Legend: i = I-type rocks; s = S-type rocks.

mafic components (reflected in MgO and FeO_{total} contents; Fig. 4B). Chemical variation between these two groups can be caused by fractionation processes. On the one hand, strongly fractionated SiO_2-rich samples occur as rhyolites or subvolcanites (Tichomirowa, 2002), and the less-fractionated SiO_2-poor samples are interpreted as intrusive rocks (Kurze, 1966; Mingram and Rötzler, 1999; Tichomirowa, 2002). On the other hand, the highly fractionated, SiO_2-rich Tremadocian rocks could result from a lower degree of anatexis due to lower temperatures (Tichomirowa, 2002).

The Late Devonian samples are granitoids, rhyolites, and rhyolitic pebbles. The latter occur in Upper Devonian conglomerates. This magmatic event is related to strike-slip tectonics during the

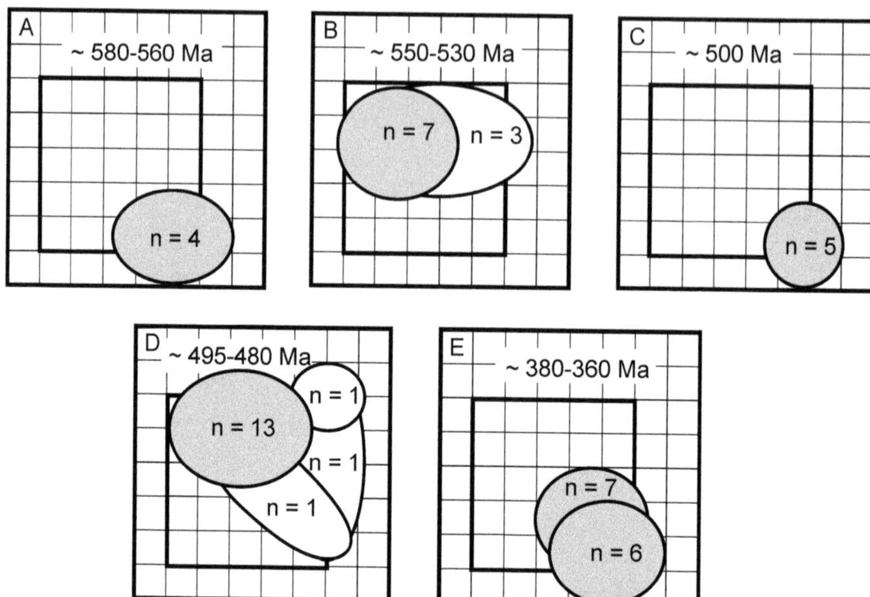

FIG. 5. Classification scheme after Pupin (1980) with typology of the investigated samples: n = number of samples. Each sample is characterized by 100 to 300 zircon crystals, the resulting field in the Pupin scheme is represented by >50% of the zircon grains per sample. A. Position of zircons of granitoids/rhyolitoids with formation ages of ~580–560 Ma. B. Position of zircons of granitoids with formation ages of ~550–530 Ma. C. Position of zircons of granitoids with formation ages of ~500 Ma. D. Position of zircons of granitoids/rhyolitoids with formation ages of ~495-480 Ma. E. Position of zircons of granitoids/rhyolitoids with formation ages of ~380–360 Ma.

oblique collision of Saxo-Thuringia with northern Europe (Gehmlich et al., 2000). The rocks are characterized by high Ga/Al and plot in the field of A-type granitoids (Fig. 4A). Due to fractionation processes, the samples show slight differences in their LREE contents. The LREE contents of the intrusive rocks are higher than those of the extrusive rocks. Gehmlich (2003) interpreted both intrusive and extrusive rocks to have formed in a within-plate environment.

According to chemical indicators (Na_2O + K_2O or Na + K / Al, the latter being used in the Pupin scheme), the samples of all age groups exhibit a calk-alkaline composition without significant differences (Kröner et al., 1995; Linnemann et al., 2000; Tichomirowa, 2002).

Results and Discussion

The granitic pebbles with ages of ~580–560 Ma contain zircon grains with a very simple morphologies. The dominant morphology of the zircon grains reveals a strong dominance of {100} prisms and {101} pyramids over {110} prisms and {211} pyramids (Fig. 5A). In many crystals, only the {110} prism and the {101} pyramid occur. Detrital zircon grains of many metagreywackes show the same shape (Tichomirowa, 2002; Gehmlich, 2003). The resulting classification within the Pupin scheme indicates a connection of these zircon crystals to dry I-type (Vavra, 1994) or subalkaline to alkaline granitoids (Pupin, 1980). Furthermore, inherited zircon is lacking within these samples.

In contrast, zircon grains in ~550–530 Ma S-type metagranitoids are characterized by dominant {211} pyramids and {110} prisms (Fig. 5B), and are typical for zircon grains in S-type granites (Pupin, 1980). Long-prismatic zircon shows a stronger dominance of the {110} prism than shorter zircon. The short-prismatic zircon may have {110} and {100} prisms of the same size. According to Vavra (1990), older cores typically occur in the zircon grains with acute pyramids, and the short-prismatic crystals commonly contain inherited cores. $^{207}Pb/^{206}Pb$ evaporation ages of core-bearing zircon grains led to the following minimal ages: ~570 Ma, 600 Ma, 680 Ma,

850 Ma, 2 Ga, 2.5 Ga, and 3 Ga (Linnemann et al., 2000; Tichomirowa, 2002).

The granitoids/rhyolitoids and pebbles with ages of ~500–480 Ma contain two completely different zircon populations. (1) The first population includes zircon grains similar to the Neoproterozoic rocks (~580–560 Ma). The zircon morphology reveals a strong dominance of {100} prisms and {101} pyramids (Fig. 5C). According to the Pupin scheme, these crystals are derived from a dry I-type (Vavra, 1994) with low contents of U, Th, and P (Benisek and Finger, 1993), a morphology that is typical for subalkaline to alkaline magmas. Furthermore, only one inherited zircon has been found and dated from these samples (~540 Ma; Drost et al., 2004). The ages of the samples with these zircon types concentrate at ~500 Ma (Fig. 2). (2) The second population with S-type morphology can be distinguished from the ~500 Ma population by a dominant {211} prism and {110} pyramid (Fig. 5D). They are similar to zircon grains in Late Cadomian granitoids (~550–530 Ma). These rocks were formed at ~495–480 Ma (Fig. 2), about 5–15 m.y. after the I-type granitoids.

The acicular, especially the short-prismatic, ~495–480 Ma zircon grains contain numerous inherited cores. $^{207}Pb/^{206}Pb$ evaporation ages of core-bearing zircon grains yield the following minimal ages: ~540 Ma, 575 Ma, 610 Ma, 1.7 Ma, 2 Ga, 2.2 Ma, 2.5 Ga, 3 Ma, and 3.3 Ga (Mohnicke et al., 1992; Linnemann et al., 2000; Tichomirowa, 2002). The youngest inherited zircon cores in both Tremadocian groups result from Late Cadomian events (~540 Ma). In the 540 Ma old granitoids, the youngest core-bearing crystals show ages of ~570 Ma (Linnemann et al., 2000). Although these are minimal ages, the reproducibility of these data suggests that the ages are a good estimate of the ages of the cores (Gehmlich, 2003), implying that the protoliths of the Tremadocian and the Late Cadomian S-type rocks are similar in age. Because the geochemical signatures of the Late Cadomian and the SiO_2-poor Tremadocian S-type rocks show no significant differences, the original materials for the various anatectic rock suites should be similar. The only difference between both rock suites seems to be the age. According to Tichomirowa (2002), the protoliths probably were Cadomian backarc sediments.

The zircon crystals in the Upper Devonian samples are very uniform in morphology. The {100} prism and the {101} pyramid are dominant and are typical of I-type (Vavra, 1994), subalkaline granitoids (Pupin, 1980). The morphologies of the zircon crystals in granitoids and rhyolitoids are very similar (Fig. 5E). $^{207}Pb/^{206}Pb$ evaporation ages of inherited cores are: ~480 Ma, 540 Ma, 560 Ma, 670 Ma, 1 Ga, and 2 Ma.

Conclusions

In five magmatic events (~580–560 Ma, ~550–530 Ma, ~500 Ma, ~495–480 Ma, and ~380–360 Ma) in Saxo-Thuringia, there is a close correspondence between zircon morphologies and geochemical signatures. Late Cambrian and Tremadocian events could not be distinguished on base of the age data alone, inasmuch as analytical uncertainty is typically only 5 to 10 m.y. (Fig. 2). Using zircon morphologies and geochemical signatures, however, allows one to distinguish between igneous rocks formed during these events.

Two major zircon types could be distinguished: (1) dominance of {100} prisms and {101} pyramids (~580–560 Ma, ~500 Ma, ~380–360 Ma); (2) dominance of {110} prisms and {211} pyramids (~550–530 Ma, ~495V480 Ma). The granitoids of the same age range and geochemical signatures have similar zircon morphologies. The observed dominance of {100} prisms and {101} pyramids in zircon grains for three magmatic events indicates dry I-type granitoids (Vavra, 1994) with a tendency toward A-type bulk-rock compositions. On the other hand, the combination of the acicular pyramids and {110} prisms as well as the abundance of rounded cores with different ages support an origin as anatectic S-type granitoids (Rottura et al., 1989). By using the Ga/Al ratios, two different groups (S-type rocks and rocks with A-type composition) can be distinguished (Fig. 4A; Whalen et al., 1987). The zircon morphology is consistent with this discrimination.

The different magmatic events reveal zircon crystals with two contrasting morphologies. During the Cadomian orogeny, a younger event of S-type granitoids (~550–530 Ma) follows after an event that produced I-type rocks (~580–560 Ma). The same sequence is observed in Cambro-Ordovician time, in which I-type rocks (~500 Ma) are followed by S-type (~495–480 Ma). The Variscan collision starts with I(A)-type granitoids (~380–360 Ma).

The events could reflect the following tectonic evolution: (1) The Cadomian magmatism at ~580–560 Ma represented by granitoid pebbles, detrital zircon grains, and ash layers (Linnemann et al., 2000; Gehmlich, 2003) was related to a backarc environment. The siliciclastic sediments contain

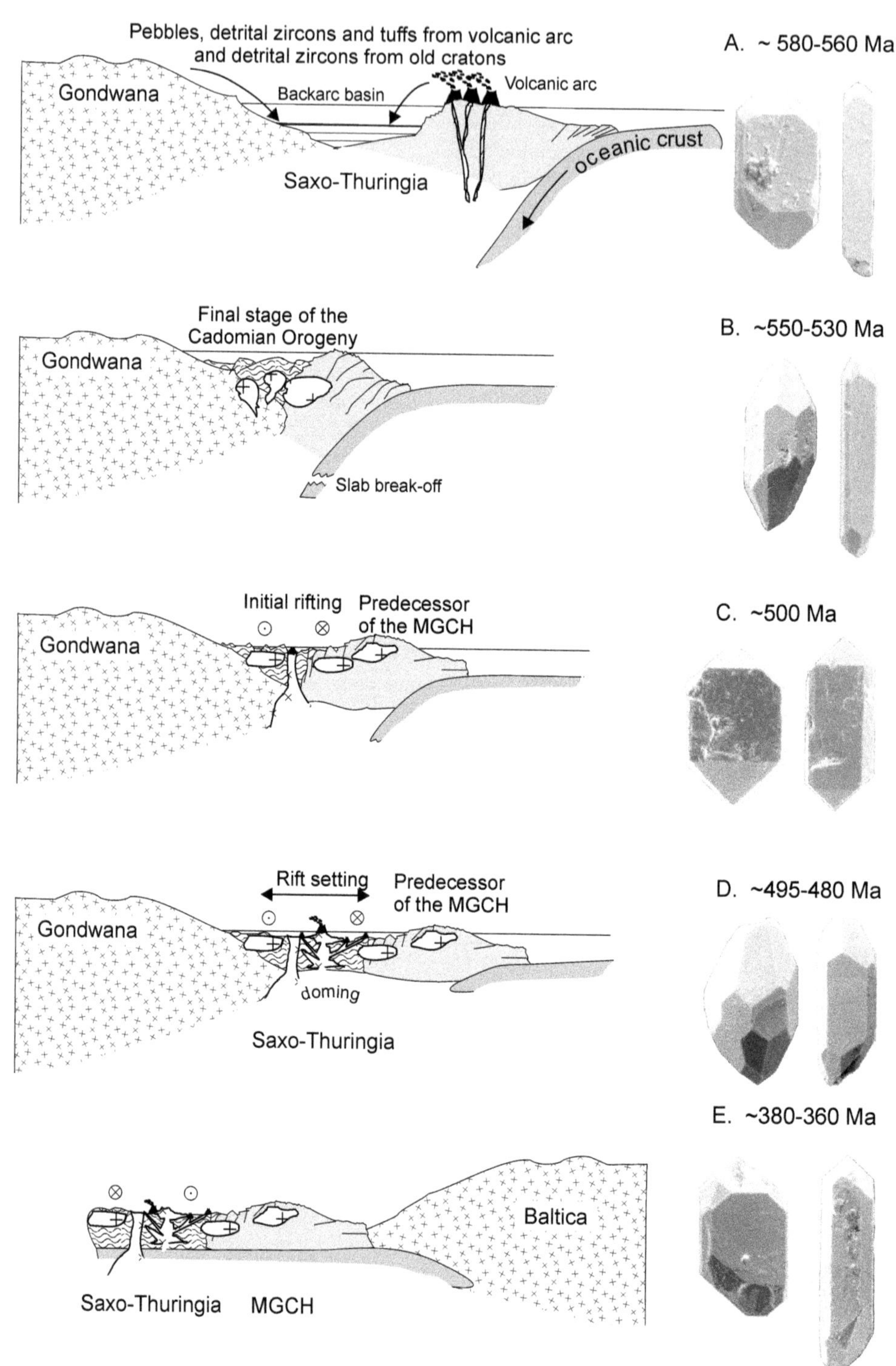

FIG. 6. Schematic tectonomagmatic model for the northern margin of Gondwana from the Cadomian orogeny to the beginning of the Variscan collision (~580–360 Ma). Typical zircon habit for each magmatic event is shown on the right. Abbreviations: MGCH = Mid-German Crystalline High.

detritus of an island arc and of an old craton of Gondwana (Fig. 6A). (2) Due to obduction of the arc, the backarc deposits were deformed and the resulting crustal thickening led to anatectic S-type granitoids (~550–530 Ma, e.g., Tichomirowa, 2002; Gehmlich, 2003) with geochemical signatures similar to the fillings of the backarc basins (Tichomirowa, 2002) (Fig. 6B). (3) I-type granitoids mainly documented by ~500 Ma granitoid pebbles (Gehmlich, 2003) could reflect rifting that led to the separation of Saxo-Thuringia from Gondwana (Kemnitz et al., 2002) (Fig. 6C). (4) As rifting proceeded, high heat flow led to anatectic melts, peaking in the Tremadocian (~495–480 Ma) (Gehmlich, 2003) (Fig. 6D). (5) The occurrence of Late Devonian granitoids, rhyolitoids, and granitoid pebbles in Upper Devonian conglomerates is connected with a coeval, large, within-plate strike-slip shear zone, transpressional horsts, and rapid uplift during the oblique collision of Saxo-Thuringia with northern Europe (Gehmlich et al., 2000) (Fig. 6E).

Acknowledgments

U. Kempe (Freiberg) is thanked for assistance with scanning electron microscopy. Furthermore, we thank U. Linnemann for a long constructive cooperation and for provision of unpublished geochemical data. The work is based on results of projects funded by the Deutsche Forschungsgemeinschaft (grants Lu 544/1 and 2, Li 521/6-1, and Li 521/14-1/2). We thank B. Murphy and the reviewers F. Finger and R. L. Romer for their constructive comments on the manuscript.

REFERENCES

Benisek, A., and Finger F., 1993, Factors controlling the development of prism faces in granite zircons: A microprobe study: Contributions to Mineralogy and Petrology, v. 114, p. 441–451.

Brätz, H., 2000, Radiometrische Altersdatierung und geochemische Untersuchungen von Orthogneisen, Graniten und Granitporphyren aus dem Ruhlaer Kristallin, Mitteldeutsche Kristallinzone: Unpubl. Ph.D. thesis, Universität Würzburg, 151 p.

Buschmann, B., Nasdala, L., Jonas, P., Linnemann, U., and Gehmlich, M., 2001: Shrimp U-Pb dating of tuff-derived and detrital zircons from Gondwanian marginal basin fragments (Neoproterozoic) in the north-eastern Saxothuringian Zone (Germany): Neues Jahrbuch für Geologie und Paläontologie, Monatsheft, v. 6, p. 321–342.

Drost, K., Linnemann, U., McNaughton, N., Fatka, O., Kraft, P., Gehmlich, M., Tonk, C., and Marek, J., 2004, New data on the Neoproterozoic–Cambrian geotectonic setting of the Teplá-Barrandian volcano-sedimentary successions: geochemistry, U-Pb zircon ages, and provenance (Bohemian Massif, Czech Republic): International Journal of Earth Sciences, v. 93, p. 742–757.

Franke, W., 2000, The mid-European segment of the Variscides: tectonostratigraphic units, terrane boundaries and plate tectonic evolution: Geological Society of London, Special Publication, v. 179, p. 9–20.

Gehmlich, M., 2003, Die Cadomiden und Varisziden des Saxo-Thuringischen Terranes–Geochronologie magmatischer Ereignisse: Freiberger Forschungshefte, v. C 500, 129 p. + appendix.

Gehmlich, M., Linnemann, U., Tichomirowa, M., Gaitzsch, B., Kroner, U., and Bombach, K., 2000, Geochronologie oberdevonischer bis unterkarbonischer Magmatite der Thüringischen und Bayerischen Faziesreihe sowie variszischer Deckenkomplexe und der Frühmolasse von Borna—Hainichen (Saxothuringisches Terrane): Zeitschrift für geologische Wissenschaften, v. 151, no. 4, p. 337–363.

Kemnitz, H., Romer R. L., and Oncken O., 2002, Gondwana breakup and the northern margin of the Saxothuringian belt (Variscides of Central Europe): International Journal of Geosciences (Geologische Rundschau), v. 91, p. 246–259.

Kröner, A., and Willner, A. P, 1998, Time of formation and peak of Variscan HP-HT metamorphism of quartz-feldspar rocks in the Central Erzgebirge, Saxony, Germany: Contributions to Mineralogy and Petrology, v. 132, p. 1–20.

Kröner, A., Willner, A. P., Hegner, E., Frischbutter, A., Hofmann J., and Bergner R., 1995, Latest Precambrian (Cadomian) zircon ages, Nd isotopic systematics and P-T evolution of granitoid orthogneisses of the Erzgebirge, Saxony and Czech Republic: Geologische Rundschau, v. 84, p. 437–456.

Kurze, M., 1966, Die Tektonisch-fazielle Entwicklung im NE-Teil des Zentralsächsischen Lineaments: Freiberger Forschungshefte, v. C 201, p. 11–89.

Linnemann, U., Gehmlich, M., Heuse, T., and Schauer, M., 1999, Die Cadomiden und Varisziden im Thüringisch-Vogtländischen Schiefergebirge (Saxothuringisches Terrane): Beiträge zur Geologie von Thüringen NF, Hft. 6, p. 9–39.

Linnemann, U., Gehmlich, M., Tichomirowa, M., Buschmann, B., Nasdala, L., Jonas, P., Lützner, H., and Bombach, K., 2000, From Cadomian subduction to Early Palaeozoic rifting: The evolution of Saxo-Thuringia at the margin of Gondwana in the light of single zircon geochronology and basin development (Central European Variszides, Germany): Geological Society of London, Special Publication, v. 179, p. 131–154.

Mingram, B., and Rötzler, K., 1999, Geochemische, petrologische, und geochronologische Untersuchungen im Erzgebirgskristallin—Rekonstruktion eines Krustenstapels: Schriftenreihe für Geowissenschaften, v. 9, p 1–80.

Mohnicke, M., Kurze, M., and Tichomirowa, M., 2001, Petrogenese und Altersstellung der Gesteine des "Coswiger Komplexes" (Elbezone): Zeitschrift für geologische Wissenschaften, v. 27, no. 5/6, p. 505–519.

Pupin, J.-P., 1980, Zircon and granite petrology: Contributions to Mineralogy and Petrology, v. 73, p. 207–220.

Rottura, A., Bargossi, G. M., Caironi, V., D`Amico, C., and Maccarone, E., 1989, Petrology and geochemistry of late Hercynian granites from the Western Central System of the Iberian Massif: European Journal of Mineralogy, v. 1, p. 667–683.

Schermaier, A., Haunschmid, B., Schubert, G., Frasl, G., and Finger, F., 1992, Diskriminierung von S-Typ und I-Typ Graniten auf der Basis zirkontypologischer Untersuchungen: Frankfurter Geowissenschaftliche Arbeiten, Serie A, Geologie-Paläontologie, v. 11, p. 149–153.

Speer, J. A., 1980, Zircon: Mineralogical Society of America, Reviews in Mineralogy, v. 5, p. 67–112.

Sylvester, P. J., 1998, Post-collisional strongly peraluminous granites: Lithos, v. 45, p. 29–44.

Taylor, S. R., and McLennan, S.M., 1985, The continental crust: Its composition and evolution: Oxford, UK, Blackwell Scientific Publications, 312 p.

Tichomirowa, M., 2002, Die Gneise des Erzgebirges—hochmetamorphe Äquivalente von neoproterozoischen Grauwacken und Granitoiden der Cadomiden: Freiberger Forschungshefte, v. C 495, 222 p.

Tichomirowa, M., Berger, H.-J., Koch, E. A., Belyatski, B. V., Götze, J., Kempe, U., Nasdala, L., and Schaltegger, U., 2001, Zircon ages of high-grade gneisses in the Eastern Erzgebirge (Central European Variszides)—constraints on origin of the rocks and Precambrian to Ordovician magmatic events in the Variscan foldbelt: Lithos, v. 56, p. 303–332.

Vavra, G., 1990, On the kinematics of zircon growth and its petrogenetic significance: A cathodoluminescence study: Contributions to Mineralogy and Petrology, v. 106, p. 90–99.

Vavra, G., 1994, Systematics of internal zircon morphology in major Variscan granitoid types: Contributions to Mineralogy and Petrology, v. 117, p. 331–344.

Whalen, J. B., Currie, K. L., and Chappell, B. W., 1987, A-type granites: Geochemical characteristics, discrimination and petrogenesis: Contributions to Mineralogy and Petrology, v. 95, p. 407–419.

The Acadian Orogeny in the Northern Appalachians

J. BRENDAN MURPHY[1]

Department of Earth Sciences, St. Francis Xavier University, Antigonish, Nova Scotia B2G 2W5, Canada

AND J. DUNCAN KEPPIE

Instituto de Geología, Universidad Nacional Autónoma de México, México D.F. 04510, México

Abstract

The Acadian orogeny, involving deposition of clastic wedges, deformation, metamorphism, magmatism, and exhumation, is limited in time to the Devonian, and in space to the northern mainland Appalachians. Conventional interpretations attribute it to collision between Laurentia and Avalon or Meguma terranes. However, advances in paleogeography indicate that the Avalon and Meguma terranes were accreted to Laurentia in the Late Ordovician and Early Silurian, which was synchronous with closure of the Iapetus Ocean. On the other hand, the Rheic Ocean remained open to the south. In this context, we propose that the Acadian orogeny developed on an Andean-type margin, and attribute it to flattening of the subduction zone as a consequence of collision of an oceanic plateau surrounding a plume. This model explains: (1) SE to NW diachronism in the onset of deformation throughout the Devonian; (2) the development of a ~400–380 Ma magmatic arc gap in Maritime Canada that was abruptly terminated in the Meguma terrane at ~380–370 Ma by (3) intrusion of voluminous felsic magmatism and plume-related lamprophyres as the plume thermally eroded the oceanic lithosphere, causing melting of the lower crust; (4) accompanying regional high-T, low-P metamorphism related to thermal anomalies above a plume; (5) emplacement of gold deposits and associated siderophile elements, possibly derived from fluid circulation above an ascending plume; and (6) rapid Late Devonian exhumation of ~10 km attributed to dynamic uplift over the plume. As the plume head migrated northward, the anomalously intense bimodal magmatism shifted into the Cobequid Highlands (Avalon terrane) at ~360 Ma, and then to the Magdalen Islands, where ~330 Ma plume-related magmatism occurred above a high-density, lower crustal lens interpreted as plume-derived underplated mafic rocks. Late Carboniferous formation of the Maritimes basin is attributed to cooling of this decapitated plume head.

Introduction

THE APPALACHIAN OROGEN extends for more than 3000 km from Newfoundland to Alabama along the eastern margin of North America, and pre-Mesozoic reconstructions imply former continuity and genetic linkages with the Caledonide and Variscan orogens of western Europe (Fig. 1). It is now generally accepted that the Appalachian-Caledonide-Variscan orogen represents the destruction of Paleozoic oceans such as the Iapetus and Rheic oceans, in which the accretion of suspect terranes to Laurentia-Baltica at various times during the Ordovician–Devonian was followed in the Permo-Carboniferous by terminal continental collision with Gondwana and the formation of Pangea (e.g., Williams, 1979; Keppie, 1985; van Staal et al., 1998 and references therein).

The destruction of these oceans is recorded by several episodes of orogenic activity that have generally been related to collision of various terranes with each other or with cratons (see Fig. 2). Thus, the earliest orogenic event, the Late Cambrian–Middle Ordovician Penobscotian ororeny, has been attributed to the amalgamation of composite terranes within the Iapetus Ocean (Keppie, 1993; Stewart et al., 1995). The early to mid-Ordovician Taconic orogeny is inferred to have been the result of collision between Laurentia and outboard oceanic arcs, backarcs, and oceanic terranes that originated in the Iapetus ocean and contain sparse Laurentian fauna (e.g., van Staal et al., 1998). The Late Ordovician–Silurian Salinic orogeny has been related to the accretion of the Gander, Avalon, Nashoba, and Carolina terranes to Laurentia during the closure of

[1]Corresponding author; email:bmurphy@stfx.ca

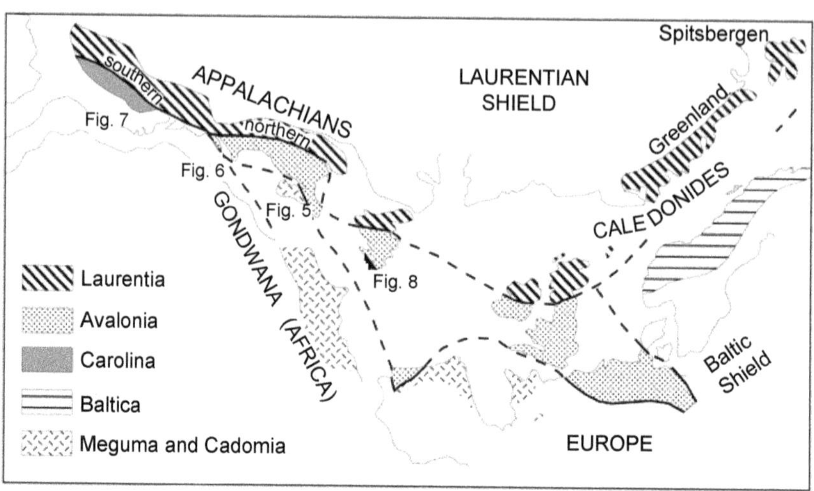

FIG. 1. Pre-Mesozoic reconstruction of the Appalachian-Caledonide-Variscan orogen, (from McKerrow et al., 2000). Locations of Figures 5 (Maritime Canada), 6 (New England) , 7 (Carolinas) , and 8 (Newfoundland) also are shown.

Iapetus (e.g., Keppie, 1985, 1993; van Staal et al., 1998; Hibbard et al., 2002). The Devonian Acadian orogeny has been attributed to collision of either Avalonia with Laurentia (Robinson et al., 1998) or to the accretion of the Meguma terrane (e.g., van Staal et al., 1998). There is general consensus that the last orogenic event, the Late Carboniferous–Permian Alleghenian orogeny, was due to terminal collision between Gondwana and Laurentia-Baltica that closed the Rheic Ocean and resulted in the formation of Pangea.

Although modern analogs record orogenic events due to collision of terranes, they also document widespread orogenic activity related to other mechanisms, such as flat-slab subduction and changes of the convergence rates and directions. Thus in the Andes, flat-slab subduction occurs in four geographically limited zones, two of which are related to collision of a mid-oceanic ridge with the trench (the Chile and Cocos ridges; Ramos and Alleman, 2001), one is inferred to result from subduction of an oceanic plateau (the Inca plateau; Gutscher, 2002), and one is genetically connected with subduction of a seamount chain (San Fernandez ridge; Ramos et al., 2002). In the Cordillera of western North America, the Late Cretaceous–Early Tertiary Laramide orogeny has been attributed to overriding of the Yellowstone hotspot (Murphy et al., 1998). These different mechanisms produce different geological records. Thus, ridge-trench collision is unique in producing a gap in the magmatic arc and slab windows that are associated with rift-related magmatism (Thorkelson and Taylor, 1989). Plume-trench collision also produces a magmatic-arc gap but is associated with a wide zone of deformation, and may be followed by extensive plume-related magmatism after the plume burns through the subducting slab and into the overlying lithosphere. Subduction of aseismic ridges and changes in the rate of convergence, although producing a magmatic-arc gap and a wide zone of deformation, are not associated with rift or plume magmatism. In the southwestern Pacific Ocean, Collins (2002) has related contractions (i.e., orogenic events) and extension to alternations of flat-slab and steep subduction that are restricted in both time and space. The difference between the southwestern and eastern Pacific may be related to different rates of convergence across the trench. However, in all of these cases, subduction is ongoing and its effects are restricted to a narrow zone, whereas flat-slab subduction is limited to short episodes with effects that have dimensions congruent with the along-strike width of the flat slab but a wide across-strike extent. In such cases, narrow and wide orogenic zones should alternate along any one active margin. This necessitates a close examination of the extent and duration of orogenic events in an orogen, rather than assuming that they may be correlated along the entire length of the orogen.

Flat-slab and convergence rate mechanisms for orogenic activity are probably more common in the

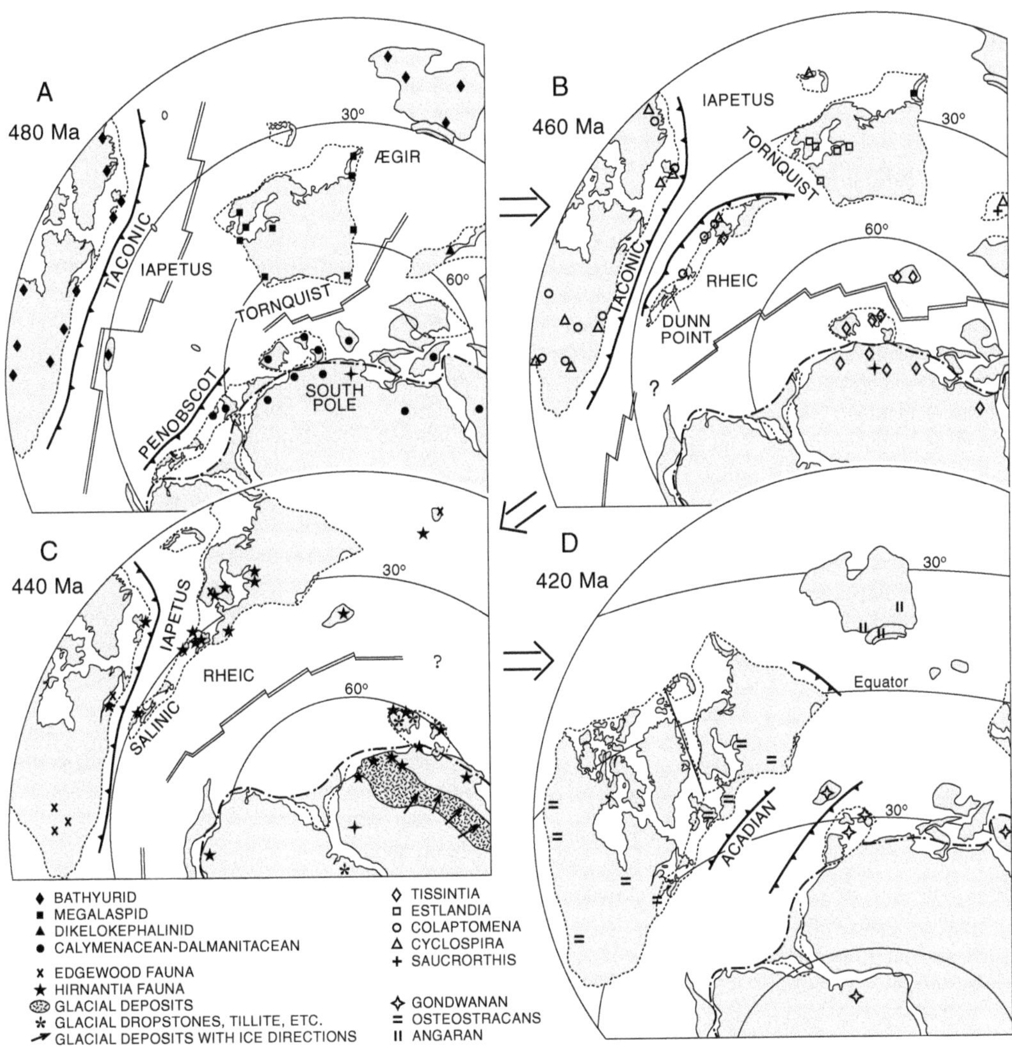

FIG. 2. Early Ordovician to Late Silurian paleogeographic reconstructions of the Circum-Atlantic region (based on Cocks and Torvik, 2002; Fortey and Cocks, 2003).

Appalachians than presently recognized. Murphy et al. (1999) and Keppie and Krogh (2000) proposed that the geological record of the Meguma and Avalon terranes in Maritime Canada during the Acadian orogeny is consistent with flat-slab subduction caused by subduction of an oceanic plateau and a related plume. This paper extends the implication of plume-trench collision as a mechanism for the Acadian orogeny to the northern Appalachians.

The mechanism deduced for the various orogenic events in the Appalachians is largely conditional upon inferred paleogeographic reconstructions. For example, the closure of ocean basins between terranes is generally related to collision orogenic events. Thus the closure of the Iapetus Ocean has been related to both the Ordovician–Silurian Salinic orogeny (e.g. Keppie, 1985, 1993; van Staal et al., 1998; Hibbard et al., 2002) and the Devonian Acadian orogeny (Robinson et al., 1998). Similarly the closure of the Rheic Ocean between the Meguma and Avalon terranes has been related to the Acadian orogeny (e.g., Schenk, 1997; van Staal et al., 1998), although some authors have suggested that the Meguma Group was deposited upon an Avalonian basement (Keppie and Dostal, 1991; Keppie et al., 1997; Murphy et al., 2004b). In view of these contrasting views, we first address the paleogeography just before the onset of the Acadian orogeny.

Silurian Paleogeography

The time of closure of the Iapetus Ocean between eastern Laurentia and Avalonia has been inferred to be either Late Ordovician–Early Silurian or Devonian; however, geological and geophysical data are accumulating in support of the former. Lithologic (O'Brien et al., 1983, 1996) and paleomagnetic data (Johnson and Van der Voo, 1986; Van der Voo, 1988) as well as faunal evidence (Pickering et al., 1988; Cocks and Fortey, 1990; Landing, 1996; Cocks and Torsvik, 2002; Fortey and Cocks, 2003) indicate that Avalonia was located along the periphery of Gondwana from the late Neoproterozoic to the Early Ordovician. Separation between Avalonia and Gondwana gradually increased during the Ordovician (Figs. 2A and 2B). Paleontological evidence supports faunal linkages of Avalonia with Baltica by the Late Ordovician (e.g., Williams et al., 1995; Fortey and Cocks, 2003). Recent geochemical and isotopic data from the ~1900 m Late Ordovician–Early Devonian sedimentary sequence of the Arisaig Group show that the clastic rocks contrast with the underlying Avalonian basement rocks, indicating that they were not derived from Avalonian basement (Murphy et al., 1996b, 2004a).

All sedimentary rocks are characterized by strongly negative ε_{Nd} (from –4.8 to –9.3) and T_{DM} ages >1.5 Ga, with an overall trend toward increasingly negative ε_{Nd} values from the base to the top of the group (Fig. 3). U-Pb detrital zircon data from Lower Silurian (Beechill Cove Formation), Middle Silurian (French River Formation), and Lower Devonian (Stonehouse Formation) strata of the Arisaig Group have similar zircon populations (Fig. 4). Some zircons are close to their respective depositional ages, suggesting that Arisaig basin formation may have been broadly coeval with active volcanism in the orogen. These rocks are also characterized by minor Ordovician and Cambrian zircons, significant populations of Neoproterozoic (550–700 Ma), Mesoproterozoic (0.95–1.3 Ga), Paleoproterozoic (2.0–2.2 Ga), and minor Archean zircons. A comparison between these data and the age of tectonothermal events in potential source areas (Fig. 4), together with regional geologic data, suggest that Arisaig Group strata were primarily derived from Baltica (Murphy et al., 2004a).

Paleomagnetic data indicate a 41°S ± 8° paleolatitude for the Avalon terrane in Nova Scotia (Dunn Point volcanics: Van der Voo and Johnson, 1985, recently dated at ~460 Ma; Hamilton and Murphy, 2004). Inasmuch as Laurentia lay at a paleolatitude of ~20°S between 460 and 440 Ma (e.g., MacNiocaill and Smethurst, 1994), this implies that Avalonia was located about 1700–2000 km south of the Laurentian margin at 460 Ma. Paleomagnetic data suggest that any paleolatitudinal separation between Avalonia and Laurentia had disappeared by the Early Silurian (Miller and Kent, 1988; Trench and Torsvik, 1992; Potts et al., 1993; MacNiocaill et al., 1997). This indicates a latitudinal convergence rate between Avalonia and Laurentia between 460 and 440 Ma of about 5.5 cm/yr (i.e., the destruction of the Iapetus Ocean). As Gondwana remained relatively stationary during this period (e.g., Torsvik et al., 1996), these data imply a northerly component of drift of Avalonia from Gondwana of about 8 cm/yr (development of the Rheic Ocean). Taken together, these data define the southern limit of Iapetus and the northern limit of the Rheic Ocean at 460 Ma. Thus these data indicate a connection between the death of the Iapetus Ocean and the Salinic orogeny, rather than with the Acadian orogeny.

In the southern and central Appalachians, paleomagnetic studies indicate minimal latitudinal separation between the Carolina terrane and Laurentia by ~455 Ma (Vick et al., 1987; Noel et al., 1988). According to Hibbard (2000), sinistral accretion of Carolina and related terranes to Laurentia began in the Middle to Late Ordovician and continued into the Early Silurian, as evidenced by $^{40}Ar/^{39}Ar$ cooling ages on micas that define the regional cleavage in the Carolina zone. Although paleomagnetic data indicate that Avalonia was at 41°S (Johnson and Van der Voo, 1990; Hamilton and Murphy, 2004) at ~460 Ma, suggesting a separation of ~2000 km, Hibbard et al. (in press) point out that because the current direct separation between sampled sites is ~1900 km, Carolina and Avalonia could represent the leading and trailing edges, respectively, of the same plate.

The Acadian orogeny has also been attributed to the collision of the Meguma and Avalona terranes and the closure of the Rheic Ocean. However, the original relationship between the Meguma and Avalon terranes is controversial (e.g., Keppie 1993; van Staal et al., 1998). Many authors have inferred

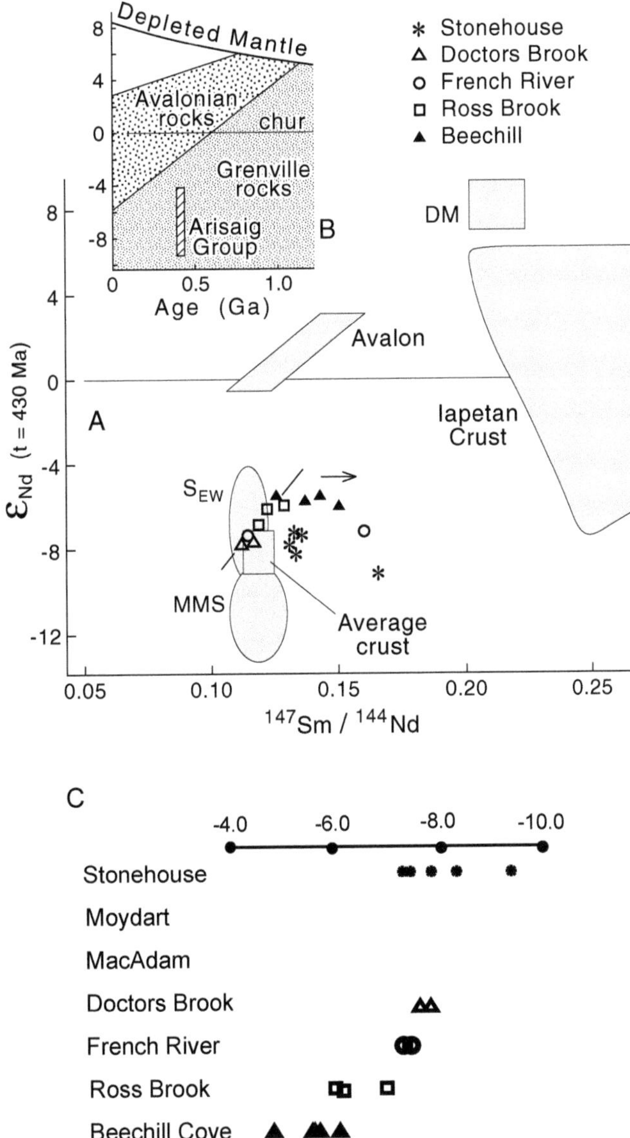

FIG. 3. A. ε_{Nd}^{t} versus $^{147}Sm/^{144}Nd$ diagram (t = 430 Ma) comparing Sm-Nd isotopic data for the Arisaig Group with typical Sm-Nd isotopic compositions of Avalonian crust (Murphy and MacDonald, 1993; Murphy et al., 1996b). The Sm-Nd isotopic characteristics for the average upper crust are bracketed between modern global average river sediment ($^{147}Sm/^{144}Nd$ = 0.114; T_{DM} = 1.52 Ga; Goldstein and Jacobsen, 1988) and the average age of sedimentary mass (Miller et al., 1986). See also Thorogood (1990). Iapetan crust includes normal and depleted island arc tholeiites, and ophiolitic complexes in Newfoundland and Norway (Pedersen and Dunning, 1997; MacLachlan and Dunning, 1998). Silurian strata of England and Wales (S_{EW}) from Thorogood et al. (1990), and Meguma terrane metasedimentary rocks (MMS) from Clarke et al. (1997). B. ε_{Nd}^{t} versus time (Ga) diagram (t = 430 Ma) comparing Sm-Nd isotopic data for the Arisaig Group with typical Sm-Nd isotopic compositions of Avalonian crust (Murphy et al., 1996b, 2000). Field for Grenville rocks after Patchett and Ruiz (1989), Dickin and McNutt (1989), Dickin et al. (1990), Daly and McLelland (1991), McLelland et al., (1993), Dickin (2000). C. Variation in ε_{Nd}^{t} (t = 430 Ma) with stratigraphic height in the Arisaig Group. Beechill Cove Formation data from Murphy et al., 1996b.

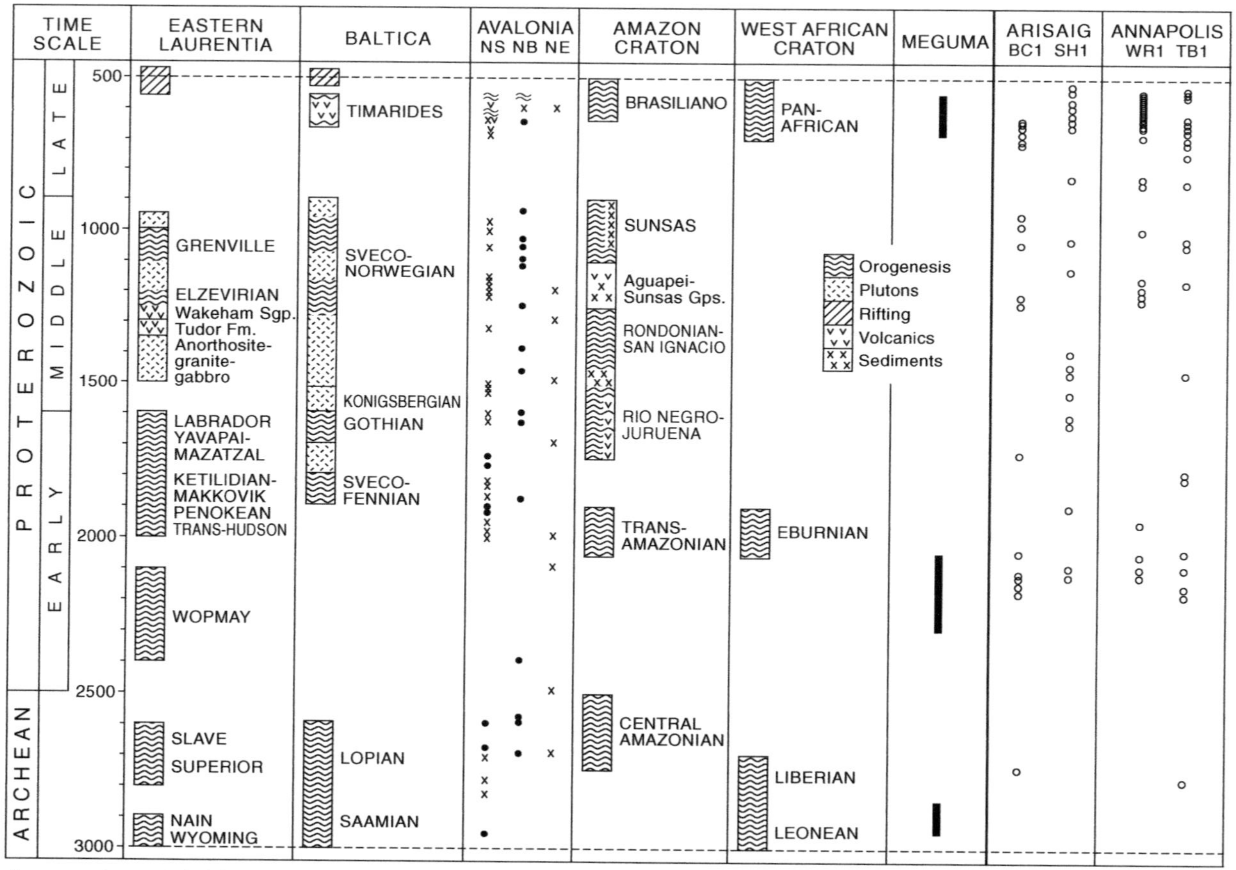

FIG. 4. Detrital zircon ages (open circles) from coeval uppermost Ordovician to Lower Devonian clastic sequences in the Avalon terrane (Arisaig Group) and Meguma terrane (Annapolis Valley), after Murphy et al. (2004b). In the Avalon terrane, BC-1 is from the Lower Silurian Beechill Cove Formation and SH-1 is from the Lower Devonian Stonehouse Formation. In the Meguma terrane, WR-1 is from the Upper Ordovician–Lower Silurian Whiterock Formation and TB-1 is from the Lower Devonian Torbrook Formation. These data are compared with detrital-zircon data from underlying uppermost Neoproterozoic–Lower Ordovician Meguma Group (Krogh and Keppie, 1990) and Neoproterozoic Avalonia (Keppie et al., 1998; Bevier et al., 1990). Symbols: x = concordant U-Pb zircon ages; filled circles = discordant $^{207}Pb/^{206}Pb$ ages. Also shown are tectonothermal events in Baltica (Gower et al., 1990; Roberts, 2003), eastern Laurentia (Cawood et al., 2001), Amazon craton (Sadowski and Bettencourt, 1996), northwest Africa (Rocci et al., 1991), and Gander (van Staal et al., 1996). Abbreviations: NS = Nova Scotia; NB = New Brunswick; NE = New England.

that the Meguma Group represents a Cambrian–Early Devonian passive margin bordering northwest Africa that was transferred to Laurentia during the Acadian orogeny (e.g., Schenk, 1997 and references therein). This was based primarily upon: (1) proposed stratigraphic correlations between the Cambro-Silurian strata in the Meguma terrane and coeval sequences in Morocco (Schenk, 1997); and (2) the Middle Devonian age of the Acadian orogeny, the oldest accretionary event recognized in the Meguma terrane. According to this model, the Avalon and Meguma terranes lay on the opposite sides of the Rheic Ocean, and were juxtaposed during the Acadian orogeny.

Alternatively, the Cambrian–Early Devonian strata of the Meguma terrane have been interpreted as a passive margin bordering the Avalon microcontinent, which would imply that the Meguma Group was deposited on Avalonian continental crust (e.g., Keppie et al., 2003 and references therein). This is based upon (1) the proposed correlation of the Upper Ordovician–Lower Devonian units in the Meguma terrane with coeval sequences in the rest of the Appalachians (Keppie and Krogh, 1999); and (2) the similarity of Nd-isotopic signatures in Late Ordovician–Early Silurian crustally derived igneous suites in the Meguma and Avalon terranes (Keppie et al., 1997). As the Silurian to Lower Devonian strata predate the Acadian orogeny, and no older deformational events are recorded, these data have been interpreted to indicate that the Meguma Group rested depositionally on Avalonian basement in the Cambro-Ordovician. In this interpretation, Avalonia and Meguma lay on the same (northern) flank of the Rheic Ocean, and together collided with Laurentia by the Early Silurian. Inasmuch as the Meguma is the most outboard terrane in the Appalachians, this latter model would imply that the eastern flank of the Appalachians would have resembled a modern Cordilleran or Andean margin.

U-Pb detrital zircon data from Ordovician and Lower Devonian clastic sedimentary rocks in the Meguma terrane (Fig. 4); (White Rock and Torbrook formations in the Annapolis Valley, respectively) are very similar to those from the Arisaig Group (Avalon terrane; Murphy et al., 2004b). In addition to abundant Cambrian–Late Neoproterozoic and Paleoproterozoic zircons, Late Ordovician–Early Devonian samples from the Meguma terrane have important Mesoproterozoic zircon populations (1.0–1.4 Ga) that are typical of Avalonia, and strongly suggest their contiguity by Late Ordovician–Early Silurian time. Because Avalonia had accreted to Laurentia-Baltica by the Late Ordovician, these data suggest that the Meguma terrane, like Avalonia, resided along the same (northern) margin of the Rheic Ocean at that time. That these terranes were juxtaposed throughout the Early Paleozoic and probably into the Neoproterozoic is also indicated by the absence of evidence for a Cambro-Ordovician accretionary event in both the Meguma and Avalonian stratigraphy, the lack of intervening suture zone ophiolitic units, and the similarity of Avalonian and Meguma basement Nd isotopic signatures in Paleozoic igneous suites (Keppie et al., 1998). In view of these recent data, the Acadian orogeny cannot be due to accretion of the Meguma and Avalon terranes.

Acadian Orogeny

The Acadian orogeny is named for Acadia, the old French name for the Maritime Provinces of Canada (Nova Scotia, New Brunswick, and Prince Edward Island). It is the term originally defined for the middle Paleozoic deformation event in the Appalachian orogen of Maritime Canada (including eastern Quebec). The term "Acadian" was introduced by Schuchert (1923) to explain Late Devonian deformation in the Chaleur Bay region of New Brunswick and Gaspé, Quebec. Several workers including Boucot et al. (1964) and Poole et al. (1970) showed that this deformation was regionally extensive in the northern Appalachians (Fig. 5). In Maritime Canada, the Acadian orogeny affected the entire width of the orogen, from the most outboard terrane (Meguma) to the Laurentian foreland. Donohoe and Pajari (1974) showed that the age of peak Acadian deformation became progressively younger northward from coastal Maine to Gaspé, varying from Gedinnian to post-Eifelian in age, a trend confirmed in a recent synthesis by Robinson et al. (1998).

The effects of the Acadian orogenic event are now widely interpreted to extend along the southeastern Laurentian margin from Pennsylvania to Newfoundland, and to correlative rocks in Western Europe (Bradley, 1989; Williams, 1979, 1993; Woodcock and Strachan, 2000) where Late Paleozoic deposits overlie, with angular unconformity, previously deformed Early Paleozoic sequences. This regional unconformity is commonly expressed by widespread Lower Devonian platformal carbonates overlying Taconic foredeep deposits, both

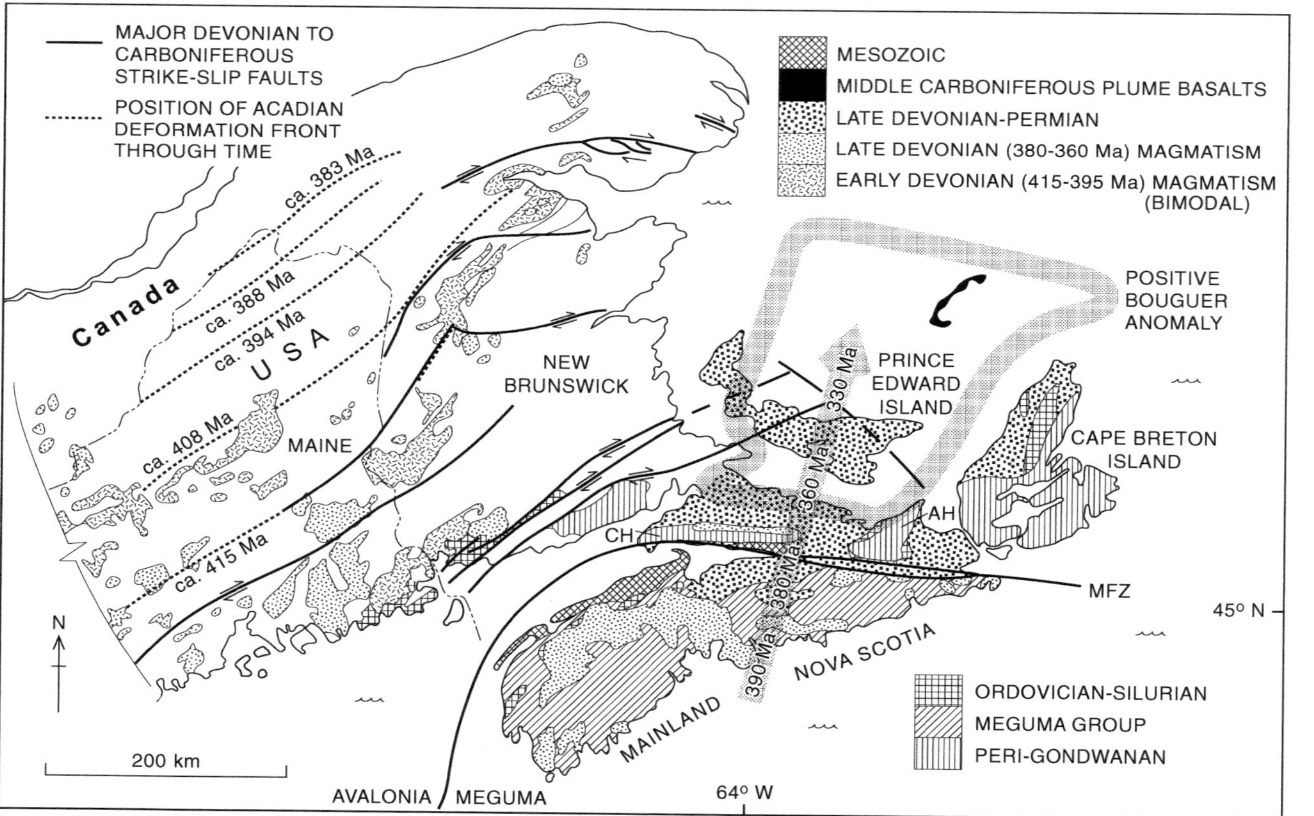

FIG. 5. Map of Appalachian orogen in Atlantic Canada and northern Maine showing Silurian–Devonian igneous rocks, the migration of Silurian–Devonian Acadian foredeep and deformation front through time (data from Robinson et al., 1998, extrapolated into New Brunswick), and Devonian–Carboniferous magmatism associated with plume activity (modified from Murphy and Keppie, 1998). Silurian arc magmatism and Early Devonian bimodal magmatism are associated with convergence of Avalonia and Meguma to Laurentia, and possibly related to resultant slab break-off (see van Staal et al., 1998). The proposed plume track from ca. 390 to 330 Ma is shown with a large arrow. The flat-slab model applies to the Acadian orogeny in Maritime Canada and adjacent Maine. Motions of upper and lower plates in a hotspot reference frame dictate that swell and plume would be overridden at different locations along the continental margin. Hence Late Devonian tectonism in New England may also be explained by this model.

traditionally recognized as clear demarcations between earlier Taconic and Acadian orogenic events. More recently, this demarcation has been related to a switch from sinistral-to dextral-dominated kinematics along major NE-trending faults (e.g., Keppie, 1993; Nance and Dallmeyer, 1993; Cawood et al., 1994; Anderson et al., 2001).

Meguma terrane

Stratigraphy. The Meguma terrane is exposed only in mainland Nova Scotia (Fig. 5), where it is separated from the Avalon terrane to the north by the Minas fault zone. The terrane is underlain by thick (~13 km) Cambrian–Ordovician turbidites (Schenk, 1997) of the Meguma Group containing a Gondwanan fauna (Pratt and Waldron, 1991) that are disconformably to conformably overlain by a 2.3 to 4.5 km thick Upper Ordovician to Lower Devonian sequence of bimodal volcanic rocks overlain by shallow-marine to continental clastic rocks with Rhenish-Bohemian fauna (Boucot, 1975). The Ordovician switch from turbidites to shallow marine deposits was accompanied by a change from a south-southwesterly to northwesterly source. Detrital zircons in the Goldenville Formation (lower unit of the Meguma Group) yielded ca. 3.0 Ga, 2.0 Ga, and 600 Ma ages and also indicate a Gondwanan (West African) source prior to separation from Gondwana (Krogh and Keppie, 1990). Near the base of the Upper Ordovician–Lower Devonian sequence, a rhyolitic tuff of the White Rock Formation yielded concordant U-Pb ages of ~440 Ma (Keppie and Krogh, 2000; MacDonald et al., 2002).

Structure. The earliest effects of the Acadian orogeny in the Meguma terrane began at ~415 Ma with regional metamorphism (predominantly greenschist-facies), polyphase deformation, and cleavage formation (Keppie and Dallmeyer, 1995) and continued to the Late Devonian (Hicks et al., 1999). On a regional scale, cleavage is axial planar to NE-trending shallowly plunging periclinal folds. Several studies propose that the onset of the Acadian orogeny at ~415 Ma was associated with oblique dextral convergence between the Meguma and Avalon terranes (Keppie and Dallmeyer, 1995). The current boundary between these terranes is defined by the E-W Minas fault zone (Fig. 5), along which several episodes of late Paleozoic dextral motion has occurred (e.g., Keppie, 1982; Murphy, 2003). Adjacent to this fault zone, regional folds are rotated clockwise, consistent with late Paleozoic dextral motion. Traced westward, this boundary swings into the NE-SW Bay of Fundy (Fig. 5). Seismic profiles in the Bay of Fundy show the major structure to be a SE-dipping Mesozoic listric normal fault inferred to be located along a late Paleozoic listric thrust zone (Keen et al., 1991). Coeval late Paleozoic (Alleghanian) deformation has been documented by Culshaw and Liesa (1997). The down-dip extension of the sole thrust either maintains its SE dip to the Moho or flattens at 15 km depth between the lower and upper crusts (Keen et al., 1991). The latter model implies that the Meguma terrane is allochthonous and that the Minas fault zone may be a lateral ramp (Keppie, 1993).

Magmatism. In the Meguma terrane, intrusion of widespread late syntectonic ~375–370 Ma granitoids and minor mafic dikes (Clarke et al., 1993, 1997) were accompanied by high T-low P metamorphism and shear zone deformation. Although these plutons clearly cross-cut early Acadian fabrics, studies by Benn et al. (1997) show that they crystallized during the latest stages of Acadian deformation and that their geometry is structurally controlled. Geophysical studies (e.g., Keen et al., 1991) show that the plutons have a broadly laccolithic geometry, and extend to depths of 5–10 km, with feeder zones extending to 25 km.

Exhumation. The earliest evidence of exhumation of Meguma terrane rocks is their occurrence as clasts in latest Devonian–Tournaisian Horton Group rocks (Martel et al., 1993), implying an uplift of between 6 and 10 km between ~370 and 360 Ma (e.g., Keppie and Dallmeyer, 1995; Jennex et al., 2000; Murphy, 2000; Murphy and Hamilton, 2000).

Avalon terrane in mainland Nova Scotia and New Brunswick

To the north of the Minas fault zone in mainland Nova Scotia, Avalonia (Fig. 5), the largest suspect terrane in the northern Appalachian orogen, occupies much of the southern flank of the Appalachians. Correlative rocks occur in Ireland, southern Britain, and adjacent parts of continental Europe, thereby providing a potential genetic linkage between orogenic events in Laurentia and Western Europe (e.g., McKerrow et al., 1991, 2000; Woodcock and Strachan, 2000).

Stratigraphy. Ordovician–Lower Devonian rocks in Avalonia of mainland Nova Scotia unconformably overlie older Neoproterozoic and Cambrian strata that contain typical Avalonian fauna (Landing and Murphy, 1991). Ordovician strata consist of ~460 Ma bimodal intracontinental volcanic rocks (Dunn

Point Formation) and interbedded red clastic sediments disconformably overlain by an 1800 m continuous stratigraphic sequence dominated by shallow-marine to continental, fossiliferous siliciclastic rocks (Arisaig Group and correlatives).

Waldron et al. (1996) interpreted the Silurian–Early Devonian subsidence history of the Arisaig Group as an initial phase of rapid subsidence and extension (30% to 60%) in the Early Llandovery, followed by thermal relaxation and slower subsidence rates in the Wenlockian and Ludlow. Vastly increased subsidence rates and accommodation space in the Pridoli (as represented by the Stonehouse Formation) was attributed to loading of the Avalonian margin due to interaction with a neighboring terrane (Waldron et al., 1996), probably the Meguma terrane.

The Arisaig Group is unconformably overlain by Middle Devonian interbedded basalts and continental rocks, and this unconformity is widely interpreted to represent the local manifestation of the Acadian orogeny.

Structure. In the Fredericton trough, the earliest Acadian deformation consists of upright, isoclinal folds with a low-grade cleavage deformed Llandovery-Gedinnian rocks before intrusion of granitoid plutons at 406 ± 7 Ma.

Magmatism. In the Cobequid Highlands of mainland Nova Scotia, Middle to Late Devonian (~360 Ma) magmatism is particularly voluminous, and is represented by bimodal volcanic and plutonic rocks (Doig et al., 1996; Piper et al., 1993).

In southern New Brunswick, the Kingston Complex (e.g., Currie, 1987) or terrane (Barr et al., 2002), is considered to be either part of Avalonia (e.g., Keppie and Dostal, 1991) or Ganderia (Barr et al., 2002). It consists of arc-related metavolcanic and metasedimentary rocks intruded by Early Silurian granite plutons and mafic dikes. The intrusive age of the mafic dikes in uncertain, but $^{40}Ar/^{39}Ar$ analyses on igneous hornblende are interpreted as a ~415 Ma cooling age (Nance and Dallmeyer, 1993). Magmatism in the Kingston Complex has been genetically related to the compositionally similar Late Silurian coastal Maine Complex.

The geochemistry of ~430–422 Ma Avalonian granitoids is typical of subduction, but a younger event (~396–367 Ma) has more mafic components and more juvenile signatures, which have been attributed to crustal delamination by Whalen et al. (1994). Silurian–Early Devonian arc-related plutonic and volcanic rocks switch off at the same time as deformation was occurring in the Central Maine Basin.

Siluro-Devonian geology in central and southern New Brunswick (north of the Kingston Complex) is dominated by relatively low grade volcanic and sedimentary rocks occurring in anticlinoria, which are affected by widespread plutonism and deformation, separated by basins that preserve a relatively continuous stratigraphic record. Deformation was accompanied by regional dextral shear (e.g., Nance and Dallmeyer, 1993; Schreckengost and Nance, 1996). Voluminous Siluro-Devonian granitoid rocks (Whalen et al., 1996) have metaluminous to slightly peraluminous, within-plate chemistry. Silurian plutons have complex histories that may involve input of a juvenile component, whereas Devonian granitoids represent partial melting of hybridized Avalonian crust. Whalen et al. (1996) attribute this magmatism to collision and lithosphere delamination accompanying and following the accretion of Avalonia to Laurentia.

Central and Northern Cape Breton Island

Important Early Silurian (~435 Ma) arc-related magmatism, metamorphism, followed by collisional orogeny in central and northern Cape Breton (Barr and Raeside, 1989; Keppie et al., 1991; see Fig. 5) may be correlative with similar-aged events in the La Poile Group of Newfoundland (e.g., Chandler et al., 1987; Dunning et al., 1990 ; O'Brien et al., 1991; Barr et al., 1998), which are attributed to the Salinic orogeny. According to Lin et al. (1994), these events collectively represented subduction followed by promontory-promontory collision between Laurentia and Avalonia during the Silurian.

Central and northern New Brunswick and neighboring Maine

Much of central and northern New Brunswick and neighboring Maine (Figs. 5 and 6) is underlain by Gondwana-derived Cambrian–Lower Ordovician strata of the Gander terrane (van Staal et al., 1998), which is unconformably overlain by a widespread Middle Ordovician volcano-sedimentary sequence interpreted as an ensialic rifted arc (van Staal, 1994) that oversteps the boundary with the Exploits subzone to the north and Avalonia to the south (Williams and Piasecki, 1990; van Staal et al., 1996) and contains peri-Gondwanan (Celtic) fauna (Neuman and Harper, 1992). Complex Late Ordovician–Silurian deformation is related to sinsistral oblique convergence and collision of peri-Gondwanan

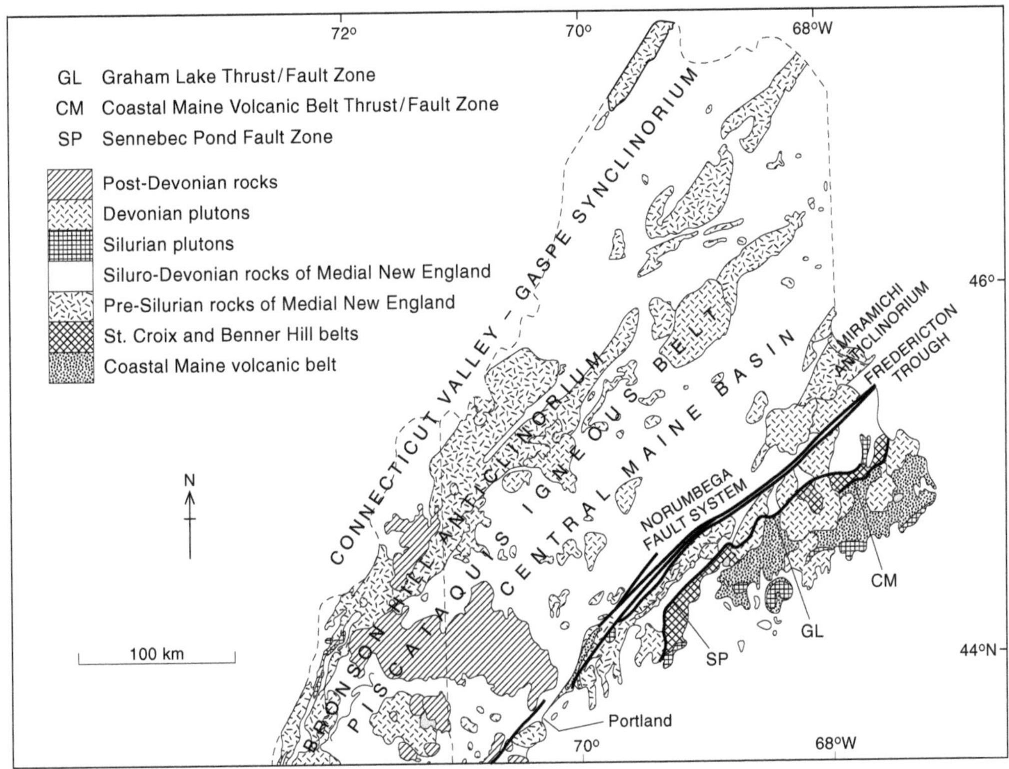

FIG. 6. Geology of northern New England (modified from Robinson et al., 1998; Tucker et al., 2001). Abbreviations: CM, GL, and SP = Coastal Maine, Graham Lake, and Sennebec Pond fault zones, respectively.

terranes against the Laurentian margin (van Staal et al., 1990). According to van Staal and de Roo (1995), several short-lived episodes of arc accretion, slab roll-back and breakup, and rapid exhumation of blueschist-facies metamorphic rocks are attributed to post-collisional extension.

A prominent feature of this region is the Silurian–Lower Devonian Tobique-Piscataquis magmatic belt and correlatives, which have been variously interpreted to represent arc (Rankin, 1968; Bradley, 1983; Thirlwall, 1988) or post-collisional magmatism (Arth and Ayuso, 1997; Ayuso and Arth, 1997) related to the closure of the Iapetus Ocean. Alternatively, they have been interpreted as intra-plate continental tholeiites intruded in an extensional environment (Dostal et al., 1989; Hon et al., 1992).

Foreland clastic wedge

In the Appalachian foreland of Quebec, Silurian–Lower Devonian sedimentary facies in the Gaspé Belt south of the Québec reentrant and the St. Lawrence promontory preserves evidence of significant syn-depositional tectonism, attributed by Malo et al. (1995) to along-strike Acadian structural variations as a consequence of collision along an irregular margin and by Bourque et al. (2000) to the Salinic "disturbance" beginning in late Llandoverian time and continuing into the Early Devonian. Llandoverian–Wenlockian regression is attributed to post-Taconian successor basin filling that culminated with extensive carbonate platform development. This was followed by extensional tectonics that produced shelf faulting and block tilting, on top of which block reefs and reef complexes developed and extended along the Gaspé-Témiscouata shelf. Early Devonian rapid subsidence in the Québec reentrant area was coeval with uplift in the St. Lawrence promontory, attributed to loading of the Laurentian margin associated with the further thrusting of Avalon and Meguma terranes over Laurentia. Lower to Middle Devonian clastic wedge deposits of the eastern Gaspé Peninsula record a gradual displacement of source areas from the southeast to the southwest.

Equivalent clastic wedge deposits in western New York state and Pennsylvania are the Middle–Upper Devonian clastic wedge (Quinlan and Beaumont, 1984) that thins to less than 30 m near Alabama (Hatcher, 1989), where subsidence curves in Middle Paleozoic strata show no effect of loading the Appalachian margin (Thomas and Whiting, 1995). The Catskill clastic wedge is widely interpreted to reflect this loading of the Laurentian margin and a westerly migrating foredeep and clastic wedge ahead of an advancing Acadian orogenic front (Bradley, 1989). Some of the source rocks for the Catskill delta deposits may have been located in the New York promontory in southern New England, where the Siluro-Devonian recumbent folds and high-grade metamorphism was followed by rapid uplift, and removal of 20 km of crust. Stewart et al. (1995) and Bradley (1989) attributed the Siluro-Devonian tectonothermal activity and westerly migration of the Catskill foredeep to the Acadian orogeny.

Post-Acadian structures in the Gaspé Peninsula affected rocks that are as young as Namurian, and such transpressive deformation, more than 1000 km away from areas of peak coeval Alleghanian metamorphism in the southeastern United States, is attributed by Jutras et al. (2003) to the far-field effects of rigid indenters during terminal continental collision. Middle-Late Devonian bimodal igneous rocks followed by Middle Carboniferous mafic rocks with plume-related geochemical signatures (Bedard, 1986) are consistent with intracontinental wrench syn-to-post Acadian tectonic regime.

In central Maine (Fig. 6), shallow-marine Silurian strata gave way to Lower Devonian deeper-water flysch deposits. The Central Maine Basin preserves a record of continuous Late Ordovician–Early Devonian deep-marine sedimentation (Berry and Osberg, 1989), with the lower strata derived from the west, and younger strata from the east. This basin has been interpreted as either a vestige of ocean between Avalonia and Laurentia (Bradley, 1989), or a post-collisional basin intracontinental regime (e.g., Hon et al., 1992).

Southern and Central Appalachians

Effects of the Acadian orogeny in the southern and central Appalachians (Fig. 7) are limited to: (1) a shear zone within the Carolina zone; (2) in the Blue Ridge province of western North Carolina, a phase of Devonian dextral strike-slip tectonics is partitioned into narrow shear zones where it is superimposed on earlier high-pressure Ordovician deformation (Willard and Adams, 1994); and (3) an upper Paleozoic ductile shear zone that defines the western limit of the Carolina zone may represent the latest stage in the movement of the zone against Laurentia. Late Devonian deformation in the central and southern Appalachians has been genetically linked with the Upper Devonian clastic wedge.

Newfoundland

The effects of the Acadian orogeny in Newfoundland are limited (Fig. 8). In central Newfoundland (Humber, Dunnage, and Gander zones), widespread Silurian (~435–415 Na) deformation, metamorphism, plutonism and exhumation are generally considered to be the climactic event following the Early Silurian, Salinic oblique sinistral collision of the Gander and Avalon terrane with Laurentia (Dunning et al., 1990; O'Brien et al., 1991; Williams, 1993; Cawood et al., 1994; Kerr, 1997; Schofield and D'Lemos, 2000) that occurred before onset of the Acadian orogeny. Devonian deformation in the Avalon terrane is limited to SE-vergent thrusting and folding adjacent to dextral faults (Keppie, 1993). An apparent hiatus in plutonism exists between the Silurian and the mid-Devonian. Devonian magmatism in Newfoundland occurs in isolated plutons in the Dunnage and Gander zones, but is most voluminous in the Avalon zone and adjacent to the Avalon-Gander boundary. The compositions of Devonian granites are also indicative of a post-collisional setting. Renewed plutonism was accompanied by a switch from sinistral kinematics of the Silurian to Devonian dextral kinematics (as exemplified by the Dover fault; Holdsworth, 1994) suggesting that the magmatism developed in response to wrench kinematics.

Summary

The data presented above indicates that the Acadian orogeny is limited in both time and space: it lasted from the Late Silurian to Late Devonian and is centered on the northern Appalachians. Outside this region, its effects are limited to strike-slip faults and sporadic magmatism. It appears to have spread diachronously from south to north with clastic wedges migrating northward and then westward, closely followed by the deformation front. In space it coincides with a gap in the magmatic arc that was

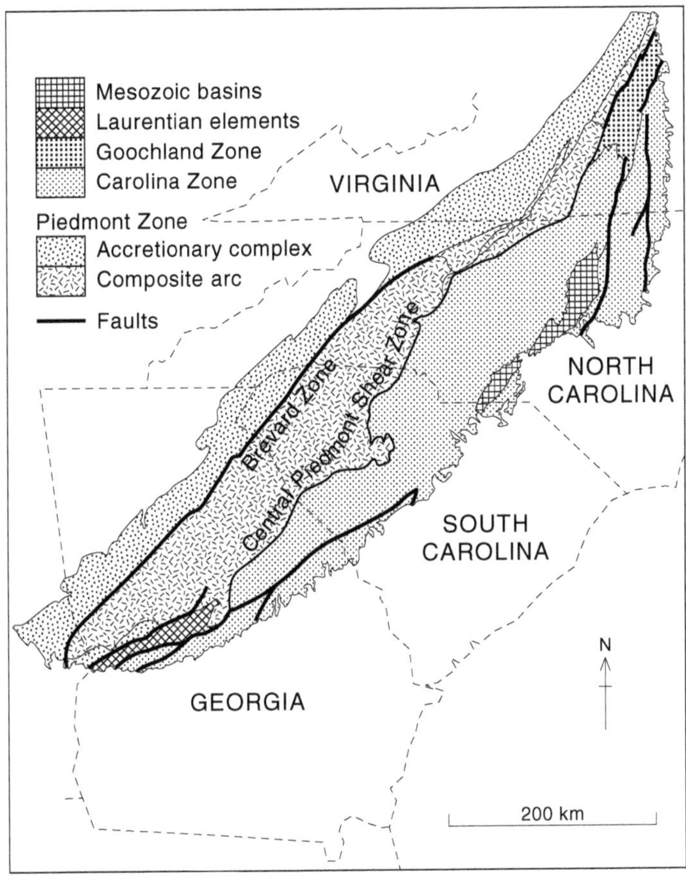

FIG. 7. Simplified geological map showing the general geology of the Southern Appalachians (after Hibbard et al., 2002; Hibbard, 2004).

subsequently interrupted by voluminous magmatism, some of which has a plume signature, and which also migrated from south to north.

Model for the Acadian Orogeny

The conclusion that the Avalon and Meguma terranes shared a common history during the Paleozoic has profound consequences for interpretation of the Acadian orogeny. Inasmuch as the Meguma is the most outboard terrane, this implies that all peri-Gondwanan terranes in the northern Appalachians were accreted prior to the mid-Silurian. The Late Silurian–Devonian Acadian orogeny, therefore, cannot be related to collision of the Meguma terrane with Avalonia or with the Laurentian margin. Instead, this scenario suggests that the Acadian orogeny occurred in a Cordilleran- or Andean-type setting.

Paleocontinental reconstructions imply continued convergence between Gondwana and Laurentia during the Late Silurian and Devonian, and consequently, subduction of the Rheic Ocean. However, over a wide region of Maritime Canada, voluminous magmatism ceased between ca. 395 and 380 Ma, and a period of relative magmatic quiescence occurred. During this period, the diachronous onset of deformation associated with the Acadian orogeny in the northern Appalachians migrated northwestward from Ludlovian in the southeast to Frasnian in the northwest (Fig. 3; Bradley, 1989; Keppie, 1993; Robinson et al., 1998), to extend across the entire width of the Appalachian orogen.

In the Meguma terrane, the period of magmatic quiescence was abruptly terminated by a voluminous ~380–370 Ma magmatic episode most dramatically represented by intrusion of the biggest

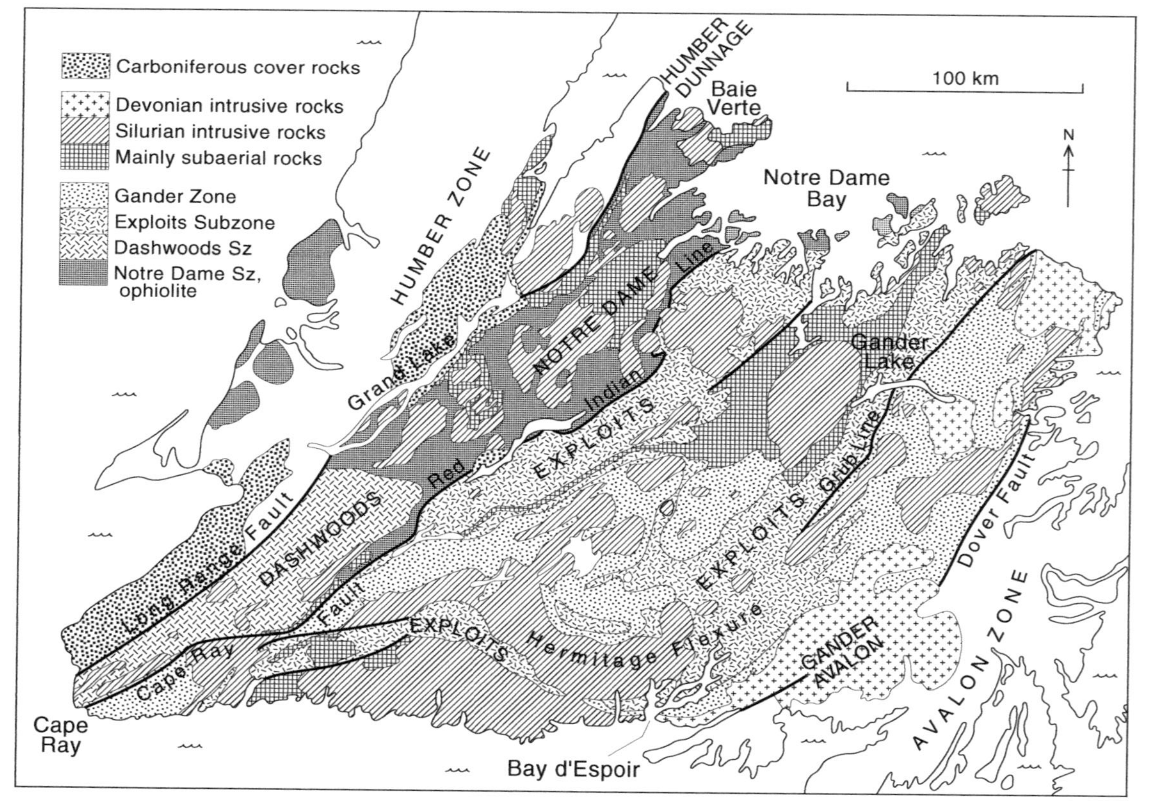

FIG. 8. Tectonostratigraphic subdivisions of central Newfoundland (after Williams and Piasecki, 1990; Williams, 1993). Age data for plutons compiled from Kerr (1997).

granitoid batholiths in the northern Appalachians and coeval lamprophyric dikes (e.g., Clarke et al., 1997). Coeval abundant gold and siderophile mineralization has an isotopic signature indicating a source within the lower crust and probably formed as a result of dehydration by mantle-derived magma and fluids (Kontak et al., 1990).

Magmatism in the Meguma terrane terminated abruptly by ~370 Ma. To the north, in the Cobequid Highlands on the Avalon terrane, voluminous ~360 Ma, bimodal magmatism commenced (Pe-Piper et al., 1989; Doig et al., 1996; Pe-Piper and Piper, 1998), with final emplacement of plutons at a shallow level, controlled by coeval dextral transpression along major E-W faults (Koukouvelas et al., 2002). Magmatism in the Cobequid Highlands terminated before ~355 Ma. Farther north, middle Carboniferous mafic rocks with plume-related geochemical signatures crop out around the periphery of the Maritimes basin and in the Magdalen Islands (Bedard, 1986). Although the Maritimes basin is poorly exposed, geophysical and borehole data indicate that ~3 km thick deposit of mainly terrestrial Carboniferous and Permian rocks. Beneath the Maritimes basin, seismic velocities of 7.2 m/s and a pronounced positive Bouguer anomaly are attributed to a 10–20 km thick Carboniferous underplating of mantle-derived magma (Marillier and Verhoef, 1989).

Murphy et al. (1999) and Keppie and Krogh (1999) proposed that the Silurian–Devonian Acadian orogeny in the type area was caused by flat-slab subduction, in a manner analogous to the modern Andes or the late Mesozoic–Cenozoic evolution of the southwestern United States (Fig. 9). This model provides a potential explanation for several anomalous aspects of the geology of the northern Appalachians that otherwise remain enigmatic. Following ~415 Ma convergence between the Avalon and Meguma terranes, the 395–380 Ma period of magmatic quiescence and the northwestward migration of the deformation front is attributed to the development of a flat slab beneath the Canadian Appalachians (Figs. 9A and 9B) and ensuing Laramide-style orogenic activity in which intermittent coupling between the subducted and overriding plates resulted in deformation that migrated to at least 600 km inboard of the continental margin. The flat-slab model is capable of explaining both the lack of 395–380 Ma arc magmatism associated with the convergence (a problem implicit in several previous models), and the diachronous migration of the deformation front and overstep sequences across the entire 500–800 km width of the orogen.

Flat-slab subduction has been attributed to several mechanisms: (1) far-field effects of terrane collision (Maxson and Tikoff, 1996); (2) overriding of an ocean plateau or dormant seamounts (e.g., Pilger, 1981; Gutchner et al., 2000); (3) subduction of young oceanic lithosphere (Gutscher, 2002) or a ridge (Yañez et al., 2002); (4) increased rates of convergence or changes in the direction of convergence (Ramos et al., 2002); or (5) overriding of a swell and an active plume (Murphy et al., 1999). All these mechanisms produce a migrating deformation front and temporarily switch off arc magmatism. However, only overriding of a plume would terminate a period of magmatic quiescence with voluminous magmatism with a linear, diachronous distribution, coupled with compositional changes from felsic- to mafic-dominated magmatism.

The Laramide orogeny is characterized by an almost complete absence of magmatism (the Paleocene magmatic gap), widespread deformation, and thick-skinned tectonics associated with basement uplifts located about 1500 km inboard of the continental margin. Most authors attribute these features to flat-slab subduction (e.g., Dickinson and Snyder, 1979; Liviccarri et al., 1981; Bird, 1988; Severinhaus and Atwater, 1990). However, origin of the flat slab is controversial, and not well constrained.

Recent geodynamic analysis of the modern flat-slab subduction zones has drawn attention to their spatial and temporal correlation with a subducting oceanic plateau (e.g., Gutscher et al., 2000; Yañez et al., 2002), suggesting a similar possibility for the Laramide orogeny. The Andean margin, for example, has several flat-slab segments, up to 500 km wide, that are each correlated with subduction of anomalously warm oceanic crust, represented by oceanic plateau (Gutscher et al., 2000). Murphy et al. (1998) speculated that the origin of flat-slab subduction in the western United States is due to the overriding by the continental margin of a mantle plume and related swell related to the ancestral Yellowstone hotspot. According to plate reconstructions, the ancestral Yellowstone hotspot would have collided with the margin at about 50 Ma. The model implies that the underlying plume would have been positioned beneath the Kula and/or Farallon oceanic plates prior to 55–50 Ma. Evidence for the existence of the plume in the oceanic realm is derived from Late Cretaceous basaltic terranes of the Coastal Ranges and from the Yukon Territory.

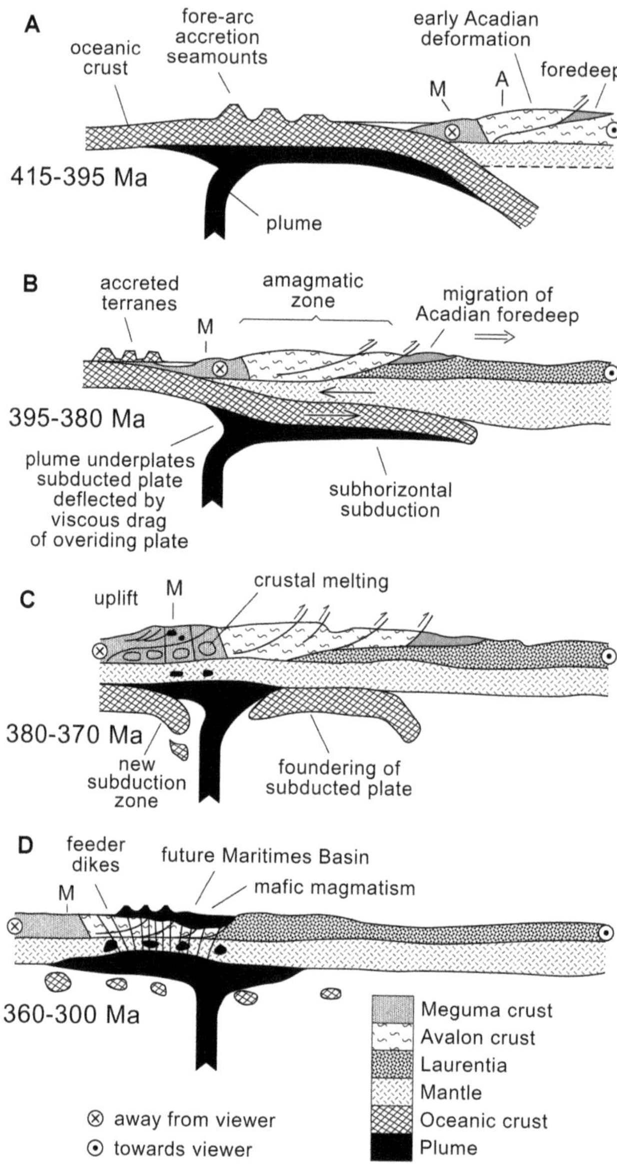

FIG. 9. Plate tectonic model for Acadian orogeny in the Maritime Appalachians. Cross-sections are coincident with the plume track in Figure 3. A. Between 415 and 395 Ma, plume resided beneath oceanic crust. Coeval magmatism and metamorphism (see Fig. 1) in Laurentia is interpreted to reflect telescoping of Avalonia (A) and Meguma (M) to Laurentia during which slab break-off may have occurred (van Staal et al., 1998). B. Between 395 and 380 Ma, the plume is overridden by the Laurentian margin, resulting in subhorizontal subduction and temporary cessation of magmatism and coeval migration of the Acadian foredeep and deformation front. C. Between 380 and 360 Ma, the plume head thermally erodes the subducted oceanic plate, causing intracrustal melting and minor mafic magmatism. D: Between 360 and 300 Ma, breakthrough of plume-related magmatism occurs. As the plume eventually dies, underplating by plume-derived magma results in the formation of a positive gravity anomaly and in the formation of the Maritimes basin. The region of flat-slab subduction may not extend to the central or southern Appalachians. Note that the process in which a continent overrides a plume and swell is considered diachronous, so that timing of events might be different in the northeastern United States.

Duncan (1982) and Wells et al. (1984) proposed that some basaltic provinces of the Coastal Ranges (such as the 60–50 Ma Crescent terrane) were seamounts generated by the Yellowstone hotspot that were accreted to western North America by the Eocene. In the Yukon Territory, the ca. 70 Ma Carmacks basaltic volcanics have plume-type geochemistry. Paleomagnetic data indicate that the Crescent and Carmacks basalts were both erupted at similar paleolatitudes to the Yellowstone hotspot (Johnston et al., 1996) although the Crescent volcanics underwent much less subsequent northward translation (Babcock et al., 1992).

The 395–380 Ma period of magmatic quiescence in Maritime Canada may be related to an incubation period in which the plume progessively assimilated or thermally eroded the overlying lithosphere (Fig. 9B), thereby exposing the continental lithosphere to the hot asthenosphere (Fig. 9C). The resulting melts led to an abrupt termination of magmatic quiescence and intrusion of voluminous ~380–370 Ma intracrustal granitoid rocks and lamprophyres of the Meguma terrane. Between 380 and 330 Ma, the site of magmatism moved northward to reside beneath the Maritimes basin by Visean times. The mafic component of the ~360 Ma magmatism in the Cobequid Highlands may represent melting of the lithospheric mantle above the plume. Visean-Westphalian mafic rocks around the Maritimes basin have plume-related geochemistry (Pe-Piper and Piper, 1998; Bedard, 1986) and are interpreted to reflect the penetration of the plume-derived magmas into the overlying continental crust (Fig. 9D).

Anomalously high heat flow associated with the plume provides a viable mechanism for regional-scale high-T, low-P metamorphism in the Meguma terrane that is broadly coeval with the intrusion of the plutons. In addition, current models for plumes involve ascent from the core-mantle boundary, a region thought to be anomalously rich in gold and siderophile elements (Brimhall, 1987), and the ascending plume may have transported these elements into the continental lithosphere (Rock and Groves, 1988). The ~20 m.y. interval between the overridding of the leading edge of the swell and its peak thermal effect provide a time window for circulating fluids to scavenge metals from the dehydrated lower crust.

The record of uplift and erosion within the Meguma and Avalon terranes can also be explained by the model. Overriding of the leading edge of the plume swell can account for the Early Devonian transition from shallow-marine to continental sedimentation recorded in both the Meguma and Avalon terranes. This change is synchronous with the earliest development of slaty cleavage in the southeastern Meguma terrane. The rapid exhumation of ~10 km in the Meguma terrane between 370 and 360 Ma (Keppie and Dallmeyer, 1995) and rapid uplift of the Cobequid Highlands at ~360 Ma (Ryan et al., 1987) may be the topographic expression of an overridden plume.

The high-density lens beneath the Maritimes basin (Marillier and Verhoef, 1989) may represent the beheaded plume, the cooling of which would induce sinking relative to surrounding uplifted areas. Such sinking could account for the geophysical anomalies and formation of the Maritimes basin (Fig. 9D). As the plume is considered stationary relative to the more mobile overlying lithosphere, there is no geophysical expression of plume activity in the Cobequid Highlands and the Meguma terrane.

The limited effects of the Acadian orogeny outside the northern mainland Appalachians may be a reflection of steepening of the subduction zone, resulting in a narrow active zone that may have become extensional rather than contractional, a factor that depends on the rate of convergence.

Discussion

In the Appalachians, several orogenic events have been recognized: Taconian, Penobscotian, Salinian, Acadian, and Alleghanian, and these have generally been correlated along the whole length of the orogen. Of these, the Salinian and Alleghenian orogenies have been interpreted as due to continent-continent collision, Laurentia-Avalonia and Laurentia-Gondwana, respectively. Thus their widespread distribution is to be expected. The Taconian and Penobscotian orogenies have been related to convergence in island arc settings, and if these island arcs have great length, their accretion to the margins may also be widespread.

The Acadian orogeny was initially interpreted to have resulted from continent-continent collision between Laurentia and Avalonia, or between the amalgamated Laurentia/Avalonia and Meguma/Africa. The former can now be discounted, inasmuch as data collected within the last 30 years indicate that Laurentia-Avalonia collision took place in the Late Ordovician–Silurian, i.e. before onset of the Acadian orogeny. On the other hand, recent detrital zircon data supports the contiguity of the Meguma and Avalon terranes during the Paleozoic, and imply that they accreted to Laurentia as a single microcontinent by the mid- to Late Silurian (~415 Ma). Hence, the collision of Avalonia or Meguma is unlikely to be the cause of the Late Devonian–Late Carboniferous tectonism and magmatism. The data support the hypothesis that the Meguma terrane is the passive margin on the southern margin of Avalonia, and that it formed along the northern (Laurentian) margin of the Rheic Ocean (Murphy et al., 2004b), which did not close until the Permo-Carboniferous.

This implies that the Acadian orogeny formed along an Andean-type margin, possibly by overriding a plume and its swell (Murphy et al., 1999; Keppie and Krogh, 1999). Arc-related magmatism and deformation related to crustal shortening occurred along most of the Andean margin. However, coeval orogeny related to flat-slab subduction, characterized by an absence of arc magmatism and deformation well inboard of the plate boundary, occurs in Peru and the Pampean, and appears to have been synchronous with changes in convergence rate and direction, and with thick-skinned tectonics related to intermittent coupling between the overridden oceanic lithosphere and the South American plate (Ramos et al., 2002). Similar variations in the dip of the subduction zone along the northern margin of the Rheic Ocean may account for the variation in the style of Acadian orogenesis from the southern Appalachians to Atlantic Canada.

Elsewhere in the circum-Pacific region, subduction-related orogenic activity appears to alternate between steeply and shallowly dipping subduction, a phenomenon called tectonic switching by Collins (2002). During periods of steep subduction, the overriding plate is generally subjected to extension, in which sedimentary basins may form, with convergent deformation limited to the trench area. An exception occurs if the convergence rate is high; then orogenic effects are more widespread. During periods of flat-slab subduction the overriding plate is subjected to convergence, causing medium- to high-grade metamorphism, polyphase deformation, and magmatism in the basins. According to Collins (2002), these tectonothermal events are often mistakenly ascribed to arc-continental collision, so tectonic switching may offer an alternative model that could resolve controversies about the timing and nature of "collisional" tectonics in the central and southern Appalachians.

Flattening of the Benioff zone is generally the result of collision of buoyant regions in the oceans such as seamounts, mid-oceanic ridges, young oceanic lithosphere, oceanic plateaus, island arc complexes, hot spots, etc. As these buoyant regions are of limited size, it follows that the associated orogenic events are limited in areal distribution. Furthermore, flattening of the subduction zone is gradual in producing diachronous orogenic effects that start at the trench and migrate inboard. On the other hand, if the flattening of the subduction zone is the result of increased convergence rates, the orogenic effects may have wider distribution.

Inasmuch as flat-slab subduction is a common feature of modern convergent margins, it should also be common in the geologic record. We believe that the Acadian orogeny in the northern mainland Appalachians may be the expression of such a process. The relative motion between the overriding plate and the plume implies that features such as the magmatic gap, plume-related magmatism, deformation fronts, and basin formation are separated in space as well as in time. As a result, these features may be difficult to recognize in ancient orogenic belts, especially where they are overprinted by polyphase deformation and dismembered by subsequent faulting. The presence of continental mafic rocks with plume-related chemistry is an important clue to the identification of such processes. Because even the strongest plumes have difficulty in penetrating continental lithosphere, the presence of plume-related volcanic rocks indicates that the plume and its related swell may have had a protracted earlier history.

Acknowledgments

We acknowledge the continuing support of the Natural Sciences and Engineering Research Council of Canada (Murphy) and Universidad Nacional Autónoma de México (Keppie) for facilitating the research. We are grateful for the thoughtful, constructive reviews by Ulf Linnemann and Rob Strachan. Contribution to IGCP (International Geological Correlation Programme) Projects 453 and 497.

REFERENCES

Anderson, S. D., Jamieson, R. A., Reynolds, P. H., and Dunning, G. R, 2001, Devonian extension in Newfoundland: $^{40}Ar/^{39}Ar$ and U-Pb data from the Mings

Bight area, Baie Verte Peninsula: Journal of Geology, v. 109, p. 191–211.

Arth, J. G., and Ayuso, R. A., 1997, The Northeast Kingdom Batholith, Vermont: Geochronology and Nd, O, Pb, and Sr isotopic constraints on the origin of Acadian granitic rocks, *in* Sinha, A. K. et al., eds., Magmatism in the Appalachians: Geological Society of America Memoir, 191, p. 1–18.

Ayuso, R. A., and Arth, J. G., 1997, The Spruce Head composite pluton: An example of mafic to silicic Salinian magmatism in coastal Maine, northern Appalachians, *in* Sinha, A. K. et al., eds., Magmatism in the Appalachians: Geological Society of America Memoir, 191, p. 19–43.

Babcock, R. S., Burmester, R. F., Engebretson, D. C., Warnock, A., and Clark, K. P., 1992, A rifted margin origin for the Crescent basalts and related rocks in the northern coast range volcanic province, Washington and British Columbia: Journal of Geophysical Research, v. 97(B5), p. 6799–6821.

Barr, S. M., and Raeside, R. P., 1989, Tectonostratigraphic zonation of Cape Breton Island, Nova Scotia: Implications for the configuration of the northern Appalachian orogen: Geology, v. 17, p. 822–825.

Barr, S. M., Raeside, R. P., and White, C. E., 1998, Geological correlations between Cape Breton Island and Newfoundland, northern Appalachian orogen: Canadian Journal of Earth Sciences, v. 35, p. 1252–1270.

Barr, S. M., White, C. E., and Miller, B. V., 2002, The Kingston terrane, southern New Brunswick, Canada: Evidence of an Early Silurian arc: Geological Society of America Bulletin, v. 114, p. 964–982.

Bedard, J. H., 1986, Pre-Acadian magmatic suites of the southeastern Peninsula: Geological Society of America Bulletin, v. 97, p. 1177–1191.

Benn, K., Horne, R. J., Kontak, D. J., Pignotta, G. S., and Evans, N. G., 1997, Syn-Acadian emplacement model for the South Mountain batholith, Meguma Terrane, Nova Scotia: Magnetic fabric and structural analyses: Geological Society of America Bulletin, v. 109, p. 1279–1293.

Berry, H. N., IV, and Osberg, P. H., 1989, A stratigraphic synthesis of eastern Maine and western New Brunswick, *in* Tucker, R. D., and Marvinney, R. G., eds., Studies in Maine geology, volume 2: Structure and stratigraphy: Augusta, ME, Maine Geological Survey, p. 1–32.

Bevier, M. L., Barr, S. M., and White, C. E., 1990, Late Precambrian U-Pb ages for the Brookville Gneiss, New Brunswick: Journal of Geology, v. 98, p. 955–968.

Bird, P., 1988, Formation of the Rocky Mountains, western United States: A continuum computer model: Science, v. 239, p. 1601–1507.

Boucot, A. J., 1975, Evolution and extinction rate controls: Developments in paleontology and stratigraphy, v. 1: Amsterdam, Netherlands, Elsevier Scientific Publication Co., 427 p.

Boucot, A. J., Field, N. T., Fletcher, R., Forbes, W. H., Naylor, R. S., and Pavlides, L., 1964, Reconnaissance bedrock geology of the Preque Isle Quadrangel, Maine: Augusta, ME, Maine Geological Survey Quadrangle Map Series, no. 2, pp. 123.

Bourque, P. A., Malo, M., and Kirkwood, D., 2000, Paleogeography and tectono-sedimentary history at margin of Laurentia during Silurian–Earliest Devonian time: The Gaspé Belt, Québec: Geological Society of America Bulletin, v. 112, p. 4–30.

Bradley, D. C., 1983, Tectonics of the Acadian orogeny in New England and adjacent Canada: Journal of Geology, v. 91, p. 381–400.

Bradley, D. G., 1989, Taconic plate kinematics as revealed by foredeep stratigraphy: Tectonics, v. 8, p. 1037–1049.

Brimhall, G. H., Jr., 1987, Preliminary fractionation patterns of ore metals through Earth history: Chemical Geology, v. 64, p. 1–16.

Cawood, P. A., Dunning, G. R., Lux, D., and van Gool, J. A. M., 1994, Timing of peak metamorphism and deformation along the Appalachian margin of Laurentia in Newfoundland: Silurian, not Ordovician: Geology, v. 22, p 399–402.

Cawood, P. A., McCausland, P. J. A., and Dunning, G. R., 2001, Opening Iapetus: Constraints from the Laurentian margin of Newfoundland: Geological Society of America Bulletin, v. 113, p. 443–453.

Chandler, F. W., Sullivan, R. W., and Currie, K. L., 1987, Springdale Group and correlative rocks: A Llandovery overlap sequence in the Canadian Appalachians: Royal Society of Edinburgh: Earth Sciences v. 78, p. 41–49.

Clarke, D. B., MacDonald, M. A., Reynolds, P. H., and Longstaffe, F. J., 1993, Leucogranites from the eastern part of the South Mountain batholith, Nova Scotia: Journal of Petrology v. 34, p. 653–679.

Clarke, D. B., MacDonald, M. A., and Tate, M. C., 1997, Late Devonian mafic-felsic magmatism in the Meguma Zone, Nova Scotia, *in*, Sinha, A. K., et al. eds., The nature of magmatism in the Appalachian orogen: Geological Society of America Memoir, 191, p. 107–127.

Cocks, L. R. M., and Fortey, R. A., 1990, Biogeography of Ordovician and Silurian faunas, *in* McKerrow, W. S., and Scotese, C. R., eds., Paleozoic Paleogeography and Biogeography: Geological Society Memoir 12, p. 97–104.

Cocks, L. R. M. and Torsvik, T. H., 2002, Earth geography from 500 to 400 million years ago: A faunal and palaeomagnetic review: Journal of the Geological Society of London, v. 159, p. 631–644.

Collins, W.J. 2002, Hot orogens, tectonic switching and creation of continental crust: Geology v. 31, p. 535–538.

Culshaw, N., and Liesa, M., 1997, Alleghanian reactivation of the Acadian fold belt, Meguma zone, southwest

Nova Scotia: Canadian Journal of Earth Sciences, v. 34, p. 833–847.

Currie, K. L., 1987, Relations between metamorphism and magmatism near Cheticamp, Cape Breton Island: Nova Scotia Geological Survey of Canada, Paper 85-23, 66 p.

Daly, J. S., and McLelland, J. M., 1991, Juvenile middle Proterozoic crust in the Adirondack Highlands, Grenville Province, northeastern North America: Geology, v. 19, p. 119–122.

Dickin, A. P., 2000, Crustal formation in the Grenville province: Nd isotopic evidence: Canadian Journal of Earth Sciences, v. 37, p. 165–181.

Dickin, A. P., and McNutt, R. H., 1989, Nd model age mapping of the southeast margin of the Archean foreland in the Grenville Province of Ontario: Geology, v. 17, p. 299–302.

Dickin, A. P., McNutt, R. H., and Clifford, P. M., 1990, A neodymium isotope study of plutons near the Grenville Front in Ontario, Canada: Chemical Geology, v. 83, p. 315–324.

Dickinson, W. R., and Snyder, W. S., 1979, Geometry of subducted slabs related to the San Andreas transform: Journal of Geology, v. 87, pp. 609–627.

Doig, R., Murphy, J. B., Pe-Piper, G., and Piper, D. J. W., 1996, U-Pb geochronology of Late Paleozoic plutons, Cobequid Highlands, Nova Scotia, Canada: Evidence for Late Devonian emplacement adjacent to the Meguma-Avalon terrane boundary: Geological Journal, v. 31, p. 179–188.

Donohoe, H.V., Jr., and Pajari, G., 1974, The age of Acadian deformation in Maine–New Brunswick: Maritime Sediments, v. 9, p. 78–82.

Dostal, J., Wilson, R. A., and Keppie, J. D., 1989, Geochemistry of the Siluro-Devonian Tobique belt in northern and central New Brunswick (Canada): Tectonic implications: Canadian Journal of Earth Sciences, v. 26, p. 1282–1296.

Duncan, R. A., 1982, A captured island chain in the Coast Range of Oregon and Washington: Journal of Geophysical Research, v. 87, p. 10,827–10,837.

Dunning, G. R., O'Brien, S. J., Colman-Sadd, S. P., Blackwood, R. F., Dickson, W. L., O'Neill, P. P., and Krogh, T. E., 1990, Silurian orogeny in the Newfoundland Appalachians: Journal of Geology, v. 98, p. 895–913.

Fortey, R. A., and Cocks, L. R. M., 2003, Palaeontological evidence bearing on global Ordovician–Silurian continental reconstructions: Earth Science Reviews, v. 61, p. 245–307.

Goldstein, S. J., and Jacobsen, S. B., 1988, Nd and Sr isotopic systematics of river water suspended material: Implications for crustal evolution: Earth and Planetary Science Letters, v. 87, p. 221–236.

Gower, C. F., Ryan, A. B., and Rivers, T., 1990, Mid-Proterozoic Laurentia–Baltica: An overview of its geological evolution and summary of the contributions by this volume, *in* Gower, C. F., Rivers, T., and Ryan, B., eds., Mid-Proterozoic Laurentia–Baltica: Geological Association of Canada Special Paper 38, p. 1–20.

Gutscher, M.-A., Spakman, W., Bijwaard, H., and Engdahl, E. R., 2000, Geodynamics of flat subduction: Seismicity and tomographic constraints from the Andean margin: Tectonics, v. 19, p. 814–833.

Gutscher, M.-A., 2002, Andean subduction styles and their effect on thermal structure and interplate coupling: Journal of South American Earth Sciences, v. 15, p. 3–10.

Hamilton, M. A., and Murphy, J. B., 2004, Tectonic significance of a Llanvirn age for the Dunn Point volcanic rocks, Avalon terrane, Nova Scotia, Canada: Implications for the evolution of the Iapetus and Rheic oceans: Tectonophysics, v. 379, p. 199–209.

Hatcher, R. D., Jr., 1989, Tectonic syntheses of the U.S. Appalachians, *in* Hatcher, R. D., Thomas, W. A., and Viele, G. W., eds., The Appalachian-Ouachita orogen in the United States: Geological Society of America, The Geology of North America, v. F-2, p. 511–535.

Hibbard, J. P., 2000, Docking Carolina: Mid-Paleozoic accretion in the southern Appalachians: Geology v. 28, p. 127–130.

Hibbard, J., 2004, The Appalachian orogen, *in* van der Pluijm, B.A., and Marshak, S., eds., Earth structure, 2nd ed.: New York, NY, Norton, p. 582–592.

Hibbard, J. P., Miller, B. V., Tracy, R., and Carter, B., in press, The Appalachian peri-Gondwanan realm: A paleogeographic perspective from the south, *in* Vaughan, A. P. M., Leat, P. L., and Pankhurst, R. J., eds., Terrane processes at the Pacific Margin of Gondwana: Geological Society of London, Special Publication, in press.

Hibbard, J. P., Stoddard, E. F., Secor, D. T., and Dennis, A. J., 2002, The Carolina Zone: Overview of Neoproterozoic to early Paleozoic peri-Gondwanan terranes along the eastern flank of the southern Appalachians: Earth Science Reviews, v. 57, p. 299–339.

Hicks, R. J., Jamieson, R. A., and Reynolds, P. H., 1999, Detrital and metamorphic $^{40}Ar/^{39}Ar$ ages from muscovite and whole-rock samples, Meguma Supergroup, southern Nova Scotia: Canadian Journal of Earth Sciences, v. 36, p 23–32.

Holdsworth, R. E., Structural evolution of the Gander-Avalon terrane boundary: A reactivated transpressional zone in the NE Newfoundland Appalachians: Journal of the Geological Society of London, v. 151, p. 629–646.

Hon, R., Fitzgerald, J. P., Sargant, S. L., Schwartz, W. D., Dostal, J., and Keppie, J. D., 1992, Silurian–Early Devonian mafic rocks of the Piscataquis volcanic belt in northern Maine: Atlantic Geology, v. 28, p. 163–170.

Jennex, L. C., Murphy, J. B., and Anderson, A. J., 2000, Post-orogenic exhumation of an auriferous terrane: The paleoplacer potential of the early Carboniferous

St. Marys Basin; Canadian Appalachians: Mineralium Deposita, v. 35, p. 776–790.

Johnson, R. J. E., and Van der Voo, R., 1986, Paleomagnetism of the Late Precambrian Fourchu Group, Cape Breton Island, Nova Scotia: Canadian Journal of Earth Sciences, v. 23, p. 1673–1685.

Johnson, R. J. E., and Van der Voo, R., 1990, Pre-folding magnetization reconfirmed for the Late Ordovician–Early Silurian Dunn Point volcanics, Nova Scotia: Tectonophysics, v. 178, p. 193–205.

Johnston, S. T., Wynne, P. J., Francis, D., Hart, C. J. R., Enkin, R. J., and Engebretson, D. C., 1996, Yellowstone in Yukon: The Late Cretaceous Carmacks Group: Geology, v. 24, p. 997–1000.

Jutras, P., Prichonnet, G., and McCutcheon, S., 2003, Alleghanian deformation in the eastern Gaspé Peninsula of Quebec, Canada: Geological Society of America Bulletin, v. 115, 1538–1551.

Keen, C. E., Kay, W. A., Keppie, J. D., Marillier, F., Pe-Piper, G., and Waldron, J. W. F., 1991, Deep seismic reflection data from the Bay of Fundy and Gulf of Maine: Tectonic implications for the northern Appalachians: Canadian Journal of Earth Sciences, v. 28, p. 1096–1111.

Keppie, J. D. 1982, The Minas geofracture, *in* St. Julien, P., and Beland, J., eds., Major structural zones and faults of the Northern Appalachians: Geological Association of Canada Special Paper 24, p. 263–280.

Keppie, J. D., 1985, The Appalachian collage, *in* Gee, D. G., and Sturt, B. eds., The Caledonide orogen, Scandinavia, and related areas: New York, NY, John Wiley and Sons, p. 1217–1226.

Keppie, J. D., 1993, Synthesis of Paleozoic deformational events and terrane accretion in the Canadian Appalachians: Geologische Rundschau, v. 82, p. 381–431.

Keppie, J. D., and Dallmeyer, R. D., 1995, Late Paleozoic collision, delamination, short-lived magmatism, and rapid denudation in the Meguma terrane (Nova Scotia, Canada): Constraints from $^{40}Ar/^{39}Ar$ isotopic data: Canadian Journal of Earth Sciences, v. 32, p. 644–659.

Keppie, J. D., Davis, D. W., and Krogh, T. E., 1998, U-Pb geochronological constraints on Precambrian stratified units in the Avalon Composite Terrane of Nova Scotia, Canada: Tectonic implications: Canadian Journal of Earth Sciences, v. 35, p. 222–236.

Keppie, J. D., and Dostal, J., 1991, Late Proterozoic tectonic model for the Avalon terrane in Maritime Canada: Tectonics, v. 10, p. 842–850.

Keppie, J. D., Dostal, J., Murphy, J. B., and Cousens, B. L., 1997, Palaeozoic within-plate volcanic rocks in Nova Scotia (Canada) reinterpreted: Isotopic constraints on magmatic source and paleocontinental reconstructions: Geological Magazine, v. 134, p. 425–447.

Keppie, J. D., and Krogh, T. E., 1999, U-Pb geochronology of Devonian granites in the Meguma terrane of Nova Scotia Canada: Evidence for hotspot melting of a Neoproterozoic source: Journal of Geology, v. 107, p. 555–568.

Keppie, J. D., and Krogh, T. E., 2000, 440 Ma igneous activity in the Meguma terrane, Nova Scotia, Canada: Part of the Appalachian overstep sequence?: American Journal of Science, v. 300, p. 528–538.

Keppie, J. D., Nance, R. D., Murphy, J. B., and Dostal, J., 1991, Northern Appalachians: The Avalon and Meguma terranes, *in* Dallmeyer, R. D., and Lecorche, J. P., eds., The Western African orogens and Circum Atlantic correlatives: Berlin, Germany, Springer Verlag, p. 315–333.

Keppie, J. D., Nance, R. D., Murphy, J. B., and Dostal, J., 2003, Tethyan, Mediterranean, and Pacific analogues for the Neoproterozoic–Paleozoic birth and development of peri-Gondwanan terranes and their transfer to Laurentia and Laurussia: Tectonophysics, v. 365, p. 195–219.

Kerr, A., 1997. Space-time composition relationships among Appalachian-cycle plutonic suite in Newfoundland, *in* Sinha, A. K. et al., eds., Magmatism in the Appalachians: Geological Society of America Memoir, 191, 193–220.

Kontak, D. J., Smith, P. K., Kerrich, R., and Williams, P. F., 1990, Integrated model for Meguma Group lode gold deposits, Nova Scotia, Canada: Geology, v. 18, p. 238–242.

Koukouvelas, I., Pe-Piper, G., and Piper, D. J. W., 2002, The role of dextral transpressional faulting in the evolution of the early Carboniferous mafic-felsic plutonic and volcanic complex: Cobequid Highlands, Nova Scotia, Canada: Tectonophysics, v. 348, p. 219–246.

Krogh, T. E., and Keppie, J. D., 1990, Age of detrital zircon and titanite in the Meguma Group, southern Nova Scotia, Canada: Clues to the origin of the Meguma Terrane: Tectonophysics, v. 177, p. 307–323.

Lambert, R. St. J., and McKerrow, W. S., 1976, The Grampian orogeny: Scottish Journal of Geology, v. 12, p. 271–293.

Landing, E., 1996, Avalon: Insular continent by the latest Precambrian, *in* Nance, R. D., and Thompson, M. D., eds., Avalonian and related peri-Gondwanan terranes of the circum-North Atlantic: Geological Society of America Special Paper 304, p. 29–63.

Landing, E., and Murphy, J. B., 1991, Uppermost Precambrian(?)–Lower Cambrian of mainland Nova Scotia: Faunas, depositional environments, and stratigraphic revision: Journal of Paleontology, v. 65, p. 382–396.

Lin, S., van Staal, C. R., and Dube, B., 1994, Promontory-promontory collision in the Canadian Appalachians: Geology, v. 22, p. 897–900.

Livaccari, R. F., Burke, K., and Sengor, A. M. C., 1981, Was the Laramide orogeny related to subduction of an oceanic plateau? Nature, v. 289, p. 276–279.

MacDonald, L. A., Barr, S. M., White, C. E., and Ketchum, J. W. G., 2002, Petrology, age, and tectonic setting of the White Rock Formation, Yarmouth area, Nova

Scotia: Canadian Journal of Earth Sciences, v. 39, p. 259–277.

MacLachlan, K., and Dunning, G., 1998, U-Pb ages and tectonomagmatic relationships of early Ordovician low-Ti tholeiites, boninites, and related plutonic rocks in central Newfoundand: Contributions to Mineralogy and Petrology, v. 133, p. 235–258.

MacNiocaill, C., and Smethurst, M. A., 1994, Palaeozoic palaeogeography of Laurentia and its margins: A reassessment of the paleomagnetic data: Geophysical Journal International, v. 116, p. 715–725.

MacNiocaill, C., van der Pluijm, B. A., and Van der Voo, R., 1997, Ordovician paleogeography and the evolution of the Iapetus Ocean: Geology, v. 25, p. 159–162.

Malo, M., Tremblay, A., Kirkwood, D., and Cousineau, P., 1995, Along-strike Acadian structural variations in the Quebec Appalachians: Consequence of a collision along an irregular margin: Tectonics, v. 14, pp. 1327–1338.

Marillier, F., and Verhoef, J., 1989, Crustal thickness under the Gulf of St. Lawrence, northern Appalachians, from gravity and deep seismic data: Canadian Journal of Earth Sciences, v. 26, p. 1517–1532.

Martel, A. T., McGregor, D. C., and Utting, J., 1993, Stratigraphic significance of Upper Devonian and Lower Carboniferous miospores from the type area of the Horton Group, Nova Scotia: Canadian Journal of Earth Sciences, v. 30, p. 1091–1098.

Maxson, J., and Tikoff, B., 1996, Hit-and-run collision model for the Laramide orogeny, western United States: Geology, v. 24, p. 968–972.

McLelland, J. M., Daly, J. S., and Chiarenzelli, J., 1993, Sm-Nd and U-Pb isotopic evidence of juvenile crust in the Adirondack lowlands and implications for the evolution of the Adirondack Mountains: Journal of Geology, v. 101, p. 97–105.

McKerrow, W. S., Dewey, J. F., and Scotese, C. R., 1991, The Ordovician and Silurian development of the Iapetus Ocean: Special Papers in Paleontology, v. 44, p. 165–178.

McKerrow, W. S., MacNiocaill, C., and Dewey, J. F., 2000, The Caledonian orogeny redefined: Journal of the Geological Society of London, v. 157, p. 1149–1154.

Miller, J. D., and Kent, D. V., Paleomagnetism of the Siluro-Devonian Andreas redbeds: Evidence of a Devonian supercontinent?: Geology, v. 16, p. 195–198.

Miller, R. G., O'Nions, R. K., Hamilton, P. J., and Welin, E., 1986, Crustal residence ages of clastic sediments, orogeny and crustal evolution: Chemical Geology, v. 57, p. 87–99.

Murphy, J. B. 2000, Tectonic influence on sedimentation along the southern flank of the Late Paleozoic Magdalen Basin in the Canadian Appalachians: Geochemical and isotopic constraints on the Horton Group in the St. Mary's Basin, Nova Scotia: Geological Society of America Bulletin, v. 112, p. 997–1011.

Murphy, J. B., 2003, Late Paleozoic formation and development of the St. Marys Basin, mainland Nova Scotia, Canada: A prolonged record of intra-continental strike-slip deformation during the assembly of Pangaea, *in* Storti, F., Holdsworth, R. E., and Salvini, F., eds., Intraplate strike-slip deformation belts: Geological Society of London, Special Publication 210, p. 185–196.

Murphy, J. B., Fernández-Suárez, J., and Jeffries, T. E., 2004a, Lithogeochemical, Sm-Nd, and U-Pb isotopic data from the Silurian–Early Devonian Arisaig Group clastic rocks, Avalon terrane, Nova Scotia: A record of terrane accretion in the Appalachian-Caledonide orogen: Geological Society of America Bulletin, v. 116, p. 1183–1201.

Murphy, J. B., Fernández-Suárez, J., Keppie, J. D., and Jeffries,T. E., 2004b, Contiguous rather than discrete Paleozoic histories for the Avalon and Meguma terranes based on detrital zircon data: Geology, v. 32, p. 585–588.

Murphy, J. B., and Hamilton, M. A., 2000, U-Pb detrital zircon age constraints on evolution of the Late Paleozoic St. Marys Basin, central mainland Nova Scotia: Journal of Geology, v. 108, p. 53–72.

Murphy, J. B., and Keppie, J. D., 1998, Late Devonian palinspastic reconstruction of the Avalon-Meguma terrane boundary: Implications for terrane accretion and basin development in the Appalachian orogen: Tectonophysics, v. 284, p. 221–231.

Murphy, J. B., Keppie, J. D., Dostal, J., and Cousins, B. L., 1996a, Repeated late Neoproterozoic–Silurian lower crustal melting beneath the Antigonish Highlands, Nova Scotia: Nd isotopic evidence and tectonic interpretations, *in* Nance, R. D., and Thompson, M. D., eds., Avalonian and related peri-Gondwanan terranes of the Circum-North Atlantic: Geological Society of America Special Paper 304, p. 109–120.

Murphy, J. B., Keppie, J. D., Dostal, J., Waldron, J. W. F., and Cude, M. P., 1996b, Geochemical and isotopic characteristics of Early Silurian clastic sequences in Antigonish Highlands, Nova Scotia, Canada: constraints on the accretion of Avalonia in the Appalachian-Caledonide orogen: Canadian Journal of Earth Sciences, v. 33, p. 379–388.

Murphy, J. B., and MacDonald, D. A., 1993, Geochemistry of Late Proterozoic arc-related volcaniclastic turbidite sequences, Antigonish Highlands, Nova Scotia: Canadian Journal of Earth Sciences, v. 30, p. 2273–2282.

Murphy, J. B., Oppliger, G. L., Brimhall, G. H., Jr., and Hynes, A., 1998, Plume-modified orogeny: An example from the western United States: Geology, v. 26, p. 731–734.

Murphy, J. B., Strachan, R. A., Nance, R. D., Parker, K. D., and Fowler, M. B., 2000, Proto-Avalonia: A 1.2–1.0 Ga tectonothermal event and constraints for the evolution of Rodinia: Geology, v. 28, p. 1071–1074.

Murphy, J. B., van Staal, C. R., and Keppie, J. D., 1999, Is the mid to late Paleozoic Acadian orogeny a plume-modified Laramide-style orogeny?: Geology, v. 27, p. 653–656.

Nance, R. D., and Dallmeyer, R. D., 1993, $^{40}Ar/^{39}Ar$ amphibole ages from the Kingston complex, New Brunswick: Evidence for Silurian–Devonian tectonothermal activity and implications for the accretion of the Avalon composite terrane: Journal of Geology, v. 101, p. 375–388.

Neuman, R. B., and Harper, D. A. T., 1992, Paleogeographic Significance of Arenig-Llanvirn Toquima-Table Head and Celtic brachiopod assemblages, *in* Webby, B. D., and Laurie, J. R., eds., Global perspectives on Ordovician geology: Rotterdam, Netherlands, and Brookfield, VT, A. A. Balkema, p. 241–254.

Noel, J., Spariosu, D., and Dallmeyer, R. D., 1988, Paleomagnetism and $^{40}Ar/^{39}Ar$ ages from the Carolina slate belt, Albemarle, North Carolina: Implications for terrane amalgamation with North America. Geology, v. 16, p. 64–68.

O'Brien, B. H., O'Brien, S. J., and Dunning, G. R., 1991, Silurian cover, late Precambrian–Early Ordovician basement, and the chronology of Silurian orogenesis in the Hermitage flexure (Newfoundland Appalachians): American Journal of Science, v. 291, p. 760–799.

O'Brien, S. J., O'Brien, B. H., Dunning, G. R., and Tucker, R. D., 1996, Late Neoproterozoic Avalonian and related peri-Gondwanan rocks of the Newfoundland Appalachians, *in* Nance, R. D., and Thompson, M. D., eds., Avalonian and related peri-Gondwanan terranes of the circum-North Atlantic: Geological Society of America Special Paper 304, p. 9–28.

O'Brien, S. J., Wardle, R. J., and King, A. F., 1983, The Avalon zone: A Pan-African terrane in the Appalachian orogen of Canada: Geological Journal, v. 18, p. 195–222.

Patchett, P. J., and Ruiz, J., 1989, Nd isotopes and the origin of the Grenville-age rocks in Texas: Implications for Proterozoic evolution of the United States mid-continent region: Journal of Geology, v. 97, p. 685–695.

Pedersen, R. B., and Dunning, G. R., 1997, Evolution of arc crust and relations between contrasting sources: U-Pb (age), Nd and Sr isotopic systematics of the ophiolite terrain of SW Norway: Contributions to Mineralogy and Petrology, v. 128, p. 1–15.

Pe-Piper, G., Murphy, J. B. and Turner, D. S., 1989, Petrology, geochemistry, and tectonic setting of some Carboniferous plutons of the Eastern Cobequid Hills: Atlantic Geology, v. 25, p. 37–49.

Pe-Piper, G., and Piper, D. J. W., 1998, Geochemical evolution of Devonian–Carboniferous igneous rocks of the Magdalen basin, eastern Canada: Pb- and Nd-isotope evidence for mantle and lower crustal sources: Canadian Journal of Earth Sciences, v. 35, p. 201–221.

Pickering, K., Basset, M. G., and Siveter, D. J., 1988, Late Ordovician–Early Silurian destruction of the Iapetus Ocean: Newfoundland, British Isles, and Scandinavia—a discussion: Transactions of the Royal Society of Edinburgh, v. 79, p. 361–382.

Pilger, R. H., 1981, Plate reconstruction, aseismic ridges, and low-angle subduction beneath the Andes: Geological Society of America Bulletin, v. 92, p. 448–456.

Piper, D. J. W., Pe-Piper, G., and Loncarevic, B. D., 1993, Devonian–Carboniferous deformation and igneous intrusion in the Cobequid Highlands: Atlantic Geology, v. 29, p. 219–232.

Poole, W. H., Sandford, B. V., Williams, H., and Kelley, D. G., 1970, Geology of southeastern Canada, *in* Douglas, R. J. W., ed., Geology and economic minerals of Canada: Geological Survey of Canada, Economic Geology Report no. 1, p. 227–304.

Potts, S., Van der Pluijm, B., and Van der Voo, R., 1993, Discordant Silurian paleolatitudes for central Newfoundland: New paleomagnetic evidence from the Springdale Group: Earth and Planetary Science Letters, v. 120, p. 1–12.

Pratt, B. R., and J. W. F. Waldron, 1991, A Middle Cambrian trilobite faunule from the Meguma Group of Nova Scotia: Canadian Journal of Earth Sciences, v. 28, p. 1843–1853.

Quinlan, G. M. and Beaumont, C., 1984, Appalachian thrusting, lithospheric flexure, and the Paleozoic stratigraphy of the Eastern Interior of North America: Canadian Journal of Earth Sciences, v. 21, p. 973–996.

Ramos, V. A., and Aleman, A. 2001, Tectonic evolution of the Andes, *in* Cordani, U. G., Thomaz Filho, A., and Campos, D. A., eds., Tectonic evolution of South America: 31st International Geological Congress, Rio de Janeiro, Brazil, p. 635–688.

Ramos, V. A., Cristallini, E. O., and Pérez, D. J., 2002, The Pampean flat-slab of the Central Andes: Journal of South American Earth Sciences, v. 15, p. 59–78.

Rankin, D. W., 1968, Volcanism related to tectonism in the Piscataquis volcanic belt, an island arc of early Devonian age in north-central Maine, *in* Zen, E-an, White, W. S., Hadley, J. B., and Thompson, J. B., Studies in Appalachian geology: Northern and maritime: New York, NY, Interscience, p. 83–94.

Roberts, D., 2003, The Scandinavian Caledonides: Event chronology, palaeogeographic settings, and likely modern analogues: Tectonophysics, v. 365, p. 283–299.

Robinson, P., Tucker, R. D., Bradley, D., Berry, H. N. V., and Osberg, P. H., 1998, Paleozoic orogens in New England, USA: Geologiska Föreningens Förhandlingar, v. 120, p. 119–148.

Rocci, G., Bronner, G., and Deschamps, M., 1991, Crystalline basement of the West African craton, *in* Dallmeyer, R. D., and Lecorche, J. P., eds., The West African orogens and circum-Atlantic correlatives: Heidelberg, Germany, Springer-Verlag, p. 31–61.

Rock, N. M. S., and Groves, D. I., 1988, Can lamprophyres resolve the genetic controversy over mesothermal gold deposits?: Geology, v. 16, p. 538–541.

Ryan, R. J., Calder, J. H., Donohoe, H. V., Jr., and Naylor, R., 1987, Late Paleozoic sedimentation and basin development adjacent to the Cobequid Highlands massif, eastern Canada, *in* Beaumont, C., and Tankard, A. J., eds., Sedimentary basins and basin-forming mechanisms: Canadian Society of Petroleum Geologists Memoir 12 and Atlantic Geoscience Society Special Publication 5, p. 311–317.

Sadowski, G. R., and Bettencourt, J. S., 1996, Mesoproterozoic tectonic correlations between eastern Laurentia and the western border of the Amazon craton: Precambrian Research, v. 76, p. 213–227.

Schenk, P. E., 1997, Sequence stratigraphy and provenance on Gondwana's margin: The Meguma zone (Cambrian-Devonian) of Nova Scotia, Canada: Geological Society of America Bulletin, v. 109, p. 395–409.

Schofield, D. I., and D'Lemos, R. S., 2000, Granite petrogenesis in the Gander Zone, NE Newfoundland: Mixing of melts from multiple sources and the role of lithospheric delamination: Canadian Journal of Earth Sciences, v. 37, p. 535–547.

Schreckengost, K. A., and Nance, R. D., 1996, Silurian–Devonian dextral reactivation near the inboard margin of the Avalon Composite Terrane: Kinematic evidence from the Kingston complex, southern New Brunswick, Canada, *in* Nance, R. D., and Thompson, M. D., eds., Avalonian and related peri-Gondwanan terranes of the circum–North Atlantic: Geological Society of America Special Paper 304, p. 165–178.

Schuchert, C., 1923, Sites and names of the North American geosynclines: Geological Society of America Bulletin, v. 34, p. 151–229.

Severinghaus, J., and Atwater, T., 1990, Cenozoic geometry and thermal state of the subducting slabs beneath North America, *in* Wernicke, B. P., ed., Basin and Range extensional tectonics near the latitude of Las Vegas, Nevada: Geological Society of America Memoir 176, p. 1–22.

Stewart, D. B., Unger, J. D., and Hutchinson, D. R., 1995, Silurian tectonic history of Penobscot Bay region, Maine: Atlantic Geology, v. 31, p. 67–79.

Thirlwall, M. F., 1988, Wenlock to mid Devonian volcanism of the Caledonian-Appalachian orogen, *in* Harris, A. L., and Fettes, D. J., eds., Caledonian-Appalachian orogen: Geological Society of London Special Publication, v. 18, p. 415–428.

Thomas, W. A., and Whiting, B. M., 1995, The Alabama promontory: An example of the evolution of the Appalachian-Ouachita thrust-belt recess at a promontory of the rifted continental margin, *in* Hibbard, J., Van Staal, C., and Cawood, P., New perspectives in the Caledonian-Appalachian orogen: Geological Association of Canada Special Paper 41, p. 7–18.

Thorkelson, D. J., and Taylor, R. P., 1989, Cordilleran slab windows: Geology, v. 17, p. 833–836.

Thorogood, E. J., 1990, Provenance of the pre-Devonian sediments of England and Wales: Sm-Nd isotopic evidence: Journal of the Geological Society of London, v. 147, p. 591–594.

Trench, A., and Torsvik, T. H., 1992, The closure of the Iapetus Ocean and Tornquist Sea: New paleomagnetic constraints: Journal of the Geological Society of London, v. 149, p. 867–870.

Torsvik, T. H., Smethurst, M. A., Meert, J. G., Van der Voo, R., McKerrow, W. S., Brasier, M. D., Sturt, B. A., and Walderhaug, H. J., 1996, Continental break-up and collision in the Neoproterozoic and Palaeozoic—a tale of Baltica and Laurentia: Earth Science Reviews, v. 40, p. 229–258.

Tucker, R. D., Osberg, P. H., and Berry, H. N., IV, 2001, The geology of a part of Acadia and the nature of the Acadian orogeny across central and eastern Maine: American Journal of Science, v. 301, p. 205–260.

Van der Voo, R., 1988, Palaeozoic paleogeography of North America, Gondwana, and intervening displaced terranes: Comparisons of palaeomagnetism within palaeoclimatology and biogeographical patterns: Geological Society of America Bulletin, v. 100, p. 311–324.

Van der Voo, R., and Johnson, R. J. E., 1985, Paleomagnetism of the Dunn Point Formation (Nova Scotia): High paleolatitudes for the Avalon terrane in the Late Ordovician: Geophysical Research Letters, v. 12, p. 337–340.

van Staal, C. R., 1994, The Brunswick subduction complex in the Canadian Appalachians: Record of the Late Ordovician to Late Silurian collision between Laurentia and the Gander margin of the Avalon: Tectonics, v. 13, p. 946–962.

van Staal, C. R., and de Roo, J. A., 1995, Mid-Palaeozoic tectonic evolution of the Appalachian Central Mobile Belt in northern New Brunswick, Canada: Collision, extensional collapse, and dextral transpression, *in* Hibbard, J., van Staal, C. R., and Cawood, P., eds., New perspectives in the Appalachian–Caledonian Orogen: Geological Association of Canada Special Paper 41, p. 367–389.

van Staal, C. R., Dewey, J. F., MacNiocaill, C., and McKerrow, W. S., 1998, The Cambrian–Silurian tectonic evolution of the Northern Appalachians and British Caledonides: History of a complex, west and southwest Pacific-type segment of Iapetus, *in* Blundell, D., and Scott, A. C., eds., Lyell: The past is the key to the present: Geological Society of London Special Publication 143, p. 199–242.

van Staal, C. R., Ravenhurst, Winchester, J. A., Roddick, J. C., and Langton, J. P., 1990, Post Taconic blueschist suture in the northern Appalachians of New Brunswick, Canada: Geology, v. 18, p. 1073–1077.

van Staal, C. R., Sullivan, R. W., and Whalen, J. B., 1996, Provenance and tectonic history of the Gander zone in the Caledonide/Appalachian orogen: Implications for the origin and assembly of Avalon, *in* Nance, R. D., and Thompson, M. D., eds., Avalonian and related peri-Gondwanan terranes of the Circum-North Atlantic: Geological Society of America Special Paper 304, p. 347–367.

Vick, H., Channell, J., and Opdyke, N., 1987, Ordovician docking of the Carolina slate belt: Paleomagnetic data: Tectonics, v. 6, p. 573–583.

Waldron, J. W. F., Murphy, J. B., Melchin, M., and Davis, G., 1996, Silurian tectonics of western Avalonia: Strain-corrected subsidence history of the Arisaig Group, Nova Scotia: Journal of Geology, v. 104, p. 677–694.

Wells, R. E., Engebretson, D. C., Snavely, P. D., Jr., and Coe, R. S., 1984, Cenozoic plate motions and the volcano-tectonic evolution of western Oregon: Tectonics, v. 3, p. 275–294.

Whalen, J. B., Jenner, G. A., Currie, K. L., Barr, S. M., Longstaffe, F. J., and Hegner, E., 1994, Geochemical and isotopic characteristics of granitoids of the Avalon Zone, southern New Brunswick: Possible evidence for repeated delamination events: Journal of Geology, v. 102, p. 269–282.

Whalen, J. B., Jenner, G. A., Longstaffe, F. J., and Hegner, E., 1996, Nature and evolution of the eastern margin of Iapetus: Geochemical and isotopic constraints from Siluro-Devonian granitoid plutons in the New Brunswick Appalachians: Canadian Journal of Earth Sciences, v. 33, p. 140–155.

Willard, R. A., and Adams, M. G., 1994, Newly discovered eclogite in the southern Appalachian orogen, northwestern Carolina: Earth and Planetary Science Letters, v. 123, p. 61–70.

Williams, H., 1979, Appalachian orogen in Canada: Canadian Journal of Earth Sciences, v. 16, p. 792–807.

Williams, H., 1993, Acadian orogeny in Newfoundland, *in* Roy, D. C., and Skehan, J. W., eds., The Acadian orogeny: Recent studies in New England, Maritime Canada, and the autochthonous foreland: Geological Society of America Special Paper 275, p. 123–133.

Williams, H., and Piasecki, M. A. J., 1990. The Cold Spring mélange and a possible model for Dunnage-Gander zone interaction in central Newfoundland: Canadian Journal of Earth Sciences, v. 27, p. 1126–1134.

Williams, S. H., Harper, D. A. T., Neuman, R. B., Boyce, W. D., and MacNiocaill, C., 1995, Lower Palaeozoic fossils from Newfoundland and their importance in understanding the history of Iapetus Ocean: Geological Association of Canada, Special Paper no. 41, p. 115–126.

Woodcock, N., and Strachan, R. A., 2000, Geological history of the British Isles: Oxford, UK, Blackwell Science.

Yañez, G., Cembrano, J., Pardo, M., Ranero, C., and Selles, D., 2002, The Challenger–Juan Fernández–Maipo major tectonic transition of the Nazca-Andean subduction system at 33–34°S: Geodynamic evidence and implications: Journal of South American Earth Sciences, v. 15, p. 23–38.

Episodic Volcanism in the Buck Creek Complex (Central British Columbia, Canada): A History of Magmatism and Mantle Evolution from the Jurassic to the Early Tertiary

J. DOSTAL,[1] J. V. OWEN,

Department of Geology, Saint Mary's University, Halifax, Nova Scotia B3H 3C3, Canada

B. N. CHURCH,

600 Parkridge Street, Victoria, British Columbia V8Z 6N7, Canada

AND T. S. HAMILTON

Camosun College, Victoria, British Columbia, V9E 2C1, Canada

Abstract

The Buck Creek volcanic complex of the Intermontane Superterrane of the Canadian Cordillera records a long history of volcanic activity from the Cretaceous through to the Eocene, when magmatic activity peaked, to the Miocene. Its basement includes primitive continental arc volcanic rocks of the Jurassic Hazelton Group (~200 Ma; mainly basalt and andesites) emplaced prior to the accretion of Stikinia during the Mesozoic. In contrast, in the post-accretionary volcanic complex, the major pulse of volcanism started with the extrusion of continental margin calc-alkaline rocks (basaltic andesites to rhyolites) of the Cretaceous Tip Top Hill Formation (~85 Ma). Their Nd-Sr isotopic compositions resemble those of the pre-accretionary Hazelton Group. However, this post-accretionary Cretaceous arc was more evolved and its crust was significantly thicker than the Hazelton. Mafic rocks of both the Hazelton and Tip Top Hill suites were generated from a spinel-bearing mantle source.

The Eocene volcanics (~50 Ma) evolved from typical high-K calc-alkaline mafic/intermediate rocks to flows that resemble intraplate tholeiitic basalts. Collectively, the Eocene rocks had a common (and comparatively deep) source, likely garnet-bearing subcontinental lithosphere. These compositional variations are probably related, in part, to an increase in the degree of partial melting. They record a gradual change from a compressional to an extensional tectonic environment, and overall represent an arc setting that matured and thickened over ca. 35 m.y., culminating in Eocene extension coincident with the cessation of compression in the foreland fold-and-thrust belt. Overlying intraplate alkali basaltic rocks of Miocene age appear to represent the final stage of this transition, as subduction-modified lithospheric mantle was replaced by a new asthenospheric mantle source. The association of Cretaceous, Eocene, and Miocene volcanic rocks in the Buck Creek complex is widespread in the region, suggesting that this model of tectono-magmatic evolution can be applied elsewhere in central British Columbia.

Introduction

THE WESTERN MARGIN of North America from southern Yukon through British Columbia to the northwestern USA records a long history of the subduction of the Farallon/Kula plate under the North American continent. Subduction was accompanied by extensive magmatism, most notably by voluminous successions of the Eocene Challis-Kamloops (CK) volcanic belt that crops out in parts of British Columbia, Idaho, and Wyoming (e.g., Ewing, 1980). In central and southern British Columbia, the CK belt, which is typically 200–400 km wide, is located about 300–500 km east of the paleotrench (Fig. 1). It is composed of a series of volcanic complexes produced by episodic igneous activity. The complexes show considerable compositional variations, with rocks ranging from calc-alkaline to alkaline. Some of the volcanic centers have been interpreted as subduction-related, whereas others appear to be back-arc or even within-plate or rift-related (Dudas, 1991; Anderson et al., 1998).

[1]Corresponding author; email: jdostal@smu.ca

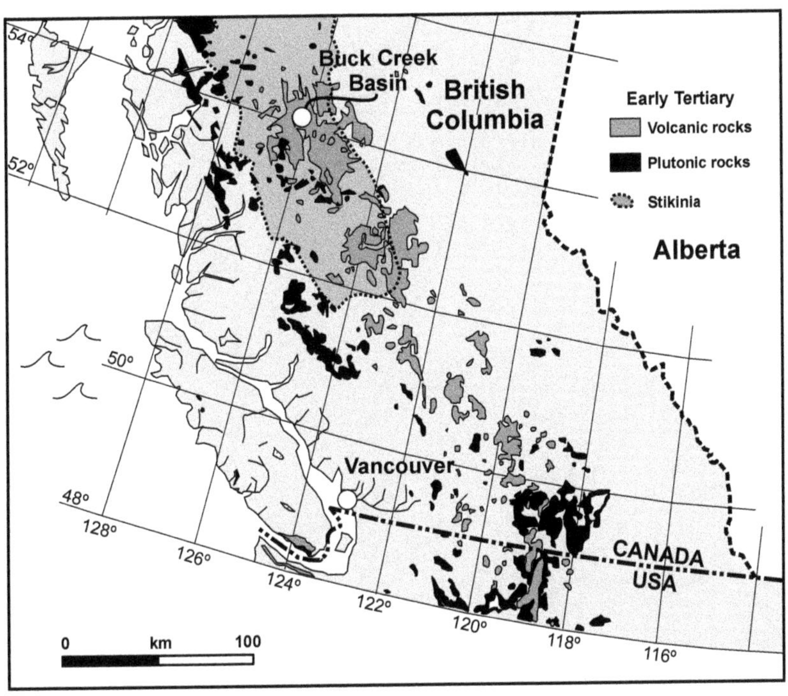

FIG. 1. Map showing the distribution of early Tertiary volcanic and plutonic rocks in southern and central British Columbia, and the location of the Buck Creek basin and the Stikine terrane of the Canadian Cordillera. Early Tertiary volcanic rocks make up the Challis-Kamloops volcanic belt.

The tectonic significance of many of these volcanic complexes, however, is still poorly understood. In fact, there are insufficient data even to characterize many of the complexes and to constrain various petrogenetic and tectonic models (Norman and Mertzman, 1991). The purpose of this paper is to present and evaluate geological and geochemical data for the Buck Creek volcanic center and related successions preserved in a local basin in west-central British Columbia (Fig. 1). This complex records a long history of volcanic activity from the Cretaceous (Tip Top Hill volcanics) through to the Eocene (Goosly Lake, Buck Creek, and Swans Lake volcanics) when magmatic activity peaked, to the Miocene (Chilcotin or Cheslatta Lake basalts) (Fig. 2). It thus documents the magmatic evolution of this part of the CK arc and constrains the changes in mantle source compositions over that period of time. The basement of the complex includes volcanic rocks of the pre-accretion Jurassic Hazelton Group and the post-accretion mid-Cretaceous Skeena Group. The Hazelton volcanic rocks can provide information on magmatic activity prior to accretion of the Intermontane superterrane to the westernmost North America. The mid-Cretaceous volcanic rocks of the Skeena Group (Rocky Ridge Formation) are relatively rare in the Buck Creek area. However, they were investigated in some detail from nearby occurrences by MacIntyre (2001) and MacIntyre and Villeneuve (2001) and, to the north of the study area, along the southern margin of the Bowser basin by Bassett and Kleinspehn (1996). The Eocene sequences were, in part, previously described by Dostal et al. (1998, 2001), whereas the Miocene volcanic rocks were studied by Dostal et al. (1996) and Anderson et al. (2001).

Field work and sampling for the present contribution were initiated during geological mapping of the area at a scale of 1:100,000 (Church, 1984; Church and Barakso, 1990). This work was prompted by the economic importance of the volcanic successions. For example, the Buck Creek complex on the Nechako Plateau, near the town of Houston, occurs in one of the more important mineral resource districts in central British Columbia. Moreover, Cretaceous rocks of the complex host two

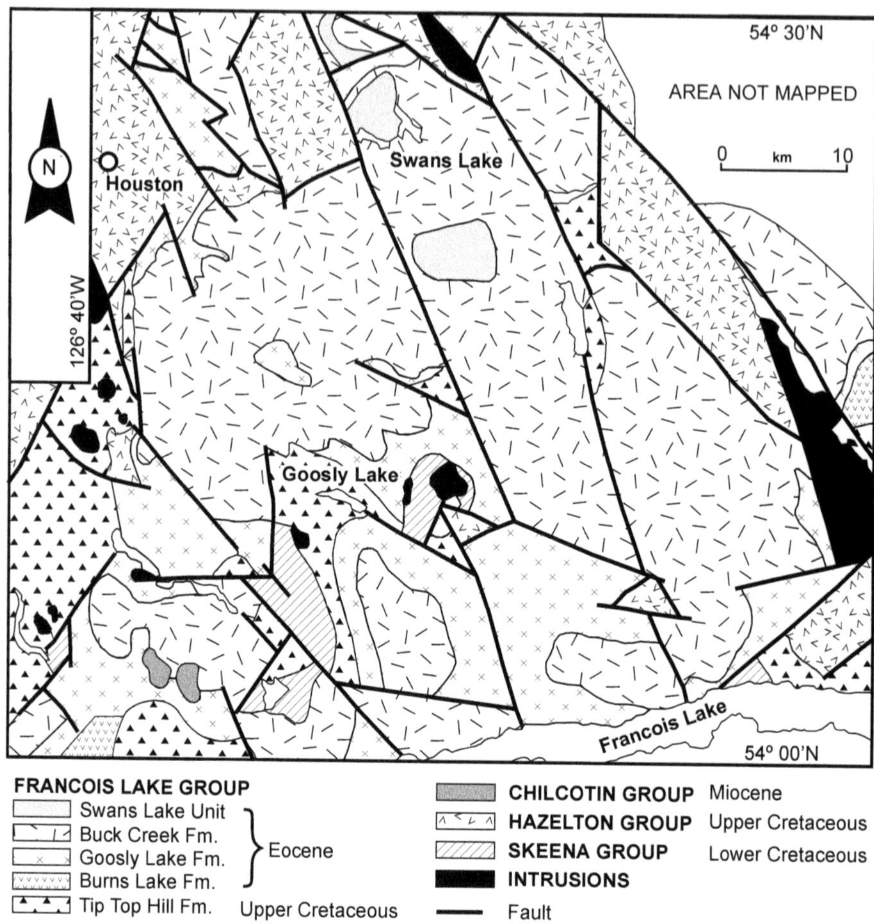

FIG. 2. Geological map of the Buck Creek basin (modified from Church, 1984). Note that the area mapped as the Tip Top Hill Formation may include coeval supracrustal rocks belonging to the Cretaceous Kasalka Group.

Au-Ag-Zn-Cu vein/replacement deposits (Equity Silver and Silver Queen, Cyr et al., 1984; Wojdak and Sinclair, 1984) (Fig. 3).

Geological Setting

The Canadian Cordillera is thought to be an assemblage of oceanic and island-arc crustal fragments or allochthonous oceanic and pericratonic terranes that were accreted to the western margin of the North America craton during the Mesozoic, but before the mid-Cretaceous (Monger and Irving, 1980; Monger and Nokelberg, 1996; Gabrielse and Yorath, 1991). Oceanic and island-arc terranes comprise the Intermontane and Insular superterranes of the Canadian Cordillera, which were accreted successively. The Buck Creek volcanic complex (Fig. 1) is located within the Stikine terrane (Stikinia) of the Intermontane superterrane (e.g., Monger and Nokleberg, 1996). Stikinia is composed of several island-arc volcanic and sedimentary assemblages and related plutonic suites ranging in age from Carboniferous to Middle Jurassic (Schiarizza and MacIntyre, 1999). The Stikine terrane was intruded by Middle Jurassic to Early Cretaceous plutons and is overlain by post-accretionary marine and nonmarine sedimentary as well as minor volcanic rocks of Late Jurassic to Late Cretaceous age. These successions are capped by Upper Cretaceous and Paleocene continental arc volcanic and associated sedimentary rocks (Struik and MacIntyre, 2001). All of these units are unconformably

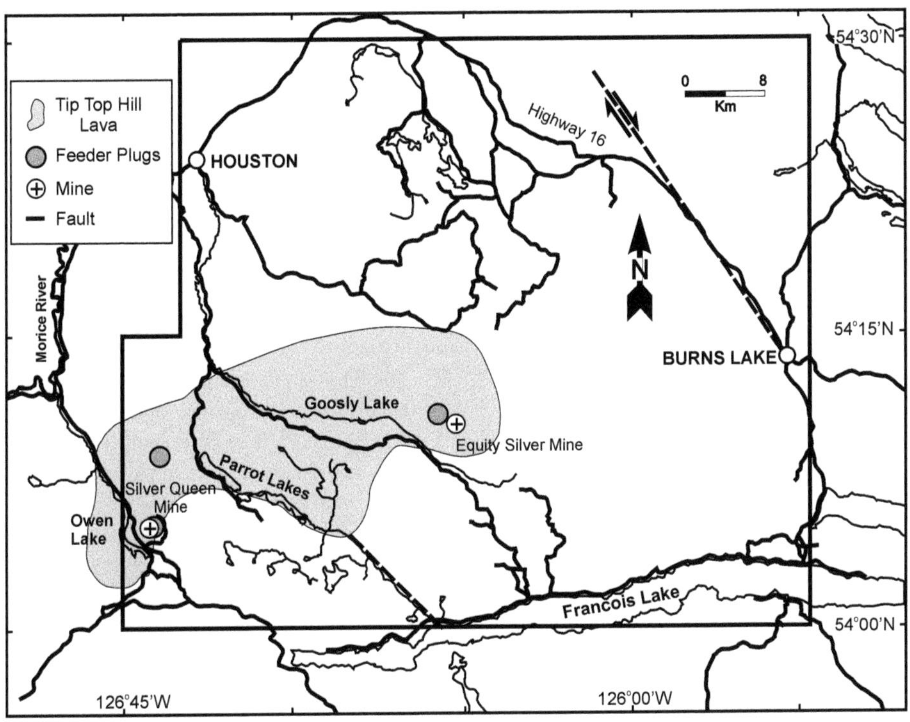

FIG. 3. Projected distribution of the Cretaceous Tip Top Hill volcanics together with the location of the feeder plugs. The inner irregular box is the area equivalent to Figure 2. Note the alignment of the feeder plugs for the volcanic rocks as well as the two deposits (the former Equity Silver and Silver Queen mines).

overlain by Eocene volcanic rocks of the CK belt. Paleomagnetic studies of the Eocene volcanic rocks within the belt show that this region has undergone little post-Middle Eocene displacement or deformation (Irving and Brandon, 1990).

The Buck Creek complex ("basin") is preserved in a block-faulted depression (~3,000 km^2 in size) controlled by a system of en echelon strike-slip faults linked by pull-aparts (Fig. 2). As elsewhere in the CK belt (Coney and Harms, 1984; Parrish et al., 1988; Armstrong and Ward, 1991), extensional faulting was broadly coincident with volcanism (Church and Barakso, 1990). The basin is filled by post-accretionary volcanic, volcaniclastic, and subordinate sedimentary rocks, which range in age from Late Cretaceous to Middle Tertiary (Fig. 2).

Buck Creek stratigraphy. The basement of the basin is poorly exposed but evidently is composed of weakly metamorphosed (greenschist-facies) volcanic and sedimentary rocks of the Jurassic Hazelton and mid-Cretaceous Skeena groups and granitoid intrusions of Jurassic to Cretaceous age (Church and Barakso, 1990; Monger et al., 1991; MacIntyre et al., 2001; MacIntyre, 2001). Most of these intrusions temporally correlate with the Topley intrusive suite of Stikinia from the nearby Babine-Takla lakes area that have been dated at 218–193 Ma (Struik and MacIntyre, 2001; MacIntyre et al., 2001).

The Lower to Middle Jurassic Hazelton Group is composed of mafic to intermediate lavas and pyroclastics, rhyolitic volcanic, and volcaniclastic and minor clastic sedimentary rocks deposited in a subaerial to shallow marine environment. The occurrence of well-preserved accretionary lapilli tuffs implies that volcanism was in part subaerial. Fossils from intercalated sedimentary units (Church and Barakso, 1990) indicate a Sinemurian–Toarcian age (~178–197 Ma; Okulitch, 1999). Unconformably overlying the Hazelton Group is the post-accretionary Skeena Group (Richards and Tipper, 1976), which consists of a ~750 m thick sequence of marine and nonmarine sedimentary strata of Hauterivian to late Albian or early Cenomanian (99–132 Ma; Okulitch, 1999) age (Church and Barakso, 1990; Bassett, 1991; Bassett and Kleinspehn, 1996; MacIntyre, 2001). The Skeena Group also includes volcanic rocks of the Rocky Ridge Formation that are in part intercalated with sedimentary rocks deposited in a fluvial-deltaic environment (Bassett, 1991). The Rocky Ridge Formation includes basaltic flows, breccias, lapilli tuffs, rhyolitic flows, pyroclastics, and domes, although in the complex and its vicinity, the formation is represented only by rhyolites. MacIntyre and Villeneuve (2001) reported U/Pb (zircon) and Ar/Ar (WR) ages of the Rocky Ridge volcanic rocks between 104 and 108 Ma. In the Buck Creek basin, the rocks of the Hazelton and Skeena groups are exposed in a series of small windows eroded through the younger overlying formations (Fig. 2).

The post-accretionary basin fill is composed of an Upper Cretaceous to Lower Tertiary stratigraphic sequence resting unconformably on the Mesozoic basement (Fig. 2). The basin sequence is made up of ~1500 m of continental volcanic and sedimentary rocks that can be divided into lower and upper segments. The lower part, preserved mainly in graben structures, consists principally of the Upper Cretaceous Tip Top Hill Formation (TTHF). The TTHF (Fig. 3) is unconformably overlain by the upper part, consisting dominantly of volcanic sequences of Early Tertiary age (Church and Barakso, 1990). There is a significant disconformity (~25–30 Ma) between the lithologically distinct lower and upper parts of the fill sequence.

The TTHF volcanics consist of lava, breccia, and tuff of andesitic to rhyolitic composition. The strata, which attain a thickness of about 600 m, were fed from several hypabyssal stocks and dikes of the Bulkley intrusions (Carter, 1981; Church and Barakso 1990; Fig. 3). A felsic and presumably comagmatic dike cross-cutting the lavas yielded a U/Pb zircon age of 84.6 ± 0.2 Ma (Leitch et al., 1992). Church (1973) reported K/Ar whole-rock ages on the TTHF volcanics ranging from 77.1 ± 2.7 to 75.3 ± 2.0. These results were corroborated by K/Ar (WR) data of Leitch et al. (1992) for the same rocks (78.7 ± 2.7 to 70.3 ± 2.5 Ma). A U/Pb zircon age of ~ 85 Ma is probably the best estimate for the TTHF rocks. The TTHF has been elsewhere in Stikinia correlated with the Kasalka Group, a prominent Cretaceous continental volcanic succession in the west-central British Columbia (MacIntyre, 1985; Armstrong, 1988; Friedman et al., 2001; MacIntyre and Villeneuve, 2001); this correlation is also being investigated here.

Lower Tertiary basin fill includes a series of volcanic, volcaniclastic, and subordinate sedimentary rocks. The Tertiary sequence begins with a conglomeratic unit (Burns Lake Formation), 50–100 m thick, which unconformably overlies the Mesozoic rocks. The Goosly Lake Formation (GLF; Church and Barakso, 1990; Dostal et al., 2001) lies atop the Burns Lake Formation. It comprises a 500 m succession of 2–6 m thick, massive to vesicular andesitic flows that are interbedded with flow breccias. These flows fan outward from three subvolcanic plugs that are upwards of 2 km in diameter, and appear to have formed (along with two ore deposits) along a single fracture system that may correspond to a reactivated Cretaceous fault zone (Fig. 4).

The GLF is conformably overlain by upwards of 400 m of subaerial volcanic rocks of the Buck Creek Formation (BCF). The BCF comprises a succession of thin, amygdaloidal, mafic to intermediate flows that are intercalated with minor volcanic breccia. In places along the margin of the basin, the BCF is overlain by a thin (15–60 m) sequence of massive mafic rocks of the Swans Lake volcanic suite (SLV; Church and Barakso, 1990). These Eocene supracrustal rocks were emplaced during two brief (<1–2 m.y.) volcanic episodes probably separated by a brief hiatus of no more than 1–2 m.y. The GLF has been dated at 52 ± 1 Ma (Ar/Ar-WR and Ar/Ar-biotite; Dostal et al., 2001). Both the BCF and SLV volcanics yield isotopic ages of 50 ± 1 Ma (Ar/Ar-WR; Dostal et al., 1998). The emplacement of these three volcanic sequences coincides with the age of the main phase of magmatic activity within the CK volcanic belt (Armstrong, 1988). The geochronological data correspond with the ages of nearby Eocene volcanic and plutonic suites of the Nechako Plateau area, as reported by Anderson et al. (1998), Grainger et al. (2001), and MacIntyre and Villeneuve (2001). Grainger et al. (2001) gave an age range between 53 and 47 Ma for these Eocene volcanics. Volcanic rocks of the Nechako Plateau include the felsic Ootsa Lake Group and mafic Endako Group. They not only overlap the ages of the Buck Creek complex, but some facies have comparable compositions.

Epithermal silver vein deposits in the Buck Creek basin (i.e., the Silver Queen and Equity Silver mines) are hosted by the TTHF, but are related to the younger Goosly Lake intrusions (Wojdak and Sinclair, 1984; Church and Barakso, 1990). These ore deposits yield K/Ar ages of 51.9 ± 1 and 51.3 ± 1.8 Ma (Leitch et al., 1992). Thus, they coincide with the emplacement of the intrusions related to the Goosly Lake volcanics. This indicates a connection

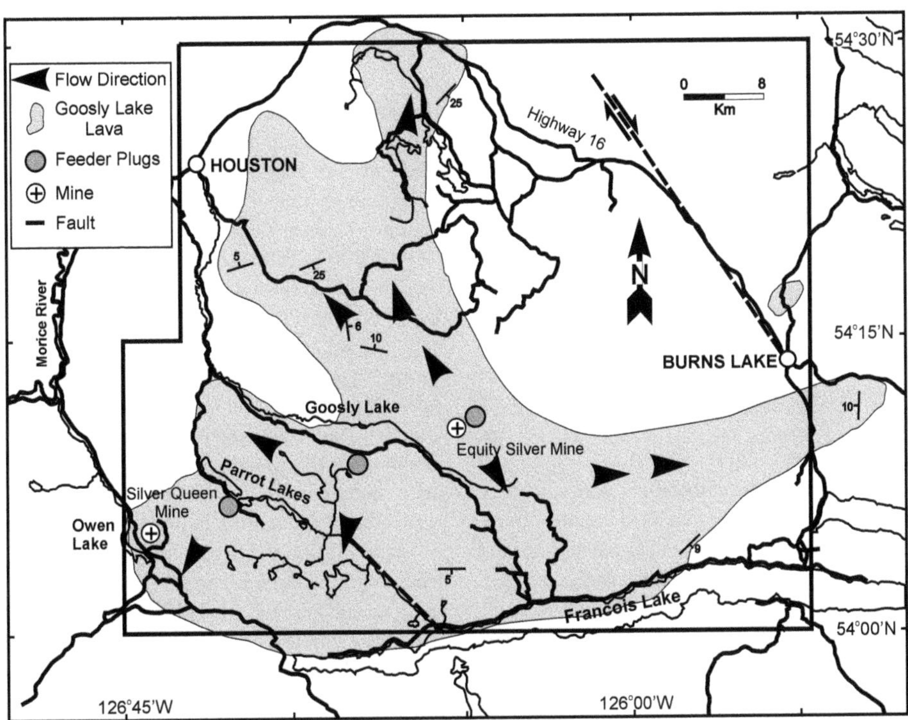

FIG. 4. Projected distribution of the Eocene Goosly Lake volcanics, together with location of the feeder plugs. Solid arrowheads indicate outflow of the lavas from source vents. Note the alignment of the feeder plugs for the Goosly volcanics as well as the two deposits (modified from Dostal et al., 2001). The inner irregular box is the area equivalent to Figure 2.

between the intrusions and silver mineralization in this area.

These Eocene volcanic sequences are unconformably overlain by Miocene Chilcotin intraplate basalts (Bevier, 1983; Dostal et al., 1996; Mathews, 1989), which has been renamed the Cheslatta Lake basalts (Anderson et al., 2001). These mafic sequences are common throughout the Intermontane belt (Souther, 1991). In the Buck Creek basin (Fig. 2), the Miocene sequence (Poplar-Buttes basalts of Church and Barakso, 1990) comprises 60 to 90 m of fresh, fine-grained olivine-bearing basalts. These rocks have yielded Early Miocene K-Ar (WR) ages (21.4 ± 1.1 Ma; Church and Barakso, 1990). The spatial association of Eocene and Miocene volcanic episodes observed in the basin is widespread in this part of British Columbia (Souther, 1991).

Petrography and Mineral Chemistry

In the Buck Creek area, the Hazelton volcanic rocks are predominantly mafic and intermediate lavas metamorphosed under greenschist-facies conditions. They are porphyritic, with 1–2 mm phenocrysts of plagioclase (now albite) and subordinate pyroxenes that have been converted to actinolite. These phenocrysts are set in a groundmass composed of fine-grained chloritized material, which includes microcrystalline actinolite, albite, magnetite, and minor ilmenite. Only traces of augite are preserved in the greenschists. A pilotaxitic texture characterizes some samples; it is defined by the subparallel alignment of microcrystalline plagioclase laths. The subordinate felsic rocks are microporphyritic with feldspar phenocrysts. The rhyolites of the Rocky Ridge Formation are typically flow-banded and feldspar-phyric.

The TTHF volcanic rocks are intermediate in composition. They are porphyritic fine- to medium-grained, and contain (micro)phenocrysts (1 to 4 mm in diameter) predominantly of zoned plagioclase (An_{39-52}), with subordinate clinopyroxene ($Wo_{28-37}En_{43-51}Fs_{19-22}$), pigeonite ($Wo_4En_{65-67}Fs_{29-30}$), hornblende, and Fe-Ti oxides. The groundmass is typically fine grained, composed of plagioclase, pyroxenes, hornblende, Fe-Ti oxides, and altered glass. Plagioclase laths may form pilotaxitic textures, whereas the pyroxenes in places occur as glomeroporphyritic clusters. The rocks show a lower grade (subgreenschist facies) of metamorphism than the Hazelton rocks. The Goosly Lake Formation includes mafic to intermediate porphyritic volcanic rocks that are coarser grained than the Hazelton and TTHF rocks. They contain large (≤1 cm) plagioclases (An_{49-58}) with subordinate clinopyroxene phenocrysts ($Wo_{39}En_{47}Fs_{14}$ to $Wo_{43}En_{42}Fs_{15}$) and minor phlogopite. Mafic to intermediate volcanic rocks of the BCF contain plagioclase (An_{43-70}), clinopyroxene (Wo_{38-44} $En_{42-47}Fs_{13-17}$), rare olivine (Fo_{75-80}) phenocrysts, and microphenocrysts of Fe-Ti oxides. The groundmass consists of altered glass, clinopyroxene, plagioclase, and Fe-Ti oxides. Similar compositions of phenocrysts characterize most of the Swans Lake volcanic suite (Cpx: $Wo_{41-44}En_{40-44}$ Fs_{13-18}; Pl: An_{55-73}; Ol: Fo_{60-80}). These rocks, however, also contain sanidine phenocrysts. The matrix consists of plagioclase, clinopyroxene, Fe-Ti oxides, and altered glass. Some flows, however, are aphyric. Clinopyroxenes of the Eocene volcanic rocks differ significantly from those of TTHF volcanics, being enriched in Ca and depleted in Fe (see Table 1 in Dostal et al., 2001). Moreover, the Eocene volcanics do not contain pigeonite. The Miocene lavas are usually microporphyritic, containing about 3–5 vol% olivine phenocrysts in a very fine grained groundmass of plagioclase and ferromagnesian minerals.

Geochemistry

Analytical techniques and alteration

Samples were selected from a suite of several hundred specimens collected during detailed mapping of the Buck Creek area by British Columbia Geological Survey (Church 1973, 1984; Church and Barakso, 1990). In addition, for a comparison with the TTHF volcanics, samples were collected from Mount Sibola in the type section of Kasalka volcanics in the Takla area (MacIntyre, 1985).

These rocks were analyzed for major and some trace elements (Rb, Sr, Ba, Ga, Zr, Y, Nb, Zn, Co, V, Ni, and Cr) by X-ray fluorescence spectrometry in the Nova Scotia Regional Geochemical Centre at Saint Mary's University, Halifax. Representative samples were selected from this set for analyses of other trace elements (rare-earth elements, Th and Nb) by ICP-MS in the Geochemical Lab of the Ontario Geological Survey in Sudbury. Precision and accuracy of the X-ray data are reported by Dostal et al. (1986) and for ICP-MS data by Ayer and Davis (1997). The analytical error of the trace element determinations is 2–10% and <5% for major elements. The chemical analyses of the Eocene volcanic rocks were reported by Dostal et al. (1998, 2001). Sixteen representative analytical results for Cretaceous volcanic rocks of the Buck Creek complex, its basement (Jurassic Hazelton volcanic rocks) as well as volcanic rocks of the Kasalka Group are presented in Tables 1–3. Seven samples were subsequently selected for Nd and Sr-isotopic analyses. Samarium and Nd abundances and Nd and Sr isotope ratios were determined by isotope dilution mass spectrometry at Carleton University. Analytical methods and quality of the data are described by Cousens (1996) and Ayers and Dostal (2000). The isotopic ratios determined for three Hazelton Group samples and four Tip Top Hill volcanics are compared with data for the Buck Creek, Swans Lake, and Goosly Lake successions (Table 4). Six repeat analyses of SRM987 standard yielded an average of $^{87}Sr/^{86}Sr$ of 0.710252 ± 18. For Nd isotopes, 34 runs of the La Jolla standard gave an average $^{143}Nd/^{144}Nd$ = 0.511875 ± 18. Data for these rocks are age corrected (50 Ma for BCF, SLV and GLF rocks, 85 Ma for Tip Top Hill rocks, and 200 Ma for the Hazelton samples).

Most of the rocks, particularly those of the Hazelton, Tip Top Hill, and Kasalka suites, exhibit petrographic evidence of deuteric alteration and (or) low-grade metamorphism. Comparison of the chemistry of these rocks with younger volcanic rocks (Gill, 1981; Thorpe, 1982) indicates that the secondary processes have not changed the main geochemical characteristics of the rocks, although concentrations of alkalis could have been modified.

Basement volcanics (Hazelton and Rocky Ridge groups)

The volcanic rocks of the Jurassic Hazelton Group from the basement of the Buck Creek volcanic complex (Fig. 5) are predominantly basalts and andesites (Table 1). The comparatively rare rhyolites in the Hazelton Group do not appear to be directly related to the mafic rocks (Fig. 6A). The mafic to intermediate rocks (Fig. 7A) are transitional between tholeiitic and calc-alkaline. Their rare-earth-element (REE) patterns (Fig. 8A) are slightly enriched in light REE (LREE) with

TABLE 1. Major and Trace Element Composition of Jurassic Volcanic Rocks of the Hazelton Group[1]

	M14B	M12	OPG-456	M1	M14A	M3	M5
SiO_2 (wt%)	47.02	47.71	54.09	56.20	56.53	75.04	75.77
TiO_2	0.89	0.94	0.92	0.93	0.88	0.26	0.09
Al_2O_3	16.74	16.48	15.83	21.93	16.94	11.77	12.28
Fe_2O_3*	10.91	11.38	11.33	6.65	8.57	3.31	0.75
MnO	0.34	0.24	0.23	0.12	0.11	0.09	0.13
MgO	9.66	6.66	5.09	2.46	5.97	0.95	0.32
CaO	6.31	9.96	2.36	0.66	4.73	1.02	2.34
Na_2O	2.92	2.33	6.52	0.15	1.63	0.20	0.93
K_2O	1.72	1.06	0.15	6.25	2.27	2.90	2.56
P_2O_5	0.15	0.14	0.19	0.20	0.12	0.07	0.02
LOI	3.51	3.40	3.89	4.46	2.40	3.71	3.90
Total	100.16	100.31	100.61	100.01	100.14	99.32	99.09
Mg#	0.64	0.54	0.47	0.42	0.58	0.36	0.46
Cr (ppm)	141	110	8	27	40	137	
Ni	58	25	7	28	16	12	10
Co	52	39	40	16	30	8	0
V	296	244	240	114	274	43	20
Zn	146	121	160	134	75	98	37
Rb	93	44	3	176	85	81	93
Ba	367	29	110	1296	198	726	1118
Sr	598	328	56	22	244	36	71
Ga	19	16	17	22	19	11	13
Nb	1.36	1.42	1.95	8.38	2.59	4.20	5.13
Hf	1.51	1.12	1.74	6.80	2.82	3.80	2.17
Zr	55	40	58	276	94	154	62
Y	20	18	21	42	29	18	9
Th	0.71	0.51	2.04	7.88	1.79	6.83	5.68
La	5.22	4.72	7.12	22.01	9.26	13.90	17.50
Ce	12.42	11.53	16.44	51.83	21.55	27.55	32.89
Pr	1.90	1.74	2.30	6.54	2.94	3.24	4.04
Nd	8.62	8.36	10.31	25.56	13.02	12.01	13.60
Sm	2.47	2.50	2.75	6.03	3.67	2.58	2.27
Eu	0.99	0.94	1.09	1.65	1.19	0.69	0.48
Gd	3.24	2.95	3.24	5.88	4.61	2.55	1.73
Tb	0.54	0.51	0.56	1.03	0.79	0.40	0.25
Dy	3.49	3.27	3.62	6.30	4.91	2.56	1.35
Ho	0.80	0.71	0.83	1.42	1.15	0.57	0.27
Er	2.22	2.01	2.39	4.27	3.42	1.66	0.81
Tm	0.34	0.31	0.37	0.68	0.51	0.29	0.13
Yb	2.04	1.86	2.41	4.36	3.43	1.92	0.87
Lu	0.34	0.31	0.39	0.70	0.55	0.36	0.14

[1]Fe_2O_3 = total Fe as Fe_2O_3; Mg# = Mg/(Mg+Fe*).

TABLE 2. Major and Trace Element Composition of Cretaceous Volcanic Rocks of the Tip Top Hill Formation[1]

	OPG-171	PS-53	NAD-51	NAD-200	G-318	OPG-743	HU-39	NAD-203
SiO_2 (wt%)	53.19	54.12	57.46	58.66	59.10	59.64	60.42	67.28
TiO_2	0.89	0.67	0.63	0.63	0.75	0.83	0.76	0.60
Al_2O_3	19.36	17.94	17.15	16.82	17.47	15.48	14.73	16.02
Fe_2O_3*	7.66	7.08	6.45	6.27	6.70	6.94	6.03	5.52
MnO	0.16	0.12	0.22	0.11	0.11	0.12	0.09	0.30
MgO	2.88	0.87	1.42	0.57	1.93	3.24	1.88	0.34
CaO	3.60	5.77	6.62	5.44	5.85	5.27	5.93	1.10
Na_2O	6.02	3.40	3.49	4.14	4.16	3.52	3.16	7.14
K_2O	2.48	3.30	2.87	2.74	2.01	2.61	2.62	1.64
P_2O_5	0.36	0.43	0.37	0.33	0.31	0.41	0.36	0.24
LOI	3.21	6.53	3.46	3.21	1.98	1.88	3.55	0.95
Total	99.80	100.23	100.13	98.92	100.37	99.94	99.53	101.13
Mg#	0.43	0.20	0.30	0.15	0.36	0.48	0.38	0.11
Cr (ppm)				13	29	74	73	13
Ni	7			29	47	34	32	4
Co	28	13	16	16	18	23	20	13
V	164	97	101	111	131	131	119	90
Zn	88	184	129	72	75	85	80	62
Rb	63	95	79	82	51	84	62	56
Ba	1786	1515	1330	1293	1065	1273	1219	1392
Sr	1043	468	1116	610	563	582	579	503
Ga	20	22	18	18	19	19	17	12
Nb	4.57	10.25	9.91	9.46	7.03	16.88	11.61	6.70
Hf	2.77	4.40	4.28	3.96	3.29	6.33	6.11	3.19
Zr	153	172	212	163	141	264	252	139
Y	20	23	27	21	14	30	29	13
Th	2.51	6.73	7.08	6.13	2.44	7.36	6.54	5.68
La	16.21	21.68	22.77	17.84	10.86	41.90	33.79	16.02
Ce	34.81	43.78	45.85	35.50	22.52	79.77	65.51	33.91
Pr	5.03	5.78	6.10	4.75	3.13	9.98	8.29	3.78
Nd	21.45	23.84	25.23	19.89	13.42	37.49	32.04	14.83
Sm	4.60	5.15	5.57	4.38	2.99	7.10	6.34	3.19
Eu	1.31	1.42	1.56	1.26	0.93	1.60	1.40	0.91
Gd	3.76	4.53	4.74	4.00	2.82	5.70	5.31	2.77
Tb	0.54	0.66	0.74	0.60	0.41	0.87	0.80	0.39
Dy	3.34	3.97	4.26	3.52	2.44	5.04	4.81	2.28
Ho	0.72	0.85	0.90	0.76	0.53	1.06	1.04	0.46
Er	2.09	2.37	2.67	2.19	1.43	3.00	2.91	1.32
Tm	0.30	0.37	0.41	0.33	0.21	0.44	0.43	0.19
Yb	1.98	2.44	2.68	2.16	1.41	2.86	2.82	1.28
Lu	0.31	0.39	0.44	0.34	0.23	0.46	0.45	0.21

[1]Fe_2O_3 = total Fe as Fe_2O_3 ; Mg# = Mg/(Mg+Fe*).

TABLE 3. Major and Trace Element Composition of Cretaceous Volcanic Rocks of the Kasalka Group[1]

	B-125	B-123	B-120	B-124	B-121
SiO_2 (wt%)	61.31	64.24	64.41	64.59	65.83
TiO_2	0.48	0.49	0.46	0.50	0.46
Al_2O_3	17.18	15.85	16.49	16.05	16.42
Fe_2O_3	4.24	4.78	2.94	4.28	3.55
MnO	0.20	0.21	0.10	0.10	0.06
MgO	3.08	1.53	1.20	1.74	0.63
CaO	2.52	4.25	3.03	3.14	3.12
Na_2O	5.01	3.38	5.32	4.32	4.78
K_2O	1.95	3.10	2.95	2.20	3.21
P_2O_5	0.21	0.22	0.20	0.22	0.21
LOI	3.40	1.50	1.50	2.20	0.80
Total	99.57	99.55	98.60	99.33	99.06
Mg#	0.59	0.39	0.45	0.45	0.26
Cr (ppm)	18	20	17	16	15
Ni	9	17	6	6	9
Co	5	8	3	3	10
V	103	86	68	77	73
Zn	137	141	46	114	52
Rb	127	110	80	102	121
Ba	459	1217	1456	916	1081
Sr	629	445	549	412	554
Ga	16	15	15	15	12
Nb	4.77	5.20	5.24	5.45	5.32
Hf	2.03	2.97	2.82	3.08	3.02
Zr	67	113	95	109	105
Y	18	18	14	16	11
Th	4.30	5.46	5.20	5.76	6.11
La	14.05	16.60	17.20	18.71	18.12
Ce	28.78	32.04	34.02	35.73	33.58
Pr	3.71	3.94	4.14	4.36	4.02
Nd	15.55	15.81	16.33	16.96	15.56
Sm	3.47	3.26	3.25	3.39	3.12
Eu	1.27	1.24	1.27	1.44	1.22
Gd	3.35	3.20	2.86	3.02	2.56
Tb	0.50	0.48	0.40	0.44	0.38
Dy	3.19	2.84	2.44	2.64	2.07
Ho	0.64	0.62	0.47	0.51	0.41
Er	1.86	1.77	1.48	1.57	1.20
Tm	0.27	0.28	0.21	0.23	0.18
Yb	1.82	1.91	1.54	1.62	1.27
Lu	0.29	0.32	0.25	0.26	0.22

[1]Fe_2O_3* = total Fe as Fe_2O_3; Mg# = Mg/(Mg+Fe*).

TABLE 4. Nd and Sr Isotopic Composition of Volcanic Rocks of the Buck Creek Area[1]

Sample no.	Nd (ppm)	Sm (ppm)	$^{143}Nd/^{144}Nd_m$	$^{147}Sm/^{144}Nd$	ε_{Nd}	Rb (ppm)	Sr (ppm)	$^{87}Sr/^{86}Sr_m$	$^{87}Sr/^{86}Sr_i$
Hazelton Group									
M14A	10.2	2.83	0.51294	0.1681	6.62	85	244	0.705894	0.703028
M14B	11.09	3.06	0.512962	0.16675	7.09	93	598	0.704719	0.70344
OPG-456	18.87	4.61	0.512864	0.1476	5.67	41	749	0.703958	0.703508
Tip Top Hill Formation									
NAD-51	22.6	4.93	0.5129	0.1319	5.81	79	1116	0.704085	0.703838
OPG-171	19.31	4.03	0.512923	0.12611	6.33	63	1043	0.703861	0.70365
G-318	16.8	3.83	0.512916	0.13769	6.06	51	563	0.703789	0.703473
H4-39	50	5.85	0.512966	0.07075	7.77	62	579	0.704362	0.703988
Goosly Lake volcanics									
HU-67	61.68	12.36	0.512796	0.12374	3.55	100	1159	0.704729	0.704552
OPG-651	54.62	9.82	0.512754	0.11099	2.81	59	1322	0.704799	0.704707
BC-77	52.97	9.42	0.512851	0.10973	4.71	90	1179	0.705017	0.70486
Buck Creek Formation									
HU-85	38.62	7.64	0.512773	0.12208	3.11	25	801	0.704496	0.704432
66	44.53	8.24	0.512736	0.11425	2.44	57	1003	0.704429	0.704312
HU-1	44.51	8.33	0.51275	0.11556	2.7	64	946	0.704853	0.704714
Swans Lake Unit									
483	24.3	5.1	0.512769	0.12965	2.98	23	632	0.704339	0.704264
379	34.58	6.5	0.512761	0.12711	2.84	16	783	0.70439	0.704348
383	31.92	6.29	0.512743	0.12168	2.53	24	760	0.704574	0.704509

[1] $^{143}Nd/^{144}Nd_m$ and $^{87}Sr/^{86}Sr_m$ are measured Nd and Sr isotopic ratios, respectively, and ε_{Nd} is the fractional difference between the $^{143}Nd/^{144}Nd$ of rock and the bulk earth at the time of crystallization. Concentrations of Nd and Sm were determined by isotope dilution, and those of Rb and Sr by X-ray fluorescence. $^{87}Sr/^{86}Sr_i$ is initial Sr isotopic ratio. ε_{Nd} and $^{87}Sr/^{86}Sr_i$ assume an age of 50 Ma for the Goosly Lake, Buck Creek, and Swan Lake volcanics, 85 Ma for Tip Top Hill, and 200 Ma for Hazelton volcanics. Values of ε_{Nd} were calculated using modern $^{143}Nd/^{144}Nd_{CHUR} = 0.512638$ and $^{147}Sm/^{144}Nd_{CHUR} = 0.1967$. Precision of concentrations of Nd and Sm is ±1%.

Age (Ma)	Volcanic sequences	Stratigraphic thickness (m)	Composition
21.4 ± 1.1	Chilcotin basalts	60 - 90	Intraplate basalts (enriched continental tholeiites)
		UNCONFORMITY	
50 ± 1	Swans Lake unit	15 - 60	Transitional basalts
50 ± 1	Buck Creek Formation	~400	High K calc-alkaline
51.9 ± 1 51.3 ± 1.8			Silver vein deposits
52 ± 1	Goosly Lake Formation (and associated intrusions)	~500	High K calc-alkaline
		UNCONFORMITY	
~85	Tip Top Hill Formation	~500	Calc-alkaline
		UNCONFORMITY	
~105	Rocky Ridge Formation		Rhyolites
		UNCONFORMITY (Cordilleran terrane accretion)	
~200	Hazelton Group		Calc-alkaline to tholeiitic

Decrease of average contents of SiO_2 and LILE

FIG. 5. Mesozoic and Tertiary igneous sequences of the Buck Creek area.

chondrite-normalized $(La/Yb)_n$ ~ 1.5–3 and $(La/Sm)_n$ ~ 1–2. Mantle-normalized patterns (Fig. 9A) are enriched in strongly incompatible trace elements and display distinct negative Nb and small Ti anomalies (Fig. 9). The Hazelton Group rocks from the Buck Creek area have high ε_{Nd} values (+ 5.7 to + 7.1) and low initial $^{87}Sr/^{86}Sr$ ~ 0.7030–0.7035 (Table 4). Rhyolites have REE patterns that are strongly enriched in light REE with $(La/Yb)_n$ ~4.4–12 and $(La/Sm)_n$ ~ 3–4.2, and without Eu anomalies (Fig. 8B). The significant difference in the shape of the REE patterns between the mafic and felsic Hazelton rocks suggests that the rhyolites are not related to the mafic rocks via fractional crystallization.

In terms of trace elements (particularly Nb vs Y; Pearce et al. 1984; MacIntyre, 2001), rhyolites of the Rocky Ridge Formation have characteristics of volcanic arc rocks. Their REE patterns show variable enrichment of LREE with $(La/Yb)_n$, ranging between 2 and 16, while $(Gd/Yb)_n$ ~ 1 (MacIntyre, 2001).

Basin volcanics

The TTHF rocks (Fig. 5) are typical calc-alkaline volcanics (Fig. 7B). They have SiO_2 (LOI-free) ranging from 55 to 72 wt% with a majority in a narrower range of 57 to 64 wt %. According to the Zr/TiO_2 versus SiO_2 diagram (Fig. 6B), these rocks are mainly subalkaline andesites. They show some systematic variations of major elements, suggesting fractionation dominated by plagioclase. Their REE patterns (Fig. 8C) are distinctly enriched in LREE with $(La/Yb)_n$ ~ 5–9. Mantle-normalized trace element patterns show negative Nb and Ti anomalies, typical of subduction-related magmas, but differ from the Hazelton volcanics by their higher abundances of strongly incompatible trace elements (Fig. 9C). In fact, their chemistry resembles the high-K calc-alkaline series (Dostal et al., 2001). The TTHF volcanic rocks have high positive values (ε_{Nd} ~ +5.8 to +7.8) and low initial $^{87}Sr/^{86}Sr$ ratios (0.7035 to 0.7040), which are similar to those of the Hazelton Group (Table 4). Overall, these rocks are similar to the Cretaceous volcanics of the Kasalka Group (Fig.

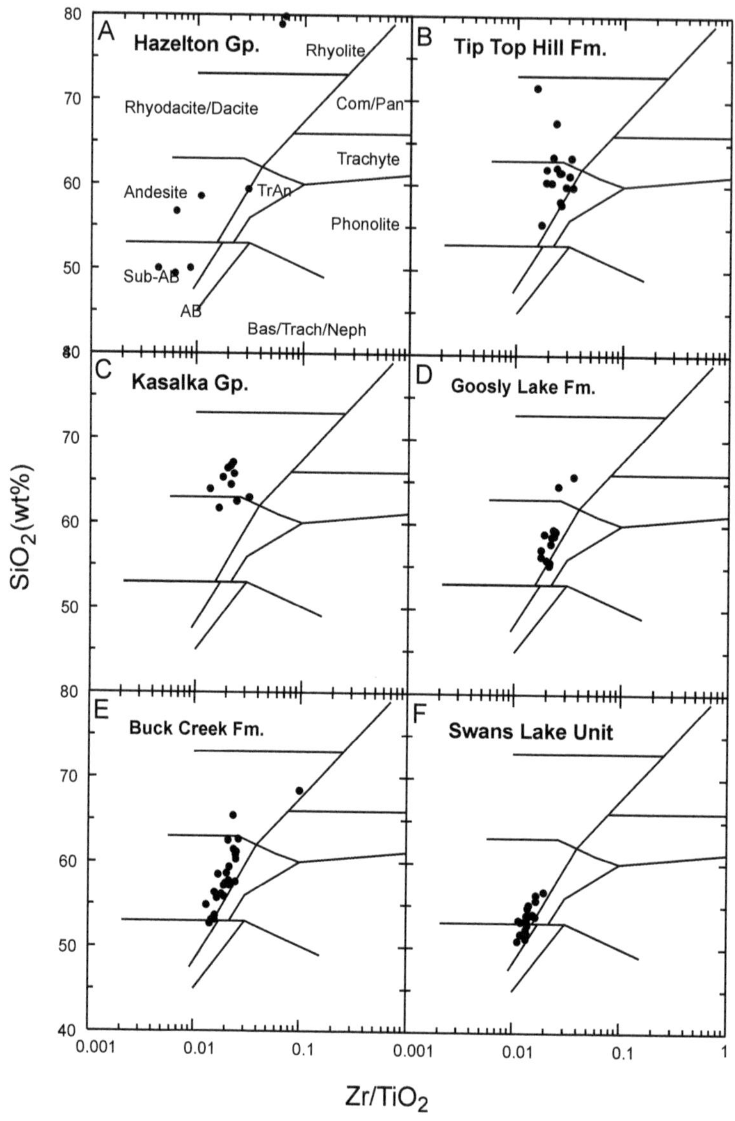

FIG. 6. Zr/TiO_2 versus SiO_2 (wt%) diagram of Winchester and Floyd (1977) for volcanic rocks of (A) Hazelton Group, (B) Tip Top Hill Formation, (C) Kasalka Group, (D) Goosly Lake Formation, (E) Buck Creek Formation, and (F) Swans Lake unit. Abbreviations: Sub-AB = subalkaline basalt; AB = alkali basalt; TrAn = trachyandesite; Bas = basanite; Neph = nephelinite; Com = comendite; Pan = pantellerite.

6C). Compared to the TTHF volcanics, however, the Kasalka rocks are mainly dacites with slightly higher SiO_2 (characteristically 63–68 wt%). For given SiO_2 values, the Kasalka samples typically have slightly lower K_2O (Tables 1–3). The shapes of their REE patterns and mantle-normalized trace element patterns are closely comparable, although on average, the Kasalka rocks have slightly lower contents of incompatible elements (Figs. 8D and 9D). The similarities between the TTHF and Kasalka volcanic rocks suggest that this type of Late Cretaceous volcanism was widespread in the central part of British Columbia.

For the most part, the Goosly Lake volcanics (Fig. 5) consist of andesites/trachyandesites enriched in K_2O and highly incompatible trace

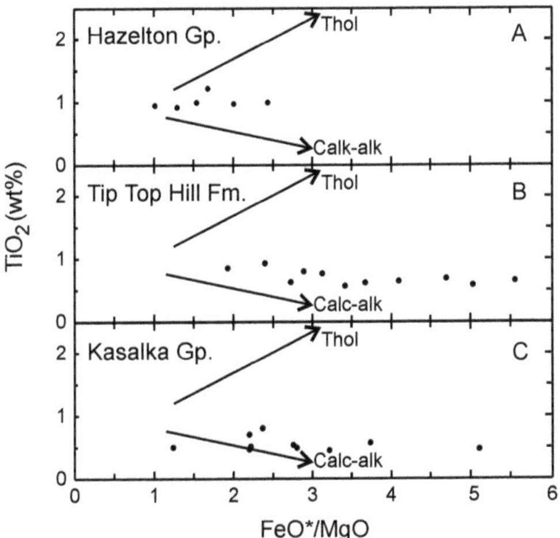

FIG. 7. FeO*/MgO versus TiO_2 (wt%) diagrams for the volcanic rocks of (A) Hazelton Group, (B) Tip Top Hill Formation, and (C) Kasalka Group. Vectors for tholeiitic (Thol) and calc-alkaline (Calc-alk) trends are after Miyashiro (1974).

elements such as Ba, Sr, and Rb. Their trace element signature is therefore characteristic of high-K rocks (Dostal et al., 1998, 2001). Chondrite-normalized REE patterns for the GLF volcanic rocks (Fig. 8E) are enriched in LREE with $(La/Yb)_n$ ~ 11–15 and $(Gd/Yb)_n$ ~ 2.0–2.6. Not surprisingly, they have mantle-normalized trace element patterns showing an overall enrichment in large-ion-lithophile elements (LILE), including Ba and LREE, relative to the heavy REE (HREE) and high-field-strength elements (HFSE) is accompanied by distinct negative anomalies for Nb and Ti (Fig. 9E). Compared to TTHF volcanics, the Goosly Lake volcanics have higher Zr, Ti, V, and higher Ti/V ratios but the same Zr/Y ratios. They have lower Zr/TiO_2 but higher Nb/Y. On average, the GLF volcanic suite has lower SiO_2 but the same K_2O and total alkalis. The GLF also has higher $(La/Yb)_n$ and $(La)_n$ compared to TTHF rocks (Fig. 8E). The initial $^{87}Sr/^{86}Sr$ ratios (0.7046–0.7049) and ε_{Nd} (+2.8 to +4.7) of the GLF rocks differ from those of the TTHF volcanics (Fig. 10).

The SLV rocks tend to have lower SiO_2 contents (47–54 wt%) than the BCF volcanics (50–67 wt% SiO_2 but mostly in the range 54–60 wt% SiO_2; Figs. 6E and 6F). Felsic rocks with >63 wt% SiO_2, however, are rare among the BCF lavas. The BCF rocks have calc-alkaline characteristics, as is typical of high-K suites, although some of the basaltic rocks have relatively high TiO_2 contents (up to 1.6 wt%). In contrast, the SLV rocks are intraplate tholeiitic to high-K calc-alkaline basalts. They show a typical tholeiitic trend of Ti enrichment with increasing FeO_{tot}/MgO ratio (Dostal et al., 1998).

Chondrite-normalized REE patterns display an enrichment of the LREE (Figs. 8F and 8G) in both suites of rocks. Their $(La/Yb)_n$ ratios range from 4 to 14 and 5 to 12 in the BCF and SLV, respectively. In addition, their mantle-normalized trace element patterns exhibit a conspicuous enrichment of the strongly incompatible trace elements, including Th and LREE, relative to the HREE, and the HFSE. The trace element patterns show negative anomalies for Nb and Ti (Figs. 9F and 9G), and thus are typical of subduction-related calc-alkaline mafic rocks (Pearce, 1983) or continental intraplate basalts that have been significantly contaminated by lithospheric mantle or crustal material (e.g., Kempton et al., 1991). The BCF and SLV have initial $^{87}Sr/^{86}Sr$ ratios and ε_{Nd} values of 0.7043–0.7047 and +2.4 to +3.1, and 0.7043–0.7045 and +2.5 to +3.0, respectively (Fig. 10). All Eocene rocks from the Buck Creek volcanic complex display restricted isotopic variations that are significantly different from those of the TTHF and Hazelton volcanics. The volcanic rocks of the Jurassic (Hazelton Group) and Cretaceous (TTHF) age have higher ε_{Nd} values and lower initial $^{87}Sr/^{86}Sr$ ratios than the Eocene rocks (Fig. 10).

The Lower Miocene Chilcotin/Cheslatta basalts in the central British Columbia (Fig. 5) are alkali olivine basalts, basanites, and transitional basalts (Mathews, 1989; Dostal et al., 1996; Anderson et al., 2001). Those from Poplar Buttes in the Buck Creek basin contain about 14% normative nepheline (Church and Barakso, 1990). The trace element abundances of the Miocene basalts (Fig. 8H) from the nearby Binta Lake volcanic center (Anderson et al., 2001) are distinctly enriched in light REE with $(La/Yb)_n$ ~ 20 and have fractionated heavy REE, suggesting derivation from a garnet peridotite source (Dostal et al., 1996). The mantle-normalized pattern of the Binta Lake basalt (Fig. 9H) displays an increase of normalized element abundances with increasing incompatibility from the heavy REE and peak at Nb, suggesting derivation from a source with an incompatible-element composition similar to that of the mantle source of oceanic island basalts (OIB). A similar OIB source has been inferred for other intraplate Chilcotin or Cheslatta Lake basalts (Dostal et al., 1996; Anderson et al., 2001).

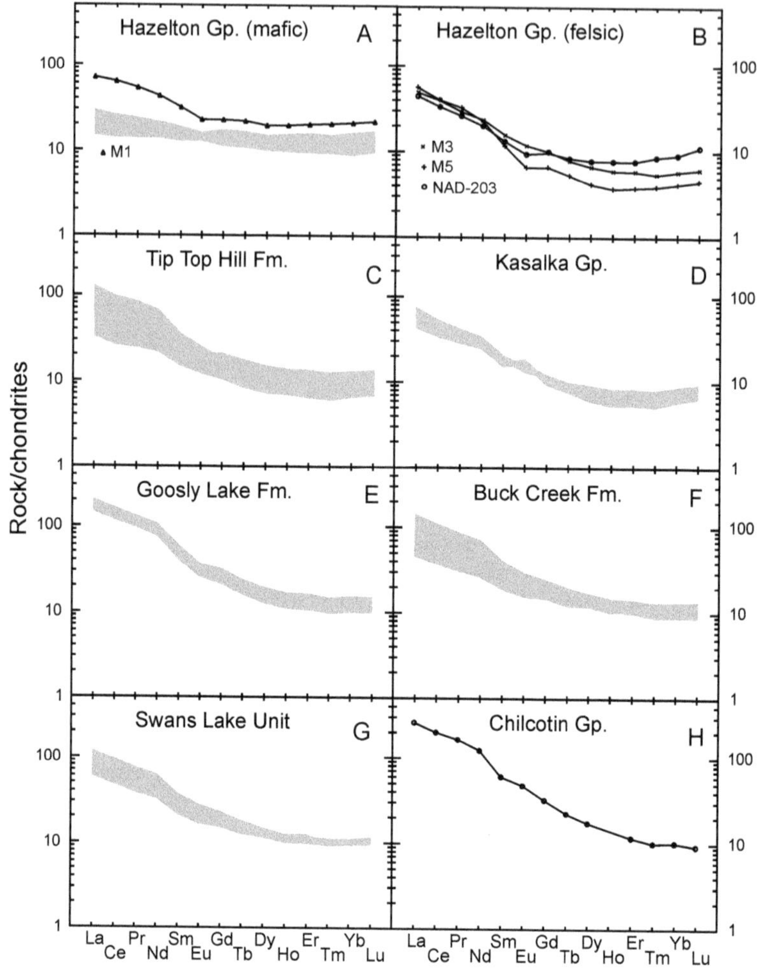

FIG. 8. Chondrite-normalized rare-earth-element abundances in the volcanic rocks of the Buck Creek volcanic complex and its basement. A. Hazelton Group mafic and intermediate rocks. B. Hazelton (samples M3 and M5) and TTHF (sample NAD-203) rhyolites. C. Tip Top Hill Formation mafic and intermediate rocks. D. Kasalka Group. E. Goosly Lake Formation. F. Buck Creek Formation. G. Swans Lake unit. H. Chilcotin/Cheslatta basalt (Binta Lake; after Anderson et al., 2001). The analyses of the Eocene rocks were taken from Dostal et al. (1998, 2001). Normalizing values are after Sun and McDonough (1989).

Petrogenetic Evolution

The long-lived and episodic volcanism of the Buck Creek basin and its basement affords a unique opportunity to assess the mantle source regions and their evolution in a single location since the Mesozoic, inasmuch as each volcanic unit is distinct with its own diagnostic features. Rocks of the Jurassic Hazelton Group (Fig. 5) are characterized by a minor enrichment in strongly incompatible trace elements, particularly LREE and Th, and display negative anomalies for Nb and Ti and relatively flat REE patterns that are typical of subduction-related suites, transitional between island-arc tholeiites and calc-alkaline rocks emplaced in a primitive continental arc setting (Crow and Condie, 1987; Hawkesworth et al., 1993). Their isotopic characteristics, particularly high positive ε_{Nd} values, suggest that they were derived predominantly from a source with a light REE depletion (e.g., the depleted mantle, which at the time of their origin had ε_{Nd} values in the range from +7.3 to + 9.8; Whalen et al., 2001). The flat HREE patterns of the Hazelton volcanic rocks

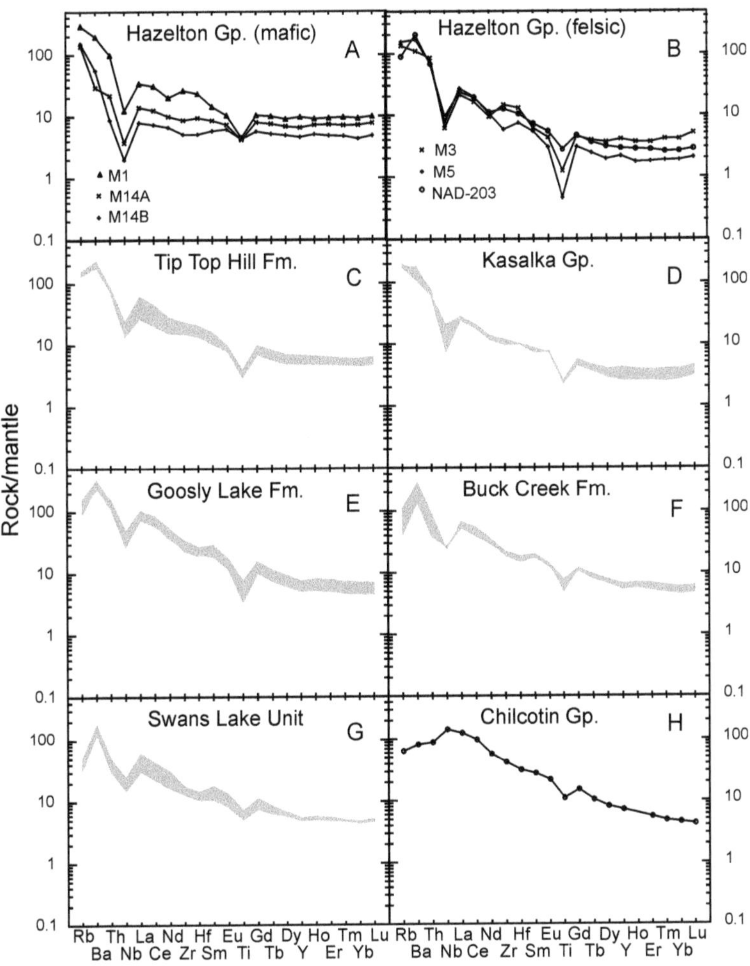

FIG. 9. Mantle-normalized incompatible trace element abundances in the volcanic rocks of the Buck Creek volcanic complex and its basement. A. Hazelton Group mafic and intermediate rocks. B. Hazelton (samples M3 and M5) and TTHF (sample NAD-203) rhyolites. C. Tip Top Hill Formation mafic and intermediate rocks. D. Kasalka Group. E. Goosly Lake Formation. F. Buck Creek Formation. G. Swans Lake unit. H. Chilcotin/Cheslatta basalt (Binta Lake; after Anderson et al., 2001). The analyses of the Eocene rocks were taken from Dostal et al. (1998, 2001). Normalizing values are after Sun and McDonough (1989).

are likely a result of melting at shallow depth, in the spinel stability field. Rare rhyolitic rocks in the area are not directly related to the mafic rocks but were probably generated by partial melting of arc crust. The same probably applies to the Rocky Ridge rhyolites (MacIntyre, 2001), which have a flat heavy REE pattern with $(Gd/Yb)_n \sim 1$, suggesting derivation from a garnet-free source.

In the Buck Creek volcanic complex, volcanism started with the extrusion of the Cretaceous TTHF volcanics (Fig. 5). These subduction-related calc-alkaline rocks have the highest average SiO_2 contents among the mafic to intermediate volcanic rocks of the basin and distinct negative Nb and Ti anomalies. Their high positive ε_{Nd} and low $^{87}Sr/^{86}Sr$ values suggest that old continental crust did not play an important role in their genesis. Instead, they were probably derived predominantly from a juvenile or DM-like source. The high SiO_2 contents and enrichment in incompatible elements indicate that although these rocks were likely derived from a juvenile source, the Cretaceous arc was more

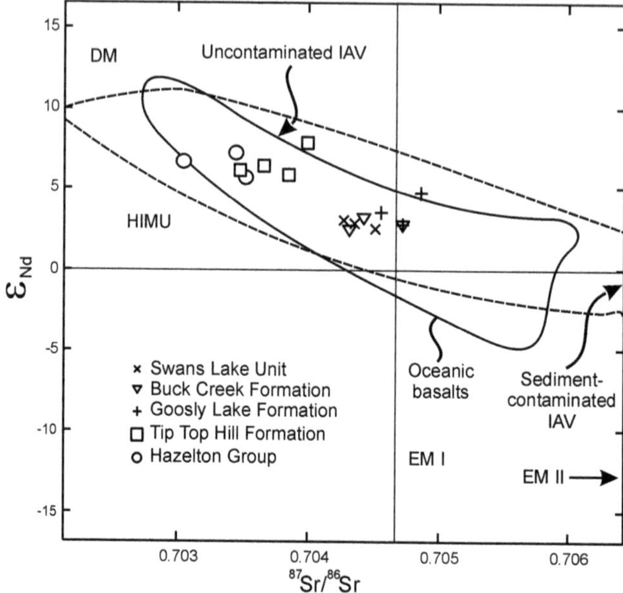

FIG. 10. ε_{Nd} versus initial $^{87}Sr/^{86}Sr$ ratio for the volcanic rocks of the Buck Creek complex, showing also for comparison the composition of oceanic basalts (Zindler and Hart, 1986) and island-arc volcanic rocks (IAV; dashed line; sediment-contaminated [e.g., Sunda arc] and uncontaminated [e.g., Aleutians, the Marianas] IAV; Samson et al. 1989). Position of mantle end-member components (DM, HIMU, EMI, and EMII) is after Zindler and Hart (1986).

evolved, its crust was significantly thicker than that of the Hazelton arc, and the rocks were probably emplaced some distance from the trench (Fig. 11). Isotopically, these rocks are similar to those of the basement of the basin, in particular to the Hazelton Group.

The GLF, BCF, and SLV volcanic rocks are genetically related (Dostal et al., 1998, 2001). The element distribution of the GLF volcanics is typical of subduction-related rocks. The distinct trace element characteristics of the GLF rocks, including an enrichment of LILE accompanied by depletion of Nb and Ti, can be attributed to a heterogeneous lithospheric mantle source enriched by subduction processes. A LREE enrichment of the REE patterns of the Eocene rocks suggests metasomatic enrichment of their mantle source (probably by a protracted subduction), whereas the low Sr isotope ratios and positive ε_{Nd} values for most samples (Fig. 10) suggest that upper-crustal contamination probably did not play an important role during their evolution. The lower ε_{Nd} and SiO_2 but higher La/Yb and La_n of the Eocene rocks compared to the Cretaceous TTHF rocks suggest that they were derived from a distinct, more LREE-enriched source. The higher enrichment of strongly incompatible trace elements in the Eocene rocks, compared to the TTHF volcanics, also suggests that during the Eocene, the continental crust was relatively thick or their mantle source was more metasomatized. The Eocene rocks have also decidedly lower ε_{Nd} (~ +2.5 to +4.7) and higher initial $^{87}Sr/^{86}Sr$ ratios (> 0.704), but still resemble oceanic basalts (Fig. 10).

Although the BCF and GLF calc-alkaline rocks appear to have similar petrogenetic histories, the lower contents of LILE and SiO_2 in BCF rocks and differences in the trace element ratios, such as lower La/Yb and La/Sm ratios, indicate that the BCF magma was derived by a relatively high degree of partial melting of mantle source rocks. The calc-alkaline BCF and tholeiitic to calc-alkaline SLV volcanic suites were derived from similar mantle source rocks, probably garnet-bearing subcontinental lithosphere. Differences between the tholeiitic and calc-alkaline suites are likely due in part to contrasts in the depth of crystallization and type of crystallizing mineral assemblages in the two suites. Evidently, calc-alkaline magmas of the BCF

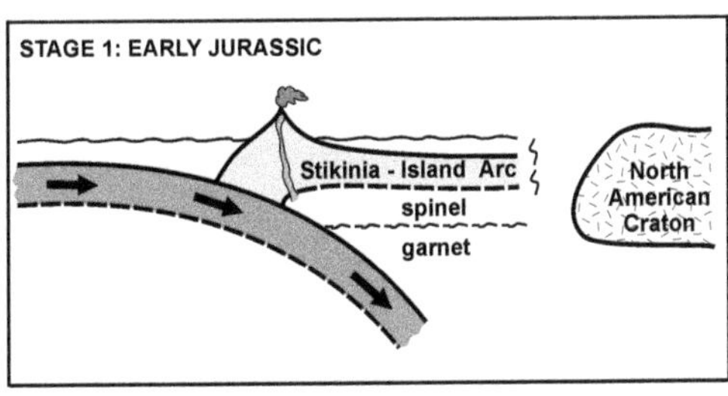

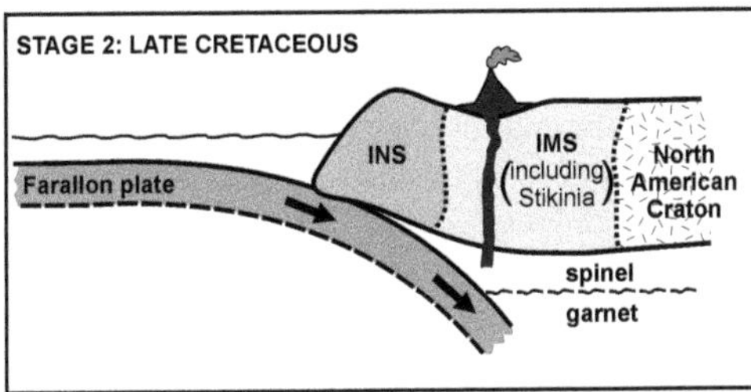

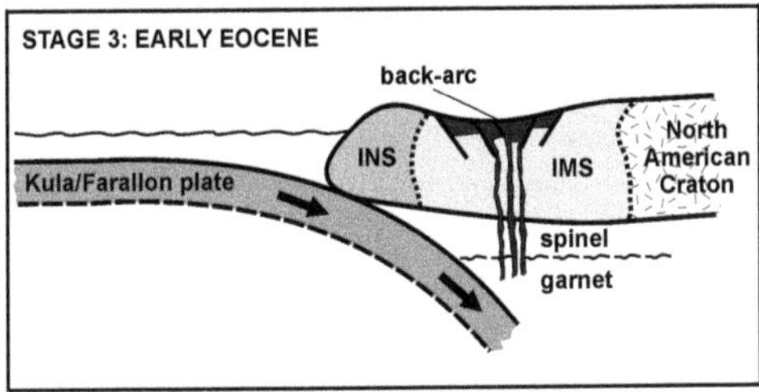

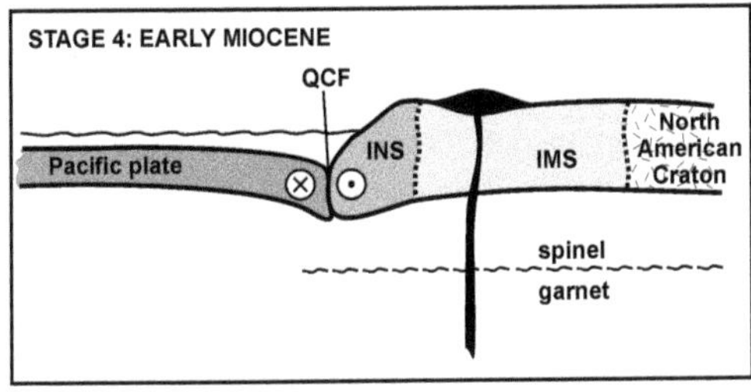

FIG. 11. Geodynamic model for the evolution of the Buck Creek basin and adjacent area (not to scale). Stage 1. Early Jurassic (~200 Ma; pre-accretion). The primitive island arc of Stikinia included Hazelton volcanics and marine sediments. The Stikine arc is an allochthonous terrane from a southerly latitude, with a shallow oceanic mantle source for the Hazelton volcanics. Stikinia's oceanic crust was relatively primitive and thin. Stage 2. Late Cretaceous (~85 Ma; post-accretion). Stikinia had been accreted to North America as a principal part of the Intermontane superterrane (IMS) by the Middle Jurassic. IMS on the sketch includes Stikinia and also the area that later became the Coast Mountains. By the Late Cretaceous, the Insular superterrane (INS) had also been accreted, moving the continental margin farther west. Rapid subduction of Farallon plate produced tectonically thickened crust with continental margin calc-alkaline rocks (basaltic andesites to rhyolites) of the Cretaceous Tip Top Hill Formation and Kasalka Group. Mafic rocks in this succession were generated from a spinel-bearing mantle source at shallow depth. Stage 3. Early Eocene (~50 Ma). The final stages of subduction of the Farallon and Kula plates under the North American continent were accompanied by crustal extention in the central part of British Columbia. Development of extensional (strike-slip and pull-apart) basins reflected a gradual transition from an arc-compressional environment to an extensional intra-arc rift setting. Volcanic activity shows a systematic change from high-K calc-alkaline rocks to rocks having many of the characteristics of intraplate tholeiites. Stage 4. Early Miocene (~20). Intraplate volcanism (Chilcotin/Cheslatta Lake basalts) was related to further thinning of Cordilleran lithosphere and a new, deeper asthenospheric mantle source that replaced the shallow subduction-modified lithospheric mantle. Melting occurred at greater depth, in the garnet stability field. Transcurrent movement along the Queen Charlotte Fault (QCF) is indicated by symbols (encircled "x" = motion is away from viewer; encircled dot = toward viewer).

underwent fractional crystallization at a greater (mid-crustal) depth than the intraplate-like SLV tholeiitic melts.

The Eocene part of the volcanic sequence shows a systematic change from typical high-K calc-alkaline rocks (GLF), to rocks having many of the characteristics of intraplate tholeiites (SLV); the BCF volcanics are intermediate between the two (Fig. 5). Average SiO_2 and LILE concentrations decline upward through the Eocene complex, although the Eocene rocks of the volcanic complex have overlapping Sr and Nd isotopic ratios, suggesting a similar mantle source. These variations are thus interpreted, in part, in terms of a higher degree of partial melting of mantle source rocks over time or in terms of a temporal decrease in magma fractionation. Flows (SLV) with intraplate characteristics probably represent a higher percentage of melting compared to the calc-alkaline flows, as has already been suggested for the BCF rocks (relative to GLF). The three formations are thus interpreted here to record a gradual change from a compressional to an extensional tectonic environment. The overlying Miocene Chilcotin (or Cheslatta Lake) intraplate alkali basaltic rocks (Mathews, 1989; Dostal et al., 1996; Anderson et al., 2001) might represent the final stage of this transition when the subduction-modified lithospheric mantle was replaced by a new asthenospheric mantle source. The distinctly fractionated heavy REE patterns of the Miocene rocks suggest their genesis at the greater depth, in the garnet stability field. In fact, the $(Gd/Yb)_n$ ratio, a measure of the slope of the heavy REE patterns (Fig. 8), shows a negative correlation with age. In mafic and intermediate rocks, it increases from Jurassic

Hazelton (~1.1) through Cretaceous TTHF and Kasalka (~1.5) to Eocene (2.0–2.4) and Miocene rocks (~3.2) suggesting a progressively deeper mantle source (McKenzie and Bickle, 1988; McKenzie and O'Nions, 1991).

Tectonic Interpretations and Implications

The Canadian Cordillera was assembled from terranes of oceanic and transitional crust prior to the Late Cretaceous (Monger and Nokelberg, 1996; Gabrielse and Yorath, 1991). The structural style of this terrane accretion is contentious, and we do not know whether the oceanic fragments were accreted together with their mantle lithosphere in trapped backarc basins or as thin-skinned, rootless flakes. Following terrane accretion, this mélange of Mesozoic and older terranes comprised the over-riding plate (as the western edge of North America) as the Farallon and Kula plates were rapidly subducted during much of the Late Cretaceous and Early Tertiary.

The Jurassic Hazelton Group is characterized by subduction-related tholeiitic to calc-alkaline mafic magmas with slight LREE enrichment and high ε_{Nd} (+5 to +6), indicative of an island-arc setting. However, the presence of rhyolitic rocks suggests that the Stikinia arc was relatively evolved and/or contained a cold, thickened transitional crust analogous to present-day subduction zones involving archipelagos of the Southwest Pacific. This arc preceded the accretion of Stikinia to North America (Fig. 11, stage 1), which took place by the Middle Jurassic (Mihalynuk et al., 1994; Monger and Nokelberg, 1996). There was a hiatus in volcanism through this crust until the initiation of Farallon plate subduction under the North American continent was accompanied by the Cretaceous volcanism of the Rocky Ridge Formation about 105 Ma and Kasalka and TTHF at about 85 Ma. Similarities in the radiogenic isotope composition of the pre-accretionary Jurassic and post-accretionary Cretaceous lavas indicate that the Stikine terrane was probably accreted as a block, with its own lithospheric mantle, and that mantle survived through Farallon subduction (Fig. 11, stage 2). However, significant differences in chemical composition between Jurassic and Cretaceous volcanic rocks, particularly in the higher abundances of incompatible trace elements and SiO_2 in the latter as well as contrasts in the shapes of their REE and mantle-normalized trace element patterns, imply different sources. Like the Jurassic volcanics, the Cretaceous mafic rocks were probably generated from a mantle source at a shallow depth in the spinel stability field, as suggested by unfractionated HREE (Fig. 8). The compositionally distinct source of the Cretaceous rocks could have been enriched by post-accretion metasomatism over the Farallon slab. These Cretaceous rocks have characteristics of continental-margin suites.

The depleted HREE pattern of the Eocene mafic and intermediate rocks indicates that magmas formed in the stability field of garnet (Fig. 8). Inasmuch as the spinel transition in the mantle occurs at a depth of about 60–80 km (Watson and McKenzie, 1991), the inferred depth of origin for the Eocene volcanic rocks must be deeper, at a greater depth than those of Cretaceous age.

The Eocene volcanism in the central part of British Columbia, which started after another 35 m.y. hiatus, is also part of the continental arc formed by the subduction of the Farallon or Kula plate under the North American continent. Volcanic activity started after the Tertiary plate reorganization in the Pacific basin. At that time, there was an abrupt change of the angle of convergence (Engebretson et al., 1985) between the continental margin of western North America and the plates of the Pacific basin. Dostal et al. (2001) suggested that decoupling of oblique plate convergence during the Eocene into a component of convergence normal to the plate boundary and a component of strike-slip faulting parallel to the plate margin within the overriding North American plate triggered crustal extension in the central part of British Columbia.

The Buck Creek volcanic complex, which is related to the development of an extensional (pull-apart) basin, probably represents a gradual transition from an arc compressional environment to one of extensional intra-arc rifting (Fig. 11, stage 3). This may mark the collapse of the Andean-style orogen. The Eocene rifting also coincided with the end of the Laramide orogeny and the cessation of thrusting in the foreland fold-and-thrust belt.

The Miocene alkali basaltic rocks, compositionally similar to OIB, were derived from asthenospheric mantle in the garnet field (Fig. 11, stage 4). Dostal et al. (1996) inferred that these rocks probably resulted from the impingement of asthenospheric mantle upwelling on lithosphere, possibly previously thinned by Early Tertiary extension. The asthenospheric upwelling also softened the lithosphere and triggered further extension and melting in the subcontinental lithosphere.

The association of Cretaceous, Eocene, and Miocene volcanic rocks as at Buck Creek is also widespread in central British Columbia. These volcanic successions record the evolution of the subcordilleran lithospheric mantle. It shows that the Stikine portion of the Cordillera evolved from (1) a primitive island arc, to (2) a post-accretion continental arc, that (3) underwent post-Laramide extension that culminated with (4) athenospheric upwelling and lithospheric mantle erosion during the Miocene. This scenario can provide a general model for the orogenic and post-orogenic evolution of the underlying mantle lithosphere in the North American Cordillera.

Acknowledgments

This research was funded by the NSERC Discovery grants to JD and JVO, the British Columbia Geological Survey Branch and a Lithoprobe grant to JD. We would like to thank Brendan Murphy and Joe English for a constructive review of the manuscript. Lithoprobe publication 1365.

REFERENCES

Anderson, R. G., Resnick, J., Russell, J. K., Woodsworth, G. J., Villeneuve, M. E., and Grainger, N. C., 2001, The Cheslatta Lake suite: Miocene mafic, alkaline magmatism in central British Columbia: Canadian Journal of Earth Sciences, v. 38, p. 697–717.

Anderson, R. G., Snyder, L. G., Wetherup, S., Struik, L. C., Villeneuve, M. E., and Haskin, M., 1998, Mesozoic to Tertiary volcanism and plutonism in southern Nechako NATMAP area, *in* Struik, L. C., and MacIntyre, D. G., eds., New geological constraints on Mesozoic to Tertiary metalogenesis and on mineral exploration in central British Columbia: Nechako NATMAP project: Short Course Notes, Cordilleran Section, Geological Association of Canada, 16 p.

Armstrong, R. L., 1988, Mesozoic and early Cenozoic magmatic evolution of the Canadian Cordillera, *in* Clark, S. P., ed., Processes in continental lithospheric deformation: Geological Society of America, Special Paper 218, p. 55–91.

Armstrong, R. L., and Ward, P. L., 1991, Evolving geographic patterns of Cenozoic magmatism in the North American Cordillera: The temporal and spatial association of magmatism and metamorphic core complexes: Journal of Geophysical Research, v. 96, p. 13,201–13,224.

Ayer, J. A., and Davis, D. W., 1997, Neoarchean evolution of differing convergent margin assemblages in the Wabigoon Subprovince: Geochemical and geochronological evidence from the Lake of the Woods greenstone belt, Superior Province, northwestern Ontario: Precambrian Research, v. 81, p. 155–178.

Ayer, J. A., and Dostal, J., 2000, Nd and Pb isotopes from the Lake of the Woods greenstone belt, northwestern Ontario: Implications for mantle evolution and the formation of crust in the southern Superior Province: Canadian Journal of Earth Sciences, v. 37, p. 1677–1689.

Bassett, K. N., 1991, Preliminary results of the sedimentology of the Skeena Group in west-central British Columbia, *in* Current Research, part A: Geological Survey of Canada, Paper 91-1A, p. 131–141.

Bassett, K. N., and Kleinspehn, K. L., 1996, Mid-Cretaceous transtension in the Canadian Cordillera: Evidence from the Rocky Ridge volcanics of the Skeena Group: Tectonics, v.15, p. 727–746.

Bevier, M. L., 1983, Implications of chemical and isotopic composition for petrogenesis of Chilcotin Group Basalt, British Columbia: Journal of Petrology, v. 24, p. 207–226.

Carter, N. C., 1981, Porphyry copper and molybdenum deposits of west-central British Columbia: British Columbia Ministry of Energy, Mines, and Petroleum Resources Bulletin 64, 150 p.

Church, B. N., 1973, Geology of the Buck Creek area, *in* Geology exploration and mining 1972: Victoria, BC, British Columbia Ministry of Energy, Mines, and Petroleum Resources, p. 353–363.

Church, B. N., 1984, Geology of the Buck Creek Tertiary outlier: Victoria, BC, British Columbia Ministry of Energy, Mines, and Petroleum Resources, map, scale 1:100,000.

Church, B. N., and Barakso, J. J.,1990, Geology, lithogeochemistry, and mineralization in the Buck Creek area, British Columbia (93L): Victoria, BC, British Columbia Ministry of Energy, Mines, and Petroleum Resources, Paper 1990-2, 95 p.

Coney, P. J., and Harms, T. A., 1984, Cordilleran metamorphic core complexes: Cenozoic extensional relics of Mesozoic compression: Geology, v. 12, p. 550–554.

Cousens, B. L., 1996, Depleted and enriched upper mantle sources for basaltic rocks from diverse tectonic environments in the northeast Pacific Ocean: The generation of oceanic alkaline vs. tholeiitic basalts, *in* Basu, A., and Hart, S. R., eds., Earth processes: Reading the isotopic code: American Geophysical Union, Geophysical Monograph 95, Washington, DC, p. 207–231.

Crow, C., and Condie, K. C., 1987, Geochemistry and origin of Late Archean volcanic rocks from the Rhenosterhoek Formation, Dominion Group, South Africa: Precambrian Research, v. 37, p. 217–229.

Cyr, J. B., Pease, R. P., and Schroeter, T. G., 1984, Geology and mineralization at Equity Silver mine: Economic Geology, v. 79, p. 947–968.

Dostal, J., Baragar, W. R. A., and Dupuy, C., 1986, Petrogenesis of the Natkusiak continental basalts, Victoria Island, N.W.T.: Canadian Journal of Earth Sciences, v. 23, p. 622–632.

Dostal, J., Church, B. N., Reynolds, P. H., and Hopkinson, L., 2001, Eocene volcanism in the Buck Creek basin, central British Columbia (Canada): Transition from arc to extensional volcanism: Journal of Volcanology and Geothermal Research, v. 107, p. 149–170.

Dostal, J., Hamilton, T. S., and Church, B. N., 1996, The Chilcotin basalts, British Columbia (Canada): Geochemistry, petrogenesis and tectonic significance: Neues Jahrbuch fur Mineralogie, Abhandlungen, v. 170, p. 207–229.

Dostal, J., Robichaud, D. A., Church, B. N., and Reynolds, P. H., 1998, Eocene Challis-Kamloops volcanism in central British Columbia: An example from the Buck Creek basin: Canadian Journal of Earth Sciences, v. 35, p. 951–963.

Dudás, F. O., 1991, Geochemistry of igneous rocks from the Crazy Mountains, Montana and tectonic models for the Montana Alkalic Province: Journal of Geophysical Research, v. 96, p. 13,261–13,277.

Engebretson, D. C., Cox, A., and Gordon, R. G., 1985, Relative motions between oceanic and continental plates in the Pacific Basin: Geological Society of America, Special Paper 206, 59 p.

Ewing, T. E., 1980, Paleogene tectonic evolution of the Pacific Northwest: Journal of Geology, v. 88, p. 619–639.

Friedman, R. M., Diakow, L. J., Lane, R. A., and Mortensen, J. K., 2001, Cretaceous magmatism and associated mineralization in the Fawnie Range, Nechako Plateau, central British Columbia: Canadian Journal of Earth Sciences, v. 38, p. 619–637.

Gabrielse, H., and Yorath, C. J., 1991., Tectonic synthesis (Chapter 18), *in* Gabrielse, H., and Yorath, C. J., eds., Geology of the Cordilleran Orogen in Canada: Geological Survey of Canada, Geology of Canada, v. 4, p. 677–705 (*also* Geological Society of America, v. G-2).

Gill, J. B., 1981, Orogenic andesites and plate tectonics: Berlin, Germany, Springer-Verlag, 390 p.

Grainger, N. C., Villeneuve, M. E., Heaman, L. M., and Anderson, R. G., 2001, New U-Pb and Ar/Ar isotopic age constraints on the timing of Eocene magmatism, Fort Fraser and Nechako River map areas, central British Columbia: Canadian Journal of Earth Sciences, v. 38, p. 679–696.

Hawkesworth, C. J., Gallagher, K., Hergt, J. M., and McDermott, F., 1993, Mantle and slab contributions in arc magmas: Annual Review Earth Planetary Sciences, v. 21, p. 175–204.

Irving, E., and Brandon, M. T., 1990, Paleomagnetism of the Flores volcanics, Vancouver Island, in place by Eocene time: Canadian Journal of Earth Sciences, v. 27, p. 811–817.

Kempton, P. D., Fitton, J. G., Hawkesworth, C. J., and Ormerod, D. S., 1991, Isotopic and trace element constraints on the composition and evolution of the lithosphere beneath the southwestern United States: Journal of Geophysical Research, v. 96, p. 13,713–13,735.

Leitch, C. H. B., Hood, C. T., Cheng, X. L., and Sinclair, A. J., 1992, Tip Top Hill volcanics: Late Cretaceous Kasalka Group rocks hosting epithermal base- and precious-metal veins at Owen Lake, west-central British Columbia: Canadian Journal of Earth Sciences, v. 29, p. 854–864.

MacIntyre, D. G., 1985, Geology and mineral deposits of the Tahtsa Lake District, west-central British Columbia: Victoria, BC, British Columbia Ministry of Energy, Mines, and Petroleum Resources, Bulletin 75.

MacIntyre, D.G., 2001, The Mid-Cretaceous Rocky Ridge Formation—a new target for subaqueous hot-spring deposits (Eskay Creek-type) in Central British Columbia: Geological Fieldwork 2000: Victoria, BC, British Columbia Ministry of Energy and Mines, Paper 2001-1, p. 253–268.

MacIntyre, D. G., and Villeneuve, M. E., 2001, Geochronology of mid-Cretaceous to Eocene magmatism, Babine porphyry copper district, central British Columbia: Canadian Journal of Earth Sciences, v. 38, p. 639–655.

MacIntyre, D. G., Villeneuve, M. E., and Schiarizza, P., 2001, Timing and tectonic setting of Stikine terrane magmatism, Babine-Takla Lakes area, central British Columbia: Canadian Journal of Earth Sciences, v. 38, p. 579–601.

Mathews, W. H., 1989, Neogene Chilcotin basalts in south-central British Columbia: Geology, ages, and geomorphologic history: Canadian Journal of Earth Sciences, v. 26, p. 969–982.

McKenzie, D., and Bickle, M. J., 1988, The volume and composition of melt generated by extension of the lithosphere: Journal of Petrology, v. 29, p. 625–679.

McKenzie, D., and O'Nions, R. K., 1991, Partial melt distributions from inversion of rare earth element concentrations: Journal of Petrology, v. 32, p. 1021–1091.

Mihalynuk, M. G., Nelson, J., and Diakow, L., 1994, Cache Creek terrane entrapment. Oroclinal paradox with the Canadian Cordillera: Tectonics, v. 13, p. 575–595.

Miyashiro, A., 1974, Volcanic rock series in island arcs and active continental margins: American Journal of Science, v. 274, p. 321–355.

Monger, J. W., and Irving, E., 1980, Northward displacement of north-central British Columbia: Nature, v. 285, p. 289–294.

Monger, J. W. H. and Nokleberg, W. J., 1996, Evolution of the northern North American Cordillera: Generation, fragmentation, displacement, and accretion of successive North American plate margin arcs, *in* Coyner,

A. R., and Fahey, P. L., eds., Geology and ore deposits of the American Cordillera: Geological Society of Nevada Symposium Proceedings, Reno-Sparks, Nevada, April 1995, p. 1133–1152.

Monger, J. W. H., Wheeler, J. O., Tipper, H. W., Gabrielse, H., Harms, T., Struik, L. C., Campbell, R. B., Dodds, C. J. Gehrels, G. E., and O'Brien, J., 1991, Cordilleran terranes, *in* Gabrielse, H., and Yorath, C. J., eds., Geology of the Cordilleran Orogen in Canada: Geological Survey of Canada, Geology of Canada, v. 4 (*also* Geological Society of America, The Geology of North America, v. G-2), p. 281–327.

Norman, M. D., and Mertzman, S. A., 1991, Petrogenesis of Challis volcanics from central and southwestern Idaho: Trace element and Pb isotopic evidence: Journal of Geophysical Research, v. 96, p. 13,279–13,293.

Parrish, R. R., Carr, S. D., and Parkinson, D. L.,1988, Eocene extensional tectonics and geochronology of the southern Omineca Belt, British Columbia and Washington: Tectonics, v. 7, p. 181–212.

Pearce, J. A., 1983, Role of subcontinental lithosphere in magma genesis at active continental margins, *in* Hawkesworth, C. J., and Norry, M. J., eds., Continental basalts and mantle xenoliths: Cambridge, MA, Shiva Publishing, p. 230–249.

Pearce, J. A., Harris, N. B. W., and Tindle, A. G., 1984, Trace element discrimination diagrams for the tectonic interpretation of granitic rocks: Journal of Petrology, v. 25, p. 956–983.

Okulitch, A. V., 1999, Geological time scale: Geological Survey of Canada, Open File 3040.

Richards, T. A., and Tipper, H. W., 1976, Smithers map-area, British Columbia (93 L): Geological Survey of Canada, Open File 351, scale 1:125,000.

Samson, S. D., McClelland, W. C., Patchett, P. J., Gehrels, G. E., and Anderson, R. G., 1989, Evidence from neodymium isotopes for mantle contributions to Phanerozoic crustal genesis in the Canadian Cordillera: Nature, v. 337, p. 705–709.

Schiarizza, P., and MacIntyre, D. G., 1999, Geology of the Babine Lake–Takla Lake area, central British Columbia (93K/11, 12, 13, 14; 93N/3, 4, 5, 6), *in* Geological Fieldwork 1998: British Columbia Ministry of Employment and Investment, Paper 1999-1, p. 36–68.

Souther, J. G., 1991, Volcanic regime, *in* Gabrielse, H., and Yorath, C. J., eds., Geology of the Cordilleran Orogen in Canada: Geological Survey of Canada, Geology of Canada, v. 4 (*also* Geological Society of America, The Geology of North America, v. G-2), p. 457–490.

Struik, L. C., and MacIntyre, D. G., 2001, Introduction to the special issue: The Nechako NATMAP Project of the central Canadian Cordillera: Canadian Journal of Earth Sciences, v. 38, p. 485–494.

Sun, S. S., and McDonough, W. F., 1989, Chemical and isotopic systematics of oceanic basalts: Implications for mantle composition and processes, *in* Saunders, A. D., and Norry, M. J., eds., Magmatism in the ocean basins: Geological Society of London, Special Publication 42, p. 313–345.

Thorpe, R. S., 1982, Andesites. Orogenic andesites and related rocks: New York, NY, John Wiley & Sons, 724 p.

Watson, S., and McKenzie, D., 1991, Melt generation by plumes: A study of Hawaiian volcanism: Journal of Petrology, v. 32, p. 501–537.

Whalen, J. B., Anderson, R. G., Struik, L. C., and Villeneuve, M. E., 2001, Geochemistry and Nd isotopes of the Francois Lake plutonic suite, Endako batholith: Host and progenitor to the Endako molybdenum camp, central British Columbia: Canadian Journal of Earth Sciences, v. 38, p. 603–618.

Winchester, J. A., and Floyd, P. A., 1977, Geochemical discrimination of different magma series and their differentiation products using immobile elements: Chemical Geology, v. 20, p. 325–343.

Wojdak, P. J., and Sinclair, A. J., 1984, Equity Silver silver-copper-gold deposit: Alteration and fluid inclusion studies: Economic Geology, v. 79, p. 969–990.

Zindler, A., and Hart, S. R., 1986, Chemical dynamics: Annual Review Earth and Planetary Science, v. 14, p. 493–571.

OROGENIC PROCESSES

Buoyancy of the Oceanic Lithosphere and Subduction Initiation

ANDREW HYNES[1]

Department of Earth and Planetary Sciences, McGill University, 3450 University Street, Montreal, Quebec, Canada H3A 2A7

Abstract

Models for melting at ridges, and the resulting mineralogy, permit precise estimates of the density of oceanic plates. The plates become denser than columns of unsegregated asthenosphere at ridges in 10–15 m.y., and than asthenosphere segregated into crust and depleted mantle almost immediately, but their mean densities never exceed that of the underlying mantle, because low-density spinel peridotite in the plates more than compensates for cooling-related density increases. Cooled oceanic plates are negatively buoyant only if melt material can flood their upper surfaces. Then the buoyancy drive for subduction is comparable to slab pull. If oceanic plates fail at new faults in compression, flooding by melt material is unlikely, and the subduction buoyancy drive is small. These observations strengthen arguments for subduction initiation at former fracture zones, where flooding of oceanic plates by melt is most likely.

Introduction

THE COOLING OF oceanic plates produces predicted density increases that are amply supported by age-depth relationships (Parsons and Sclater, 1977). There is widespread acceptance that the density increases lead to negative buoyancy of the plates, which are therefore unstable and remain at the surface only because of work required to bend them, or to overcome resistance to thrusting into the mantle (e.g., McKenzie, 1977; Mueller and Phillips, 1991). Negative buoyancy of the plates has, however, been defined in different ways. Oxburgh and Parmentier (1977) compared the density of the thermal boundary layer with that of the underlying mantle, and suggested that plates become negatively buoyant in 40 to 50 m.y. Davies (1992) also defined buoyancy by comparing the density of the plate with that of the underlying mantle. Cloos (1993) compared the density of oceanic plates with that of an asthenospheric column at the ridge, and argued that plates become negatively buoyant as young as 10 Ma. Mueller and Phillips (1991) generally equated negative buoyancy to ridge push, which would make all oceanic plates negatively buoyant. Using a calculation like that of Oxburgh and Parmentier (1977), however, they suggested oceanic plates become negatively buoyant at about 26 Ma.

A light object resting in a more dense fluid has no tendency to sink and is therefore buoyant. A dense object in a less dense fluid is typically negatively buoyant because the fluid flows over it as it sinks toward isostatic equilibrium. It therefore never achieves equilibrium, and continues to sink. A boat, however, can be positively buoyant despite being denser than the water in which it lies, if it is designed so that it displaces enough water to balance its excess weight. In these circumstances, the boat is buoyant provided water cannot flow in over the sides of the boat. An analogous situation exists for oceanic plates. Cooling increases the density and thickness of oceanic plates as they move away from spreading ridges. This cooling results in subsidence as the oceanic plates maintain isostasy. If the subsided plates can be flooded by melts from rising underlying asthenosphere, they will become negatively buoyant due to the excess weight of the melts. Thus, if melt flooding is possible, oceanic plates are negatively buoyant as soon as they have cooled at all, and the negative buoyancy force may be calculated from the weight of the melts that flood them. If they cannot be flooded, however, they are not effectively negatively buoyant. In this case, negative buoyancy develops only as plates are thrust into the underlying mantle, and as spinel peridotite is converted to garnet peridotite and basalt is converted to eclogite. There are, therefore, significant differences in the potential negative buoyancy of oceanic plates that depend on whether or not the oceanic plates can be flooded by mantle-derived melts.

In this paper, I calculate the density variation in oceanic plates as a function of depth in the plate and age of the plate. I then use these density variations to quantify the potential negative buoyancy of

[1]Email: andrew.hynes@mcgill.ca

the oceanic plates, and discuss implications of the results for sites at which subduction may be initiated.

Composition of Oceanic Plates

In their early analysis of ocean-floor buoyancy, Oxburgh and Parmentier (1977) estimated the decrease in intrinsic density of the oceanic crust and upper mantle due to extraction of the crust from the mantle and the resulting loss of dense aluminous phases in the residual mantle. I broadly follow this procedure, but use better-constrained models for melting processes at the ridge to estimate the variation in composition and the resulting mineralogy of oceanic plates as a function of depth. Only a general description of the methods used is included here; a more detailed description may be found in the Appendix.

The oceanic plates are assumed to consist of undepleted peridotite at depth, peridotite that has been depleted by partial melting near the surface, and crust of broadly basaltic composition that is the product of the partial melting. Melting at oceanic ridges is widely believed to result from adiabatic decompression of mantle peridotite as it rises to fill the gap created at the ridge by sea-floor spreading. Contributions to the melt are derived from all parts of the rising column of mantle that cross the peridotite solidus (McKenzie and Bickle, 1988). Inasmuch as shallower parts of the mantle column overstep the solidus to greater degrees than deeper parts, proportionately more of the melt that gives rise to the oceanic crust is derived from them. As a result, the mantle in the oceanic plates is probably highly depleted in basaltic components at shallow depths, where it has been melted more extensively, and progressively less depleted at greater depths, until a depth is reached at which the solidus is not crossed by the rising mantle, below which there is no depletion (cf. Langmuir et al., 1993).

For this paper, the solidus of McKenzie and Bickle (1988) is assumed, and melt is derived from all parts of the mantle that cross it, producing the crust. The composition of the underlying mantle is determined by the amount of melt that has been extracted from it. Knowledge of the mantle melting profile is limited by uncertainties about the solidus and the amount of melting necessary for efficient melt extraction. These uncertainties largely affect the nature of small melt fractions at depth, however, rather than the melt fractions that produce most of the crust and the major characteristics of the depleted-mantle layer. They are not therefore critical to the problem treated here.

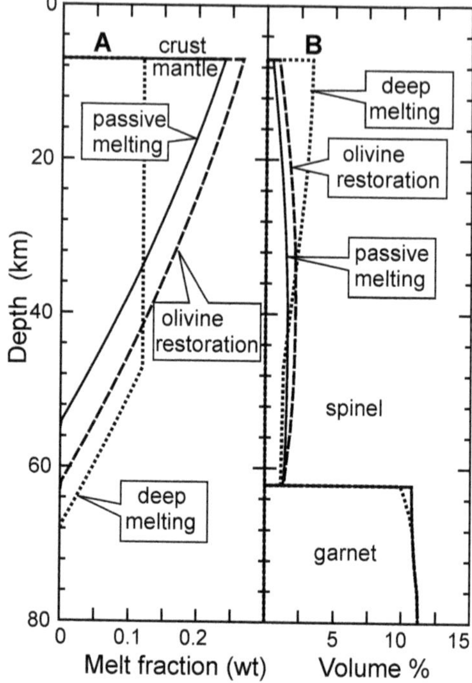

FIG. 1. A. Melt fraction extracted from mantle at spreading ridges as a function of depth, for a passive-melting model, an olivine-restoration model, and a deep-melting model. B. Vol% spinel and garnet in residue from melting, for the same three models, for 50 Ma oceanic lithosphere, using melting parameterization of Niu and Batiza (1991).

A simple model for melting is that of McKenzie and Bickle (1988), in which the amount of melt is proportional to the temperature difference between the rising peridotite and the solidus. This model is here termed the "passive melting" model (cf. Shen and Forsyth, 1995). With this model, a total melt thickness of 7 km can be produced from mantle with a potential temperature of 1307°C. The melt is extracted from a mantle column 47 km long (Fig. 1A). Depletion of the oceanic mantle declines with depth, to zero at 54 km (7 km crust + 47 km depleted mantle).

Two other models are also considered. In the first, the "olivine restoration" model, olivine is restored to the depleted mantle in proportion to the amount of melt first extracted, as described by Niu

(1997) to account for features of abyssal peridotites (e.g., Elthon, 1992; Niu, 1997; Baker and Beckett, 1999). Because the olivine is restored from the melt, the net melt production is smaller than for the passive-melting model. To produce a crustal thickness of 7 km, the mantle temperature must thus be raised to 1336°C, resulting in melting beginning approximately 10 km deeper, and in larger melt fractions at all depths than for the passive-melting model (Fig. 1A).

A second model, the "deep melting" model, limits the total amount of melting at depths less than 47 km; i.e., the melt fraction is uniform between 47 km and the base of the crust. This has been suggested to accommodate trace-element characteristics of ocean-ridge basalts consistent with the presence of garnet in their source regions for (e.g., Shen and Forsyth, 1995). More recent partitioning studies indicate that garnet may in fact not be necessary to explain the trace-element features (Blundy et al., 1998); there is little support for a significant role of garnet from a recent survey of the chemistry of mantle peridotites (Canil, 2004), but the model is included here for completeness. To produce 7 km of crust, a model of this kind requires a mantle temperature of 1356°C, and results in melting that starts at 68 km (Fig. 1A).

The depth variation in mantle chemistry and the crustal composition are different for each model, and were determined from a parameterization of the melting chemistry. I estimated melt compositions assuming batch melting, with the melting parameterization of Niu and Batiza (1991), which is based on experiments with peridotite MPY90. Details of the melting parameterization are given in the Appendix. Undepleted mantle was therefore assigned the composition of MPY90. The composition of the mantle residue, calculated by mass balance, varies markedly through the depleted mantle, due to the variation in melt fraction (Table 1). I treated crustal composition as uniform, and consisting of the weighted aggregate of melt fractions calculated at 1 km intervals through the melting column. To test the sensitivity to the melting parameterization, I also used the parameterization of Langmuir et al. (1993) with Al, Ca, and Si contents of the melts after Hirose and Kushiro (1993). In this case, undepleted peridotite was assigned the composition of KLB1, as in the experiments used for the parameterization. Resulting variations with depth in the chemistry of oceanic lithosphere are given in Table 1.

Density of Oceanic Plates

The density profile of an oceanic plate is determined by its thermal structure and its mineralogy. I used the infinite-series expression of Parsons and Sclater to determine the thermal structure of the plates (Eq. 5 in Parsons and Sclater, 1977; see Appendix). I estimated crustal mineralogy with a CIPW norm procedure. In the mantle, aluminum was assigned to spinel or garnet, depending on pressure, and to pyroxenes. The pressure of the spinel/garnet transition is uncertain (e.g., Herzberg and O'Hara, 1998). Recent experimental studies with MORB pyrolite are consistent with pressures near 2.6 GPa at 1400°C (Robinson and Wood, 1998). For this paper, the transition was assumed to occur at 2.5 GPa (25 kbar), but some calculations were made for a transition pressures of 2 GPa and 3 GPa. I used the thermodynamic models of Wood (1987) to estimate Al concentrations in pyroxenes coexisting with spinel or garnet (see Appendix). Norms were then calculated assuming equal values for clinoenstatite/clinoferrosilite, enstatite/ferrosilite, forsterite/fayalite and spinel/hercynite, with pyrope/almandine/grossular assigned in proportions 55/36/9, the normative proportions used by Asimow and Ghiorso (1998) in their initial guesses for subsolidus norms in the program MELTS, in the garnet zone.

The marked depletion of the uppermost mantle results in the near absence of spinel in all melting models. Spinel abundance increases with depth until undepleted mantle is reached, or until it is replaced by garnet at the spinel-to-garnet transition (e.g., Fig. 1B).

Densities were calculated using partial molar volumes for components of all phases, corrected for P and T (see Appendix). Crustal densities are consistently low compared with those in the mantle, due to the abundance of feldspar. In the mantle, densities increase with depth at very young ages, with a jump at the spinel-to-garnet transition (Fig. 2). As cooling penetrates from the surface, however, densities increase faster nearer the surface than they do at greater depths, and minimum mantle densities occur near 30 km. In the depleted mantle, densities are consistently lower than for undepleted peridotite at the same P and T (the dashed lines on Fig. 2), reflecting lower Al and Fe concentrations, but the differences rarely exceed 20 kg/m^3, even in the most depleted mantle.

The general effects of varying the melting models, or the melting parameterization, or of permitting

TABLE 1. Compositional Variation with Depth

Depth, km	X melt	SiO_2	TiO_2	Al_2O_3	FeO	MgO	CaO	Na_2O	K_2O
Crustal composition, wt%:		50.84	0.82	17.21	6.45	10.4	11.37	2.71	0.21
				Niu and Batiza					
7	0.24	42.71	0.03	0.70	8.18	47.73	0.65	0.00	0.00
12	0.22	43.09	0.03	0.99	8.08	46.85	0.96	0.00	0.00
17	0.20	43.46	0.04	1.30	7.98	45.95	1.28	0.00	0.00
22	0.18	43.79	0.04	1.64	7.89	45.03	1.61	0.00	0.00
27	0.15	44.10	0.05	2.01	7.80	44.10	1.94	0.00	0.00
32	0.13	44.37	0.06	2.40	7.73	43.16	2.27	0.01	0.00
37	0.10	44.61	0.07	2.82	7.67	42.20	2.58	0.05	0.00
42	0.07	44.80	0.09	3.26	7.62	41.24	2.88	0.10	0.00
47	0.05	44.96	0.11	3.72	7.60	40.28	3.15	0.18	0.01
52	0.01	45.07	0.15	4.17	7.60	39.32	3.37	0.31	0.01
55	0.00	44.74	0.17	4.37	7.55	38.57	3.38	0.40	0.03
Crustal composition, wt%:		53.09	0.65	15.93	6.54	10.49	11.37	1.77	0.15
				Langmuir					
7	0.24	41.84	0.01	0.11	8.71	48.46	0.85	0.01	0.00
12	0.22	42.36	0.01	0.37	8.63	47.51	1.10	0.02	0.00
17	0.20	42.84	0.01	0.66	8.54	46.54	1.38	0.02	0.00
22	0.18	43.28	0.01	0.97	8.47	45.57	1.68	0.02	0.00
27	0.15	43.67	0.02	1.31	8.40	44.59	1.99	0.03	0.00
32	0.13	44.01	0.03	1.67	8.33	43.61	2.30	0.05	0.00
37	0.10	44.30	0.04	2.06	8.28	42.64	2.61	0.07	0.00
42	0.07	44.55	0.05	2.47	8.23	41.69	2.90	0.10	0.00
47	0.05	44.77	0.07	2.92	8.20	40.74	3.16	0.15	0.00
52	0.01	44.95	0.10	3.38	8.16	39.80	3.37	0.23	0.00
55	0.00	44.88	0.12	3.59	8.10	39.22	3.44	0.30	0.02

no Al substitution into pyroxenes, are shown in Figure 3. With the exception of moving the spinel-to-garnet transition to 2 GPa or 3 GPa, which results in major density changes in the depth zone between these two pressures, the changes in these various assumptions have little effect on overall density profiles.

The density difference between asthenosphere and an oceanic plate segregated into crust and depleted mantle is small relative to the density changes due to cooling (Fig. 2). The cooling effect therefore rapidly overcomes the density decrease due to the segregation (cf. Cloos, 1993). Despite this, oceanic plates remain less dense than the mantle that underlies them. When cooling from the surface has not penetrated to the depth of the garnet zone, this is due both to the amount of mantle in the spinel zone and to the density increase with depth in the garnet zone. At greater ages, when cooling has penetrated sufficiently, density no longer increases with depth in the garnet zone (e.g., see the 200 Ma curve on Fig. 2) so that density beneath the plate is greater than in the lowermost plate, but the mean plate density remains lower than the density beneath it. For example, if the plate is taken as the region cooler than 1200°C, a 200 Ma plate is

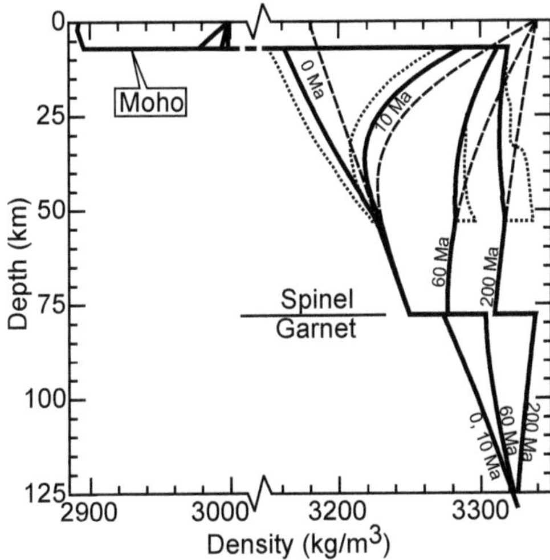

FIG. 2. Density-depth profiles in oceanic plates at 0, 10, 60, and 200 Ma, using passive-melting model and melting parameterization of Niu and Batiza (1991). Dashed lines = equivalent profiles for unsegregated asthenosphere; dotted lines = effect of addition of 10% mafic rock.

represented by the top 112 km of the profile of Figure 2. The density of the garnet-zone mantle in the plate is higher than that beneath the plate, but the difference is insufficient to offset the effect of the overlying spinel-zone mantle and the crust.

The melting model adopted assumes all melts from the melting region contribute to the crust formed at the ridge axis. It is possible, however, that some melt generated far from the axis may rise directly into the overlying lithosphere. Support for such a process is supplied by mafic dikes in peridotites of ophiolite complexes (e.g., Ceuleneer and Rabinowicz, 1992). Addition of 10% mafic rock to the depleted layer increases its density (dotted lines, Fig. 2). Only in old plates where mafic rocks deep in the depleted layer have passed fully through the eclogite transition is the increase sufficient, however, to make the plate more dense than the underlying mantle, and 10% is a very significant overestimate of the amount of melt possible at these depths.

Thus, oceanic plates never develop mean densities greater than the underlying asthenosphere. This observation is surprising given the amount of cooling, and runs counter to a widely held belief. As already discussed, however, it has little direct bearing on buoyancy.

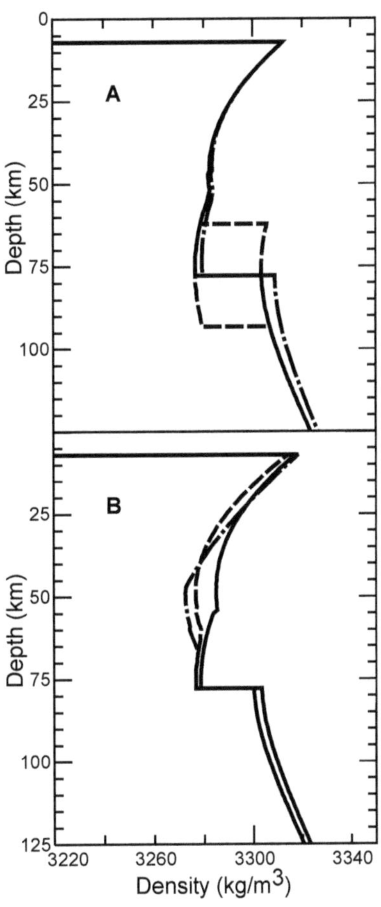

FIG. 3. Depth profiles of density in 60 Ma oceanic plate, for various melting models and assumptions. A. Passive melting using melting parameterization of Niu and Batiza (1991). Solid line = for spinel-garnet transition at 2.5 GPa; dashed lines = for spinel-garnet transition at 2 GPa and 3 GPa; dashed-dotted line = for no Al substitution permitted in pyroxenes. B. Solid line = passive melting using melting parameterization of Langmuir et al. (1993); dashed line = olivine restoration model with melting parameterization of Niu and Batiza (1991); dashed-dotted line = deep melting with melting parameterization of Niu and Batiza (1991).

Buoyancy Forces on Oceanic Plates

Whether an oceanic plate will sink depends on the force balance on it. The top of a cooled, isostatically equilibrated oceanic plate rides lower than the ridge, but the downward force due to the weight of the plate is balanced by the pressure in the underlying asthenosphere (Fig. 4A). If melts from rising asthenosphere can break through the plate and flood it, there is a resulting unbalanced downward force

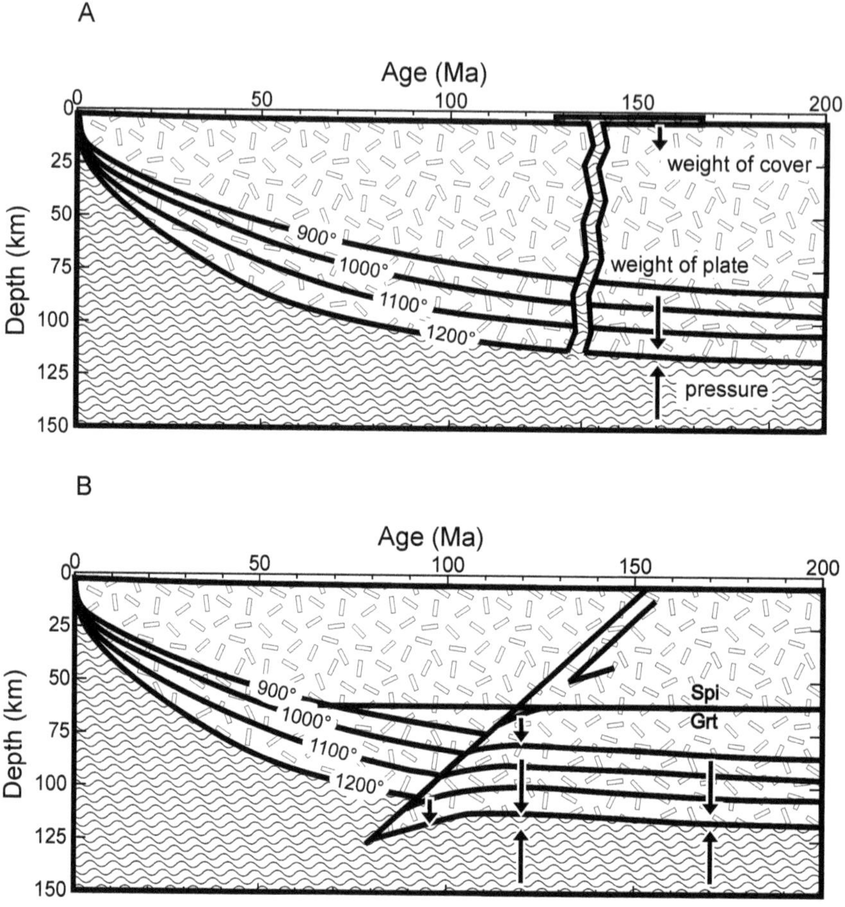

FIG. 4. A. Forces on an oceanic plate flooded by melts derived from the rising asthenosphere. B. Forces on an oceanic plate that develop as it is thrust into the underlying mantle. The base of the plate is taken as the 1200°C isotherm.

equal to the weight of the new material on the plate minus that of the displaced water (Fig. 4A). The new material that covers the plate can have a thickness equal to the difference in elevation between the ridge and the cooled and subsided plate and the force due to it is easily quantified.

The force is little affected by the choice of melting model or melting parameterization (Fig. 5). For all models, the top of an oceanic plate rides above an adjacent column of undifferentiated asthenosphere until the plate is 10 to 15 m.y. old, so that the potential downward force due to flooding of the plate by melt is negative (cf. Cloos, 1993). The delay results from comparison of unsegregated asthenosphere with oceanic plate segregated into crust and depleted mantle. The unsegregated asthenosphere, because it has higher intrinsic density, rides lower than its segregated equivalent, and the 10 to 15 m.y. is the time required for cooling to compensate for the difference. A more pertinent result arises from comparing the cooled oceanic plate with asthenosphere that has been segregated into crust and depleted mantle at the ridge, since arrival of unsegregated asthenospheric material at the surface seems unlikely. This calculation (lower curves on Fig. 5) shows the magnitude of the potential unbalanced downward force on an oceanic plate, and resulting negative buoyancy, which starts to develop as soon as the plate begins to cool.

Oceanic plates are only potentially negatively buoyant once they start to cool. The downward forces of Figure 5 occur only if rising melt material can cover the upper surface of the depressed plate. This is typically not possible, because the material

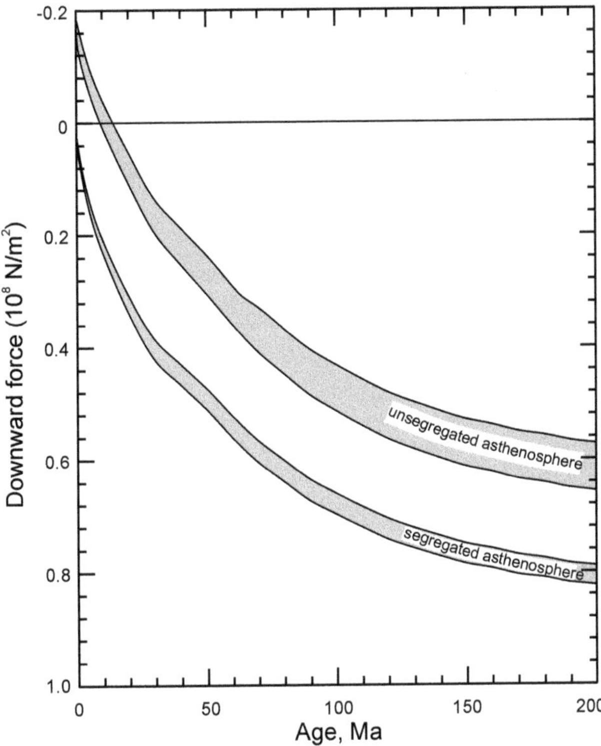

FIG. 5. Downward force on plate flooded by asthenospheric material as function of age. Upper shaded region encompasses results for all models treated in Figure 3, assuming the rising asthenosphere does not segregate into crust and depleted mantle. Lower shaded region is the same, but for asthenospheric material segregated into crust and depleted mantle.

freezes as it reaches the surface. One can, however, envisage a situation in which an older, cool plate experiences extensional failure, with melt derived from the resulting upwelling asthenosphere flowing out on top of the plate (cf. Kemp and Stevenson, 1996). Then the downward forces active on an oceanic plate reach magnitudes of 0.8×10^8 N/m^2 in old oceans, and are half that magnitude in only 25 m.y. (Fig. 5).

Implications for Subduction Initiation

Although negative buoyancy forces could provide one of the principal drives for subduction, they are not generally treated explicitly in assessing conditions for subduction initiation in compression. Instead, cooling of the plate is used to calculate a ridge-push force, which is then compared with resistance to subduction due to motion on plate-penetrating faults and bending of the descending plate (e.g., Mueller and Phillips, 1991; Toth and Gurnis, 1998). Flooding of a cool, old oceanic plate by ridge material provides an attractive drive for subduction initiation (Fig. 6). As illustrated by Figure 5, the drive is significant. The magnitude of ridge-push forces is on the order of 3×10^{12} N/m in old oceans (Parsons and Richter, 1980). With a force of 0.8×10^8 N/m^2 due to flooding by asthenosphere, only 37.5 km^2 of ocean floor must be flooded for each line-km of potential new trench to produce a negative buoyancy force as large as ridge push. Such negative buoyancy forces might develop in oceanic plates failing in extension, at leaky transform faults (Stern and Bloomer, 1992) or in plates failing in compression if failure occurred at former transform faults or fracture zones, in which case there could be upward flow of asthenosphere at corners between the faults and ridge segments.

In compressional environments in which subduction initiation occurred on new faults, rise of asthenospheric material through the plate would be unlikely. The newly generated fault would not be an effective pathway through which underlying asthenosphere could rise to the surface. There would therefore be no negative buoyancy, and it would have to develop as subduction proceeded.

If a plate is severed by a thrust, and the tip of the plate is thrust into the adjacent asthenosphere, negative buoyancy forces develop for two reasons. First, the underthrust plate displaces warmer material, less dense than the lowermost plate, so that pressure at the base of the underthrust plate increases less than the weight of the depressed part of the plate (isotherms are displaced downward at the thrust; Fig. 4B). Secondly, parts of the plate pass from the spinel zone to the garnet zone, increasing the density of the plate (Fig. 4B). In Figure 7, a thrust with a constant dip of 30° is depicted. Horizontal convergence through a distance *c* moves the thrust tip distance *c* into the asthenosphere, and a zone in the lower plate, delimited by the dashed line, is deformed to conserve area. With a simple geometry like this, and a starting density profile, the magnitude of downward forces developing with convergence may be calculated. These forces increase from zero at the thrust tip to a maximum, and then decline again to zero outside the deformation zone, with perturbations where the thrust breaks surface because of double crustal thicknesses there (Fig. 8). The forces increase progressively with increasing *c*. For 50 Ma oceanic plates and a thrust dip of 45°, forces reach values of up to 0.23×10^8 N/m^2 after 100 km of convergence. Magnitudes as high as 0.3×10^8 N/m^2 are reached if the age of the oceanic plate is increased.

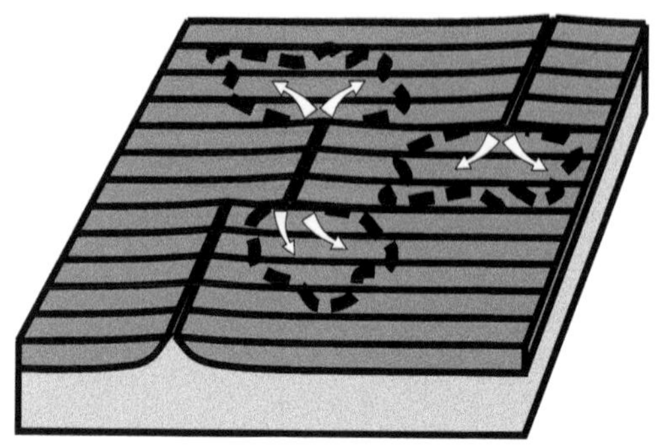

FIG. 6. Flooding of older, depressed parts of oceanic plates due to melt derived from oceanic ridges where they abut leaky transform faults.

The net force on an underthrust plate may be determined by integrating the force along the slab (Fig. 9). Underthrusting of roughly 60 km is required before the effective negative buoyancy of the plate becomes comparable with the force due to ridge push, and even for 100 km of underthrusting, the net force is still lower than that achieved by flooding of the plate over a width of 100 km. Thus, not only does the work required to underthrust and deform the plate work against the development of new failure planes in a compressional environment, but the negative buoyancy of the plate itself also provides little assistance until subduction is well under way.

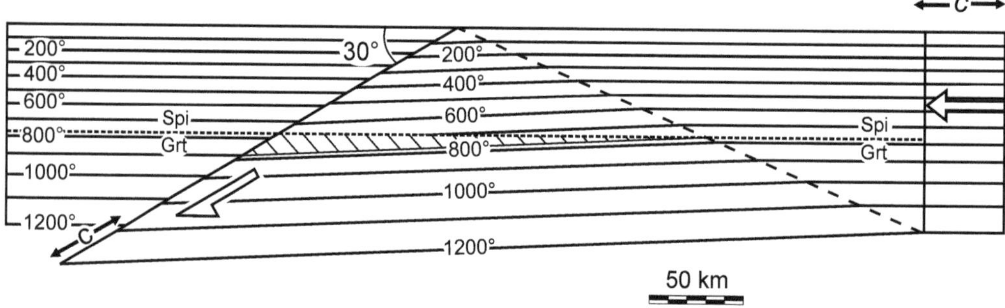

FIG. 7. Schematic illustration of conditions for negative buoyancy with compressional failure of oceanic plate. Plate, identified by isotherms to basal temperature of 1200°C, has been underthrust a distance *c* on a fault with constant dip. Dashed line = right-hand limit of region of deformation in descending plate; dotted line = depth of spinel-to-garnet transition; cross-hatched region = where plate has gone through spinel-to-garnet transition.

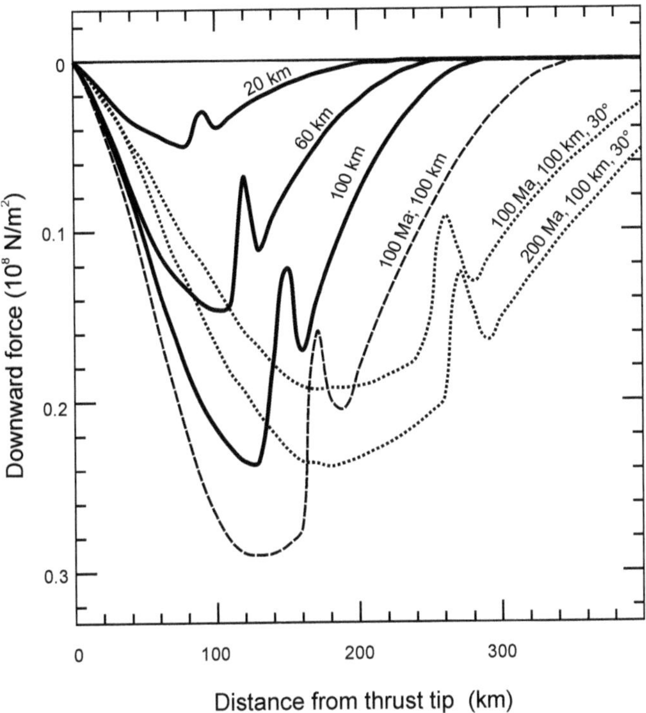

FIG. 8. Downward force on underthrust plate as a function of distance from thrust tip. Solid lines = 50 Ma plates with thrust dip of 45° and different amounts of convergence (*c* of Fig. 7). Dotted lines = plates with thrust dip of 30°, convergence of 100 km, and various plate ages; dashed line = 100 Ma plate, with thrust dip of 45° and convergence of 100 km. All calculations use passive-melting model, melting parameterization of Niu and Batiza (1991), and spinel-to-garnet transition at 2.5 GPa.

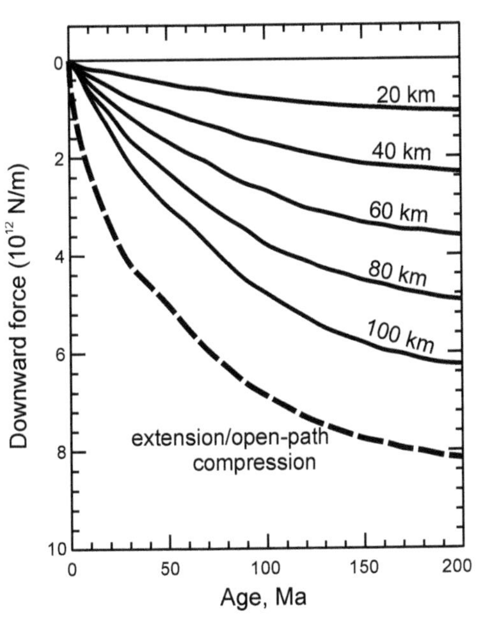

Conclusions

Difficulties with the initiation of subduction are well known (cf. McKenzie, 1977). The analysis presented here shows that negative buoyancy forces can play a significant role in overcoming these difficulties provided a pathway exists whereby asthenospheric mantle can reach the surface. In the absence of such a pathway, negative buoyancy forces are very much smaller, and develop only as underthrusting proceeds.

FIG. 9. Total downward force on the depressed region of an underthrust plate as a function of age, for various magnitudes of convergence (*c*) with a thrust angle of 45°. Also shown (dashed line) is the age-dependent negative-buoyancy force in extension for magma flooding extending 100 km from the zone of failure (cf. Fig. 5). All calculations use the passive-melting model, the melting parameterization of Niu and Batiza (1991), and a spinel-to-garnet transition at 2.5 GPa.

In both extensional and compressional environments, the favorable place at which to initiate subduction would be at a transform fault or fracture zone. There the oceanic plate is cool and depressed, but still relatively young and therefore weak. In an extensional environment, an asthenospheric pathway is directly available due to the extension; in a compressional environment the pathway may be supplied by a re-entrant at a former ridge/transform intersection.

The idea that subduction zones may be initiated at transform faults or fracture zones was one of the major conclusions of Mueller and Phillips (1991), and has been advocated for several places (e.g., Scholl et al., 1975; Hilde et al., 1977; Hegarty et al., 1982; Stern and Bloomer, 1992; Collot et al., 1995; Lebrun et al., 2000). In this paper, I have shown that the successful implementation of negative buoyancy, rather than just the weakness of the oceanic plate, may play a major role in localizing subduction initiation at these places.

Acknowledgments

This research was supported by the National Science and Engineering Research Council of Canada. I thank unidentified reviewers of an earlier version of this manuscript, whose comments clarified my thinking on the problems discussed, and Dante Canil, who made several constructive suggestions that have improved the current version.

REFERENCES

Asimow, P. D., and Ghiorso, M. S., 1998, Algorithmic modifications extending MELTS to calculate subsolidus phase relations: American Mineralogist, v. 83, p. 1127–1132.

Baker, M. B., and Beckett, J. R., 1999, The origin of abyssal peridotites: A reinterpretation of constraints based on primary bulk compositions: Earth and Planetary Science Letters, v. 171, p. 49–61.

Blundy, J. D., Robinson, J. A. C., and Wood, B. J., 1998, Heavy REE are compatible in clinopyroxene on the spinel lherzolite solidus: Earth and Planetary Science Letters, v. 160, p. 493–504.

Canil, D., 2004, Mildly incompatible elements in peridotites and the origins of mantle lithosphere: Lithos, v. 77, p. 375–393.

Ceuleneer, G., and Rabinowicz, M., 1992, Mantle flow and melt migration beneath oceanic ridges: Models derived from observations in ophiolites, *in* Morgan, J., Blackman, D. K., and Sinton, J. M., Mantle flow and melt generation at mid-ocean ridges: Washington, DC, American Geophysical Union Monographs, v. 71, p. 123–154.

Cloos, M., 1993, Lithospheric buoyancy and collisional orogenesis: Subduction of oceanic plateaus, continental margins, island arcs, spreading ridges, and seamounts: Geological Society of America Bulletin, v. 105, p. 715–737.

Collot, J.-Y., Lamarche, G., Wood, R. A., Delteil, J., Sosson, M., Lebrun, J.-F., and Coffin, M. F., 1995, Morphostructure of an incipient subduction zone along a transform plate boundary: Puysegur Ridge and trench: Geology, v. 23, p. 519–522.

Davies, G. F., 1992, On the emergence of plate tectonics: Geology, v. 20, p. 963–966.

Elthon, D., 1992, Chemical trends in abyssal peridotites: Refertilization of depleted oceanic mantle: Journal of Geophysical Research, v. 97, p. 9015–9025.

Hegarty, K. A., Weissel, J. K., and Hayes, D. E., 1982, Convergence at the Caroline-Pacific plate boundary: Collision and subduction, *in* Hayes, D. E., ed., The tectonic and geologic evolution of Southeast Asian seas and Islands, part 2: Washington, DC, American Geophysical Union Monographs, v. 27, p. 326–348.

Herzberg, C., and O'Hara, M. J., 1998, Phase equilibrium constraints on the origin of basalts, picrites, and komatiites: Earth Science Review, v. 44, p. 39–79.

Hilde, T. W. C., Uyeda, S., and Kroenke, L., 1977, Evolution of the western Pacific and its margin: Tectonophysics, v. 38, p. 145–165.

Hirose, K., and Kushiro, I., 1993, Partial melting of dry peridotites at high pressure: Determination of compositions of melts segregated from peridotites using aggregates of diamonds: Earth and Planetary Science Letters, v. 114, p. 477–489.

Holland, T. J. B., and Powell, R., 1990, An enlarged and updated internally consistent thermodynamic dataset with uncertainties and correlations: the system K_2O-Na_2O-CaO-MgO-MnO-FeO-Fe_2O_3-Al_2O_3-TiO_2-SiO_2-C-H_2-O_2: Journal of Metamorphic Geology, v. 8, p. 89–124.

Kemp, D. V., and Stevenson, D. J., 1996, A tensile, flexural model for the initiation of subduction: Geophysical Journal International, v. 125, p. 73–94.

Langmuir, C. H., Klein, E. M., and Plank, T., 1993, Petrological systematics of mid-ocean ridge basalts: Constraints on melt generation beneath ocean ridges, *in* Morgan, J., Blackman, D. K., and Sinton, J. M., Mantle flow and melt generation at mid-ocean ridges: Washington, DC, American Geophysical Union Monographs, v. 71, p. 183–280.

Lebrun, J.-F., Lamarche, G., Collot, J.-Y., and Delteil, J., 2000, Abrupt strike-slip fault to subduction transition: The Alpine Fault–Puysegur Trench connection, New Zealand: Tectonics, v. 19, p. 688–706.

McKenzie, D., 1977, The initiation of trenches: A finite amplitude instability, *in* Maurice Ewing Series, v. 1:

Washington, DC, American Geophysical Union, p. 57–61.

McKenzie, D., and Bickle, M. J., 1988, The volume and composition of melt generated by extension of the lithosphere: Journal of Petrology, v. 29, p. 625–679.

Mueller, S., and Phillips, R. J., 1991, On the initiation of subduction: Journal of Geophysical Research, v. 96, p. 651–665.

Niu, Y., 1997, Mantle melting and melt extraction processes beneath ocean ridges: Evidence from abyssal peridotites: Journal of Petrology, v. 38, p. 1047–1074.

Niu, Y., and Batiza, R., 1991, An empirical method for calculating melt compositions produced beneath mid-ocean ridges: Applications for axis and off-axis (seamounts) melting: Journal of Geophysical Research, v. 96, p. 21,753–21,777.

Oxburgh, E. R., and Parmentier, E. M., 1977, Compositional and density stratification in oceanic lithosphere—causes and consequences: Journal of the Geological Society of London, v. 133, p. 343–355.

Parsons, B., and Richter, F. M., 1980, A relation between the driving force and the geoid anomaly associated with mid-ocean ridges: Earth and Planetary Science Letters, v. 51, p. 445–450.

Parsons, B., and Sclater, J. G., 1977, An analysis of ocean floor bathymetry and heat flow with age: Journal of Geophysical Research, v. 83, p. 803–827.

Robinson, J. A. C., and Wood, B. J., 1998, The depth of the spinel to garnet transition at the peridotite solidus: Earth and Planetary Science Letters, v. 164, p. 277–284.

Scholl, D. W., Buffington, E. C., and Marlow, M. S., 1975, Plate tectonics and the structural evolution of the Aleutian–Bering Sea region: Geological Society of America, Special Paper 131, p. 1–31.

Shen, Y., and Forsyth, D. W., 1995, Geochemical constraints on initial and final depths of melting beneath mid-ocean ridges: Journal of Geophysical Research, v. 100, p. 2211–2237.

Stern, R. J., and Bloomer, S. H., 1992, Subduction zone infancy: Examples from the Eocene Izu-Bonin-Mariana and Jurassic California arcs: Geological Society of America Bulletin, v. 104, p. 1621–1636.

Toth, J., and Gurnis, M., 1998, Dynamics of subduction initiation at preexisting fault zones: Journal of Geophysical Research, v. 103, p. 18,053–18,067.

Wood, B. J., and Holloway, J. R., 1984, A thermodynamic model for subsolidus equilibria in the system $CaO\text{-}MgO\text{-}Al_2O_3\text{-}SiO_2$: Geochimica et Cosmochimica Acta, v. 48, p. 159–176.

Wood, B. J., 1987, Thermodynamics of multicomponent systems containing several solid solutions, *in* Carmichael, I. S. E., and Eugster, H. P., eds., Thermodynamic modeling of geological materials: Minerals, Mineralogical Society of America, Reviews in Mineralogy, v. 17, p. 71–96.

Appendix

Physics of melting

The solidus for mantle peridotite was assumed to obey the equation

$$P = (T_S - 1100)/136 + 4.968 \times 10^{-4} \exp[1.2 \times 10^{-2}(T_S - 1100)],$$

following McKenzie and Bickle (1988), where P is the pressure in GPa and T_S is the solidus temperature in °C. $T_S(P)$ in this expression was determined by numerical iteration. The liquidus temperature, T_L, was determined, again following McKenzie and Bickle (1988), from

$$T_L = 1736.2 + 4.343\ P + 180 \tan^{-1}(P/2.2169).$$

Batch melting was assumed, and the melt fraction (by weight) extracted from the rising peridotite was assumed to be proportional to the difference in temperature between T_S at the pressure of interest and the temperature of the peridotite.

Chemistry of melting

The chemical compositions of melts were determined mostly using the parameterization of Niu and Batiza (1991) for peridotite MPY-90. This parameterization expresses the bulk solid-liquid partition coefficient, D_i, for each major oxide component as a function of the fraction of melt produced, F, using the equation

$$D_i = e + f\,F + g/F + h\,P + i\,P/F,$$

where F is in weight percent, P is in kbar, and e, f, g, h, and i are coefficients that depend on the major oxide. The values for the coefficients are given in Appendix Table 1.

Melt fractions were extracted as batch melts, with compositions calculated at kilometer intervals in the asthenospheric column, and the composition of the residual peridotite was determined by mass balance. The composition and thickness of the crust were determined by summing all the melts derived from the melt column, and treating the crust as having uniform composition.

Mineralogy

The mineralogy of the crust was estimated with a conventional CIPW norm procedure. Calculation of peridotite mineralogy was conducted as follows. All Ti was assigned to ulvospinel, all K was assigned to potassium feldspar and all Na was assigned to jadeite (clinopyroxene component).

APPENDIX TABLE 1. Coefficients for Melting Parameterization

Oxide	e	f	g	h	i
SiO_2	0.8480	−0.0022	0.00	0.0055	0.00
TiO_2	0.0910	−0.0020	0.00	0.00	0.00
$Al2O_3$	0.1960	−0.0065	−0.0250	0.0021	0.00
FeO	1.4720	0.00	0.2730	−0.0350	−0.0130
MgO	5.6230	−0.0451	0.00	−0.0810	0.00
CaO	0.3270	−0.0120	0.3071	0.0005	0.00
Na_2O	0.0509	−0.0038	0.00	0.00	0.00
K_2O	0.0099	−0.0002	0.00	0.00	0.00

Aluminum was assigned to clinopyroxene and orthopyroxene in accordance with the relationship

$$T_k \ln(W) = a + b\,P_{kbar} + c\,T_k + d\,T_k^3,$$

where T_k is temperature (K), W is the weight percent Al_2O_3 in the pyroxene, P_{kbar} is the pressure (kbar). The coefficients a, b, c, and d (Appendix Table 2) were determined by regressing the equilibrium concentrations of Al in clinopyroxene and orthopyroxene coexisting with spinel or garnet calculated at 100°C and 1 kbar intervals over the range of conditions crossed by oceanic geotherms, using the thermodynamic models of Wood (1987) and the composition of peridotite KLB1. The form chosen for the equation reflects the thermodynamically reasonable linear relationship of T ln(W) to P and T and an additional T-dependence evident in the results (cf. curves in Wood and Holloway, 1984).

Clinopyroxene compositional data were constrained by the expression

$$D = a + b\,P_{kbar} + c\,T_k,$$

where D is molar Di/(CFs+CEn), and a, b and c are coefficients (Appendix Table 2) derived by regression of results from the same calculations as for Al in pyroxenes. Amounts and compositions of clinopyroxene, orthopyroxene, olivine, and either spinel or garnet were assigned by an iterative routine, respecting the values of D, W (clinopyroxene) and W (orthopyroxene) dictated by the regressions and

APPENDIX TABLE 2. Constants and Correlation Coefficients for Regressions of wt% Al_2O_3 in Pyroxene, and Di/(CFs + CEn) in Clinopyroxene

	a	b	c	d	CC
Al_2O_3 in Opx (Spi)	–3.6500E + 03	–7.6130E + 00	5.0280E + 00	–1.5430E – 07	1.0000
Al_2O_3 in Cpx (Spi)	–1.1600E + 03	–2.8480E + 01	1.0230E + 00	1.1730E – 06	0.9872
Al_2O_3 in Opx (Grt)	–4.0890E + 02	–9.3420E + 01	2.3540E + 00	3.0660E – 07	0.9979
Al_2O_3 in Cpx (Grt)	1.0150E + 03	–1.2170E + 02	–1.4860E–01	1.2790E – 06	0.9840
Di/(CFs + Cen) (Spi)	5.1440E + 01	2.0100E–01	–3.6160E–02		0.9722
Di/(CFs + Cen) (Grt)	5.6270E + 01	1.8030E–01	–3.9110E–02		0.9665

the values of P and T. No hedenbergite was assigned to clinopyroxenes. Clinoenstatite/clinoferrosilite, enstatite/ferrosilite and forsterite/fayalite were all assigned the same values (i.e., uniform Fe/Mg). For spinel-zone calculations, spinel/hercynite was also assigned the same value. For garnet-zone calculations, pyrope/almandine/grossular were assigned in the ratios 55/36/9.

APPENDIX TABLE 3. Molar Volumes, Coefficients of Thermal Expansion, and Compressibilities of Normative Minerals, Used in Density Calculations[1]

Mineral	Formula	Volume, cm^3	a x 105, K^{-1}	b x 10^6, bar^{-1}
Clinopyroxene	$NaAlSi_2O_6$	60.40	2.66	0.75
	$CaMgSi_2O_6$	66.10	2.92	0.82
	$Mg_2Si_2O_6$	62.64	2.92	1.01
	$Fe_2Si_2O_6$	65.98	3.89	1.01
	$CaAl_2SiO_6$	63.62	2.70	0.80
Orthopyroxene	$CaFeSi_2O_6$	67.88	3.89	0.82
	$MgAl_2SiO_6$	58.93	2.92	1.01
	Mg_2Si2O_6	62.64	2.92	1.01
	$Fe_2Si_2O_6$	65.98	3.89	1.01
Garnet	$Mg_3Al_2Si_3O_{12}$	114.00	2.57	0.47
	$Fe_3Al_2Si_3O_{12}$	116.00	2.44	0.53
	$Ca_3Al_2Si_3O_{12}$	125.00	2.34	0.54
Olivine	Mg_2SiO_4	43.79	4.14	0.79
	Fe_2SiO_4	46.39	3.19	0.91
Spinel	$MgAl_2O_4$	39.71	2.75	0.49
	$FeAl_2O_4$	40.75	2.54	0.49
	Fe_2TiO_4	46.82	4.10	0.49
Feldspar	$KAlSi_3O_8$	108.80	1.89	0.59

[1]Molar volumes at 1 bar and 298 K. Values are from Wood (1987), except for ulvospinel, for which the values are from Holland and Powell (1990).

Calculation of density

The density of the crust and mantle were calculated using partial molar volumes for components of all phases in the norms, corrected for pressure and temperature. Thus, partial molar volumes of mineral components were calculated from

$$V_i = V_i^0[1 + a_i\,(TK - 298)](1 - b_i\,P),$$

where V_i^0 is the partial molar volume at 1 bar and 298 K, a_i is the coefficient of thermal expansion, and b_i is the compressibility for each component. Values used are given in Appendix Table 3. Molar volumes were then calculated for each phase, assuming ideal mixing, and used, with the norm, to calculate the density of the mineral aggregate.

Oceanic geotherms

Following Parsons and Sclater (1977), Eq. 5, oceanic geotherms were calculated from:

$$T' = (1 - z') + \sum_{n = 1, \infty} a_n \sin n\pi z \exp(-\beta_n x'),$$

where $\beta_n = [(R^2 + n^2\pi^2)^{1/2} - R]$ and $a_n = 2(-1)^{n+1}/(n\pi)$. In these expressions, $T' = T/T_1$, where T_1 is the temperature at the base of the plate, taken as 1350°; $z' = z/a$, where z is height, and a is the maximum plate thickness, taken as 128 km; $x' = x/a$, where x is the horizontal distance from the oceanic ridge; and $R = u\,a\,/\,[2(k/\rho C_P)]$ is the thermal Reynolds number, where u is the half-rate of spreading, k is the thermal conductivity (3.1 J/°C m s), ρ is the density (3330 kg/m^3), and C_P is the specific heat (1.17×10^3 J/kg °C).

The Laramide Orogeny: What Were the Driving Forces?

JOSEPH M. ENGLISH[1] AND STEPHEN T. JOHNSTON

School of Earth and Ocean Sciences, University of Victoria, P.O. Box 3055 STN CSC, Victoria, British Columbia, V8W 3P6 Canada

Abstract

The Laramide orogeny is the Late Cretaceous to Paleocene (80 to 55 Ma) orogenic event that gave rise to the Laramide block uplifts in the United States, the Rocky Mountain fold-and-thrust belt in Canada and the United States, and the Sierra Madre Oriental fold-and-thrust belt in east-central Mexico. The Laramide orogeny is believed to post-date the Jurassic and late Early Cretaceous accretion of the terranes that make up much of the North American Cordillera, precluding a collisional origin for Laramide orogenesis. Instead, the deformation belt along much of its length likely developed 700–1500 km inboard of the nearest convergent margin. The purpose of this paper is to show, through a review of proposed mechanisms for producing this inboard deformation (retroarc thrusting, "orogenic float" tectonics, flat-slab subduction and Cordilleran transpressional collision), that the processes responsible for orogeny remain enigmatic.

Introduction

THE LARAMIDE OROGENY is the Late Cretaceous to Paleocene (80 to 55 Ma) orogenic event that gave rise to the Laramide block uplifts in the United States, as well as the Rocky Mountain fold-and-thrust belt in Canada and the United States, and the Sierra Madre Oriental fold-and-thrust belt in east-central Mexico (Fig. 1). The Laramide orogeny is believed to post-date the Jurassic and late Early Cretaceous accretion of the terranes that make up much of the North American Cordillera (e.g., Dickinson and Snyder, 1978; Monger et al., 1982; Monger and Nokleberg, 1996; Dickinson and Lawton, 2001); a collisional origin for Laramide orogenesis has therefore been ruled out. Thus, along much of its length, the deformation belt is thought to have developed 700–1500 km inboard of the nearest convergent margin. The purpose of this paper is to illustrate, through the review of proposed mechanisms for producing this inboard deformation, that the processes responsible for orogeny remain enigmatic. Proposed mechanisms reviewed here are retroarc thrusting, "orogenic float" tectonics, flat-slab subduction, and Cordilleran transpressional collision.

Retroarc Thrusting

Retroarc thrusting, during which thrust faulting is antithetic with respect to subduction at the plate margin, has been suggested for backthrusting of the Cordillera onto the adjacent continental interior (e.g., Price, 1981; Fig. 2) due to strong regional compressive stresses. The zone of backthrusting represents a locus of continuing relative motion between the Cordillera and the continental interior; its location may be controlled by inherited weaknesses in the lithosphere produced by a craton-rifted margin boundary or by thermal weakening during an earlier plutonic/metamorphic episode (e.g., Omineca Belt in Canada). In this model, the rate of western advancement of the North American plate exceeds the rate of slab rollback in the subducting oceanic plate, and the continent essentially "collides" with the oceanic slab due to its inability to actively override the slab. A similar mechanism has been proposed for Andean deformation (Russo and Silver, 1996). This mechanism has the advantage of being of continental-scale; along-strike variations in deformational style may be produced by variations in crustal heterogeneity.

The retroarc thrusting model requires that stress, attributable to the upper-plate versus lower-plate collision at the subduction interface, was transmitted across the entire Cordillera without causing significant deformation within the internal portions of the orogen. This requires that: (1) the differential stress attributable to collision be small enough in order to avoid failure and shortening proximal to the subduction zone: and (2) that the highly faulted accretionary complex that comprises the central portion of the Cordillera consists of a coherent,

[1]Corresponding author; email: englishj@uvic.ca

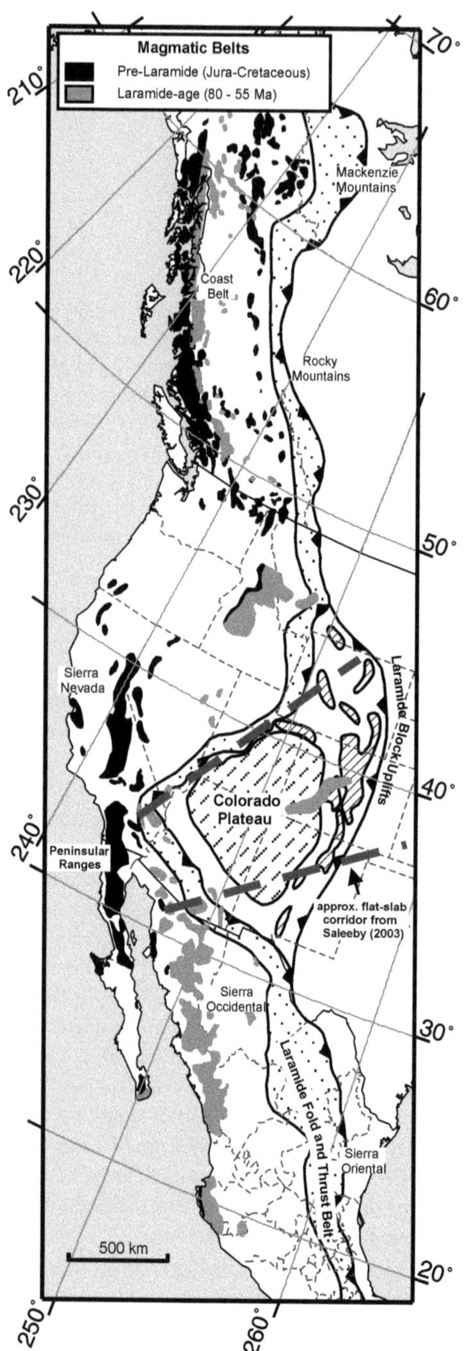

FIG. 1. Map showing areas of pre-Laramide and Laramide magmatism, the extent of the Laramide thin-skinned fold-and-thrust belt and thick-skinned block uplifts, and the approximate location of the Laramide flat slab according to Saleeby (2003). Cenozoic extension has not been restored in this figure.

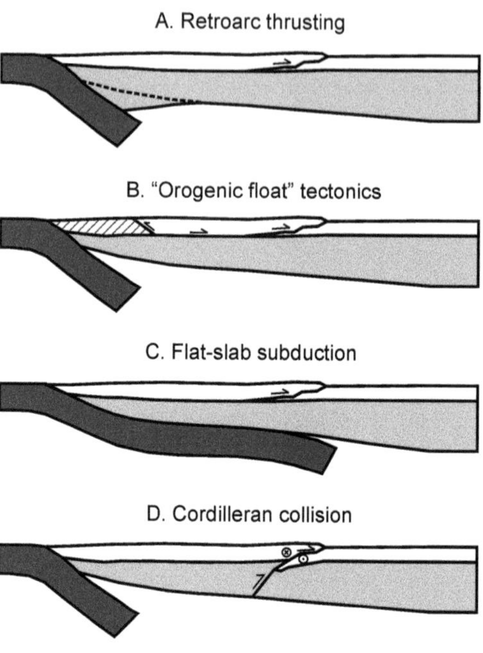

FIG. 2. Schematic diagrams illustrating the various models that have been proposed for producing the inboard deformation belts of the Laramide orogeny.

strong block despite the presence of numerous older crustal-scale faults. It is also not apparent in this model how the shortening was accommodated in the mantle lithosphere.

"Orogenic Float" Tectonics

The addition of "orogenic float" (Oldow et al., 1990) at the convergent margin may have driven far-field strain in the foreland region of the Cordillera. In this model, faults in the foreland are linked to a collision zone at the plate boundary by a major through-going deep-crustal basal detachment, so that the entire crustal section "floats" on its underlying lithosphere (Fig. 2). In convergent margins characterized by large-scale terrane collision and accretion, the addition of mass (i.e., terranes) above the basal detachment exerts a horizontal normal compressive stress on the continental crust and drives deformation in the foreland. There is no contraction within the Cordilleran mantle lithosphere, and a lithospheric root is not developed; the mantle formerly attached to the accreted terranes must be removed by subduction. This concept has been used to relate current deformation in the Mackenzie

Mountains of northern Canada to collision of the Yakutat block in southeastern Alaska (Mazzotti and Hyndman, 2002). There are, however, no known terrane accretion events coeval with Laramide orogeny. In Canada, inboard terranes within the central part of the orogen (e.g., Quesnellia and Stikinia) are stitched to pericratonic crust of presumed North American affinity by Jurassic plutons, whereas the outboard terranes (e.g., Wrangellia) are stitched to the inboard terranes by mid-Cretaceous and possibly older plutons. The Eocene accretion of the Crescent terrane outboard of Wrangellia post-dates Laramide orogeny and was responsible for nearby deformation, including buckling of Wrangellia and the formation of a fold-and-thrust belt (Johnston and Acton, 2003). Therefore, there was no driving force present to apply a horizontal compressive stress above the proposed deep-crustal basal detachment, and this model can be effectively ruled out for the Laramide orogeny.

Flat-Slab Subduction

In the flat-slab subduction model for Laramide deformation (e.g., Coney and Reynolds, 1977; Dickinson and Snyder, 1978; Bird, 1988), the oceanic slab subducting along the western margin of the continent remained in contact with the upper plate for a distance of > 700 km inboard of the trench (Fig. 2). Flat-slab subduction is postulated to have formed in response to the subduction of the Kula-Farallon spreading center, or in response to the subduction of a buoyant oceanic plateau (e.g., Livaccari et al., 1981; Murphy et al., 2003; Saleeby, 2003). Stress coupling of the upper plate with the flat slab could have transmitted stresses eastward and caused basement-cored block uplifts in the foreland of the U.S. Cordillera (e.g. Dickinson and Snyder, 1978; Bird, 1988). Deformation is inferred to have developed above the zone along which the subducting slab eventually steepened and descended into the deep mantle. Basement-cored uplifts in Wyoming, Colorado, and New Mexico are comparable to structures in the Sierra Pampeanas in Argentina (Jordan and Allmendinger, 1996), where deformation has been linked to a period of flat-slab subduction (Kay and Abbruzzi, 1996). However, the development of a magmatic arc within 300 km of the trench in Canada and Mexico during Laramide time refutes the existence of flat-slab subduction in these regions (Fig. 1; English et al., 2003). Therefore, if we accept that the Laramide orogeny in the United States is the product of flat slab subduction, then the coeval fold-and-thrust belt formation to the north in Canada, and to the south in Mexico, must be attributed to some other process. While possible, it is not clear why such disparate regions should all have synchronously undergone contractional deformation if the processes responsible for shortening differed along the length of the orogen.

Cordilleran Transpressional Collision

Paleomagnetic studies of Cretaceous layered sedimentary and volcanic rocks within the Cordilleran orogen have been interpreted to show that much of the Canadian Cordillera originated 2000–3000 km to the south relative to autochthonous North America (e.g., Irving et al., 1996; Wynne et al., 1998; Enkin et al., 2002). Timing constraints indicate that northward translation must have occurred between 85 and 55 Ma, coeval with the duration of Laramide orogenesis. Therefore, it is possible that Laramide orogenesis occurred in response to a dextral transpressional collision between a northward migrating ribbon continent and autochthonous North America (Johnston, 2001). Northern pinning and buckling of this ribbon continent is proposed to have led to the formation of the Alaskan oroclines during collision with North America in Laramide time. In this model, the Rocky Mountain fold-and-thrust belt may have been produced by offscraping of passive margin sediments during subduction of the North American plate beneath the northward migrating ribbon continent (Fig. 2). However, foreland basin formation and extensive westerly derived sedimentation on the North American craton from at least Early Cretaceous time (e.g. Stott, 1993), and the absence of an inner Late Cretaceous arc magmatic belt, cast doubt on the existence of an intervening oceanic basin between the Cordillera and the continental interior during Late Cretaceous–Paleocene time.

Discussion

Jarrard (1986) proposed that four variables (convergence rate, slab age, intermediate slab dip–average dip to 100 km, and absolute motion of the upper plate) determine the strain regime of the overriding plate at a convergent margin. Convergence rate and slab age affect the strain regime by changing the level of coupling between the plates. For example, an increased convergence rate increases the horizontal compressive stress applied to the upper plate, whereas an increase in slab age may reduce this stress due to increased gravitational pull on an older slab. Intermediate slab dip may affect the strain regime of the upper plate by changing the surface area over which coupling occurs at the plate boundary.

If the convergence rate, intermediate slab dip, or slab age were important factors for driving Laramide deformation, then orogenesis was ultimately driven by the subducting oceanic slab. Plate reconstruction models indicate that the Kula and Farallon oceanic plates, separated by a spreading center, were subducting beneath western North America during Laramide time (e.g., Engebretson et al., 1985; Kelley, 1993), although the location of the Kula–Farallon–North America triple junction during Laramide time remains a matter of much debate (e.g., Breitsprecher et al., 2003). It is, therefore, likely that the age of the subducting slab varied significantly along strike. The convergence rate is estimated to have been high during Laramide time (average trench-normal component of 10–15 cm/yr for the Farallon and North American plates, and somewhat less for the Kula–North American plates). However, given that Laramide-age deformation occurred along the entire length of North America regardless of which plate was subducting beneath it, it seems unlikely that slab age or convergence rate were the principal driving force of orogenesis. It is possible that a ~100 km eastward migration of arc magmatic belts in western Canada and Mexico during Laramide time (e.g., van der Heyden, 1992; Friedman and Armstrong, 1995; Fig. 1) may reflect a shallowing of the subducting slab, and hence enhanced coupling across the plate boundary. However, it is not apparent why this increased horizontal compressive stress would be transferred up to 1000 km inboard to drive foreland deformation. Nor does it explain why slab shallowing should simultaneously occur along the length of the continent spanning two subducting plates. Flat-slab subduction beneath the southwestern U.S. Cordillera may have locally influenced the structural style and location of orogenesis in this region. This may be analogous to the modern Andes where variations in subduction angle have been interpreted to exert a significant influence on the anatomy of the orogen; one significant difference is that the Laramide deformation belt along the entire length of the North American Cordillera lies at great distances (700–1500 km) from the current plate margin. Thus, post-Laramide extension alone cannot account for the inboard location of this orogenic belt.

Alternatively, the absolute motion of the overriding plate may have been important for driving Laramide orogenesis (as in the retroarc thrusting model). The high rate of western advancement of the North American plate must have been driven by ridge-push from the Atlantic spreading-center or by basal drag caused by mantle convection. Basal drag is not considered to be an important factor in driving plate motions (e.g. Stüwe, 2002); this model, therefore, requires that horizontal normal compressive stresses applied by ridge-push from the mid-Atlantic spreading center must have been sufficient to drive Laramide orogenesis. However, it is not apparent why ridge-push would only have driven orogenesis during the Laramide, as spreading rates were not anomalously high during this time. Although Africa and North America began to separate at ~180 Ma, the main phase of drift between Eurasia and North America did not occur until ~80 Ma; separation varied from 5.0 to 4.0 cm/yr until 53 Ma, and then dropped off to ~2 cm/yr (Pitman and Talwani, 1972). It remains a possibility that pre-Laramide and Laramide-age spreading rates in the North Atlantic were sufficient to culminate in a greater ridge-push force on the North American continent during Laramide time. Subsequent slowing of the spreading rates during the Eocene may have resulted in a decline in the ridge-push force and absolute motion of the North American continent.

None of the proposed mechanisms for driving Laramide orogenesis satisfactorily explain the geometry, timing, or extent of this inboard, continental-scale orogeny. In addition to our assessment of available models for Laramide orogenesis, we offer three major questions that require resolution.

1. Are the paleomagnetic data implying northward translation correct, and if so, how expansive was the Cordilleran entity that underwent > 2000 km of northward translation during Laramide time? Existing data suggest that much of the Alaskan and Canadian Cordillera underwent northward translation, but the extension of this Cordilleran entity into the United States Cordillera is currently unconstrained.

2. What was the relationship between northward translation, dextral transpression, and fold-and-thrust belt formation? Does the fold-and-thrust belt represent the product of a collisional event following closure of an intervening ocean basin, or does the fold and thrust belt root into an oblique dextral

transpressional system that assisted in accommodating northward translation? Attributing Laramide orogeny to collision with a northward-translating ribbon continent leaves unexplained the presence of pre–Late Cretaceous westerly-derived molasse deposited on North America.

3. Why did the Laramide strain zone form so far inboard (~1000 km) of the inferred plate boundary along the entire length of the continent?

Acknowledgments

This research was supported by a University of Victoria Fellowship to JE, and an NSERC Discovery Grant to STJ. Figure 1 was prepared with the GMT software (Wessel and Smith, 1995). We would like to thank John Geissman, John Bartley, Lucinda Leonard, Kaesy Gladwin, William Dickinson, Damian Nance, Andrew Hynes, Brendan Murphy, and three anonymous reviewers for constructive comments.

REFERENCES

Bird, P., 1988, Formation of the Rocky Mountains, western United States: A continuum computer model: Science, v. 239, p. 1501–1507.

Breitsprecher, K., Thorkelson, D. J., Groome, W. G., and Dostal, J., 2003, Geochemical confirmation of the Kula-Farallon slab window beneath the Pacific Northwest in Eocene time: Geology, v. 31, p. 351–354.

Coney, P. J., and Reynolds, S. J., 1977, Cordilleran Benioff zones: Nature, v. 270, p. 403–406.

Dickinson, W. R., and Lawton, T. F., 2001, Carboniferous to Cretaceous assembly and fragmentation of Mexico: Geological Society of America Bulletin, v. 113, p. 1142–1160.

Dickinson, W. R., and Snyder, W. S., 1978, Plate tectonics of the Laramide Orogeny, *in* Matthews, V., ed., Laramide folding associated with basement block faulting in the western United States: Geological Society of America Memoir 151, p. 355–366.

Engebretson, D. C., Cox, A., and Gordon, R. G., 1985, Relative motions between oceanic and continental plates in the Pacific Basin: Geological Society of America Special Paper, 59 p.

English, J. M., Johnston, S. T., and Wang, K., 2003, Thermal modelling of the Laramide orogeny: Testing the flat-slab subduction hypothesis: Earth and Planetary Science Letters, v. 214, p. 619–632.

Enkin, R. J., Mahoney, J. B., Baker, J., Kiessling, M., and Haugerud, R. A., 2002, Syntectonic remagnetizations in the southern Methow block: Resolving large displacements in the southern Canadian Cordillera: Tectonics, v. 21 [10.1029/2001TC001294].

Friedman, R. M., and Armstrong, R. L., 1995, Jurassic and Cretaceous geochronology of the southern Coast Belt, British Columbia, 49° to 51°N, *in* Miller, D. M., and Busby, C., eds., Jurassic magmatism and tectonics of the North American Cordillera: Geological Society of America Special Paper 299, p. 95–139.

Irving, E., Wynne, P. J., Thorkelson, D. J., and Schiarizza, P., 1996, Large (1000–4000 km) northward movements of tectonic domains in the northern Cordillera, 83 to 45 Ma: Journal of Geophysical Research, v. 101, p. 17,901–17,916.

Jarrard, R. D., 1986, Relations among subduction parameters: Reviews of Geophysics, v. 24, p. 217–284.

Johnston, S. T., 2001, The great Alaskan terrane wreck: Reconciliation of paleomagnetic and geological data in the northern Cordillera: Earth and Planetary Science Letters, v. 193, p. 259–272.

Johnston, S. T., and Acton, S., 2003, The Eocene southern Vancouver Island orocline—a response to seamount accretion and the cause of fold and thrust belt and extensional basin formation: Tectonophysics, v. 365, p. 165–183.

Jordan, T. E., and Allmendinger, R. W., 1986, The Sierras Pampeanas of Argentina: A modern analogue of Rocky Mountain foreland deformation: American Journal of Science, v. 286, p. 737–764.

Kay, S. M., and Abbruzzi, J. M., 1996, Magmatic evidence for Neogene lithospheric evolution of the central Andean "flat-slab" between 30°S and 32°S: Tectonophysics, v. 259, p. 15–28.

Kelley, K., 1993, Relative motions between North America and oceanic plates of the Pacific basin during the past 130 million years: Unpubl. M.Sc. thesis, Western Washington University, 89 p.

Livaccari, R. F., Burke, K., and Sengor, A. M. C., 1981, Was the Laramide orogeny related to subduction of an oceanic plateau?: Nature, v. 289, p. 276–278.

Mazzotti, S., and Hyndman, R. D., 2002, Yakutat collision and strain transfer across the northern Canadian Cordillera: Geology, v. 30, p. 495–498.

Monger, J. W. H., and Nokleberg, W. H., 1996, Evolution of the northern North American Cordillera: Generation, fragmentation, displacement, and accretion of successive North American plate margin arcs, *in* Geology and ore deposits of the American Cordillera: Reno, NV, Geological Society of Nevada, p. 1133–1152.

Monger, J. W. H., Price, R. A., and Tempelman-Kluit, D. J., 1982, Tectonic accretion and the origin of the two major metamorphic and plutonic welts in the Canadian Cordillera: Geology, v. 10, p. 70–75.

Murphy, J. B., Hynes, A. J., Johnston, S. T., and Keppie, J. D., 2003, Reconstructing the ancestral Yellowstone plume from accreted seamounts and its relationship to flat-slab subduction: Tectonophysics, v. 365, p. 185–194.

Oldow, J. S., Bally, A. W., and Ave' Lallemant, H. G., 1990, Transpression, orogenic float, and lithospheric balance: Geology, v. 18, p. 991–994.

Pitman, W. C., and Talwani, M., 1972, Sea-floor spreading in the North Atlantic: Geological Society of America Bulletin, v. 83, p. 619–646.

Price, R. A., 1981, The Cordillera foreland thrust and fold belt in the southern Canadian Rocky Mountains, *in* McClay, K., ed., Thrust and nappe tectonics: Geological Society of London Special Publication 9, p. 427–448.

Russo, R. M., and Silver, P. G., 1996, Cordillera formation, mantle dynamics, and the Wilson cycle: Geology, v. 24, p. 511–514.

Saleeby, J., 2003, Segmentation of the Laramide slab—evidence from the southern Sierra Nevada region: Geolgical Society of America Bulletin, v. 115, p. 655–668.

Stott, D. F., 1993, Evolution of Cretaceous foredeeps: A comparative analysis along the length of the Canadian Rocky Mountains, *in* Caldwell, W. G. E., and Kauffman, E. G., eds., Evolution of the Western Interior Basin: Geological Association of Canada, p. 131–150.

Stüwe, K., 2002, Geodynamics of the lithosphere: An introduction: Berlin, Germany and New York, NY, Springer-Verlag, 463 p.

van der Heyden, P., 1992, A Middle Jurassic to Early Tertiary Andean-Sierran arc model for the Coast Belt of British Columbia: Tectonics, v. 11, p. 82–97.

Wessel, P., and Smith, W. H. F., 1995, New version of the Generic Mapping Tools released: EOS (Transactions of the American Geophysical Union, v. F359.

Wynne, P. J., Enkin, R. J., Baker, J., Johnston, S. T., and Hart, C. J. R., 1998, The big flush—paleomagnetic signature of a 70 Ma regional hydrothermal event in displaced rocks of the northern Canadian Cordillera: Canadian Journal of Earth Sciences, v. 35, p. 657–671.

Do Supercontinents Turn Inside-in or Inside-out?

J. BRENDAN MURPHY[1]

Department of Earth Sciences, St. Francis Xavier University, Antigonish, Nova Scotia, B2G 2W5 Canada

AND R. DAMIAN NANCE

Department of Geological Sciences, 316 Clippinger Laboratories, Ohio University, Athens, Ohio 45701

Abstract

Supercontinent amalgamation and dispersal has occurred repeatedly since the Archean. However, the mechanisms responsible for these events are unclear. Following supercontinent breakup, two geodynamically distinct oceans may be distinguished: an *interior ocean* formed between the dispersing continents, whose lithosphere is younger than the time of supercontinent breakup (T_R); and an *exterior ocean* that surrounded the supercontinent prior to breakup, and consequently is dominated by lithosphere that is older than the time of breakup. In order to evaluate geodynamic models for supercontinent formation, it is essential to determine which of these two types of ocean is consumed during supercontinent amalgamation. Although much of the evidence needed is destroyed by subduction, vestiges of oceanic lithosphere are preserved in mafic complexes accreted to continental margins prior to terminal collision. Because the age contrast between interior and exterior oceans diminishes as the continents drift apart, the ages of the earliest accreted complexes are the most diagnostic of the ocean in which they formed. Constraints on the age of the mantle lithospheric sources (T_{DM}) that give rise to these accreted complexes can be derived from Sm-Nd isotope systematics. In the case of Pangea, for example, the North American Cordillera represents an accretionary orogen along the leading edge of a dispersing supercontinent. Within this orogen, the oldest accreted oceanic terranes, characterized by high ε_{Nd} values close to contemporary depleted mantle values, show similar crystallization and T_{DM} model ages that imply crustal formation and arc activity during the lifespan of Pangea, that is, within the exterior Panthalassa ocean (i.e., $T_{DM}>T_R$). This example suggests that a similar approach applied to older orogens may constrain the relationship between continental margins and their accreted mafic complexes.

Pangea was formed by closure of Paleozoic oceans (e.g., Iapetus and Rheic) that were formed after the ca. 550 Ma breakup of Pannotia. Uncontaminated mafic rocks from both oceans that have ε_{Nd} values close to depleted mantle values at their respective times of emplacement show closely matching crystallization and depleted mantle model ages that do not exceed the age of rifting (i.e., $T_{DM} \leq T_R$). This indicates that the oceanic lithospheric source of these suites was generated after the rifting of Pannotia, such that Pangea was formed by the closure of interior oceans (introversion). In contrast, mafic terranes accreted in orogens that terminated in the formation of the Late Neoproterozoic supercontinent Pannotia have Sm-Nd T_{DM} model ages between ca. 1.2 and 0.71 Ga, implying that much of the oceanic lithosphere that was subducted and recycled to yield these complexes was formed *before* the ca. 755 Ma breakup of the supercontinent Rodinia (i.e. $T_{DM} > T_R$). These mafic complexes are therefore vestiges of oceanic lithosphere that formed within the peri-Rodinian ocean, such that Pannotia was formed by the closure of an exterior ocean (extroversion). This analysis suggests that Pangea and Pannotia were assembled by fundamentally distinct geodynamic processes. Hence, the "supercontinent cycle" may have a more complex origin than previously considered.

Introduction

GROWING EVIDENCE exists that the repeated amalgamation and dispersal of supercontinents have had a profound effect on the Earth's evolution since the end of the Archean (e.g., Worsley et al., 1984; Nance et al., 1986; Hoffman, 1991; Dalziel, 1992; Windley, 1993; Rogers, 1996; Condie, 1998; Dalziel et al., 2000). The amalgamation of Wegener's supercontinent Pangea at the end of the Paleozoic (ca. 350–250 Ma) was preceded by that of Pannotia (Stump, 1992; Dalziel, 1997) at the end of the

[1]Corresponding author; email: bmurphy@stfx.ca

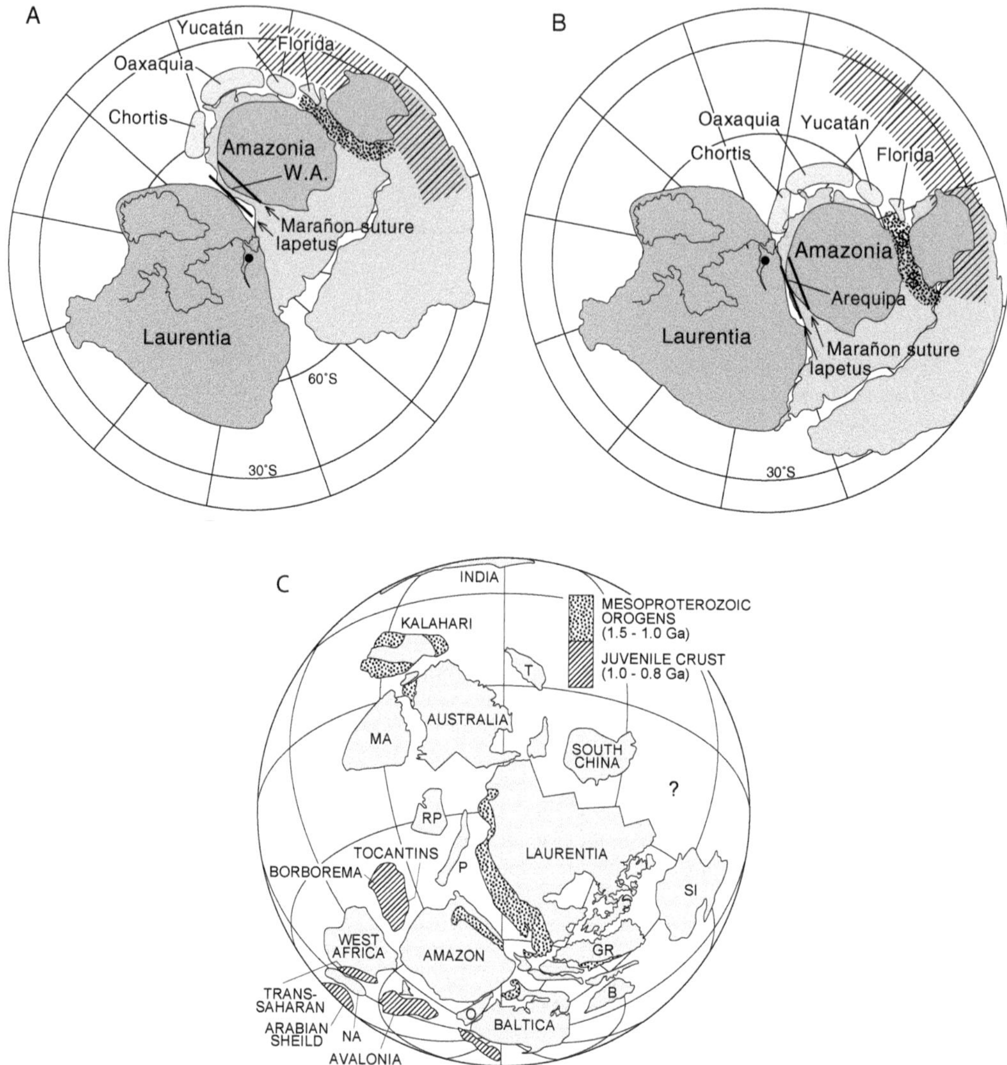

FIG. 1. Supercontinents before Pangea. Two versions of the Late Neoproterozoic supercontinent Pannotia: A. Modified after Stump (1992) and Dalziel (1997). B. Modified after Hoffman (1991) and Weil et al. (1998). C. Rodinia at ca. 1.1–0.76 Ga (modified after McMenamin and McMenamin, 1990; Pisarevsky et al., 2003; Murphy et al., 2004). D (facing page). Columbia (after Rogers and Santosh, 2002; Zhao et al., 2002). E (facing page). *Partial* reconstruction of Kenorland (after Williams et al., 1991; Heaman, 1997). In (C), the conventional configuration for Baltica (e.g., Gower et al., 1990) is used instead of the geographically inverted reconstruction (Torsvik and Rehnström, 2001; Hartz and Torsvik, 2002).

Neoproterozoic (ca. 650–550 Ma; Figs. 1A and 1B) and, although its configuration is controversial, there is general acceptance of the existence of the supercontinent Rodinia (McMenamin and McMenamin, 1990) at the end of the Mesoproterozoic (ca. 1.1 Ga to 750 Ma) (Fig. 1C). Other proposed supercontinents include Nuna (ca. 1.8–1.5 Ga; Hoffman, 1989) or Columbia (ca. 1.8–1.5 Ga; Rogers and Santosh, 2002; Zhao et al., 2002; Fig. 1D), Kenorland (ca. 2.7–2.5 Ga; Williams et al., 1991, Heaman, 1997) (Fig. 1E), and Ur (ca. 3 Ga; Rogers, 1996). Following these intervals of supercontinent

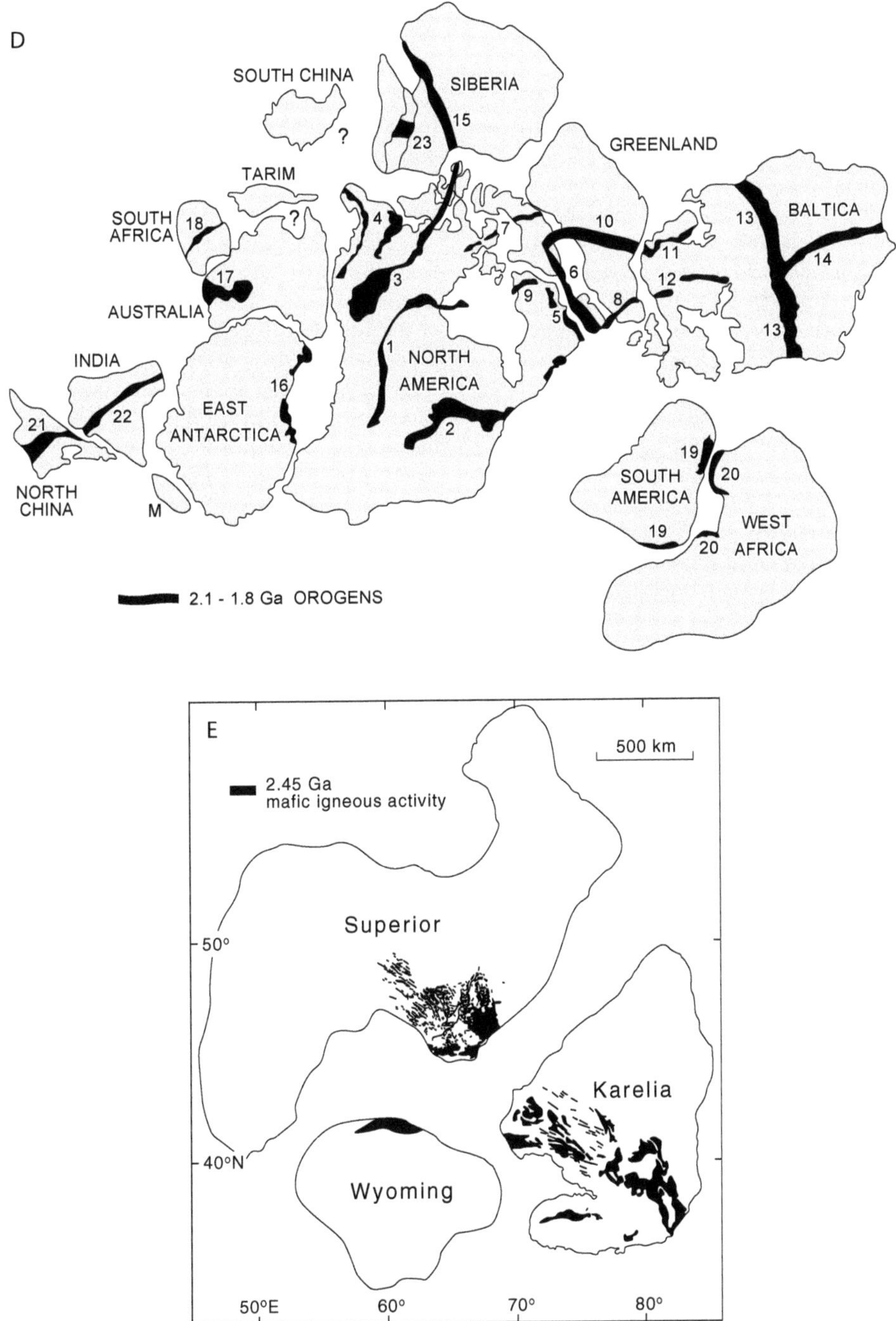

FIG. 1. D. and E.

amalgamation are periods of supercontinent breakup and dispersal that are recorded by mafic dike swarms, passive margin development, and the formation of oceanic lithosphere. Continental convergence, ultimately leading to renewed supercontinent amalgamation, is represented sequentially by subduction initiation, arc development, the accretion of outboard terranes to continental margins, and continent-continent collisions.

First encapsulated in the concept now known as the Wilson cycle (Wilson, 1966), the breakup and dispersal of continents is necessarily accompanied by the generation of new oceanic lithosphere between the dispersing continental blocks. The evolution of the Paleozoic Appalachian-Caledonide orogen is generally held to be the type example of a Wilson cycle (e.g., Dewey and Bird, 1971). Here, rifting between Laurentia and Gondwana in the latest Neoproterozoic and Early Cambrian resulted in the generation of a Laurentian passive margin flanking a widening Iapetus Ocean. The subsequent subduction of oceanic crust generated between Laurentia and Gondwana started in the Late Cambrian–Early Ordovician, and led first to foundering of the passive margin, arc development, ophiolite obduction, and terrane accretion, and culminated in collision between Laurentia and Gondwana to form Pangea (Williams, 1979; Keppie, 1985; van Staal et al., 1998).

To date, our understanding of the geodynamic forces that drive these global-scale events remains incomplete. However, over the past 30 years, two end-member models have emerged (Fig. 2). The Wilson cycle, which as originally proposed involves the symmetrical opening and closing of ocean basins, forms one end-member (e.g., Worsley et al., 1984; Collins, 2003). Here, the terminal collisions of supercontinent assembly are produced by the destruction of the oceanic lithosphere generated by the breakup and dispersal of the previous supercontinent, such that the supercontinent turns "inside in," or "introverts." The interior continental margins of the former supercontinent, produced during the dispersal stage, consequently become the interior orogenic belts of the next supercontinent (Figs. 2A–2C). Mechanisms for subduction initiation are controversial, although recent modeling suggests that preexisting fracture systems play an important role (e.g., Toth and Gurnis, 1998; Hall et al., 2003). But in general, subduction is more likely to initiate near an ocean's margins, where its lithosphere is oldest, because oceanic lithosphere becomes more negatively buoyant with age.

In the second end-member (Figs. 2A–2D), supercontinent dispersal is considered to reflect dissipation of mantle thermal upwelling, such that the subsequent supercontinent amalgamates above areas of mantle downwelling that are sites of subduction (e.g., Gurnis, 1988). In contrast to the Wilson cycle, the supercontinent in this scenario turns "outside in" or "extroverts"—that is, the leading or exterior continental margins of the supercontinent during the dispersal stage become the interior orogenic belts of the next supercontinent. In this model, the supercontinent traps mantle heat, which ultimately leads to its breakup and dispersal toward areas of mantle downwelling where the next supercontinent assembles. The model of Hoffman (1991) for the assembly of Pannotia—in which the breakup of the supercontinent Rodinia "turned Gondwanaland inside out"—exemplifies such extroversion (Fig. 3). According to this model, the ca. 0.755 Ga rifting from Western Laurentia of what would later become East Gondwana was followed at ca. 0.6–0.55 Ga by the collision of its exterior margin with the exterior margin of West Gondwana to form the supercontinent Pannotia. Hence, in the context of the resulting supercontinent, it is the interior oceans that close during introversion, whereas it is the exterior ocean that closes during extroversion (Murphy and Nance, 1991).

In a companion paper (Murphy and Nance, 2003), we showed that these end-member models can be distinguished by comparing the Sm-Nd crust-formation ages of accreted mafic complexes in the collisional orogens of supercontinent assembly with the breakup age of the previous supercontinent. For supercontinents generated by introversion, these mafic complexes will yield crust-formation ages that postdate the age of rifting (T_R) of the previous supercontinent (Fig. 2C). But for supercontinents generated by extroversion, portions of the oceanic lithosphere consumed during reassembly will predate the breakup of the previous supercontinent such that the crust-formation ages of the earliest accreted mafic complexes will be older than the age of rifting (Fig. 2D). In the Paleozoic Appalachian–Caledonide, Variscan, and Uralide orogens, for example, the oldest crust-formation ages of accretionary mafic complexes respectively postdate the formation of the Iapetus (ca. 615–540 Ma; Meert et al., 1998; Cawood et al., 2001), Rheic (ca. 510–470 Ma; Giese and Buehn, 1994; Cocks et al., 1997; Kemnitz et al., 2002), and paleo-Uralian (ca. 540–480 Ma; Leech et al., 2002; Bogolepova, 2003) oceans, whose closures they record, suggesting that the supercontinent Pangea was assembled by introversion. By contrast, the Neoproterozoic East African and Brasiliano orogens, which formed during the amalgamation of Pannotia, contain mafic

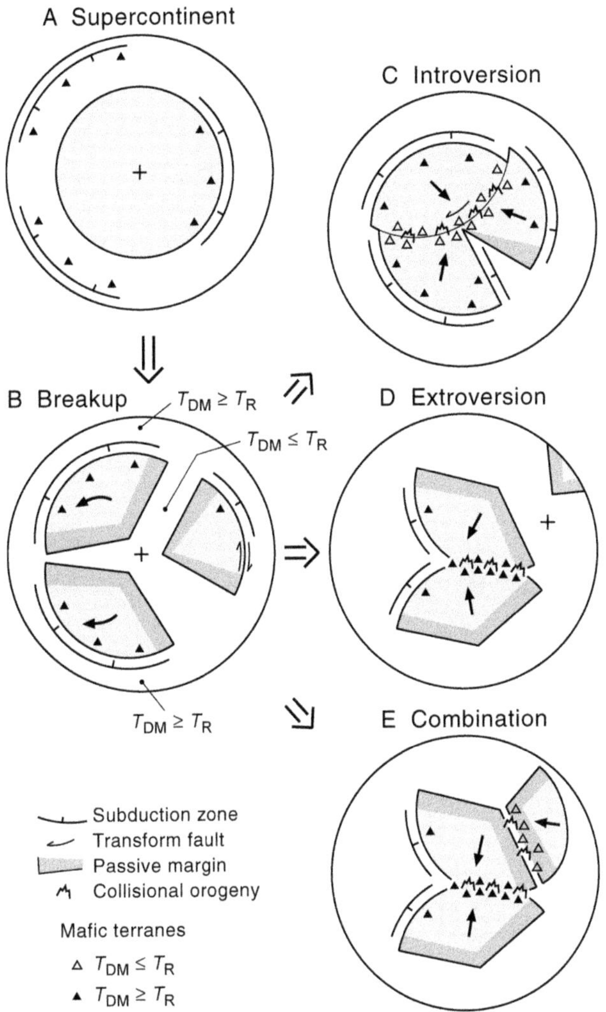

FIG. 2. A. Supercontinent flanked by an exterior ocean with subduction zones and mafic complexes (e.g., de Wit et al., 1988). B. Breakup of supercontinent at T_R (time of rifting) creates an interior ocean with Sm-Nd depleted-mantle model ages (T_{DM}) ≤ T_R. Exterior ocean has $T_{DM} \geq T_R$. C. If supercontinent is formed by introversion, then accreted oceanic lithosphere in collisional orogens has $T_{DM} \leq T_R$. D. If supercontinent is formed by extroversion, then accreted oceanic lithosphere in collisional orogens is primarily derived from the exterior ocean with $T_{DM} \geq T_R$. E. An intermediate case in which one ocean is closed by introversion and the other by extroversion.

complexes with ca. 0.75–1.2 Ga crust-formation ages that predate the ca. 755 Ma breakup of Rodinia (e.g., Wingate and Giddings. 2000). Hence, these complexes must have formed from lithosphere in the exterior ocean that surrounded Rodinia, implying that it was this ocean that was consumed during the amalgamation of Pannotia. If so, the supercontinent Pannotia must have formed by extroversion. The implication of these data, that Pangea and Pannotia were formed by introversion and extroversion, respectively, suggests that supercontinents can be assembled by fundamentally distinct geodynamic processes.

In this paper, we develop these ideas and examine their implications for the "supercontinent cycle." For each model, there should be examples of each stage of their development. To test this observation, we look at the western margin of North

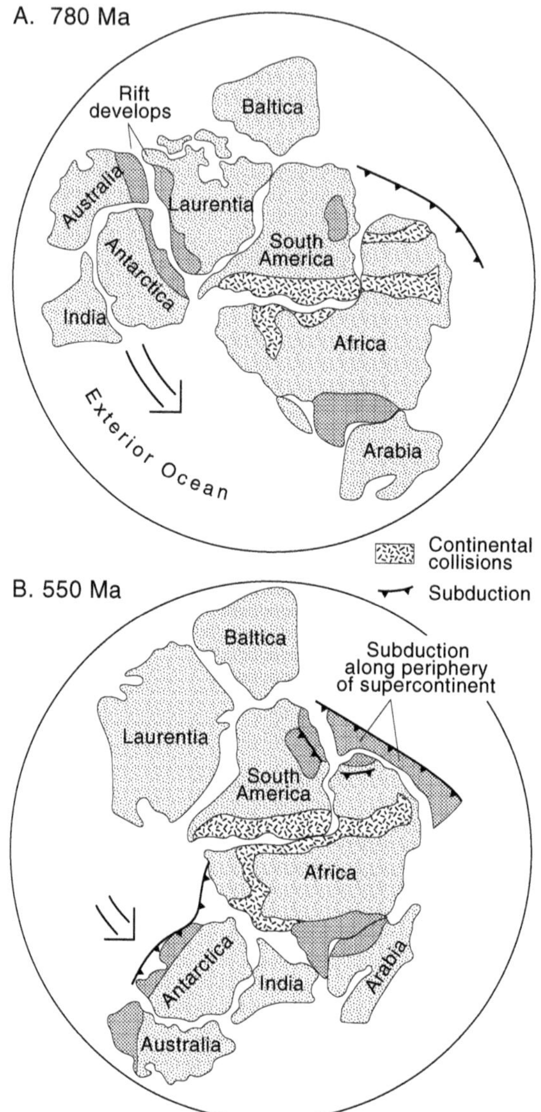

FIG. 3. Schematic diagram showing the extroversion implied by the reconstructions of Hoffman (1991) and Dalziel (1992). A. The breakup of Rodinia at 0.780 Ga initially recorded by rifting of Antartica–Australia–South China from Laurentia to form the Pacific Ocean. B. Closure of a portion of the exterior peri-Rodinian ocean resulted in collision between East and West Gondwana to form a late Neoproterozoic supercontinent. For a more recent reconstruction of Rodinia, see Figure 1C (modified from Pisarevsky et al., 2003).

America, where Late Paleozoic mafic complexes formed within Panthalassa were accreted to the Laurentian continental margin in the Mesozoic following the breakup of Pangea. We then review evidence for the origin of Pangea by introversion and that of Pannotia by extroversion. Finally, we extend this approach into the Mesoproterozoic and the Paleoproterozoic by examining the signature of accreted mafic complexes in the Grenville and Trans-Hudson orogens, respectively. In these older orogens, interpretation is hindered by uncertainties in the timing of collisional and rifting events, and by our lack of understanding of continental configurations. Nevertheless, we show how this approach can help constrain the geodynamics of specific orogenic belts. We emphasize that the amalgamation of most supercontinents likely reflects situations that are intermediate between these two end-member models. Nevertheless, we feel the approach offers insights into the development of orogens by focusing attention on the relationships between the consumption of oceans by subduction and the geodynamic forces responsible for such consumption.

Sm-Nd Isotope Data

Although vestiges of oceanic lithosphere are preserved during ocean closure in ophiolite complexes and other mafic terranes that are accreted to continental margins prior to terminal collision, subduction of oceanic lithosphere during continental convergence generally destroys much of the primary evidence that would distinguish between the two end-member models for supercontinent amalgamation. In addition, some mafic terranes are generated by the subduction process itself, such that their crystallization ages postdate the onset of subduction and, hence, cannot be used to distinguish between the two end-member models. In this scenario, crust-formation ages, rather than crystallization ages, are the key to distinguishing between the two models.

Constraints on the age of the mantle sources that give rise to these accreted complexes can be derived from Sm-Nd isotope systematics (Fig. 4A). Variations in Sm/Nd ratios in crustal rocks are insensitive to intracrustal processes and are largely inherited from the depleted mantle, which preferentially retains samarium over neodymium. The decay of ^{147}Sm to ^{143}Nd produces $^{143}Nd/^{144}Nd$ ratios that, over time, have become higher in the depleted mantle than in the bulk Earth (represented by the "chondrite uniform reservoir" or CHUR), the $^{143}Nd/^{144}Nd$ ratio of which is, in turn, higher than those in the crust. Upon melting, a magma acquires the $^{143}Nd/^{144}Nd$ ratio of its source, known as its initial ratio $(^{143}Nd/^{144}Nd)_0$. Differences between the initial ratio and CHUR are expressed as ε_{Nd} values (DePaolo, 1988), which are usually calculated for the age of the rock, and can be used to distinguish between igneous rocks derived from juvenile crust (such as mid-oceanic ridges, plumes, oceanic arcs, and backarcs) and those derived from the recycling of ancient crust.

In some samples, the age at which this crust was first extracted from the mantle (T_{DM}) may be derived by extrapolating the ε_{Nd} growth line backward to the depleted-mantle curve (DePaolo, 1988). However, for samples involving mixtures of ancient recycled crust and new mantle material, this depleted-mantle model age reflects only an average crustal residence time and so has no real geologic meaning (e.g., Arndt and Goldstein, 1987). Because of this, extreme caution must be exercised in the use of T_{DM} model ages, and their interpretation as mantle extraction ages must be restricted to those cases where the possibility of crustal contamination can be eliminated. Large-scale contamination is generally identifiable from field observations of mixing or mingling, from abundant and dismembered xenoliths, and from geochemical features that show linear trends between end-member compositions. Smaller amounts of crustal contamination, which may not be obvious in the field or in many geochemical variation diagrams, can be detected by comparing ε_{Nd} values with indices of contamination such as Nb anomalies (Nb*) or typical crustal values of $^{147}Sm/^{144}Nd$. In contaminated suites ε_{Nd} values will decrease with increasing Nb* and decreasing $^{147}Sm/^{144}Nd$. To minimize the effects of crustal contamination, we concentrate on the Sm-Nd isotope signature of mafic complexes with no geochemical evidence of contamination. For such complexes, ε_{Nd} values will lie close to contemporary depleted mantle values such that their T_{DM} model ages will differ little from their ages of crystallization. In cases where the time interval between mantle extraction and crystallization is short, these data can provide important constraints on the way the crust was formed.

Interior and Peripheral Orogens

Because supercontinents produced by introversion result from the closure of interior oceans whereas those that form by extroversion are the product of exterior ocean closure (Fig. 2), a comparison between the mantle extraction ages of mafic complexes accreted within the collisional orogens of supercontinent amalgamation and those that record the breakup of the preceding supercontinent provides a potential means of distinguishing between the two processes of assembly. However, inasmuch as supercontinents are generally bordered by active continental margins, not all of the orogenic belts associated with supercontinents mark the sites of

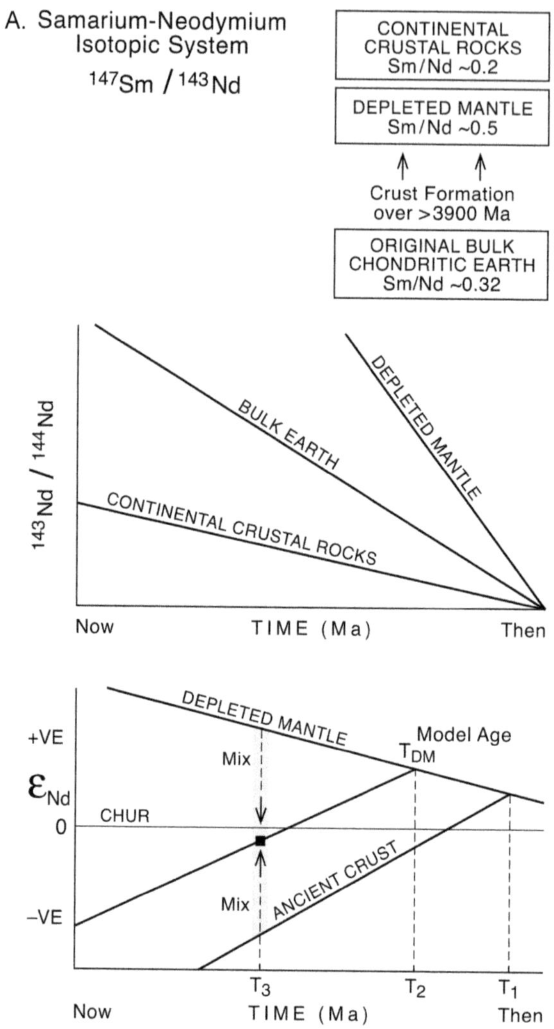

FIG. 4. Schematic representation of Sm-Nd isotopic system showing Nd isotopic reservoirs, their evolution through time, and the interpretation of Nd isotopic data. A. Depleted mantle preferentially retains Sm over Nd, so that its Sm/Nd ratio (~0.5) is greater than that of bulk earth (CHUR) (~0.32) and the continental crust (~0.2). As a result of the variations in Sm/Nd and the decay of ^{147}Sm to ^{143}Nd, the ^{143}Nd/^{144}Nd initial ratio increases more rapidly in the depleted mantle than in CHUR and in continental crustal source rocks. Differences between the ^{143}Nd/^{144}Nd initial ratio of depleted mantle, CHUR, and crustal rocks are expressed as ε_{Nd} values relative to CHUR. The Sm/Nd ratio and the ε_{Nd} value of the sample (calculated for the age of the rock) can be used to produce a growth line that intersects the depleted mantle curve. This intersection yields a depleted mantle (T_{DM}) model age. This model age is open to various interpretations. It could reflect the time (T_2) at which the crust was derived from the depleted mantle. Alternatively, the model age may reflect a mixing age of more ancient (e.g., with a T_{DM} model age of T_1) with more juvenile (with a T_{DM} model age of T_3) crust.

former ocean closure. Instead, orogenic belts will include both the collisional orogens of supercontinent assembly (interior orogens of Murphy and Nance, 1991) and the accretionary orogens of any peri-supercontinental subduction systems (peripheral orogens of Murphy and Nance, 1991); these orogens must be distinguished before our approach can be applied.

Because supercontinental assembly occurs as the result of ocean closure, accretionary orogenesis

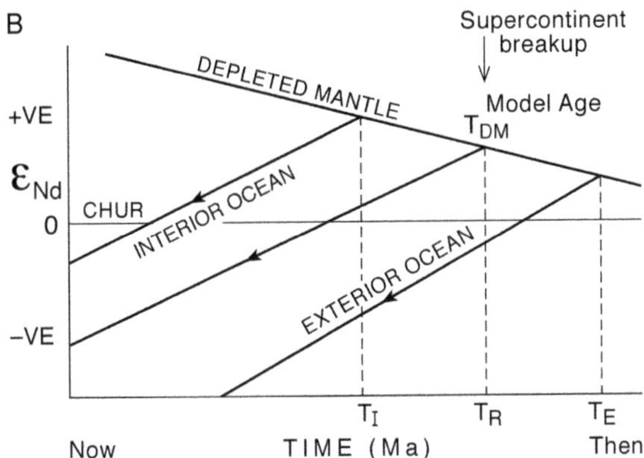

FIG. 4. B. Schematic representation of the Sm-Nd isotopic evolution of oceanic lithosphere from the interior and exterior oceans. Depleted mantle model ages for the interior ocean (T_I) are younger than the time of supercontinent breakup, whereas the depleted mantle model ages for the exterior ocean (T_E) are older than the time of supercontinent breakup.

invariably precedes the collisional (interior) orogenesis of supercontinent amalgamation. During introversion, however, accretionary orogenesis entirely postdates the accreting supercontinent because it reflects the closure of oceans created by former supercontinent breakup. During extroversion, on the other hand, the onset of accretionary orogenesis will predate supercontinent breakup inasmuch as the orogens first develop as peripheral orogens along the margins of the former supercontinent. Hence, the peripheral orogens of one supercontinent become the interior orogens of the next during extroversion, whereas, during introversion, this is not the case. Because the age contrast between interior and exterior oceans is greatest just after supercontinent breakup and diminishes as the continents drift apart, it is the ages of the earliest accreted complexes that are most diagnostic of the ocean in which they formed. To illustrate this distinction and its relevance to the T_{DM} model ages of accreted mafic complexes (Fig. 4B), we first turn to the Phanerozoic to compare the accretionary orogenic record of the North American Cordillera with the Paleozoic collisional orogens of Pangea assembly.

Application to the Phanerozoic

North American Cordillera

The North American Cordillera (Fig. 5A) is characterized by a collage of suspect terranes accreted to the western margin of Laurentia in the Mesozoic and Cenozoic (Coney et al., 1980). Although the details are complex, accretion of terranes to this continental margin was geodynamically related to the westerly migration of North America following the breakup of Pangea. Thus the North American Cordillera is an accretionary orogen that formed along the leading edge of a dispersing continent, and that evolved in response to peri-supercontinental subduction systems (peripheral orogen of Murphy and Nance, 1991).

In northern British Columbia, southern Yukon, Oregon, and northern California, a belt of dismembered Late Paleozoic oceanic terranes is preserved west of ancestral North America and the adjacent pericratonic Omineca terrane. These terranes (Slide Mountain, Quesnel, Stikine, Klamath Mountains, and correlatives) preserve vestiges of Late Paleozoic oceanic crust that must have surrounded Pangea prior to supercontinent breakup (Monger et al., 1982, 1991). The Slide Mountain terrane is exposed in a series of allocthonous sheets that contain Mississippian to Upper Permian volcanic and sedimentary rocks sitting unconformably on Upper Devonian–Lower Carboniferous strata (Monger, 1977). Upper Pennsylvanian mafic volcanics have geochemical features similar to N-MORB with ε_{Nd} values (t = 300 Ma) of +7.7 to +10.2, which are among the highest for Paleozoic basalts, and are thought to have been derived from depleted mantle (Smith and Lambert, 1995; Fig 5B). Recent data, which identify N-MORB and E-MORB components, have been used to attribute the magmatism to hot

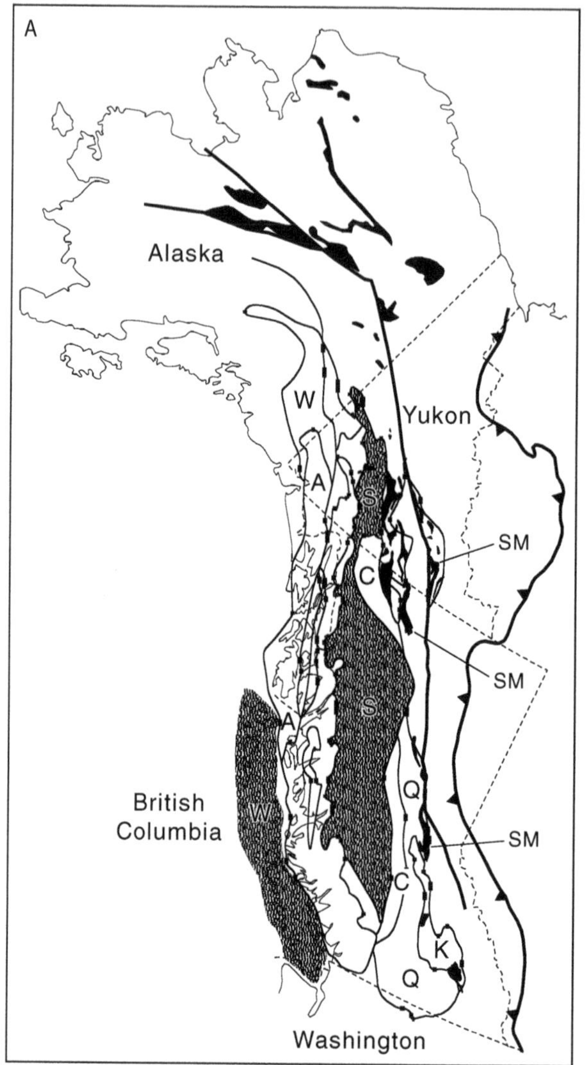

FIG. 5. A. Summary tectonic map of the Canadian and Alaskan Cordillera (based on Monger, 1997; Gehrels and Kapp, 1998) showing some of the accreted mafic terranes (in black). Abbreviations: W = Wrangellia; S = Stikinia; YT = Yukon-Tanana; K = Kootenay; SM = Slide Mountain; C = Cache Creek; Q = Quesnel terranes.

spot and related plume activity (Tardy et al., 2003). The Klinkit Group, which has been interpreted as having been deposited within the Slide Mountain basin, contains Permian calc-alkalic basalts that show no geochemical evidence for crustal contamination, and have ε_{Nd} values (t = 281 Ma) ranging from +6.7 to +7.4 (Simard et al., 2003), close to depleted mantle values.

The Jurassic (ca. 200 Ma) Hazelton Group, which is basement to the Buck Creek volcanic complex in central British Columbia, consists of mafic to felsic volcanic rocks emplced prior to the accretion of Stikinia to North America in the Mesozoic. Mafic rocks in this group have ε_{Nd} values (t = 200 Ma) ranging from +5.7 to +7.1 (Dostal et al., in press), which are typical of depleted mantle values.

Mid- to Upper Triassic flood basalts of the Wrangellia terrane (Alaska and British Columbia) overlie an arc sequence that was active until the Late Pennsylvanian and Early Permian. The

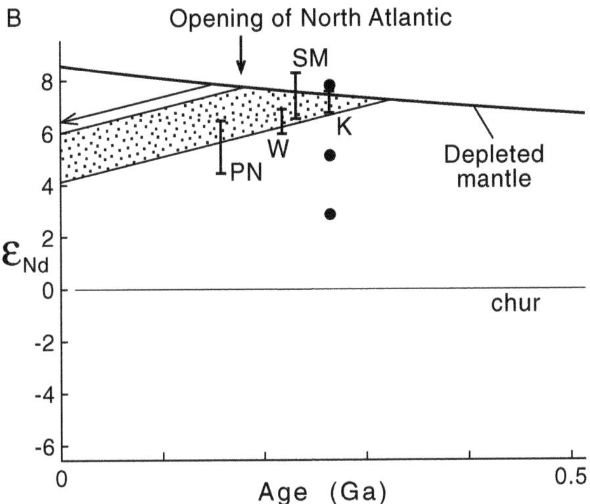

FIG. 5. B. Summary diagram of $(\varepsilon_{Nd})_t$ vs. time (Ga) for late Paleozoic and Mesozoic mafic complexes accreted to the western margin of North America. Examples from Klamath Mountains include Yellow Dog (dots; after Ernst et al., 1991) and post-Nevadan plutons (PN; after Barnes et al., 1992). Examples from northern Cordillera include Wrangellia (W; Lassiter et al., 1995), the Klingit Group (K; Simard et al., 2003), and Slide Mountain (SM; Smith and Lambert, 1995; Tardy et al., 2003).

Wrangellia terrane was accreted to North America by the Early Cretaceous and is stitched to inboard portions of the orogen by mid-Cretaceous plutons (Nokleberg et al., 1985). The flood basalts have ε_{Nd} values (t = 230 Ma) of +4 to +7 with the lower values attributed to contamination, whereas the higher values (+6 to + 7) are considered to be representative of the plume source (Lassiter et al., 1995).

The Klamath Mountain province of California and Oregon consists of a tectonic collage of polydeformed terranes characterized by Permian to Early Cretaceous metavolcanic and plutonic rocks that originated in an oceanic setting. The province consists of westward-younging, easterly dipping belts that were accreted to North America in discrete episodes of predominantly eastward-dipping subduction during the Mesozoic (Irwin, 1981; Ernst, 1987; Ernst et al., 1991). Some igneous rocks in the province are characterized by juvenile isotopic characteristics, and are thought to have been derived directly from the mantle (Farmer and DePaolo, 1983). Others are intra-crustal melts or have sustained variable contamination from continental crust or subducted sediments (e.g., Barnes et al., 1992). Older complexes, such as the ca. 265 Yellow Dog complex, have variable ε_{Nd} values of + 3.2 to + 7.9 (Ernst et al., 1991). Younger complexes, such as the Jurassic to Early Cretaceous plutons, have widely varying characteristics, explained by assimilation-fractionation during cooling. A suite of tonalite-trondheimite plutons have geochemical characteristics consistent with partial melting of underthrust ophiolitic rocks. They have ε_{Nd} values within the mantle array, ranging from + 4.65 to + 6.47, with T_{DM} varying from 0.28 to 0.48 Ga (Barnes et al., 1992), suggesting derivation from mid- to late Paleozoic oceanic lithosphere.

Taken together, the accreted mafic terranes in the Canadian and U.S. Cordillera exemplify terranes accreted at the leading edge of a supercontinent during its dispersal. The similarity of the ε_{Nd} values of these late Paleozoic terranes to contemporary depleted mantle values, and the close correspondence of crystallization and T_{DM} model ages, implies crustal formation and arc activity during the lifespan of Pangea, that is, within the Panthalassa Ocean (Fig. 5B).

Appalachian-Caledonide orogen

Following the amalgamation of Pannotia in the Late Neoproterozoic, ocean formation between Gondwana and Laurentia occurred in several discrete episodes (Fig. 6A; Cawood et al., 2001). A late Neoproterozoic–Early Cambrian rifting event resulted in the formation of the Iapetus Ocean. Subsequently, during the Late Cambrian to Middle

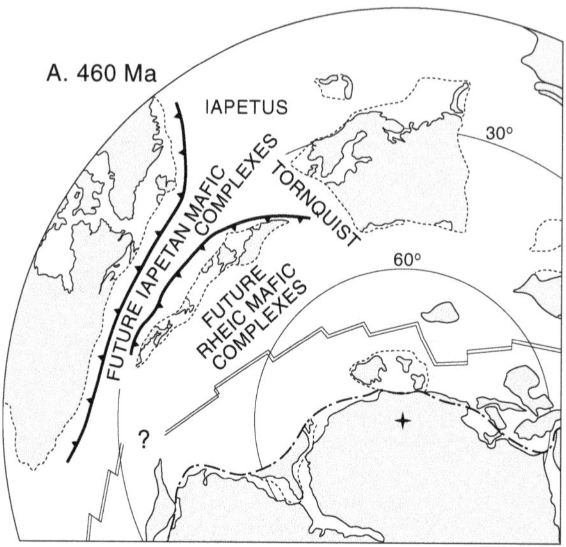

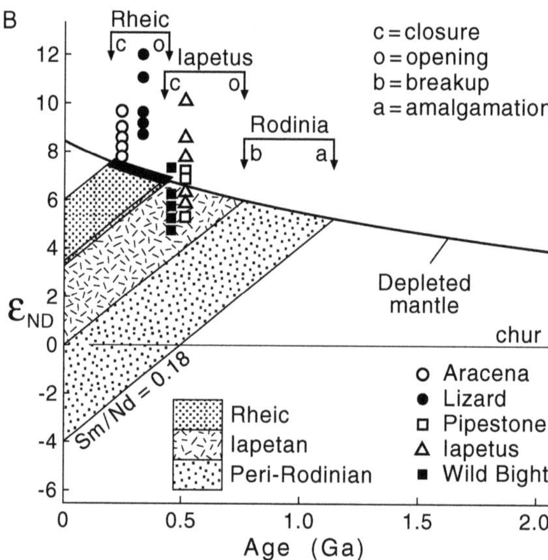

FIG. 6. A. Ordovician paleogeographic reconstruction of the circum-Atlantic region (based on Cocks and Torvik, 2002; Fortey and Cocks, 2003) showing paleogeography of future Iapetan and Rheic ocean mafic complexes. Cambrian rift-related volcanic rocks of Laurentia are detailed by Cawood et al. (2001). B. Diagrams of $(\varepsilon_{Nd})_t$ vs. time (in Ga) showing compilation of Sm-Nd isotope compositions from Iapetan and Rheic ocean complexes in the Paleozoic Appalachian-Caledonide orogen. Field for peri-Rodinian oceanic lithosphere defined by time of amalgamation (a), and breakup (b) of Rodinia. Field for oceanic lithosphere of Iapetus and Rheic oceans are respectively defined between their times of initial opening (o) and closure (c). Isotope evolution of peri-Rodinian, Iapetan, and Rheic oceanic lithosphere calculated by assuming a typical Sm/Nd ratio of 0.18. Data from Camiré et al. (1995), Bedard and Stevenson, (1999), Swinden et al. (1990, 1997), Jenner and Swinden (1993), MacLachlan and Dunning (1998), Pin and Paquette (1997), Nutman et al. (2001), Sandeman et al. (2000), and Castro et al. (1996).

Ordovician, the Rheic Ocean opened when a number of terranes (e.g., Avalonia, Carolina) rifted from the Amazonian–West African margin of Gondwana (e.g., Giese and Buehn, 1994; Prigmore et al., 1997; Keppie and Ramos, 1999). More controversially, a third ocean, known as Paleotethys (e.g., Stampfli and Borel, 2002) or Prototethys (e.g., Winchester et al., 2002), is defined on paleomagnetic grounds as the ocean formed by the rifting of Armorica (Iberia–northwestern France–Bohemia] from the Gondwanan margin in the Silurian. However, its existence does not affect the first-order arguments presented and it is not considered further here.

Closure of the Iapetus Ocean resulted in the collision of the Laurentian margin with Baltica, as well as with Avalonia, Carolina, and other peri-Gondwanan terranes (e.g., van Staal et al., 1998). Closure of the Rheic Ocean produced terminal collision in the Ouachita-Alleghanian-Variscan orogen, which, together with the Uralide orogen formed by closure of the paleo-Uralian ocean, was one of the principal events in the late Paleozoic assembly of the supercontinent Pangea. Inasmuch as the Ouachita-Alleghanian-Variscan orogen was located in the interior of Pangea after supercontinent amalgamation, the orogen is classified as an interior orogen (Murphy and Nance, 1991).

Vestiges of Iapetus. The rift-to-drift and convergence histories of the Iapetus Ocean are well preserved along the eastern margin of North America. Rifting associated with the generation of the Iapetus Ocean produced several late Neoproterozoic to Early Cambrian mafic complexes along the Laurentian margin, including the Long Range dikes of Newfoundland, and the Caldwell Group and Tibbit Hill lavas of southern Quebec (Camiré et al, 1995; Bedard and Stevenson, 1999). Several of these suites show evidence of crustal contamination. For example, the Tibbit Hill volcanics have ε_{Nd} values (t = 550 Ma) ranging from +3.5 to +6.5 that correlate with indicators of crustal contamination such as $^{147}Sm/^{144}Nd$ and negative Nb anomalies. However, the Caldwell Group has N-MORB chemistry, shows little crustal contamination, and has ε_{Nd} values (t = 600 Ma) of +6.9 to +10.0 that are close to, or above, the depleted mantle curve (Fig 6B). These data are interpreted by Bedard and Stevenson (1999) to reflect derivation from depleted asthenosphere and emplacement during the final stages of rifting.

Vestiges of Iapetus are also preserved in the ophiolitic complexes of Newfoundland and Quebec that were obducted during ocean closure (e.g., Williams, 1979). The mafic rocks in these complexes range widely in chemistry and include low-Ti tholeiites, boninites, and suites with alkalic affinities related to the rifting of arc complexes. Many of these suites show evidence of contamination with slab components or subducted sediments, principally reflected in negative correlations with $^{147}Sm/^{144}Nd$ or by Nb* anomalies (e.g., Swinden et al., 1990). However, mafic suites unaffected by these processes have ε_{Nd} values (t = 480 Ma) that are typical of depleted mantle (Fig. 6), and range from about +5.6 to +7.7 (e.g., Swinden et al., 1990, 1997; Jenner and Swinden, 1993; MacLachlan and Dunning, 1998).

Uncontaminated mafic rocks from both the rifting and convergent stages of Iapetus consequently have compositions that lie at or very close to depleted mantle values at their respective times of emplacement, such that their model ages closely match their ages of crystallization, which do not exceed the age of rifting (i.e., $T_{DM} \leq T_R$). This indicates that the oceanic lithospheric source of these suites was generated after the rifting event, and must therefore have formed in an interior ocean (Fig. 4B).

Vestiges of the Rheic Ocean. Vestiges of the Rheic Ocean include rift-related complexes associated with its formation and arc-related high- to low-pressure (eclogite, amphibolite, and greenschist facies) complexes associated with its subduction. These suites, which were widely dispersed by the subsequent breakup of Pangea, are well preserved along the contact between the Central Iberian and Ossa Morena zones in Iberia (e.g., Quesada, 1990, 1997; Sanchez-Garcia et al., 2003), in the Massif Central (Pin and Marini, 1993; Pin and Paquette, 1997), in an isolated occurrence in southern Britain (Lizard Complex; Davies, 1984), and in the Acatlán Complex of Mexico (Keppie and Ramos, 1999; Nance et al., 2004). Unfortunately Sm-Nd isotopic data from these complexes are scarce. The best data for the rifting stage are from the Massif Central, where ε_{Nd} values (t = 480 Ma) show evidence of mixing between a LREE-depleted mantle component with values of ca. +6.0, and a crustal rhyolite with ε_{Nd} of –6.3 (Fig. 4; Pin and Marini, 1993). The mantle source for these basalts is similar to the depleted mantle value at t = 480 (ca. +6.8).

Mafic complexes that record the convergent stage are preserved as high-grade assemblages. In southwestern Spain, amphibolites from the Aracena metamorphic belt have MORB-like geochemical characteristics and ε_{Nd} values (t = 350 Ma) close to or above the depleted mantle curve at + 7.9 to +9.2 (Castro et al., 1996). In the northern Massif Central, ca. 360 Ma mafic bodies have OIB affinities and ε_{Nd} values (t = 360 Ma) ranging from + 6.1 to + 8.0 (Pin and Paquette, 1997). In southern Britain, the ca. 390 Ma Lizard Complex (Nutman et al., 2001; Sandeman et al., 2000) has MORB-like geochemistry and ε_{Nd} values (t = 390 Ma) ranging from + 9.0 to +11.8, again reflecting derivation from depleted mantle of equivalent (i.e., ca. 390 Ma) age (Fig. 6B). As with Iapetus, the correspondence of ε_{Nd} data with contemporary depleted mantle values, and thus the match between T_{DM} and crystallization ages, suggests that the oceanic lithospheric sources of these suites were generated during or after Rheic rifting and, hence, were formed within an interior ocean (Fig. 4B).

Uralides

The paleo-Uralian ocean along the eastern (present coordinates) margin of Baltica is thought to have opened as a result of rifting in the Cambrian–Ordovician (e.g., Leech et al., 2002; Bogolepova, 2003). The Uralides are a late Paleozoic collisional orogen in which the continental margin of Baltica was subducted eastward (current coordinates) beneath the Magnitogorsk and Tagil island arcs in the Late Devonian and Early Carboniferous, leading to accretion of volcanic arcs, oceanic crust, and possible continental fragments (e.g., Brown and Spadea, 1999). This was followed in the Late Carboniferous to Early Permian by terminal collision with the Kazakhstan and Siberian plates and closure of the paleo-Uralian ocean (e.g., Zonenshain et al., 1984, 1990; Puchkov, 1997).

Sm-Nd isotopic data are not currently available for the mafic components of these accreted terranes, but in those showing little crustal contamination, T_{DM} model ages would be expected to be Paleozoic if terrane accretion records the closure of an interior ocean as appears to be the case for the Appalachian-Caledonide orogen.

Application to Neoproterozoic Orogens

As with the Phanerozoic assembly of Pangea, two predominant end-member types of orogenic belts are associated with the Neoproterozoic assembly of Pannotia (e.g., Murphy and Nance, 1991). Orogenic belts whose histories culminate in continental collision (e.g., Borborema, Trans-Sahara) lay in the interior of Pannotia after its assembly, and are classified as interior orogens (Fig. 1C). Orogenic belts that lay along the periphery of Pannotia have histories of subduction and terrane accretion, but do not terminate in or record a terminal continental collision (e.g., Avalonia) and are classified as peripheral orogens (Fig. 1C). Some belts contain elements of collision and terrane accretion, such as the 5000 km long system that includes the Mozambique belt in the south, which reflects the collision between East and West Gondwana, and the Arabian-Nubian shield to the north, which is dominated by accretionary tectonics. Whether or not the collisional belts of Pannotia assembly record the closure of interior oceans, as was the case for the assembly of Pangea, depends on the age of their oceanic lithosphere (T_{DM}) relative to that (T_R) of the breakup of the previous supercontinent, Rodinia, which is dated at ca. 755 Ma (Wingate and Giddings, 2000).

Interior orogens

Neoproterozoic collisional fold belts associated with the amalgamation of Pannotia as exemplified by the Borborema and Tocantins provinces of northeast and central Brazil (Brito Neves, 1983; Teixeira et al., 1989), and the Trans-Saharan orogenic belt of western Africa are examples of interior orogens (Murphy and Nance, 1991). Borborema Province lies between the West African–São Luis and São Francisco–Congo–Kasai cratons (e.g., de Wit et al., 1988), and was deformed and metamorphosed by collision between these cratons during the 500 to 600 Ma Pan-African Brasiliano orogeny. However, several fold belts (including the Sergipano and Oros belts) pre-date the Brasiliano orogeny, whereas Pre-Brasiliano supracrustal sequences have ages of 2.0–1.9, 1.8–1.7, 1.1 to 1.0, and ~ 0.6 Ga. Archean and Paleoproterozoic basement gneisses in the southern part of the province are thought to represent tectonic slivers of the São Francisco craton. Metavolcanic and granitoid units of ca. 1.0 Ga age have ε_{Nd} values of +0.6 to –10.1 and T_{DM} ages that range from 1.17 to 1.77 Ga (van Schmus et al., 1995). These Sm-Nd isotopic characteristics are interpreted to reflect mixing between ancient crust (> 2.1 Ga) and a juvenile (ca. 1.0 Ga) component (Fig. 7A) thought to have originated during a major crust-forming event.

Tocantins Province lies between the Amazon, São Francisco, and possibly the Parana cratons, and likewise consists of a number of fold belts (Fig. 1C; Teixeira et al., 1989; Pimental et al., 1997). The Paraguai belt along the Amazonian margin

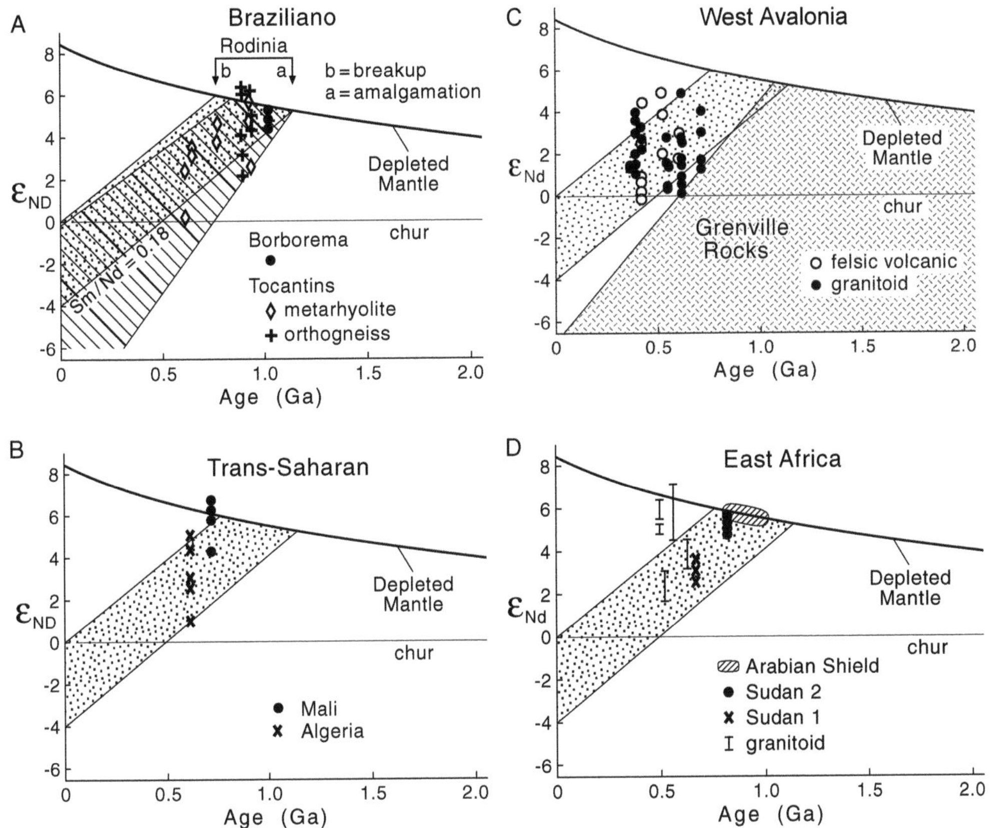

Fig. 7. Summary diagrams of $(\varepsilon_{Nd})_t$ vs. time (Ga) for inferred peri-Rodinian crust preserved in Neoproterozoic orogens. A. Borborema and Tocantins provinces of the Brasiliano orogenic belt, Brazil (Pimental and Fuck, 1992; van Schmus et al., 1995). B. Trans-Saharan belt of West Africa (Mali from Caby et al., 1989; Algeria from Dostal et al., 2002). C. West Avalonia (Murphy et al., 1996, 2000; Murphy and Nance, 2002). D. East African orogens (Duyverman et al., 1982; Stern et al., 1991; Kröner et al., 1992; Küster and Liégeois, 2001; Stern, 2002, Stein, 2003). In each diagram, the evolution of ε_{Nd} with time for peri-Rodinian oceanic lithosphere is shown in stipples and is defined by assuming a depleted mantle composition for the oceanic lithosphere formed between the time of amalgamation (a) and breakup (b) of Rodinia at ca. 1.0 Ga and 0.75 Ga, respectively. The evolution assumes a typical Sm/Nd ratio of 0.18. In (a), the stippled region is the Sm-Nd isotopic envelope for the Brasiliano orogens. In (c), the field for Grenvillian rocks (shaded) is compiled from Dickin and McNutt (1989), Patchett and Ruiz (1989), Marcantonio et al. (1990), Daly and McLelland (1991), Dickin et al. (1990), and Dickin (1998, 2000).

predominantly consists of Neoproterozoic and Eocambrian platformal metasedimentary rocks. Along the western flank of the São Francisco craton, the Uruaña and Brasilia belts expose ca. 800 Ma polydeformed supracrustal rocks that contain terranes with ophiolitic affinities (Pimental et al., 1991). The Goias Massif, in the central portion of the province, is characterized by metavolcanic and metasedimentary assemblages that are juxtaposed against orthogneisses along steeply dipping mylonite zones (Pimental and Fuck, 1987). The massif is dominated by ca. 950–850 Ma mafic igneous protoliths that evolved within a primitive ensimatic arc-setting, and younger (ca. 760–600 Ma) arc-related rocks of more felsic composition (Pimental and Fuck, 1992; Pimental et al., 1997). Sm-Nd isotopic constraints for this province suggest derivation from juvenile sources. Initial ε_{Nd} values (calculated for the age of crystallization) range from +0.2 to +6.9, and T_{DM} depleted mantle model ages lie between 0.9 and 1.2 Ga (Pimental and Fuck, 1992; Fig. 7A).

The Trans-Saharan orogenic belt (Fig. 1C; e.g., Cahen et al., 1984; Caby et al., 1989; Black et al., 1994; Dostal et al., 1994) was formed by collision between the passive margin of the West African craton and the active eastern margin of the East Saharan craton. The Tuareg shield, which lies between these cratons, consists of a number of displaced arc-related terranes with variable tectonostratigraphic, metamorphic, and magmatic characteristics. Separated by subvertical shear zones, these terranes were accreted to the East Saharan craton at various times between 750 and 550 Ma (Black et al., 1994). Thus, the Trans-Saharan belt sequentially underwent arc-arc, arc-continent, and continent-continent collisional orogenesis in the Neoproterozoic.

In contrast to the West African craton, the Tuareg shield contains Neoproterozoic igneous rocks, indicating that convergence between these cratons was accompanied by easterly directed subduction (Caby, 1994). In Mali, a ca. 720 Ma complex of intra-oceanic arcs, known as the Tilemsi belt, formed above such an east-dipping subduction zone, and was accreted to the continental margin of West Africa during Pan-African collisional orogenesis. The arcs are characterized by mafic to intermediate volcanic and plutonic rocks with calc-alkalic and island arc tholeiitic affinities (Dostal et al., 1994). The volcanics are interbedded with metagreywackes. Available Sm-Nd isotopic data indicate that the plutonic rocks and metagreywackes have high ε_{Nd} values (+6.3 to +6.6 and +4.4 to +5.8 at t = 730 Ma) with model ages of 710 to 760 Ma and 840 to 940 Ma, respectively. These data indicate that the plutonic rocks were derived from depleted mantle, whereas the metagreywackes, although predominantly derived from juvenile crust, had a limited contribution from sialic crust (Caby et al., 1989). To the north, in southwestern Algeria and southern Morocco, inliers of Trans-Saharan mafic rocks include shoshonites with arc-related geochemical signatures and high ε_{Nd} values (+1.0 to +5.0) that yield model ages of 950 Ma to 1.20 Ga (Dostal et al., 2002; Fig. 7B).

On the basis of these data, each of these interior orogens can be shown to contain accreted mafic complexes with T_{DM} model ages that exceed the ca. 755 Ma breakup age of Rodinia (i.e. $T_{DM}>T_R$; Fig. 4B). This suggests that the oceanic lithospheric sources of these suites were generated prior to the breakup of Rodinia and, hence, were formed in the exterior, peri-Rodinian ocean.

Peripheral orogens

On the basis of faunal, lithostratigraphic, geochemical, and paleomagnetic data, a group of terranes collectively known as peri-Gondwanan are traditionally interpreted to represent segments of the northern (Amazonian and West African) Gondwanan margin of Pannotia during the Neoproterozoic and early Paleozoic (Fig. 1C; Theokritoff, 1979; Cocks and Fortey, 1982, 1990; Keppie, 1985; van der Voo, 1988; Murphy and Nance, 1989; McNamara et al., 2001, Mac Niocaill et al., 2002; Murphy et al., 2002; Keppie et al., 2003). Each of these terranes is characterized by Neoproterozoic magmatism that records a history of subduction beneath the Gondwanan continental margin (e.g., O'Brien et al., 1983; Keppie, 1985; Murphy et al., 1990; Nance et al., 1991; Egal et al., 1996; Linnemann et al., 2000; Linnemann and Romer, 2002; von Raumer et al., 2002).

Some of these terranes (e.g., Avalonia-Carolina) rifted from Gondwana in the early Paleozoic to form the Rheic Ocean (Fig. 6A), and were subsequently involved in Paleozoic orogenesis; thus they are now preserved as a collection of suspect terranes in the Caledonian-Appalachian orogen (O'Brien et al., 1983; Murphy and Nance, 1989, 1991; Keppie et al., 1991; Tucker and Pharoah, 1991; Gibbons and Horák, 1996; Hibbard et al., 2002). The Paleozoic history of other peri-Gondwanan terranes is less clear. Paleomagnetic data have been interpreted to reflect the Silurian rifting of Cadomia (Iberia-Armorica-Bohemia) from Gondwana with the opening of Paleotethys (e.g., Stampfli and Borel, 2002) or Prototethys (e.g., Winchester et al., 2002), and its subsequent collision with Baltica during the Devonian. Another view is that these terranes remained attached to the West African portion of the Gondwanan margin until they were involved in collisional orogenesis between Gondwana and Baltica in the Devono-Carboniferous (e.g., Robardet, 2003).

Neoproterozoic tectonothermal activity in some of these peri-Gondwanan terranes begins with an early phase of ca. 760–660 Ma arc-related magmatism and is followed in West Avalonia (e.g., Krogh et al., 1988; Swinden and Hunt, 1991; Bevier et al., 1993; Doig et al, 1993; O'Brien et al., 1996), East Avalonia (Gibbons and Horák, 1996; Tucker and Pharoah, 1991, Strachan et al., 1996), and Cadomia (Egal et al., 1996) by ca. 660–650 Ma medium to high-grade metamorphism. The early arc-related complexes are thought to have formed outboard of the Gondwanan margin, and metamorphism is interpreted to have occurred during their subsequent accretion (Murphy et al., 2000; Nance et al., 2002).

At 635 Ma, the main phase of arc-related activity commenced broadly synchronously along the Gondwanan margin. Arc-related rocks typically include voluminous calc-alkaline mafic to felsic volcanics, coeval plutons, and pull-apart sedimentary basin deposits that contain detritus derived from the arc. The termination of subduction, on the other hand, occurred diachronously between 590 Ma and 540 Ma, and was accompanied by a transition to an intracontinental wrench regime. This reorganization is interpreted to record ridge-trench collision in a manner analogous to the Oligocene collision between western North America and the East Pacific Rise, which led to the diachronous initiation of the San Andreas transform margin (Murphy et al., 1999; Nance et al., 2002).

Sm-Nd isotopic data from crustally derived rocks in both the early arc and main arc phases of West and East Avalonia yield similar ranges in initial ε_{Nd} values (–2.5 to +5.0) and T_{DM} model ages of 0.8–1.2 Ga in Atlantic Canada and 1.0–1.3 Ga in the British Midlands (Murphy et al., 1996; 2000; Fig 7C). Initial ε_{Nd} values of +0.5 to +5.9 and T_{DM} model ages of 0.7–1.1 Ga also have been reported from the main-phase Virgilina sequence of Carolina (Samson et al., 1995; Wortman et al., 2000).

Inasmuch as no record of tectonothermal activity exists in Avalonia or Carolina between the oldest T_{DM} ages of ca. 1.2 Ga and the onset of early-phase arc magmatism at ca. 760 Ma, the isotopic signature of these terranes is considered to represent a relatively simple evolution involving recycling of ca. 1.0 to 1.2 Ga crust. If so, the depleted-mantle model ages represent a genuine tectonothermal event that produced juvenile crust that was recycled by subsequent Neoproterozoic tectonothermal activity (Murphy et al., 2000). These data are therefore interpreted to reflect the evolution of primitive Avalonian and Carolinian crust (proto-Avalonia-Carolina) at 1.2–1.0 Ga in the peri-Rodinian ocean, followed by the development of mature oceanic arcs at 750–650 Ma (early Avalonian magmatism). The accretion of these oceanic arcs to Gondwana at ca. 650 Ma was then followed by Andean-style arc development at 635–570 Ma (main Avalonian-Carolinian magmatism), which was predominantly generated by recycling pre-existing ca. 1.2–1.0 Ga crust.

Sm-Nd isotopic data for Cadomia, however, suggest that the relationship of this terrane with respect to Gondwana was different. ε_{Nd} values from main-phase arc rocks in northern France and the Channel Islands are predominantly negative (+1.9 to –9.9), and yield T_{DM} model ages ranging from 1.0 to 1.9 Ga. These data are interpreted to reflect the mixing of material derived from the mantle at ca. 600 Ma with continental basement represented by the 2.1 Ga Icartian Gneiss (D'Lemos and Brown, 1993; Nance and Murphy, 1994, 1996). In contrast to Avalonia and Carolina, evidence of the significant influence of a juvenile crustal component in Cadomia is lacking. Instead, Cadomia appears to have originated as an ensialic arc built upon crust like that of the West African craton.

Margins with interior and peripheral characteristics

Stretching for 5000 km along the East African margin, a succession of Neoproterozoic orogenic belts, ranging from ca. 950 to 550 Ma in age, show a complete spectrum of the varying styles of Pan African orogenesis (Kröner, 1984; Stern, 1994). To the north, the Arabian-Nubian shield (Fig. 1C) consists of a collage of ca. 850–650 Ma arcs and Paleoproterozoic–late Archean continental microplates that accreted at various times during the Pan-African orogeny (Kröner, 1985; Blasband et al., 2000). Although models vary in detail, accretion appears to have occurred in a number of phases, including the closure of fore-arc (Greenwood et al., 1980) or back-arc (Kröner et al., 1987) basins, followed by the docking of younger arcs to the shield's eastern margin (present coordinates). Arc terranes are separated by ophiolitic complexes that represent sutures between the oceanic microplates (Kröner, 1985; Pallister et al., 1988; Kröner et al., 1992, 2000). In a review of crustal evolution in the East African Orogen, Stern (2002) pointed out that crystallization ages and T_{DM} model ages for the Arabian shield both show a tight clustering around 750 Ma (Fig. 7D), suggesting that these complexes are vestiges of juvenile lithosphere. Several ca. 700–550 Ma calc-alkaline volcanics and granites produced during Pan-African orogenesis have similar isotopic characteristics that are consistent with recycling of this juvenile lithosphere during late Neoproterozoic subduction (Duyverman et al., 1982; Stein, 2003).

To the south, the Mozambique belt is a zone of highly deformed and metamorphosed rocks that have been variously interpreted to be predominantly juvenile, or to be reworked Archean crust that potentially forms the basement to the Arabian-

Nubian shield (Burke and Şengör, 1986; Shackleton, 1986). However, the presence of highly dismembered ophiolites in suture zones that can be traced southward from the Arabian-Nubian Shield (e.g., Behre, 1990), suggests that the Mozambique belt was generated by collisional tectonics (Stern and Dawoud, 1991).

A transition zone between the Arabian-Nubian shield and the Mozambique belt occurs in Yemen (Windley et al., 1996; Whitehouse et al., 2001), Somalia (Lenoir et al., 1994), and Sudan (Stern and Dawoud, 1991; Küster and Liégeois, 2001). In Yemen, microcontinental blocks consisting of late Archean and Proterozoic gneisses are separated by obducted island arc complexes. To the south, in Somalia, comparable island arc terranes contain Neoproterozoic and early Paleozoic granitoids, some of which were derived from juvenile crust (Lenoir et al., 1994).

Geochemical and isotopic data from voluminous Neoproterozoic, predominantly mafic volcanic and interbedded metasedimentary rocks in the Bayuda Desert of Sudan (Küster and Liégeois, 2001) record an oceanic island arc or backarc environment. These rocks were metamorphosed at about 700 Ma during an event attributed to collision with the East Saharan craton. The Bayuda sequence has two components, a juvenile ca. 800 Ma mafic and sedimentary sequence characterized by high ε_{Nd} values (+3.6 to +5.2 at t = 800 Ma) and T_{DM} ages of 0.78 to 0.90 Ga (Fig. 7D), and a metasedimentary sequence derived from an older continental source with T_{DM} ages of about 2.1 Ga. To the south, ca. 740 Ma granulites and charnockites are interpreted to have formed as a result of collisional orogenesis followed by exhumation (Behre, 1990). Available ε_{Nd} data for these high-grade rocks (Stern et al., 1991) range from +2.9 to +3.4 (t = 740 Ma) and yield T_{DM} ages of 0.96 to 0.98 Ga, suggesting derivation from a juvenile mantle source. As with the other interior orogens, the existence of mafic complexes with T_{DM} model ages that predate the breakup of Rodinia (i.e., $T_{DM}>T_R$) (Fig. 4B) suggests that the ocean whose closure resulted in orogenesis was part of the exterior, peri-Rodinian ocean.

Application to Older Orogens

Application of this approach to older supercontinents is hindered by uncertainties in paleocontinental reconstructions. But while global-scale interpretations are consequently difficult, there are examples where the approach yields testable hypotheses. Such is the case for mafic complexes accreted to the Superior craton of North America in the Paleoproterozoic and Mesoproterozoic. These occur in each of two orogens, the Grenville belt to the east and the Trans-Hudson belt to the west.

Of the Mesoproterozoic continental collisions (ca. 1.2 to 1.0 Ga) that resulted in the formation of the supercontinent Rodinia, the Grenville belt of eastern North America is arguably the best studied. It consists of SE-dipping stacked allochthons of Neoarchean, Paleoproterozoic, and Mesoproterozoic rocks (Ludden and Hynes, 2000; White et al., 2000) on the eastern flank the Superior craton. The Grenville terminal collision was associated with the generation of a wide range of igneous complexes that have geochemical and isotopic signatures consistent with recycling of ancient continental crust. For example, Sm-Nd isotopic signatures are overwhelmingly negative with T_{DM} model ages ranging up to 2.7 Ga. These model ages are predominantly mixing ages and have no geological meaning. However, a number of accretionary orogenic events (the 1.9 to 1.6 Ga Penokean, the 1.68–1.67 Ga Labradorian, the 1.50–1.45 Ga Pinwarian, and the 1.25–1.19 Ga Elzeverian) occurred along the eastern flank of the Superior province (Fig. 8A) prior to ca. 1.19 and 0.98 Ga terminal continental collision (Rivers and Corrigan, 2000), suggesting that the Superior craton and its accreted terranes faced an open ocean to the east (present coordinates) for nearly 0.8 Ga.

The oldest of these events, the 1.9–1.6 Penokean orogeny, is attributed to the accretion of predominantly mafic arc complexes (Dickin and McNutt, 1989; Dickin, 2000), whereas the younger orogens are attributed to the accretion of ensialic arcs and backarc basins (e.g., Gower et al., 1990; Gower, 1996; Rivers and Corrigan, 2000). According to Dickin (1998, 2000), the Penokean and Labradorian arcs are juvenile and did not incorporate significant crustal material. They have crystallization ages of 1.74 Ga, ε_{Nd} values of + 1 to + 3, and T_{DM} model ages of ca. 1.9 Ga (Fig. 8B). Similar Sm-Nd isotopic signatures have been documented in Scandinavia, Greenland, Scotland, and Colorado (DePaolo, 1981; Patchett and Bridgewater, 1984; Patchett and Kuovo, 1986; Marcantonio et al., 1990), and are interpreted by Dickin and McNutt (1989) to reflect large-scale accretion of arc material in response to a change in plate geometry. Given a breakup age for the supercontinent (Nuna/Columbia) that preceded Rodinia of ca. 1.4–1.6 Ga (Rogers and Santosh,

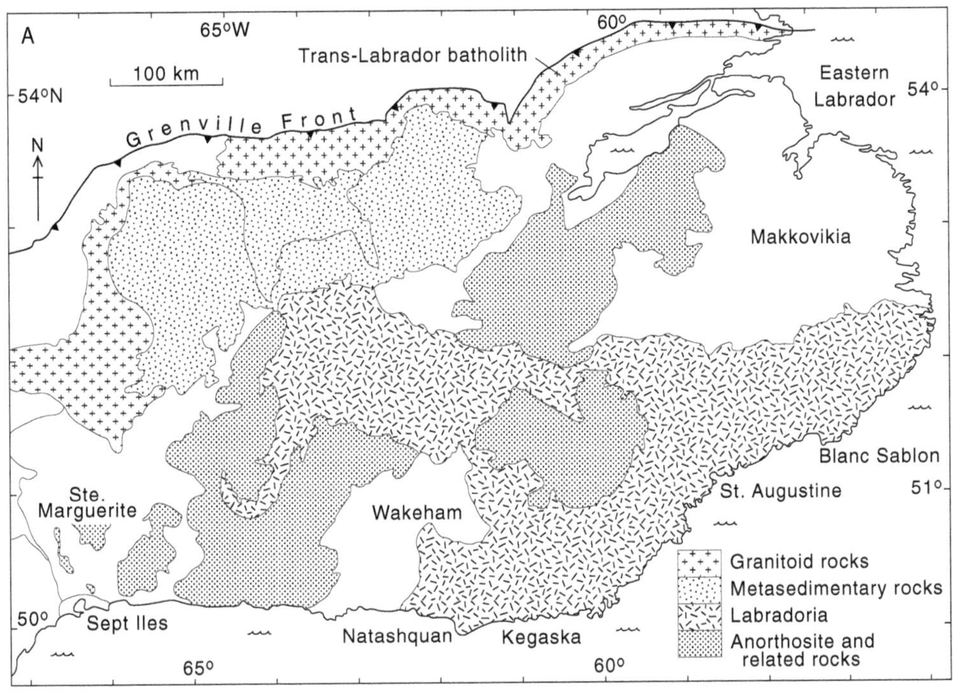

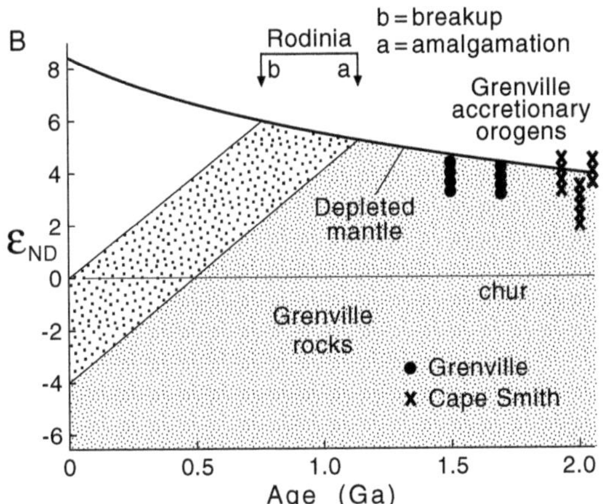

Fig. 8. A. Crustal formation map (after Dickin, 2000) for Makkovia and Labradoria within the Grenville belt. Distribution of anothosites and related rocks after Davidson (1998). According to Dickin (2000), the Wakeham terrane is underlain by 1.68–1.67 Ga Labradorian crust that was reworked in the 1.50–1.45 Pinwarian. B. Summary diagrams of $(\varepsilon_{Nd})_t$ vs. time (Ga) for inferred peri-Rodinian crust preserved in Proterozoic orogens. The evolution of ε_{Nd} with time for peri-Rodinian oceanic lithosphere is shown in stipples (see Fig. 7). The shaded area shows the evolution of ε_{Nd} with time of the oceanic lithosphere prior to the amalgamation of Rodinia. Cape Smith data from Scott et al. (1992), Dunphy et al. (1995), and Thériault et al. (2001). Grenville data from Dickin and McNutt (1989) and Dickin (1998, 2000).

2002), these model ages and the long time interval over which the eastern flank of the Superior craton faced an open ocean, suggest that the Grenville belt records the closure of an exterior ocean.

On the northern and western margins of the Superior province, the Trans-Hudson orogen consists of Archean (3.22–2.7 Ga) autochthonous basement gneisses, an imbricated 2.04–1.92 Ga rift-drift sequence (St. Onge et al., 2000), a 2.0 Ga ophiolite (Watts Group; Scott et al., 1992), and a 1.86–1.82 Ga magmatic arc, which reflects the convergence that culminated in continental collision (Dunphy and Ludden, 1998). The Watts ophiolite has two distinct suites, an older one with MORB-like geochemistry and ε_{Nd} values (t = 2.0 Ga) of +4.0 to + 5.3 (Fig 8C), similar to depleted mantle values, and a younger one derived from a more enriched source with OIB chemistry and ε_{Nd} values (t = 2.0 Ga) of +2.5 to +3.6 (Scott et al., 1992; Dunphy et al., 1995; Thériault et al., 2001). According to St.-Onge et al. (2000, 2001), the Trans-Hudson and correlative Ungava orogens represent a Wilson cycle in which 3.2 to 2.74 Ga Superior basement was rifted between 2.04 and 1.92 Ga to form a passive margin flanking oceanic crust. Arc development beginning at 1.86 Ga represents the onset of convergence (Dunphy and Ludden, 1998) with terminal collision at about 1.8 Ga. The Watts and Purtuniq ophiolites are interpreted as vestiges of the oceanic crust involved in the collision. Because the depleted mantle model ages of the mafic complexes are similar to their crystallization ages and are coeval or younger than the rifting event (i.e., $T_{DM} \leq T_R$; Fig. 4B), the ocean whose closure the Trans-Hudson orogen records was likely to have been an interior ocean.

Interestingly, rifting along the western Superior cratonic margin was coeval with accretion of mafic terranes along its eastern margin in a manner analogous to the Cenozoic breakup of Pangea and the accretion of mafic terranes in the North American Cordillera. We speculate that a similar geodynamic relationship may have existed for the Superior Province in the Paleoproterozoic.

Discussion

Taken together, the examples selected represent various stages and locations of supercontinent assembly, amalgamation, and dispersal. Because the age contrast between interior and exterior oceans diminishes as the continents drift apart, the ages of the earliest accreted complexes are the most diagnostic of the ocean in which they formed.

Accretion of terranes in the North American Cordillera began in the late Paleozoic and continued during the breakup of Pangea and the westerly drift of North America. Thus the North American Cordillera is an example of terrane accretion at the leading edge of a continent during supercontinent amalgamation, breakup, and dispersal. Accreted late Paleozoic and early Mesozoic mafic terranes (Klamath Mountains, and correlatives) represent vestiges of the Panthalassa Ocean that surrounded Pangea. Although accreted Mesozoic–Cenozoic mafic terranes are complexes that formed after the breakup of Pangea, those complexes with Paleozoic T_{DM} ages were developed from Panthalassa oceanic lithosphere, so their origin can be traced back to the exterior ocean.

As modern examples of exterior ocean terranes accreted to the leading edge of a dispersing continent following supercontinent breakup, the origin and age distribution of those in the North American Cordillera are instructive in interpreting similar environments created during the dispersal of previous supercontinents. First, inasmuch as oceanic lithosphere is being continuously created, accretion of terranes with $T_{DM} > T_R$ will be succeeded by terranes with $T_{DM} < T_R$ as accretion proceeds. Should this accretionary continental margin eventually become involved in collisional orogenesis, the latter category of terranes will have isotopic characteristics similar to those of an interior ocean. This implies that a range of mafic complexes must be studied before the oceanic domain (exterior or interior) can be identified. It is conceivable that such convergent continental margins may not become involved in terminal collisional orogenesis for some time, and that during this interval significant changes may occur in the tectonic regime within the oceanic domain outboard from the margin. Changes in the tectonic regime should be suspected where a significant time gap exists in the tectonothermal evolution of a margin between its histories of accretion and terminal collision. This is the case in the Cordillera where, after >200 Ma of accretionary tectonics, there remains little to suggest any impending terminal continental collision.

Accreted mafic terranes in the Paleozoic Appalachian-Caledonide-Variscan belt represent stages of convergence that led to the eventual amalgamation of Pangea. Avalonia and related peri-Gondwanan terranes document subduction processes along the periphery of Pannotia during and after its assembly. The available data from the Appalachian-Caledonide belt indicate that mafic complexes representing vestiges of the oceans that were consumed to produce Pangea are younger than the Neoproterozoic–Early Cambrian rifting of Laurentia from Gondwana (i.e., $T_{DM} < T_R$). It appears, therefore, that neither vestiges of pre-Iapetan oceanic lithosphere nor mafic complexes recycled from that lithosphere are preserved in the Appalachian-Caledonide orogen. The lack of pre-Iapetan oceanic lithosphere implies that the consumption of Iapetus is entirely an introversion event (see Collins, 2003), and that Avalonia was tectonically isolated from the pre-Iapetan oceanic lithosphere that would have lain outboard of the terrane in the Late Neoproterozoic and Early Cambrian. This scenario is consistent with the predominantly clockwise rotation of Laurentia relative to Gondwana implied during the breakup of Pannotia by most continental reconstructions. Assuming Gondwana behaved as a coherent entity during the Cambrian, the clockwise rotation of Laurentia requires that half-spreading rates in the southern Iapetus Ocean were higher than those in the north.

Mafic terranes in the Boborema-Tocantins of Brazil, the Trans-Saharan belts of West Africa, and the Mozambique belt of East Africa were accreted in orogens that terminated in collisional orogenesis and in the formation of Pannotia. These terranes have Sm-Nd T_{DM} model ages between ca. 1.2 and 0.71 Ga, implying that much of the oceanic lithosphere that was subducted and recycled to yield these complexes was formed *before* the ca. 755 Ma breakup of the 1.1–0.75 Ga supercontinent Rodinia (i.e., $T_{DM}>T_R$). Hence, these mafic complexes could not have been derived from oceanic lithosphere formed after the breakup of Rodinia. Instead, they are vestiges of oceanic lithosphere that formed within the peri-Rodinian ocean, and hence represent fragments of an exterior ocean. Data for the amalgamation of Pannotia are therefore consistent with extroversion. Unlike Pangea, whose interior orogens were formed by the subduction of interior oceans, Pannotia's interior orogens were formed by subduction of an exterior ocean. This analysis indicates that Pangea and Pannotia were formed by geodynamically distinct processes. Consequently, the origin of the supercontinent cycle must be more complex than any previous model has considered.

The late Neoproterozoic evolution of the Arabian-Nubian shield and the peri-Gondwanan terranes are examples of subduction and acccretionary tectonics along the periphery of Pannotia during its assembly. Sm-Nd isotopic signatures of peri-Gondwanan terranes distinguish Neoproterozoic arc terranes derived from recycling juvenile crust (e.g., Avalonia, Carolina) from those that recycle continental crust (e.g., Cadomia). Sm-Nd data further suggest that the lithosphere involved or recycled in the Arabian-Nubian shield and Avalonia-Carolina represents vestiges of 0.75 to 1.1 Ga oceanic lithosphere. This lithosphere must therefore have coexisted with the supercontinent Rodinia, and so must have been part of the Panthalassa-type ocean surrounding Rodinia.

Just as the amalgamation and dispersal of Pangea exerted a first-order influence on tectonothermal events in the Phanerozoic, global-scale tectonics during the middle to late Proterozoic were profoundly influenced by the amalgamation of the supercontinent Rodinia at ca. 1.2–1.0 Ga, its subsequent breakup beginning at 0.75 Ga, and the amalgamation of Pannotia at 0.65 to 0.55 Ga. Because magmatism associated with Grenvillian collisional orogenesis is predominantly derived from recycled crust, T_{DM} model ages are generally significantly older than the orogenic event itself (e.g., Daly and McLelland, 1991; Dickin, 2000). During the lifespan of Rodinia, however, ocean ridge processes, ensimatic subduction, oceanic plateau generation, and plume activity resulted in the formation of juvenile oceanic crust with ca. 1.2–0.75 Ga Sm-Nd depleted mantle (T_{DM}) model ages in the peri-Rodinian ocean surrounding this supercontinent. Vestiges of this crust are preserved in Late Neoproterozoic orogenic belts such as Avalonia-Carolina and the Arabian-Nubian shield as terranes that were accreted to the leading margins of continents following the breakup and dispersal of Rodinia, and in Neoproterozoic collisional (interior) orogens such as in the Borborema and Tocantins provinces of Brazil, and West Africa in which arc accretion preceded terminal collision, which would have stabilized and cratonized this crust.

A schematic paleocontinental reconstruction of Rodinia at ca. 0.8 Ga, showing the inferred location of 1.0–0.8 Ga juvenile crust generated in the peri-Rodinian ocean, is shown in Figure 1C. For collisional orogens, vestiges of juvenile crust are shown between the converging cratons. For accretionary orogens, the juvenile crust is shown adjacent to the craton to which it accreted. Vestiges of the peri-Rodinian oceanic crust can be identified by: (1) crystallization and T_{DM} model ages that coincide with or pre-date Rodinia's existence; and (2) strongly positive (i.e., mantle-derived) ε_{Nd} isotopic signatures.

During the late Neoproterozoic, portions of the northern Gondwanan margin of Pannotia evolved from western to eastern Pacific–style subduction (e.g., Avalonia, Carolina). Although previously accreted juvenile crust was recycled by subsequent orogenic activity, its existence can be identified from the ε_{Nd} signatures in younger, crustally derived igneous rocks that indicate a crustal source that was itself extracted from the mantle while Rodinia existed as a coherent supercontinent.

Taken together, the formation and Neoproterozoic evolution of peri-Rodinian oceanic lithosphere is geodynamically linked to the amalgamation and breakup of Rodinia. Thus it provides constraints on the configuration of Rodinia, and the timing of the major tectonothermal events involved in its growth and subsequent breakup (e.g., Murphy et al., 2000).

Acknowledgments

We acknowledge the continuing support of the Natural Sciences and Engineering Research Council of Canada (Murphy) and National Science Foundation grant EAR 0308105 (Nance), and the thoughtful, constructive reviews by Steve Johnson and Cecilio Quesada. Contribution to IGCP (International Geological Correlation Programme) Projects 453 and 497.

REFERENCES

Arndt, N. T., and Goldstein, S. L., 1987, Use and abuse of crust formation ages: Geology, v. 15, p. 893–895.

Barnes, C. G., Petersen, S. W., Kistler, R. W., Prestvik, T., and Sundvoll, B., 1992, Tectonic implications of isotopic variation among Jurassic and Early Cretaceous plutons, Klamath Mountains: Geological Society of America Bulletin, v. 104, p. 117–126.

Bedard, J. H., and Stevenson, R., 1999, The Caldwell Group lavas of southern Quebec: MORB-like tholeiites associated with the opening of the Iapetus Ocean: Canadian Journal of Earth Sciences, v. 36, p. 999–1019.

Behre, S. M., 1990, Ophiolites in northeast and East Africa: Implications for Proterozoic crustal growth: Journal of the Geological Society, London, v. 147, p. 41–57.

Bevier, M. L., Barr, S. M., White, C. E., and Macdonald, A. S., 1993, U-Pb geochronologic constraints on the volcanic evolution of the Mira (Avalon) terrane, southeastern Cape Breton Island, Nova Scotia: Canadian Journal of Earth Sciences, v. 30, p. 1–10.

Black, R., Latouche, L., Liegeois, J. P., Caby, R., and Bertrand, J. M., 1994, Pan-African displaced terranes of the Tuareg shield (central Sahara): Geology, v. 22, p. 641–644.

Blasband, B., White, S., Brooijmans, P., De Boorder, H., and Visser, W., 2000, Late Proterozoic extensional collapse in the Arabian-Nubian Shield: Journal of the Geological Society, London, v. 157, p. 615–628.

Bogolepova, O. K., 2003, The post-Timanian evolution of the northeastern and eastern margin of Baltica: Facies and biogeography—a review, *in* Albanesi, G. L., Beresi, M. S., and Peralta, S. H., eds., Ordovician from the Andes. Instituto Superior de Correlación Geológica (INSUGEO), Serie Correlación Geológica, no. 17.

Brito Neves, B. B., 1983, O mapa geológico de Nordeste oriental do Brasil, escala 1/1000000. Unpublished livre docencia thesis, Instituto de Geosciencias, University de São Paulo, 177 p.

Brown, D., and Spadea, P., 1999, Processes of forearc and accretionary complex formation during arc-continent collision in the southern Urals: Geology, v. 27, p. 649–652.

Burke, K., and Şengör, A. M. C., 1986, Tectonic escape in the evolution of the continental crust: American Geophysical Union, Geodynamic Series, v. 14, p. 41–53.

Caby, R., 1994, First record of Precambrian coesite from northern Mali: Implications for Late Proterozoic plate tectonics around the West African craton: European Journal of Mineralogy, v. 6, p. 235–244.

Caby, R., Andreopoulos-Renaud, U., and Pin, C., 1989, Late Proterozoic arc/continent and continent/continent collision in the Trans-Saharan belt: Canadian Journal of Earth Sciences, v. 26, p. 1136–1146.

Cahen, L., Snelling, N. J., Delhal, J., and Vail, J. R., 1984, The geochronology and evolution of Africa: Oxford, UK, Clarendon Press, 512 p.

Castro, A., Fernandez, C., de la Rosa, J., Moreno-Ventas, I., and Rogers, G., 1996, Significance of MORB-derived amphibolites from the Aracena metamorphic belt, southwest Spain: Journal of Petrology, v. 37, p. 235–260.

Camiré, G., Laflèche, M. R., and Jenner, G. A., 1995, Geochemistry of pre-Taconian mafic volcanism in the Humber Zone of the northern Appalachians: Chemical Geology, v. 119, p. 55–77.

Cawood, P. A., McCausland, P. J. A., and Dunning, G. R., 2001, Opening Iapetus: Constraints from the Laurentian margin in Newfoundland: Geological Society of America Bulletin, v. 113, p. 443–453.

Cocks, L. R. M., and Fortey, R. A., 1982, Faunal evidence for oceanic separations in the Palaeozoic of Britain: Journal of the Geological Society, London, v. 139, p. 465–478.

Cocks, L. R. M., and Fortey, R. A., 1990, Biogeography of Ordovician and Silurian faunas, *in* McKerrow, W. S., and Scotese, C. R., eds., Paleozoic paleogeography and biogeography: Geological Society of London Memoir 12, p. 97–104.

Cocks, L. R. M., McKerrow, W. S., and van Staal, C. R., 1997, The margins of Avalonia: Geological Magazine, v. 134, p. 627–636.

Cocks, L. R. M., and Torsvik, T. H., 2002, Earth geography from 500 to 400 million years ago: A faunal and palaeomagnetic review: Journal of the Geological Society, London, v. 159, p. 631–644.

Collins, W. J., 2003, Slab pull, mantle convection, and Pangaean assembly and dispersal: Earth and Planetary Science Letters, v. 205, p. 225–237.

Coney, P. J., Jones, D. L., and Monger, J. W. H., 1980, Cordilleran suspect terranes: Nature, v. 288, p. 329–333.

Condie, K., 1998, Episodic continental growth and supercontinents: A mantle avalanche connection: Earth and Planetary Science Letters, v. 163, p. 97–108.

Daly, J. S., and McLelland, J. M., 1991, Juvenile middle Proterozoic crust in the Adirondack Highlands, Grenville Province, northeastern North America: Geology, v. 19, p. 119–122.

Dalziel, I., 1992, On the organization of American plates in the Neoproterozoic and the breakout of Laurentia: GSA Today, v. 2, no. 11, p. 238–241.

Dalziel, I., 1997, Overview: Neoproterozoic–Paleozoic geography and tectonics: Review, hypotheses, and environmental speculations: Geological Society of America Bulletin, v. 109, p. 16–42.

Dalziel, I. W. D., Mosher, S., and Gahagan, L. M., 2000, Laurentia–Kalahari collision and the assembly of Rodinia: Journal of Geology, v. 108, p. 499–513.

Davidson, A., 1998, Geological map of the Grenville Province, Canada and adjacent parts of the United States of America: Geological Survey of Canada, Map 1947A, scale 1:2,000,000.

Davies, G. R., 1984, Isotopic evolution of the Lizard Complex: Journal of the Geological Society, London, v. 141, p. 3–14.

DePaolo, D. J., 1981, Neodymium isotopes in the Colorado Front range and crust-mantle evolution in the Proterozoic: Nature, v. 291, p. 193–196.

DePaolo, D. J., 1988, Neodymium isotope geochemistry: An introduction: New York, NY, Springer-Verlag, 187 p.

de Wit, M., Jeffrey, M., Bergh, H., and Nicolaysen, L., 1988, Geological map of sectors of Gondwana reconstructed to their disposition ca. 150 Ma: Tulsa OK, American Association of Petroleum Geologists, scale 1:10,000,000.

Dewey, J. F., and Bird, J. M., 1971, Origin and emplacement of the ophiolites suite: Appalachian ophiolites in Newfoundland: Journal of Geophysical Research, v. 76, p. 3179–3206.

Dickin, A. P., 1998, Nd isotopic mapping of a cryptic continental suture, Grenville Province of Ontario: Precambrian Research, v. 91, p. 445–454.

Dickin, A. P., 2000, Crustal formation in the Grenville province: Nd isotopic evidence: Canadian Journal of Earth Sciences, v. 37, p. 165–181.

Dickin, A. P., and McNutt, R. H., 1989, Nd model age mapping of the southeast margin of the Archean foreland in the Grenville Province of Ontario: Geology, v. 17, p. 299–302.

Dickin, A. P., McNutt, R. H., and Clifford, P. M., 1990, A neodymium isotope study of plutons near the Grenville Front in Ontario, Canada: Chemical Geology, v. 83, p. 315–324.

D'Lemos, R. S., and Brown, M., 1993, Sm-Nd isotope characteristics of late Cadomian granite magmatism in northern France and the Channel Islands: Geological Magazine, v. 130, p. 797–804.

Doig, R., Murphy, J. B., and Nance, R. D., 1993, Tectonic significance of the Late Proterozoic Economy River Gneiss, Cobequid Highlands, Avalon composite terrane, Nova Scotia: Canadian Journal of Earth Sciences, v. 30, p. 474–479.

Dostal, J., Dupuy, C., and Caby, R., 1994, Geochemistry of the Neoproterozoic Tilemsi belt of Iforas (Mali, Sahara): A crustal section beneath an oceanic island arc: Precambrian Research, v. 65, p. 55–69.

Dostal, J., Caby, R., Keppie, J. D., and Maza, M., 2002, Neoproterozoic magmatism in Southwestern Algeria (Sebkha el Melah inlier): A northerly extension of the Trans-Saharan orogen: Journal of African Earth Sciences, v. 35, p. 213–225.

Dostal, J., Owen, V., Church, B. N., and Hamilton, T. S., in press, Episodic volcanism in the Buck Creek Complex of Central British Columbia: A history of magmatism and mantle evolution from the Jurassic to the Upper Tertiary: International Geology Review, in press.

Dunphy, J. M., and Ludden, J. N., 1998, Petrological and geochemical characteristics of a Paleoproterozoic magmatic arc (Narsajuaq terrane, Ungava Orogen, Canada) and comparisons to Superior Province granitoids: Precambrian Research, v. 91, p. 109–142.

Dunphy, J. M., Ludden, J. N., and Francis, D. F., 1995, Geochemistry of mafic magmas from the Ungava orogen of Quebec, Canada and implications for mantle reservoir compositions at 2.0 Ga: Chemical Geology, v. 120, p. 361–380.

Duyverman, H. J., Harris, N. B. W., and Hawkesworth, C. J., 1982, Crustal accretion in the Pan-African: Nd and Sr isotope evidence from the Arabian Shield: Earth and Planetary Science Letters, v. 59, p. 315–326.

Egal, E., Guerrot, C., Le Goff, D., Thieblemont, D., and Chantraine, J., 1996, The Cadomian orogeny revisited in northern France, *in* Nance, R. D., and Thompson, M. D., eds., Avalonian and related peri-Gondwanan terranes of the circum-North Atlantic: Geological Society of America Special Paper 304, v. 281–318.

Ernst, W. G., 1987, Mafic meta-igneous arc rocks of apparent komatiitic affinities, Sawyers Bar area, central Klamath Mountains, northern California: Special Publication of the Geochemical Society, v. 1, p. 191–208.

Ernst, W. G., Hacker, B. R., Barton, M. D., and Sen, G., 1991, Igneous petrogenesis of magnesian metavolcanic rocks from the central Klamath Mountains, northern California: Geological Society of America Bulletin, v. 103, p. 56–72.

Farmer, G. L., and DePaolo, D. J., 1983, Origin of Mesozoic and Tetiary granite in the western United States and implications for pre-Mesozoic crustal structure 1: Nd and Sr isotopic studies in the geocline of the northern Great Basin: Journal of Geophysical Research, v. 88, p. 3379–3401.

Fortey, R. A., and Cocks, L. R. M., 2003, Palaeontological evidence bearing on global Ordovician–Silurian continental reconstructions: Earth Science Reviews, v. 61, p. 245–307.

Gehrels, G. E., and Kapp, P.A., 1998, Detrital zircon geochronology and regional correlation of metasedimentary rocks in the Coast Mountains, southeastern Alaska: Canadian Journal of Earth Sciences, v. 35, p. 269–279.

Gibbons, W., and Horák, J. M., 1996, The evolution of the Neoproterozoic Avalonian subduction system: Evidence from the British Isles, *in* Nance, R. D., and Thompson, M. D., eds., Avalonian and related peri-Gondwanan terranes of the circum-North Atlantic: Geological Society of America Special Paper 304, p. 269–280.

Giese, U., and Buehn, B., 1994, Early Paleozoic rifting and bimodal volcanism in the Ossa-Morena aone of south-west Spain: Geologische Rundschau, v. 83, p. 143–.

Gower, C. F., 1996, The evolution of the Grenville Province in eastern Labrador, Canada, *in* Brewer, T. S., ed., Precambrian crustal evolution in the North Atlantic Region: Geological Society of London Special Publication 112, p. 197–218.

Gower, C. F., Ryan, A. B., and Rivers, T., 1990, Mid-Proterozoic Laurentia-Baltica: An overview of its geological evolution and summary of the contributions by this volume, *in* Gower, C. F., Rivers, T., and Ryan, A. B., eds., Mid-Proterozoic Laurentia-Baltica: Geological Association of Canada Special Paper 38, p. 1–20.

Greenwood, W. R., Hadley, D. G., Anderson, R. E., Fleck, R. J., and Roberts, R. J., 1980, Precambrian geologic history and plate tectonic evolution of the Arabian shield: Saudi Arabian Directorate General of Mineral Resources Bulletin 24, 35 p.

Gurnis, M., 1988, Large-scale mantle convection and the aggregation and dispersal of supercontinents: Nature, v. 322, p. 695–699.

Hall, C. E., Gurnis, M., Sdrolias, M., Lavier, L. L., and Müller, R. D., 2003, Catastrophic initiation of subduction following forced convergence across fracture zones: Earth and Planetary Science Letters, v. 212, p. 15–30.

Hartz, E. H., and Torsvik, T. H., 2002, Baltica upside down: A new plate tectonic model for Rodinia and the Iapetus Ocean: Geology, v. 30, p. 255–258.

Heaman, L. M., 1997, Global mafic magmatism at 2.45 Ga: Remnants of an ancient large igneous province?: Geology, v. 25, p. 299–302.

Hibbard, J. P., Stoddard, E. F., Secor, D. T., and Dennis, A. J., 2002, The Carolina zone: Overview of Neoproterozoic to Early Paleozoic peri-Gondwana terranes along the eastern flank of the southern Appalachians: Earth Science Reviews, v. 57, p. 299–339.

Hoffman, P. F., 1989, Speculations on Laurentia's first gigayear (2.0–1.0 Ga): Geology, v. 17, p. 135–138.

Hoffman, P. F., 1991, Did the breakout of Laurentia turn Gondwana inside out?: Science, v. 252, p. 1409–1412.

Irwin, W. P., 1981, Tectonic accretion of the Klamath Mountains, *in* Ernst, W. G., ed., The geotectonic development of California (Rubey volume 1): Englewood Cliffs, NJ, Prentice Hall, p. 29–49.

Jenner, G. A., and Swinden, H. S., 1993, The Pipestone Pond Complex, central Newfoundland: Complex magmatism in an eastern Dunnage ophiolite: Canadian Journal of Earth Sciences, v. 30, p. 434–448.

Kemnitz, H., Romer, R. L., and Oncken, O., 2002, Gondwana break-up and the northern margin of the Saxothuringian belt (Variscides of Central Europe): International Journal of Earth Sciences, v. 91, p. 246–259.

Keppie, J. D., 1985, The Appalachian Collage, *in* Gee, D. G., and Sturt, B., eds., The Caledonide orogen, Scandinavia, and related areas: New York, NY, John Wiley and Sons, p. 1217–1226.

Keppie, J. D., Nance, R. D., Murphy, J. B., and Dostal, J., 1991, Northern Appalachians: Avalon and Meguma terranes, *in* Dallmeyer, R. D., and Lécorché, J. P., eds., The West African orogens and circum-Atlantic correlatives. New York, NY, Springer-Verlag, p. 315–334.

Keppie, J. D., Nance, R. D., Murphy, J. B., and Dostal, J., 2003, Tethyan, Mediterranean, and Pacific analogues for the Neoproterozoic–Paleozoic birth and development of peri-Gondwanan terranes and their transfer to Laurentia and Laurussia: Tectonophysics, v. 365, p. 195–219.

Keppie, J. D., and Ramos, V. A., 1999, Odyssey of terranes in the Iapetus and Rheic oceans during the Paleozoic, *in* Ramos, V. A., and Keppie, J. D., eds., Laurentia-Gondwana connections before Pangea: Geological Society of America Special Paper 336, p. 267–276.

Krogh, T. E., Strong, D. F., O'Brien, S. J., and Papezik, V. S., 1988, Precise U-Pb zircon dates from the Avalon terrane in Newfoundland: Canadian Journal of Earth Sciences, v. 25, p. 442–453.

Kröner, A., 1984, Late Precambrian tectonics and orogeny: A need to re-define the term Pan-African, *in* Klerkx, J., and Michot, J., eds., African geology: Tervuren, Belgium, Royal Museum for Central Africa, p. 23–26.

Kröner, A., 1985, Ophiolites and the evolution of tectonic boundaries in the Late Proterozoic Arabian-Nubian shield of northeast Africa and Arabia: Precambrian Research, v. 27, p. 277–300.

Kröner, A., Greiling, R., Reischman, T., Hussein, I. M., Stern, R. J., Durr, S., Kruger, J., and Zimmer, M., 1987, Pan-African crustal evolution in the Nubian segment of northeast Africa, *in* Kröner, A., ed., Proterozoic lithospheric evolution: American Geophysical Union, Geodynamic Series, v. 15, p. 235–257.

Kröner, A., Hegner, E., Collins, A. S., Windley, B. F., Brewer, T. S., Razakamanana, T., and Pidgeon, R. T., 2000, Age and magmatic history of the Antananarivo block, central Madagascar, as derived from zircon geochronology and Nd isotopic systematics: American Journal of Science, v. 300, p. 251–288.

Kröner, A., Pallister, J. S., and Fleck, R. J., 1992, Age of initial oceanic magmatism in the late Proterozoic Arabian shield: Geology, v. 20, p. 803–806.

Küster, D., and Liégeois, J.-P., 2001, Sr, Nd isotopes and geochemistry of the Bayuda Desert high-grade metamorphic basement (Sudan): An early Pan-African oceanic convergent margin, not the edge of the East Saharan ghost craton: Precambrian Research, v. 109, p. 1–23.

Lassiter, J. C., DePaolo, D. J., and Mahoney, J. J., 1995, Geochemistry of the Wrangellia flood basalt province: Implications for the role of continental and oceanic lithosphere in flood basalt genesis: Journal of Petrology, v. 36, p. 983–1009.

Leech, M. L., Metzger, E. P., Wooden, J. L., Jones, R. E., Schwartz, C. L., and Beane, R. J., 2002, New eclogitization and protolith ages for the Maksyutov Complex (south Ural Mountains) based on U-Pb zircon SHRIMP data [abs.]: EOS (Transactions of the American Geophysical Union), v. 83, no. 47, Fall Meeting Supplement, Abstract V51B-1263.

Lenoir, J.-L., Küster, D., Liégeois, J.-P., Utke, A., Haider, A., and Matheis, G., 1994, Origin and regional significance of late Precambrian and early Paleozoic granitoids in the Pan-African belt of Somalia: Geologische Rundschau, v. 83, p. 624–641.

Linnemann, U., Gehmlich, M., Tichomirowa, M., Buschmann, B., Nasdala, L., Jonas, P., Lützner, H., and Bombach, K., 2000, From Cadomian subduction to Early Palaeozoic rifting: The evolution of Saxo-Thuringia at the margin of Gondwana in the light of single-zircon geochronology and basin development (Central European Variscides, Germany), *in* Franke,W., Altherr, R., Haak, V., Oncken, O., and Tanner, D., eds., Orogenic processes—quantification and modelling in the Variscan Belt of Central Europe: Geological Society of London Special Publication 179, p. 131–153.

Linnemann, U., and Romer, R. L., 2002, The Cadomian Orogeny in Saxo-Thuringia, Germany: Geochemical and Nd-Sr-Pb isotopic characterization of marginal basins with constraints to geotectonic setting and provenance: Tectonophysics, v. 352, p. 33–64.

Ludden, J., and Hynes, A. J., 2000, The lithoprobe Abitibi-Grenville transect: Two billion years of crust formation and recycling in the Precambrian shield of Canada: Canadian Journal of Earth Sciences, v. 37, p. 459–476.

MacLachlan, K., and Dunning, G., 1998, U-Pb ages and tectonomagmatic relationships of early Ordovician low-Ti tholeiites, boninites, and related plutonic rocks in central Newfoundand: Contributions to Mineralogy and Petrology, v. 133, p. 235–258.

Mac Niocaill, C., van der Pluijm, B. A., Van der Voo, R., and McNamara, A. K., 2002, Discussion and reply: West African proximity of Avalon in the latest Precambrian: Geological Society of America Bulletin, v. 114, p. 1049–1052.

Marcantonio, F., Dickin, A. P., McNutt, R. H., and Heaman, L. M., 1990, Isotopic evidence for the crustal evolution of the Frontenac Arch in the Grenville Province of Ontario, Canada: Chemical Geology, v. 83, p. 297–314.

McMenamin, M. A., and McMenamin, D. L., 1990, The emergence of animals: The Cambrian breakthrough. New York, NY: Columbia University Press, 217 p.

McNamara, A. K., Mac Niocaill, C., van der Pluijm, B. A., and Van der Voo, R., 2001, West African proximity of Avalon in the latest Precambrian: Geological Society of America Bulletin, v. 113, p. 1161–1170.

Meert, J. G., Dahlgren, S., Eide, E. A., and Torsvik, T. H., 1998, Tectonic significance of the Fen Province, S. Norway: Constraints from geochronology and paleomagnetism: Journal of Geology, v. 106, p. 553–564.

Monger, J. W. H., 1977, Upper Paleozoic rocks of western Canadian Cordillera and their bearing on Cordillera evolution: Canadian Journal of Earth Sciences, v. 14, p. 1832–1858.

Monger, J., 1997, Plate tectonics and northern Cordilleran geology: An unfinished revolution: Geoscience Canada, v. 24, p. 189–198.

Monger, J. W. H., Price, R. A., and Templeman-Kluit, D., 1982, Terrane accretion and the origin of two metamorphic and plutonic welts in the Canadian Cordillera: Geology, v. 10, p. 70–75.

Monger, J. W. H., Wheeler, J. O, Tipper, H. W., Gabrielese, H., Harms, T., Struik, L. C., Campbell, R. B., Dodds, C. J., Gehrels, G. E., and O'Brien, J., 1991, Part B, Cordilleran terranes, *in* Gabrielse, H., and Yorath, C. J., eds., Geology of the Cordilleran orogen in Canada: Geological Survey of Canada, Geology of Canada, p. 287–321 (also Geological Society of America, Geology of America, v. G-2).

Murphy, J. B., Dostal, J., Nance, R. D., and Keppie, J. D., 2004, Grenville-aged juvenile crust development in the peri-Rodinian ocean, *in* Tollo, R. P., Corriveau, L., McLelland, J., and Bartholomew, M. J., eds., Proterozoic tectonic evolution of the Grenville orogen in North America: Geological Society of America Memoir 197, in press.

Murphy, J. B., Keppie, J. D., Dostal, J., and Cousins, B. L., 1996, Repeated late Neoproterozoic–Silurian lower crustal melting beneath the Antigonish Highlands, Nova Scotia: Nd isotopic evidence and tectonic interpretations *in* Nance, R. D., and Thompson, M. D., eds., Avalonian and related peri-Gondwanan terranes of the circum-North Atlantic: Geological Society of America Special Paper 304, p. 109–120.

Murphy, J. B., Keppie, J. D., Dostal, J., and Hynes, A. J., 1990, Late Precambrian Georgeville group: A volcanic arc rift succession in the Avalon terrane of Nova Scotia: Geological Society of London Special Publication 51, p. 383–393.

Murphy, J. B., Keppie, J. D., Dostal, J., and Nance, R. D., 1999, Neoproterozoic–early Paleozoic evolution of Avalonia, *in* Ramos, V. A., and Keppie, J. D., eds., Laurentia-Gondwana connections before Pangea: Geological Society of America Special Paper 336, p. 253–266.

Murphy, J. B., Keppie, J. D., and Nance, R. D., 2002, West African proximity of the Avalon terrane in the latest Precambrian by McNamara, A. K., Mac Niocaill, C., van der Pluijm, B. A., and Van der Voo, R.: Discussion and reply: Geological Society of America Bulletin, v. 114, p. 1049–1052.

Murphy, J. B., and Nance, R. D., 1989, Model for the evolution of the Avalonian-Cadomian belt: Geology, v. 17, p. 735–738.

Murphy, J. B., and Nance, R. D., 1991, A supercontinent model for the contrasting character of Late Proterozoic orogenic belts: Geology, v. 19, p. 469–472.

Murphy, J. B., and Nance, R. D., 2002, Sm-Nd isotopic systematics as tectonic tracers: An example from West Avalonia in the Canadian Appalachians: Earth Science Reviews, v. 59, p. 77–100.

Murphy, J. B., and Nance, R. D., 2003, Do supercontinents introvert or extrovert?: Sm-Nd isotopic evidence: Geology, v. 31, p. 873–876.

Murphy, J. B., Strachan, R. A., Nance, R. D., Parker, K. D., and Fowler, M. B., 2000, Proto-Avalonia: A 1.2–1.0 Ga tectonothermal event and constraints for the evolution of Rodinia: Geology, v. 28, p. 1071–1074.

Nance, R. D., Keppie, J. D., and Miller, B. V., 2004, Record of the closure of the Rheic Ocean in the Acatlán Complex of Southern Mexico [abs.]: Geological Society of America, Abstracts with Programs, v. 36, no. 2, in press.

Nance, R. D., and Murphy, J. B., 1994, Contrasting basement isotopic signatures and the palinspastic restoration of peripheral orogens: Example from the Neoproterozoic Avalonian-Cadomian belt: Geology v. 22, p. 617–620.

Nance, R. D., and Murphy, J. B., 1996, Basement isotopic signatures and Neoproterozoic paleogeography of Avalonian-Cadomian and related terranes in the circum-North Atlantic, *in* Nance, R. D., and Thompson, M. D., eds., Avalonian and related peri-Gondwanan terranes of the circum-North Atlantic: Geological Society of America Special Paper 304, p. 333–346.

Nance, R. D., Murphy, J. B., and Keppie, J. D., 2002, Cordilleran model for the evolution of Avalonia: Tectonophysics, v. 352, p. 11–31.

Nance, R. D., Murphy, J. B., Strachan, R. A., D'Lemos, R. S., and Taylor, G. K., 1991, Late Proterozoic tectonostratigraphic evolution of the Avalonian and Cadomian terranes: Precambrian Research, v. 53, p. 41–78.

Nance, R. D., Worsley, T. R., and Moody, J. B., 1986, Post-Archean biogeochemical cycles and long-term episodicity in tectonic processes: Geology, v. 14, p. 514–518.

Nokleberg, W. J., Jones, D. L., and Silbererg, N. J., 1985, Origin and tectonic evolution of the Maclaren and Wrangellia terrane, eastern Alaska Range, Alaska: Geological Society of America Bulletin, v. 96, p. 1251–1270.

Nutman, A. P., Green, D. H., Cook, C. A, Styles, M. T., and Holdsworth, R. E., 2001, U-Pb dating of the exhumation of the Lizard peridotite and its exhumation over crustal rocks: Journal of the Geological Society, London, v. 158, p. 809–820.

O'Brien, S. J., O'Brien, B. H., Dunning, G. R., and Tucker, R. D., 1996, Late Neoproterozoic Avalonian and related peri-Gondwanan rocks of the Newfoundland Appalachians, *in* Nance, R. D., and Thompson, M. D., eds., Avalonian and related peri-Gondwanan terranes of the circum-North Atlantic: Geological Society of America Special Paper 304, p. 9–28.

O'Brien, S. J., Wardle, R. J., and King, A. F., 1983, The Avalon zone: A pan-African terrane in the Appalachian orogen of Canada: Geological Journal, v. 18, p. 195–222.

Pallister, J. S., Stacey, J. S., Fischer, L. B., and Premo, W. R., 1988, Precambrian ophiolites of Arabia: Geologic settings, U-Pb geochronology, Pb isotopic characteristics, and implications for continental accretion: Precambrian Research, v. 38, p. 1–54.

Patchett, P. J., and Bridgewater, D., 1984, Origin of continental crust of 1.9–1.7 b.y. age defined by Nd isotopes in the Ketilidian terrane of South Greenland: Contributions to Mineralogy and Petrology, v. 87, p. 311–318.

Patchett, P. J., and Kuovo, O., 1986, Origin of continental crust of 1.9–1.7 Ga age: Nd isotopes and U-Pb zircon ages in the Svecokarelian terrain of South Finland: Contributions to Mineralogy and Petrology, v. 92, p. 1–12.

Patchett, P. J., and Ruiz, J. 1989, Nd isotopes and the origin of the Grenville-aged rocks in Texas: Implications for Proterozoic evolution of the United States mid-continent region: Journal of Geology, v. 97, p. 685–695.

Pimental, M. M., and Fuck, R. A., 1987, Late Proterozoic granitic magmatism in southwestern Goias, Bazil: Revisita Brasiliera Geosciencias, v. 17, p. 415–425.

Pimental, M. M., and Fuck, R. A., 1992, Neoproterozoic crustal accretion in central Brazil: Geology, v. 20, p. 375–379.

Pimental, M. M., Heaman, L., Fuck, R. A., and Marini, O. J., 1991, U-Pb zircon geochronology of Precambrian tin-bearing continental-type acid magmatism in central Brazil: Precambrian Research, v. 52, p. 321–335.

Pimental, M. M., Whitehouse, M. J., das G. Viana, M., Fuck, R. A., and Machado, N., 1997, The Mara Rosa arc in the Tocantins province: Further evidence for Neoproterozoic crustal accretion in central Brazil: Precambrian Research, v. 81, p. 299–310.

Pin, C., and Marini, F., 1993, Early Ordovician continental break-up in Variscan Europe: Nd-Sr isotope and trace element evidence from bimodal igneous associations of the southern Massif Central, France: Lithos, v. 29, p. 177–196.

Pin, C., and Paquette, J.-L., 1997, A mantle-derived bimodal suite in the Hercynian belt: Nd isotope and trace element evidence for a subduction-related rift origin of the Late Devonian Brevenne metavolcanics, Massif Central (France): Contributions to Mineralogy and Petrology, v. 129, p. 222–238.

Pisarevsky, S. A., Wingate, M. T. D., Powell, C. McA., Johnson, S., and Evans, D. A. D., 2003, Models of Rodinia assembly and fragmentation, *in* Yoshida, M., Windley, B., and Dasgupta, S., eds., Proterozoic East Gondwana: Supercontinent assembly and breakup: Geological Society of London Special Publication 206, p. 35–55.

Prigmore, J. K., Butler, A. J., and Woodcock, N. H., 1997, Rifting during separation of Eastern Avalonia from Gondwana: Evidence from subsidence analysis: Geology, v. 25, p. 203–206.

Puchkov, V. N., 1997, Structure and geodynamics of the Uralian orogen, *in* Burg, J.-P., and Ford, M., eds., Orogeny through time: Geological Society of London Special Publication 121, p. 201–236.

Quesada, C., 1990, Precambrian terranes in the Iberian Variscan foldbelt, *in* Strachan, R. A., and Taylor, G. K., eds., Avalonian and Cadomian geology of the North Atlantic: London, UK, Blackie, p. 109–133.

Quesada, C., 1997, Evolución geodinámica de la zona Ossa-Morena durante el ciclo Cadomiense, *in* Livro de Homenagem ao Prof. Francisco Goncalves: Estudio sobre a Geologia de zona de Ossa-Morena (Macico Ibérico): Évora, Portugal, Univ. Évora, p. 205–230.

Rivers, T., and Corrigan, D., 2000, Convergent margin on southeastern Laurentia during the Mesoproterozoic: Tectonic implications: Canadian Journal of Earth Sciences, v. 37, p. 359–383.

Robardet, M., 2003, The Armorican "microplate": fact or fiction? Critical review of the concept and contradictory palaeobiogeographical data: Palaeogeography, Palaeoclimatology, Palaeoecology, v. 195, p. 125–148.

Rogers, J. J. W., 1996, A history of continents in the past three billion years: Journal of Geology, v. 104, p. 91–107.

Rogers, J. J. W., and Santosh, M., 2002, Configuration of Columbia, a Mesoproterozoic supercontinent: Gondwana Research, v. 5, p. 5–22.

St.-Onge, M. R., Scott, D. J., and Lucas, S. B., 2000, Early partitioning of Quebec: Microcontinent formation in the Paleoproterozoic: Geology, v. 28, p. 323–326.

St.-Onge, M. R., Scott, D. J., and Wodicka, N., 2001, Terrane boundaries within Trans-Hudson orogen (Quebec–Baffin segment), Canada: Changing structural and metamorphic character from foreland to hinterland: Precambrian Research, v. 107, p. 75–91.

Samson, S. D., Hibbard, J. P., and Wortman, G. L., 1995, Nd isotopic evidence for juvenile crust in the Carolina terrane, southern Appalachians: Contributions to Mineralogy and Petrology, v. 121, p. 171–184.

Sánchez-García, T., Bellido, F., and Quesada, C., 2003, Geodynamic setting and geochemical signatures of Cambrian–Ordovician rift-related igneous rocks (Ossa-Morena Zone, SW Iberia): Tectonophysics, v. 365, p. 233–255.

Sandeman, H. A., Clark, A. H., Scott, D. J., and Malpas, J. G., 2000, The Kennack gneiss of the Lizard Peninsula, Cornwall, S.W. England: Commingling and mixing of mafic and felsic magmas during Givetian continental incorporation of the Lizard ophiolite: Journal of the Geological Society, London, v. 157, p. 1227–1242.

Scott, D. J., Helmstaedt, H., and Bickle, M. J., 1992, Purtuniq ophiolite, Cape Smith belt, northern Quebec, Canada: A reconstructed section of Early Proterozoic oceanic crust: Geology, v. 20, p. 173–176.

Shackleton, R. M., 1986, Precambrian collision tectonics in Africa, *in* Coward, M. C., and Ries, A. C., eds., Collision tectonics: Geological Society of London Special Publication 19, p. 329–349.

Simard, R.-L., Dostal, J., and Roots, C. F., 2003, Late Paleozoic development of volcanic arcs in the Canadian Cordillera: An example from the Klinkit Group of northern British Columbia and southern Yukon: Canadian Journal of Earth Sciences, v. 40, p. 907–924.

Smith, A. D., and Lambert R. St. J., 1995, Nd, Sr, and Pb isotopic evidence for the contrasting origin of Late Paleozoic volcanic rocks from the Slide Mountain and Cache Creek terranes, south-central British Columbia: Canadian Journal of Earth Sciences, v. 32, p. 447–459.

Stampfli, G. M., and Borel, G. D., 2002, A plate tectonic model for the Paleozoic and Mesozoic constrained by dynamic plate boundaries and restored synthetic oceanic isochrones: Earth and Planetary Science Letters, v. 196, p 17–33.

Stein, M., 2003, Tracing the plume material in the Arabian-Nubian Shield: Precambrian Research, v. 123, p. 223–234.

Stern, R. J., 1994, Arc assembly and continental collision in the Neoproterozoic East African orogen: Implications for the consolidation of Gondwanaland: Annual Reviews of Earth and Planetary Sciences, v. 22, p. 319–351.

Stern, R. J., 2002, Crustal evolution in the East African orogen: A neodymium isotopic perspective: Journal of African Earth Sciences, v. 34, p. 109–117.

Stern, R. J., and Dawoud, A. M., 1991, Late Precambrian (740 Ma) charnockite, enderbite, and granite from Moya, Sudan: A link between the Mozambique belt and the Arabian-Nubian shield: Journal of Geology, v. 99, p. 648–659.

Stern, R. J., Kröner, A., and Rashwan, A. A., 1991, A late Precambrian ~710 Ma high volcanicity rift in the southern Eastern Desert of Egypt: Geologische Rundschau, v. 80, p. 155–170.

Strachan, R. A., D'Lemos, R. S., and Dallmeyer, R. D., 1996, Late Precambrian evolution of an active plate margin: North Armorican Massif, France, *in* Nance, R. D., and Thompson, M. D., eds., Avalonian and related peri-Gondwanan terranes of the circum-North Atlantic: Geological Society of America Paper 304, p. 319–332.

Stump, E., 1992, The Ross orogen of the Transantarctic Mountains in the light of the Laurentian-Gondwana split: GSA Today, v. 2, no. 2, p. 25–27 and 30–33.

Swinden, H. S., and Hunt, P. A., 1991, A U-Pb zircon age from the Connaigre Bay Group, southwestern Avalon zone, Newfoundland: Implications for regional correlations and metallogenesis, *in* Radiometric age and isotopic studies, Report 4: Geological Survey of Canada Paper 90-2, p. 3–10.

Swinden, H. S., Jenner, G. A., Fryer, B. J., Hertogen, J., and Roddick, J. C., 1990, Petrogenesis and paleotectonic history of the Wild Bight Group, an Ordovician rifted island arc in central Newfoundland: Contributions to Mineralogy and Petrology, v. 105, p. 219–241.

Swinden, H. S., Jenner, G. A., and Szybinski, Z. A., 1997, Magmatic and tectonic evolution of the Cambro-Ordovician Laurentian margin of Iapetus: Geochemical and isotopic constraints from the Notre Dame subzone, Newfoundland, *in* Sinha, A. K., Hogan, J. P., and Whalen, J. B., eds., Magmatism in the Appalachians: Geological Society of America Memoir 191, p. 337–365.

Tardy, M., Lapierre, H., Bosch, G., Cadeux, A., Narros, A., Struik, L. C., and Brunet, P., 2003, Le terrane de Slide Mountain, Cordillères canadiennes: Une lithosphere océanique marqueée par des points chaudes: Canadian Journal of Earth Sciences, v. 40, p. 833–852.

Teixeiera, W., Tassinari, C. C. G., Cordani, U. G., and Kawashita, K., 1989, A review of the geochronology of the Amazon craton: Tectonic implications: Precambrian Research v. 42, p. 213–227.

Theokritoff, G., 1979, Early Cambrian provincialism and biogeographic boundaries in the North Atlantic region: Lethaia, v. 12, p. 281–295.

Thériault, R. J., St.-Onge, M. R., and Scott, D. J., 2001, Nd isotopic signature of the Paleoproterozoic Trans-Hudson orogen, southern Baffin Island: Implications for the evolution of eastern Laurentia: Precambrian Research, v. 108, p. 113–138.

Torsvik, T. H., and Rehnström, E. F., 2001, Cambrian palaeomagnetic data from Baltica: Implications for true polar wander and Cambrian palaeogeography: Journal of the Geological Society, London, v. 158, p. 321–329.

Toth, J., and Gurnis, M., 1998, Dynamics of subduction initiation at preexisting fault zones: Journal of Geophysical Research, v. 103, p. 18,053–18,067.

Tucker, R. D., and Pharoah, T. C., 1991, U-Pb zircon ages of late Precambrian rocks in southern Britain: Journal of the Geological Society, London, v. 148, p. 435–443.

van der Voo, R., 1988, Paleozoic paleogeography of North America, Gondwana, and intervening displaced terranes: Comparisons of paleomagnetism with paleoclimatology and biogeographical patterns: Geological Society of America Bulletin, v. 100, p. 311–324.

van Schmus, W. R., de Brito Neves, B. B., Hackspacher, P., and Babinski, M., 1995, U/Pb and Sm/Nd geochronologic studies of the eastern Borborema Province, Northeastern Brazil: Initial conclusions: Journal of South American Earth Sciences, v. 8, p. 267–288.

van Staal, C. R., Dewey, J. F., Mac Niocaill, C., and McKerrow, W. S., 1998, The Cambrian–Silurian tectonic evolution of the Northern Appalachians and British Caledonides: History of a complex, west and southwest Pacific-type segment of Iapetus, *in* Blundell, D., and Scott, A., eds., Lyell: The past is the key to the present: Geological Society of London Special Publication 143, p. 199–242.

von Raumer, J. F., Stampfli, G. M., Borel, G., and Bussy, F., 2002, Organization of pre-Variscan basement areas at the north-Gondwanan margin: International Journal of Earth Sciences, v. 91, p. 35–52.

Weil, A. B., Van der Voo, R., Mac Niocaill, C., and Meert, J. G., 1998, The Proterozoic supercontinent Rodinia: Paleomagnetically derived reconstruction for 1100 to 800 Ma: Earth and Planetary Science Letters, v. 154, p. 13–24.

White, D. J., Forsyth, D. A., Asudeh, I., Carr, S. D., Wu, H., Easton, R. M., and Mereu, R. F., 2000, A seismic-based cross-section of the Grenville Orogen in southern Ontario and western Quebec: Canadian Journal of Earth Sciences, v. 37, p. 183–192.

Whitehouse, M. J., Windley, B. F., Stoeser, D. B., Al-Khirbash, S., Ba-Bttat, M. A. O., and Haider, A., 2001, Precambrian basement character of Yemen and correlations with Saudi Arabia and Somalia: Precambrian Research, v. 105, p. 357–369.

Williams, H., 1979, Appalachian orogen in Canada: Canadian Journal of Earth Sciences, v. 16, p. 792–798.

Williams, H., Hoffman, P. F., Lewry, J. F., Monger, J. W. H., and Rivers, T., 1991, Anatomy of North America: Thematic geologic portrayals of the continent: Tectonophysics, v. 187, p. 117–134.

Wilson, J. T., 1966, Did the Atlantic Ocean close and then re-open?: Nature, v. 211, p. 676–681.

Winchester, J. A., Pharaoh, T. C., and Verniers, J., 2002, Palaeozoic amalgamation of Central Europe: An introduction and synthesis of new results from recent geological and geophysical investigations, *in* Winchester, J. A., Pharaoh, T. C, and Verniers, J., eds., Palaeozoic amalgamation of Central Europe: Geological Society of London Special Publication 201, p. 1–18.

Windley, B. F., 1993, Uniformitarianism today: Plate tectonics is the key to the past: Journal of the Geological Society, London, v. 150, p. 7–19.

Windley, B. F., Whitehouse, M. J., and Ba-Btatt, M. A. O., 1996, Early Precambrian gneiss terranes and Pan-African island arcs in Yemen: Crustal accretion of the eastern Arabian shield: Geology, v. 24, p. 131–134.

Wingate, M. T. D., and Giddings, J. W., 2000, Age and palaeomagnetism of the Mundine Well dyke swarm, Western Australia: Implications for an Australia-Laurentia connection at 755 Ma: Precambrian Research, v. 100, p. 335–357.

Worsley, T. R., Nance, R. D., and Moody, J. B., 1984, Global tectonics and eustasy for the past 2 billion years: Marine Geology, v. 58, p. 373–400.

Wortman, G. L., Samson, S. D., and Hibbard, S. D., 2000, Precise U-Pb zircon constraints on the earliest magmatic history of the Carolina terrane: Journal of Geology, v. 108, p. 321–338.

Zhao, G., Cawood, P. A., Wilde, S. A., and Sun, M., 2002, Review of global 2.1–1.8 Ga orogens: Implications for a pre-Rodinia supercontinent: Earth Science Reviews, v. 59, p. 125–162.

Zonenshain, L. P., Korinevsky, V. G., Kazmin, V. G., Pechersky, D. M., Khain, V. V., and Mateveenkov, V. V., 1984, Plate tectonic model of the south Urals development: Tectonophysics, v. 109, p. 95–135.

Zonenshain, L. P., Kuzmin, M. I., and Natapov, L. M., 1990, Uralian foldbelt, *in* Page, B. M., ed., Geology of the USSR: A plate-tectonic synthesis: American Geophysical Union, Geodynamics Series, v. 21, p. 27–54.

INDEX

A

B

C

D

E

F

G

H

I

N

O

P

Q

R

S

T

U

V

W

X

Y

Z